DEVELOPMENTS IN GEOTECHNICAL ENGINEERING, 36

RHEOLOGICAL FUNDAMENTALS OF SOIL MECHANICS

Further titles in this series:

1. G. SANGLERAT — THE PENETROMETER AND SOIL EXPLORATION
2. Q. ZÁRUBA AND V. MENCL — LANDSLIDES AND THEIR CONTROL
3. E.E. WAHLSTROM — TUNNELING IN ROCK
4. R. SILVESTER — COASTAL ENGINEERING, 1 and 2
5. R.N. YONG AND B.P. WARKENTIN — SOIL PROPERTIES AND BEHAVIOUR
6. E.E. WAHLSTROM — DAMS, DAM FOUNDATIONS, AND RESERVOIR SITES
7. W.F. CHEN — LIMIT ANALYSIS AND SOIL PLASTICITY
8. L.N. PERSEN — ROCK DYNAMICS AND GEOPHYSICAL EXPLORATION
   Introduction to Stress Waves in Rocks
9. M.D. GIDIGASU — LATERITE SOIL ENGINEERING
10. Q. ZÁRUBA AND V. MENCL — ENGINEERING GEOLOGY
11. H.K. GUPTA AND B.K. RASTOGI — DAMS AND EARTHQUAKES
12. F.H. CHEN — FOUNDATIONS ON EXPANSIVE SOILS
13. L. HOBST AND J. ZAJÍC — ANCHORING IN ROCK
14. B. VOIGHT (Editor) — ROCKSLIDES AND AVALANCHES, 1 and 2
15. C. LOMNITZ AND E. ROSENBLUETH (Editors) — SEISMIC RISK AND ENGINEERING DECISIONS
16. C.A. BAAR — APPLIED SALT-ROCK MECHANICS, 1
    The In-Situ Behavior of Salt Rocks
17. A.P.S. SELVADURAI — ELASTIC ANALYSIS OF SOIL-FOUNDATION INTERACTION
18. J. FEDA — STRESS IN SUBSOIL AND METHODS OF FINAL SETTLEMENT CALCULATION
19. Á. KÉZDI — STABILIZED EARTH ROADS
20. E.W. BRAND AND R.P. BRENNER (Editors) — SOFT-CLAY ENGINEERING
21. A. MYSLIVEC AND Z. KYSELA — THE BEARING CAPACITY OF BUILDING FOUNDATIONS
22. R.N. CHOWDHURY — SLOPE ANALYSIS
23. P. BRUUN — STABILITY OF TIDAL INLETS
    Theory and Engineering
24. Z. BAŽANT — METHODS OF FOUNDATION ENGINEERING
25. Á. KÉZDI — SOIL PHYSICS
    Selected Topics
26. H.L. JESSBERGER (Editor) — GROUND FREEZING
27. D. STEPHENSON — ROCKFILL IN HYDRAULIC ENGINEERING
28. P.E. FRIVIK, N. JANBU, R. SAETERSDAL AND L.I. FINBORUD (Editors) — GROUND FREEZING 1980
29. P. PETER — CANAL AND RIVER LEVÉES
30. J. FEDA — MECHANICS OF PARTICULATE MATERIALS
    The Principles
31. Q. ZÁRUBA AND V. MENCL — LANDSLIDES AND THEIR CONTROL
    Second completely revised edition
32. I.W. FARMER (Editor) — STRATA MECHANICS
33. L. HOBST AND J. ZAJÍC — ANCHORING IN ROCK AND SOIL
    Second completely revised edition
34. G. SANGLERAT, G. OLIVARI AND B. CAMBOU — PRACTICAL PROBLEMS IN SOIL MECHANICS AND FOUNDATION ENGINEERING, 1 and 2
35. L. RÉTHÁTI — GROUNDWATER IN CIVIL ENGINEERING
37. P. BRUUN (Editor) — DESIGN AND CONSTRUCTION OF MOUNDS FOR BREAKWATERS AND COASTAL PROTECTION
38. W.F. CHEN AND G.Y. BALADI — SOIL PLASTICITY
    Theory and Implementation

DEVELOPMENTS IN GEOTECHNICAL ENGINEERING, 36

RHEOLOGICAL FUNDAMENTALS OF SOIL MECHANICS

SERGEI S. VYALOV

*Gersevanov Research Institute for Bases and Underground Structures,
6, Vtoraya Institutskaya Ulitsa, 109389 Moscow, U.S.S.R.*

Translated from the Russian
by
O.K. SAPUNOV

ELSEVIER
Amsterdam — Oxford — New York — Tokyo 1986

ELSEVIER SCIENCE PUBLISHERS B.V.
Sara Burgerhartstraat 25
P.O. Box 211, 1000 AE Amsterdam, The Netherlands

Distributors for the United States and Canada:

ELSEVIER SCIENCE PUBLISHING COMPANY INC.
52, Vanderbilt Avenue
New York, N Y 10017

Library of Congress Cataloging-in-Publication Data

Vialov, Sergeĭ Stepanovich.
  Rheological fundamentals of soil mechanics.

  (Developments in geotechnical engineering ; 36)
  Bibliography: p.
  Includes index.
  1. Soil mechanics. 2. Rheology. I. Title.
II. Series.
TA710.V4973 1986   624.1'5136   85-27586
ISBN 0-444-42223-4

ISBN 0-444-42223-4 (Vol. 36)
ISBN 0-444-41662-5 (Series)

© Elsevier Science Publishers B.V., 1986

All rights reserved. No part of this publication may be reproduced, stored in a retrieval system or transmitted in any form or by any means, electronic, mechanical, photocopying, recording or otherwise, without the prior written permission of the publisher, Elsevier Science Publishers B.V./Science & Technology Division, P.O. Box 330, 1000 AH Amsterdam, The Netherlands.

Special regulations for readers in the USA — This publication has been registered with the Copyright Clearance Center Inc. (CCC), Salem, Massachusetts. Information can be obtained from the CCC about conditions under which photocopies of parts of this publication may be made in the USA. All other copyright questions, including photocopying outside of the USA, should be referred to the publishers.

PREFACE

Soil rheology is a branch of soil mechanics investigating the origin of, and the time-dependent changes in, the stressed and strained state of soil. However, the words "rheological fundamentals" in the title of the book are interpreted by the author in a wider sense. He considers rheology as the science concerned with how the state of stress and strain is formed and altered in a body on the one hand, and with the particulars of the body's behaviour failing to fit the traditional concepts of elasticity and plasticity on the other. Following Academician Sedov[*1] one may say that the objective of rheology is to produce novel models describing the behaviour of media other than idealized bodies. Soils belong to media of this kind.

As is known, soil mechanics examines and predicts the behaviour of soils in response to either external or internal forces. Engineering problems are solved by examining the physical conditions of the soil, which are determined by its geological origin, and by employing continuum mechanics mathematics. However, unlike this branch of science, proceeding from the concept of continuity of bodies, soil mechanics treats soil as a dispersed porous medium capable of irreversibly changing its volume, i.e. of compacting.

Modern soil mechanics is based on the following three principles.

(1) The pressure and volumetric changes in a soil, i.e., the changes in void ratio, are assumed to be in direct proportion as are the shear stresses and changes in shape.

(2) The compaction of soil with time (consolidation) takes place due to the migration of water through the voids according to the laws of percolation.

(3) Soil, as a dispersed medium, displays not only inter-particle cohesion but internal resistance as well; these properties determine its resistance to failure.

The role of the above principles in developing soil mechanics can hardly be overestimated. They have been instrumental in formulating the theories of linear soil deformation, percolation consolidation, and limiting equilibrium. Moreover, a wide range of engineering problems have been solved on the basis of the above principles by using mathematics of the theories concerned.

At the same time, the principles mentioned have somewhat schematized the behaviour of soil; in fact, it appears to be more complex when the soil sustains a load. For example, the deformation of clay soils is significantly influenced by time-dependent phenomena such as creep, relaxation and the deterioration of strength due to a long-term load application. In other words, it may be said that clay soils are capable of changing their state of stress and strain with time.

[*1] For the references see Chapter 1.

Another noteworthy feature of real soils is the non-linearity of the stress–strain relation, especially if the strain is developing with time. Furthermore, one must keep in mind that internal friction — a cardinal property of soils — manifests itself in both limiting and subliminting conditions, influencing the manner in which the deformation is developing. It will be shown that this particular aspect of soil behaviour is attributed to the difference in the resistance which the soil offers to compressional and tensional deformation. This gives rise to such anomalies as changes in volume under a shear stress (dilatancy) and changes in shape due to all-around pressure, etc.

Thus, the actual behaviour of soil differs substantially from schematized concepts. Notwithstanding the fact that schematization has frequently been helpful in arriving at results acceptable in engineering practice, many cases are recorded where ignoring all aspects of soil behaviour has led to significant departures from reality. A few examples to that end are: structural deformations induced by long-term creep; failure of slopes and retaining walls because they have been designed in terms of instantaneous strength rather than for the long term; and actual settlements being a far cry from computed values because consideration of their non-linearity was neglected.

By taking into account all peculiarities of soil deformation we can obtain precise knowledge of the soil properties. We can then improve our analytical predictions so that they approach actual soil behaviour.

From what is said here it appears appropriate that all particulars of soil behaviour be investigated and systematized. All the more so because the Building Codes and Regulations, 1972 (BC and R II-A.10.71, Structures and Foundations; Main Aspects of Design) make it a point that "the physical non-linearity, plastic and rheological properties of materials and soils be taken into account whenever necessary in analysing structures and the soil below foundations". It is exactly these problems that are tackled in this book.

To assist students in comprehending the subject, some general aspects of continuum mechanics, including well-known notions from soil mechanics, are considered in this book along with the main theme. It was the author's intention to present this material in an optimum combination with recent findings in the field of rheology and soil mechanics (including some theoretical aspects of non-linear deformation), which may also be of interest to the reader with an advanced knowledge of the subject. Accordingly, the book may be subdivided into three parts in ascending order of complication and novelty.

By referring to "rheological fundamentals" in the title of the book, the author wishes to stress the impact of rheology on soil mechanics in general.

In considering the basic concepts of classical soil mechanics, the author followed the guidelines in the work by Tsytovich (1976), a generally recognized textbook. The problems of soil physics are discussed in the light of the well-known treatise by E.M. Sergeev et al. (1961).

The three parts into which the book may be subdivided are as follows. The first part (Chapters 1–4) deals with basic rheological concepts and terms, the physics of soil, principles of stress–strain theory, elasticity, plasticity and viscosity — all cardinal rheological properties.

The second part (Chapters 5–10) explains the rheological processes taking place in soils, such as creep and long-term strength, which are examined by the author with allowance for non-linear deformation. Along with the known phenomenological theories, attention is paid to the novel kinetic (physical) theory of deformations and long-term strength.

The third part (Chapters 11 and 12) outlines the generalized theory of soil deformation. It explains why soil offers different resistances to tensional and compressional deformations and derives the generalized rheological equation of state, enabling the effect of the three stress tensor invariants on the changes in shape and volume to be taken into account.

Chapter 13 exemplifies solutions, from the standpoint of the theory discussed, of some problems facing soil mechanics.

Chapter 14, specially written for this particular edition, reviews mathematical models representing the actual behaviour of a soil under load and provides for engineering problems numerical solutions obtained with the aid of these models on a computer.

The author wishes to express his deep satisfaction with the fruitful cooperation in examining the problems outlined of Yu.K. Zaretsky, S.E. Gorodetsky, N.K. Pekarskaya, R.V. Maximyak and other colleagues from the Laboratory of Frozen Soil Mechanics, of which he is in charge, at the Foundation Research Institute. The author expresses his gratitude to Prof. G.A. Geniev, D.Sc. (Eng.), and Prof. R.S. Ziangirov, D.Sc. (Geol. Min.), for reviewing the book and also to Prof. Sergeev, Corresponding Member of the USSR Academy of Science, Head of the Department of Pedology and Engineering Geology, Moscow State University, who read the manuscript and made valuable comments. The author is also thankful to T.P. Vyalova, M.E. Slepak and V.P. Razbegin for their assistance in preparing the manuscript for printing.

S.S. VYALOV

CONTENTS

Preface v

Chapter 1. RHEOLOGICAL PROPERTIES OF SOILS 1
1.1 Peculiarities of soil deformation 1
1.2 Basic concepts, rheology defined 4
1.3 Research into the rheology of soils 9
1.4 Structural deformations encountered in engineering practice 15
1.5 References 23

Chapter 2. STRUCTURE OF SOIL AND STRUCTURAL BONDS 27
2.1 Composition and structure of soil 27
2.2 Soil components 29
2.3 Interaction between solid and liquid components of soil 32
2.4 Forces of interaction between solid particles in soil 35
2.5 Interparticle bonds in soil 40
2.6 Soil structure 43
2.7 Soil texture 44
2.8 Anisotropy in soil 50
2.9 References 53

Chapter 3. STRESSES AND STRAINS 57
3.1 Stresses and strains at a point 57
3.2 Changes in volume and shape 63
3.3 Tensors of stress, strain and strain rate 67
3.4 Stress tensor invariants 73
3.5 Strain tensor invariants 80
3.6 Basic equations of plasticity theory 84
3.7 Deformation theories 88
3.8 References 95

Chapter 4. ELASTICITY, PLASTICITY AND VISCOSITY 97
4.1 Elasticity and plasticity 97
4.2 Limiting state of stress 100
4.3 Non-linear deformation 107
4.4 Viscosity 116
4.5 Non-linear viscosity and Bingham flow 124
4.6 After-effect and relaxation 128
4.7 Viscoelastic–plastic properties of soil 134

4.8	Thermodynamic relations	137
4.9	References	144

Chapter 5. CREEP OF SOILS . . . 147
5.1	Mechanism of creep	147
5.2	Time-dependent stress–strain relation	151
5.3	Equations of creep	156
5.4	Equations of flow	160
5.5	Experimental data	169
5.6	Similarity of the patterns of soil deformation	181
5.7	References	189

Chapter 6. METHODS OF EXPERIMENTAL DATA HANDLING . . . 193
6.1	Testing soils for creep	193
6.2	Choice of empirical formula	195
6.3	Practical selection of formula	201
6.4	Comparison of the results of computations	209
6.5	Testing formulae for closeness of fit	212
6.6	References	219

Chapter 7. THEORIES OF CREEP . . . 221
7.1	The theory of linear viscoelastic deformation	221
7.2	Mechanical models of soil	229
7.3	The theory of hereditary creep	235
7.4	Kernels of the integral equations of hereditary creep	241
7.5	Engineering theories of creep	247
7.6	The molecular theory of flow	257
7.7	References	265

Chapter 8. THE THEORY OF SOIL CONSOLIDATION . . . 267
8.1	Volumetric creep	267
8.2	Percolation consolidation of soils	272
8.3	Primary and secondary consolidation of soil	277
8.4	References	283

Chapter 9. LONG-TERM STRENGTH OF SOILS . . . 285
9.1	Creep and long-term strength	285
9.2	Experimental data	289
9.3	Loss of shear strength by soil	299
9.4	"Peak" and residual soil strength	302
9.5	Criteria of long-term failure	306
9.6	Equations of long-term strength	311
9.7	Methodology of experimental data handling	315
9.8	Effect of loading conditions	328
9.9	Techniques of testing soil for long-term strength	332
9.10	References	337

Chapter 10. THE KINETIC THEORY OF SOIL STRENGTH AND SOIL CREEP 339
10.1 Soil deformation as a thermoactivated process 339
10.2 Changes in the micro-structure of soil due to deformation 342
10.3 Kinetic nature of long-term strength 352
10.4 Physical meaning of long-term strength parameters 359
10.5 Taking account of variable loading 363
10.6 Kinetic nature of creep in soils 369
10.7 Equation of deformation 373
10.8 Experimental data handling 378
10.9 References 381

Chapter 11. THE THEORY OF COHESIVE SOIL DEFORMATION 383
11.1 Particulars of soil deformation in combined stress state 383
11.2 Generalized rheological equation of state 388
11.3 Effect of mean normal stress 390
11.4 Equation of soil creep allowing for the effect of mean normal stress . . 399
11.5 Equation of viscoplastic flow of soil allowing for the effect of mean normal stress 409
11.6 Kinetic equation of soil deformation allowing for the effect of mean normal stress 411
11.7 Equation of long-term soil strength allowing for the effect of mean normal stress 417
11.8 References 420

Chapter 12. SOME PECULIARITIES OF SOIL DEFORMATION IN COMBINED STRESS STATE 421
12.1 Dilatancy 421
12.2 Effect of the arrangement of stress 430
12.3 Effect of loading conditions 436
12.4 References 439

Chapter 13. THE THEORY OF NON-LINEAR CREEP: PROBLEMS AND SOLUTIONS . . . 441
13.1 Generalized equation of soil deformation 441
13.2 Axi-symmetrical problem 443
13.3 Approximate methods of allowing for non-linear stress–strain relation in settlement computations 447
13.4 Effect of concentrated force on the soil below a foundation 451
13.5 Determining the settlement of the soil below a foundation and the reaction pressure of soil 459
13.6 Experimental data 470
13.7 References 478

Chapter 14. NON-LINEAR SOIL MODELS AND NUMERICAL SOLUTIONS OF PROBLEMS 481
14.1 Original relationships 481
14.2 Loading surface and limiting surface 486

14.3 Allowing for arrangement of stress, loading path and geometrical non-linearity 494
14.4 Equations of deformation based on the deformation theory of plasticity and the theory of plastic flow 503
14.5 Soil models survey 511
14.6 References 552

APPENDIX. 557

SUBJECT INDEX. 559

Chapter 1

RHEOLOGICAL PROPERTIES OF SOILS

1.1 PECULIARITIES OF SOIL DEFORMATION

Non-linear stress–strain relation

An all-around pressure applied to a soil irreversibly changes its volume, whereas a shear stress displaces mineral soil particles and aggregated lumps relative to each other, distorting their shape.

In the general case, a volumetric strain changes non-linearly with load, as does a shear strain (Fig. 1-1). The shear strain continuously increases with the stress, but the intensity of the volumetric strain decreases while the load increases and the strain itself approaches a certain limit at which final compaction of the soil occurs.

Real and approximate settlement curves

It is known that the real relation between a load and the settlement of a foundation is represented by a curve having two characteristic points. One indicates a critical load, p_c, under which the soil compaction phase comes to an end and the shear phase begins. The other point corresponds to an ultimate load, p_u, at which the zone of limiting state develops to its full extent and the bearing capacity of the foundation becomes completely exhausted.

The existing methods of foundation engineering are based on the assumption that a soil subjected to a load less than p_c (or, to be precise, less than R, which is the design load close to p_c) can be treated as a linearly deforming medium obeying Hooke's law. When the load reaches the value of p_u, the soil behaviour is described by the theory of limiting equilibrium and the soil can be treated in accordance with the Mohr-Coulomb law of the limiting state of stress.

Thus, the real "settlement versus load" curve illustrated in Fig. 1-2 can be approximated by dashed line Obc. This involves the following assumptions. In the subliminiting state, we consider segment Oa limited by the value p_c rather than the entire curve Oac. In fact, the soil is capable of effectively sustaining a load over the range $p_c \leqslant p \leqslant p_u$, commonly disregarded during analysis, providing its bearing capacity is not exhausted. Neglecting to take the above loading into consideration is neither reasonable nor logical. Although analysis in the subliminiting state of stress is aimed at limiting the settlement, S, which should not exceed a specified ultimate value, $S \leqslant S_u$, this does not imply that S_u is correlated with p_c. The ultimate settlement is dependent upon the allowable deformation of the structure, the load causing this settlement

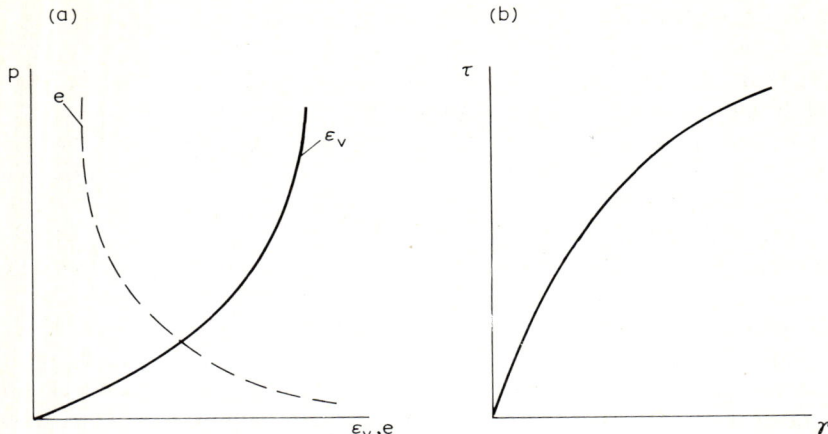

Fig. 1-1. Stress–strain relationship in soil: (a) volumetric strain; (b) shearing strain.

being either smaller or greater than p_c. Ironically, we calculate deformation operating with loads within the confines of the linear load–settlement relation. Taking the non-linear dependency of soil settlement upon load into account apparently creates the prospect of calculating settlements over the entire range of stresses.

We have approximated above the real settlement-versus-load curve by the dashed line of Fig. 1-2. This is assuming we differentiate soil behaviour. Segment Oa is represented by a model of a Hookean solid whose properties are expressed in terms of deformation. In this case, such a cardinal property as internal friction is ignored in spite of the unquestionable fact that frictional resistance is inherent in both limiting and sublimiting states of stress. This can be readily proved by a simple shear test in a conventional tester, using a load smaller than the ultimate one. The shear strain

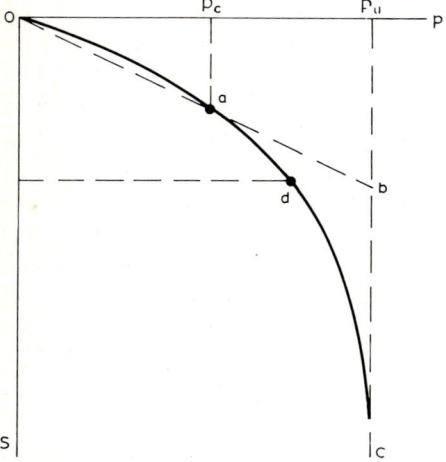

Fig. 1-2. Real and approximated settlement versus load curves.

resulting from the application of a changing normal stress will decrease with an increase in the stress. This is the manifestation of frictional resistance.

Segment bc, characterizing soil behaviour in the limiting state, is represented by a model of a Mohr-Coulomb body whose properties are expressed in terms of strength with due regard for the frictional resistance. The strength characteristics involved have no bearing on the deformation characteristics serving as the criterion of soil behaviour in the sublimiting state, although the strength and deformation properties of a soil are apparently correlated, being a manifestation of the same forces of soil particle interaction. Studying the sublimiting state involves ascertaining the fields of stress and strain, while investigating the limiting state requires estimating the stresses only. The problem in this latter case is thus reduced to evaluating the ultimate load, p_u, deformations being neglected. However, in many situations it is necessary to determine not only the ultimate load but also the resulting deformation, as is the case in assessing the behaviour of soils with explicit rheological properties; when exposed to an ultimate load they display a non-attenuating creep rather than the loss of stability or rapid failure.

One of the problems facing the rheology of soils is to develop a comprehensive soil model which is helpful in describing the pattern of deformation over the entire range of stresses up to the limiting value. Such a model should take account of the frictional resistance in the sublimiting state, providing thereby a correlation between the strength and deformation characteristics. All transient phenomena should also be taken into consideration while designing the model, since the process of soil deformation is accompanied by viscous drag of the particles surrounded by a film of water.

Difference between the strength of soil in tension and in compression

To take into account the frictional resistance of a soil is the equivalent of taking into consideration the difference between the strength the soil exhibits in tension and in compression.

In effect, the parameters of the equation of limiting state according to Mohr-Coulomb — cohesion and the angle of internal friction — are the parameters of a tangent to the circles representing the stresses at failure in compression and tension; the slope of tangent, which determines the angle of internal friction, indicates the difference existing between the strength of soil in compression and in tension.

It will be shown that the same holds for the sublimiting state. A soil compression under a load is smaller than the soil extension due to the same load acting in the opposite direction, and the modulus of compression will be, consequently, greater than the modulus of elongation. This point is often neglected, the modulus of compression alone being introduced into the calculations.

To give due consideration to the difference between soil compression and elongation, not only in the limiting state but in the sublimiting as well (or, which is the same, to take account of the internal frictional resistance), is equivalent to introducing into the equation of state an expression defining the shear strain as a function of all-around pressure and another expression giving the volumetric strain as a function

1.2 BASIC CONCEPTS, RHEOLOGY DEFINED

Rheological processes

The rheological properties of a body manifest themselves in the form of creep, relaxation, and deterioration of strength due to long-term load application.

The creep of soil is understood to be a process of deformation progressing in time even under a constant load. The ability to deform with time is inherent in many substances, from colloid systems and polymers to metals and from suspensions to rocks. Although all real bodies are in principle apt to creep, this property can be observed only if, in addition to a certain load and temperature, an adequate interval of time is available. The period required to perceive the flow of a liquid is a few seconds or minutes, whereas in the case of ice it amounts to several days or months. Soils and metals can flow under a high temperature, the associated displacement becoming perceptable within hours or months. The flow of glass can be perceived not earlier than in the course of centuries, and the rocks in the earth's crust flow at a speed measurable in terms of geological time.

Thus, it seems to be quite natural that the science which studies the flow of substance is called rheology, this term stressing the truth of Heraclitus' illustrious expression "$\pi\alpha\nu\tau\alpha\ \varrho\varepsilon\iota$" (everything flows).

A convincing proof of the flow of solids is provided by the disturbed outline of the walls of ancient Mexican temples repeating in our days the undulated contour of the terrain. Some time in the past the ground was even and the skyline of Aztecan temples straight, but the topographical changes taking place century after century have bent the walls without any failure.

Rheological phenomena in soils and rocks can be observed everywhere. Solifluction, mud flows, landslides, glacier movement, etc., are all rheological processes of a duration ranging between a few hours or days and a century or more. They cause various tectonic disturbances, such as the folding and bending of strata. Sometimes these disturbances result from a slow flow of the rock yielding to gravitational pressure in the course of an extremely long period measured by the millennium.

More frequently, however, rheological processes are triggered by the interaction of soil or rock with engineering structures and manifest themselves within periods of time shorter than the structure's life.

Many cases are known where the creep of clay soil below foundations has led to long-term and differential settlements of structures, tilting of retaining walls, instability of slopes, etc. Landslides, incurring considerable losses, are a frequent occurrence also caused by soil creep.

The phenomenon of creep and the associated stress redistribution is often observed in underground structures and mine workings drifted in rock. Rock pressure manifestation, subsidence of ground at tunnel construction sites, deterioration of roof strength and rock displacement in mines are all evidence of rock creep. Explicit forms of creep are observed in permafrost and in ice-soil retaining structures built by the artificial freezing technique in order to sink or drift workings in loose soils.

The above enumeration, although far from being a comprehensive one, provides ample evidence that none of the basic problems facing soil and rock mechanics can be tackled unless the rheological properties of soils are given due consideration. The requirement to be fulfilled in analysing the limiting state with allowance for these properties is as follows: the analysis in terms of deformation should be aimed at estimating the load which induces a deformation not exceeding an allowable limit during a given interval of time (say, the useful life of the structure).

Whenever a non-attenuating creep is present, impairing the resistance of soil, the analysis of the limiting state in terms of the bearing capacity can be reduced to ascertaining the load causing stresses in the soil which reach the long-term ultimate strength at a given moment.

Objective of rheology

In the narrow sense of the word, rheology is the science studying the flow of various substances. However, in recent years this term has acquired a broader meaning.

According to the classical theories of elasticity and plasticity, the state of stress and strain of a body is explicitly defined by the magnitude of the applied load and the manner of its application. If the load remains unchanged, the resulting stresses and strains remain also unchanged. In real bodies, the stress–strain behaviour changes with time and is dependent upon the history of the preceding load. As a result, the stress–strain relation is many-valued rather than single-valued, for even if one of the values — be it stress or strain — is constant the other one is changing with time. Investigations of the mechanism of the state of stress and strain and of the ways this state is changing with time is the objective of rheology.

In examining simple idealized bodies, the classical theory of elasticity and plasticity postulates that the patterns of deformation due to combined and simple stress are identical. However, as far as real bodies are concerned, the relation between stress, strain and rate of strain is non-linear and, moreover, dependent upon the arrangement of stresses and loading conditions. The study of these problems, which are outside the scope of elasticity and plasticity theory, is also rheology's concern. In other words, rheology must give the answer to the question what stresses and strains will come about at a given point depending on any kind of relation between their components and time.

Macro- and microrheology

The study of those processes in real bodies which are externally perceivable, i.e. can be observed with the aid of conventional measuring means (e.g. the increase in deformations and stresses), is the domain of rheology. It neglects structural particulars and treats any body as a homogeneous continuum possessing certain idealized properties. Chief considerations in this respect are elasticity, plasticity and viscosity, both individually and in their various combinations.

The manner in which the properties of a body influence its behaviour in response to external actions is studied in macrorheology by employing the so-called phenomenological approach. It neglects the physical processes taking place in the body and provides, on the basis of a macroexperiment, mathematical description of the way these processes manifest themselves externally. Relying on experiment, the macroscopic phenomenological theories yield solutions which, albeit with some reservations, are in good agreement with practice. In other words, these theories are an adequately effective tool for solving engineering problems.

However, a more general approach calls for establishing the mechanism of macroprocesses by considering their physical essence, i.e. by probing into the macroprocesses occurring in real bodies. This comes within the scope of microrheology investigating particulars of the framework of a body and the interrelation of the elemental structural particles.

Historical outline

Rheology is one of the oldest branches of knowledge; it dates back to ancient Egypt, where an hourglass, invented in 1540 B.C., was capable of measuring time by the trickle of water from a bulb with due allowance for the changes in viscosity during the warm and cool spells of a day.

The scientific fundamentals of rheology were laid in the 16th century due to Newton's law (stating that the resistance of a liquid to flow is proportional to the rate of shear), Poiseuille's law, and Stokes' equation.

Rheology has developed as a science by virtue of the rapid strides in engineering chemistry, which has given birth to a variety of man-made materials such as polymers, plastics and cellulose. They all must have controllable properties, plasticity and elasticity ranking first. The studies of these properties have materialized in numerous treatises on rheology which made their appearance from the end of the 19th century. By that time Kelvin (1875) and Voigt (1890) had formulated their equations of the deformed viscous solid and later Poynting and Thomson (1902), and Burgers (1935) developed mechanical models of that solid.

The final shaping of rheology into a self-contained branch of science is associated with E. Bingham. In his famous *Fluidity and Plasticity*, published in 1922, he suggested the idea of setting up a rheological society, which has, consequently, come into being — since 1928. The coining of the term "rheology" dates back to the same time. Numerous other works on rheology have appeared since then, a good deal of them being included in a three-volume treatise by Eirich (1956). Reiner (1949, 1958, 1960) is another author credited with writing a number of well-known books on rheology.

In the Soviet Union, much research into the rheology of colloid systems has been performed by M.P. Volarovitch (1949–1958) and, in particular, by P.A. Rebinder

(1937–1968) — the masterminds of physicochemical mechanics, a new branch of knowledge aimed at producing man-made materials possessing mechanical properties and a structure specified in advance.

Creep. The deformation of solids in time has been the object of studies beginning in the 19th century, simultaneously with the investigation of the flow of viscous media. First to observe the development of deformation in a solid were Vicate (1834) and Weber (1835). Experimenting with quartz glass filaments and silk threads, Weber noted that their elastic deformation, set up instantaneously on the application of a tensile load, was followed by further elongation progressing in time. The recovery of specimens from the deformation on removing the load was also a time-dependent process. The phenomenon of developing a delayed deformation has been termed elastic after-effect.

Proceeding with Weber's experiments, Kohlrausch (1847) has shown that the stress, too, may change with time if the set up deformation is maintained unchanged. This phenomenon has been called the after-effect of stress. The attenuation in time of the stress required to retain a permanent deformation has become known as the relaxation of stress. The theory of relaxation as developed by Maxwell was outlined in his well-known book *On the Dynamic Theory of Gases* (1868).

Earlier decades of this century have seen the inception of studies into the behaviour of metals at elevated temperatures. Experiments have proved that flow is an inherent property of metals as well. Thus, the phenomenon of a slow flow of solids has been termed creep.

Creep investigations seem to have been pioneered by Andrade (1910), who has described experiments revealing creep in copper, lead, and steel. The subsequent intensive probing into the creep of metals has been conducive to formulating the theory of creep — a branch of continuum mechanics.

The most comprehensive description of the state of affairs in modern rheology can be found in a fundamental monograph by Rabotnov (1966), as well as in the works of Goldenblat and Nikolaenko (1960), Kachanov (1960), Malinin (1968), Nadai (1950), Oding et al. (1959) and Rzhanitsin (1968).

Rheology and the theory of creep

Consider these terms from the standpoint of their similarity and difference. The theory of creep is a branch of the mechanics of solids treating the entire range of the phenomena resulting from changes in the state of stress and strain of a body as a function of time. Accordingly, it appears that the problems facing the theory of creep are identical with those dealt with in rheology. Yet the former term is used in the mechanics of solids, apparently due to its familiarity, and the latter one is more frequently employed in connection with viscous media.

In recent years the concept of rheology has acquired a broader meaning, being applicable to all bodies. We are going to use the term "rheology" exactly in this context. The term "creep" will be reserved for use in those cases when its direct meaning as the process of time-dependent deformation is implied, and this process will be considered as a manifestation of the rheological properties of the body. Their other manifestations are stress relaxation and the lessening of the resistance to failure with time.

Creep and flow

The term "flow", as applied to solids, is a loan-word from the theory of viscous liquid, where it denotes an unceasing and unconfined change in shape. In other words, flow is thought of as a shear deformation progressing in time at a constant rate. Typical in this respect is the flow of a perfectly viscous (Newtonian) liquid. Eventually the term flow has also been applied to the processes of slowly progressing deformation in solids.

Flow is also a term frequently used to describe the secular deformation of rock mass, glacier movements, etc. It is also popular in soil mechanics, being used there as a synonym of the term "creep". However, if we use creep to denote, as settled above, the time-dependent deformation of all kinds, then, strictly speaking, flow is a special case of creep, referring to that stage when deformation is developing at a constant rate.

The term "plastic flow" is also in use. In the theory of plasticity it denotes an unconfined development of plastic deformation when the load reaches a certain limit (yield point). In the theory of creep, plastic flow is understood to be a viscous flow induced by a load exceeding a certain limit (the so-called Bingham limit of plasticity); quite frequently it is referred to as viscoplastic flow.

The graph in Fig. 1-3 depicts the way deformation is developing in time. Straight line *1*, characterizing the development of deformation at a constant rate, represents a viscous flow; curve *3* corresponds to an attenuating creep at a decreasing rate, and curve *2* indicates a non-attenuating creep. Since the rate of deformation inside

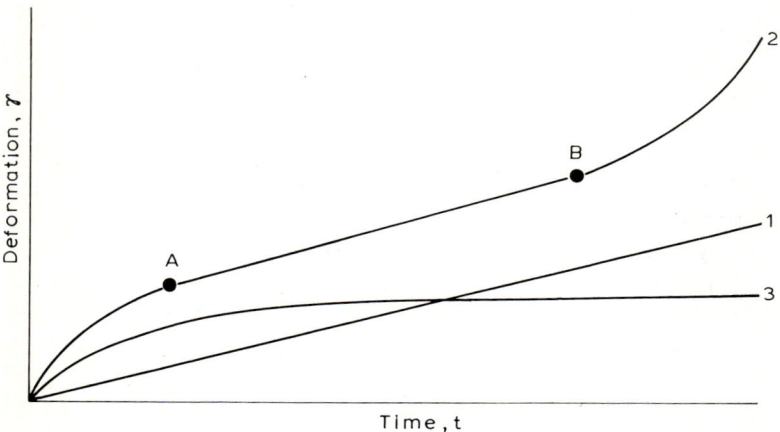

Fig. 1-3. Development of deformation in a body with time: *1* = viscous flow; *2* = non-attenuating creep; *3* = attenuating creep.

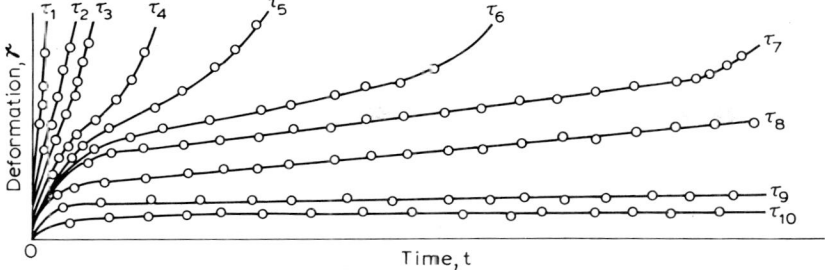

Fig. 1-4. Curves of creep in soil for various constant loadings.

segment AB of this latter curve is almost constant, this stage of the process can be regarded, by analogy with straight line 1, as one at which the flow is of a viscoplastic nature.

The configuration of creep curves depends on the magnitude of load.

Fig. 1-4 illustrates a family of creep curves. Each of them corresponds to a certain loading, an increase of which leads to more serious deformation developing at a faster rate; the process of creep, hitherto an attenuating one, turns into the non-attenuating, and the soil fails, the earlier the higher the load.

1.3 RESEARCH INTO THE RHEOLOGY OF SOILS

Studies on rheology of frozen soils

Containing ice along with non-freezing water, frozen soil can be regarded as a classical example of a rheological body. Therefore, most of the headway made in rheology is due to frozen-soil mechanics.

The presence of flow in frozen soils was discovered in the 1930s by N.A. Tsytovich and his fellow workers. The probing into long-term strength of frozen soils was pioneered by M.N. Goldstein between 1941 and 1948 on the basis of Maxwell-Schwedoff's theory of relaxation.

Studies which the author of this book undertook at the Igarka Permafrost Research Station between 1950 and 1955 had proved that frozen soils are apt to display creep in all its classical forms. Inherent in these soils are both attenuating (at low loads) and non-attenuating (at high loads) deformations, including a non-steady state, a viscoplastic flow, and a progressive creep.

These studies established the criteria of long-term failure and proved the necessity of introducing the concept of ultimate long-term strength into engineering practice. The term denotes a stress below which no failure is likely to occur, whereas higher stress leads to progressive creep culminating in failure. This strength, as far as frozen soils are concerned, has appeared to be between 1/5 and 1/15 of the resistance to rapid failure referred to as the hypothetically instantaneous strength. This fact necessitated a radical revision of the strength specifications of the building codes used in permafrost foundation engineering; they remain in force up to now, being only slightly modified.

The above investigations created the basis for the formulation of the postulates of the rheology of frozen soils (holding, in general, for non-frozen soils as well) which are outlined in a monograph, *Rheological Properties and Bearing Capacity of Frozen Soils*, by Vyalov (1959). These postulates have received further development in other works by the author and his colleagues (Vyalov et al., 1962, 1966).

A series of works on the rheology of frozen soils was published by K.F. Voitkovsky, S.E. Gorodetsky, S.E. Grechishchev, V.V. Dokuchaev, A.G. Zatsarnaya, Yu.K. Zaretsky, N.K. Pekarskaya, R.V. Maximyak, Yu.S. Mirenburg, D.R. Sheinman. E.P. Shusherina and many other authors between 1960 and 1970. Much attention was devoted to the problem by N.A. Tsytovich (1975).

During the same period, i.e. in the 1960s and, particularly, in the 1970s, a number of works on the rheology of frozen soils appeared in Canada and the U.S.A. (O. Andersland, A. Assur, L. Gold, C. Kaplar, B. Ladany, N. Morgenstern, J. Nixon, F. Sayles, E. Chamberlain, D. Anderson). They were reviewed, along with other things, in a state-of-the-art report presented by Anderson and Morgenstern (1973) at the 2nd International Conference on Permafrost, Yakutsk, and in collective monographs edited by Andersland and Anderson (1978) and by H. Johnston (1981). Problems of the rheology of artificially frozen soils, arising in drifting or sinking mine workings, were discussed in a publication by Jessberger (1979). The rheology of frozen soils received considerable attention at the International Conferences on Permafrostology (Lafayette, U.S.A., 1963; Yakutsk, U.S.S.R., 1978; Edmonton, Canada, 1978) as well as at the International Symposia on Ground Freezing (Bochum, B.R.D., 1977; Trondheim, Norway, 1980).

Studies on rheology of clay soils

Although the rheology of clay soils has sprouted into a self-contained branch of knowledge quite recently, the rheological properties of these soils have been given very careful consideration for a long time. Referring to the data on long-term tests of clay, Terzaghi (1925, 1943) has pointed out that soils explicity exhibit the properties of elastic after-effect (denoted as creep).

The importance of taking into account the property of flow in soils was stressed by Puzyrevsky (1934). Gersevanov (1937) wrote that soil is likely to develop Bingham's flow which is a state in which a body, subjected to a certain stress, begins to unceasingly change its shape, passing into a viscous condition like a viscous fluid.

Studies of the rheological properties of soils require two lines of enquiry. On the one hand, the creeping flow in the form of long-term landslides due to shear is being examined to solve the problem of long-term stability of dams, artificial slopes, retaining walls, etc. These studies are particularly significant in connection with the sliding of natural slopes, which is a problem of paramount importance. On the other hand, the phenomenon of creep is investigated in the context of the volumetric deformations in soil developing with time due to the creep of its skeleton. This phenomenon is behind long-term settlements of the clay soil below foundations.

Studies of soils as viscoplastic bodies

The rheological properties of clay soils were considered in the course of their studies from ever-changing points of view.

At an early stage attempts were made to apply the Newtonian mechanism of ideally viscous flow to soils. For example, Hvorslev (1937, 1939) treated soils on these lines. In his fundamental treatises on the creep of soil, which were among the first to appear on the subject, he described long-term creep tests of soils in the ring shear apparatus — Hvorslev's invention in wide-spread use nowadays.

In the 1940s, Haefeli and Schaerer (1946) also employed the Newtonian mechanism of viscous flow to analyse the data acquired from creep tests of soils in ring shear, uni- and triaxial compression. However, they pointed out that in the general case the flow rate–stress relation is non-linear in soils. Approximately at the same time, Casagrande and Wilson (1950) succeeded in establishing the fact of clay failure due to long-term creep from uniaxial compression.

The conclusion drawn from almost all the early rheological studies was that, unlike an ideal viscous liquid flowing in response to any load, soils set out to flow only when stress exceeds a certain limit. This implies that the Bingham theory of viscoplastic flow suits soils better than the theory of Newtonian viscous flow.

First to formulate this, as early as 1937, was Gersevanov (1937), a relevant passage from his book being referred to above. First to prove experimentally the applicability of the Bingham law to soils was Maslov (1936), credited with carrying out one of the pioneer tests of clay for creep in shear in order to predict the long-term stability of the water-retaining structures of the Swir hydro-power station in 1933–1936. In 1952 he suggested classifying soils into plastic and latently plastic types, the former obeying Newton's law of viscous flow and the latter being amenable to Bingham's law of viscoplastic flow.

The applicability of Bingham's law to soils was also demonstrated by Geuze and Tan (1954) in their papers on the torsional tests of hollow cylinders presented at the 2nd International Congress on Rheology as well as in a number of later publications.

Experimental proofs of the applicability of the law governing the viscoplastic Bingham flow to soils were furnished in the works of Sotnikov (1960), Kisiel (1967), Sorokina (1965), and others.

A modification of Bingham's law which takes into account the non-linear relation between the rate of flow and stress was suggested by Vyalov (1959).

A point to be noted is that unlike the theory of viscoplastic flow describing a steady-state process developing at a constant rate, the process of creep in soils is characterized by a variable rate of flow (see Fig. 1-3); moreover, this latter process is associated not only with viscous deformations but also with elastic ones.

This has given a reason for treating soils on the basis of the theory of elasto-viscoplastic deformation, employing various rheological models. Among the first works in this field were the experiments by Geuze and Tan (1954), revealing plastic and viscous deformations in soils. The authors described the process of deformation, using a mechanical model due to Burger. The patterns of soil deformation have eventually been described more exhaustively by other authors, using other models.

The notion that the pattern of viscous deformation is applicable to soils owes its origin to the theory of viscous media. This can be easily understood, for flow manifests itself in the most conspicuous way in wet, plastic soils behaving like viscous media. Moreover, the engineering geologist and pedologist are more familiar with the concepts of physics and chemistry of dispersed media, on which the theory of viscosity is based, than with anything else.

Further studies have revealed that soil sustaining a long-term load behaves on lines resembling the creep of solids, for, in the general case, all phases of this latter process — attenuating creep, flow at a roughly constant speed, and progressive flow — were observed in soils.

These data were acquired between 1950 and 1970 by A. Casagrande and S. Wilson, S.S. Vyalov, M.N. Goldstein, A.M. Skibitsky, M. Saito and Uezawa, S. Murayama, T. Shibata and D. Karube, A. Bishop and G. Lovenbery, J. Feda and B. Kamenov, W. Liam Finn, J. Biarez, and Tjong-Kie Tan and others, experimenting with various soils, from plastic to consolidated clays, frozen ground and even rock.

Much research had been done by Meschyan (1967, 1978), who generalized his findings in two monographs. The problems of soil rheology were also discussed in the works of Folque (1961), Kisiel and Lysik (1966), and Push (1979).

Goldstein's three-volume treatise (1979) was written with due regard to rheological concepts and the aspects of rheology were not omitted from the well-known textbook on soil mechanics by Tsytovich (1976).

Studies of long-term stability of artificial and natural slopes simultaneously undertaken with investigations into creep were also a significant contribution to soil rheology. Noteworthy in this respect are the works of Terzaghi (1950), Skempton (1964), Skempton and Hutchinson (1969), and in particular, the monographs by Maslov (1968a,b), as well as the books by Ter-Stepanyan (1961, 1975) and Šuklje (1969).

Milestones in soil rheology

The shaping of rheology of soils as a new branch of knowledge was started at the 3rd International Conference on Soil Mechanics and Foundation Engineering, Zurich, 1953, by presenting and discussing several papers. R. Haefeli, reporting on the creep of soils, snow and ice, brought forward the point that further progress in soil mechanics would be inseparable from the success scored in investigating creep, because the deformation of creep was active, explicitly or implicitly, in all the processes with which soil mechanics was concerned.

The flow of papers on the rheology of soils steadily increased during the 4th to 10th Conferences convened between 1957 and 1981. The state-of-the-art reports by Scott and Hon-Yim-Ko (1969) and De Mello (1969) at the 7th Mexico City Conference in 1969 dealt, for the best of their parts, with this subject.

An IUTAM Symposium on Rheology and Soil Mechanics was assembled in Grenoble in 1964, where papers on various aspects of rheology were considered (IUTAM, 1966). The Soviet Union was the host country for symposia on the rheology of soils in 1966 and 1979 (Leningrad) and in 1973 and 1976 (Erevan) as well as for symposia on the rheology of rocks (Dnepropetrovsk, 1966, 1969).

Investigations into the process of soil consolidation

Irrecoverable volumetric changes (consolidation) developing with time in response to load are a salient feature of soil systems. One may say that the theory of consolidation (or the theory of soil mass) developed by K. Terzaghi and N.M. Gersevanov in the 1920s and 1930s furnished the basis for advancing soil mechanics (its fundamentals were laid by Coulomb in the 1680s) to the ranks of an independent science.

At an early stage of study, it was thought that consolidation resulted from the forcing of water out of the pore spaces in the ground by pressure and that the migration of water obeys Darcy's law of laminar percolation. Subsequent examinations of the effect of some additional factors, such as changes in the coefficient of permeability and compressibility of the air dissolved in the water, contributed to the development of the percolation theory of consolidation. Significant progress was made in the late 1930s and early 1940s by virtue of the works of Biot (1941) and Florin (1959) suggesting that the effect of pore water pressure on soil skeleton, manifesting itself in the form of body forces, should be taken into consideration.

Primary and secondary consolidation

In the middle of the 1930s, Gray (1936) and Buisman (1936) found that changes in the volume of soil resulted not only from expelling water but also from the creep of soil skeleton. Buisman's experiments have shown that the process of volumetric creep develops over a long period of time and can be described by a logarithmic law. It is not irrelevant to point out that in 1934, i.e. somewhat earlier, a similar law was deduced by Pokrovsky proceeding from energy considerations (Pokrovsky and Nekrasov, 1934).

Consolidation brought about by the percolation of pore water has been termed primary consolidation and that resulting from the creep of soil skeleton has been called secondary consolidation (Fig. 1-5).

In the early 1950s, Florin (1959) developed a generalized theory of consolidation by considering that the primary and secondary consolidations occurred simultaneously, the latter in the form of hereditary creep.

Other theories describing the simultaneous process of primary and secondary consolidations have proceeded from the concept of secondary consolidation as an elastoplastic volumetric deformation represented by a certain rheological model (Taylor and Merchant, Tjong-Kie Tan, R. Gibson, K. Lo and others).

The theory of consolidation in its most general form, which takes into account the percolation of gas-laden fluid and the presence of structural bonds influencing the

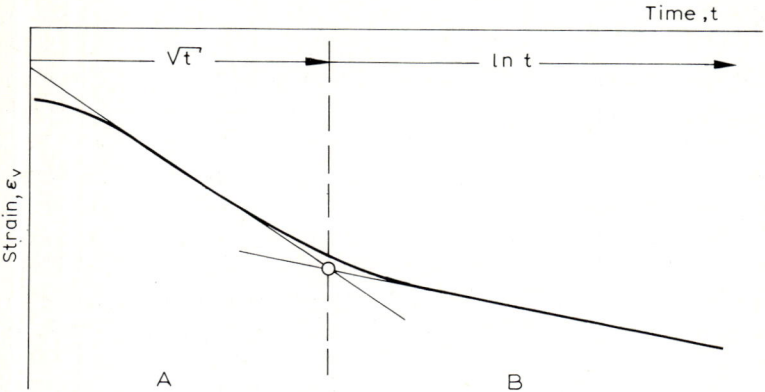

Fig. 1-5. Consolidation of soil, after B. Hansen. A. Primary; B. Secondary.

initial pore pressure and the way this pressure dissipates, has been developed by Tsytovich et al. (1967). Applying the above provisions to the Biot-Florin model of body forces, Zaretsky (1967) suggested a general solution of the three-dimensional problem and examined the problem of consolidation as a whole.

Šuklje's monograph referred to above, appearing in Russian in 1973, was a comprehensive disclosure of the effects of rheological properties of soil on consolidation and long-term strength of slopes, both natural and artificial, and retaining walls. It also described a semigraphical method of assessing consolidation referred to as the method of isotachs. (According to the author, isotachs are curves representing the relation between void ratio and effective pressure for a constant rate of consolidation.)

Studies of rheological properties of rocks

The study of creep in hard rock was initiated in the 1930s at the time of the advance in the field of the rheology of loose rocks. Stocke (1937) and Griggs (1939) apparently pioneered the research, discovering the deformation of creep in sandstones, argillites and siltstones when these were subjected to a load between 12.5 and 80% of the load at failure. Creep behaviour of hard rocks has been corroborated by other authors at later dates.

Significant experimental and theoretical studies were undertaken in Alma-Ata by Erzhanov (1964) and his fellow workers (Erzhanov et al., 1970). The authors went into the phenomenon of creep in rock, and the findings have enabled them to solve certain problems in mining engineering (rock movement, estimating non-steady state rock pressure in mine openings, etc.) as well as to describe such geophysical phenomena as the folding of rock mass. Another centre of research into the rheology of rock is the Institute of Geotechnical Mechanics, Dnepropetrovsk, where V.T. Glushko (1970), M.I. Rosovsky and others are at work.

Rheology and geophysics

A rheological approach to geophysical phenomena is practiced, by no means to a lesser extent, in dealing with engineering problems. This is exemplified by some of the problems discussed by Nadai (1950), such as the sinking and postglacial rising of the earth's surface, the mechanism of Weber's continental drift, salt dome intrusions, plastic deformation of rock strata, etc

Rocks subjected to normal compressive loads deform little. However, if the pressure imposed by the overlaying strata and the temperature are high, even a slight difference between the vertical and horizontal components of pressure may create conditions under which the resulting impulse of a duration measurable in terms of geological time will cause a hard rock to flow.

In particular, this may provide an explanation of the fact that old rock overlies strata of more recent origin (e.g., the salt domes occurring at the Gulf of Mexico, in the north German plain, below the Alps, around the Carpathians and elsewhere). The salt, originally deposited in horizontal strata as the retreating sea evaporated, was eventually buried under loose sediments turning into hard rock due to compaction with time. Discontinuities and weak spots in the rock provided sufficient outlets through which the slow salt flow was squeezed to the surface.

Rheological processes are behind such phenomena as, for example, the postglacial thrust fault along the Canadian side of the Great Lakes and Alpine deposits (their origins were the overthrust faults of older strata flowing bodily along folds in the earth's crust due to gravity). It is believed that the continental masses of the earth's crust themselves are in a state of viscous flow.

The vertical movements of the earth's surface during the glacial and postglacial periods are thought of as the process of bending of a viscoplastic plate floating on a semifluid, yet more dense, base. The well-known Wegener hypothesis stating that Africa and South America, as well as Europe and North America, a single continent some time in the past, were torn apart by a cataclysmic crack triggered by the globe's rotation is based on an assumption that the continents have separated since then due to their drift over viscous abyssal rocks.

Similar points were raised in the paper on the problems of geomechanics by Tsytovich (1973b).

1.4 STRUCTURAL DEFORMATIONS ENCOUNTERED IN ENGINEERING PRACTICE

Deformations of dams and bridges

Numerous cases are on record when the consequences of rheological process were disastrous. Reporting at the 2nd ICSMFE (Rotterdam, 1948), Peck disclosed that in the U.S.A. long-term deformations of slopes were the cause of complete failures of 18% of the retaining walls built to hold them back. In 53 cases the walls were in the

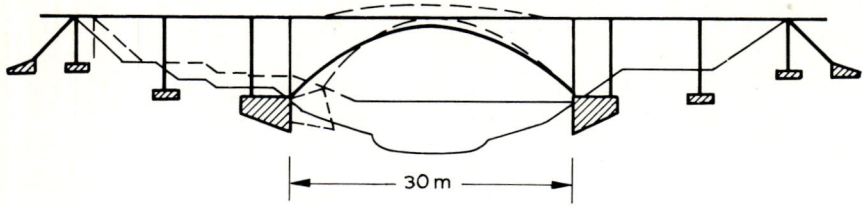

Fig. 1-6. Displacement of a viaduct abutment in Switzerland due to creep of soil (dashed lines).

state of progressive displacement, 4% of the walls inspected showed insignificant displacement and only 11% were in the state of stabilized displacement.

Noteworthy were the failures of two dams in France. One of them, a 520 m × 22 m structure resting on a 6-m bed of argillaceous sandstone collapsed 11 years after its commissioning in 1895, being incapable of resisting a progressive displacement induced by soil creep. The other, measuring 550 m × 28.3 m and built in 1938, was standing on compact fissured clay at a depth between 2 and 10 m. The dam suffered appreciable deformation from the same origin (Maslov, 1968b).

Numerous similar failures of engineering structures caused by soil creep were reported at the 3rd ICSMFE. Peterson (1953) told about a number of clay embankment failures due to the creep of their slopes during a period between 6 months and 4 years after placing. Geuze disclosed a long-term displacement of a bridge pier resting on a 1.5-m bed of clay and peat (Geuze et al., 1954).

A paper by R. Haefeli, Ch. Schaerer and G. Amberg dealt with the deformation of an r.c. arched viaduct in Switzerland resulting from the creep of the slope supporting an abutment (Fig. 1-6). The abutment had given in immediately on completion of the bridge and went on shifting at a steady rate of roughly 37 mm per year. In thirteen years the displacement amounted to 480 mm, whereas the settlement was only 14.9 mm although the layer of soil affected by the creep was 12 m deep. The bridge has been reconstructed (Haefeli, 1953).

A point to be noted is that bridge abutment deformation due to creep is a frequent occurrence. Luga (1964) described a case when the creep of stiff and hard clay soils had moved three abutments of a railway bridge through horizontal distances ranging from 9.3 to 46 cm in 38 years (1916–1954).

Land abutments of bridges over the Danube in Budapest, as well as those of the Reichsbrücke, etc., have been known to suffer from long-term displacements.

Structural deformations in London Clay

A paper presented by Henkel (1957) at the 4th ICSMFE dwelled on failures of retaining walls due to the creep of slopes. Fissured clays of marine origin heavily compacted by an 150- to 200-m overlying deposit were completely eroded eventually. The clays had a liquid limit W_L = 70–90%, a plastic limit W_P = 24–32% and a natural moisture content slightly exceeding W_P.

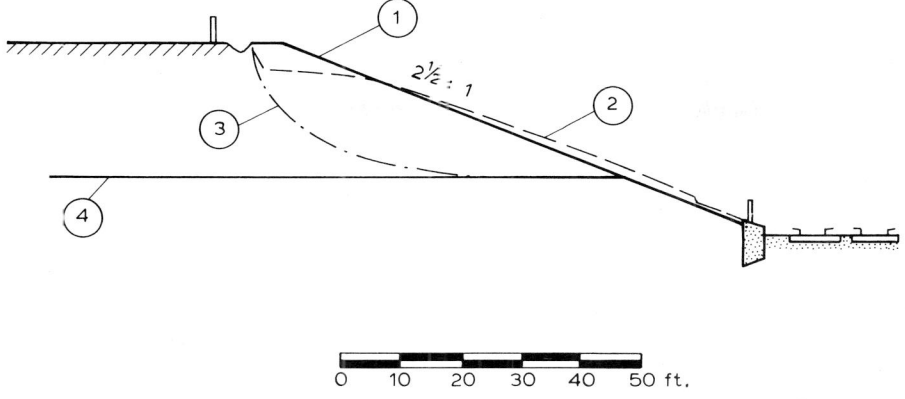

Fig. 1-7. Landslide at Northolt Station: *1* = ground surface before slide; *2* = ground surface after slide; *3* = slip surface; *4* = lower boundary of London brown clay stratum.

One of the slope failures in a 10-m excavation occurred at Northolt Station. The excavation, made in 1936, developed cracks at the upper end of the slope by 1955 and collapsed during the same year (Fig. 1-7). The moisture content appeared to be 44% in the zone of the slide and 30% above and below the zone.

Another London Clay slide occurred at Wood Green Station, where a retaining wall 4 m high was erected in 1893 to hold back the soil at a cutting. By 1948, i.e., in 55 years, the wall displaced through 90 cm and cracked.

Skempton described other slope failures in London Clay. A retaining wall erected in Kensal Green in 1912 collapsed in 1941 after a slope slide. The annual rate of slide increased from 6 mm in 1929, when the first observations were made, to 457 mm short of the failure. The batter of a wall erected at Sudbury Hill collapsed in 1949 after 49 years of service (Skempton and Hutchinson, 1969).

Other cases of slope failures

Also at the 4th ICSMFE Peynircioglu (1957) gave details of a displacement involving a 6.3-m retaining wall erected in Devonian clays to hold back a steep slope in the Bosporus region. The slope, cut off by the wall, started to slide 2 years after its construction and went on displacing at a daily rate of 1 mm, until the wall suffered the damage shown in Fig. 1-8.

An intricate case of deformation suffered by the intake structure of the Dzora hydropower project, Armenia, due to a slow downgrade movement of a soil mass was described by Lomize (1945). The structure was erected next to the steep slope of a bank, some 60 m high, deposited from crushed andesite and dacite. Although the deposition rested on a tuffaceous bedrock, a layer of clay soil of eluvial origin occurred between the bed and rock. The bed also had a gradient of 8 or 9 degrees. Soon after being placed, the bed started to slide at a rate of 2 cm/year, damaging the

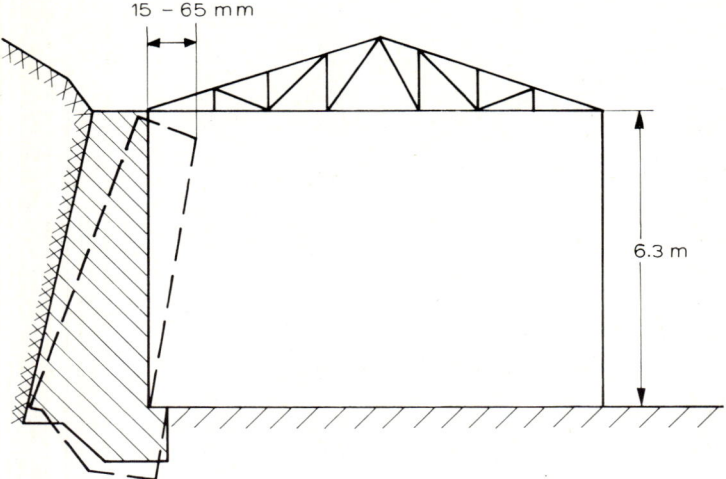

Fig. 1-8. Displacement of a retaining wall due to soil creep in the Bosporus region.

intake structure. The landslide was provoked by the deformation of creep of the clay layer.

Another case of long-term deformation resulting from soil creep, which was described by Mogilevskaya (1968), occurred at the Farkhad hydropower station. A high frontal wall of the forebay erected on clay loam and sandy loam, as were the rest of the structures, displaced through 140 mm between 1947 and 1966 and eventually cracked. Reconditioning was performed by a special technique.

Deformations of sheet pile walls (bulkheads)

A few cases illustrating the behaviour of sheet pile walls erected in soils subjected to creep were examined by Budin (1974). His field tests have proved the vulnerability of these structures to creep deformation which, if neglected, may lead to damage. The creep of soil causes a relaxation of the reaction stresses at the bottom of the pile walls with the result that the bending moment applied to the wall increases.

A.Yu. Budin also carried out, in Leningrad, a comparative study of the behaviour of a cantilever sheet pile retaining wall, erected in soft ribbon clay with a moisture content of 36%, and that of a tied wall (Fig. 1-9), set up in stiff boulder clay loam with a moisture content of 10.2%. In the former case, creep caused the top of the piling to displace through 95 mm in 4.5 years at approximately the same rate (Fig. 1-10). In the latter case, the displacement was less pronounced, varying between 7 and 21.5 mm, but the r.c. piles suffered ever-increasing bending. The stress in the land ties of the wall also increased with time. In another similar case, the creep-induced bending of the piles resulted in failure of the bulkhead.

A method of creep control by lessening the loading was described by Andreev. A quay built in 1960 in Leningrad clays overlying boulder clay loam consisted of an r.c.

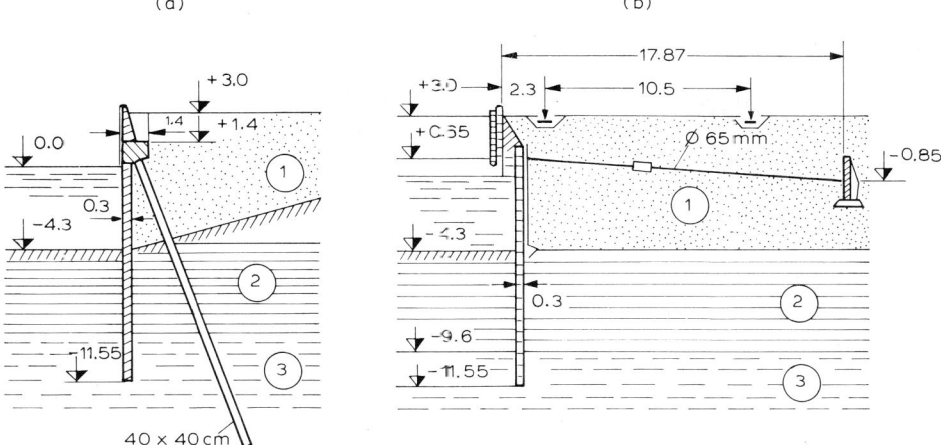

Fig. 1-9. Sheet pile retaining walls in clay soil (Leningrad). (a) Cantilever type. (b) Tied wall. *1* = sand fill; *2* = clay; *3* = clay loam.

pier on a raft foundation with metal sheet piles of the cantilever type. A quay load of $1.66 \cdot 10^5$ Pa provoked a disastrous deformation involving a 33-cm sidewise displacement and a 16-cm settlement of the wall in 10 days of service. A reduction of the load to $1.0 \cdot 10^5$ Pa failed to stop the deformation, although slowing down its rate. Only after a further lessening of the load to $0.6 \cdot 10^5$ Pa did the deformation attenuate, stabilizing in the course of the next 3 years. The total horizontal displacement, as recorded between 1960 and 1964, was 38 cm and the maximum settlement amounted to 17 cm (Andreev et al., 1968).

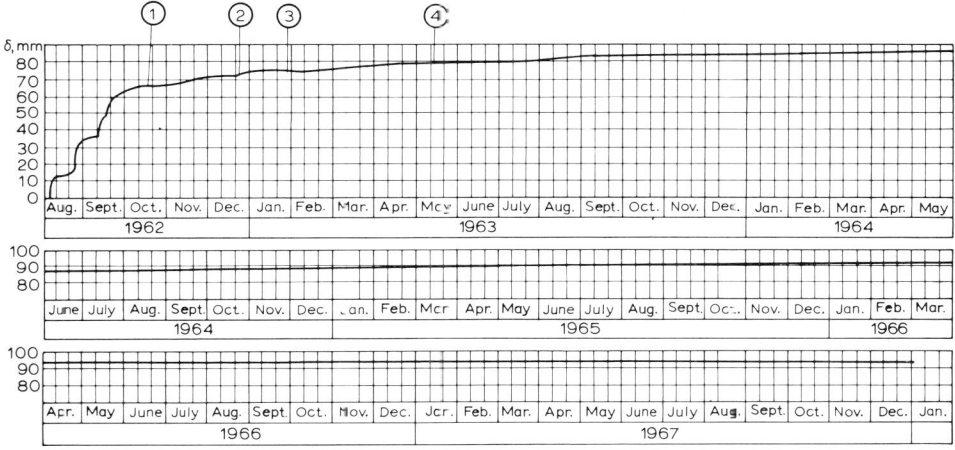

Fig. 1-10. Displacements of cantilever sheet pile wall at the top. *1*: time of completing project; quay loadings: *2*: $p = 0.16 \cdot 10^5$ Pa; *3*: $p = 0.42 \cdot 10^5$ Pa; *4*: $p = 0.5 \cdot 10^5$ Pa.

The case disclosed is of particular value in that it proves that creep may vary in its nature, being either non-attenuating or attenuating, depending on the loading.

Long-term soil settlement under a load

The examples considered above all refer to cases when the deformation of creep is induced by horizontal shearing forces. However, there are cases known where a disastrous deformation resulted from vertical loads. We have in mind long-term structural deformation attributed to the squeezing out of a layer of clay soil overlying an incompressible (rocky) bed. The layer behaves like a plastic-viscous strip drawn apart to the sides by a punch thrust into it.

A settlement on these lines was disclosed at the 5th ICSMFE, Paris, 1961. A 6-m layer of silt clay appeared to underlie an embankment supported by a bed rock (see Fig. 1-11). Sidewise transport of the clay below the embankment caused the settlement, which resulted in the embankment sinking 120 cm in 4 years. Laboratory tests of the silt clay from the site had shown that its creep began under a shearing load equal to 60% of the standard strength determined by a quick test. At lower stresses, the creep bore an attenuating nature, and this value was adopted as the allowable load in reconditioning the embankment.

Creep and sidewise transport of the clay in the form of lenticles occurring in the depths of the sand soil below the famous Leaning Tower of Pisa seems to be a plausible explanation of the structure's settlement offered by many scientists, although secondary consolidation also seems to be involved in this case. It will be recalled that the tower settled by 50 cm in the course of its erection between 1174 and 1350 and has carried on subsiding at a rate of 2 mm per annum throughout the 600 years elapsed since then. At present, the settlement is 1.5 m and the displacement of the top of the tower from the vertical is 1.5 m (resulting from non-uniform settlement).

Facts illustrating long-term settlement of clayey silts supporting the plain earth dams of the Kakhovka hydro-power station on the Dnieper are given by Karpyshev et al. (1972). The moisture content of the silts is 77% and the liquid limit is 68%. However, their behaviour is one of a soft plastic body, not a fluid, apparently due to structural cohesion. The compressibility of the silts under a load of $3.5 \cdot 10^5$ Pa varies

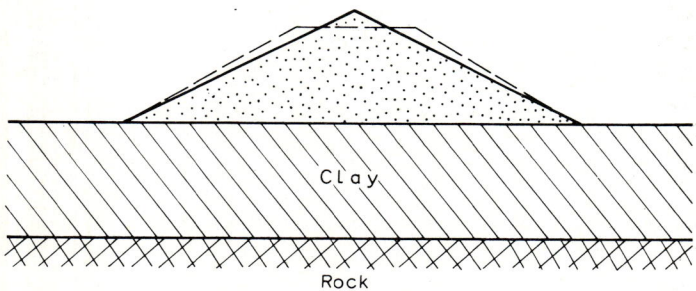

Fig. 1-11. Long-term settlement of embankment resulting from sidewise transportation of clay soil.

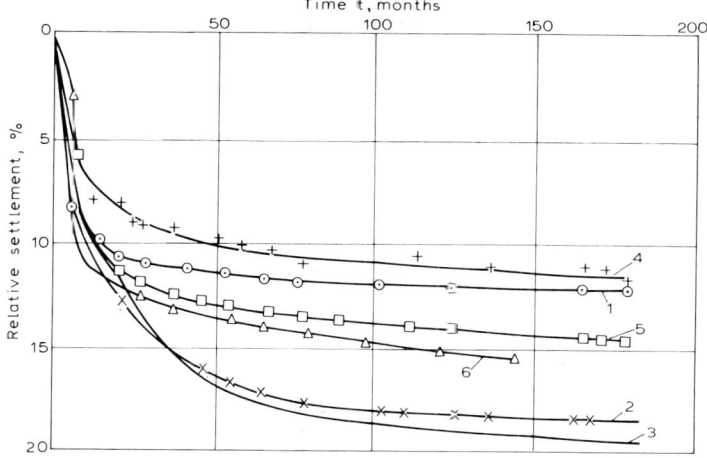

Fig. 1-12. Settlement of the silts below the Kakhovka Dam.

	Right-hand bank			Left-hard bank		
Curves:	1	2	3	4	5	6
Loading, 10^5 Pa	1.62	2.72	3.10	1.95	2.62	3.10
Depth of layer, mm	3.5	4.0	4.8	1.0	2.4	2.95

between 150 and 200 mm/m. Occurring in a bed at a depth between 1.0 and 4.0 m, the silts are overlain by an 8-m stratum of lean fine-grained sands. The settlement of the silts is 970 mm, whereas the aggregate settlement of the soil below the dam, including the deposits of sand and gravel, amounts to 1120 mm. The settlement is a long-term process, being still in progress here and there even after 15 years (Fig. 1-12).

Deformation of mine openings

One of the manifestations of creep properties of rock is the distortion of mine openings along the contour occurring in the course of long periods due to rock pressure. Fig. 1-13 depicts in-situ observations of contour distortions of a Karaganda coal mine opening, with a clear cross-section of 14 m^2, made by Erzhanov et al. (1970) during a period of 264 days. The recorded displacement, at two points at the walls of the heading drifted in sandstone at a depth of 527 m below the surface, amounted to 5.5 mm and 8.2 mm. The roof displacement was 25 mm over the same period.

Theoretically derived soil models and engineering practice

The generalized rheological model of soil mentioned above, which takes into account the specific features of a soil system, substantially narrows the gap between

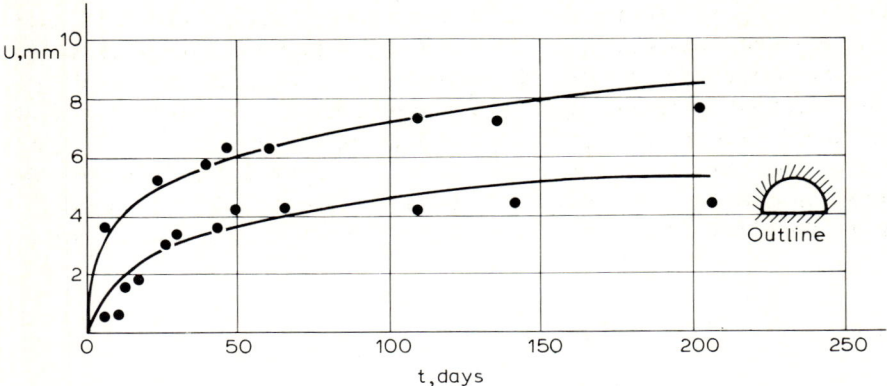

Fig. 1-13. Displacement (*u*) of rock around the contour of a mine heading, Karaganda.

analytical solutions and the behaviour of soil in practice. However, the use of a more sophisticated model than available at present seems to be doubtful, for this will complicate calculations and invite difficulties in acquiring the parameters required to obtain a solution (the effort involved may be unjustified owing to the spread in the values of soil characteristics).

Electronic computers, various alignment charts, tabulations, etc., minimize the difficulties encountered in calculations. The fact that numerous factors are to be incorporated into the constitutive equations does not imply that all of them are required for solving an engineering problem. The factors are considered for the sole purpose of assessing their respective effects. If this effect is negligible, the factor in question can be disregarded. However, prudence is required in this case, for neglecting to take a factor into consideration without good reason, except for lack of knowledge, will cause disagreement between the computed and actual data. The generalized equation of rheological state is called upon to bring these data more closely to each other and thus provide a more comprehensive allowance for the real properties of soil. This may be simplified after evaluating the effect of each parameter entering the equation.

The role of the experiment is highly important in this connection, particularly the one conducted in the field, and the in-situ observations of the behaviour of the structure. These investigations alone can provide a final criterion by which we can judge whether or not the model adopted is trustworthy and the solution correct. Likewise, to cope with the ever-increasing number of soil characteristics to be taken into consideration, new, more reliable, methods of assessing them should be developed, particularly for use in the field. Finally, no accurate assessment of the behaviour of a soil either supporting, surrounding, or used as the material of, a structure can be assured unless the particulars of its geological formation are taken into consideration. This is obvious because the conditions of origin of the soil which is the object of our studies, have a bearing on its properties.

In conclusion, a few words about engineering intuition in analytical computations. In dealing with soil, an intricate and heterogeneous medium when one fully realizes the vagueness of the data providing the basis for engineering calculations, there is little confidence that the results will be well in agreement with the real behaviour of the soil. In this context, an engineering intuition based on practical experience becomes a factor whose importance can hardly be overestimated. However, the assessment made will be reasonable and valuable only if based on the knowledge of the real pattern of soil deformation and failure, rather than on sheer empiricism.

1.5 REFERENCES

Andersland, O.B. and Anderson, D.M. (Editors), 1978. Geotechnical Engineering for Cold Regions. McGraw-Hill, New York, N.Y.

Anderson, D.M. and Morgenstern, N.R., 1973. Physics, chemistry and mechanics of frozen ground. A review. Permafrost. 2nd Int. Conf. North Am. Contrib., Washington, D.C.

Andrade, E.N., 1910. On the viscous flow of metals and allied phenomena. Proc. R. Soc. Lond., 84(567).

Andreev, G.A., Zlatoverkhnikov, N.F and Karpov, V.M., 1968. On the problem of evaluating long-term deformation of the soil below a quay. Proc. Coordinating Conf. on Hydraulic Engineering, Issue 38, Studies on Rheological Properties of Soils. Energiya, Leningrad (in Russian).

Biot, M.A., 1941. General theory of three-dimensional consolidation. J. Appl. Phys., 12.

Budin, A.Ya., 1974. Thin Retaining Walls. Stroiizdat, Leningrad (in Russian).

Buisman, K.A.S., 1936. Results of long duration settlement tests. Proc. Int. Conf. Soil Mech. Found. Eng., 1st, Cambridge, Vol. 2.

Burgers, I.M., 1935. First Report on Viscosity and Plasticity. Nordemann, Amsterdam.

Casagrande, A. and Wilson, S.D., 1950. Effect of Rate of Loading on the Strength of Clays and Shales of Constant Water Content. Harvard Soil Mech. Series, No. 39, Harv. Univ., Cambridge, Mass.

Eirich, F.R. (Editor), 1956. Rheology Theory and Application. Acad. Press, New York, N.Y.

Erzhanov, Zh.S., 1964. The Theory of Creep of Rocks and its Application. Nauka, Alma-Ata (in Russian).

Erzhanov, Zh.S., Samkov, A.S., Gumeniuk, T.N., Vexler, Yu.A. and Nesterov, G.A., 1970. Creep of Sedimentary Rocks. Nauka, Alma-Ata (in Russian).

Florin, V.A., 1959. Fundamentals of Soil Mechanics, vol. 1, 1959; vol. 2, 1961. Gosstroiizdat, Moscow (in Russian).

Folque, J.B., 1961. Reologia de Solos Nao Saturados. Laboratorio Nac. de Eng. Civil., Lisboa, Mem. No. 176.

Gersevanov, N.M., 1937. Fundamentals of Dynamics of Soil Mass, V. 2. Stroivoenmorizdat, Moscow (in Russian).

Geuze, E.C.W.A., 1953. Discussion Session 2, Proc. 3rd ICSMFE, v. III, Zürich.

Geuze, E.C.W.A. and Tan, Tjong-Kie, 1954. The mechanical behaviour of clays. Proc. 2nd Int. Congress of Rheology, Oxford, 1953. London.

Glushko, V.T., Usachenko, I.I. and Vaganov, I.I., 1970. Study of rheological properties of rocks in Donbass. In: Problems of Rheology of Soils. Naukova Dumka Publ., Kiev (in Russian).

Goldenblat, I.I. and Nikolaenko, H.A., 1960. The Theory of Creep of Building Materials and its Application. Gosstroiizdat, Moscow (in Russian).

Goldstein, M.N., 1979. Engineering Properties of Soil. Stroiizdat, Moscow, 3 vols.: 1971, 1973, 1979 (in Russian).

Gray, H., 1936. Proc. 1st ICSMFE. Cambridge, 2, 138.

Griggs, D.T., 1939. Creep of rocks. J. Geol., 47.

Haefeli, R. and Schaerer, Ch., 1946. Der Triaxial Apparat: ein Instrument der Boden- und Eismechanik zur Prüfung von Verformungs- und Bruchzuständen. Schweiz. Bauzeitung, Bd. 128(I).

Haefeli, R., Schaerer, Ch. and Amberg, G., 1953. The behaviour under the influence of soil creep pressure of the concrete bridge built at Klosters by Rhaetiln Railway Company, Switzerland. Proc. 3rd ICSMFE, v. II, Zurich.

Henkel, D.J., 1957. Investigation of two long-term failures in London Clay slopes at Wood Green and Northolt. Proc. 4th ICSMFE, v.II, London.

Hvorslev, M.J., 1937. Über die Festigkeitseigenschaften gestörter bindiger Böden. Thesis, Copenhagen.

Hvorslev, M.J., 1939. Torsion shear tests and their plays on the determination of the shearing resistance of soils. Am. Soc. Testing Mater., v. 39.

IUTAM Symposium, Grenoble, 1966. Rheology and Soil Mechanics. Springer Verlag, Berlin, New York, 1966.

Jessberger, H.L. (Editor), 1979. Grundbau und Bodenmechanik, Serie Grundbau, Heft I. Ruhr Universität, Bochum.

Johnston, G.H. (Editor), 1981. Permafrost. Engineering Design and Construction. John Wiley and Sons, Toronto, New York, Chichester, Brisbane.

Kachanov, L.M., 1960. Theory of Creep. Fizmatizdat, Moscow (in Russian).

Karpyshev, E.S., Molokov, L.A., Tulinov, R.G. and Kalmykova, N.I., 1972. Deformation of clay soils below dams: an engineering and geological analysis. In: Engineering and Geological Properties of Clay Soils and Processes Therein. Moscow State University, Issue 2.

Kelvin (Sir W. Thomson), 1875. Elasticity. In: Encyclopædia Britannica, 9th ed. 1875; Papers 3, 1890, London.

Kisiel, J., 1964. On determining rheological constants of weak soils. Proc. Seminar on Soil Mech. and Found. Eng., Lodz.

Kisiel, J., 1967. Reologia w budownictwie. Warszawa.

Kisiel, J. and Lysik, R., 1966. Zarys reologii gruntow. Dzialanie obciazenia ztalycznego na grunt, Warszawa.

Lomize, G.M., 1945. Deformation of Intake Structure, Dzora Hydro-Power Project. Gruznitostroiteli Publ. (in Russian).

Luga, A.A., 1964. The Effect of Creep in Clay Soils on the Displacement of Bridge Abutments. Moscow (in Russian).

Malinin, I.N., 1968. Applied Theory of Plasticity and Creep. Mashinostroenie, Moscow (in Russian).

Maslov, N.N., 1936. To the problem of estimating settlement and deformation of structures. In: Swir'stroi Project. Issue X, Leningrad (in Russian).

Maslov, N.N., 1968a. Fundamentals of Soil Mechanics and Engineering Geology. Vysshaya Shkola, Moscow (in Russian).

Maslov, N.N., 1968b. Long-Term Stability and the Displacement Deformation of Retaining Walls. Energiya, Moscow (in Russian).

Maxwell, I., 1868. On the dynamical theory of gases. Phil. Mag., (4): 35, 129, 135.

De Mello, V.F.B., 1969. Foundations of buildings in clay. State-of-the-art report. Proc. 7th ICSMFE, State-of-the-Art Volume, Mexico City.

Meschyan, S.R., 1967. The Creep of Clay Soils. Acad. Sci. Armenian SSR, Erevan (in Russian).

Meschyan, S.R., 1978. Initial and Long-Term Strength of Soils. Nedra, Moscow (in Russian).

Mogilevskaya, S.E., 1968. Results of studies of creep in loess soil in connection with deformation of water-development works. In: Studies of Rheological Properties of Soil. Energiya, Leningrad (in Russian).

Nadai, A., 1950. Theory of Flow and Fracture of Solids. McGraw-Hill, New York, N.Y., v. 1, 2.

Oding, I.A., Ivanova, V.S., Burduksky, V.V. and Geminov, V.N., 1959. Theory of Creep and Long-Term Strength of Metals. Metalurgizdat, Moscow (in Russian).

Peterson, R., 1953. Discussion. In: Proc. 3rd ICSMFE, v. III, Zürich.

Peynircioglu, H., 1957. Earth movement investigations in a landslide area on the Bosporus. Proc. 4th ICSMFE, v. II, London.

Pokrovsky, G.I. and Nekrasov, A.A., 1934. Statistical theory of gases, VIA-6 News. In: Foundation Engineering, No. I (in Russian).

Poynting, J.H. and Thomson, I.I., 1902. Properties of Matter. London.

Push, R., 1979. Creep of Soils. Ruhr-Universität, Bochum.
Puzyrevsky, N.P., 1934. Foundations. ONTI, Gosstroiizdat, Leningrad-Moscow (in Russian).
Rabotnov, Yu.N., 1966. Creep of Structural Elements. Nauka, Moscow (in Russian).
Reiner, M., 1949. Twelve Lectures on Theoretical Rheology. Amsterdam.
Reiner, M., 1958. Rheology. Springer Verlag, Berlin.
Reiner, M., 1960. Deformation, Strain and Flow. Lewis, London.
Rzhanitsin, A.R., 1968. Theory of Creep. Stroiizdat, Moscow (in Russian).
Scott, R.F. and Ko, H.Y., 1969. Stress-deformation and strength characteristics, State-of-the-art report. Proc. 7th ICSMFE, v. I, Mexico City.
Sedov, L.I., 1973. Continuum Mechanics. Nauka, Moscow, Vol. 1 (Vol. 2, 1976) (in Russian).
Sergeev, E.M., Golodkovsky, G.A. et al., 1961. Fedology. Moscow State University, Moscow (in Russian).
Skempton, A.W., 1964. Long-term stability of clay slopes (4th Rankine lecture). Géotechnique, 14(2).
Skempton, A.W. and Hutchinson, J., 1969. Stability of Natural Slopes and Embankment Foundations, State-of-the-Art Volume. Proc. 7th ICSMFE. Mexico City.
Sorokina, G.V., 1965. Building Properties of Marine Silts. In: Proc. All-Union Conf. Construction in Weak Saturated Clay Soils, Tallin (in Russian).
Sotnikov, S.N., 1960. Mechanism of Creep in Shear, Scientific Report, Leningrad Civil Engineering Institute, 1960 (in Russian).
Stocke, K., 1937. In: Z. Berghütten Salinenwerke, 2.
Šuklje, L., 1969. Rheological Aspects of Soil Mechanics. Wiley-Interscience, London.
Ter-Stepanyan, G.I., 1961. On Long-Term Stability of Slopes. Acad. Sci. Armenian SSR, Erevan (in Russian).
Ter-Stepanyan, G.I., 1975. The Theory of Progressive Failure in Rocks (in Russian).
Terzaghi, K., 1925. Erdbaumechanik und bodenphysikalische Grundlage. Wien.
Terzaghi, K., 1943. Theoretical Soil Mechanics. New York.
Terzaghi, K., 1950. Mechanism of Landslides. Application of Geology to Engineering Practice. Geol. Soc. Am., Berkeley Volume.
Tsytovich, N., 1973. Problems of soil and rock mechanics in Geomechanics. Proc. 8th ICSMFE, v. 2, Moscow.
Tsytovich, N., 1975. The Mechanics of Frozen Ground. McGraw-Hill, New York, N.Y.
Tsytovich, N., 1976. Soil Mechanics. Mir, Moscow.
Tsytovich, N., Zaretsky, Yu.K., Malyshev, M.V., Ableev, M.Yu. and Ter-Martirosyan, Z.G., 1967. Forecasting the Rate of Settlement of the Soil Below Foundations. Gosstroiizdat, Moscow (in Russian).
Voigt, W., 1890. Ueber die innere Reibung der festen Körper insbesonder Kristalle. Abh. Akad. Wiss. Königsberg, Math. Kl., v. 36; Ann. Phys., 47, 671.
Vyalov, S.S., 1959. Rheological Properties and Bearing Capacity of Frozen Soils. USSR Acad. Sci., Moscow; translated into English by USA Cold Regions Res. Eng. Lab., Transl. 74, Hanover, NH, 1965.
Vyalov, S.S. et al., 1962. The Strength and Creep of Frozen Soils and Calculations for Ice-Soil Retaining Structures. Translated into English by USA Cold Regions Res. Eng. Lab., Transl. 75, Hanover, NH.
Vyalov, S.S., Gorodetsky, S.E., Ermakov, V.F., Zatsarnaya, A.G. and Pekarskaya, N.K., 1966. Methods of Determining Creep, Long-Term Strength and Compressibility Characteristics of Frozen Soils. Translated into English by Natl. Res. Counc. Can., Tech. Transl. 1364, Ottawa, Ont., 1969.
Zaretsky, Yu.K., 1967. The Theory of Soil Consolidation. Nauka, Moscow (in Russian).

Chapter 2

STRUCTURE OF SOIL AND STRUCTURAL BONDS

2.1 COMPOSITION AND STRUCTURE OF SOILS

Structure and mechanical properties of soils

Soil structure and structural bonds have been examined in numerous investigations. Referring those who wish to obtain detailed information to Sergeev et al. (1977), Larionov (1966) and Nerpin and Chudnovsky (1967), we shall confine the discussion to a brief review of the modern concepts of soil structure, which are essential for understanding the physical aspects of the theory of soil deformation considered in the chapters which follow.

A point to be noted before all is that the physical meaning of the mechanical processes taking place in soils is determined by the structure and structural bonds alone. If these processes are considered from the standpoint of microrheology, it justifies introducing the concept of "structural soil mechanics", which studies the mechanisms of soil behaviour in response to a load in the light of the intrinsic dependence of the mechanical properties of a soil on the structure and on the structural changes occurring due to deformation.

Structural bonds in soils

Whatever rock is taken, it consists of either crystals or individual mineral particles interlinked by bonds formed during the geological process of rock formation. Depending on the nature of the bonds, distinction is made between rocks with rigid cementing bonds and those devoid of such bonds. Falling under the former category are rocks of massive crystalline character (igneous and metamorphic) as well as consolidated sedimentary rocks. Belonging to the latter category are loose sedimentary rocks, both cohesive (clayey, silty) and cohesionless (fragmented and sandy).

Soil as a multicomponent system

In rocks of the massive crystalline kind, the individual crystals are joined to one another over the entire surface of contact. These rocks are of the single-component type and can be treated in mechanics as solids.

In loose rocks, the particles contact each other only point-wise. Thus, a porous structure is formed, the pores being filled with either water or air. Clay soils represent three-component systems, whose constituents are solid, liquid and gaseous components. The term "three-phase system", used quite frequently, is inaccurate from the point of view of thermodynamics, for the phases are understood to be different states of the same matter. For example, an equilibrium system consisting of ice, water and steam is, according to Gibbs, a single-component three-phase system.

For cohesionless soils in which the constituents (mineral grains, water, air) are not bonded into a single whole, the concept of "multicomponent system" is inapplicable; these soils are treated as a granular medium.

Water-saturated clay soils are considered to be a dispersed system consisting of a continuum (dispersed medium) and individual fine particles (dispersed matter) distributed over the medium. The dispersed medium is a liquid phase and the dispersed matter is represented by mineral particles (and air bubbles). The degree of dispersion is determined by the size of the particles. Those of the colloidal size form a colloidal solution. When the soil particles are suspended in water without being interlinked the system is termed suspension.

Colloidal particles are apt to stick to each other and grow in size; this process is known as coagulation. The process of disintegration of aggregations and larger particles is referred to as dispersion.

Elements of soil structure

Individual parts of a soil, the adjacent particles of which are separated from one another in a certain way, are termed elements of soil structure (Larionov, 1966). In massive hard rock they are represented by mineral grains or cemented fragments interlinked with one another and in cohesionless loose rocks they consist of separate grains, particles and fragments; the elements of structure of fine clayey rock are mineral particles, stacks and aggregations of particles, as well as individual grains. Strictly speaking, the above elements comprise the soil skeleton; in a broader sense, also coming under the category of structural elements are adsorbed air, single layers of water, voids, etc.

Distinction is made between primary elements — individual mineral particles, stacks of particles — and secondary elements, i.e. aggregates.

Aggregation is a salient feature of clay soils. The smaller a particle the greater its specific surface and, consequently, the stronger the molecular forces of attraction which cause the colloidal and clayey particles to stick to each other to form strong ultra- and micro-aggregates (the former less than 0.0002 mm and the latter between 0.0002 and 0.1 mm). The ultra- and micro-aggregates, in their turn tend to join into structural units of a higher order — aggregates exceeding 0.1 mm. These join to form blocks and domains. In addition, there are semiaggregates, i.e. particles of silt or sand surrounded by a film of colloidal particles.

Cohesionless soils consist predominantly of primary elements, the bulk of cohesive clay soils is composed of secondary elements (this does not apply to fresh deposits after precipitation).

Structure and texture of soils

Factors such as (a) size, shape and character of the surface of the elements comprising the soil, (b) quantitative relation between these elements, and (c) the nature of their mutual bonds decide the soil structure. Widening the scope of the term "structure" so that the nature of interparticle bonds is embraced is justifiable, for such a connotation — differing from the narrow one describing structure from the standpoint of the particulars of mutual arrangement of particles only — is in line with the modern concept of physical and chemical nature of soil (Sergeev et al., 1971; Larionov, 1966).

Soil texture is determined by the spatial arrangement of the elements comprising the soil, irrespective of their size. The texture can be an orderly one, characterized by a certain orientation of structural elements with respect to any of the solid axes. Alternatively, the texture can be of the random type, when the structural elements are positioned in a chaotic manner. The degree of orientation of particles is expressed in terms of an index C, which is the ratio of the number of structural elements oriented along a given axis and that of the elements arranged at right angles to it. The deviation of particles from this axis is defined by an angle α.

Depending on the particle size, distinction is made between micro-, meso-, and macro-structures. Microstructure defines the combination of mineral particles, ultra- and micro-aggregates less than 1–5 μm. The spatial arrangement of the microelements determines the microstructure.

The mesostructure reflects the spatial arrangement of microaggregates, separate microblocks and inclusions in the form of grains of silt and sand. The elements of the mesostructure commonly vary between 1 and 5 mm, yet this is not their upper size limit.

The macrostructure defines details of soil structure visible by the naked eye. Macrostructural elements (blocks, grains, fragments, etc.) are in the millimetre- to metre-size ranges.

2.2 SOIL COMPONENTS

The solid constituent

The solid constituent of soil consists of mineral particles, aggregates and fragments having different chemical composition and mineralogical nature.

All rock minerals are, as a rule, of the crystalline structure characterized by a lattice of regular shape. They are subdivided into the following groups: primary silicates, in which covalent bonds are prevalent, simple salts, in which ionic bonds are mainly present, clay minerals with bonds of various kinds, none of which are explicitly dominant, and organic matter.

Silicates are the main building blocks of igneous, metamorphic and some sedimentary rocks. Widely represented in other sedimentary rocks are simple salts such as halides, sulphates and carbonates. Their salient feature is solubility in water, ascribed to the ionic character of intercrystalline bonds.

Clay minerals, formed from rocks of the silicate group predominantly due to the chemical effects of weathering, are characterized by high dispersity and a structure of the sheet- and stackwise nature. Constituting the bulk of dispersed sedimentary deposits, they are classified — depending on the crystalline structure and chemical composition — into kaolinites, montmorillonites and hydrated micas; less frequently clay minerals occur in the form of chlorites, vermiculites, and the like.

Grading of soils

The size of the particles a soil is composed of is commonly defined by their minimum dimension. Therefore, falling under the category of fragments (gravel, broken rock, cobbles) are particles greater than 2 mm. Particles between 2.0 and 0.05 mm are called sand, those of 0.05–0.01 mm are referred to as silt, and the smaller particles come under the category of clay. Clay particles less than 0.00025 mm are termed colloids.

A quantitative relation between the fractions of different grain size defines the granulometric composition of a soil; this composition is determined by taking into account the primary elements only. Whenever the primary as well as the secondary elements are taken into consideration, the relation thus obtained is called the microaggregate composition. The two compositions materially differ from one another. For example, in kaoline clay the fractions with a particle size less than 0.001 mm account for 35% of the granulometric composition and for only 15% of the microaggregate composition.

Colloidal clay particles. Clay minerals are characterized by a large specific surface; this, in its turn, results from the high dispersity, water-receptivity, capacity for absorbtion and ion exchange of clay particles.

Colloidal clay particles display an activity second to none among other components of dispersed soils and, due to that, significantly influence the strength and deformation characteristics of soils. The degree of this influence varies with the specific surface of the particles which, in its turn, varies with their mineralogical composition. The specific surface of montmorillonite amounts to $800 \, m^2/g$, that of illite is $80 \, m^2/g$, whereas for kaolinite it is $10 \, m^2/g$ only. This implies that the colloidal activity (i.e. the degree by which a soil approaches a colloidal system with regard to its properties) may be regarded as being a maximum in montmorillonite and a minimum in kaolinite.

Organic matter. This is met with in many soils, not only peat and top soil, predominantly in the form of vegetable remnants either decomposed (humus) or not. All organic matter, and humus in particular, is characterized by high water receptivity and reactivity, which makes it active in various physical and chemical processes occurring in soils (e.g. oxidizing and reduction processes).

The organic matter present in a soil influences its mechanical properties, rendering the soil more plastic and deformable, particularly with regard to volumetric changes due to compression. Humus also imparts good water resistance to soil.

Water in soil

This may occur in a variety of forms such as vapour, bound water (firmly or loosely) and free water (capillary and gravitational). Water may also be present in solid form (applies to frozen solids), in crystalline form and as a chemically bound water.

Since the water contained in soil is saturated with various salts, it is reasonable to treat it as a solution. The chemical ingredients of the solution control, to a considerable extent, the forces of interaction between the particles and, consequently, the strength of the interparticle bonds.

Bound water is a substance of very intricate nature. A salient feature of this water is an orderly orientation of the molecules with respect to the surface of the particle it surrounds, the degree of orientation diminishing with distance from the particle. At a certain distance, the molecules become oriented at random and, consequently, the water is transformed from the bound state into a free state.

The properties of bound water materially differ from those of a free one, approaching the properties of a solid. For example, the density of firmly bound water is above normal, amounting to $1.2–2.4 \text{ g/cm}^3$. The molecules display less mobility than in free water. As a result, firmly bound water has elasticity and resists shear; its cohesion is also sufficiently high. The density of loosely bound water differs little from that of free water, the same being true for the rest of properties.

The maximum quantity of firmly bound water approximately corresponds to the maximum amount of hygroscopicity; this gives good reason to refer to the firmly bound water as hygroscopic or adsorbed water, as is often the case. The total content of firmly and loosely bound water in a soil corresponds to the maximum molecular moisture capacity of the soil, i.e. to the amount of water held fast by the forces of surface tension.

The total amount of water of all kinds contained in a soil determines its moisture content, an absolute one, W, when the weight of water is referred to the weight of dry soil, or a relative water content, S_r, when the volume of water is referred to that of voids. Gases are present in soils in either free or dissolved form If dissolved, they are always contained in ground water, influencing the composition and properties of solutions and responding to the changes in temperature and pressure in an appreciable way.

Some of the free gases can communicate with the atmosphere and are termed free gases proper. Others are locked in the soil and are referred to as trapped gases. These constitute the bulk, filling the voids, cracks, etc. Clay soils contain trapped gases.

Porosity

Porosity is one of the main structural features of soil. Denoted as n, it is the ratio of the volume of voids to the total volume of the soil. The void ratio, e, is determined by the ratio of the volume of voids to the volume of soil solids.

Larionov (1966) has suggested differentiating: (a) ultramicropores having a size of less than 1 μm, which are located between microparticles, stacks and run through ultra-aggregates; (b) interparticle pores (1–100 μm), which occur between coarser particles and aggregates or inside aggregates; and (c) wide pores of a size in excess of 100 μm.

The porosity contributing to the bulk of volume change in soil being compacted is called active porosity. The ratio of active and total porosity is the main consideration in evaluating soil consolidation.

Apart from pores, micro- and macro-cracks running between or inside aggregates are also present in soil.

2.3 INTERACTION BETWEEN SOLID AND LIQUID COMPONENTS OF SOIL

Electric nature of the interaction

The interaction of solid and liquid constituents of soil bears an electric nature and manifests itself in the form of ion exchange. It will be recalled that ions are atoms or groups of atoms having an excess or deficiency of electrons compared with the neutral atoms.

Inside a solid particle, the oppositely charged ions exist in an equilibrium. At the surface, where such an equilibrium is unfeasible, the particle behaves as an electrically charged body. Since the dielectric constant of rocks is lower than that of water, the solid particles of a dispersed water-saturated soil acquire, according to Köhn's rule, a negative charge at the surface. Theoretically, the charge must be uniformly distributed over the surface. However, it has been noted that the edges of clay particles are positively charged, i.e. opposite to the charge on the surface.

The dipoles of water surrounding the particle are acted upon by the electrical field, become oriented and are attracted to its surface. This, being negatively charged, attracts positive ions (cations) of water; negative ions (anions) will be present in a limited number at some distance from the surface.

The distribution of ions depending on distance from the surface of a particle is represented in Fig. 2-1. Note that ion concentration decreases with distance (Lambe, 1958).

Ion exchange

Because of the oppositively charged surface of a particle and molecular layer of bound water, the solid and liquid components of the soil exchange cations. When the positively charged ions (cations) of the layer become attracted by the negatively charged surface, they enter into the reaction of exchange with the surface molecules and, furthermore, expel the cations of the particle lattice into the liquid. It will be noted that the ione exchange takes place not only at the interface (in the diffuse layer) but also inside the particle, i.e. in its lattice.

The exchange (adsorbing) capacity of soil depends on the chemical and mineralogical composition of clay particles. In kaolinite, the particles of which have a rigid

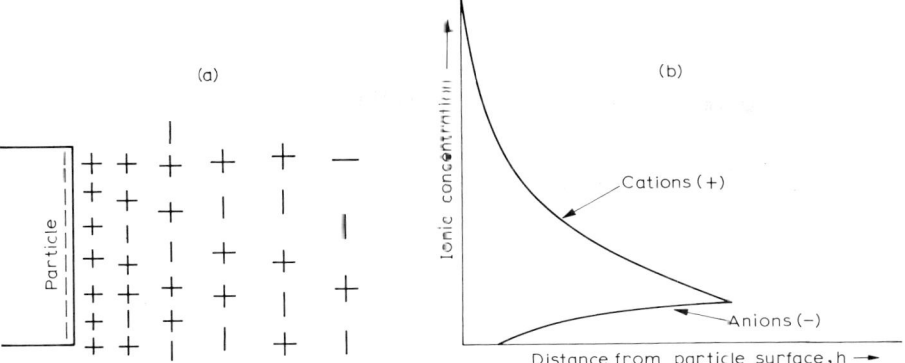

Fig. 2-1. Distribution of ions at the surface of a mineral particle. (a) Distribution of positive and negative ions. (b) Curves of ionic concentration versus distance from particle. (After Lambe, 1958).

lattice, the possibility of ion exchange in the space between the stacks is excluded and the exchange reactions are localized at the end faces of the particles only. Therefore, the adsorbing capacity is only 3–15 me./100 g. In illinite, also composed of particles with a rigid lattice, the ion exchange spreads to the outside basal surfaces of the stack and the adsorbing capacity consequently increases to 10–40 me./100 g. In montmorillonite, the process of ion exchange is also inside the basal surface so the adsorbing capacity reaches 80–150 me /100 g.

The mineral particle and hydrated envelope

A colloidal particle in conjunction with the bound water enveloping it forms a micelle (Fig. 2-2). Next to the surface where the forces of attraction are rather high (amounting to hundreds of thousands of newtons*[1] per square centimetre), the molecules of water are firmly attached to the particle. The oriented molecular layer so formed takes no part in the movement of fluid. This layer of firmly-bound (adsorbed) water, representing a boundary phase, is regarded as constituting a single whole with the mineral particle. The thickness of the boundary phase corresponds to that of a molecule or a few molecules.

The oriented molecules of loosely bound (lysisorbed) water located at some distance from the surface form a diffuse layer which takes part in the movement of fluid, flowing along the solid particle in the course of deformation.

The adsorbed and lysisorbed water form a double layer, a few hundred ångström thick (1 Å = 10^{-4} μm = 10^{-8} cm). Outside the double layer, i.e. outside the range of the forces of molecular attraction, free water is located. The whole of the hydrated envelope surrounding the mineral particle is called the lyosphere.

The double electric layer results from the opposing charges at the surface of the solid particle (negatively charged) and the contiguous layer of bound water, the ions of

---

*[1] SI units and their conversion into units of other systems are given in the Appendix, p. 557.

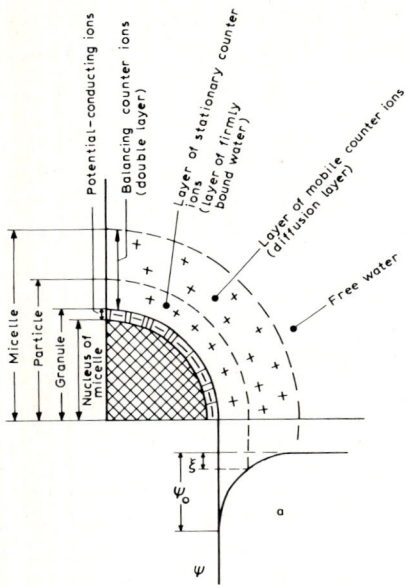

Fig. 2-2. Electromolecular forces of a mineral particle and the hydrated envelope.

which are positively charged. This layer is characterized by an electric potential, ψ (Fig. 2-2), which is at a maximum ψ_0 at the surface of the particle, diminishing as distance from the particle increases. The minimum value of the potential is in the free water, being determined by the concentration of the electrolyte. Thus, the point at which the potential difference is zero determines the boundary between the bound and free water.

Distinction is made between thermodynamic and electrokinetic potentials. The former, ψ, defines the aggregate bonding forces in any layer of the micelle, the latter, ξ, characterizes the bonds in the diffuse layer of water; the thinner this layer the higher the value of ξ.

The theory of double electric layer was formulated by G. Gouy in 1910. D.L. Chapman developed it in 1913, establishing an analytical relation between the electric potential and the distance to the surface of the particle. However, this theory describes the interaction of individual idealized particles and, if applied to real dispersed soils, is suitable from the qualitative point of view only.

On forces of electromolecular attraction

Being formed between the mineral particle and the bound water due to the attraction of water dipoles to the electrically charged surface, these forces act at right angles to the surface and determine the resistance which should be overcome in order to pull the molecules of water off the surface of the particle. Coexisting with the electromolecular forces are tangential forces which, being formed at the interface between the solid and fluid phases, resist the motion of water along the surface of the particle. Although far weaker than the pull forces, the tangential forces play an important part in rheological processes.

Since bonding between the mineral particle and the surrounding water is always weaker than the strength of a direct contact, it would be logical to treat the hydrated envelope as a kind of link between the solid particles rather than a cementing bond. The effect of adsorbed water is indirect, manifesting itself by changes in the stresses transmitted to the direct contacts.

2.4 FORCES OF INTERACTION BETWEEN SOLID PARTICLES IN SOIL

Fields of energy

In the preceding section we examined the interaction between the solid particle and the surrounding hydrated envelope. Consider the interaction between solid particles separated by a film of water.

Each such particle is acted upon by a system of external and internal forces as well as by fields of energy induced by the forces. External fields are set up by the applied load and gravitational forces. Internal fields are produced by interparticle forces arising between the soil components themselves. In general, the following inter- and intra-particle forces occur: forces of a chemical nature, molecular forces, ionic-electrostatic forces, capillary-electrostatic (Coulomb) forces and magnetic forces.

The bonding forces of a chemical nature are intermolecular forces set up inside mineral particles and between crystals of crystalline rocks and loose rocks with rigid cementing bonds. These bonds provide a strength approaching that of mineral grains. The rest of the bonds are of the intermolecular (interparticle) form, serving as a link between the particles in a clay soil. Sometimes they are referred to as water-colloidal bonds. The strength of the interparticle bonds is far inferior to that of the intra-molecular ones. Granular soils have no bonds at all, their strength being determined by the gravitational forces alone.

Bonding forces of a chemical nature. Because they originate from the electrical interaction of atoms, these forces may exist in either ionic or covalent forms.

Intramolecular forces are characterized by a high bond energy (i.e. the energy liberated when a given bond is being formed), between 400 and 1250 J/mole[*1] (100–300 kcal./mol), and a short radius of effective action (0.5–3.5 Å). Thus, they appear to be short-range forces.

Also falling under the category of chemical bonds are hydrogen bonds resulting from the interaction of a hydrogen located between two negatively charged atoms in water-containing compounds (crystal hydrates, ice). Their nature is similar to that of ionic bonds.

[*1] Mole is a unit of the quantity of matter equal to its molecular weight in grams.

Molecular or Van der Waals–London forces are produced by molecular interactions of the following type: oriented interaction of polar (axially symmetric) molecules, inductive interaction due to the polarization of non-dipolar molecules in a field produced by dipolar molecules, and dispersive orientation resulting from the interaction of electrons in a molecule. The energy of the interaction between a pair of molecules, U, and the force of their mutual attraction, f, are assumed to equal:

$$U = \frac{s}{h^6} \qquad f = -\frac{dU}{dh} = \frac{6s}{h^7} \tag{2-1}$$

where h is the distance between the two molecules; s is a constant dependent upon the dipole moment of the molecules, their polarizability, frequency of oscillations and temperature.

The forces of molecular interaction also come into play between two solid particles. According to Livshits (1954) they may be determined by the formula $f = B/h^n$, where $n = 3$ for $h \leqslant \lambda$ or $n = 4$ for $h > \lambda$ (λ is a spectral characteristic of the atoms in the interacting particles).

Note that expression 2-1 refers to the interaction of two individual molecules. Actually, if the mutual influence of a multitude of molecules is taken into account, the resultant force of the interaction between the two molecules will be described by a more complex relation.

The bond energy due to molecular forces is far smaller than that produced by intramolecular forces, being at maximum 40 J/mol (9 kcal./mol). On the other hand, the range of action of the molecular forces is considerably longer than is the case with intramolecular forces, varying between a few (some 5 to 10) and several thousand ångströms. Consequently, the molecular forces are regarded to be of a long-range nature.

When the distance is very short (2 Å or less), the forces of mutual attraction between adjacent particles are transformed into forces of repulsion, the Brownian movement of electron shells being behind this effect. Increasing as the distance between the molecules shortens, the forces of repulsion are inversely proportional to this distance raised to the power of $n \geqslant 9$. Since, according to expression 2-1, $n = 7$ for forces of attraction, the action of the molecular forces of repulsion is influenced more by distance than are molecular forces of attraction.

In spite of their comparatively low magnitude, molecular forces significantly affect the strength characteristics of clay deposits, especially at the early stages of lithogenesis such as sedimentation (coagulation and settling of deposits) and diagenesis. Thus, in freshly settled deposits (or artificially remoulded clays) the strength-imparting bonds owe their origin mainly to the molecular or Van der Waals forces. The subsequent compaction of clay deposits gives rise to predominantly ionic-electrostatic forces.

Ionic-electrostatic forces. The mineral particles of a water-saturated clay soil interacting with exchange cations acquire an electric charge. When one charged particle approaches another, the cations interact with both of them simultaneously and, as a result, an ionic-electrostatic interparticle bond is formed. This bonding is

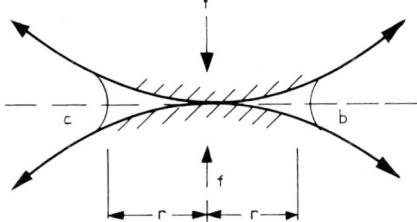

Fig. 2-3. Schematic illustration of capillary forces in action.

effective when particles are separated by a distance of a few ångströms and attains the peak of strength in an absolutely dry clay soil.

Capillary forces, first studied by K Terzaghi, owe their origin to the capillary pressure build up at the interface between the fluid and gaseous phases of a soil system (Fig. 2-3). They may reach an appreciable strength determined by the relationship:

$$f = 2\pi rs \tag{2-2}$$

where r is the radius of the particle and s is the surface tension of the fluid.

Electrostatic forces result from the accumulation of electric charges at the surface of the particles. Those charged oppositely attract each other, while particles bearing a charge of the same sign are repulsed; in other words, the charges interact according to the Coulomb pattern. The magnitude of electrostatic forces is low. The energy of interaction and the force of interaction are given, respectively, by:

$$U = \frac{1}{\varepsilon}\frac{q_1 q_2}{h} \qquad f = \frac{1}{\varepsilon}\frac{q_1 q_2}{h^2} \tag{2-3}$$

where ε is the relative permittivity of the fluid phase surrounding the particles, h is the distance between the particles; q_1 and q_2 are the charges of the particles; in like charges $f > 0$ and in unlike charges $f < 0$.

Since the particles are negatively charged at the sides and positively at the edges, their side-to-side contact gives rise to the electrostatic forces of repulsion; a side-to-edge contact produces forces of attraction. The effect of electrostatic forces is felt at the early stages of lithogenesis (in young clay deposits).

Magnetic forces. Investigations carried out at the Moscow State University have revealed that forces of a magnetic nature may be formed in finely dispersed systems along with the forces considered above. The origin of these forces are ferromagnetic substances (hematite, goethite, hydrometite) which occur in clayey soils in the form of a surface film varying in thickness between 0.05 and 0.5 μm. The rigid magnetic moment of the film transmits the effect of coagulation to the adjacent particles. The magnitude of magnetic forces is low, and they are a factor at the stage of sedimentation only (Sergeev et al., 1971).

TABLE 2-I

Strength of bonds in clay soils

Type of bond	Strength of an individual contact, 10^5 N (dyne)	Strength of soil system as a whole (Pa)
Chemical	Up to 10^2	(10^7–10^8) (hundreds of kg/cm^2)
Molecular (Van der Waals)	10^{-4}–10^{-3}	Up to 10^4 (up to 0.1 kg/cm^2)
Ionic-electrostatic	10^{-3}–$4 \cdot 10^{-1}$	Up to 10^6 (up to 10 kg/cm^2)
Capillary	10^{-2}	Up to $4 \cdot 10^5$ (up to 4 kg/cm^2)
Electrostatic (Coulomb)	10^{-5}	10^3–10^4 (a few g/cm^2)
Magnetic	10^{-10}	10^2–10^3 (a fraction of g/cm^2)

Strength of bonds

The forces discussed above, producing fields of energy, form bonds between the particles of a dispersed system. The strength of the bonds varies over a wide range (up to 20 orders of magnitude) depending on the type of prevailing bonding. It must be stressed that distinction is made between the strength of a single bond (individual contact) and the strength of a soil system as a whole. The relevant values are given in Table 2-I compiled by Osipov (1979).

Forces of attraction and repulsion

As already pointed out, the forces of particle interaction may cause either attraction or repulsion of the particles in a clay soil. Attraction results from the action of the molecular Van der Waals forces when the distance between the molecules is over 2 Å; the same effect is achieved by the ionic-electrostatic, capillary and electrostatic (Coulomb) forces, provided the particles contact each other with their oppositely charged surfaces, and also by the magnetic and hydrogen forces. The electrostatic (Coulomb) forces arising from a contact between the sides of two particles bearing like charges are repulsive forces and so are the molecular (Born) forces acting at a short range.

Wedging forces. These owe their origin to the hydrated envelopes surrounding the mineral particles. As stated above, the hydrated envelope consists of a layer of firmly bound adsorbed water and a diffuse layer of loosely bound water (see Figs. 2-1 and 2-2). In the former, the water molecules are oriented, in the latter they are not. When the particles contact each other so that their surfaces are separated at the point of contact by a distance not greater than the size of the layer of firmly bound water, the water film exerts a wedging action. This phenomenon has been detected by Deryagin

(1956). The wedging force is:

$$f = A/h^3 \qquad (2\text{-}4)$$

where A is the Van der Waals constant equal to 10^{-9} J (10^{-2} erg) and h is the distance between the particles.

The above expression is similar to the one suggested by Livshits (1954) to calculate the forces of molecular attraction; this implies that the wedging forces obey the same law as the forces of molecular interaction.

If the contact between the particles is effected through the diffuse layers, the wedging action of the water films will be attributed, on the one hand, to the electrostatic forces of repulsion and, on the other, to the osmotic forces. Correspondingly, the total repulsive force will be:

$$f_r = f_e + f_o \qquad (2\text{-}5)$$

where

$$f_e = \varepsilon E^2/8\pi^2 \quad \text{and} \quad f_o = Re\,\theta\,(n_i - n_\infty) \qquad (2\text{-}6)$$

Here E is the electrostatic field strength; Re is the Reynolds number; θ is the thermodynamic temperature, in Kelvin; n_i is the concentration of ions in the bound water between the particles; and n_∞ is the concentration of ions in the free water.

The theory of osmotic pressure considers the condition of equilibrium of two solid particles separated by a layer of fluid. The concept of the wedging action exerted by a layer of water according to B.V. Deryagin is based on examining the condition of equilibrium of the fluid located between two solid particles. Nevertheless, both these approaches yield similar results.

Swelling of soils

The wedging effect of the film of water brings about swelling of clay soils. If the particles are separated by a distance not greater than twice the thickness of the layer of oriented water, the swelling will be of the intracrystalline type only. This effect is characteristic of soils with a water content less than their maximum hygroscopicity. For a macroswelling to begin, the water content must correspond to the lower limit of plasticity and increase to the upper limit in order to enable the swelling to progress.

Investigations carried out by Mustafaev and his colleagues (1976) have revealed that clay soils swell after some time, rather than immediately on being wetted, in a manner analogous to the phenomenon of after-effect. The resulting increase in the volume of the soil appears to be far greater than the volume of water added. This gives a good reason to conclude that swelling results not only from the shifting of the clay particles to the sides due to an increase in the thickness of the water film, but also from the intracrystalline volumetric changes occurring in the particles themselves after the addition of water.

2.5 INTERPARTICLE BONDS IN SOIL

Interaction of adjacent particles

Consider two parallel particles of a clay soil separated by a layer of water having a thickness h (Fig. 2-4). These particles are acted upon by the forces of attraction f_a and those of repulsion f_r, the latter, being of the ionic-electrostatic nature f'_r, exerting their action over a distance exceeding a given value h_a; moreover, when $h < h_a$, the contact, Born's forces of repulsion f''_r come into existence. The particles are subjected to the action of a resultant force:

$$F = f_r - f_a \tag{2-7}$$

where $f_r = f'_r$ for $h > h_a$, and $f_r = f'_r + f''_r$ for $h < h_a$.

If the forces of repulsion are greater than those of attraction, the resultant force acquires the sign "+", if the opposite is true, the sign is "−". Thus, the resultant force F varies with h depending on the relation between the component forces f_a and f_r.

Possible patterns of dependence according to Nerpin and Chudnovsky (1967) are depicted in Fig. 2-4. Curve 1 corresponds to the case when only the forces of attraction are at work, $F = f_a$, and curve 4 represents the solitary action of the forces of repulsion, $F = f_r$. Curves 2 and 3 refer to the instances when attractive and repulsive forces act simultaneously. Curve 3 illustrates the case when the repulsive forces

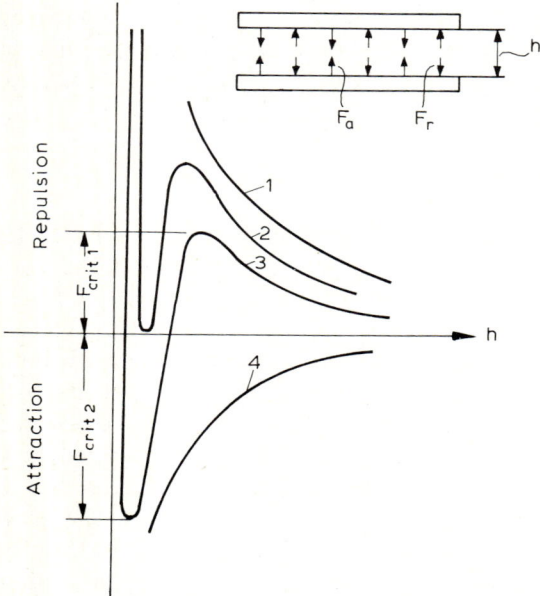

Fig. 2-4. Relationship between the forces of particle interaction, F, and distance between particles, h. For further explanation see text.

predominate, whatever the distance h, and the resultant force is positive, and curve 2 exemplifies a situation in which the resultant force F reverses its sign. The two extreme values of this force are denoted on the graph by $F_{\text{crit.1}}$ and $F_{\text{crit.2}}$, respectively.

The shape of the "force F versus distance h" curves depends on the mineralogical nature and chemical composition of the particles. Thus, curve 1 is characteristic of kaoline, in which the stacks of particles are in contact at their oppositely charged sides so that the electrostatic forces of attraction come into existence in addition to the Van der Waals forces. Montmorillonite, in which the stacks are linked by like charges, is represented by curves located between curves 3 and 4, and illite by those confined between curves 2 and 3. The repulsive forces increase with concentration of the electrolyte. Therefore, curve 4 is characteristic of marine deposits with a high electrolyte content and curve 2 of soils with a low concentration of salts in the pore solution.

Condition of the equilibrium of particles

Two parallel particles separated by a distance h, at which the resultant force of their interaction F is repulsive, remain in equilibrium only as long as an external load $N = \sigma's$ is applied, where σ' is the effective pressure (transmittable to the mineral particles) and s is the surface:

$$N - F = 0 \tag{2-8}$$

If the external force is increased by ΔN, the equilibrium can be restored by drawing the particles closer to each other through a distance Δh, thereby increasing the repulsive force by ΔF. The work performed when the particles approach each other, $\Delta A = F\Delta h$, changes the potential bond energy U. The curves depicting the dependence of this energy upon the distance h between the particles are identical to those of Fig. 2-4.

Assessing the forces of interaction

Geuze and Tan (1954) made an attempt to assess the forces of particle interaction, investigating the case when the particles are in an "edge-to-face" contact (see Fig. 2-5) and make an angle of 45° with one another. Prevalent in this case are the forces of attraction, $F = f_a$, the components of f_a being the Van der Waals and negative electrostatic forces. The relationship between the total force of attraction and the edge-to-face distance h between the particles is defined by a curve of the same configuration as curve 1 in Fig. 2-4. This curve may be described by the expression:

$$F = -B\frac{\delta l}{h^4} \tag{2-9}$$

where F is the force of particle interaction; B is the coefficient of attraction; δ is the thickness of a particle; and l is its unit width.

Taking $B = 10^{-24}\,\text{N cm}^2$, $\delta = 10^{-7}\,\text{cm}$, $l = 1\,\text{cm}$ and $h = 10\,\text{Å}$, we obtain

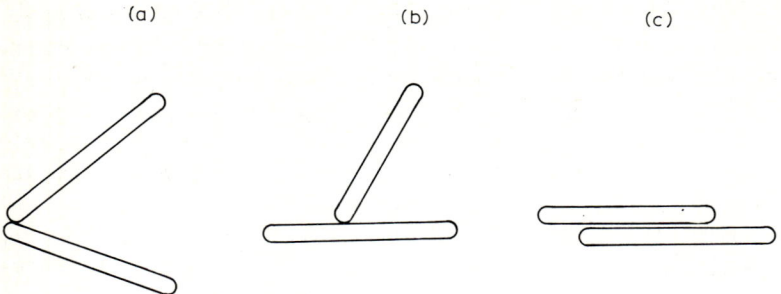

Fig. 2-5. Modes of particle association: (a) edge-to-edge; (b) edge-to-face; (c) face-to-face.

$F = -10^{-31} \cdot 10^{28} = -10^{-3}$ N. If this force is referred to as the width of a particle equal to between 100 and 1000 Å rather than to its unit width of 1 cm, we arrive at values of F anywhere between 10^{-9} and 10^{-8} N. An increase in the distance h from 10 to 20 Å reduces the value of F by 6.25%.

On assessing bond energy

The interparticle forces discussed above decide the interaction of two idealized particles of soil. However, this idealized approach is of no avail in analytical computations of the resultant force of interaction F, for the components of this force are of such an intricate nature that a conventional arithmetic summation is inapplicable. We must examine not an individual particle but an assemblage of particles of different size and shape along with aggregations.

It is expedient to assess the forces of interaction between particles in terms of the averaged potential energy $\langle U \rangle$ of bonding between the arbitrary static particles in a characteristic elemental volume of soil, using integration.*[1] In proceeding thus, it seems feasible to describe the relation between the potential energy and the statistically averaged distance between the particles in the first approximation by a phenomenological formula of the type:

$$\langle U \rangle = \frac{a}{h^n} - \frac{b}{h^m} \qquad (2\text{-}10)$$

where the first term defines repulsion and the second attraction.

Another method of assessing the strength of interparticle bonds in a soil system consists of determining R, the strength, as a product of the strength F_i of an individual contact and the number of contacts λ in a unit volume of the soil system:

$$\langle R \rangle = \lambda F_i \qquad (2\text{-}11)$$

For example, if a particle in the form of a sphere with a radius r is taken, $\lambda = 1/4r^2 n^2$, where n is the porosity.

---

*[1] The symbol $\langle \ \rangle$ denotes the averaging of the energy in a characteristic volume of soil, which is a very small fraction of the volume under consideration.

The value of F_i varies between 10^{-12} and 10^{-3} N (or 10^{-7} to 10^2 dynes). If the value of R is known from a macroexperiment, the strength of an individual contact may be approximated accurately to within an order of magnitude, thus shedding light on the nature of the bond.

2.6 SOIL STRUCTURE

Typical structural bonds

The electrical nature of interparticle bonds in soil was discussed above. Now consider the way these bonds are influenced by physical and chemical conditions of rock formation. Studying the problem, Rebinder (1956) has evolved a classification of soil systems into two basic classes: coagulation–thixotropy structures and condensation–crystallization structures.

Coagulation–thixotropy structures are characterized by the presence of water-colloidal bonds, i.e. by interparticle bonds of the molecular, ionic and electrostatic nature discussed above. Since these bonds are formed at the instant of formation of sedimentary rock, they are referred to — taking into account diagenesis — as primary adhesion. In dispersed soils, the bonds in question develop during settling, when the clay deposit consists of primary nonaggregated particles each surrounded by a film of bound water. Since the thickness of the films and, consequently, the magnitude of intermolecular forces are a function of soil density, the primary adhesion increases with compaction of the soil. These bonds have a low strength, high mobility and re-form if broken, i.e. they are capable of gaining strength in a thixotropic way. Remoulded clay is a typical representative of the coagulation–thixotropy structure.

In soils of *condensation–crystallization structure*, the particles are interlinked directly, without separating water films. Condensation bonds are thought of as dry contacts between particles; such contacts are likely to be established when the particles approach one another, as is the case during evaporation, etc. Bonds of chemical nature formed due to cementation at the stage of diagenesis are referred to as crystallization bonds; they are rigid and break under high stress without re-forming.

In crystalline rocks, crystallization bonds are formed during cooling of magma and metamorphic recrystallization, i.e. their origin concurs with diagenesis. Consequently, they are classified as primary bonds. In sedimentary rocks, crystallization bonds take shape due to chemical, physical and biochemical processes at a later stage of formation. Therefore, in sedimentary rocks these bonds come, according to Denisov's classification (1956), under the category of secondary bonds.

On the author's suggestion (Vyalov, 1959), a third form of adhesion has been singled out in frozen soils and is referred to as ice-cementation adhesion. It owes its origin to the cementation bonds between ice crystals and mineral particles established by way of a film of unfrozen water. Ice-cementation adhesion varies with soil temperature and ceases to exist as soon as the soil thaws.

Soil classification by the nature of structural bonds

Rebinder's classification was developed by I.M. Gor'kova (1966). Taking into account that particle interaction is mainly influenced by the aggregation factor, i.e. by the ratio between the particle content in terms of granulometric and microaggregate compositions, she suggested using this factor as the basis for the classification of soil structures.

Stabilization structures with particles of less than 1 μm have an aggregation factor equal to unity. They are formed in finely dispersed soils when a hydrophilic stabilizing agent, preventing the particles sticking to one another, is present at the surface. The agent augments the layer of adsorbed water surrounding the particle and, consequently, impairs soil strength. Stabilization structures are characteristic of sedimentary rocks formed in an alkaline medium.

Coagulation structures occur most frequently in soils with a low (1.5% or less) electrolyte content, causing structural coagulation and the formation of structural skeleton. The aggregation factor is 4.5–5 for particles less than 2 μm and 1 when the particles are less than 5 μm. The strength of interparticle bonding in coagulation structures is significantly larger than in stabilization structures.

Plasticization–coagulation structures are characterized by a coagulation factor of 3 or less for particles under 1 μm and of 2 for those under 5 μm. The structure also originates from coagulation processes, provided organic plasticizing agents are present in an amount of at least 0.7%, before all calcium carbonates (up to 50%). The electrolyte concentration varies over a wide range between 0.3 and 10%.

Mixed coagulation–crystallization or coagulation–condensation structures are characterized by an aggregation factor of 6–38 when the particles are less than 1 μm and 2–30 for particles less than 5 μm. Bonds are of the coagulation–cementation type resulting, as indicated above, from the joint action of the intramolecular forces of chemical origin and the interparticle forces. Inherently strong, brittle and water-resistant, bonds are formed during the cementation of rocks by amorphous silica, iron oxides, etc.

2.7 SOIL TEXTURE

Mutual arrangement of particles

Clay particles may contact one another in various ways: edge-to-edge, edge-to-face, face-to-face (Fig. 2-5). In the first instance, the interaction is between surfaces bearing like positive charges; in the second instance the surfaces involved are with unlike charges; and in the third instance, those charged similarly but negatively. In fresh deposits, the mechanism of particle interaction depends on the conditions of settling. Particles of colloidal size weighing little may stay suspended for a long period if acted upon by repulsive forces. Consequently, a scattered (dispersed) system of the parallel pattern illustrated in Fig. 2-6(a) is formed. If the weight is not as light as is necessary

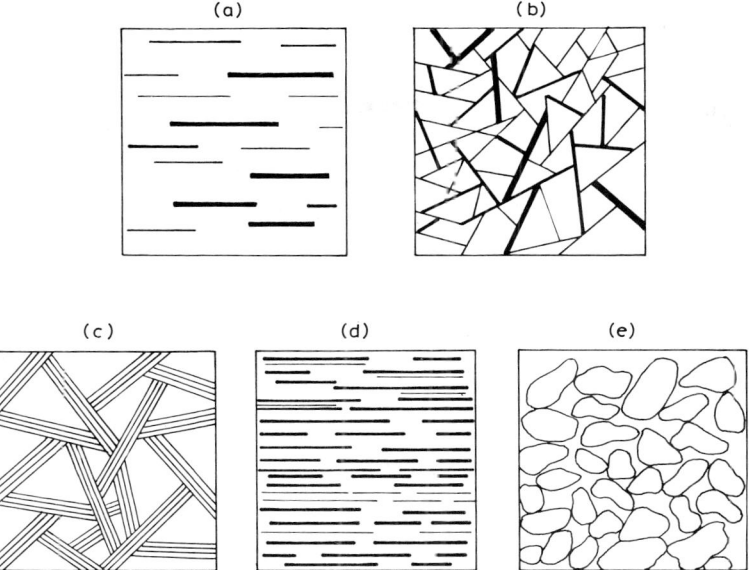

Fig. 2-6. Patterns of particle arrangement: (a) scattered; (b) flocculated ("card house"); (c) flocculated and aggregated ("book house"); (d) packed and oriented; (e) granular.

to enable the particles to stay suspended, they will settle and, contacting each other, form a structure shown in Fig. 2-6(b). It is known as the card house, a term coined by Rosenqvist (1959).

The settling in quiet water of bigger clay particles (over $0.25\,\mu m$) produces the so-called (after Filatov, 1936) ordinary cellular system. Salts, which may be present in water, impair the magnitude of the repulsive forces and bring about flocculation (the growing of particles in size due to the action of attractive forces). Unlike coagulation, resulting from changes in the double electric layer of ions at the surface of a particle, flocculation accrues from weakness of the molecular bonds between the particles and the fluid. Fresh deposits settling in sea water consequently form a flocculated system.

Particles (such as those of silt or dust) of a weight sufficient to counteract the interparticle forces settle without bonding, forming a granular structure (Fig. 2-6(e)).

The patterns of particle arrangement depicted in Fig. 2-6(a) and (b), are of a rather hypothetical meaning; in any case, as far as real soils are concerned, they may be met with only in freshly laid, thickening deposits and suspensions. Aggregations of particles are characteristic of the overwhelming majority of normally compacted clay soils. Therefore, it seems more appropriate to use the term "book house" (Fig. 2-6(c)) instead of card house, highlighting the fact that not individual particles but their aggregations are in contact.

There is evidence that the strength of soils is significantly influenced by the history and the type of texture of the soils. Kulkarny (1973) examined the ultimate short-term strength, long-term strength under the conditions of creep and Bingham's yield limit of marine clay in the light of the above factors. Three series of specimens were subjected to testing. In one, the specimens were prepared from clay having an initial moisture content $W = 2W_L = 230\%$ (silts). The other series, consisted of specimens prepared from a suspension with an initial moisture content of 100% reduced eventually to 80% by compaction with the retention of the initial structure. The salt content was 26 g/l in each case. In the third series were specimens of the above two kinds with a salt content artificially reduced to 1.5–2 g/l. The three series of specimens were classified by the author as having dispersed, flocculated and leached systems, respectively. Preparatory to the tests, all specimens were brought into the state of consolidation (at various values of W) and over-consolidation by compaction. Testing was done on a triaxial compression apparatus, and the results were presented in the form of Mohr shearing diagrams.

It was found that cohesion and the angle of internal friction increased with density during the test for short-term strength, the angle of friction of the dispersed and flocculated systems being greater than that of the leached system. A decrease in W brought about an increase in strength, the yield limit at a given W being lowest for the leached system, somewhat higher for the dispersed system and at a maximum for the flocculated system. The effect of the electrolyte content (salinity) was more pronounced the greater the variance in the history of soil origin.

Shaping of texture. A random microtexture is characteristic of freshly settled deposits in most cases. Orderly texture, if met with in natural conditions, is, as a rule, a secondary texture formed from randomly oriented systems when these undergo changes under a compacting pressure imposed by the ever-increasing dead weight of the deposit. Under pressure, the particles of a random initial structure assume parallel orientation and those arranged in an initial parallel pattern form a more closely packed system. Texture may also arise from a long-term shearing deformation of the soil.

Microtexture. All the systems discussed above (except, of course, the pattern in Fig. 2-6(e)) come under the category of ultramicrotextures, characterizing mutual arrangement of individual colloidal clay particles and ultramicroaggregates of particles.

However, as pointed out earlier, real clay soils are composed of aggregates and blocks of various sizes between which is the clay mass. Then, it is logical to regard the aggregates interlinked by the clay mass rather than the individual particles as the structural elements of the cellular and flocculated systems. Such systems found in marine- and fresh-water deposits are depicted in Fig. 2-7. The former are characterized by coarse aggregates and wide pores; the latter display, owing to the strong forces of attraction, a pattern of considerably smaller dimensions.

Mesotexture. In assessing mesotexture, the mutual arrangement of the particles, aggregates and blocks formed by particles must be taken into account. Numerous possible combinations are classified, according to Raitburd et al. (1968) as follows: random texture — one with over 90% of the particles located at random; random texture with elements of order — some 10–20% of the particles and aggregates display an orderly (axial) orientation; orderly texture — over 90% of the particles are in axial

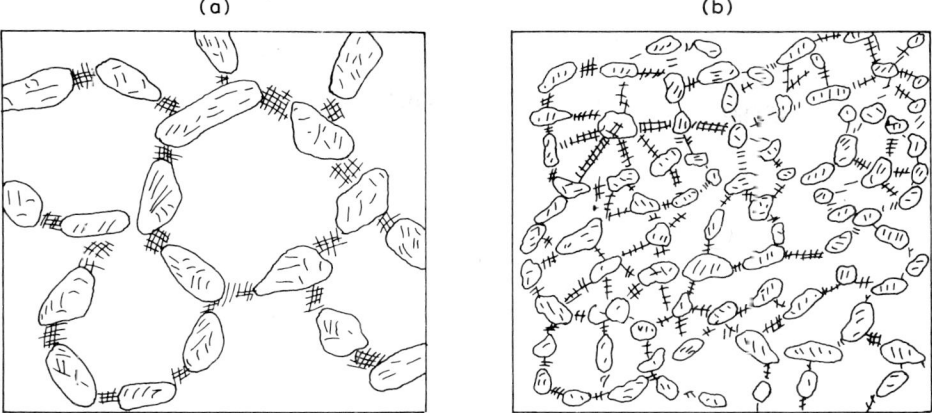

Fig. 2-7. Flocculated and cellular system of clay soils after aggregation. (a) Marine deposits. (b) Freshwater deposits.

orientation; layerwise orderly texture — the degree of orientation and the angle of deviation varies from layer to layer in a stratified soil.

Reorientation of particles in response to deformation. The initial orientation of clay particles may alter when soil is deformed either in natural conditions or in the laboratory.

Among the pioneers of research into the process of particle reorientation were Filatov (1836) and Popov (1944, 1956) studying, by optical means, the changing microstructure of clays subjected to compression. Ter-Stepanyan, too, investigated the shear-induced changes in microstructure (1948).

Later, the changing microstructure of clay soils suffering various deformations stirred the interest of other researchers both in this country and abroad. The findings of Bondarik et al. were published in 1968 and 1970. Among the researchers working abroad were Lambe (1958), Skempton (1964), Morgenstern and Tchalenko (1967), Ingles (1968), Barden et al. (1970), Green-Kelly et al. (1970), Barden (1972), Collins and McGown (1974), Morgenstern (1969), Push (1979). Matsuo and Kaman (1977), studying under an electronic microscope changes in the microstructure of soil due to consolidation and shear, have suggested a relationship between particle reorientation and loading in the form:

$$M = 0.13 \log P + 0.26$$

where:

$$M = \frac{1}{\sum n_i} \sqrt{\left(\sum n_i \sin 2\bar{\theta}\right)^2 + \left(\sum n_i \cos 2\bar{\theta}\right)^2}$$

n_i is the number of observations in each group; $\bar{\theta}$ is the azimuth of the group of observations.

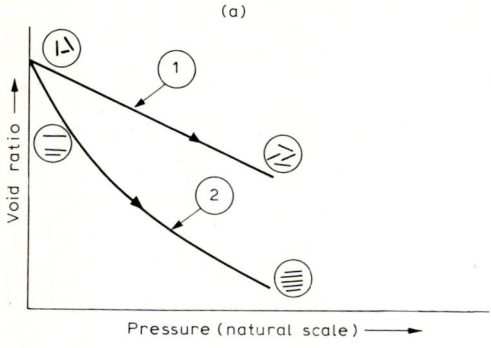

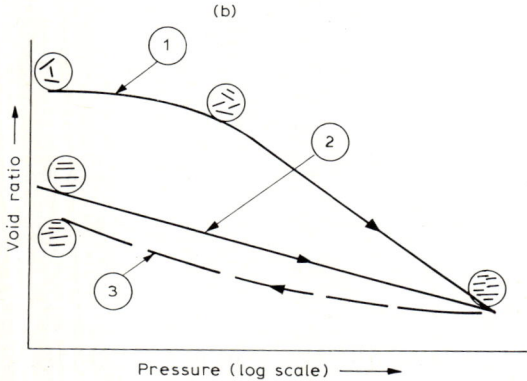

Fig. 2-8. Changes in soil structure due to compaction. (a) Compaction by low pressure. (b) Compaction by high pressure. *1* = specimen with moisture content less than the optimum one or specimen with undisturbed structure; *2* = specimen with moisture content higher than the optimum one or specimen from remoulded clay; *3* = unloading curve for both specimens.

Reorientation of particles in response to compaction of soil. Changes in the texture of clay soils resulting from compaction are explicitly illustrated in the graphs of Fig. 2-8 plotted by Lambe (1958).

Both curves 1 correspond to specially prepared specimens with a water content less than the optimum value and to samples with an undisturbed original structure of a highly random kind. Curves 2 represent specimens with a water content in excess of the optimum and samples of remoulded soil; their original structure approached the orderly type with a parallel arrangement of particles.

Compaction of type 1 specimens by a comparatively light load (Fig. 2-8(a)) results in partial reorientation without attaining an absolutely parallel packing. In type 2 specimens, having an initial oriented structure, the same loading brings the parallel particles closer to one another. An increase in the compacting load intensifies both processes. So, in particular, the totally random structure of type 1 specimens assumes an oriented arrangement and, under a certain loading, the structural difference

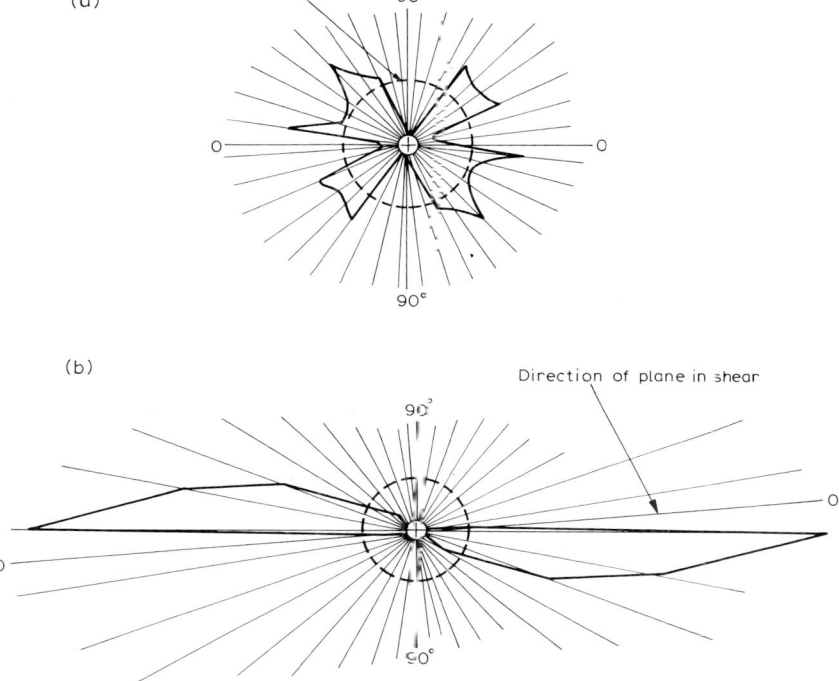

Fig. 2-9. Reorientation of particles in clay soil in response to shearing deformation: (a) circle diagram representing orientation of particles before shear; (b) ditto after shear. (After A.Ya. Turovskaya.)

between the specimens of both types ceases to exist, the particles being packed strictly into a parallel arrangement in either case. During unloading, this arrangement remains almost intact except for a slight increase in the distance between the particles. This indicates that the process of reorientation is an irreversible one.

Reorientation of particles during the process of shear. A more pronounced structure-orienting response is observed in soils subject to shearing forces. In this case, the soil particles tend to set their basal planes parallel to the direction of shear.

The way particles reorient themselves is exemplified in Fig. 2-9, which shows the circle diagrams of particle orientation obtained by Goldstein and Turovskaya (1964) during shearing tests of kaoline clay. The elongated shape of the diagram extending in the direction of the shear indicates that the particles appear to be oriented in the given direction after deformation.

Similar textural changes are observed in natural conditions, particularly during slides. Skempton (1964) noted that the microstructural investigations of the soil samples collected in the zone of slide prove that the surface of natural slip consists of continuous streaks, inside which the clay particles are strongly oriented in the direction of shear. Other researchers have arrived at similar results.

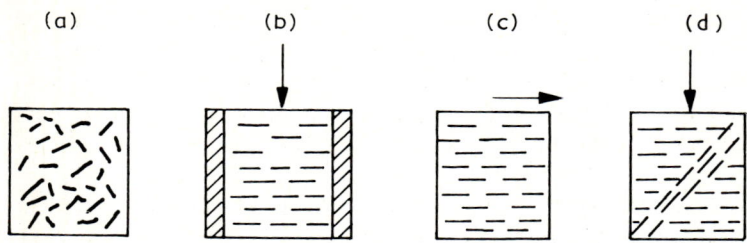

Fig. 2-10. Reorientation of soil particles depending on the arrangement of stress: (a) original condition; (b) confined compression; (c) shear; (d) uniaxial compression.

Reorientation of particles depending on the arrangement of stress in soils. Let us summarize the results of the studies considered above. A deformed soil undergoes structural and textural changes consisting of particle rearrangement and reorientation. If at the onset of deformation the particles have a random rearrangement or are oriented orderly but different to the external force direction, then in the course of deformation they tend to assume an orderly arrangement consistent with the minimum free energy.

Particle reorientation on the above lines results from deformation of any kind (except uniform all-around compression) but the direction of the reoriented particles depends on the arrangement of stress (Fig. 2-10). Thus, compression without provision for sidewise expansion (confined compression) causes the particles to take up a position in which their basal planes are at right angles to the normal compacting force. Reorientation begins even under a light load, getting more pronounced as the load increases. Shear deformation causes the particles to orient themselves parallel to the direction of shear.

Under uniaxial compression (with provision for sidewise expansion), the basal planes of particles are set perpendicular to the force direction but, with an increase in load and progress of shear deformation, the particles appearing in the zone of shear turn in the direction of the maximum tangential stress.

2.8 ANISOTROPY IN SOIL

Anisotropic mechanical properties of soil

In an orderly soil structure with parallel orientation of particles, the mechanical properties are anisotropic — the resistance of the soil to deformation and failure varying with respect to the coordinate axes. Distinction is made between anisotropic deformation and anisotropic strength. The former manifests itself irrespective of the pattern of orientation, which may be between or within aggregations. The latter poses a problem.

More than once it has been experimentally proved, that the shear strength of an oriented clay in a direction parallel to the strata is inferior to that at right angles to the basal surfaces. Therefore, according to Larinov (1968), the ultimate long-term strength of Tortonian clay composed of montmorillonite and mica hydrates appeared to be $7 \cdot 10^5$ Pa (7 kg/cm^2) and 10^6 Pa (10 kg/cm^2), respectively. Morgenstern and Tchalenko (1967) have found that the shear strength of remoulded kaoline with an orderly texture decreases by 20% in those cases when the shear force is applied parallel to the strata.

However, there are also somewhat different data. Barden (1972) believes that soils with an orderly texture, despite deforming anisotropically may have a strength which is — contingent on the structure — either isotropic or anisotropic. He attributes the failure of clay soils to weakening of the forces of attraction between the macro-aggregates and the blocks the soil is composed of. Hence, an anisotropic strength is inherent only in soils with an orderly macrotexture. If, however, the blocks are arranged at random, the soil strength will be isotropic even if the microtexture of the blocks is of the orderly kind. Experiments with kaoline of an orderly texture have yielded cohesion and friction data not influenced by the force direction, whereas the deformation has been anisotropic.

The problem of anisotropy in soil and its effect on mechanical properties of the material was discussed in a number of papers presented at the 7th ICSMFE. Sergeev and Osipov (1977) pointed out that the shear strength ratios of specimens with perpendicular and parallel orientation varied between 0.8 and 2.2 in either case depending on the degree of orientation during the test. Starzewski and Thomas (1977), testing compacted clay for triaxial compression, discovered that the modulus of deformation, E, and Poisson's ratio, v, varied appreciably in their values during deformation in the vertical and horizontal (with respect to the natural occurrence) directions. They suggested equations (linear) of axial, radial and volumetric deformations allowing for the difference mentioned. Equations of shearing and volumetric deformations accounting for stratification and anisotropy in soil were also derived by Sobotka et al. (1977) from shear tests of specimens in which the layers of sand and clay were inclined in various directions.

Effect of particle orientation

Most of the researchers (Goldstein, Larionov, Tan, Lambe, Skempton) believe that the shear strength of soils having an oriented texture decreases in the direction of orientation. The explanation is that repulsive Coulomb forces arise between each pair of parallel particles, whereas in the case of disorderly texture the contact between unlike charges produces attractive forces. It is also believed that the film of bound water increasing in thickness exerts a wedging effect on particles with parallel arrangement.

Other authors oppose this point of view. They think that a parallel arrangement of particles adds to the strength of soil, for the contact area and number of contacts between particles increase in this case. These were the results of experiments with shear-strain apparatus at Moscow State University, obtained, in particular, by Shibakova (Shibakova, 1970; Shibakova et al., 1972). The values of cohesion, c, and

TABLE 2-II

Shear test of remoulded clay with oriented and random structure

Soil	Oriented structure		Random structure	
	c (10^5 Pa)	ϕ	c (10^5 Pa)	ϕ
Kaoline	0.01	9	0.005	7.2
Montmorillonite	0.09	7.4	0.20	2.7

the angle of internal friction, ϕ, obtained from testing kaoline ($W = 47\%$) and montmorillonite ($W = 86\%$) specimens with random and orderly texture (in this latter case the shear has been applied in the direction of particle orientation) are given in Table 2-II. As can be seen from that table, the angle of friction, ϕ, is higher in soils with oriented texture. Cohesion of the montmorillonite having an oriented texture is lower than in the case of random texture; the cohesion of the kaoline approaches zero in both cases.

The shear tests of clays with undisturbed structure of various textures have yielded the following values of shear strength: (1) soil with block-like oriented texture ($W = 72.5\%$), $\tau = 0.3 \cdot 10^5$ Pa; (2) soil with oriented texture ($W = 93.5\%$), $\tau = 0.22 \cdot 10^5$ Pa; (3) soil with random texture ($W = 94\%$), $\tau = 0.17 \cdot 10^5$ Pa. Thus, it is apparent that the strength of clay soil with an oriented texture is higher than that of soil having a random texture.

At the same time, a point to be noted in comparing the strength of soils with different textures is that parallel packing adds to the density of soil and influences its strength. If the particles come so close together that the forces of attraction predominate, they by far exceed the forces of interaction of the randomly arranged particles. Hence, it may be postulated that in the above examples the key factor is the difference in soil density rather than the difference in orientation.

Difference in the resistance a soil is offering to compression and tension

One of the basic properties of soil is that the resistance it offers to compression is different from that observed in tension. This not only influences soil behaviour with regard to strength and deformation but, as will be shown later, has a pronounced effect on the stress–strain state of a soil.

The nature of this phenomenon is as follows. The water-colloidal bonds linking the fine particles of a dispersed soil are weaker in resisting a tearing-off action than in opposing compression. In this latter case, the water films become thinner, the number of contacts increases, including the so-called "dry" contacts having a maximum strength, and, as a result, the soil skeleton is included in the operation. This all adds to the compressive strength of soil and impedes its deformation under a given loading.

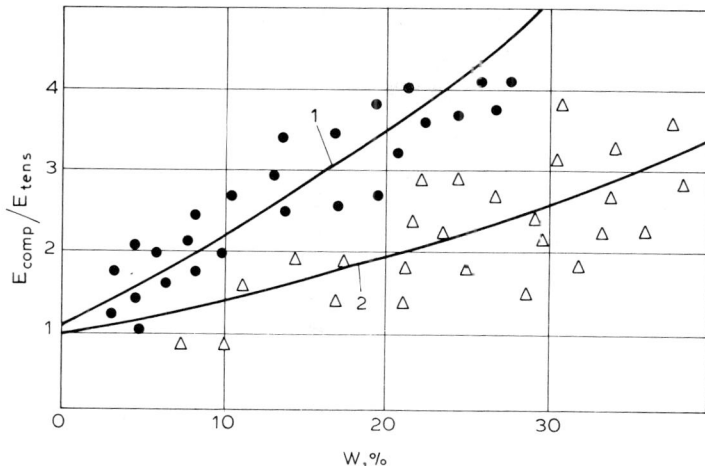

Fig. 2-11. Relationship between the moduli of deformation in compression, E_{comp}, and tension, E_{tens}, depending on the moisture content of soil: *1* = clay loam; *2* = clay. (After I.V. Lushnikov et al., 1973.)

Fig. 2-11 depicts the results of the uniaxial compression and tensile tests of specially prepared specimens in clay loam and clay having various water contents, which were undertaken by Lushnikov et al. (1973) at the Urals Polytechnic. The curves represent the way the ratio of the modulus of deformation due to compression, E_{comp}, and that due to tension, E_{tens}, are changing with water content. It can be seen that for sandy clay the ratio is higher than for sand, being $E_{comp}/E_{tens} = 4$ in the former case and 3 in the latter. This is natural for the interparticle forces and, consequently, the strength of bonds are far weaker in sandy clay than in clay. Apparently, the difference between the moduli E_{comp} and E_{tens} will increase with a decrease in cohesion, for in cohesionless granular soils the ratio E_{comp}/E_{tens} is infinity (because in this case $E = \sim 0$).

Another feature represented by the curves is an increase in the ratio of E_{comp} and E_{tens} with water content. This is also quite logical, for a higher water content augments the thickness of water films and drastically reduces the tensile strength of interparticle bonds. Inside the limits specified for suspensions, E_{comp}/E_{tens} also tends towards infinity, for, in this case, $E_{tens} = 0$.

The above tests have revealed a difference not only between the values of the moduli of deformation but also between those of the proportional limit. Therefore, the proportional limit in compression was $1.5 \cdot 10^5$–$2.5 \cdot 10^5$ Pa against $0.2 \cdot 10^5$–$0.5 \cdot 10^5$ Pa in tension, the former case being 7.5 times the latter.

The strength and deformation characteristics of crystalline rock in compression and tension are also different owing to macro- and micro-cracks as well as other structural flaws.

2.9 REFERENCES

Barden, L., Sides, G.R. and Karunaratue, J.P., 1970. A microscopic study of aspects of clay structure. Proc. 2nd Southeast Asian Conf. Soil Eng., Singapore.

Barden, L., 1972. The influence of structure on deformation and failure in clay soils. Geotechnique, 22 (1).
Bondarik, G.K., Berezkina, G.M., Tsareva, A.M., Raitburd, T.M., Ponomarev, V.V. et al., 1968. Modern Methods of Studying Engineering Properties of Rocks. All-Union Res. Inst. Geol., Moscow (in Russian).
Bondarik, G.K., Berezkina, G.M., Tsareva, A.M., Raitburd, T.M., Ponomarev, V.V. et al., 1970. Problems of Engineering Geology. Reports of Soviet scientists, Int. Congr. Eng. Geol., Moscow, All-Union Inst. Sci. Tech. Information (in Russian).
Collins, K. and McGown, A.M., 1974. The form and function of microfabric features in a variety of natural soils. Géotechnique, 24.
Denisov, N.Yu., 1956. Engineering Properties of Clay Soils and Their Use in Hydraulic Engineering. Gosenergoizdat Publ., Moscow, Leningrad (in Russian).
Deryagin, B.V., 1956. The theory of thin water layers as an explanation of the properties of clay soils. Proc. Conf. Eng. Prop. Rocks and Methods of Their Study, 1, USSR Acad. Sci., Moscow (in Russian).
Filatov, M.M., 1936. On microscopic structure of soils in connection with their deformation under load. In: Physics of USSR Solids. Proc. Conf. Int. Assoc. Pedologists, V. 5, No. 3, Moscow (in Russian).
Geuze, E.C.W.A. and Tan, Tjong-Kie, 1954. The mechanical behaviour of clays. Proc. 2nd Int. Congress of Rheology, Oxford, 1953.
Goldstein, M.N., Turovskaya, A.Ya. and Babitskaya, S.S., 1964. On the strength of weak saturated clays. Proc. 7th ICSMFE, Mexico City, V. 1.
Gor'kova, I.M., 1966. Theoretical Principles of Assessing Sedimentary Rocks for Engineering and Geological Purposes. Nauka Publ., Moscow (in Russian).
Green-Kelly, K. and Mackney, D., 1970. Preferred orientation of clays in soils. The effect of drying and wetting. In: D.A. Osmord et al. (Editors), Micromorphological Techniques and Applications. Agric. Res. Council Soil Surv., Tech. Monogr., No. 2.
Ingles, O.G., 1968. Soil chemistry relevant to the engineering behaviour of soils. In: I. Lee (Editor), Soil Mechanics. Selected Topics. Butterworth's, London.
Kulkarni, R., 1973. Effect of structure on properties of marine clay. Proc. 8th ICSMFE, 1 (1): 217–220.
Lambe, R.W., 1958. The structure of compacted clay. J. Soil Mech. Found Div., Proc. ASCE, V. 84, No. SM4.
Larionov, A.K., 1966. Engineering and Geological Study of the Structure of Soft Sedimentary Rocks. Nedra Publ., Moscow (in Russian).
Larionov, A.K., 1968. To the problem of the nature of creep in clay soils. In: Studies of Rheological Properties of Soils. Proc. Coordinating Conf. Hydrol. Eng., Issue 38, Energiya Publ. (in Russian).
Livshits, E.M., 1954. The theory of molecular forces of attraction between condensed bodies. Proc. USSR Acad. Sci., V. 97, No. 4 (in Russian).
Lushnikov, V.V., Vulis, P.D. and Litvinov, B.M., 1973. On the Relationship between the Moduli of Deformation in Compression and Tension of Soils. SMFE, No. 6 (in Russian).[*1]
Matsuo, S. and Kaman, M., 1977. Microscopic study on deformation and strength of clays. Proc. 9th ICSMFE, V. 1.
Morgenstern, N.K. and Tchalenko, J.S., 1967. Microstructural observations on shear zones from slips in natural clays. Proc. Geotech. Conf., V. 1, Oslo.
Morgenstern, N.K., 1976. Structural and physico-chemical effects on the properties of clays. Proc. 7th ICSMFE, Spec. Section, V. 3, Mexico City.
Mustafaev, A.A., 1969. Fundamentals of Settling Soil Mechanics. Stroiizdat Publ., Moscow (in Russian).
Nerpin, S.V. and Chudnovsky, A.F., 1967. The Physics of Soils. Nauka Publ., Moscow (in Russian).
Osipov, V.I., 1979. The Nature of Strength and Deformation Properties of Soils. Moscow State University, Moscow (in Russian).
Popov, I.V., 1944. Creep structure of clays during deformation. Proc. USSR Acad. Sci., V. 45, No. 4 (in Russian).
Popov, I.V., 1956. The effect of crystalline structure in minerals of clay soils on their properties. Proc. Conf. Eng. Geol. Properties of Rocks and Methods of Study, V. I, USSR Acad. Sci., Moscow (in Russian).

[*1] Here and in what follows "SMFE" stands for "Soil Mechanics and Foundation Engineering", a monthly publication appearing in the USSR and translated into English by Plenum Publ. Corp., New York.

Push, R., 1979. Creep of Soils. Ruhr-Universität, Bochum.

Raitburd, M., Tsareva, A.M. and Ponomarev, V.V., 1968. Methodology of studying the structure of clay soils. In: Modern Methods of Studying Physical and Mechanical Properties of Rocks. Moscow (in Russian).

Rebinder, P.A., 1956. Structural and mechanical properties of clay soils and modern concept of chemical physics of colloids. Proc. Conf. Eng. Geol. Properties of Rocks and Methods of Study, V. 1, USSR Acad. Sci., Moscow (in Russian).

Rosenqvist, I.T., 1959. The physico-chemical properties of soils: soil–water systems. J. Soil. Mech. Found. Div., Proc. ASCE, SM2, V. 85.

Sergeev, E.M. and Osipov, V.I., 1977. Structural aspects of shear resistance of clays. Proc. 9th ICSMFE, V. 1, Tokyo.

Sergeev, E.M., Golodkovsky, G.A. et al., 1971. Pecology. Moscow State University, Moscow (in Russian).

Shibakova, V.S., 1970. Effect of texture on shearing resistance of clays. In: Problems of Engineering Geology, Reports of Soviet Scientists, Int. Congr., VINITI Publ. House (in Russian).

Shibakova, V.S. and Polunovsky, A.G., 1972. Clay Texture as a Factor in Assessing Engineering and Geology Properties. In: Engineering and Geological Properties of Clays and Processes Thereinto. Proc. Int. Symp., Issue I, Moscow State University (in Russian).

Skempton, A.W., 1964. Long-term stability of clay slopes (4th Rankine Lecture). Géotechnique, V. 14, No. 2.

Sobotka, Z., Kamenov, B. and Pruška, L., 1977. Stress–strain relations and shear strength of soils. Proc. 9th ICSMFE, V. 1, Tokyo.

Starzewski, K. and Thomas, S.P., 1977. Anisotropic behaviour of an overconsolidated clay. Proc. 9th ICSMFE, V. 1, Tokyo.

Ter-Stepanyan, G.I., 1948. On the effect of the shape and arrangement of particles on the process of shear in soils. News Acad. Sci. Armenian SSR, Section Phys., Math., Nat., Eng. Sci., I (2) (in Russian).

Vyalov, S.S., 1959. Rheological Properties and Bearing Capacity of Frozen Soils. USSR Acad. Sci., Moscow. (Translated into English by USA Cold Regions Res. Eng. Lab., Transl. 74, Hanover, NH 1965.)

Chapter 3

STRESSES AND STRAINS

3.1 STRESSES AND STRAINS AT A POINT

Soil as a continuum

The ultimate goal of continuum mechanics is to provide a mathematical description of the motion of deformed bodies.

Under the concept of "continuum" we understand a model of a material body (solid, liquid or gaseous) filling a part of the space in a continuous way in spite of being made up of individual particles — atoms and molecules. The structural continuity of such an idealized body is retained during the process of deformation.

The hypothesis of a continuum of a material body provides the basis for examining the state of stress and strain of the body as a whole by considering the stress and strain of infinitesimal volumes and applying the techniques of differential calculus.

Soil, an assemblage of individual particles separated by voids filled with either air or water, is essentially a discrete medium rather than a continuous one. Therefore, a model of real soil will be one providing the description of the interaction between individual particles in terms of statistics with allowance for the physical nature of the existing bonding.

Attempts to produce a soil model of this kind are known. For example, Muler (1962) and Kandaurov (1966) considered the mechanism of stress distribution in a discrete cohesionless medium in terms of a statistical analysis of the forces of interaction between particles as they contact or pass one another.

A widely-known model by Rowe (1962) describes the form of a medium consisting of spheres whose contacts give rise to frictional forces. Similar and different models of discrete media have been suggested by other authors. However, a new and comprehensive theory of soil deformation is a matter for the future.

Now, the mechanism of the stressed and deformed state of a soil is successfully described in terms of continuum mechanics. The applicable principles appear to be valid because the elements of soil structure are of a size which is a small fraction of the least volume of soil under consideration.

It should be noted that continuum mechanics does not examine the microprocesses taking place in real bodies and disregards their structural details; this is the domain of theoretical physics. However, if probing into microprocesses results in establishing the stress–strain relation, one can describe macroprocesses in terms of continuum mechanics.

This chapter deals with the basic formulae used in the theory of stress and strain of a continuum, with emphasis on the data which will be helpful in examining the combined stress–strain state outlined in later chapters.

Components of stress and strain

A stressed and deformed state of a medium at a given point is comprehensively defined by nine components of stress and nine components of deformation at three mutually perpendicular planes. The components of stress (Fig. 3-1) are three normal stresses, σ_x, σ_y, σ_z, and three pairs of reciprocally equal tangential stresses, $\tau_{xy} = \tau_{yx}$, $\tau_{yz} = \tau_{zy}$, $\tau_{zx} = \tau_{xz}$. The components of strain are three linear strains (unit elongations), ε_x, ε_y, ε_z, and three pairs of reciprocally equal angular distortions (unit shears), $\gamma_{xy} = \gamma_{yx}$, $\gamma_{yz} = \gamma_{zy}$, $\gamma_{zx} = \gamma_{xz}$.

The above notation is accepted in engineering literature. In continuum mechanics preference is given to the notation σ_{ij} and ε_{ij}, where $i, j = x, y, z$ or $1, 2, 3$. Like values of subscripts $i = j$ correspond to the normal stresses and strains, the unlike ones, $i \neq j$, represent the tangential stresses and strains. Thus, instead of σ_x, σ_y, σ_z and ε_x, ε_y, ε_z, we obtain σ_{xx}, σ_{yy}, σ_{zz} and ε_{xx}, ε_{yy}, ε_{zz} or σ_{11}, σ_{22}, σ_{33}, and ε_{11}, ε_{22}, ε_{33}, whereas τ_{xy}, τ_{yz}, τ_{zx} and γ_{xy}, γ_{yz}, γ_{zx} are replaceable by σ_{xy}, σ_{yz}, σ_{zx} and ε_{xy}, ε_{yz}, ε_{zx} or σ_{12}, σ_{23}, σ_{31} and ε_{12}, ε_{23}, ε_{31}. Although this system is more compact, we shall use the notation familiar to the engineer and adopted in textbooks on civil engineering.

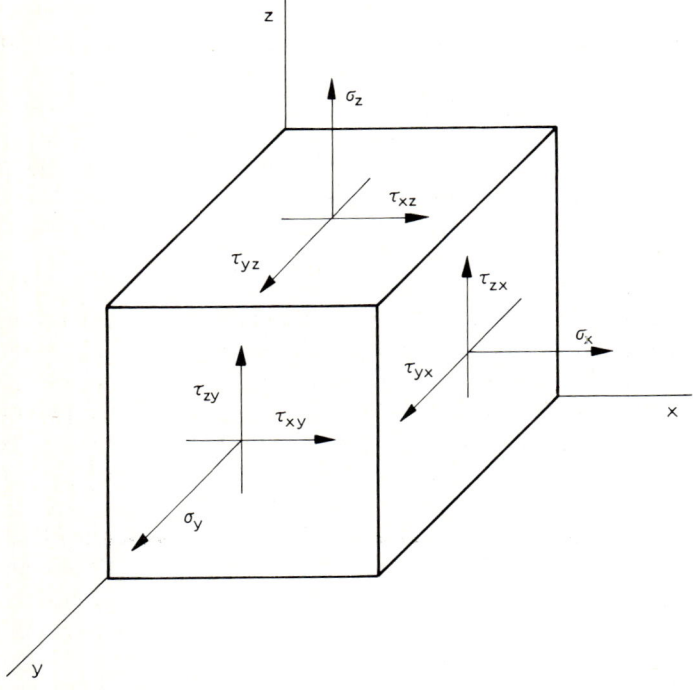

Fig. 3-1. Stress components.

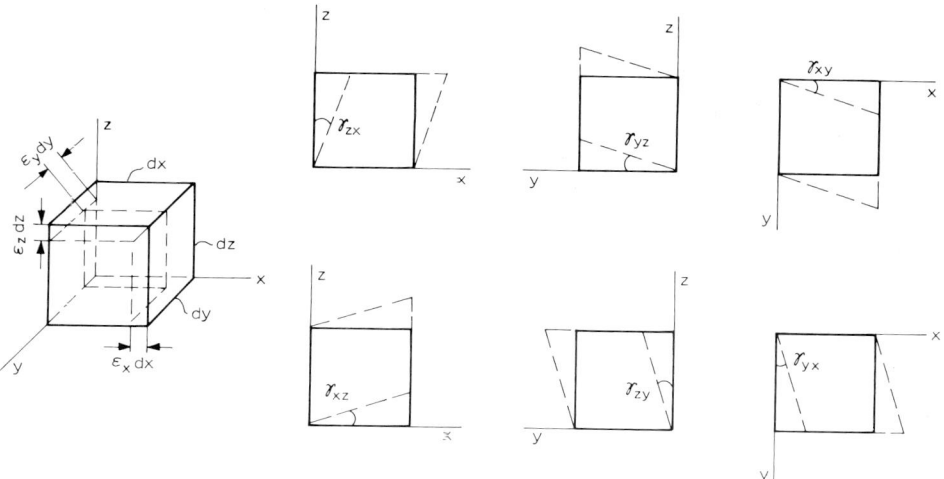

Fig. 3-2. Strain components.

The deformed state of an elemental volume dx by dz acted upon by normal and tangential stresses is depicted in Fig. 3-2. The linear strains ε_x, ε_y, ε_z represent unit elongations of the edges of an elemental parallelepiped. Linear strains producing shortenings are commonly taken in soil mechanics as being positive and those which result in extensions, as negative.

The unit shears $\gamma_{xy} = \gamma_{yx}$, $\gamma_{zy} = \gamma_{yz}$, $\gamma_{zx} = \gamma_{xz}$ define the amount by which the angles between the faces of the parallelepiped are distorted. It is assumed that a reduction of the angle between the positive direction of the axes corresponds to a positive shear and an increase in the angle indicates that the shear is negative.

Principal stresses and principal strains

Consider the state of stress at a point of an elemental parallelepiped provided that all nine components of the stresses acting on its faces are known.

Through the given point draw an oblique plane with a normal v, arbitrarily inclined with respect to the faces of the parallelepiped (Fig. 3-3). Denote the direction cosines of the angles the normal v makes with the x-, y-, z-axes as $\cos(x, v) = l$, $\cos(y, v) = m$, $\cos(z, v) = n$, respectively, keeping in mind that $l^2 + m^2 + n^2 = 1$.

From the equation of equilibrium of the tetrahedron we obtain the components of stress at the oblique plane:

$$\begin{aligned} p_{xv} &= \sigma_x l + \tau_{xy} m + \tau_{xz} n \\ p_{yv} &= \tau_{xy} l + \sigma_y m + \tau_{yz} n \\ p_{zv} &= \tau_{zx} l + \tau_{xy} m + \sigma_z n \end{aligned} \qquad (3\text{-}1)$$

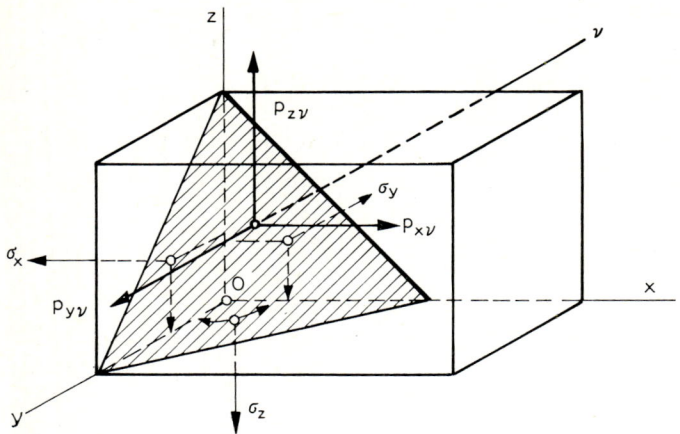

Fig. 3-3. Stresses at a given point.

The total stress acting on the oblique plane is then given by:

$$p_v^2 = p_{xv}^2 + p_{yv}^2 + p_{zv}^2 \tag{3-1'}$$

On resolving the vector of the total stress into its components along the normal and tangent to it, we can compute the normal and tangential stress set up at the plane under consideration. Thus, the normal stress is:

$$\sigma_v = p_{xv}l + p_{yv}m + p_{zv}n$$

or:

$$\sigma_v = \sigma_x l^2 + \sigma_y m^2 + \sigma_z n^2 + 2\tau_{xy}lm + 2\tau_{yz}mn + 2\tau_{xz}nl \tag{3-2}$$

The tangential stress is given by the formula:

$$\sigma_v^2 + \tau_v^2 = p_v^2 \tag{3-3}$$

Apparently, among the multitude of possible positions of the arbitrarily drawn plane there is one in which the tangential stress $\tau_v = 0$ and the normal stress coincides with the total stress in both magnitude and direction $\sigma_v = p_v$. Through any point one can draw at least three mutually perpendicular planes called principal planes; their axes 1, 2, 3 are referred to as principal axes.

Accordingly, the normal stresses acting on these planes are termed the principal normal stresses, $\sigma_1, \sigma_2, \sigma_3$, and the linear strains on similar planes are known as the principal linear strains, $\varepsilon_1, \varepsilon_2, \varepsilon_3$. The principal stresses represent the maximum and minimum values of all the stresses at the given point; the axes are commonly designated so that $\sigma_1 \geqslant \sigma_2 \geqslant \sigma_3$.

Through any point one can also draw three planes on which the tangential, rather than normal, stresses are at maximum. These maximum stresses are called the principal tangential stresses:

$$\tau_1 = \frac{\sigma_2 - \sigma_3}{2} \qquad \tau_2 = \frac{\sigma_3 - \sigma_1}{2} \qquad \tau_3 = \frac{\sigma_1 - \sigma_2}{2} \tag{3-4}$$

The angular distortions (or principal shears) on similar planes are:

$$\gamma_1 = \varepsilon_2 - \varepsilon_3 \qquad \gamma_2 = \varepsilon_3 - \varepsilon_1 \qquad \gamma_3 = \varepsilon_1 - \varepsilon_2 \tag{3-5}$$

Mohr circles

Should the plane be rotated, the values of the normal and tangential stresses will apparently change depending on the angle the normal, n, to the plane makes with the axes 1, 2, 3. In accordance with expressions 3-2 and 3-3 and taking $n = v$, these stresses are given by:

$$\sigma_n = \sigma_1 l^2 + \sigma_2 m^2 + \sigma_3 n^2$$

$$\sigma_n^2 + \tau_n^2 = \sigma_1^2 l^2 + \sigma_2^2 m^2 + \sigma_3^2 n^3$$

where $l = \cos(1, n)$, $m = \cos(2, n)$, $n = \cos(3, n)$.

Assuming that the values of $\cos(1, n)$, $\cos(2, n)$ and $\cos(3, n)$ are in constant succession, we can construct three circles of stress, using the points $(\sigma_2 + \sigma_3)/2$, $(\sigma_1 + \sigma_3)/2$, $(\sigma_1 + \sigma_2)/2$ as the centres and $(\sigma_2 - \sigma_3)/2$, $(\sigma_1 - \sigma_3)/2$, $(\sigma_1 - \sigma_2)/2$ as the radii.

The circles obtained (Fig. 3-4) are the Mohr circles for the three-dimensional stress state. The points between the confines of the large and two small circles define the

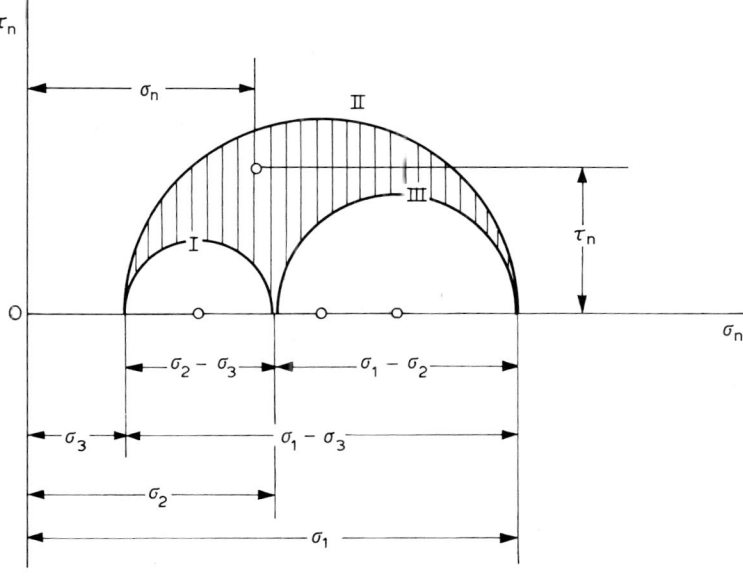

Fig. 3-4. Mohr circles.

Stresses and strains on the octahedral plane

The plane drawn through a given point can be located so as to be uniformly inclined with respect to the three planes of the principal stresses (Fig. 3-5). This plane, displaying certain specific properties, is referred to as the octahedral plane.

Since the direction cosines on the octahedral plane are all the same ($l = m = n = 1/\sqrt{3}$), the total stress on this plane is given according to formula 3-1' by:

$$p_v^2 = \tfrac{1}{3}(\sigma_1^2 + \sigma_2^2 + \sigma_3^2) \tag{3-6}$$

The normal and tangential stresses, in their turn, are given in accordance with formulae 3-2 and 3-3 by:

$$\sigma_{oct} = \tfrac{1}{3}(\sigma_1 + \sigma_2 + \sigma_3) \tag{3-7}$$

$$\tau_{oct} = \sqrt{p_v^2 + \sigma_{oct}^2} = \tfrac{1}{3}\sqrt{(\sigma_1 - \sigma_2)^2 + (\sigma_2 - \sigma_3)^2 + (\sigma_3 - \sigma_1)^2}$$
$$= \tfrac{2}{3}\sqrt{\tau_{12}^2 + \tau_{23}^2 + \tau_{31}^2} \tag{3-8}$$

By referring the stress components to a system of coordinates x, y, z, expressions 3-7 and 3-8 take the form:

$$\sigma_{oct} = \tfrac{1}{3}(\sigma_x + \sigma_y + \sigma_z) \tag{3-7'}$$

$$\tau_{oct} = \tfrac{1}{3}\sqrt{(\sigma_x - \sigma_y)^2 + (\sigma_y - \sigma_z)^2 + (\sigma_z - \sigma_x)^2 + 6(\tau_{xy}^2 + \tau_{yz}^2 + \tau_{zx}^2)} \tag{3-8'}$$

The above σ_{oct} and τ_{oct} are commonly termed octahedral normal stress and octahedral tangential stress, respectively.

A plane drawn through a given point in a body can be uniformly inclined with respect to the three principal planes of strain. In classical mechanics, it is assumed that this plane coincides with the octahedral plane in terms of stresses; possible deviations

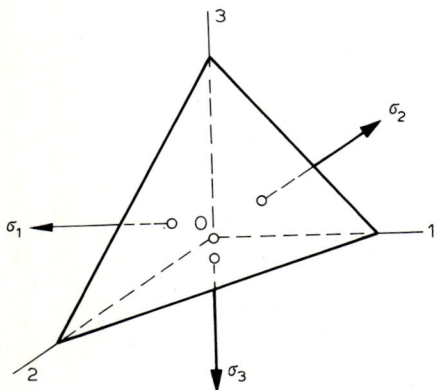

Fig. 3-5. Octahedral plane.

from this assumption will be discussed later in this book. A unit elongation normal to the octahedral plane (octahedral elongation) will be given by:

$$\varepsilon_{oct} = \tfrac{1}{3}(\varepsilon_1 + \varepsilon_2 + \varepsilon_3) = \tfrac{1}{3}(\varepsilon_x + \varepsilon_y - \varepsilon_z) \qquad (3\text{-}9)$$

and an angular distortion on the same plane, i.e. an octahedral angular distortion, by:

$$\begin{aligned}\gamma_{oct} &= \tfrac{2}{3}\sqrt{(\varepsilon_1 - \varepsilon_2)^2 + (\varepsilon_2 - \varepsilon_3)^2 + (\varepsilon_3 - \varepsilon_1)^2} \\ &= \tfrac{2}{3}\sqrt{(\varepsilon_x - \varepsilon_y)^2 + (\varepsilon_y - \varepsilon_z)^2 + (\varepsilon_z - \varepsilon_x)^2 + \tfrac{3}{2}(\gamma_{xy}^2 + \gamma_{yz}^2 + \gamma_{zx}^2)}\end{aligned} \qquad (3\text{-}10)$$

3.2 CHANGES IN VOLUME AND SHAPE

Volume changes

An elemental parallelepiped acted upon by three mutually equal compression stresses is said to be in the state of all-around or hydrostatic compression:

$$p = \sigma_x = \sigma_y = \sigma_z \qquad (3\text{-}11)$$

The tangential stresses at any point of the parallelepiped will equal zero.

The applied normal stresses bring about unit elongations (shortenings) of the parallelepiped edges $\varepsilon_x = \varepsilon_y = \varepsilon_z$. Assuming that the initial dimension of every edge is unity (the volume being $V = 1 \times 1 \times 1$), the fractional change in volume of such a soil is:

$$\varepsilon_V = \frac{\Delta V}{V} = 1 - (1 - \varepsilon_x)(1 - \varepsilon_y)(1 - \varepsilon_z)$$

If the elongations can be ignored compared with unity, the above expression, on being developed, takes the form (the quantities of the second and third order are neglected):

$$\varepsilon_V = \varepsilon_x + \varepsilon_y + \varepsilon_z = \varepsilon_1 + \varepsilon_2 + \varepsilon_3 \qquad (3\text{-}12)$$

It is thus evident that three mutually equal unit elongations (shortenings) resulting from an all-around pressure induce a volumetric strain ε_V on the parallelepiped without distorting its shape.

Distortion of shape

Consider now the effect of tangential stresses and of the angular distortions they produce. When the angle of shear is small compared with unity, it may be assumed that no elongation of the edges of the parallelepiped is produced and the only deformation it suffers in this case is skewing and rotation. A deformation caused by the action of two mutually equal tangential stresses $\tau_{ij} = \tau_{ji}$ is referred to as simple shear.

Denoting the angle through which the parallelepiped skews under a stress by α and

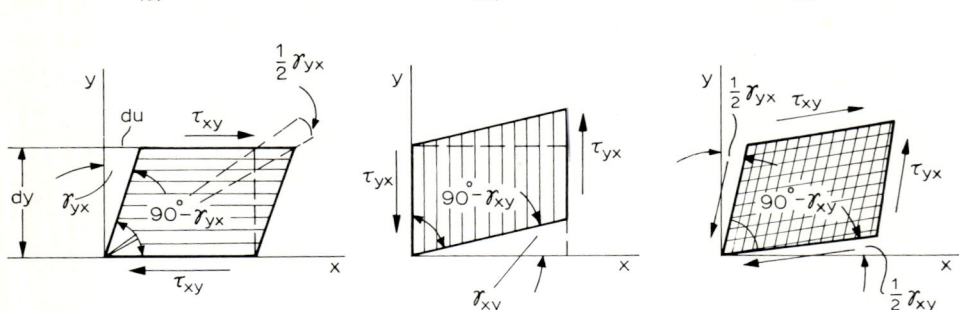

Fig. 3-6. (a), (b) Simple shear. (c) Pure shear resulting from the superposition of the simple shears.

the displacement along the axis by u (Fig. 3-6(a)), the angular distortion can be expressed as the gradient of displacement $\gamma_{yx} = \tan \alpha = (du/dy)$.

Taking quantity as infinitesimal, we can write $\gamma_{yx} = \alpha$, approximately. Note that simultaneously with skewing, the parallelepiped is rotated through an angle $\gamma_{yx}/2$. Apparently, a similar angular distortion will occur due to a stress τ_{yx} (Fig. 3-6(b)).

Superposition of two simple shears with the same angles of distortion, $\gamma_{xy} = \gamma_{yx}$, at the adjacent edges results in pure shear. In other words, the shape of the parallelepiped is distorted without changing its volume or rotating about its axes (Fig. 3-6(c)). Pure shear may also result from the action of two equal stresses of opposite signs, $\sigma_1 = -\sigma_3$, formed in the parallelepiped.

It is thus evident that deformation of a body is distortion of its shape due to the action of tangential stresses and a volumetric change resulting from an all-around pressure.

A different approach to the concept of deformation is necessary for studying the patterns of deformations which may obey different laws. Therefore, in elastic and viscous bodies, volumetric strain varies directly with all-around pressure, whereas their resistance to distortion of shape is sharply different: elastic bodies changing their shape directly with shear stress but viscous bodies being incapable of retaining shape and resisting shear.

Strains and displacements

Let a point M, with the coordinates x, y and z, displace, due to the body's deformation, into a position M_1, defined by the coordinates $x + \Delta x$, $y + \Delta y$, $z + \Delta z$ (Fig. 3-7). Denote the components of the total displacement vector $\overrightarrow{MM_i}$, called the displacement components, by u_x, u_y, u_z (sometimes the displacement components are denoted by u, v, w). Then the total displacement is $\delta = \sqrt{u_x^2 + u_y^2 + u_z^2}$, being a continuous function of the coordinates. This latter condition fully conforms with the hypothesis of structural continuity of the body.

Around point M cut out an elemental parallelepiped so that its edges are parallel to the coordinate axes. The deformation of the parallelepiped can be expressed in

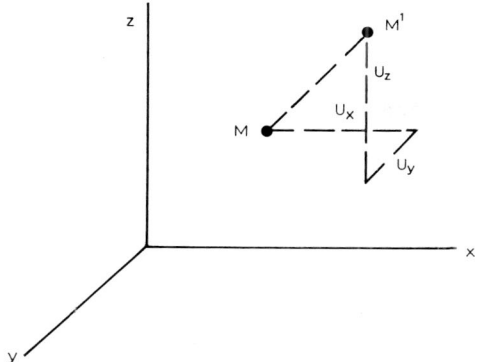

Fig. 3-7. Displacement of point M in a body due to deformation.

terms of displacements as:

$$\varepsilon_x = \frac{\partial u_x}{\partial x} + \frac{1}{2}\left[\left(\frac{\partial u_x}{\partial x}\right)^2 + \left(\frac{\partial u_y}{\partial x}\right)^2 + \left(\frac{\partial u_z}{\partial x}\right)^2\right]$$

. . . .

$$\gamma_{xy} = \frac{\partial u_x}{\partial y} + \frac{\partial u_y}{\partial x} + \left[\frac{\partial u_x}{\partial x}\frac{\partial u_x}{\partial y} + \frac{\partial u_y}{\partial x}\frac{\partial u_y}{\partial y} + \frac{\partial u_z}{\partial x}\frac{\partial u_z}{\partial y}\right] \qquad (3\text{-}13)$$

Continuum mechanics commonly examines small deformations, i.e. those which are far less than unity. In this case, the expressions $(\partial u_x/\partial x)^2$, $(\partial u_x/\partial x)(\partial u_x/\partial y)$, etc., can be omitted from formulae 3-13 which, consequently, then take the form:

$$\varepsilon_x = \frac{\partial u_x}{\partial x} \qquad \varepsilon_y = \frac{\partial u_y}{\partial y} \qquad \varepsilon_z = \frac{\partial u_z}{\partial z}$$

$$\gamma_{xy} = \frac{\partial u_x}{\partial y} + \frac{\partial u_y}{\partial x} \qquad \gamma_{yz} = \frac{\partial u_y}{\partial z} + \frac{\partial u_z}{\partial y} \qquad \gamma_{zx} = \frac{\partial u_z}{\partial x} + \frac{\partial u_z}{\partial z} \qquad (3\text{-}14)$$

Expressions 3-13 and 3-14 correspond to different definitions of deformation or the so-called measures of deformation. Expression 3-14 is obtained if the fractional change in the length of the parallelepiped edge is taken as the measure of deformation. The unit elongation along the x-axis is:

$$\varepsilon_x = \frac{l_0 - l}{l_0} = \frac{\Delta l_0}{l_0} = 1 - \frac{l}{l_0} \qquad (3\text{-}15)$$

where $l_0 = dx$ is the original length of the edge and l is its final length.

Denoting the displacement of the parallelepiped along the x-axis by u_x and the elongation (absolute) of its edge by $(\partial u_x/\partial y)\,dy$ (accurately to within the infinitesimals of the first order of magnitude), we obtain $l = dx + (\partial u_x/\partial y)\,dy$ or $l = dx + du_x$.

Substituting into expression 3-15, we arrive at expressions 3-14, which are called Cauchy's equations. The method of determining deformation by expression 3-15 is termed Cauchy's measure of deformation. Continuum mechanics deals precisely with such small deformations.

Finite strains

Eqs. 3-13 are called Green's equations; Green's measure of strain corresponding to these equations is given by:

$$\varepsilon_x^* = \left(1 - \frac{l}{l_0}\right) - \frac{1}{2}\left(1 - \frac{l}{l_0}\right)^2 = \frac{1}{2}\left[1 - \left(\frac{l}{l_0}\right)^2\right]$$

(3-16)

Omitting the infinitesimals of the second order of magnitude, the above expressions take the form of eq. 3-15.

Expression 3-13 finds application in determining finite strains, i.e. those strains which are commensurate with unity. Another measure of finite strain used more frequently is Hencky's measure.

Consider a small element of body having length x. A change of its length is dx and a unit elongation is dx/x. The total strain of the body will then be equal to the sum of the strains of the individual elements:

$$\varepsilon_x^* = \int_{l_0}^{l} \frac{dx}{x} = \ln \frac{l}{l_0}$$

(3-17)

where l_0 is the original length and l is the final length of the body.

Green's measure and Hencky's measure

Novozhilov (1951, 1952) has shown that the problems of finite deformations can be solved, in terms of Green's measure, by determining the strain components directly from relations 3-13 and the stress components, corresponding to these strains, by the formulae:

$$\sigma_x^* = \sigma_x \sqrt{\frac{(1 + 2\varepsilon_x)(1 + 2\varepsilon_x) - \gamma_{yz}^2}{1 + 2\varepsilon_x}}$$

$$\tau_{xy}^* = \tau_{xy} \sqrt{\frac{(1 + 2\varepsilon_y)(1 + 2\varepsilon_x) - \gamma_{yz}^2}{1 + 2\varepsilon_x}}$$

(3-18)

The above approach, impeccable as it is with regard to mathematics, may seem to be objectionable because of the shear modulus of elasticity, G, the bulk modulus, k, and Poisson's ratio, v, losing their physical meaning.

If Hencky's measure is used, the stresses are taken in their true values and the strains as defined by formula 3-17. Hence, the elastic constants acquire a quite definite physical meaning: $G = \tau_{oct}/\gamma^*_{oct}$ and $k = \tau_{oct}/\varepsilon^*_{oct}$.

The asterisk serves to denote that the strains ε_1, ε_2, and ε_3 entering the expressions for γ^*_{oct} and ε^*_{oct} are finite values determined by formula 3-17. The values of τ_{oct} and σ_{oct} are adopted in accordance with formulae 3-7 and 3-8.

The finite volumetric strain, $\varepsilon^*_V = 3\varepsilon^*_{oct}$, can be determined directly from the following relation, similar to eq. 3-17:

$$\varepsilon^*_V = \ln(V/V_0) \tag{3-19}$$

In fact, considering the volumetric strain of a cube with edges of length $l_{1(0)} = l_{2(0)} = l_{3(0)} = l_0$ changing due to deformation to $l_1 = l_2 = l_3$, we obtain:

$$\varepsilon^*_V = 3\varepsilon^*_{oct} = \ln\frac{l_1}{l_{1(0)}} + \ln\frac{l_2}{l_{2(0)}} + \ln\frac{l_3}{l_{3(0)}} = \ln\left(\frac{l}{l_0}\right)^3 = \ln\frac{V}{V_0}$$

The possibility of summing logarithmic strains in order to obtain the logarithms of the total strain is one of the assets of Hencky's measure of deformation.

3.3 TENSORS OF STRESS, STRAIN AND STRAIN RATE

Stresses at a point

Tensor analysis is widely used in continuum mechanics, providing a means of expressing the relations between all components of stress and strain in a compact form. Referring the reader who wishes to acquire detailed knowledge to special literature, e.g. Il'yushin (1963) and Sedov (1983), we shall concentrate here on the fundamentals of tensor analysis used in further discussion.

Consider the state of stress at a point. Let a plane ΔF in the vicinity of this point with a normal v be subjected to the action of a force ΔP. The stress at the point will then be given by a vector:

$$p_v = \lim_{\Delta F \to 0} \frac{\Delta P}{\Delta F}$$

The above stress may be resolved into three components (parallel to the coordinate axes) expressed in the form of a matrix:

$$p_v = [p_{xv} \quad p_{yv} \quad p_{zv}]$$

If the direction of v is changed, the stress components acquire other values. Assuming that this direction coincides with the x-axis, the components of the stress

$p_v = p_x$ will be a normal stress, σ_x, and two tangential stresses, τ_{xy} and τ_{xz}. Likewise, $p_v = p_y$ is resolved into the stresses σ_y, τ_{yx}, τ_{yz}, and $p_v = p_z$ into σ_z, τ_{zx}, τ_{zy}.

Thus, the stresses formed at the faces of an elemental parallelepiped whose edges are dx, dy, and dz will be:

$$p_x = [\sigma_x \quad \tau_{xy} \quad \tau_{xz}]$$
$$p_y = [\tau_{yx} \quad \sigma_y \quad \tau_{yz}]$$
$$p_z = [\tau_{zx} \quad \tau_{zy} \quad \sigma_z]$$

This implies that the state of stress at a point is defined by the vectors p_x, p_y, p_z which, in their turn, are resolved each into a normal component and two tangential components.

Stress tensor

The state of stress at a point in a body considered above can be defined by the matrix:

$$T_\sigma = \begin{bmatrix} p_x \\ p_y \\ p_z \end{bmatrix} = \begin{bmatrix} \sigma_x & \tau_{xy} & \tau_{xz} \\ \tau_{yx} & \sigma_y & \tau_{yz} \\ \tau_{zx} & \tau_{zy} & \sigma_z \end{bmatrix} \tag{3-20}$$

The quantity T_σ written in the form of the above matrix is called a stress tensor. It is a tensor of second rank representing a set of quantities $A_{ij}a_j$ satisfying the condition that if a is an arbitrary vector then $A_{ij}a_j$ and $A_{ji}a_i$, in their turn, are components of the vector.

In matrix 3-20, normal stresses are arranged diagonally and tangential stresses of the same magnitude are located symmetrically with respect to the diagonal. Consequently, the stress tensor 3-20 is called a symmetrical tensor. When referred to the principal axes, this tensor may be represented in the form of the matrix:

$$T_\sigma = \begin{bmatrix} \sigma_1 & 0 & 0 \\ 0 & \sigma_2 & 0 \\ 0 & 0 & \sigma_3 \end{bmatrix} \tag{3-20'}$$

Strain tensor and rate of strain tensor

The above reasoning may be repeated with respect to strains and rates of strain. The strain tensor is defined by the matrix:

$$T_\varepsilon = \begin{bmatrix} \varepsilon_x & \tfrac{1}{2}\gamma_{xy} & \tfrac{1}{2}\gamma_{xz} \\ \tfrac{1}{2}\gamma_{yx} & \varepsilon_y & \tfrac{1}{2}\gamma_{yz} \\ \tfrac{1}{2}\gamma_{zx} & \tfrac{1}{2}\gamma_{zy} & \varepsilon_z \end{bmatrix} \quad \text{or} \quad \begin{bmatrix} \varepsilon_1 & 0 & 0 \\ 0 & \varepsilon_2 & 0 \\ 0 & 0 & \varepsilon_3 \end{bmatrix} \tag{3-21}$$

Likewise, the rate of strain tensor may be expressed as:

$$T_{\dot{\varepsilon}} = \begin{bmatrix} \dot{\varepsilon}_x & \tfrac{1}{2}\dot{\gamma}_{xy} & \tfrac{1}{2}\dot{\gamma}_{xz} \\ \tfrac{1}{2}\dot{\gamma}_{yx} & \dot{\varepsilon}_y & \tfrac{1}{2}\dot{\gamma}_{yz} \\ \tfrac{1}{2}\dot{\gamma}_{zx} & \tfrac{1}{2}\dot{\gamma}_{zy} & \dot{\varepsilon}_z \end{bmatrix} \quad \text{or} \quad \begin{bmatrix} \dot{\varepsilon}_1 & 0 & 0 \\ 0 & \dot{\varepsilon}_2 & 0 \\ 0 & 0 & \dot{\varepsilon}_3 \end{bmatrix} \tag{3-22}$$

where $\dot{\varepsilon}_x, \ldots, \dot{\gamma}_{xy}, \ldots$, are the components of the strain rates.

The state of stress and strain at a given point appears to be distinctly defined if the stress tensor and the strain tensor or the rate of strain tensor are known.

Resolution of stress tensor

A stressed state at a point may be imagined as a sum of two stressed states (Fig. 3-8). One of them is due to the action of three pairs of equal normal stresses equalling the mean values:

$$\sigma_m = \frac{\sigma_x + \sigma_y + \sigma_z}{3} = \frac{\sigma_1 + \sigma_2 + \sigma_3}{3} \tag{3-23}$$

The quantity σ_m, referred to as mean normal stress, equals the all-around (hydrostatic) pressure $\sigma_m = p$ and brings about a volume change only.

The other component of the stressed state owes its origin to the tangential stresses $\tau_{xy} = \tau_{yx}$, $\tau_{yz} = \tau_{zy}$, $\tau_{zx} = \tau_{xz}$ and to the difference between the normal stresses and their mean values:

$$s_x = \sigma_x - \sigma_m, \quad s_y = \sigma_y - \sigma_m, \quad s_z = \sigma_z - \sigma_m \tag{3-24}$$

This state of stress may also be expressed in terms of the principal stresses: $s_1 = \sigma_1 - \sigma_m$, $s_2 = \sigma_2 - \sigma_m$, $s_3 = \sigma_3 - \sigma_m$.

Thus, the stresses s_x, s_y, s_z determine how much a given state of stress differs from all-around compression or tension. It can easily be seen that:

$$s_x + s_y + s_z = s_1 + s_2 + s_3 = 0 \tag{3-25}$$

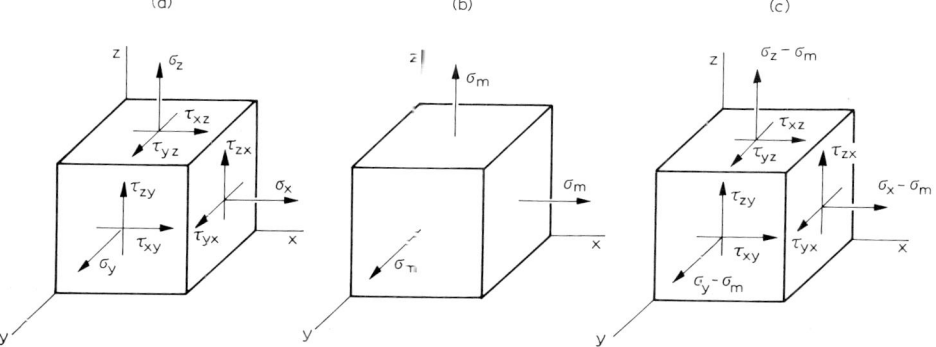

Fig. 3-8. (a) Stress tensor, resolved into (b) spherical stress tensor; (c) stress deviator.

This indicates that in the second state of stress the volumetric strain is zero and the stresses s_x, s_y, s_z (or s_1, s_2, s_3) cause distortion of shape only.

The stress tensor of the first state is called a spherical strain tensor and is presented in the form of the matrix:

$$T_\sigma^{(0)} = \begin{bmatrix} \sigma_m & 0 & 0 \\ 0 & \sigma_m & 0 \\ 0 & 0 & \sigma_m \end{bmatrix} \quad (3\text{-}26)$$

The stress tensor of the second state, termed stress deviator, is expressed in the following form:

$$D_\sigma = \begin{bmatrix} s_x & \tau_{xy} & \tau_{xz} \\ \tau_{yx} & s_y & \tau_{yz} \\ \tau_{zx} & \tau_{zy} & s_z \end{bmatrix} \quad (3\text{-}27)$$

Hence, the resolution of the state of stress illustrated in Fig. 3-8 is equivalent to the resolution of the stress tensor into a spherical strain tensor and a stress deviator:

$$T_\sigma = T_\sigma^0 + D_\sigma \quad (3\text{-}28)$$

The spherical stress tensor causes volumetric change and the stress deviator brings about distortion of shape.

Resolution of strain tensor

A strained state of an elemental volume may be imagined as the sum of two strained states. One of them results from three pairs of equal linear strains whose mean values are:

$$\varepsilon_m = \frac{\varepsilon_x + \varepsilon_y + \varepsilon_z}{3} = \frac{\varepsilon_1 + \varepsilon_2 + \varepsilon_3}{3} \quad (3\text{-}29)$$

The quantity ε_m is referred to as mean linear strain. Apparently, ε_m does not distort the shape of an elemental parallelepiped, defining the volumetric strain alone, $\varepsilon_m = \varepsilon_V/3$.

The second state of strain is dependent upon the angular distortions $\gamma_{xy} = \gamma_{yx}$, $\gamma_{yz} = \gamma_{zy}$, $\gamma_{xz} = \gamma_{zx}$ and the difference between the linear strains:

$$e_x = \varepsilon_x - \varepsilon_m \qquad e_y = \varepsilon_y - \varepsilon_m \qquad e_z = \varepsilon_z - \varepsilon_m \quad (3\text{-}30)$$

or:

$$e_1 = \varepsilon_1 - \varepsilon_m \qquad e_2 = \varepsilon_2 - \varepsilon_m \qquad e_3 = \varepsilon_3 - \varepsilon_m \quad (3\text{-}31)$$

which determines how much a given state of strain differs from all-around uniform compression (or tension).

It follows from expression 3-30 that:

$$e_x + e_y + e_z = e_1 + e_2 + e_3 = 0 \tag{3-32}$$

and, consequently, $\varepsilon_v = 0$. This indicates that the second state of strain does not involve a change in volume and that deformation consists of distortion of shape only.

The resolution of the state of strain described above is equivalent to the resolution of the strain tensor into a spherical strain tensor and a strain deviator:

$$T_\varepsilon = T_\varepsilon^0 + D \tag{3-33}$$

The spherical strain tensor and the strain deviator can be represented in the form of matrices similar to matrices 3-26 and 3-27. The spherical strain tensor describes a change in volume and the strain deviator a distortion of shape.

Resolution of the rate of strain tensor

The rate of strain may be determined by formulae similar to formulae 3-29 through 3-33. The quantity:

$$\dot{\varepsilon}_m = \frac{\dot{\varepsilon}_x + \dot{\varepsilon}_y + \dot{\varepsilon}_z}{3} = \frac{\dot{\varepsilon}_1 + \dot{\varepsilon}_2 + \dot{\varepsilon}_3}{3} = \tfrac{1}{3}\dot{\varepsilon}_v \tag{3-34}$$

is referred to as the mean rate of linear strain defining the rate of volumetric strain.

The quantities $\dot{e}_x = \dot{\varepsilon}_x - \dot{\varepsilon}_m$, etc. (or $\dot{e}_1 = \dot{\varepsilon}_1 - \dot{\varepsilon}_m$, etc.), and $\dot{\gamma}_{xy} = \dot{\gamma}_{yx}$, etc., determine the rate of distortion of shape.

Thus, the rate of strain tensor may be resolved into a spherical tensor of the rate of strain and a rate of strain deviator

$$T_{\dot{\varepsilon}} = T_{\dot{\varepsilon}}^0 + D_{\dot{\varepsilon}} \tag{3-35}$$

The quantities $T_{\dot{\varepsilon}}^0$ and $D_{\dot{\varepsilon}}$ may be written as matrices similar to matrices 3-26 and 3-27.

In the above expressions, the rates of strain are denoted by $\dot{\varepsilon}_x$, $\dot{\gamma}_{xy}$, etc. The dot above the symbol indicates that the quantity is differentiated in time so that $\dot{\varepsilon}_x$, $\dot{\gamma}_{xy}$, ..., can, apparently, be regarded as:

$$\dot{\varepsilon}_x = \frac{d\varepsilon_x}{dt} \qquad \dot{\gamma}_{xy} = \frac{d\gamma_{xy}}{dt} \quad \ldots$$

However, for higher accuracy, the quantities $\dot{\varepsilon}_x$, $\dot{\gamma}_{xy}$, ... should be expressed in terms of the components of displacement rates.

Let a point in a body be displaced in the x-, y-, and z-directions during an infinitesimal interval of time dt by the amounts du_x, du_y and du_z. The rates of these displacements will then be:

$$v_x = \dot{u}_x = \frac{du_x}{dt} \qquad v_y = \dot{u}_y = \frac{du_y}{dt} \qquad v_z = \dot{u}_z = \frac{du_z}{dt} \tag{3-36}$$

In accordance with eq. 3-14, the rates of unit strains take the form:

$$\dot{\varepsilon}_x = \frac{\partial \dot{u}_x}{\partial x} \qquad \dot{\varepsilon}_y = \frac{\partial \dot{u}_y}{\partial y} \qquad \dot{\varepsilon}_z = \frac{\partial \dot{u}_z}{\partial z}$$

$$\dot{\gamma}_{xy} = \frac{\partial \dot{u}_x}{\partial y} + \frac{\partial \dot{u}_y}{\partial x} \qquad \dot{\gamma}_{yz} = \frac{\partial \dot{u}_y}{\partial z} + \frac{\partial \dot{u}_z}{\partial y} \qquad \dot{\gamma}_{zx} = \frac{\partial \dot{u}_x}{\partial z} + \frac{\partial \dot{u}_z}{\partial x} \qquad (3\text{-}37)$$

The above expressions coincide with Cauchy's measure of deformation. If use is made of the measures of deformation according to Green or Hencky, it is appropriate to employ formulae of the form given in eqs. 3-13 and 3-17.

Other notation for components of stress and strain

The tensors of stress, strain and rate of strain represented in the form of matrices similar to 3-20, 3-21 amd 3-22 may be replaced by a compact notation frequently used in the theory of plasticity:

$$T_\sigma = \{\sigma_{ij}\} \qquad T_\varepsilon = \{\varepsilon_{ij}\} \qquad T_{\dot\varepsilon} = \{\dot\varepsilon_{ij}\}$$

where $i, j = 1, 2, 3$. The quantities σ_{ij}, ε_{ij} and $\dot\varepsilon_{ij}$ denote the general components of the tensors of stress, strain and rate of strain, respectively; individual tensor components are obtained by substituting 1, 2, 3 for the subscripts i, j.

Likewise, the deviators of stress, strain and rate of strain, as determined from matrices 3-27, etc., are then written as:

$$D_\sigma = \{s_{ij}\} \qquad D_\varepsilon = \{e_{ij}\} \qquad D_{\dot\varepsilon} = \{\dot e_{ij}\}$$

where s_{ij}, e_{ij} and $\dot e_{ij}$ are the general components of the deviators of stress, strain and rate of strain; individual components of the deviators are obtained by substituting 1, 2, 3 for the subscripts i, j.

From eqs. 3-28, 3-33 and 3-35 it follows that:

$$\sigma_{ij} = 3\sigma_{kk}\delta_{ij} + s_{ij} \qquad \varepsilon_{ij} = 3\varepsilon_{kk}\delta_{ij} + e_{ij} \qquad \dot\varepsilon_{ij} = 3\dot\varepsilon_{kk}\delta_{ij} + \dot e_{ij}$$

where δ_{ij} is the Kronecker symbol ($\delta_{ij} = 1$ when $i = j$, and $\delta_{ij} = 0$ when $i \neq j$); $\sigma_{kk} = \sigma_m/3$, $\varepsilon_{kk} = \varepsilon_m/3$; $\dot\varepsilon_{kk} = \dot\varepsilon_m/3$.

Forestalling events, the invariants of the stress tensor, as represented by expressions 3-39, may then be written as:

$$I_1(T) = \sigma_{ij} \qquad I_2(T) = \sigma_{ij}\sigma_{ji}/2 \qquad I_3(T) = \sigma_{ij}\sigma_{jk}\sigma_{ki}/3$$

The invariants of the stress deviator given by the expressions of 3-41 will, in this case, take the form $I_1(D) = 0$; $I_2(D) = s_{ij}s_{ji}/2$; $I_3(D) = s_{ij}s_{jk}s_{ki}/3$.

The invariants of strain tensor and rate of strain tensor as well as those of strain deviator and rate of strain deviator may be represented in a similar form.

3.4 STRESS TENSOR INVARIANTS

In examining the state of stress and strain of a body, stress tensors are commonly replaced by stress tensor invariants, i.e. such combinations of the components of stress and strain tensors which do not change their values after rotation of the axes.

It may be demonstrated that the principal stresses σ_1, σ_2 and σ_3 are the roots of a cubic equation:

$$\sigma^3 - I_1(T)\sigma^2 - I_2(T)\sigma - I_3(T) = 0 \qquad (3\text{-}38)$$

where I_1, I_2, I_3 are coefficients.

Since the principal stresses σ_1, σ_2, σ_3 are independent of the axes of coordinates selected, the coefficients I_1, I_2, I_3 of the cubic equation appear to be invariant as well. Referred to as the first (linear), and second (square) and the third (cubic) invariants of the stress tensor, respectively, they are expressed as follows:

$$\begin{aligned} I_1(T) &= \sigma_1 + \sigma_2 + \sigma_3 = 3\sigma_m \\ I_2(T) &= -(\sigma_1\sigma_2 + \sigma_2\sigma_3 + \sigma_3\sigma_1) \\ I_3(T) &= \sigma_1\sigma_2\sigma_3 \end{aligned} \qquad (3\text{-}39)$$

By analogy with expressions 3-39, we may obtain three invariants of the spherical stress tensor:

$$\begin{aligned} I_1(T^0) &= 3\sigma_m = \sigma_1 + \sigma_2 + \sigma_3 \\ I_2(T^0) &= 3\sigma_m^2 = -\tfrac{1}{3}(\sigma_1 + \sigma_2 + \sigma_3)^2 \\ I_3(T^0) &= \sigma_m^3 = \tfrac{1}{27}(\sigma_1 + \sigma_2 + \sigma_3)^3 \end{aligned} \qquad (3\text{-}40)$$

and three invariants of the stress deviator:

$$\begin{aligned} I_1(D) &= s_1 + s_2 + s_3 = (\sigma_1 - \sigma_m) + (\sigma_2 - \sigma_m) + (\sigma_3 - \sigma_m) = 0 \\ I_2(D) &= -(s_1 s_2 + s_2 s_3 + s_3 s_1) \\ &= \tfrac{1}{6}[(\sigma_1 - \sigma_2)^2 + (\sigma_2 - \sigma_3)^2 + (\sigma_3 - \sigma_1)^2] \\ I_3(D) &= s_1 s_2 s_3 = (\sigma_1 - \sigma_m)(\sigma_2 - \sigma_m)(\sigma_3 - \sigma_m) \end{aligned} \qquad (3\text{-}41)$$

The quantities $I_{1,2,3}(T)$, $I_{1,2,3}(T^0)$ and $I_{1,2,3}(D)$ may also be expressed in terms of the stress components σ_x, σ_y, σ_z, τ_{xy}, τ_{yz}, τ_{zy}.

In analysing the state of stress and strain of a body, the main considerations are the first invariant of the stress tensor and the second invariant of the stress deviator. As can be seen from expressions 3-39 and 3-40, the first invariant of the stress tensor is identical with the first invariant of the spherical stress tensor and both these values are identically equal to three times the mean normal stress:

$$I_1(T) = I_1(T^0) = 3\sigma_m$$

The mean normal stress, in its turn, is identical with the octahedral normal stress: $\sigma_m = \sigma_{oct} = (\sigma_1 + \sigma_2 + \sigma_3)/3$. The value of σ_m is commonly used in the analysis.

As far as the second invariant of the stress deviator $I_2(D)$ is concerned, it is more convenient to use not the invariant itself but a certain quantity proportional to the square root thereof, which is thus also invariant in itself; this quantity is called the stress intensity.

The *stress intensity*, as given by:

$$\sigma_i = \sqrt{3}\sqrt{I_2(D)} \qquad (3\text{-}42)$$

is referred to as the intensity of normal stresses; given by:

$$\tau_i = \sqrt{I_2(D)} \qquad (3\text{-}43)$$

it is called the intensity of tangential stresses (this quantity is also frequently denoted by σ_i).

Taking into account expressions 3-41, the above relations can be expressed in the following form:

$$\sigma_i = \frac{1}{\sqrt{2}}\sqrt{(\sigma_1 - \sigma_2)^2 + (\sigma_2 - \sigma_3)^2 + (\sigma_3 - \sigma_1)^2}$$

$$= \frac{1}{\sqrt{2}}\sqrt{(\sigma_x - \sigma_y)^2 + (\sigma_y - \sigma_z)^2 + (\sigma_z - \sigma_x)^2 + 6(\tau_{xy}^2 + \tau_{yz}^2 + \tau_{zx}^2)} \qquad (3\text{-}44)$$

$$\tau_i = \frac{1}{\sqrt{6}}\sqrt{(\sigma_1 - \sigma_2)^2 + (\sigma_2 - \sigma_3)^2 + (\sigma_3 - \sigma_1)^2}$$

$$= \frac{1}{\sqrt{6}}\sqrt{(\sigma_x - \sigma_y)^2 + (\sigma_y - \sigma_z)^2 + (\sigma_z - \sigma_x)^2 + 6(\tau_{xy}^2 + \tau_{yz}^2 + \tau_{zx}^2)} \qquad (3\text{-}45)$$

Comparing the value of $I_2(D)$ as determined by formula 3-41 with the value of the octahedral tangential stress τ_{oct} given by expression 3-8, we can see that:

$$\tau_{oct} = \frac{2}{\sqrt{6}}\sqrt{I_2(D)} \qquad (3\text{-}46)$$

It follows from expressions 3-42, 3-43 and 3-46 that the three values — normal stresses intensity σ_i, tangential stresses intensity τ_i, and octahedral tangential stress τ_{oct} — equal the square root of the second invariant of the stress deviator, accurate to within a constant factor, and differ from one another by the value of the factor. This may be written as the equality:

$$I = k\sqrt{I_2(D)} \qquad (3\text{-}47)$$

where $I = \sigma_i$ for $k = \sqrt{3}$; $I = \tau_i$ for $k = I$ and $I = \tau_{oct}$ for $k = 2/\sqrt{6}$.

Consider the advantages that any of the above quantities may offer in practice. It may be shown that the octahedral tangential stress is of a magnitude approaching that of the maximum tangential stress at the same point, i.e.:

$$0.94\tau_{max} > \tau_{oct} > 0.816\tau_{max} \qquad (3\text{-}48)$$

where $\tau_{max} = (\sigma_1 - \sigma_2)/2$.

However, the above expression is of limited practical use, for it is sometimes a problem to determine which of the principal stresses will be at maximum and which at minimum. The alternative is to use the expanded formula 3-8 or the similar formulae 3-44 and 3-45. The two last-named formulae are given preference in the literature on the subject because the values of σ_i and τ_i correspnd to a certain simple stress — an obvious practical advantage. These values are sometimes referred to as generalized stresses.

τ_{oct}, σ_i and τ_i values depending on the stressed state

Consider the values acquired by the generalized stresses τ_{oct}, σ_i and τ_i in analysing the states of stress most commonly met in the laboratory. Referring to Fig. 3-9, the states of stress mentioned are as follows.
(1) Uniaxial compression:

$$\sigma_1 > 0; \quad \sigma_2 = \sigma_3 = 0 \quad \text{and} \quad \varepsilon > 0; \quad \varepsilon_2 = \varepsilon_3 = -\nu\varepsilon_1$$

The coefficient of lateral strain, ν, is treated here not as an elastic constant but as a parameter which, in the general case, is a variable dependent on the amount of deformation. For an incompressible body ($\varepsilon_V = \varepsilon_1 + \varepsilon_2 + \varepsilon_3 = 0$), $\nu = 0.5$.
(2) Pure shear:

$$\sigma_1 = -\sigma_3 = \tau; \quad \sigma_2 = 0 \quad \text{and} \quad \varepsilon_1 = -\varepsilon_3 = \tfrac{1}{2}\gamma; \quad \varepsilon_2 = 0$$

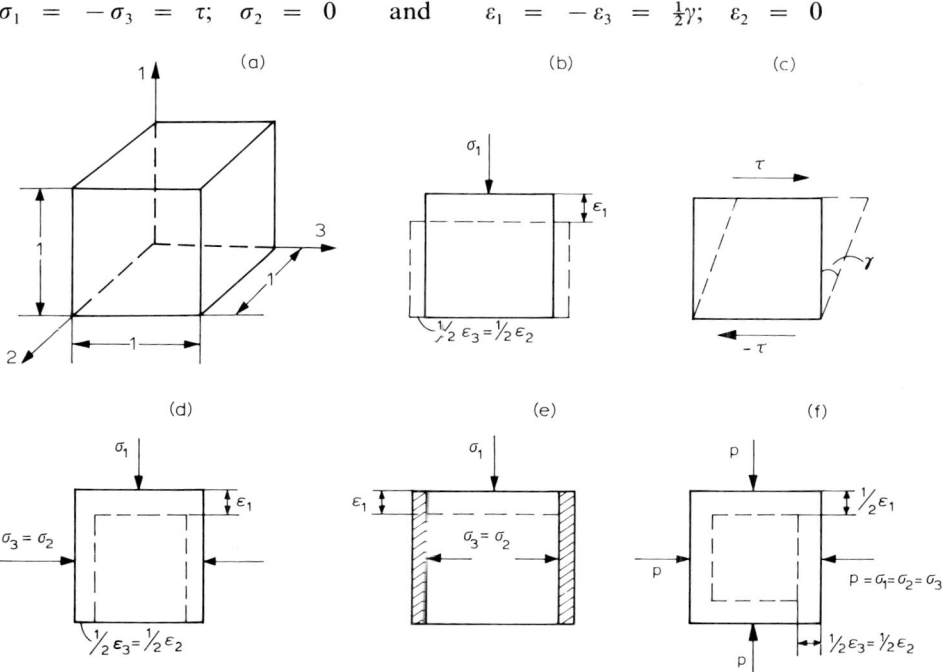

Fig. 3-9. Arrangement of stress: (a) designation of principal axes; (b) uniaxial compression; (c) triaxi-symmetrical compression; (d) laterally confined compression; (e) all-around uniform compression.

(3) Triaxi-symmetrical compression:

$$\sigma_1 > \sigma_2 = \sigma_3 \quad \text{and} \quad \varepsilon_1 > 0, \quad \varepsilon_2 = \varepsilon_3 < 0$$

(4) Compression without provision for lateral expansion (confined compression):

$$\sigma_1 > 0, \quad \sigma_2 = \sigma_3 = \frac{v}{1-v}\sigma_1 \quad \text{and} \quad \varepsilon_1 > 0, \quad \varepsilon_2 = \varepsilon_3 = 0$$

(5) Uniform all-around compression:

$$\sigma_1 = \sigma_2 = \sigma_3 = p \quad \text{and} \quad \varepsilon_1 = \varepsilon_2 = \varepsilon_3 = \varepsilon_v/3$$

Substituting the values of σ_1, σ_2 and σ_3 defining the above states of stress into expressions 3-38, 3-44 and 3-45, we arrive at the generalized stresses τ_{oct}, σ_i, and τ_i also defining these states. The results are presented in Table 3-I. Also tabulated here are the values of mean normal stress σ_m for the given states. The lower part of the table presents the values of generalized strains which will be considered below.

As can be seen from Table 3-I, the value of normal stress intensity σ_i equals the stress due to the uniaxial compression and the value of tangential stress intensity τ_i equals the stress formed during pure shear. Since the bulk of the soil resistance is in shear, we shall use the value of τ_i.

Angle of stress arrangement and Lode parameter

Consider the third invariant of the stress deviator $I_3(D)$. Defining the arrangement of stress, a subject of subsequent discussion, this quantity plays an important role in continuum mechanics. However, quite frequently I_3 is substituted by other invariant quantities such as the angle of stress arrangement ω_σ connected with $I_3(D)$ by the relationship:

$$-\cos 3\omega_\sigma = \frac{3\sqrt{3}I_3(D)}{2[I_2(D)]^{3/2}} \tag{3-49}$$

If the value of ω_σ is expressed in terms of the principal normal stresses, we obtain the relationship:

$$\tan \omega_\sigma = \frac{(\sigma_1 - \sigma_2)\sqrt{3}}{\sigma_1 + \sigma_2 - 2\sigma_3} \tag{3-50}$$

where $\sigma_1 > \sigma_2 > \sigma_3$.

Another frequently used invariant characteristic defining the stressed state is known as the Nadai–Lode parameter (for stresses):

$$\mu_\sigma = \frac{2\sigma_2 - \sigma_1 - \sigma_3}{\sigma_1 - \sigma_3} \tag{3-51}$$

This parameter is connected with the angle ω_σ by the relationship:

$$\mu_\sigma = \sqrt{3} \cot(\omega_\sigma + \pi/3) \tag{3-52}$$

Generalized stresses and strains for various arrangements of stress

Invariant formulae defining state of stress	Arrangements of stress:				
	uniaxial compression	shear	triaxi-symmetrical compression	confined compression	all-around pressure
Stresses					
$I_2(D) = \frac{1}{6}[(\sigma_1-\sigma_2)^2 + (\sigma_2-\sigma_3)^2 + (\sigma_3-\sigma_1)^2]$	$\frac{1}{3}\sigma_1^2$	τ^2	$\frac{1}{3}(\sigma_1-\sigma_3)^2$	$\frac{1}{3}\left(\frac{1-2v}{1-v}\right)^2\sigma_1^2$	0
$\tau_{oct} = \frac{2}{\sqrt{6}}\sqrt{I_2(D)}$	$\frac{\sqrt{2}}{3}\sigma_1$	$\frac{2}{\sqrt{6}}\tau$	$\frac{\sqrt{2}}{3}(\sigma_1-\sigma_3)$	$\frac{\sqrt{2}}{3}\left(\frac{1-2v}{1-v}\right)\sigma_1$	0
$\sigma_i = \sqrt{3}\sqrt{I_2(D)}$	σ_1	$\sqrt{3}\tau$	$\sigma_1-\sigma_3$	$\frac{1-2v}{1-v}\sigma_1$	0
$\tau_i = \sqrt{I_2(D)}$	$\frac{1}{\sqrt{3}}\sigma_1$	τ	$\frac{\sigma_1-\sigma_3}{\sqrt{3}}$	$\frac{1}{\sqrt{3}}\frac{1-2v}{1-v}\sigma_1$	0
$\sigma_m = \frac{\sigma_1+\sigma_2+\sigma_3}{3}$	$\frac{\sigma_1}{3}$	0	$\frac{\sigma_1+2\sigma_3}{3}$	$\frac{1}{3}\left(\frac{1-2v}{1-v}\right)\sigma_1$	P
Strains					
$J_2(D) = \frac{1}{6}[(\varepsilon_1-\varepsilon_2)^2 + (\varepsilon_2-\varepsilon_3)^2 + (\varepsilon_3-\varepsilon_1)^2]$	$\frac{1}{3}(1+v)^2\varepsilon_1^2$ $[\frac{3}{4}\varepsilon_1^2]$	$\frac{1}{4}\gamma^2$	$\frac{1}{3}(\varepsilon_1+\varepsilon_3)^2$	$\frac{1}{3}\varepsilon_1^2$	0
$\gamma_{oct} = \frac{4}{\sqrt{6}}\sqrt{J_2(D)}$	$\frac{2\sqrt{2}}{3}(1+v)\varepsilon_1$ $[\sqrt{2}\varepsilon_1]$	$\frac{2}{3}\gamma$	$\frac{2\sqrt{2}}{3}(\varepsilon_1+\varepsilon_3)$	$\frac{2\sqrt{2}}{3}\varepsilon_1$	0
$\varepsilon_i = \frac{2}{\sqrt{3}}\sqrt{J_2(D)}$	$\frac{2}{3}(1+v)\varepsilon_1$ $[\varepsilon_1]$	$\frac{1}{\sqrt{3}}\gamma$	$\frac{2}{3}(\varepsilon_1+\varepsilon_3)$	$\frac{2}{3}\varepsilon_1$	0
$\gamma_i = 2\sqrt{J_2(D)}$	$\frac{2}{\sqrt{3}}(1+v)\varepsilon_1$ $[\sqrt{3}\varepsilon_1]$	γ	$\frac{2}{\sqrt{3}}(\varepsilon_1+\varepsilon_3)$	$\frac{2}{\sqrt{3}}\varepsilon_1$	0
$\varepsilon_m = \frac{\varepsilon_1+\varepsilon_2+\varepsilon_3}{3}$	$\frac{1-2v}{3}\varepsilon_1$ $[0]$	0	$\frac{\varepsilon_1+2\varepsilon_3}{3}$	$\frac{1}{3}\varepsilon_1$	$\frac{1}{3}\varepsilon_v$

Note: The values of generalized strain for $v = 0.5$ are between brackets.

In the case of simple stress, the parameter μ_σ and the angle ω_σ assume the following values:

(1) Uniaxial compression ($\sigma_1 > 0, \sigma_2 = \sigma_3 = 0$) and triaxi-symmetrically stressed state when $\sigma_1 > \sigma_2 = \sigma_3 > 0$:

$$\mu_\sigma = -1; \qquad \omega_\sigma = \pi/3 \tag{3-53}$$

(2) Uniaxial tension ($\sigma_1 = \sigma_2 = 0; \sigma_3 < 0$) or triaxi-symmetrically stressed state when $\sigma_1 = \sigma_2 > \sigma_3$:

$$\mu_\sigma = +1; \qquad \omega_\sigma = 0 \tag{3-54}$$

(3) Pure shear when $\sigma_1 = -\sigma_3 = \tau, \sigma_2 = 0$ or for $\sigma_2 = 0.5(\sigma_1 + \sigma_3) > 0$:

$$\mu_\sigma = 0; \qquad \omega_\sigma = \pi/6 \tag{3-55}$$

Thus, the parameters μ_σ and ω_σ vary over the limits:

$$-1 \leq \mu_\sigma \leq +1; \qquad \pi/3 \geq \omega_\sigma \geq 0 \tag{3-56}$$

Stressed states are said to be similar if the parameters μ_σ (or the angles ω_σ) of the states are equal. This can be readily verified by constructing stress circles for two different states of stress. If the values of σ_1, σ_2 and σ_3 defining each of the states are such that the parameters μ_σ are the same, the stress circles will appear to be similar in both cases.

Novozhilov (1952) has found that specifying an angle of stress arrangement ω_σ is equivalent to specifying the relation between the maximum tangential stress τ_{max} and the tangential stress intensity τ_i.

In conclusion it is necessary to say that the state of stress at any point in a body is completely defined if, at this point, the mean normal stress σ_m, the tangential stress intensity τ_i and the arrangement of stress as determined by either the parameter μ_σ or the angle ω_σ are known.

Geometrical representation

Consider a geometrical representation of the stressed state of an isotropic body in a stress space ($\sigma_1 - \sigma_2 - \sigma_3$) (Fig. 3-10(a)).

Any point M of this space with the coordinates σ_1, σ_2, and σ_3 represents a certain state of stress defined by the principal stresses σ_1, σ_2, and σ_3. The points representing simple tension or compression along the coordinate axes are located on the axes, the points representing the state of plane stress lie on the coordinate planes and the origin corresponds to a point where no stresses are present in the body. A straight line making the same angle α ($\cos \alpha = 1/\sqrt{3}$) with the three coordinate axes, called space diagonal, defines the position of the points $\sigma_1 = \sigma_2 = \sigma_3 = p$ corresponding to the hydrostatic state.

For a geometrical representation of the values of σ_m and τ_i, draw a plane through point M so that the space diagonal (hydrostatic axis) is normal to it. The projection of point M on the normal along this plane gives point M'. The distance of this point

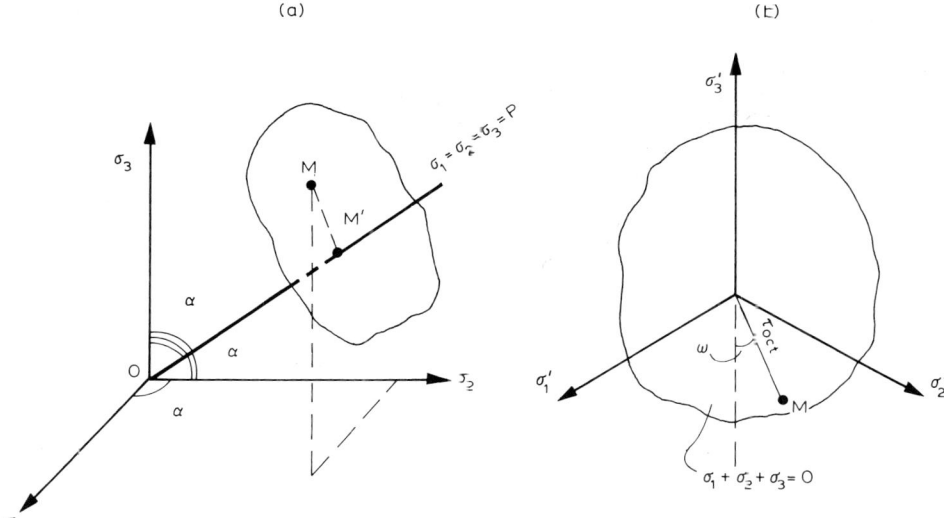

Fig. 3-10. (a) Stress space $\sigma_1, \sigma_2, \sigma_3$. (b) Deviatoric plane $\sigma_1 + \sigma_2 + \sigma_3 = 0$.

on the hydrostatic axis from the origin is:

$$\overline{OM'} = \sqrt{3}\,\frac{\sigma_1 + \sigma_2 + \sigma_3}{3} = I_2(T)/\sqrt{3} = \sqrt{3}\sigma_{oct} = \sqrt{3}\sigma_n$$

The radial distance between the points M and M', in its turn, is:

$$\overline{MM'} = \sqrt{2[I_2(D)]^{1/2}} = \sqrt{3}\tau_{oct} = \sqrt{2}\tau_i$$

Thus, the hydrostatic axis also appears to be the axis of the octahedral normal stress and the plane, to which this axis is the normal, becomes the octahedral plane. It is also called the deviatoric plane, for any vector lying in this plane defines the stress deviator $I(D)$ of any state of stress $M(\sigma_1, \sigma_2, \sigma_3)$.

For a geometrical representation of the angle ω, draw a deviatoric plane through the origin for $\sigma_1 = \sigma_2 = \sigma_3 = 0$ (Fig. 3-10(b)), and denote the projections of the axes σ_1, σ_2 and σ_3 on this plane by σ'_1, σ'_2 and σ'_3. The equation satisfying the plane drawn is $\sigma_1 + \sigma_2 + \sigma_3 = 0$. Constructing a projection of the distance MM' on this plane, we can see that line $M'M = OM$ forms an angle ω with the axis $O\sigma'$. The angle formed is given by the formula:

$$\tan\omega = \sqrt{3}\,\frac{\sigma_1 - \sigma_2}{\sigma_1 + \sigma_2 - 2\sigma_3}$$

which corresponds to expression 3-50. This implies that the angle ω is the angle of stress arrangement, $\omega = \omega_\sigma$.

3.5 STRAIN TENSOR INVARIANTS

Strain tensor invariants

The statements regarding the stress tensor invariants in Section 3.4 can be repeated with respect to the strain tensor invariants. By analogy with expression 3-39, the three invariants of strain tensor (linear, square and cubic) can be written in the form:

$$\begin{aligned} J_1(T) &= \varepsilon_1 + \varepsilon_2 + \varepsilon_3 = 3\varepsilon_m \\ J_2(T) &= -(\varepsilon_1\varepsilon_2 + \varepsilon_2\varepsilon_3 + \varepsilon_3\varepsilon_1) \\ J_3(T) &= \varepsilon_1\varepsilon_2\varepsilon_3 \end{aligned} \quad (3\text{-}57)$$

The invariants of spherical strain tensor will take the form:

$$\begin{aligned} J_1(T^0) &= 3\varepsilon_m = \varepsilon_1 + \varepsilon_2 + \varepsilon_3 \\ J_2(T^0) &= -3\varepsilon_m^2 = -\tfrac{1}{3}(\varepsilon_1 + \varepsilon_2 + \varepsilon_3)^2 \\ J_3(T^0) &= \varepsilon_m^3 = \tfrac{1}{27}(\varepsilon_1 + \varepsilon_2 + \varepsilon_3)^3 \end{aligned} \quad (3\text{-}58)$$

and the invariants of strain deviator can be written as:

$$\begin{aligned} J_1(D) &= (\varepsilon_1 - \varepsilon_m) + (\varepsilon_2 - \varepsilon_m) + (\varepsilon_3 - \varepsilon_m) = 0 \\ J_2(D) &= \tfrac{1}{6}[(\varepsilon_1 - \varepsilon_2)^2 + (\varepsilon_2 - \varepsilon_3)^2 + (\varepsilon_3 - \varepsilon_1)^2] \\ J_3(D) &= (\varepsilon_1 - \varepsilon_m)(\varepsilon_2 - \varepsilon_m)(\varepsilon_3 - \varepsilon_m) \end{aligned} \quad (3\text{-}59)$$

The first invariants in expressions 3-57 and 3-58 define the quantity:

$$J_1(T) = J_1(T^0) = 3\varepsilon_m \quad (3\text{-}60)$$

where $\varepsilon_m = (\varepsilon_1 + \varepsilon_2 + \varepsilon_3)/3$ is the mean linear strain equalling one third of the volumetric strain ε_v; this quantity also equals the unit elongation on the octahedral plane, $\varepsilon_m = \varepsilon_{oct} = \varepsilon_v/3$.

Strain intensity

In the theory of plasticity, much importance is attached to the second invariant of strain deviator, $I_2(D)$ — an inclusive characteristic defining distortion of the shape of a body element. However, the usual practice is to replace $J_2(D)$ by another invariant quantity such as the linear strain intensity:

$$\varepsilon_i = \frac{2}{\sqrt{3}}\sqrt{J_2(D)} \quad (3\text{-}61)$$

or the angular strain intensity (the intensity of shear strain):

$$\gamma_i = 2\sqrt{J_2(D)} \quad (3\text{-}62)$$

Using the second expression of 3-59, the above quantities may be written as:

$$\varepsilon_i = \frac{\sqrt{2}}{3}\sqrt{(\varepsilon_1 - \varepsilon_2)^2 + (\varepsilon_2 - \varepsilon_3)^2 + (\varepsilon_3 - \varepsilon_1)^2}$$

$$= \frac{\sqrt{2}}{3}\sqrt{(\varepsilon_x - \varepsilon_y)^2 + (\varepsilon_y - \varepsilon_z)^2 + (\varepsilon_z - \varepsilon_x)^2 + \tfrac{2}{3}(\gamma_{xy}^2 + \gamma_{yz}^2 + \gamma_{zx}^2)} \quad (3\text{-}63)$$

$$\gamma_i = \sqrt{\tfrac{2}{3}}\sqrt{(\varepsilon_1 - \varepsilon_2)^2 + (\varepsilon_2 - \varepsilon_3)^2 + (\varepsilon_3 - \varepsilon_1)^2}$$

$$= \sqrt{\tfrac{2}{3}}\sqrt{(\varepsilon_x - \varepsilon_y)^2 + (\varepsilon_y - \varepsilon_z)^2 + (\varepsilon_z - \varepsilon_x)^2 + \tfrac{3}{2}(\gamma_{xy}^2 + \gamma_{yz}^2 + \gamma_{zx}^2)} \quad (3\text{-}63')$$

Comparing the value of $I_2(D)$, as determined by formula 3-59, with the value of octahedral angular strain γ_{oct}, given by formula 3-10, we can see that:

$$\gamma_{oct} = \frac{4}{\sqrt{6}}\sqrt{J_2(D)} \quad (3\text{-}64)$$

Thus, the three values of γ_{oct}, ε_i and γ_i are proportional to the square root of the second invariant of strain deviator:

$$J = k\sqrt{J_2(D)} \quad (3\text{-}65)$$

The above values differ from one another only by the value of a constant factor k; $J = \varepsilon_i$ for $k = 2/\sqrt{3}$, $J = \gamma_i$ for $k = 2$ and $J = \gamma_{oct}$ for $k = 4/\sqrt{6}$.

γ_{oct}, ε_i and γ_i values depending on the stressed state

Consider the values acquired by the generalized strains γ_{oct}, ε_i and γ_i depending on the existing state of stress. The various states of stress and the conditions of deformation defining them were examined in Section 3.4. It will be recalled that the cases of uniaxial compression, pure shear, triaxi-symmetrical compression, compression without provision for lateral expansion, and all-around compression were considered.

Substituting the values of strain for the cases under consideration into formulae 3-10, 3-63 and 3-63', we obtain the values of γ_{oct}, ε_i and γ_i. These values are entered in Table 3-I above, along with the values of ε_m.

Note that the parameter v in the formulae given in Table 3-I can be a variable, defining the relation between vertical and lateral strains under a given condition of loading and at a given instance; the quantity $v = 0.5$ corresponds to the condition of volumetric incompressibility of a body.

From Table 3-I we can see that the value of linear strain intensity ε_i coincides (for $v = 0.5$) with the deformatin of uniaxial compression and the value of shear strain intensity γ_i corresponds to the deformation of pure shear. The value γ_i is used hereafter.

Angle of strain arrangement and Lode parameter

The arrangement of strain is defined by a quantity ω_ε, which is called the angle of strain arrangement. This quantity is connected with the third invariant of strain deviator by the relationship:

$$-\cos 3\omega_\varepsilon = \frac{3\sqrt{3} I_3(D)}{2[J_2(D)]^{3/2}} \tag{3-66}$$

The value of ω_ε is connected with the strain components by the expression:

$$\tan \omega_\varepsilon = \frac{1}{\sqrt{3}} \frac{e_1 - e_2}{e_1 + e_2} = \sqrt{3} \frac{\varepsilon_1 - \varepsilon_2}{\varepsilon_1 + \varepsilon_2 - 2\varepsilon_3} \tag{3-67}$$

where $e_{1,2,3} = \varepsilon_{1,2,3} - \varepsilon_m$, provided $\varepsilon_1 \geqslant \varepsilon_2 \geqslant \varepsilon_3$.

The arrangement of strain can be defined by another quantity which is the Lode–Nadai parameter given by:

$$\mu_\varepsilon = 3\frac{e_3 + e_1}{e_3 - e_1} = \frac{2\varepsilon_2 - \varepsilon_1 - \varepsilon_3}{\varepsilon_1 - \varepsilon_3} \tag{3-68}$$

The parameters ω_ε and μ_ε are correlated by the expression:

$$\mu_\varepsilon = \sqrt{3} \cot (\omega_\varepsilon + \pi/3) \tag{3-69}$$

In the case of simple strain, ω_ε and μ_ε assume the following values:

(1) Uniaxial compression ($\varepsilon_1 > 0$, $\varepsilon_2 = \varepsilon_3 = -\nu\varepsilon_1$) and symmetrical triaxial compression when $\varepsilon_1 > 0$, $\varepsilon_2 = \varepsilon_3 < 0$:

$$\mu_\varepsilon = -1, \quad \omega_\varepsilon = \pi/3 \tag{3-70}$$

(2) Uniaxial tension ($\varepsilon_3 < 0$, $\varepsilon_1 = \varepsilon_2 = -\nu\varepsilon_3$) and axially symmetrical deformation when $\varepsilon_1 = \varepsilon_2 > 0$ and $\varepsilon_3 < 0$:

$$\mu_\varepsilon = 1, \quad \omega_\varepsilon = 0 \tag{3-71}$$

(3) Pure shear ($\varepsilon_1 = -\varepsilon_3 = \gamma/2$, $\varepsilon_2 = 0$):

$$\mu_\varepsilon = 0, \quad \omega_\varepsilon = \pi/6 \tag{3-72}$$

Thus, the parameters μ_ε and ω_ε vary between the same limits as the parameters μ_σ and ω_ε:

$$-1 \leqslant \mu_\varepsilon \leqslant +1, \quad \frac{\pi}{3} \geqslant \omega_\varepsilon \geqslant 0 \tag{3-73}$$

States of strain are said to be similar if the parameters μ_ε (or ω_ε) are equal.

In concluding it is necessary to emphasize that the state of strain at a given point in a body is completely defined if, for this point, the mean linear strain ε_m, as determined by formula 3-29, the intensity of angular distortion γ_i, as given by formula 3-63′, and the strain arrangement as defined by either the parameter μ_ε or the angle ω_ε are known.

A geometrical representation of a state of strain in the ($\sigma_1 - \sigma_2 - \sigma_3$) stress space can be made on the same lines as illustrated in Fig. 3-10, provided the axes σ_1, σ_2 and σ_3 are replaced by the axes ε_1, ε_2 and ε_3.

Invariants of the rate of strain tensor

The above expressions defining strains also hold for the rates of strain. Therefore, by analogy with expression 3-60, the first invariant of the rate of strain tensor and the first invariant of the rate of strain spherical tensor will be equal:

$$\dot{J}_1(T) = \dot{J}_1(T^0) = 3\dot{\varepsilon}_m \tag{3-74}$$

where $\dot{\varepsilon}_m$ is the rate of mean linear strain, $\dot{\varepsilon}_m = (\dot{\varepsilon}_1 + \dot{\varepsilon}_2 + \dot{\varepsilon}_3)/3 = \dot{\varepsilon}_v/3 = \dot{\varepsilon}_{oct}$.

The notation $\dot{\varepsilon}_m$, $\dot{\varepsilon}_i$ and $\dot{\gamma}_i$ is adopted for the sake of symbol identification. Strictly speaking, the intensity of shear rate is the first-order derivative of the intensity of shear strain increments rather than the total derivative of shear strain intensity; the equality $\dot{\gamma}_i = d\gamma_i/dt$, etc., holds in so far as the principal axes of strain deviators and of the rate of strain deviators coincide.

The rate of the octahedral angle distortion $\dot{\gamma}_{oct}$, the intensity of the rate of linear strain $\dot{\varepsilon}_i$ and the intensity of the rate of shear strain $\dot{\gamma}_i$ are correlated with the second invariant of the rate of strain $J_2(D)$ by a relationship similar to 3-65:

$$\dot{J} = k\sqrt{\dot{J}_2(D)} \tag{3-75}$$

For $k = 2/\sqrt{3}$, $\dot{J} = \dot{\varepsilon}_i$; for $k = 2$, $\dot{J} = \dot{\gamma}_i$; and for $k = 4/\sqrt{6}$, $\dot{J} = \dot{\gamma}_{oct}$. The quantity $\dot{\gamma}_i$, which is the intensity of the rate of shear strain, will be used hereafter. It is given by the formula:

$$\dot{\gamma}_i = \sqrt{\tfrac{2}{3}} \sqrt{(\dot{\varepsilon}_1 - \dot{\varepsilon}_2)^2 + (\dot{\varepsilon}_2 - \dot{\varepsilon}_3)^2 + (\dot{\varepsilon}_3 - \dot{\varepsilon}_1)^2}$$

$$= \sqrt{\tfrac{2}{3}} \sqrt{(\dot{\varepsilon}_x - \dot{\varepsilon}_y)^2 + (\dot{\varepsilon}_y - \dot{\varepsilon}_z)^2 + (\dot{\varepsilon}_z - \dot{\varepsilon}_x)^2 + \tfrac{2}{3}(\dot{\gamma}_{xy}^2 + \dot{\gamma}_{yz}^2 + \dot{\gamma}_{zx}^2)} \tag{3-76}$$

In the case of simple strain, we adopt the values of $\dot{\gamma}_{oct}$, $\dot{\gamma}_i$ and $\dot{\varepsilon}_i$ given in Table 3-I. Formulae of the type 3-66 through 3-73, defining the angle of strain arrangement ω_ε and the Lode–Nadai parameter μ_ε, also hold for strain rates.

Relation between principal stresses and invariants

Solving relationships 3-23, 3-44 and 3-51 for σ_1, σ_2 and σ_3, we arrive at the following relation between the principal normal stresses and the invariants τ_m, τ_i and μ_σ:

$$\sigma_{1,3} = \sigma_m \pm \frac{3 \mp \mu_\sigma}{3} \sqrt{\frac{3}{3 + \mu_\sigma^2}} \tau_i$$

$$\sigma_2 = \sigma_m + \frac{2\mu_\sigma}{3} \sqrt{\frac{3}{1 + \mu_\sigma^2}} \tau_i \tag{3-77}$$

Likewise, the principal strains ε_1, ε_2 and ε_3 or the rates of principal strains $\dot{\varepsilon}_1$, $\dot{\varepsilon}_2$ and $\dot{\varepsilon}_3$ may be expressed in terms of the invariants ε_m, γ_i, μ_ε or $\dot{\varepsilon}_m$, $\dot{\gamma}_i$, $\mu_{\dot{\varepsilon}}$.

3.6 BASIC EQUATIONS OF PLASTICITY THEORY

Equations of equilibrium

Consider the equilibrium of an elemental parallelepiped (Fig. 3-11) with edges dx, dy and dz.

Acting on the invisible sides of the parallelepiped are the forces $P_x = \sigma\, dy\, dz$, $T_{xy} = \tau_{xy}\, dy\, dz$ and $T_{xz} = \tau_{xz}\, dy\, dz$ and on the visible sides, the forces $P_x + (\partial P_x/\partial y)\, dy$, $T_{xy} + (\partial T_{xy}/\partial x)\, dx$ and $T_{xz} + (\partial T_{xz}/\partial x)\, dx$. In addition, body forces, whose projections are denoted by X, Y, Z, are also at work. Projecting all the forces in action on the relevant planes, we obtain the following equations of equilibrium of the parallelepiped:

$$\frac{\partial \sigma_x}{\partial x} + \frac{\partial \tau_{yx}}{\partial y} + \frac{\partial \tau_{zx}}{\partial z} + X = 0$$

$$\frac{\partial \tau_{xy}}{\partial x} + \frac{\partial \sigma_y}{\partial y} + \frac{\partial \tau_{zy}}{\partial z} + Y = 0 \qquad (3\text{-}78)$$

$$\frac{\partial \tau_{xz}}{\partial x} + \frac{\partial \tau_{yz}}{\partial y} + \frac{\partial \sigma_z}{\partial z} + Z = 0$$

When the body is in a state of motion, the body forces will have a component equal to the product of mass and acceleration taken with the "minus" sign (d'Alembert's

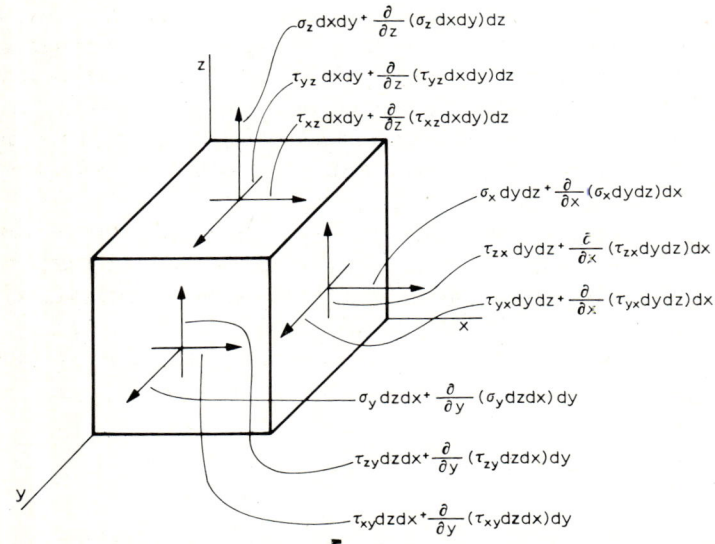

Fig. 3-11. Equilibrium conditions of an elemental parallelepiped.

inertial forces), i.e.:

$$X = \varrho\left(F_x - \frac{\partial^2 u_x}{\partial t^2}\right) \ldots \tag{3-79}$$

where ϱ is density; $F_{x,y,z}$ are the components of the body force referred to as unit mass; and $u_{x,y,z}$ are the components of the displacements.

If the body is at rest, the accelerations $(\partial^2 u_{x,y,z}/\partial t^2)$ are zero and $X = \varrho F_x$, $Y = \varrho F_y$, $Z = \varrho F_z$ will represent the projections of the body forces.

Boundary conditions

To make use of the equations of equilibrium, one must know the relation between the external load components applied along the contour of the body and the components of internal stress formed on the planes adjacent to the contour. This relation is called boundary condition.

Static boundary conditions are determined by eq. 3-1. Then, the quantities $p_x v$, $p_y v$, $p_z v$ entering the equations are treated as the projections of the external forces acting on the unit area of the external surface and the quantities l, m, n are regarded as the cosines of the angles the normals v make with the coordinate axes.

Kinematic boundary conditions are expressed in terms of displacement components (or the rates of displacements). The possibility that mixed conditions are given at a boundary, being expressed partly in terms of load and partly in terms of displacement, is not excluded.

Compatibility equations

Eliminating the displacement components u_x, u_y, u_z from eq. 3-14, we obtain equations of compatibility (or continuity) of deformations:

$$\begin{aligned}
\frac{\partial^2 \gamma_{xy}}{\partial x \partial y} &= \frac{\partial^2 \varepsilon_x}{\partial y^2} + \frac{\partial^2 \varepsilon_y}{\partial x^2} \\
\frac{\partial^2 \gamma_{yz}}{\partial y \partial z} &= \frac{\partial^2 \varepsilon_y}{\partial z^2} + \frac{\partial \varepsilon_z}{\partial y^2} \\
\frac{\partial^2 \gamma_{zx}}{\partial z \partial x} &= \frac{\partial^2 \varepsilon_z}{\partial x^2} + \frac{\partial^2 \varepsilon_x}{\partial z^2} \\
2\frac{\partial^2 \varepsilon_x}{\partial y \partial z} &= \frac{\partial}{\partial x}\left(-\frac{\partial \gamma_{yz}}{\partial x} + \frac{\partial \gamma_{zx}}{\partial y} + \frac{\partial \gamma_{xy}}{\partial z}\right) \\
2\frac{\partial^2 \varepsilon_y}{\partial z \partial x} &= \frac{\partial}{\partial y}\left(\frac{\partial \gamma_{yz}}{\partial x} - \frac{\partial \gamma_{zx}}{\partial y} + \frac{\partial \gamma_{xy}}{\partial z}\right) \\
2\frac{\partial^2 \varepsilon_z}{\partial x \partial y} &= \frac{\partial}{\partial z}\left(\frac{\partial \gamma_{yz}}{\partial x} + \frac{\partial \gamma_{zx}}{\partial y} - \frac{\partial \gamma_{xy}}{\partial z}\right)
\end{aligned} \tag{3-80}$$

The above equations represent, in mathematical form, one of the basic postulations of continuum mechanics: that a deformed body and deformation are continuous.

The condition of compatibility in eq. 3-80 must also hold if not deformations but their rates are considered. To that end, the components of deformation rates $\dot{\varepsilon}_x$, $\dot{\gamma}_{xy}$, etc., must be substituted into eq. 3-80.

Simple and combined loading

Since the stress–strain state of a body is influenced by the way the loading is applied, distinction is made between simple and combined loading. According to Il'yushin (1963), we have a simple loading when the components of a stress deviator increase in proportion to only one of the parameters. When this condition is not fulfilled, the loading is regarded as being complex.

In particular, a simple loading is the loading of a body in a homogeneous state of stress (possible only if the body forces are disregarded), which results from an increase in the external forces in proportion to a parameter. In this case, the state of strain will also be homogeneous, the equations of equilibrium (eq. 3-78) and those of compatibility of deformations (eq. 3-80) being identically satisfied. Consequently, an increase in loading obeying the law βP will cause an increase in the stress deviator components $(\sigma_x - \sigma_m)$, τ_{xy}, etc., in proportion to the factor β. Note, that β can be time, i.e. a time-dependent loading may be a simple one.

Il'yushin has shown that a simple loading is possible when, for example, the relation between the intensities of stresses and strains obeys a power law, $\tau_i = A\varepsilon_i^m$ (under the conditions of incompressibility). However, simple loading may also exist in conjunction with other nonlinear stress–strain relations. Recourse to simple loading makes problem solutions significantly easier.

Active and passive deformation

Distinction is made between active deformation, i.e. one resulting from loading a body, and passive deformation, which is formed when the body is being unloaded. In the case of simple stress, these concepts are satisfactorily defined.

When a body is in the state of combined stress, the deformation at each given instant of time is said to be active, provided the tangential stress intensity τ_i at this instant is greater than any of the preceding values. If the stress τ_i is less than at least one of the preceding values, the deformation is passive (Il'yushin, 1963). Active deformation is characterized by an increase in both the elastic and plastic deformation; in the case of passive deformation, the plastic component remains constant while the elastic deformation decreases.

Thus, when the load increases, the stress intensity increases as well, $d\tau_i > 0$; if the load decreases, the stress intensity diminishes, $d\tau_i < 0$; $d\tau_i = 0$ corresponds to the so-called neutral deformation.

If the process of deformation entails both a distortion of shape and a change in volume, being also a function of time, the active deformation is defined by the expressions:

$$\frac{d\tau_i}{dt} > 0 \quad \frac{d\sigma_m}{dt} > 0$$

or:

$$\frac{d\tau_i}{dt} \geq 0 \quad \frac{d\sigma_m}{dt} > 0$$

Since the case $d\tau_i = 0$ introduces some ambiguity, there is a more strict way of determining the active and passive deformation (Meschyan, 1967) when the strain tensor T_ε is expressed in the form of a vector space of six coordinates. The process of loading is then expressed as the motion of the tip of the vector along a curve called the deformation path. A plane for $\gamma_i =$ const. can be drawn through any point of this space, corresponding to a shear–strain intensity γ_i.

If the increment dT_ε forms an acute angle with the normal v to the above plane and, consequently, $d\gamma_i > 0$, extra plastic deformation will occur, indicating that the loading increases. If the angle between dT_ε and v is greater than $\pi/2$, unloading takes place. Finally, when the vector dT_ε is located in a plane tangential to the surface $\gamma_i =$ const., a neutral deformation will be observed.

Rheological equation of state

The three equations of equilibrium in 3-78 and the six equations of the compatibility of deformation in 3-80 contain a total of fifteen unknown quantities (six stress components, six strain components and three displacement components). Consequently, to solve a problem, one needs six more equations, unless some other unknowns enter these equations.

Eqs. 3-78 and 3-80 hold for any continuum without reflecting its physical properties. This implies that they must be defined by the six extra equations, describing the relation or the rate of strain components.

The above equations can be written in the general form as a single equation, establishing a relation between the stress tensor and strain tensor. Such an equation is termed the equation of state.

When time-dependent processes are considered, the above equation must also include time (in the explicit or implicit form). This equation is called the rheological equation of state. It should provide a correlation between the stress tensor T_σ, strain tensor T_ε, rate of loading tensor $T_{\dot\sigma}$, rate of strain tensor $T_{\dot\varepsilon}$ and time t:

$$R(T_\sigma, T_\varepsilon, T_{\dot\sigma}, T_{\dot\varepsilon}, t) = 0 \tag{3-81}$$

When soils having a variable moisture content W and variable temperature θ are involved, these quantities should also be included in the equation of state.

All in all, we have three equations of equilibrium, six equations of continuity, and six equations of state. The equations of equilibrium are static equations of continuum mechanics holding for any point in a body. The equations of the compatibility of deformations are geometrical equations of continuum mechanics fulfilling the condition that the body under consideration is continuous and remains as such on being deformed. The equation of state represents (in a microscopic form) the physical properties of the body.

Equations of changes in shape and volume

As already pointed out, when considering the deformation of any body, distinction must be made between the distortion of its shape γ_i, produced by the tangential stresses intensity τ_i, and the change in the volume $\varepsilon_m = \varepsilon_v/3$, resulting from the mean normal (all-around) pressure σ_m; the laws which the changes in shape and volume obey can, however, be different. Taking this into account, the rheological equation of state is commonly represented by two equations:

$$R(\tau_i, \gamma_i, \dot{\tau}_i, \dot{\gamma}_i, t) = 0 \tag{3-82}$$

$$R^*(\sigma_m, \varepsilon_m, \dot{\sigma}_m, \dot{\varepsilon}_m, t) = 0 \tag{3-83}$$

The classical theory of elasticity and plasticity assumes that the two equations are independent of one another. However, we shall show that as far as soils are concerned this assumption is inapplicable and the two equations appear to be correlated.

3.7 DEFORMATION THEORIES

Theories of deformation define the stresses and deformations at various points in a body under given boundary conditions. Serving this purpose are the equations of equilibrium (3-78), the equations of the compatibility of deformations (3-80) and the rheological equations of state (3-82 and 3-83).

The equation of state in its general form incorporates functions providing a correlation between the components of stress, strain and rate of strain. In use are two theories of plasticity or, to be exact, two groups of theories. One of them, called deformation theory, considers the relation between the components of stress and strain; the other, referred to as the theory of plastic flow, is concerned with the strain increments rather than with strains proper.

The term "flow" has been introduced because plastic deformation increments are identified with the plastic flow of material. However, this term has no association with the time-dependent viscous flow.

Both theories of plasticity disregard time in its explicit form. The effect of time is taken into account by the theory of creep discussed below.

Basic assumptions of plasticity theory

The theories of plasticity mentioned above are based on the following basic assumptions:

(1) A distortion of shape (or an increment in this distortion) is regarded as being brought about by a stress deviator and is not influenced by the spherical stress tensor. A change in volume (or an increment in volume change) results from the spherical stress tensor and is not influenced by the stress deviator.

(2) The relation between the components of stress and strain (or their increments) remains unchanged whatever the state of stress.

(3) The state of stress and that of strain in a body are regarded as being similar.

The extent within which these assumptions are valid in the case of soils will be outlined below.

Deformation theory of plasticity

Most popular among the deformation theories is the theory of small elastoplastic deformations by Il'yushin (1963). Assumptions (1) and (2) are formulated for this theory in the form of the relations:

$$\gamma_i = \phi(\tau_i) \qquad (3\text{-}84)$$

$$\varepsilon_m = \sigma_m/k \qquad (3\text{-}85)$$

Relationship 3-84 represents the condition in which the shear strain intensity is a function of the shear stress intensity. Relationship 3-85 states that the volumetric strain is a function of the mean normal stress only; it is also assumed that volumetric changes are elastic so that the function to be adopted must be, consequently, of the linear type. When the plastic deformations develop to a considerable extent, the elastic volumetric strains may be neglected. Hence, condition 3-85 is replaced by the conditions of incompressibility of the material:

$$\varepsilon_m = 0 \qquad (3\text{-}86)$$

Since γ_i is a definite function of τ_i and τ_i alone, the curve of the $\gamma_i - \tau_i$ relationship (Fig. 3-12) will be of the invariant nature whatever the state of stress, thus satisfying assumption (2). This implies that the curve can be plotted on the basis of data acquired from any test: shear, compression, tensile, combined stress.

In any case, the $\gamma_i - \tau_i$ graph obtained from substituting the generalized stresses τ_i and strains γ_i for the test data must be of identical shape. By way of illustration, if a curve defining a $\sigma_z - \varepsilon_z$ relation is plotted from a uniaxial compression test, a curve in the $\tau_i - \gamma_i$ coordinates can be obtained by reducing the test data in accordance with Table 3-I, i.e. by taking $\tau_i = \sigma_z/\sqrt{3}$ and $\gamma_i = 2\varepsilon_z(1 + v)/\sqrt{3}$. In the case of pure shear test, the $\tau - \gamma$ and $\tau_i - \gamma_i$ graphs will coincide totally.

Assumption (3) is represented by the relationship:

$$D_\varepsilon = \chi D_\sigma \qquad (3\text{-}87)$$

indicating that the strain deviator is proportional to the stress deviator, the factor χ — sometimes referred to as the modulus of plasticity — being a function of the stress invariants.

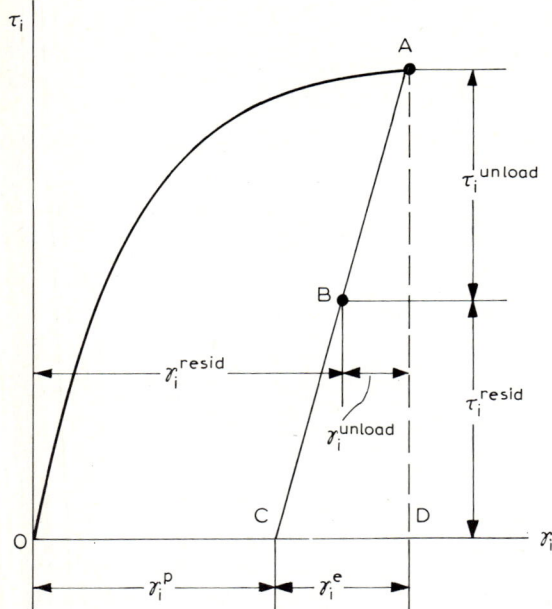

Fig. 3-12. Curve of relationship $\gamma_i - \tau_i$.

From relationship 3-87 it follows that:

$$\frac{e_1}{s_1} = \frac{e_2}{s_2} = \frac{e_3}{s_3} = \chi \qquad \frac{\gamma_1}{\tau_1} = \frac{\gamma_2}{\tau_2} = \frac{\gamma_3}{\tau_3} = \chi \qquad (3\text{-}88)$$

where $e_1 = \varepsilon_1 - \varepsilon_m, \ldots;\ s_1 = \sigma_1 - \sigma_m, \ldots$.

Relationships 3-88, in their turn, imply that the principal values of the stress deviator components $s_{1,2,3}$ and those of the strain deviator components $e_{1,2,3}$ are proportional and coincide in direction; the same applies to the principal tangential stresses $\tau_{1,2,3}$ and principal shears $\gamma_{1,2,3}$.

Relationships 3-88 may be reduced to the form:

$$\frac{(\sigma_2 - \sigma_1) - (\sigma_2 - \sigma_3)}{\sigma_1 - \sigma_3} = \frac{(\varepsilon_2 - \varepsilon_1) + (\varepsilon_2 - \varepsilon_3)}{\varepsilon_1 - \varepsilon_3}$$

from which it is apparent that the above assumptions are equivalent to the condition:

$$\mu_\sigma = \mu_\varepsilon \qquad (3\text{-}89)$$

stating that the state of stress and that of strain are similar.

Hencky's equations

Relationships 3-88 are readily transformed into expressions establishing a link between the components of stress and strain (Hill, 1950):

$$\begin{aligned}
\varepsilon_x - \varepsilon_m &= \chi(\sigma_x - \sigma_m) & \gamma_{xy} &= 2\chi\tau_{xy} \\
\varepsilon_y - \varepsilon_m &= \chi(\sigma_y - \sigma_m) & \gamma_{yz} &= 2\chi\tau_{yz} \\
\varepsilon_z - \varepsilon_m &= \chi(\sigma_z - \sigma_m) & \gamma_{zx} &= 2\chi\tau_{zx}
\end{aligned} \quad (3\text{-}90)$$

In the general case, we shall have $T_\varepsilon = \chi T_\sigma$.

The above expressions, known as Hencky's equations, hold (under the assumptions adopted) for any deformed body. They can be rendered fit to represent the physical properties of a body if the type of χ function is determined experimentally. Substituting formulae 3-45 and 3-63 into equalities 3-88, we obtain:

$$\chi = \frac{\gamma_i}{2\tau_i} \quad (3\text{-}91)$$

Elastic and plastic deformations

In the case of elastic deformations, $\chi = 1/2G$, where G is the shear modulus. Taking into account that $G = E/2(1 + \nu)$ and $\varepsilon_m = \sigma_m/k$, where $k = E/(1 - 2\nu)$, eqs. 3-90 are transformed into the following equations from the theory of elasticity:

$$\begin{aligned}
\varepsilon_x^e &= \frac{1}{E}[\sigma_x - \nu(\sigma_y + \sigma_z)] \ldots \\
\gamma_{xy}^e &= \frac{1}{G}\tau_{xy} \ldots
\end{aligned} \quad (3\text{-}92)$$

When the deformation consists of elastic and plastic constituents:

$$\chi = \frac{1}{2G} + \lambda \quad (3\text{-}93)$$

where $\lambda = \gamma_i^p/2\tau_i$ is a function defining the plastic constituent of the deformation. Hence, relationships 3-90 can be written as:

$$\varepsilon_x = \varepsilon_x^e + \varepsilon_x^p \ldots; \; \gamma_{xy} = \gamma_{xy}^e + \gamma_{xy}^p \ldots \quad (3\text{-}94)$$

where the superscripts e and p indicate the elastic and plastic constituents of the deformation; ε_x^e and γ_{xy}^e are determined from expressions 3-92; $\varepsilon_x^p = \lambda(\sigma_x - \sigma_m)$ and $\gamma_{xy}^p = 2\lambda\tau_{xy}$.

Unloading

When a body is being unloaded, the intensity of shear strains decreases (Fig. 3-12):

$$\gamma_i^{unload} = \gamma_i - \gamma_i^{resid}$$

where γ_i^{unload} is the strain resulting from unloading and γ_i^{resid} is the residual strain.

It is assumed that the decrease in strain intensity during the process of load relief is directly proportional to the decrease in stress intensity, i.e. $\tau_i^{unload} = G\gamma_i^{unload}$. Correspondingly, relationships 3-92 hold for the process of unloading.

Theory of plastic flow

Developed by Prandtl and Reuss, this theory treats increments in strain as the sum of the increments in elastic and plastic strains (Hill, 1950):

$$d\gamma_i = d\gamma_i^e + d\gamma_i^p \quad \text{and} \quad d\varepsilon_m = d\varepsilon_m^e + d\varepsilon_{an}^2 \tag{3-95}$$

The increments in elastic strains are given by:

$$d\gamma_i^e = \frac{1}{G}d\tau_i \quad d\varepsilon_m^e = \frac{1}{k}d\sigma_m \tag{3-96}$$

and the increments in plastic strains by:

$$d\gamma_i^p = \phi(\tau_i)\,d\tau_i \quad d\varepsilon_m^p = 0 \tag{3-97}$$

Statements 3-96 and 3-97 represent a formulation of assumption (1): that the increment in shear strains is a function of tangential stress intensity and the increment in volumetric strain is a function of mean normal stress, provided the volumetric strains are elastic and change directly with mean stress.

Assumption (2) and the consequent conditions in eqs. 3-96 and 3-97 imply, by analogy with deformation theory, that the $\gamma_i^p - \tau_i$ curve is invariant with respect to the arrangement of stress.

Assumption (3) may be represented in the form:

$$D_{d\varepsilon}^p = d\lambda D_\sigma \tag{3-98}$$

which implies that the deviator of plastic strain increment is proportional to the stress deviator, the proportionality factor $d\lambda$ being a function of stress.

From relationship 3-98 it follows that:

$$\frac{de_1^p}{s_1} = \frac{de_2^p}{s_2} = \frac{de_3^p}{s_3} = d\lambda \quad \frac{d\gamma_1}{\tau_1} = \frac{d\gamma_2}{\tau_2} = \frac{d\gamma_3}{\tau_3} = d\lambda \tag{3-99}$$

where $e_1 = \varepsilon_1 - \varepsilon_m \ldots; s_1 = \sigma_1 - \sigma_m \ldots$.

Relationships 3-99, in their turn, may be readily reduced to the equality:

$$\mu_\sigma = \mu_{d\varepsilon} \tag{3-100}$$

which implies that the field of stresses and the field of strain increments (rate of strain increments) are similar.

Saint-Venant–von Mises' equations

Substituting relationships 3-96 and 3-97 into equalities 3-95 and taking into account relations 3-99, we may obtain the following relations, establishing a connection between increments in strain and stress components:

$$d\varepsilon_x = \frac{1}{E}[d\sigma_x - v(d\sigma_y + d\sigma_z)] + d\lambda(\sigma_x - \sigma_m)$$

. . . .

$$d\gamma_{xy} = \frac{1}{G}d\tau_{xy} + 2\,d\lambda(\tau_{xy}) \quad (3\text{-}101)$$

. . . .

where $d\lambda = d\gamma_i^p/2\tau_i$.

If a body is strained only elastically, $d\lambda = 0$; when the strains are only of the plastic nature, $G = E = \infty$.

Considering the last-named case and dividing both parts of eqs. 3-101 by dt, we obtain the equations of Saint-Venant, Levy and von Mises correlating the components of plastic strain rates with the components of stress:

$$\dot{\varepsilon}_x^p = \frac{\dot{\gamma}_i^p}{2\tau_i}(\sigma_x - \sigma_m) \qquad \dot{\gamma}_{xy}^p = \frac{\dot{\gamma}_i^p}{\tau_i}\tau_{xy}$$

$$\dot{\varepsilon}_y^p = \frac{\dot{\gamma}_i^p}{2\tau_i}(\sigma_y - \sigma_m) \qquad \dot{\gamma}_{yz}^p = \frac{\dot{\gamma}_i^p}{\tau_i}\tau_{yz} \quad (3\text{-}102)$$

$$\dot{\varepsilon}_z^p = \frac{\dot{\gamma}_i^p}{2\tau_i}(\sigma_z - \sigma_m) \qquad \dot{\gamma}_{zx}^p = \frac{\dot{\gamma}_i^p}{\tau_i}\tau_{zx}$$

In the general form, it may be written $T_{\bar{\varepsilon}} = d\lambda T_\sigma$.

Although, strain rates are entering eqs. 3-102, time in its explicit form is omitted because the ratios of rates are equivalent to the ratios of strain increments:

$$\frac{d\varepsilon_x^p}{dt} : \frac{d\gamma_i^p}{dt} = d\varepsilon_x^p : d\gamma_i^p, \text{ etc.}$$

Plasticity potentials

Generalizing the theory of plastic flow, von Mises (1928) has introduced the notion of plasticity potential. This is thought of as a stress function, a partial derivative of which varies with the plastic strain increment:

$$d\gamma_i^p = d\lambda \frac{\partial f}{\partial \tau_i} \quad (3\text{-}103)$$

If expanded, the above statement takes the form:

$$d\varepsilon_x^p = d\lambda \frac{\partial f}{\partial \sigma_x} \ldots \quad d\gamma_{xy}^p = d\lambda \frac{\partial f}{\partial \tau_{xy}} \qquad (3\text{-}104)$$

Assuming, as von Mises did, that the plasticity potential coincides with the loading function, expressions 3-104 will be equivalent to the second terms in the right-hand side of equalities 3-101 defining the plastic strain. Hence, expressions 3-104 may be written:

$$\frac{\partial f}{\partial \sigma_x} = (\sigma_x - \sigma_m) \ldots \quad \frac{\partial f}{\partial \tau_{xy}} = 2\tau_{xy} \ldots \qquad (3\text{-}105)$$

Thus, it follows, that the partial derivatives of plasticity potential equal (under the assumption referred to above) the stress deviator components.

The above relationships are referred to as the associated law of flow, for relations 3-103 are connected (associated) under the above von Mises assumption with the law of plastic flow.

The plastic potential can also be expressed in terms of tangential stress intensity. Finding the values of $d\tau_i^2/d\sigma_x \ldots$ from expression 3-45 and comparing them with relations 3-105, we obtain:

$$f = \tau_i^2 \qquad (3\text{-}106)$$

Now, if we construct a surface $\tau_i^2 = $ const. on the σ_1, σ_2 and σ_3 coordinates (similar to Fig. 3-10), we shall obtain a circular cylinder. Its axes will extend at right angles to the deviatoric plane and the vector of plastic strain increment will be directed along the radius of the cylinder. The surface obtained is called the surface of plastic potential. It is assumed, under the associated law of flow, that this surface and the limiting surface to be discussed in Chapter 4 (see Fig. 4-3(b)) coincide in shape.

Comparison of plasticity theories

In the case of a simple loading, solutions yielded by deformation theory and the theory of plastic flow are identical. When a combined loading is involved, solutions obtained on the basis of the theory of plasticity coincide more or less with the experimental results, whereas the deformation theory appears to be, strictly speaking, inapplicable. However, when departure from the conditions of a simple loading is small, particularly when the deformation path is represented in the $(\varepsilon_1 - \varepsilon_2 - \varepsilon_3)$ strain space by a line close to the straight one, the solutions arrived at through the use of both theories will differ little. Thus, the theory of plastic flow operating with strain increments rather than with the strains proper is more versatile than the deformation theory. On the other hand, this latter theory is simpler and is, consequently, used more widely in solving engineering problems, particularly those pertaining to soil mechanics.

3.8 REFERENCES

Hill, R., 1950. The Mathematical Theory of Plasticity. Clarendon Press, Oxford.

Il'yushin, A.A., 1963. Plasticity. USSR Acad. Sci., Moscow (in Russian).

Kandaurov, I.I., 1966. Mechanics of Granular Medium and Its Application in Construction. Moscow (in Russian).

Meschyan, S.R., 1967. The Creep of Clay Soils. Acad. Sci. Armenian SSR, Erevan (in Russian).

Muler, R.A., 1962. To the statistical theory of stress distribution in granular foundation. SMFE, 4: 4–6 (in Russian).

Novozhilov, V.V., 1951. On stress–strain relations in a nonlinearly elastic medium. Appl. Math. Mech., 15 (2) (in Russian).

Novozhilov, V.V., 1952. On the physical meaning of stress invariants. Appl. Math. Mech., 16 (5) (in Russian).

Rowe, R.W., 1962. Stress–dilatancy relation for static equilibrium of an assembly of particles in contact. Proc. R. Soc., A 269.

Sedov, L.I., 1983. Continuum Mechanics. Nauka, Moscow, Vol 1 (Vol. 2, 1984) (in Russian).

von Mises, R., 1928. Mechanik der plastischen Formänderung von Kristallen. Z. Angew. Math. Mech., 8: 161–185.

Chapter 4

ELASTICITY, PLASTICITY AND VISCOSITY

4.1 ELASTICITY AND PLASTICITY

Idealized solids and liquids

Continuum mechanics deals with idealized bodies possessing specified properties. The extreme examples of these bodies are the Euclidian (non-deformable) solid and the Pascalian (perfect) liquid.

The rheological equation of state of the Euclidian solid takes the form:

$$\gamma_i \equiv 0 \quad \varepsilon_m \equiv 0 \tag{4-1}$$

The Pascalian liquid is incapable of resisting shear, yet it can withstand an all-around (hydrostatic) pressure without experiencing volumetric strain. The rheological equation of state of this liquid can be written as:

$$\tau_i \equiv 0 \quad \varepsilon_m \equiv 0 \tag{4-2}$$

The remaining idealized bodies represent the transition between perfect solids and perfect liquids.

Elasticity

This is the property of a solid to restore its shape and volume, or of a liquid or gas to restore its volume, as soon as external forces are removed. Correspondingly, an elastic deformation is one which completely disappears after the load has been withdrawn. Since the disappearance of deformation implies that the body has been reinstated to its original dimensions, such a deformation is referred to as 'recoverable' or 'reversible'.

In a perfectly elastic body, deformations are set up immediately on applying a load and cease to exist as soon as the load is removed. The elasticity of solids is attributed to the forces of interaction (attraction and repulsion) between the atoms.

Elastic deformation can be either linear, changing directly with the stress, or nonlinear, in which case we have to deal with non-linear elasticity. The rheological equation of state of a perfectly elastic linearly deforming body — a term which refers to the Hookean solid — is:

$$\tau_i = G\gamma_i \quad \sigma_m = k\varepsilon_m \tag{4-3}$$

Relationships 4-3 refer to the state of combined stress, defined by the tangential stress intensity τ_i, shearing strain intensity γ_i, mean normal stress σ_m and mean unit elongation ε_m.

Since for pure shear we know (according to Table 3-I) that $\tau_i = \tau$ and $\gamma_i = \gamma$, the first of expressions 4-3 may be represented in the form:

$$\tau = G\gamma \tag{4-3'}$$

where τ and γ are, respectively, the tangential stress and angular distortion in pure shear.

Similarly, the remainder of the relations describing the state of combined stress may be employed in the case of pure shear simply by substituting τ and γ for τ_i and γ_i.

For uniaxial compression or tension we have (according to Table 3-I) $\tau = \sigma_z/\sqrt{3}$ and $\gamma_i = \varepsilon_z 2(1 + v)/\sqrt{3}$; hence $\sigma_z = 2G(1 + v)\varepsilon_z = E\varepsilon_z$.

The second of expressions 4-3 may be written in the form:

$$p = k\varepsilon_v/3 \tag{4-3''}$$

where $3p = \sigma_1 + \sigma_2 + \sigma_3$ is the all-around pressure and $\varepsilon_v = \varepsilon_1 + \varepsilon_2 + \varepsilon_3$ is the volumetric strain.

Elastic constants. The parameters G and k used above, known as the shear modulus (or the modulus of lateral elasticity) and the bulk modulus, respectively, are connected by the relation:

$$k = 2G\frac{1 + v}{1 - 2v} = \frac{E}{1 - 2v} \tag{4-4}$$

where $E = 2G(1 + v)$ is the modulus of longitudinal elasticity (Young's modulus) and $v = \varepsilon_x/\varepsilon_z$ is the coefficient of lateral strain (Poisson's ratio).

The parameter v, in turn, may be expressed in terms of the constants G and k, i.e. $v = (k - 2G)/2(G + k)$. The coefficient v varies between 0 and 0.5.

For $v = 0$, deformation occurs only in the z-direction ($\varepsilon_x = 0$ and $k = E = 2G$); for $v = 0.5$, a change in shape takes place without a change in volume (the material is said to be incompressible); in this case, $k \to \infty$ and $E = 3G$.

The constants G, E and k are measured in units of stress; the constant v is dimensionless. Since these constants are interrelated we only need to determine any two of them, the other two then being readily obtained by computation. Occasionally, use is made of the so-called Lamé constant:

$$\lambda = 2vG/(1 - 2v)$$

Since the moduli of elasticity of most real bodies are apt to decrease (relax) under a load, their values depend upon the rate of load application. Distinction is commonly made between a static modulus of elasticity, as determined from static tests at a standard rate of load application, and a dynamic modulus of elasticity corresponding to a rate of loading equal to the speed of sound. This latter modulus is always greater than its static counterpart.

In an anisotropic medium the moduli of elasticity acquire different values, i.e.

$E_x \neq E_y \neq E_z$ and $v_x \neq v_y \neq v_z$. For rock in natural occurrence it is typical when $E_z \neq E_x = E_y$.

Moduli of elasticity of rocks. Below are given the values of static moduli of linear deformation (in 10^5 Pa) of some rocks (10^5 Pa = 1 kg/cm^2):

Gypsum	1300
Sandstone	5000 and upwards
Granite	6000
Limestone	8500
Basalt	9700
Quartzite	10 000
Corundum	52 000
Igneous rocks	3000–13 000

The physical meaning of the constants E, G and k is somewhat different for soils than for solids. The stress–deformation relation in soils is assumed to be linear over a short range of the stresses applied, rendering Hooke's law (eq. 4-3) applicable to soils over this range. The resulting deformations are, however, partly recoverable and partly irrecoverable, so that eq. 4-3 holds for the process of loading only and, in contrast to the law of elasticity, is known as the law of linear deformation. Similarly, the parameters E, G and k are called the moduli of linear deformation.

Applicable building codes and regulations recommend for use the following standard values of the modulus of linear deformation (in 10^5 Pa):

Sand (medium- and coarse-grained)	300–500
Sand (fine-grained and silty)	110–480
Clay soils	50–400

The values of E corresponding to the elastic, recoverable part of the deformation (as obtained, for example, from settlement plate tests with unloading) appear to be considerably higher and, according to Barkan (1948), are as follows (in 10^5 Pa):

Sand (medium-grained)	830
Clay loam	310–2950
Loess	1000–1300

Plasticity

This is the property of a body to change its shape irreversibly without failure under the action of external forces. A perfectly plastic body is said to be in the plastic state when the maximum tangential stress τ_{max} attains a certain ultimate value τ_s, called the yield point in shear. It is assumed that the material that is in the plastic state is incompressible. Accordingly, the equation of state may be written in this case as:

$$\tau_{max} = \tau_s = \text{const.} \quad \varepsilon_v = 0 \tag{4-5}$$

Eq. 4-5 is referred to as the Tresca–Saint-Venant condition of plasticity.

A body which does not deform when subjected to a shearing stress less than the yield limit, $\tau < \tau_s$, but which suffers from an unconfined plastic deformation when $\tau = \tau_s$, is classified as a Saint-Venant rigid–plastic body (Fig. 4-1(a)). On the other

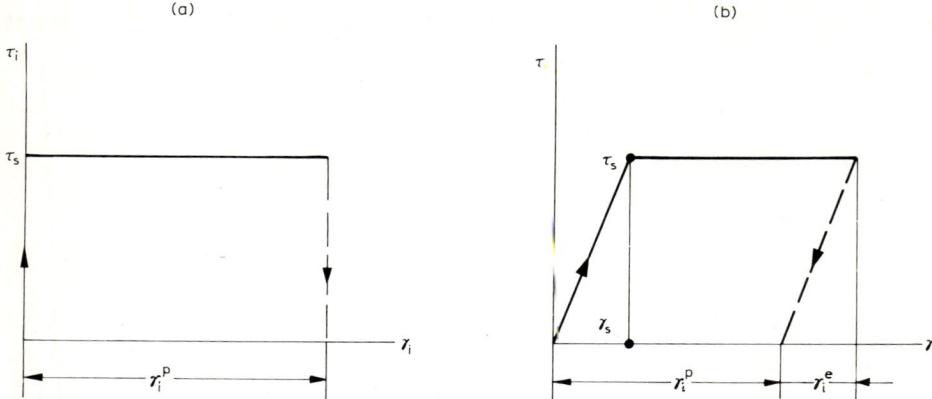

Fig. 4-1. Diagrams of deformation: (a) rigid–plastic body; (b) elastoplastic body.

hand, a body which deforms elastically (according to Hooke's law, $\tau = G\gamma$) under a stress less than the yield limit, $\tau < \tau_s$, and which deforms plastically when $\tau = \tau_s$, is called a Prandtl elastoplastic body (Fig. 4-1(b)).

It will be recalled that an unconfined plastic deformation under a steady load is often referred to as 'flow', apparently by analogy with the flow of a liquid. This notion is not identical with the notion of viscous flow which will be discussed later.

A rigid–plastic body relieved of load fails to recover from deformation, and is thus totally plastic, $\gamma = \gamma^p$.

An elastoplastic body recovers from deformation only partly after a load relief and, as a result, the aggregate shearing deformation is the sum of an elastic deformation γ^e and a plastic deformation γ^p:

$$\gamma = \gamma^e + \gamma^p \tag{4-6}$$

4.2 LIMITING STATE OF STRESS

Theory of limiting equilibrium

Plastic soil deformations require careful consideration, for, if progressing, they may lead to a loss of stability and failure of soil mass.

Depending on the load applied, soil can be in one of these two states of stress: subliminiting or limiting. The former is defined by quite definite deformations which may change due to an increase in the level of stress or to some time-dependent process (e.g. creep, consolidation). The latter results from a certain critical combination of stresses in which the external loading and the internal forces of soil resistance appear to be in limiting equilibrium. In such a case, even the slightest increase in the forces acting on the soil will destroy the interparticle forces, trigger cracks and crevices, provoke an irreversible shear along the surfaces of slip, and lead eventually to a loss of stability in the soil.

Strength in a broad sense of this word is understood to be the property of a material to stand up to failure or to a build-up of big plastic deformations which distort the shape of the body beyond a tolerable limit. The state of limiting stress and the state of plasticity may be thought of as the conditions under which the granular cohesive and plastic bodies, respectively, retain their strength. The three terms referred to are sometimes used as synonyms.

Studying the limiting state of stress of a cohesive medium is the object of the theory of limiting equilibrium, the fundamentals of which were formulated by Ch. Coulomb in 1773 and developed by W. Rankine in 1857 and L. Prandtl in 1920. A purely mathematical treatment of this theory, together with the general solution of a wide range of problems, are given in the works of Sokolovsky (1960) and others.

Limiting state

The limiting equilibrium of a body may be brought about by a combination of stresses. The distinction between limiting states or failure theories as they are sometimes called, depends on the stress or stresses involved. Some of these states are of a historical interest only; however, as far as soils are concerned, two basic states or theories are applicable. One is the Mohr–Coulomb condition of failure, stating that the limiting state results from a certain relation between the tangential and normal stresses acting on the same surface. The other is the von Mises–Schleicher condition according to which the limiting condition is the outcome of a certain relation between the tangential stress intensity and the mean normal stress.

The Mohr–Coulomb theory

Concerning the limiting equilibrium of a granular medium, Coulomb had shown that the maximum tangential stress τ_n acting on a surface of possible slippage (the normal to which is n) is directly proportional to the normal stress σ_n on the same surface:

$$|\tau_n| = \sigma_n \tan \phi \qquad (4\text{-}7)$$

where ϕ is the angle of internal friction.

For a cohesive medium displaying both friction and cohesion, c, the τ_n–σ_n relation takes the form:

$$|\tau_n| = c + \tau_n \tan \phi = (\sigma_n + H) \tan \phi \qquad (4\text{-}8)$$

where $H = c/\tan \phi$ is the resistance to all-around tension.

Eq. 4-7 due to Coulomb and eq. 4-5 due to Tresca–Saint-Venant were later generalized by Mohr who formulated a new condition which stated that in the limiting state the maximum tangential stress on a surface with a normal n is a certain function

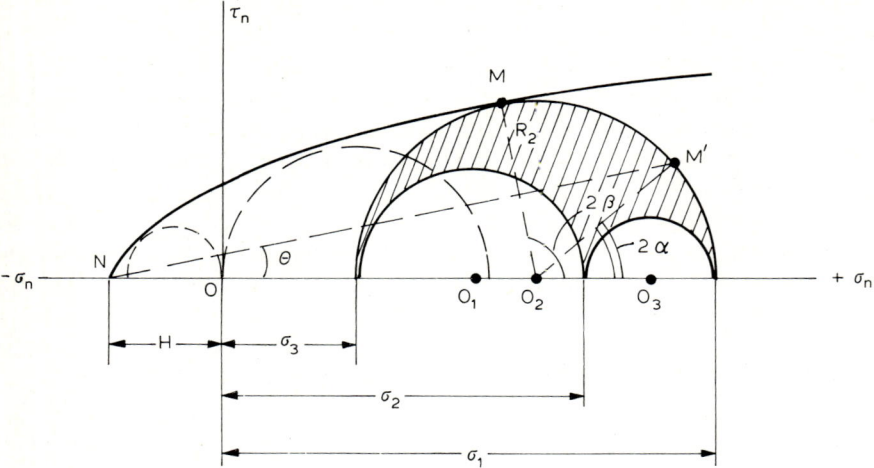

Fig. 4-2. Mohr-circle diagram of limiting stress.

of the normal stress acting on the same surface, i.e.:

$$|\tau_n| = f(\sigma_n) \tag{4-9}$$

Eq. 4-8, called the Mohr–Coulomb condition at failure, is a specific case of eq. 4-9.

The elemental surfaces satisfying condition 4-9 are called the fields of slip (although it seems to be more appropriate to call them the fields of limiting equilibrium). There are two such fields, and they are inclined to form an angle of $\pm(\pi/2 - \phi)$.

Mohr diagram of stresses at failure. The limiting state of stress at a given point of a medium can be visualized in the form of a Mohr diagram constructed for the stresses at failure on the τ_n–σ_n plane (Fig. 4-2). The limiting state, as represented on the diagram, will be defined by three circles of limiting stresses with their centres at $OO_1 = (\sigma_2 + \sigma_3)/2$, $OO_2 = (\sigma_1 + \sigma_3)/2$, and $OO_3 = (\sigma_1 + \sigma_2)/2$.

The points corresponding to the stresses τ_n and σ_n on the plane are located inside, and only inside, a curvilinear triangle (hatched in the diagram) confined between the semicircles. The points located on the biggest of the limiting semicircles (e.g. point M') correspond to the stresses τ'_n and σ'_n on a plane inclined with respect to principal plane 1 through an angle α.

The relation $\tau'_n/(H + \sigma'_n) = \tan\theta$ defines the so-called angle of obliquity, θ, which the total stress p on a given plane makes with a normal n. Apparently, the angle of obliquity changes with the angle α, reaching a certain maximum at a point M where $\alpha = \beta$. This point will satisfy the condition:

$$\tan\theta_{max} = \max\left|\frac{\tau_n}{H + \sigma_n}\right| = f \tag{4-10}$$

where f is a function.

The above expression can be written as $\tau_n = f(H + \sigma_n)$, which is equivalent to eq. 4-8. Thus, the angle β determines the inclination of the field of slip, and the point M corresponds to the limiting stresses τ_n and σ_n on this field.

Limiting semicircles can be drawn on the τ_n–σ_n diagram in an infinite number of families, each corresponding to various values of σ_1, σ_2 and σ_3. An envelope of the largest stress circles, contacting these at point M, is the limiting curve on which all points satisfy eq. 4-8.

Some of the semicircles (shown by dashed lines on the diagram), of which the envelope is the limiting curve, represent the cases of uniaxial compression and tension when $\sigma_1 \gtrless 0$ and $\sigma_2 = \sigma_3 = 0$. Apparently, if the compression and tensile strengths are the same, the envelope is transformed into a straight horizontal line. If the strengths are different, the envelope will depart from such a line.

Mohr's eq. 4-8 does not take the intermediate principal stress σ_2 into consideration. In fact, the point M, representing the limiting state of the body, is located on a large semicircle. To fulfil eq. 4-8, the radius R_2 of the principal circle should be a function of the position of the centre O_2 of this circle, i.e.:

$$\frac{\sigma_1 - \sigma_3}{2} = f\left(\frac{\sigma_1 + \sigma_3}{2}\right) \tag{4-11}$$

It can be seen that the intermediate principal stress σ_2 is omitted from the above expression.

Approximating the envelope (Fig. 4-2) by an inclined straight line, we fulfil Mohr–Coulomb's eq. 4-9 which is then expressed in terms of principal stresses by the equation:

$$\sin \phi = \frac{\sigma_1 - \sigma_3}{\sigma_1 + \sigma_3 + 2H} \tag{4-12}$$

This formula is used to assess the behaviour of soils possessing cohesion and friction. However, in the case of real soils at low — let alone negative — values of σ_m, the envelope is almost invariably of curvilinear shape. The replacement of this curve by an inclined straight line is an approximation; thus the quantities c and ϕ entering into condition 4-9 should not be treated as physical characteristics but as parameters of an approximated shear diagram.

For perfectly granular, cohesionless soils relationship 4-12 takes the form:

$$\sin \phi = \frac{\sigma_1 - \sigma_3}{\sigma_1 + \sigma_3} \tag{4-13}$$

which corresponds to Coulomb's eq. 4-7.

For perfectly cohesive soils displaying cohesion only and devoid of friction, the envelope (Fig. 4-2) is transformed into a horizontal straight line satisfying Saint-Venant's or Tresca's eq. 4-5. To describe the principal stresses, this condition may be written as:

$$\sigma_1 - \sigma_3 = 2c = 2\tau_s \tag{4-14}$$

As mentioned above, it is assumed in the case under consideration that the yield stresses in compression and tension are the same, $+\sigma_s = -\sigma_s = \sigma_s$. According to Saint-Venant's theory the value of σ_s is related to the yield stress in shear by the relation:

$$\tau_s = c = \sigma_s/2 \tag{4-15}$$

A non-linear envelope of the circle of limiting stresses in soils may be approximated, for example, by the power equation:

$$|\tau_n| = \tau_0\left(1 + \frac{\sigma_n}{H}\right)^\lambda \tag{4-16}$$

or by the equation of a cycloid:

$$\sigma_n + H = \tfrac{1}{2}K(4\beta - \sin 4\beta) \quad \tau_n = \tfrac{1}{2}K(1 - \cos 4\beta) \tag{4-17}$$

where $1 \geqslant \lambda \geqslant 0.5$, τ_0 is the shearing strength in the case of pure shear; 2β is the angle a line normal to the cycloid makes with the x-axis ($\pi/2 \leqslant \beta_2 \leqslant \pi$), and K is the diameter of the cycloid's generating circle.

The envelope may also be described by a combination of a cycloid and a straight line, using expression 4-15 to describe the curvilinear segment of the envelope and expression 4-8 to describe the straight segment.

Von Mises' theory

This theory states that in the limiting state of a material the intensity of tangential stresses is constant:

$$\tau_i = \tau_s = \text{const.} \tag{4-18}$$

Expressing the value of τ_i in terms of the principal normal stresses, we obtain the following condition for the limiting state of stress:

$$(\sigma_2 - \sigma_3)^2 + (\sigma_3 - \sigma_1)^2 + (\sigma_1 - \sigma_2)^2 = 6\tau_s^2 \tag{4-19}$$

The values of τ_s and σ_s are also correlated by the expression:

$$\tau_s = \sigma_s/\sqrt{3} \tag{4-20}$$

which differs by a permanent factor from condition 4-15 describing the theory of Saint-Venant. However, the main difference between von Mises' and Saint-Venant's equation is that the former takes account of all the three principal stresses.

Applying eq. 4-19 generally, we obtain the von Mises–Schleicher condition according to which the tangential stress intensity is a funtion of the mean normal stress when the material is in the limiting state:

$$\tau_i = f(\sigma_m) \tag{4-21}$$

A linear form of the above relation has been suggested by Botkin (1939, 1940a, b) for soils:

$$\tau_i = (H + \sigma_m) \tan \psi \qquad (4\text{-}22)$$

where $H = \tau_s/\tan \psi$ and ψ are the parameters of the τ_i–σ_m straight line, H being regarded as the ultimate strength under conditions of all-around tension and ψ as the angle of friction on the octahedral plane.

Relation 4-22 may be represented in the form:

$$\sqrt{(\sigma_1 - \sigma_2)^2 + (\sigma_2 - \sigma_3)^2 + (\sigma_3 - \sigma_1)^2} = \sqrt{2/3}(\sigma_1 + \sigma_2 + \sigma_3 + 3H) \tan \psi \qquad (4\text{-}23)$$

Energy proof of von Mises' theory. Beltrami in 1885 voiced an opinion that a solid may be destroyed only by overcoming the forces of molecular interaction. This requires work which should be used as a criterion of strength.

Hubert in 1904 provided a more exact definition by stating that the failure of a material is ascribed to that part of the work which is used to change the shape. This work, as will be shown later, is given by:

$$A_D = \frac{\tau_i^2}{2G} = \frac{1+v}{6E}[(\sigma_1 - \sigma_2)^2 + (\sigma_2 - \sigma_3)^2 + (\sigma_3 - \sigma_1)^2]$$

Since, in the case of uniaxial compression or tension, $A_D = \sigma_s^2(1 + v)/3E$, the above expression may be written as follows:

$$(\sigma_1 - \sigma_2)^2 + (\sigma_2 - \sigma_3)^2 + (\sigma_3 - \sigma_1)^2 = 2\sigma_s^2$$

Taking eq. 4-20 into account, one can see that the above expression is identical with von Mises' eq. 4-19 and for this reason it is known as the von Mises–Hubert condition.

Condition of limiting state in stress space σ_1–σ_2–σ_3

The condition of a limiting state of stress may be presented geometrically in the principal stress space $(\sigma_1$–σ_2–$\sigma_3)$ (see Fig. 3-10) as a limiting surface:

$$\Phi(\sigma_1, \sigma_2, \sigma_3) = 0$$

Saint-Venant's condition 4-14 of the limiting state is represented in this space as a hexagonal prism with the axis $\sigma_1 = \sigma_2 = \sigma_3 = p$ (Fig. 4-3(a)). The area obtained when the prism is being cut by the deviatoric plane $\sigma_1 + \sigma_2 + \sigma_3 = 0$ is a regular hexagon; the radii of the circles inscribed into, and described about, the hexagon are, respectively:

$$R_1 = \sqrt{2}\tau_s = \frac{1}{\sqrt{2}}\sigma_s$$

$$R_2 = 2\frac{\sqrt{2}}{3}\tau_s = \frac{\sqrt{2}}{3}\sigma_s$$

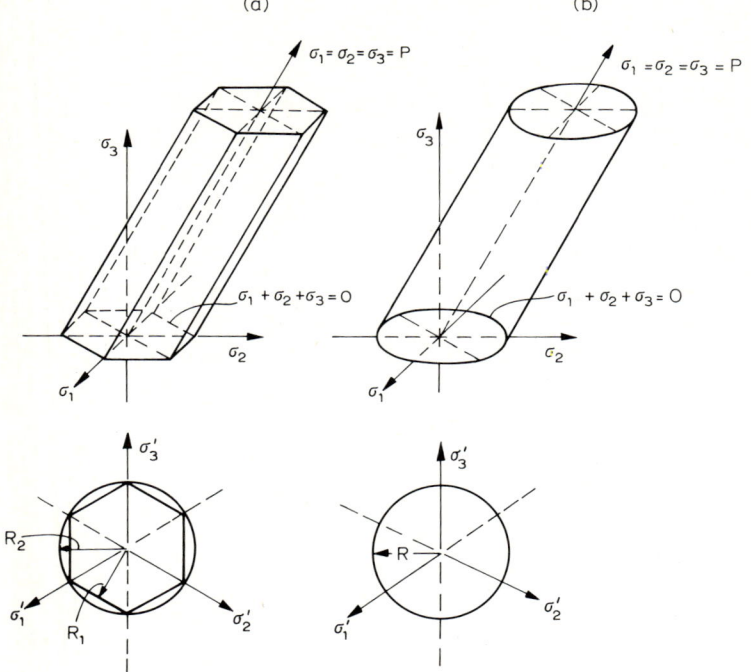

Fig. 4-3. Limiting surface $\sigma_1, \sigma_2, \sigma_3$ and its contour on the deviatoric plane $\sigma_1 + \sigma_2 + \sigma_3 = 0$: (a) for the Saint-Venant condition; (b) for the von Mises condition.

Mohr–Coulomb's eq. 4-12 is represented in the form of a hexagonal pyramid (Fig. 4-4(a)), which has its apex at a point defined by the coordinates $\sigma_1 = \sigma_2 = \sigma_3 = H$, and has a hexagonal contour on the deviatory plane. The nearest and farthest vertexes of the hexagon are distant from the centre at:

$$R_{1,2} = \frac{2\sqrt{6}H \sin \phi}{3 \pm \sin \phi}$$

Von Mises' eq. 4-19 is represented in the stress space $(\sigma_1-\sigma_2-\sigma_3)$ by a circular cylinder with its axis $\sigma_1 = \sigma_2 = \sigma_3$ (see Fig. 4-3). The area obtained due to the intersection of the cylinder with the deviatoric plane $\sigma_1 + \sigma_2 + \sigma_3 = 0$ is a circle of radius:

$$R = \sqrt{2}\tau_s = \sqrt{2/3}\sigma_s$$

The condition of von Mises–Schleicher–Botkin given by eq. 4-23 is represented by a cone (Fig. 4-4(b)) with the apex at a point defined by $\sigma_1 = \sigma_2 = \sigma_3 = H$.

Comparing diagrams (a) and (b) of Fig. 4-3, we can visualize the difference between the limiting conditions at failure according to Saint-Venant and von Mises. Diagrams (a) and (b) of Fig. 4-4 provide the same information about the conditions of

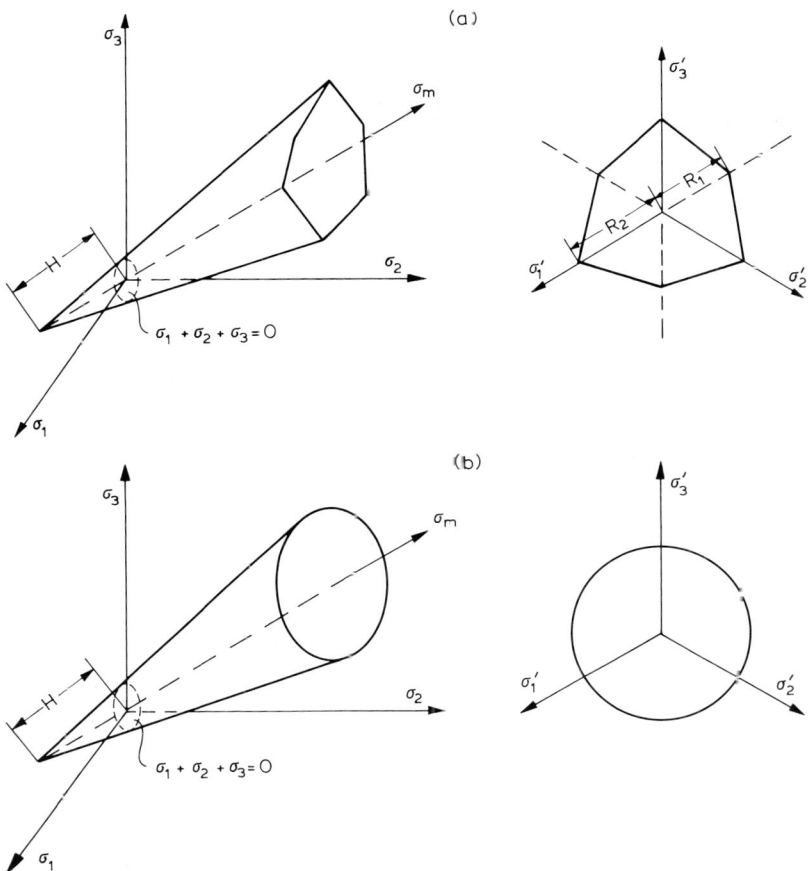

Fig. 4-4. Limiting surface σ_1, σ_2, σ_3 and its contour on the deviatoric plane $\sigma_1 + \sigma_2 + \sigma_3 = 0$: (a) for the Mohr–Coulomb condition; (b) for the von Mises–Botkin condition.

Mohr–Coulomb and von Mises–Botkin, respectively. Thus, it is evident that von Mises' circle can be described around Saint-Venant's hexagon, and the stresses at failure according to von Mises–Botkin appear to be higher (not only in the zone of compression) than those defined by the Mohr–Coulomb condition.

4.3 NON-LINEAR DEFORMATION

Non-linear shear strain

For many materials, including soils, the stress–strain relation is non-linear. In the case of a state of combined stress, this non-linearity is given by:

$$\tau_i = \phi(\gamma_i) \tag{4-24}$$

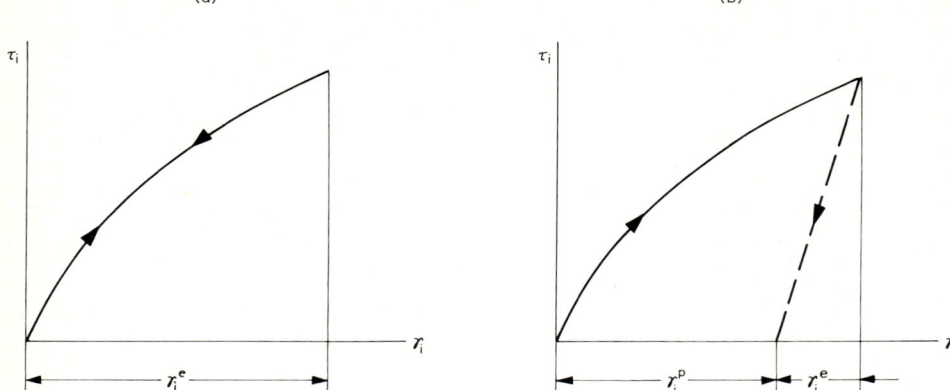

Fig. 4-5. Deformation curves of non-linearly deforming bodies: (a) non-linearly elastic body; (b) plastic body capable of strain hardening.

A similar expression may be used in the case of pure shear:

$$\tau = \phi(\gamma) \qquad (4\text{-}24')$$

Occasionally, we may use the statement:

$$\tau_i = G(\gamma_i)\,\gamma_i \qquad (4\text{-}25)$$

where the quantity $G(\gamma_i)$ is treated as a variable shear modulus dependent upon deformation.

When a non-linear deformation is fully recoverable (Fig. 4-5(a)), the body is said to be non-linearly elastic. However, in most cases the body only partly recovers from non-linear deformation, so that the deformation actually consists of an elastic deformation and a plastic deformation, i.e.:

$$\gamma_i = \gamma_i^e + \gamma_i^p, \quad \text{where} \quad \gamma_i^e = \tau_i/G \quad \text{and} \quad \gamma_i^p = f(\tau_i)$$

The deformation occurring in the above expression is sometimes referred to as plastic deformation with a strain hardening effect. This latter term owes its origin to the fact that $d\tau_i/d\gamma_i > 0$, so that it is necessary to increase the stress in order to obtain an increment in deformation. In other words, the material gains in strength compared with the perfectly plastic state ($dG = 0$).

Non-linear volumetric strain

The classical theory of plasticity postulates that volumetric strains are always elastic, i.e. obeying Hooke's law (eq. 4-3), and are fully recoverable. However, in soils, volumetric strains appear to be non-linear and quite often are irrecoverable:

$$\sigma_m = \phi^*(\varepsilon_m) \quad \text{or} \quad \sigma_m = k(\varepsilon_m)\,\varepsilon_m \qquad (4\text{-}26)$$

where $k(\varepsilon_m)$ is a variable bulk modulus. The aggregate volumetric strain consists of a recoverable deformation and a residual deformation, $\varepsilon_m = \varepsilon_m^e + \varepsilon_m^p$. The curve of a volumetric strain will be concave towards the axis of stress.

Physical and geometrical non-linearity

A stress-, strain- or time-dependent variation in the value of physical characteristics of a medium such as, for example, the modulus of elasticity, Poisson's ratio or the coefficient of viscosity, is called 'physical non-linearity'. However, non-linear relations may also exist when these characteristics are constant. This applies to situations when finite strains are set up, because — according to eqs. 3-16 and 3-17 — the relation between elongations and finite strains is non-linear.

Stress–strain curve

The stress–strain curve describing the behaviour of real bodies is of a more complex nature than the idealized curves of Fig. 4-1 or 4-5.

Deformation curves typical of solids (e.g. metals) are presented in Fig. 4-6. Referring to curve (a), depicting the effect of a tensile load, the initial linear segment OA defines the elastic behaviour of the body. the flat BC corresponds to the plastic flow when deformation progresses without an increase in stress, and segment CF represents the effect of plastic strain hardening.

Point A on the curve indicates the proportional limit, $\sigma_{p.l.}$, which is the stress beyond which the stress–strain relation departs from linearity. The flat BC is termed the 'yield flat' and the respective stress, the 'yield limit' (denoted by σ_y or σ_s).

The maximum stress the material is capable of sustaining before failure (point F) is known as the 'ultimate strength', σ_u. Beyond this point a stress attenuation is

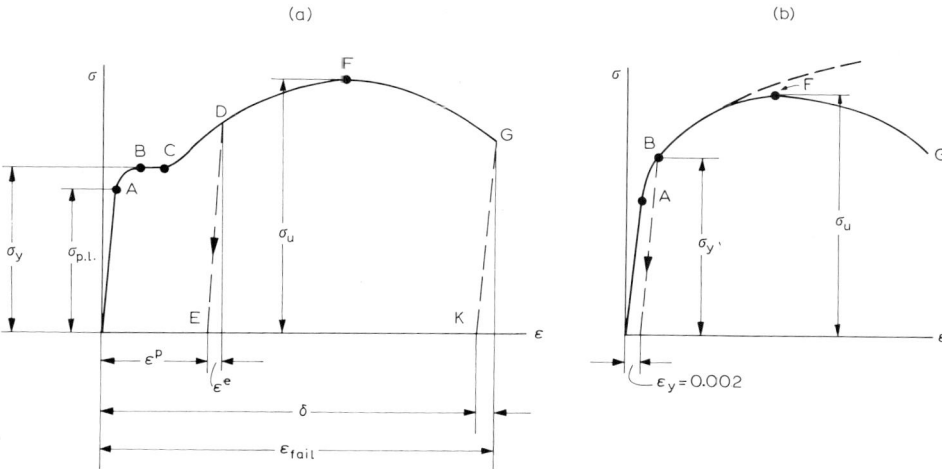

Fig. 4-6. Stress–strain curves: (a) in tension; (b) in compression.

observed, being particularly noticeable during a tensile test due to necking, which weakens the cross-sectional area of the specimen. As long as $\sigma < \sigma_{p.l.}$, the deformations are recoverable. When $\sigma > \sigma_{p.l.}$ and the load is removed, the stress decreases along the straight line DE and the body recovers from deformation but only partly as the total deformation consists of an elastic deformation ε^e and a residual, plastic, deformation ε^p.

The total deformation at failure, ε_{fail}, may also be subdivided into an elastic deformation ε^e_{fail} and a plastic one $\varepsilon^p_{fail} = \delta$. The value of δ is commonly used as a yardstick of the plasticity of a material.

Some materials, among which are many soils, display no explicit yield flat, as can be seen from Fig. 4-6(b). The yield limit to be used in such a case is the stress producing a plastic (residual) strain of a certain magnitude; e.g. for metals, $\varepsilon_y = 0.002$.

Compression curve

Most homogeneous materials respond to any loading — tensile, compression, in shear — by deformation curves of identical configurations. Therefore, in the general case, the curve of Fig. 4-6(a) appears to be correct also for compression. For other materials, including soils, the deformation curves in compression differ in configuration from those in tension. Thus, compression appears to be expressed more explicitly in Fig. 4-6 by curve (b) than by curve (a).

Certain materials whose failure is likely to be of a viscous nature (for example, plastic clays) may not fail and may only flatten when a specimen is being compressed. Their deformation curve has no extremum at point F and indicates a steady increase in stress, which is shown in Fig. 4-6(b) by the dashed line. Appropriate for use in conjunction with such materials is the concept of conventional yield limit which is understood to be a stress due to which the strain attains a certain reasonably high level. The limiting strain recommended for use in soils is 20%.

Fictitious and true tensile-compression curves. The stress applied during a tensile-compression test is given by $\sigma = P/f_0$, where f_0 is the original cross-sectional area of the specimen. The original area changes, however, during the test, either decreasing due to necking when a tensile load is applied or increasing if the material flattens under the compression. Consequently, a curve plotted on the basis of the above stress is called a 'fictitious' curve. To plot a true curve, the change in the specimen's cross-sectional area must be taken into account. To that end, the stress should be determined by the formula $\sigma_{true} = P/(f_0 \pm \Delta f_0)$, where Δf_0 is the increment in cross-sectional area. The increment is determined by measuring the lateral deformations, or may be approximated (assuming that, for large deformations, $v = 0.5$) by the formula:

$$\Delta f = f_0 \frac{\varepsilon}{1 + \varepsilon}$$

where $\varepsilon = \Delta l/l = (l/l_0) - 1$ (l_0 is the original height of the specimen).

The true strain can be determined from the expression $\varepsilon = \ln(f/f_0)$ or $\varepsilon = \int_{l_0}^{l} dl/l = \ln(l/l_0)$, where l_0 and l are the original and current lengths or heights of the specimen during the tensile-compression test. Neglecting any changes in volume, we arrive at $Al = A_0 l_0$, implying that the two expressions above for true strain are identical.

A true tensile curve will always be located above the fictitious curve, and a true compression curve will always be located below it. Consequently, the true ultimate strength will be higher than the fictitious curve in the former case and lower in the latter.

In the case of pure shear, the cross-sectional area remains constant and, consequently, the true and fictitious curves coincide.

True deformation curves are represented in a simpler schematic form in most cases. Thus, the curves depicted in Fig. 4-6 are replaced by either straight lines (linear strain hardening) or a combination of a straight line and a curve (non-linear strain hardening) or by jagged curves or, lastly, by just one curve.

Il'yushin's strain function. The equation of non-linear strain (4-24) may be written, as suggested by Il'yushin (1963), in the form:

$$\tau_i = G\gamma_i(1 - \omega) \qquad (4-27)$$

where ω is a dimensionless function indicating how far the deformation curve departs from a straight line. It varies between $0 \leqslant \omega \leqslant 1$, where $\omega = 0$ for elastic deformation and $\omega = 1$ for a perfectly plastic deformation.

Power relation between stress and strain

Most frequently, a non-linear deformation curve is approximated by a single curve (Fig. 4-7). The curve in question is described by various empirical formulae, but preference is given to Bach's power expression (Fig. 4-7(a)):

$$\tau_i = A\gamma_i^m \qquad (4-28)$$

where A is a deformation factor (in Pa) and $m \leqslant 1$ is a strain-hardening factor.

In the case of pure shear, eq. 4-28 remains unaltered (see Table 3-I):

$$\tau = A\gamma^m \qquad (4-29)$$

For uniaxial compression (or tension), the above relation takes the form:

$$\sigma_z = A_z \varepsilon_z^m \qquad (4-30)$$

where $A_z = 3^{(1-m)/2}[2(1 + v)]^m A$ or, alternatively, for $v = 0.5$, $A_z = 3^{(m+1)/2} A$, in which A is the modulus of deformation for pure shear and A_z is the modulus of deformation for uniaxial compression (or tension).

Relations 4-28 and 4-29 may also be expressed according to eq. 4-25 in terms of the variable moduli of deformation:

$$G(\gamma_i) = G(\gamma) = A\gamma_i^{m-1} \qquad E(\varepsilon_x) = A_z \varepsilon_z^{m-1} \qquad (4-31)$$

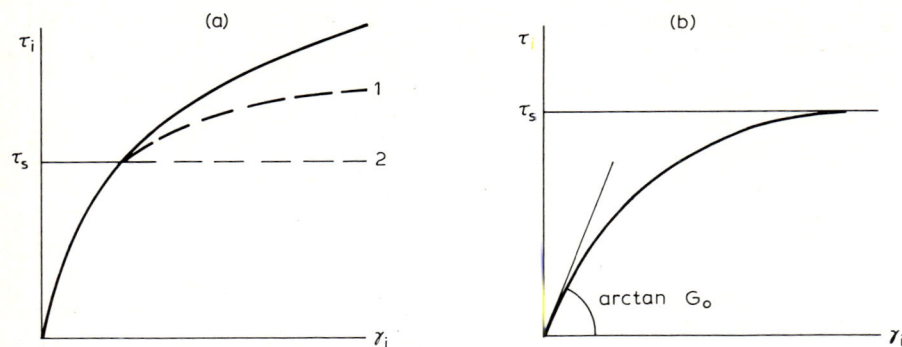

Fig. 4-7. Stress–strain relationships: (a) power; (b) linear rational.

At the same time, the moduli $G(\gamma_i)$ and $E(\varepsilon_z)$ remain to be connected by relation 4-4:

$$E(\varepsilon_z) = 2G(\gamma_i)[1 + \nu] \tag{4-32}$$

For $m = 1$, $G(\gamma_i) = $ const. and relation 4-28 is transformed into Hooke's law.

Strictly speaking, the power relation is not free of all limitations. These were pointed out by Reiner (1960, see Section 1.5) in a study of the pattern of flow of viscous media. As far as relations 4-28 to 4-31 are concerned, the limitations are detailed below.

Firstly, the dimensionless quantity m lacks any physical meaning. Secondly, for $\gamma_i \to 0$, the derivative $d\tau_i/d\gamma_i$ tends to infinity, which implies that for $\gamma_i \to 0$ the modulus of deformation is $G(\gamma_i) = d\tau_i/d\gamma_i \to \infty$, i.e. the body does not deform as long as it is not loaded, but deforms on being loaded. In the case of an unconfined strain, $\gamma_i \to \infty$, the derivative $d\tau_i/d\gamma_i$ does not become zero (i.e. no state of flow is set up) and the body deforms under conditions of unconfined strain hardening. Behaviour on these lines is possible only under conditions of viscous compression, but not shear, let alone compression.

However, simplicity and a satisfactory agreement with experimental data for a wide range of materials, including soils, make the power relation the most popular method of describing non-linear deformation. One must only define the scope of application of the power law.

Combined power relations between stress and strain

The field of application of the power law can be expanded if it is supplemented by some other relation. Thus, for example, we can combine the power law with a linear relation of the form:

$$\gamma_i = \frac{\tau_i}{G} + \left(\frac{\tau_i}{A}\right)^{1/m} \tag{4-33}$$

where it is sometimes assumed that $1/m = 2$.

In other cases, the stress–strain curve may be described by two power relations having different exponents (curve 1 in Fig. 4-7(a)). The relevant expression for $\tau_i < \tau_s$ is:

$$\tau_i = A_1 \gamma_i^{m_1}$$

and for $\tau_i > \tau_s$ is:

$$\tau_i = A_2 \left[\frac{A_1}{A_2} \gamma_s^{m_1} + (\gamma_i - \gamma_s)^{m_2} \right] = A_2 \gamma_i^{m_2} \text{ approx.} \tag{4-34}$$

where γ_s is the strain corresponding to the limiting stress τ_s. Here, and in what follows, γ_s and τ_s are considered (unless another meaning is reserved) to be the limiting strain, $\gamma_s = \gamma_{i(s)}$, and the limiting stress, $\tau_s = \tau_{i(s)}$, in the state of combined stress.

It may also be assumed that a body deforms non-linearly as long as the stress is below the yield limit and assumes a state of unconfined flow at this limit (curve 2 in Fig. 4-7(a)). Thus, for $\tau_i < \tau_s$:

$$\tau_i = A \gamma_i^m \tag{4-35}$$

and for $\tau_i = \tau_s = \text{const.}$:

$$\gamma_i \to \infty$$

Linear–fractional relation

Another form of stress–strain relation (Fig. 4-7(b)) in wide-spread use is called the linear–fractional (hyperbolic) relation and is given by:

$$\tau_i = \frac{G_0 \tau_s}{\tau_s + G_0 \gamma_i} \gamma_i \tag{4-36}$$

where τ_s and G_0 (both in Pa) are parameters whose physical meaning will be explained below. Rewriting eq. 4-36 in the form:

$$\frac{\tau_i}{\gamma_i} = \frac{G_0 \tau_s}{\tau_s + G_0 \gamma_i}$$

we obtain, for $\gamma_i \to 0$, that $(\tau_i/\gamma_i) \to G_0$. Hence, G_0 is the initial modulus of shear corresponding to an infinitesimal strain.

Assuming in relation 4-36 that $\gamma_i \to \infty$, and evaluating the indetermined form, we obtain $\tau_i \to \tau_s$. Thus, it appears that τ_s is a limiting value of the stress (yield point) which is set up when the strain develops unconfined. This latter assumption appears to be acceptable in a number of cases.

Relation 4-36, as set forth for the state of combined stress, remains the same in the case of pure shear:

$$\tau = \frac{G_0 \tau_s}{\tau_s + G_0 \gamma} \gamma \tag{4-37}$$

where τ_s is the ultimate strength (yield limit) in pure shear.

In the case of uniaxial compression (or tension), relation 4-36 takes the form:

$$\sigma_z = \frac{E_0 \sigma_s}{\sigma_s + E_0 \varepsilon_z} \varepsilon_z \tag{4-38}$$

where E_0 is the initial modulus of compression and τ_s is the yield limit (ultimate strength) under compression. These parameters are connected by the well-known relations

$$E_0/G_0 = 2(1 - v_0) \quad \text{and} \quad \sigma_s/\tau_s = \sqrt{3}$$

The first of these relations is identical with the relation between the moduli of linear deformation E and G. This is understandable if we take into account the meaning of the parameters G_0 and E_0 considered above. The second relation is identical with the relation between the yield limits of a perfectly plastic body — a fact in agreement with the meaning of the parameters σ_s and τ_s in eqs. 4-37 and 4-38.

Expressing relations 4-36 through 4-38 in terms of the variable moduli of deformation $\tau_i = G(\gamma_i)\gamma_i$ and $\sigma_z = E(\varepsilon_z)\varepsilon_z$, we obtain:

$$G(\gamma_i) = \frac{G_0 \tau_s}{\tau_s + G_0 \gamma_i}; \quad E(\varepsilon_z) = \frac{E_0 \sigma_s}{\sigma_s + E_0 \varepsilon} \tag{4-39}$$

For $\tau_s = \infty$, $G(\gamma_i) = G_0$, indicating that relation 4-36 is transformed in Hooke's law.

Thus, the linear–fractional law of eq. 4-36, which also contains deformation (G_0) and strength (τ_s) characteristics, creates the prospect of describing both the sublimiting and limiting ($\gamma_i \to \infty$) state of a material by a single curve.

Alternatively, relation 4-28 may be used in combination with relation 4-36, provided the quantity γ_i (or τ_i) in formula 4-36 is raised to the power of m.

Relations of other kinds

Apart from the power and linear–fractional stress–strain relations, the function of the $\phi(\gamma_i)$ type may be approximated in some other way. For example, use is made of the relations:

$$\phi(\gamma_i) = a \operatorname{arctanh} \gamma_i \qquad \phi(\gamma_i) = b \operatorname{arcsinh} \gamma_i \qquad \phi(\gamma_i) = \tau_s(1 - e^{-\alpha \gamma_i})$$

Whatever expression is used, a point to be noted is that the relations derived directly from the experiment are phenomenological and, consequently, which of them should be given preference may be a subjective decision. The criteria of the right choice are: (a) maximum agreement with experimental data; (b) maximum simplicity of the form of the relation; (c) maximum conformity to the conditions of the problem and convenience of use. It may appear, for example, that relation 4-28 is more suitable for use in conjunction with a narrow range of stresses, whereas a wider range requires that preference is given to relation 4-36.

In general, the power relation is employed (probably owing to its simplicity) more often than the linear–fractional relation despite the latter's advantages considered above.

Stress–volumetric strain relation

A non-linear change in volume is described by the relation:

$$\sigma_m = \phi * (\varepsilon_m) \qquad (4\text{-}40)$$

which indicates the attenuating nature acquired by the volumetric strain ε_m with a rise in all-around pressure σ_m (Fig. 4-8).

A volumetric strain obeying the power law may be determined by the formula:

$$\sigma_m = D\varepsilon_m^x \qquad (4\text{-}41)$$

where $x \geqslant 1$.

The linear–fractional law defining volumetric strain is given by

$$\sigma_m = \frac{k_0 \varepsilon_s}{\varepsilon_s - \varepsilon_m} \varepsilon_m \qquad (4\text{-}42)$$

where k_0 is the initial ($\varepsilon_m \to 0$) bulk modulus and ε_s is the limiting value of volumetric strain induced by the stress $\sigma_m \to \infty$ when the material attains maximum density.

Expressing relations 4-41 and 4-42 in terms of the variable bulk modulus, $\sigma_m = k(\varepsilon_m) \varepsilon_m$, we obtain:

$$k(\varepsilon_m) = A\varepsilon_m^{x-1}$$

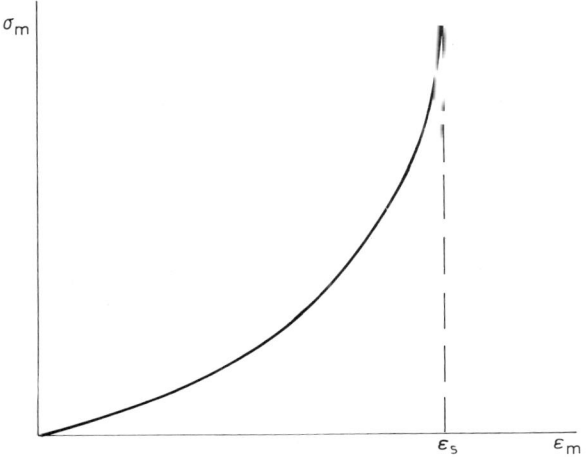

Fig. 4-8. Relationship between all-around pressure σ_m and volumetric strain ε_m.

and:

$$k(\varepsilon_m) = \frac{k_0 \varepsilon_s}{\varepsilon_s - \varepsilon_m} \qquad (4\text{-}43)$$

The exponential relation suggested by Grigoryan (1960) and Meschyan (1957) is also applicable in this case:

$$\varepsilon_m = \varepsilon_s (1 - e^{-b\sigma_m}) \qquad (4\text{-}44)$$

Coefficient of lateral expansion

It follows from eq. 4-4 that the coefficient of lateral expansion is connected with the shear and bulk moduli by the relation:

$$v = \left[\frac{k(\varepsilon_m)}{2G(\gamma_i)} - 1\right] : \left[\frac{k(\varepsilon_m)}{G(\gamma_i)} + 1\right] \qquad (4\text{-}45)$$

It can be seen that the coefficient v is variable when the deformation is non-linear. It acquires a constant value either when the material is incompressible, $[k(\varepsilon_m) = \infty]$ and $v = 0.5$, or when the ratio of the moduli $k(\varepsilon_m)$ and $G(\gamma_i)$ is constant over the entire range of stress. This can be the case when the functions $\phi(\gamma_i)$ and $\phi(\varepsilon_m)$, describing the shear and volumetric strains, respectively, are identical. For example, if we use the power relations 4-29 and 4-41, v will be constant only when the exponents in both relations are numerically equal, $m = x$.

One must take into account that the condition of identity of the functions $\phi(\gamma_i)$ and $\phi^*(\varepsilon_m)$ cannot be fulfilled for real materials because their deformation curves in shear and when subjected to all-around pressure have different shapes, i.e. if $m \leqslant 1$, then $x \geqslant 1$. This means that the above condition may be employed as an assumption simplifying computations rather than as a physical postulate.

4.4 VISCOSITY

Perfectly viscous liquids

Viscosity is the property of liquids (and gases) to resist the motion of elemental particles with respect to one another. Viscosity is also thought of as internal friction, thus stressing the point that frictional forces come into play when two layers of liquid move relative to each other. Some scientists, Reiner (1958) in particular, regard it erroneous that internal friction should be used as a term denoting viscosity.

In contrast to internal friction, an external or contact friction is the interaction of solids, at their point of contact, which interferes with the relative motion of the bodies. The external friction between two moving bodies is called 'kinematic friction' and that between two static bodies is referred to as 'the friction of rest'.

The friction of rest manifests itself in that an external force $F > F_0$ must be applied

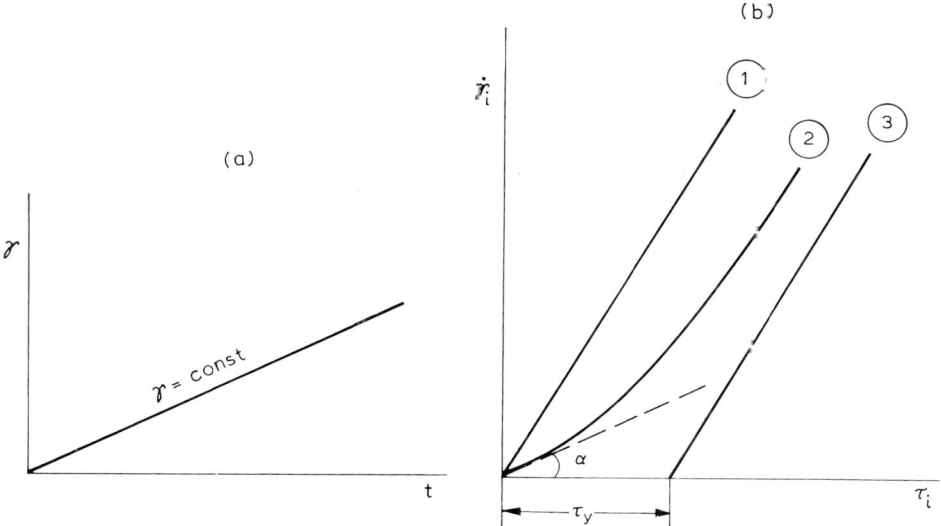

Fig. 4-9. Rheological curves: (a) development of flow in a viscous liquid with time; (b) relationship between shearing stress and flow rate. *1* = Newtonian viscous liquid; *2* = non-linearly viscous liquid; *3* = plastic-viscous Bingham liquid.

to displace two bodies relative to one another (F_0 is the ultimate force of the friction of rest). In its turn, $F = fN$, where N is the normal component force and f is the coefficient of sliding friction, $f = \tan \phi$, in which ϕ is the angle of friction. Thus, the notion "the angle of internal friction", as used in soil mechanics, should in fact be referred to as the angle of contact friction.

Newton (1687) was the first to investigate viscosity. He found that the resistance offered by a flowing liquid resulted from poor slippage of the particles, and that this resistance was proportional to the shear velocity at which the particles were displaced relative to one another.

Viscous flow is induced by any shear stress greater than zero, and progresses at a constant velocity $\dot{\gamma} = d\gamma/dt = $ const. which is directly proportional to the stress (Fig. 4-9); the deformation of viscous flow is totally irrecoverable. Any fluid fulfilling these conditions is called a Newtonian perfectly viscous liquid.

A viscous liquid acted upon by an all-around, hydrostatic pressure behaves as a perfectly elastic body, its density changing inversely with the volume.

Newton's law. The rheological equation of state of a Newtonian liquid is given by:

$$\tau_i = \eta \dot{\gamma}_i \tag{4-46}$$

$$p = k\varepsilon_m \tag{4-47}$$

where τ_i and $\dot{\gamma}_i$ are, respectively, the shear stress intensity and the intensity of shear strain rate, $p = \sigma_m$ is the all-around pressure equalizing the mean normal stress,

$\varepsilon_m = \varepsilon_v/3$ is the mean strain equalling one-third of the volumetric strain, k is the bulk modulus, and η is the coefficient of viscosity (in shear) or simply the viscosity measured in $N\,s/m^2$. In the C.G.S. system, the unit of viscosity is the poise (P). It will be recalled that 1 poise = 1 dyne s/cm^2 or 1 poise = $10^{-1} N\,s/m^2$ = $0.012\,g\,s/cm^2$ (in the engineering system).

The viscosity η is also called dynamic viscosity, which is distinct from kinematic viscosity, $v = \eta/\varrho$, where ϱ is the density of the liquid (or gas). The unit of v is m^2/s or, in the C.G.S. system, stokes; $1\,St = 10^{-4}\,m^2/s$.

Under pure shear, expression 4-46 retains the form:

$$\tau = \eta \dot{\gamma} \tag{4-48}$$

The reciprocal of viscosity, $\phi = 1/\eta$, is called fluidity. This is the property of a body to change its shape in an unconfined way, without changing volume, due to minimum effort.

When $k = \infty$ and $\eta = 0$, a viscous liquid becomes a Pascalian liquid.

A viscous liquid is capable of resisting compression (and tension) and in this case expression 4-46 takes the form:

$$\sigma_s = \lambda \dot{\varepsilon}_z \tag{4-49}$$

where λ is the coefficient of viscosity in compression (or tension), sometimes referred to as Trouton's factor.

Deriving equation of flow

Consider a two-dimensional laminar flow of a viscous liquid confined between two parallel solid plates, one of which is static and the other is being displaced at a velocity v_0 due to a tangential force F (Fig. 4-10). The molecules of the layer of liquid in contact with the moving plate adhere to it and, displacing at the same velocity v_0, entrain the molecules of the next layer. These, in turn, induce movement in the molecules of the next layer below, and so on. The molecules in the layer contiguous with the lower plate remain static, as if they were stuck there.

Every liquid layer is displaced in a plane parallel to the plates (like playing-cards shifted in a pack) at a velocity v_j of its own changing with the distance y_j from the plane. Thus, the velocity gradient will be:

$$\frac{v_j - v_{j-1}}{y_j - y_{j-1}} = \frac{dv}{dy}$$

The shear stresses τ (internal friction) set up between the liquid layers as they interact with one another bring, at balance, a force F, i.e. $\tau = F/s$, where s is the area of shear. The stress τ will, apparently, change as the velocity gradient:

$$\tau = \eta \frac{dv}{dy} \tag{4-50}$$

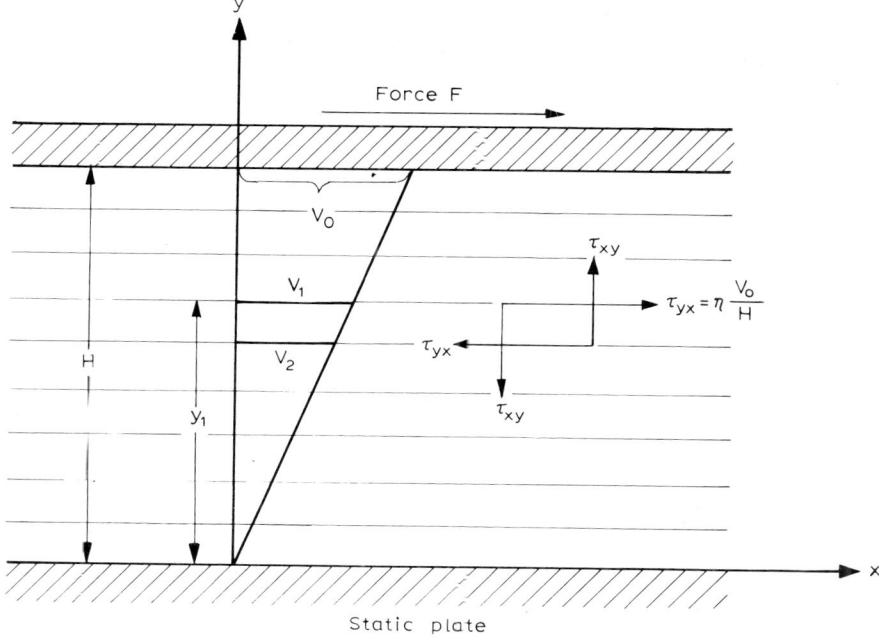

Fig. 4-10. Forces in a two-dimensional laminar flow.

The velocity gradient $dv/dy = v_0/H$ may be written as $(1/H)(du_0/dt)$ where u_0 is the displacement of the liquid in the x-direction. Assuming that the depth H is constant, and taking into account that $u_0/H = \gamma$, we arrive at the following expression for the velocity gradient:

$$\frac{1}{H}\left(\frac{du}{dt}\right) = \frac{d}{dt}\left(\frac{u_0}{H}\right) = \frac{d\gamma}{dt}$$

In other words, $dv/dy = d\gamma/dt$, i.e. eqs. 4-48 and 4-50 are identical.

From eq. 4-48 one can visualize the physical meaning of the coefficient of friction η. Thus, this is a quantity numerically equal to the frictional force produced between two layers of liquid with an area equalling unity when the velocity gradient is unity.

Navier–Stokes equations of a viscous liquid flow

Reducing conditions 4-46 to the form:

$$\sigma_x - p = 2\eta(\dot{\varepsilon}_x - \dot{\varepsilon}_m) \ldots \qquad (4\text{-}51)$$

$$\tau_{xy} = \eta \dot{\gamma}_{xy} \ldots$$

and substituting into eqs. 3-78, we obtain:

$$\frac{dv_x}{dt} = F_x - \frac{1}{\varrho}\frac{\partial p}{\partial x} + \frac{\eta}{\varrho}\left(\frac{\partial^2 v_x}{\partial x^2} + \frac{\partial^2 v_x}{\partial y^2} + \frac{\partial^2 v_x}{\partial z^2}\right) + \frac{1}{\varrho}(\eta_v + \eta)\frac{\partial \varepsilon_m}{\partial x}$$

$$\frac{dv_y}{dt} = F_y - \frac{1}{\varrho}\frac{\partial p}{\partial y} + \frac{\eta}{\varrho}\left(\frac{\partial^2 v_y}{\partial x^2} + \frac{\partial^2 v_y}{\partial y^2} + \frac{\partial^2 v_y}{\partial z^2}\right) + \frac{1}{\varrho}(\eta_v + \eta)\frac{\partial \varepsilon_m}{\partial y} \quad (4\text{-}52)$$

$$\frac{dv_z}{dt} = F_z - \frac{1}{\varrho}\frac{\partial p}{\partial z} + \frac{\eta}{\varrho}\left(\frac{\partial^2 v_z}{\partial x^2} + \frac{\partial^2 v_z}{\partial y^2} + \frac{\partial^2 v_z}{\partial z^2}\right) + \frac{1}{\varrho}(\eta_v + \eta)\frac{\partial \varepsilon_m}{\partial z}$$

where $\varepsilon_m = (1/3)(\varepsilon_x + \varepsilon_y + \varepsilon_z)$, η and η_v are the shear and volumetric viscosities, ϱ is the density, $v_{x,y,z}$ are the components of displacement rates, p is the pressure, and $F_{x,y,z}$ are the projections of the external force vector referred to a unit mass of the liquid. The notions of shear and volumetric viscosities will be discussed below in formulae 4-54 and 4-67.

The left-hand sides of eqs. 4-52 may be written in the form:

$$\frac{\partial v_x}{\partial t} + v_x\frac{\partial v_x}{\partial x} + v_y\frac{\partial v_x}{\partial y} + v_z\frac{\partial v_x}{\partial z} + \ldots$$

Eqs. 4-52 or, as they are called, the Navier–Stokes equations are the basic equations of the mechanics of a viscous liquid. When a liquid is incompressible, $\varepsilon_m = 0$ and the third terms of the equations vanish. When $\eta_v = \eta = 0$, eqs. 4-52 become Euler's equations for an ideal liquid. Eqs. 4-52 are supplemented by an equation of state, correlating pressure with density and temperature, and by an equation of continuity.

Viscosity of soils

The viscosity of various media varies over a wide range: $1.8 \cdot 10^{-4}$ P for air, 10^{-2} P for water, 0.5–10 P for various oils, and $5 \cdot 10^{22}$ P for the earth's crust.

As far as soils are concerned, the available experimental data on viscosity have a spread varying between 10^6 and 10^{17} P. Thus, the plastic viscosity of a soft clay-type silt is, according to Sorokina (1965), 10^6 P and, according to Roza (1954), between $0.6 \cdot 10^{12}$ and $10 \cdot 10^{12}$ P.

The viscosity of stiff Maikop clay has been found to be between 10^{12} and 10^{14} P (according to Karaulova, 1968) and that of clay with a moisture content of 25%, $5 \cdot 10^{14}$ P (according to Denisov, 1956). Mogilevskaya (1968) has found that the viscosity of loess silts is between $2 \cdot 10^{13}$ and $4 \cdot 10^{14}$ P.

Ermolaeva et al. (1968) at the Hydraulic Engineering Research Institute have conducted tests in a shear box with hinged end pieces which tilt during shear. These tests have yielded the following data on the plastic viscosity of clays: remoulded Cambrian clay ($W = 24$–27%), between $1.5 \cdot 10^9$ and $8 \cdot 10^{12}$ P; remoulded loessial clay loam ($W = 13$–21%), between $3.6 \cdot 10^{10}$ and $2.1 \cdot 10^{14}$ P; remoulded Khvalynsk clay ($W = 38\%$), between $1.5 \cdot 10^7$ and $1.3 \cdot 10^{10}$ P; the same clay with undisturbed structure ($W = 30\text{–}40\%$), between $6.8 \cdot 10^9$ and $2.8 \cdot 10^{12}$ P.

According to the experimental data acquired by Gor'kova (1965), the viscosity of sedimentary rocks (at low stresses inducing the flow of rock with intact structure) is as follows: silty sands having a disturbed structure ($W = 27\text{–}42\%$), 10^7 P; Black Sea silts and lacustrine clays ($W = 48\text{–}106\%$), between 10^8 and 10^9 P; Caspian Sea silts and postglacial marine clays ($W = 49\text{–}70\%$), between 10^9 and 10^{10} P; Maikop clays ($W = 12\text{–}24\%$), 10^{11} P; Jurassic clays of Moscow ($W = 32\text{–}45\%$), between $2.5 \cdot 10^{11}$ and $5.0 \cdot 10^{11}$ P.

Maslov (1968) recommends the use of the following averaged viscosities of clay soils, as obtained from an analysis of his own tests and the test data acquired by other authors: soft clay, $10^{10}\text{–}10^{11}$ P; firm clay, $10^{12}\text{–}10^{13}$ P; stiff clay $10^{14}\text{–}10^{15}$ P; hard clay, $10^{15}\text{–}10^{17}$ P.

The viscosity of rapid landslides is estimated to be of the order of 10^{11} P; in slowly flowing slopes it is between 10^{13} and 10^{14} P.

The viscosity of some rocks (at the stage of steady flow), as determined in the laboratory by Maximov et al. (1969), is 10^{17} to 10^{18} P for clay shale and $10^{18}\text{–}10^{19}$ P for sandstone and sandy shale.

The viscosity of the earth's crust taking part in rheological processes, for a duration measured in geological time, is estimated to be between 10^{20} and 10^{25} P.

Viscosity of ice. This may vary with temperature, structure and load over the range $10^{10}\text{–}10^{15}$ P (Voitkovsky, 1960). However, field observations of glacier movement by some investigators have shown that the spread in the viscosity data for polycrystalline ice at temperatures close to zero narrows to between $1.4 \cdot 10^{13}$ and $2.3 \cdot 10^{14}$ P (averaging $6.5 \cdot 10^{13}$ P). According to Haefeli (1953), the viscosity of ice is $1.7 \cdot 10^{14}\text{–}2.5 \cdot 10^{14}$ P.

On methods of determining viscosity of soils

The significant spread in soil viscosity data that have been obtained from tests does not only result from the diversity of soil properties. The diversity of methods employed to determine the coefficient of viscosity, together with different definitions of this quantity, are also reasons of no less importance. The notion that viscosity is a constant, $\eta = \tau/\dot{\gamma}$, is correct only for a perfectly viscous Newtonian medium. Soils, however, obey other laws. The relation between the stress and rate of flow is non-linear in soils, and the flow itself is induced by the difference $\tau - \tau_y$ (where τ_y is the yield limit) rather than by the total stress τ. Therefore, the coefficient of viscosity is a variable which is dependent upon the applied load. Investigators who neglect to take this into account may obtain casual results.

Likewise, one must bear in mind that the notion of Newtonian viscosity is applicable to the process of viscous flow at a constant rate. Soils, however, like most real bodies, deform at a variable rate. Only at a certain stage of the process is the rate of deformation constant. Taking into account what has been said, the notion of Newtonian viscosity is applicable to this stage only. In all other cases, viscosity is to be treated as a time-dependent variable which may significantly change, say, by a factor of thousands. Thus, if at the beginning of a process the viscosity is between 10^9 and 10^{10} P, its final value can be 10^{13} to 10^{14} P. This point will be considered in more detail.

Finally, one must take into consideration the method of viscosity measurements. Some researchers use a standard direct shear apparatus. Although widely employed in determining strength parameters, it is of little value in studying deformation as it can introduce gross errors into the viscosity measurements. The inaccuracy results, firstly, from the non-uniform state of stress produced (stress concentrations are set up at the edges of the specimen) and, secondly, because the area of shear in the specimen is small and indeterminate, making accurate determination of the shear strain γ a problem. Variations in the active area of specimen, confined deformation and a fluctuating clearance between the upper and lower parts of the container are other disadvantages of the direct shear apparatus. The same apparatus, but with a shear box with hinged end pieces which tilt during shear, produces better results, although this also produces a confined deformation and an indeterminate state of stress.

The most suitable viscosity-measuring instrument is the torsion test apparatus. This creates conditions of pure shear during the test and allows an almost unconfined deformation to be set up, with no change in the active area of the specimen. For hard and stiff soils, one may use apparatus applying torque to an end face of a cylindrical specimen, whereas in the case of plastic and liquid soils a viscometer of some type (e.g. with two coaxial cylinders) can be used.

Viscosity measurements in the state of combined stress may be carried out on the same torsion test apparatus while simultaneously applying a vertical load or an all-around pressure (or both). A triaxial compression test apparatus is a suitable alternative.

In conclusion, let us say a few words about viscosity measurements by means of the falling-ball viscometer. In this case, viscosity is calculated from the rate of penetration of the soil by a ball due to gravity and an applied load, using Stokes' formula:

$$\eta_{St} = \frac{\varrho_b - \varrho_s}{18v} d^2$$

where ϱ_b is the density of the ball with due regard for the load applied, ϱ_s is the density of the soil, d is the ball diameter, and v is the constant rate of ball penetration, in cm/s.

The falling-ball technique is widely used in studying the viscous behaviour of liquids, and must be applied to soils with care as the Stokes formula has been deduced for a Newtonian liquid having a constant coefficient of viscosity. In the case of a non-linearly viscous medium, such as soil, the values of η_{St} obtained will be influenced by the load applied to the ball. This technique is also inapplicable to soils that have a yield point (Bingham medium) because the Stokes formula is invalid in this case.

Bulk viscosity

As already mentioned, classical continuum mechanics postulates that volumetric deformation is present in all idealized bodies, solid and liquid alike, and is instantaneous and fully recoverable. Consequently, the patterns of volumetric strain in an elastic (eq. 4-3) and a viscous (eq. 4-47) body are identical. However, real bodies depart from this pattern, as their volumetric strains may be partly irrecoverable and develop in time. In such a case the body is said to have bulk viscosity.

For a body displaying both elastic and viscous volumetric strain the equation of volumetric deformation (eq. 4-47) takes the form:

$$p = k\varepsilon_m + \eta_v \dot{\varepsilon}_m \quad \text{or} \quad \varepsilon_m = \varepsilon_m^e + \varepsilon_m^v \tag{4-53}$$

where k is the coefficient of bulk elasticity and η_v is the coefficient of bulk viscosity.

For a body experiencing viscous volumetric strain only, the equation of volumetric deformation may be written as:

$$p = \eta_v \dot{\varepsilon}_m \tag{4-54}$$

If a body has volumetric deformations that are described by formula 4-54, then the coefficient of viscosity in shear η and in compression (tension) λ, and the coefficient of bulk viscosity η_v, are connected by the relation:

$$\lambda = \frac{3\eta\eta_v}{\eta + \eta_v} \tag{4-55}$$

Coefficient of lateral strain of viscous flow

Since this coefficient refers to irrecoverable deformations which increase continuously, its physical meaning is somewhat different from Poisson's ratio defining elastic properties of a body. However, from a formal point of view, the coefficient of lateral strain and Poisson's ratio can be treated as identical quantities, defining the ratio of lateral and longitudinal strains. Accordingly, the coefficient of lateral strain of a viscous flow, v_v, may be presented in the form:

$$v_v = \dot{\varepsilon}_x / \dot{\varepsilon}_y \tag{4-56}$$

It is obvious, therefore, that the coefficient of viscous flow v_v is constant only when the rates of lateral and longitudinal strains are either constant or changing with time according to the same law.

If relation 4-55 connecting the parameters of viscous flow is expressed in terms of the coefficient v_v, we obtain an expression similar to eq. 4-4:

$$\eta_v = \frac{\lambda}{1 - 2v_v} = 2\eta \frac{1 + v_v}{1 - 2v_v} \tag{4-57}$$

In the absence of bulk viscosity ($\eta_v = \infty$), the coefficient v_v becomes equal to 0.5 and expression 4-57 takes the form $\lambda = 3\eta$.

4.5 NON-LINEAR VISCOSITY AND BINGHAM FLOW

Non-linear viscosity

Many a real viscous body departs from Newton's law (eq. 4-46) describing the flow of a perfectly viscous liquid. This departure, called anomalous viscosity, manifests itself as variations in the coefficient of viscosity η depending on the magnitude and duration of the load applied.

The dependence of the parameter η upon loading is equivalent to a non-linear relation between the rate of flow $\dot{\gamma}_i$ and the stress τ_i (see curve 2 in Fig. 4-9(b)). This flow is called 'non-linear viscosity' and the body in such a state of flow is referred to as a non-linearly viscous body or a non-Newtonian liquid.

The rheological equation of state of a non-linearly viscous body may be presented in the form:

$$\tau_i = \phi(\dot{\gamma}_i) \quad \text{or} \quad \dot{\gamma}_i = f(\tau_i) \tag{4-58}$$

The above expression also holds for pure shear:

$$\tau = \phi(\dot{\gamma}) \quad \text{or} \quad \dot{\gamma} = f(\tau) \tag{4-58'}$$

Likewise, bulk viscosity may be expressed as a non-linear relation:

$$\sigma_m = \phi^*(\dot{\varepsilon}_m) \quad \text{or} \quad \dot{\varepsilon}_m = f^*(\sigma_m) \tag{4-59}$$

Non-linear dependence (eq. 4-58) may be written as:

$$\tau_i = \eta(\tau_i)\dot{\gamma}_i \tag{4-60}$$

where $\eta(\tau_i)$ is the variable coefficient of viscosity decided by the level of stress. The reciprocal of this quantity (the coefficient of fluidity), $\phi = 1/\eta$, is defined as the tangent an angle α (curve 2 in Fig. 4-6(b)) makes with the axis of the abscisses (with due regard for the scale of the graph).

With an increase in stress, fluidity ϕ also increases and viscosity decreases, varying over the range $\eta_0 \geq \eta(\tau) \geq \eta_m$, where η_0 is the initial or maximum viscosity for $\tau_i \to 0$ and η_m is the final or minimum viscosity for $\tau_i \to \infty$.

Flow rate–stress relation. Consider the possible forms of relation 4-60. The most frequently used is the power relation derived by De Waele and, independently, by Ostwald (1925) for liquid media and by Norton (1929) and Bailey (1933) for solid media:

$$\dot{\gamma}_i = a\tau_i^{1/m} \tag{4-61}$$

or, more accurately:

$$\dot{\gamma}_i = \dot{\gamma}^* \left(\frac{\tau_i}{\tau^*}\right)^{1/m} \tag{4-61'}$$

where $m \leq 1$ and a, τ^*, $\dot{\gamma}^*$ are parameters, with the quantity τ^* being introduced to observe dimensions (it may be taken as unity).

In a more complex form, the power relation 4-61 may be written as:

$$\dot{\gamma}_i = a\tau_i + b\tau_i^{1/m} \tag{4-62}$$

where, on occasions, $1/m = 2$; alternatively, we may write:

$$\dot{\gamma}_i = a\tau_i + b\tau_i^2 + c\tau_i^3 \tag{4-63}$$

The following expressions are widely used as representations of the stress–flow rate relation:

$$\dot{\gamma}_i = \dot{\gamma}^* \sinh \frac{\tau_i}{\tau^*} \tag{4-64}$$

$$\dot{\gamma}_i = \dot{\gamma}^* e^{\tau_i/\tau^*} \tag{4-65}$$

The physical meaning of these expressions will be examined later. Note that eq. 4-65 is a particular case of eq. 4-64 for $\sinh(\tau_i/\tau^*) = 1/2 \exp(\tau_i/\tau^*)$ approx. when $\tau_i/\tau^* \gg 1$.

Relations 4-61, 4-64 and 4-65 may be represented in the form of relation 4-60 if the coefficient of viscosity is assumed to be equal, respectively, to:

$$\eta(\tau) = \left[\frac{\dot{\gamma}^*}{\tau^*} \left(\frac{\tau_i}{\tau^*} \right)^{1-m/m} \right]^{-1} \tag{4-66}$$

$$\eta(\tau) = \left[\frac{\dot{\gamma}^*}{\tau_i} \sinh \frac{\tau_i}{\tau^*} \right]^{-1} \tag{4-66'}$$

$$\eta(\tau) = \left[\frac{\dot{\gamma}^*}{\tau_i} e^{\tau_i/\tau^*} \right]^{-1} \tag{4-66''}$$

Structural viscosity

Ostwald (1926), a pioneer of research into anomalously viscous media, came to the conclusion that those media which possess structure depart from the pattern of Newtonian flow. The explanation is that the structure changes in the course of flow and, as a result, so does viscosity. Accordingly, this variable viscosity is called 'structural viscosity'.

Referring to Fig. 4-9(b), the rheological curve (i.e. the curve of τ–γ relation) for a non-linearly viscous body is of a smooth configuration having no inflections. However, this is true only over a certain range of stresses. In the general case, the rheological curve of a non-Newtonian liquid does have inflection points, as can be seen in Fig. 4-11.

Mikhailov and Rebinder (1955) have classified structuralized media into quasi-liquid and quasi-solid varieties. In a quasi-liquid medium, the flow is induced by any stress, but in this case, unlike a Newtonian liquid, there are two critical points corresponding to the critical stresses τ_r and τ_f (Fig. 4-11(a)).

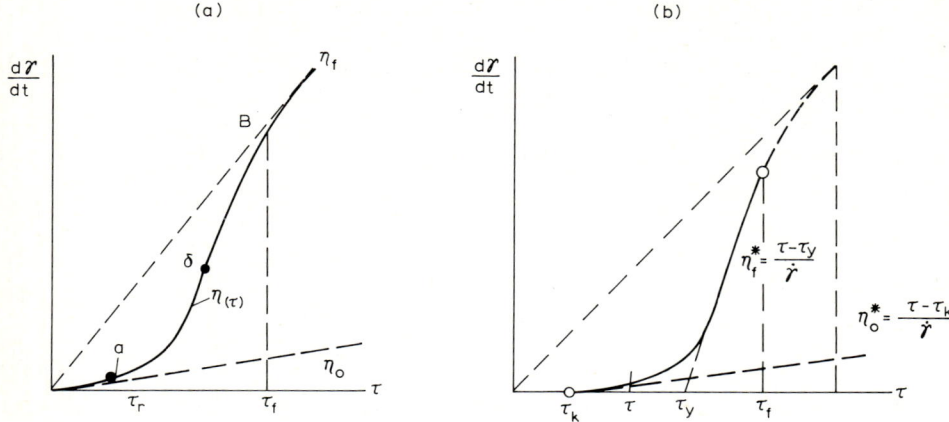

Fig. 4-11. (a) Rheological curve of a liquid. (b) Rheological curve of a solid.

When the shearing stress is low, $\tau < \tau_r$, the rheological curve approaches a straight line — i.e. at an early stage, ending at point a, the flow develops under the conditions of constant maximum viscosity η_0. The straight line is succeeded by a curvilinear segment aB inflecting at δ, which represents a changeable or so-called 'effective' viscosity, $\eta = \eta(\tau)$.

With an increase in shearing stress, $\tau > \tau_f$, the rheological curve again approaches a straight line (beyond point B) whereas the flow progresses with a constant minimum viscosity η_f. While the viscosity is η_0, the structure is undisturbed by the flow which, accordingly, obeys Newton's law. When the viscosity changes to η_f, the structure suffers maximum damage; at this stage, the effect of structural changes ceases to exist and the flow again assumes the Newtonian pattern, but at minimum viscosity.

In quasi-solid media (Fig. 4-11(b)), the flow is induced only on exceeding a certain limit τ_k. Below this point, the deformations are purely elastic or are developing at a slow rate which may be neglected. In other respects the rheological curve of a quasi-solid medium is analogous to that of a quasi-liquid medium, except that it is moved along the x-axis by the value of τ_k. A segment of this curve between the limits $\tau_k < \tau < \tau_r$ may be treated as linear at a constant viscosity (the Schwedoff viscosity) $\eta_0^* = (\tau - \tau_k)/\dot{\gamma}$.

When $\tau > \tau_r$ the structure begins to break down and the viscosity becomes variable. The process ends in a flow that has an utterly disturbed structure and an extremely low viscosity or in a complete loss of continuity.

Another limit τ_y may be introduced on the assumption that within the interval $\tau_y < \tau < \tau_f$ the curve is approximated by a straight line and the flow proceeds at a constant viscosity $\eta_f^* = (\tau - \tau_y)/\dot{\gamma}$. A similar approximation may, apparently, be adopted for the $d\gamma/dt$ versus τ curve (Fig. 4-11(a)).

Viscoplasticity. The rheological curve of a quasi-solid body (Fig. 4-11(b)) may be represented schematically in the form of a straight line (see graph 3 in Fig. 4-9(b)).

This implies that the body begins to flow only when the stress exceeds the yield τ_y (also referred to as the limiting shearing stress).

The departure from the Newtonian flow pattern has been detected by Bingham (1919, 1922) in studying the viscous flow of oil colours and clay suspensions. An oil colour is known to be a dispersion of particles of a pigment in oil. The colours should meet two requirements: ease of application to the work (requires the highest-possible fluidity) and freedom from runs (adequate viscosity facilitates this).

If a suspension is treated as a Newtonian liquid, the two requirements contradict one another. Bingham and Green, laying down the fundamentals of rheology in their work in 1919, indicated that the colours, fulfilling the two requirements simultaneously, were totally new bodies, viz. viscoplastic bodies.

These authors have introduced the notion of a viscoplastic model, which is now known as a Bingham body.

The equation of flow of a Bingham body is written in the form:

$$\tau_i = \tau_y + \eta_p \dot{\gamma}_i \tag{4-67}$$

where η_p is the coefficient of plastic viscosity (also called plastic viscosity).

Note that Schwedoff's equation deduced in 1890 (which will be considered in the next section) may be reduced to the same form. Consequently, relation 4-67 is often referred to as the Schwedoff–Bingham law.

The equations of state of a Bingham body may be written in general form as:

$$\begin{aligned} \tau_i &= G\gamma_i \quad \text{for} \quad \tau_i < \tau_y \\ \tau_i &= \tau_y + \eta_p \dot{\gamma}_i \quad \text{for} \quad \tau_i \geqslant \tau_y \\ p &= k\dot{\varepsilon}_m \end{aligned} \tag{4-68}$$

The above statement implies that a body subjected to a stress less than the limiting value τ_y deforms elastically in accordance with Hooke's law, but begins to flow at a constant rate proportional to the stress difference $(\tau_i - \tau_y)$ on reaching τ_y. The volumetric strains are, however, elastic, while the condition of incompressibility, $k = \infty$, is a special case.

It can be seen that the first term τ_y in eqs. 4-68 is similar to Saint-Venant's condition of plasticity for a solid, given by eq. 4-5; the second term defines the Newtonian viscous flow (eq. 4-46). Eq. 4-68 thus describes the behaviour of a combinative body possessing the properties of ideal plasticity and ideal viscosity. Note that the flow of such a body is called plastic flow, stressing the analogy with the plastic deformation that occurs when the stress reaches the limiting level.

The rheological curve of a Bingham body is depicted in Fig. 4-9(b) by curve 3. The plot explicitly illustrates the difference between the coefficients η and η_p. The former is the cotangent of the slope of the straight line 1 extending from the origin, and the latter is the cotangent of the slope of the straight line 3 making $\tau_i = \tau_y$ intercept on the axis of abscissas.

If the coefficient of viscosity of Bingham flow is formally defined as $\eta = \tau_i/\dot{\gamma}_i$, then this quantity is variable since the slope of a straight line interconnecting the origin and a point given on the Bingham straight line is also variable. This is referred to as 'apparent' viscosity, η'.

For $\tau_y = 0$, a Bingham body is transformed into a Newtonian liquid; for $\eta_{pl} = 0$, it becomes a Saint-Venant body; and for $G \to \infty$, the outcome is a rigid–plastic Prandtl body.

4.6 AFTER-EFFECT AND RELAXATION

Viscoelasticity

Elasticity, viscosity and plasticity are cardinal rheological properties of a continuum. Idealized bodies possessing these properties have been considered above and are represented graphically in Fig. 4-12. The curves in (a), (b) and (c) refer to solids; the curves in (d), (e) and (f) apply to viscous bodies.

The behaviour of solids in response to a load is defined by a stress–strain relation. In viscous bodies, the deformation develops with time and, as a result, the stress–strain relation is not single-valued. The behaviour of such bodies sustaining a load is described by a relation between the stress and the strain rate.

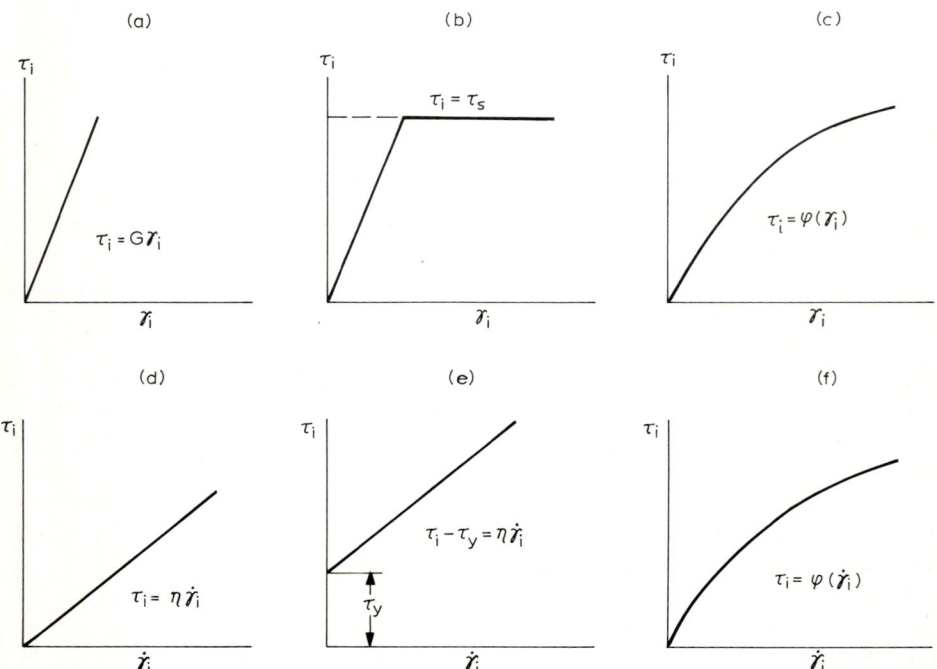

Fig. 4-12. Deformation curves of various bodies: (a) elastic; (b) elastoplastic; (c) non-linearly elastic; (d) viscous; (e) viscoplastic; (f) non-linearly viscous.

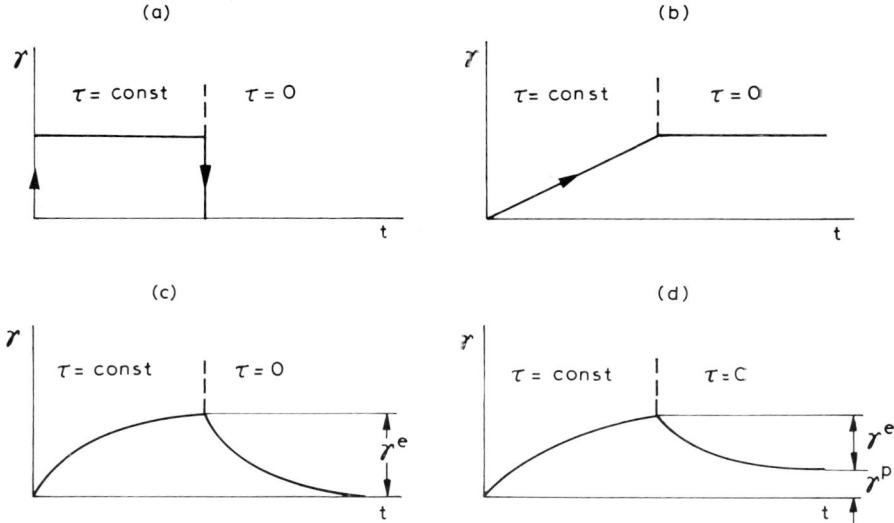

Fig. 4-13. Development of deformation with time in response to loading (τ = const.) and unloading ($\tau = 0$) in: (a) an ideally elastic body; (b) an ideally viscous body; (c) a viscoelastic body; (d) a viscoelastic–plastic body.

Note the analogy existing between the deformation curves of solid and viscous bodies. This creates the prospect of solving the problems of viscous flow in terms of the theory of elasticity and plasticity: linear in the former case and non-linear in the latter case. The solution may be applied (under certain restrictions) in a formal way by substituting the strain rate $\dot{\gamma}_i$ for the strain γ_i.

Elasticity is a property characteristic of solids, and viscosity is a salient feature of liquids. However, many real bodies display both these properties, and are consequently known as viscoelastic bodies. Elasticity manifests itself by recovery from deformation after removal of a load; and the evidence of viscosity is the development of deformation with time in response to loading.

A typical manifestation of viscoelastic properties is the elastic after-effect. However, in many bodies the recoverable (elastic) deformations, and also the residual (plastic) deformations, are time-dependent; this phenomenon is termed the plastic after-effect and the medium where the process is developing is referred to as a viscoelastic–plastic body.

It will be recalled that the bodies subjected to an elastic or viscous deformation before the stress reaches a limiting value, and develops a plastic flow on attaining this limit, are classified as elastoplastic and viscoplastic bodies.

Referring to Fig. 4-13, we can see the way deformation is developing with time in various bodies. The deformation of an elastic body (Fig. 4-13(a)) is induced and recovered instantaneously without changing in time. In the case of a viscous body (Fig. 4-13(b)), the deformation steadily increases with time at a constant rate and fails to recover after unloading.

The deformation of a viscoelastic body (Fig. 4-13(c)) develops with time but is of an attenuating nature, increasing at an ever-decreasing rate. After unloading, this deformation totally recovers, but with time. The deformation of a viscoelastic–plastic body (Fig. 4-13(d)) is time-dependent, is of a non-attenuating nature, and only partly recovers.

Kelvin's equation of after-effect

The rheological equation of state of a viscoelastic medium (Fig. 4-13(c)) was considered by Kelvin in his well-known treatise *Elasticity* (1890) and, almost simultaneously, by Voigt. This equation is of the form:

$$\tau_i = G\gamma_i + \eta\dot{\gamma}_i \tag{4-69}$$

where G is the shear modulus of elasticity and η is the coefficient of viscosity.

Solving eq. 4-69 for γ_i (for $\tau_i = $ const.), we obtain:

$$\gamma_i = \frac{\tau_i}{G}(1 - e^{-t/T_a}) \tag{4-70}$$

where $T_a = \eta/G$ is the period of after-effect.

Expression 4-70 describes the process of a delayed elastic deformation (after-effect). At an initial instant, the deformation is zero; at $t \to \infty$ it reaches its limiting magnitude $\gamma_i = \tau_i/G = \gamma_i(\infty)$. When the load is removed, the deformation recovers in accordance with the pattern defined by:

$$\gamma_i = \gamma_{i(0)} e^{-t/T_a} \tag{4-71}$$

and ceases to exist at $t \to \infty$.

According to Kelvin, the phenomenon of delayed deformation in a solid is attributed to the forces of internal friction (viscosity). He wrote that molecular friction existing in an elastic solid may be referred to quite reasonably as the viscosity of the solid, for, reflecting the internal resistance to a change in shape and being dependent upon the rate of the change, it must come under the same category as the molecular friction in liquids known as viscosity.

The body to which Kelvin had referred was an elastic porous medium whose voids were filled with a viscous liquid. Some clay is exactly such a medium; it was quite logical that K. Terzaghi and N.M. Gersevanov (Gersevanov and Polshin, 1948) employed a model of the Kelvin body to describe the process of soil consolidation.

Relaxation

Further evidence of the viscoelastic properties of a body is shown by the relaxation (lessening) of stresses.

Suppose a beam in a viscoelastic material is put to a bending test (Fig. 4-14). A constant force F applied to the beam brings about a deflection f which increases with time, obeying a certain law $f(t)$, i.e. the law of after-effect. If at a given instant

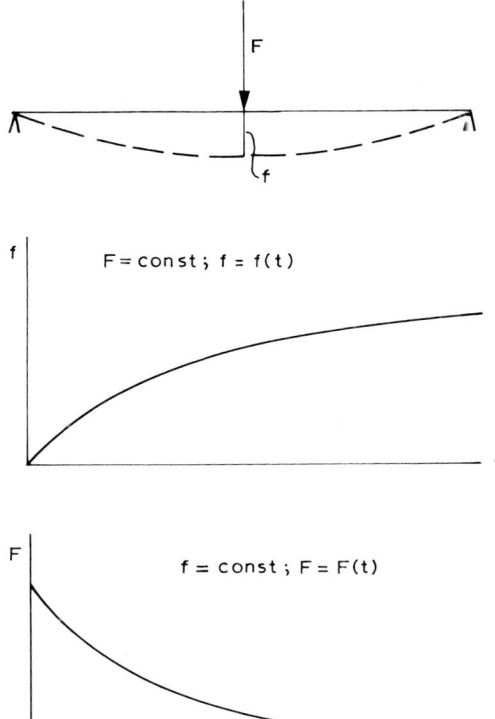

Fig. 4-14. After-effect and relaxation curves.

(assuming that this is zero, $t = 0$) the deflection must be checked and maintained at a constant level (f_0 = const.), then the force F should apparently be reduced in accordance with the law $F(t)$. The lessening of the stress in time in order to maintain a constant deformation is called stress relaxation.

Relaxation results from a redistribution of elastic and plastic deformations. In fact, the aggregate deformation of a viscoelastic body is the sum of elastic deformation (e) and plastic deformation (p):

$$\gamma_i = \gamma_i^e + \gamma_i^p$$

Since the deformation γ_i^p increases with time, the condition $\gamma_{i(0)}$ = const. may be fulfilled if the value of γ_i^e decreases. Noting that $\gamma_i^e = \tau_i/G$, this condition then takes the form:

$$\tau_{i(0)} = \tau_i/G + \gamma_i^p = \text{const.} \tag{4-72}$$

Hence, it follows that the constancy of a deformation, $\gamma_{i(0)}$ = const., is ensured by reducing the stress in time, $\tau_i = \tau_i(t)$.

A relaxation test consists of setting up an initial deformation $\gamma_i(0)$ in a specimen corresponding to an initial stress $\tau_{i(0)}$ and of measuring the stress as this changes with time while the deformation is maintained constant.

Maxwell's equation of relaxation

Interpreting the viscosity of gases as the relaxation of elastic stresses, Maxwell pointed out in his dynamics studies that an elastic force τ inducing a deformation γ in a relaxing medium tends to fade, rather than to remain constant, at a rate $d\tau/dt$ which depends upon the magnitude of the force and the material.

At a later date, the phenomena of after-effect and relaxation were examined by Thomson (1888). Boltzmann (1876), too, developed a mathematical theory of the stress and strain after-effect, making due allowance for the history of loading. This theory eventually became the keystone of Volterra's (1913, 1931) theory of hereditary media.

Relaxation, as considered from the standpoint of statistical physics, is a process of establishing a statistical equilibrium in a physical system at which the microscopic quantities defining the state of the system (i.e. stresses) asymptotically approach their equilibrating values.

The equation of a relaxing viscoelastic (Maxwell) body takes the form:

$$\dot{\gamma}_i = \frac{1}{\eta}\tau_i + \frac{1}{G}\dot{\tau}_i \qquad (4\text{-}73)$$

Assuming that the deformation is constant, $\gamma_i = \gamma_{i(0)} = \text{const.}$, we arrive at Maxwell's law of relaxation:

$$\tau_i = \tau_{i(0)}\, e^{-t/T_r} \qquad (4\text{-}74)$$

where $\tau_{i(0)} = G\gamma_{i(0)}$ is the initial stress and $T_r = \eta/G$ is the period of relaxation.

As can be seen, at $t \to \infty$ the stress set up in a Maxwell body becomes zero.

Experimenting some time later with gelatin, F.N. Schwedoff in 1890 discovered that some liquid media are anomalous in point of viscosity and elasticity of shape. Also, the stresses relax not to zero but to some finite value $\tau_{i(k)}$. Accordingly, he introduced a correction into Maxwell's equation (eq. 4-74), presenting it in Schwedoff's form as:

$$\tau_i = \tau_{i(k)} + [\tau_{i(0)} - \tau_{i(k)}]\, e^{-t/T_r} \qquad (4\text{-}75)$$

Hence, the equation of state of a Schwedoff body is:

$$\dot{\gamma}_i = \frac{1}{\eta}[\tau_i - \tau_{i(k)}] + \frac{1}{G}\dot{\tau}_i \qquad (4\text{-}76)$$

For $\gamma_i = \text{const.}$ the above equation is transformed into the equation of relaxation (eq. 4-75) and for $\tau_i = \text{const.}$ it becomes the equation of viscoplastic Bingham flow (eq. 4-67), provided it is assumed that $\tau_{i(k)} = \tau_y$.

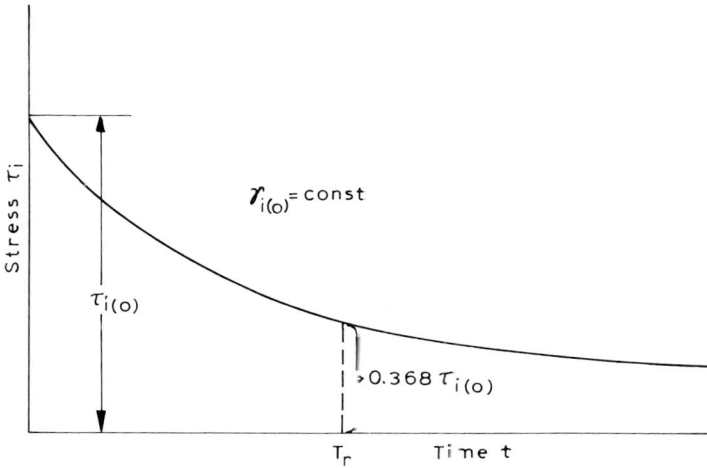

Fig. 4-15. Stress relaxation curve.

Formally, the parameters T_r (period of relaxation) and T_p (period of after-effect) both equal η/G. However, they differ appreciably in point of physical meaning.

A viscoplastic Kelvin body satisfying eq. 4-69 in the state of after-effect may be treated as an elastic solid possessing the property of viscosity but devoid of the property of relaxation. For γ_i = const. the law expressed by eq. 4-69 is transformed into Hooke's law of elasticity.

A viscoelastic Maxwell body satisfying eq. 4-73 may be treated as a viscous liquid possessing elastic properties. However, it is incapable of displaying the after-effect and, when γ_i = const., flows at a constant rate. In such a case relation 4-73 becomes Newton's law, which is obeyed by perfectly viscous liquids.

The period of relaxation T_r is one of the most essential rheological parameters. To determine its formal value, consider an instant $t = T_r$. In this case, eq. 4-74 takes the form $\tau_i = \tau_{i(0)}/e$. Hence, the period of relaxation t_r may be defined as a period during which the stress τ_i decreases by a factor of $t = 2.718$ (Fig. 4-15) compared with the initial stress $\tau_{i(0)}$.

The period of relaxation, as far as its physical meaning is concerned, corresponds to the so-called period of settled life of molecules in a state of transient equilibrium. In other words, the period of relaxation defines the "mobility" of a material. The period of "settled life" of molecules in a liquid is 1/100,000,000 that in crystalline solids. As a result, the smaller the value of T_r, the closer the material approaches a liquid state and, similarly, the greater the value of T_r, the more solid is the body.

Generally speaking, all real bodies possess elasticity (characteristic of solids) and viscosity (characteristic of liquids). These properties manifest themselves according to the period of load application (or the period of observation) or, to be precise, according to the relation between the period of observation and the period of relaxation. In fact, let us present formula 4-73 as:

$$\gamma_i = \frac{\tau_i}{G} + \frac{1}{GT_r} \int_0^t \tau_i \mathrm{d}t$$

or:

$$\tau_i = \eta \dot{\gamma}_i - T_r \dot{\tau}_i$$

If the period of observation is shorter than the period of relaxation ($t \ll T_r$), the body in question behaves as a Hookean solid, $\gamma_i = \tau_i/G$. If $t \gg T_r$, the body has the properties of a Newtonian liquid, $\tau_i = \eta \dot{\gamma}_i$. By way of illustration, the period of relaxation of rocks in the earth's crust is measured by the millennium, being 10^{10} s for limestone; in glass this period lasts for centuries, in ice it amounts to several hundreds of seconds, and in air and water is, respectively, $1.96 \cdot 10^{-10}$ s and 10^{-11} s.

Accordingly, water acted upon by a force lasting less than 10^{-11} s will behave elastically. On the other hand, rocks which sustain a pressure over periods measured in terms of geological time may develop a viscous flow — a fact convincingly proved by geology.

Typical in this respect is ice. During the period of relaxation, which is 10^2 to 10^3 s, ice behaves like an elastic body, failing, for example, in its brittle form if rapidly struck. A long-term load causes ice to flow as a viscous liquid, as occurs in glaciers. Similar behaviour — i.e. brittle failure under a rapid load application and viscous flow due to a long-term load — can be observed in frozen soils.

Stress accumulation

Suppose a loaded viscoelastic body is relieved of its load. In this case, deformation will recover with time (see Fig. 4-13(c)). Let us interrupt the process of recovery, adopting for this instant a zero reading of $t = 0$, and let us maintain the recovering deformation at some constant level, γ_0^c = const. To this end, it is necessary to apply a stress τ^c capable of checking the recovery, and then increase the stress in time according to a law $\tau^c = \tau^c(t)$. The accretion of stresses in time to a level that enables a recoverable deformation to remain constant is called 'stress accumulation'. As can be seen, this process acts as a reciprocal of relaxation: in correlation with it as a direct after-effect in the case of loading and as a reverse after-effect in the case of unloading.

4.7 VISCOELASTIC–PLASTIC PROPERTIES OF SOIL

Soil as a viscoelastic–plastic medium

Soil is to be treated as a compressible viscoelastic–plastic medium with a non-linear behaviour.

Non-linearity shows itself by the changes in intensity of strain developed with each increase in stress: elasticity becomes evident due to recoverable deformation of the soil, plasticity manifests itself in the development of irrecoverable deformation, and viscosity is represented by the capacity of the deformation to develop with time. Irrecoverable deformation is evident during shear and during a volumetric change in the soil.

Shearing deformation of a soil results from the relative displacement and rearrangement of mineral particles surrounded by a hydrate film. Volumetric changes arise due to the rearrangement of particles into a more compact structure accompanied by a shrinkage in the void volume, reduction of free water and gas, and dimensional changes of the micelles and particles proper.

Recoverable soil deformation. Elastic shear deformation of the soil results from a corresponding deformation of the soil skeleton (bending of flaky mineral particles, etc.) and a reversible displacement of the mineral particles.

Recoverable volumetric changes are attributed mainly to the dimensional changes of the micelles (the key factor), volumetric changes in the soil skeleton, and recoverable volumetric changes of trapped gas bubbles.

An increase in load brings the particles closer to one another and reduces the thickness of the water film. After unloading, the forces of repulsion cause the particles to move farther apart, which provides an explanation for the recoverable part of volumetric deformation. Recoverable volumetric deformation is possible only then when the loading is high enough to bring the particles so close to each other that they appear to be within the range of repulsive forces.

The reversible contraction of micelles has been proved experimentally more than once. Lambe (1958), reporting on experiments with illite and montmorillonite, indicated that repetitive loading and unloading cycles yielded a totally reversible compression curve, and a change in the composition of the solution (obtained by adding some salt) altered the configuration of the curve. He also succeeded in obtaining a reversible compression curve by changing the temperature of the soil — an increase in the temperature reducing the thickness of the water film, and a decrease causing the film to swell.

The two facts are in accord with the theory of double layers which, as pointed out earlier, provides a qualitative explanation for certain phenomena occurring in soil without furnishing quantitative relationships.

More recent findings state that the elastic deformation of soil is not attributed to variations in the thickness of the water film but results from an elastic deformation of the soil skeleton itself. Thus, on tests on soil soaked in water and dried at 150°C for compression, Larionov (1970) detected no decrease in elastic deformation. Indeed, during some of the tests an increase in elastic deformation was observed.

In studying elastic deformation, one must distinguish between the behaviour of soils with either water-colloidal or crystallization bonds.

Water-colloidal bonds, originating from an electrical interaction between the particles, are controlled by the interparticle distance and re-form, consequently, on removal of the load. However, owing to their weakness, even a light load may induce not only a recoverable but also an irrecoverable deformation entailing particle rearrangement. Therefore, the elastic part of the deformation suffered by soils with water-colloidal bonds is small: it increases with an increase in soil density and is of considerable magnitude only in heavily compacted soils.

Crystallization bonds are rigid and fail to re-form on being broken. Therefore, the soils with crystallization bonds deform in a recoverable way only when they are subjected to a load that is below the fail point. Consequently, the elastic part of the deformation in such soils sustaining a low load may be of appreciable magnitude.

Irrecoverable deformation arises from an irreversible displacement and rearrangement of the particles. Irreversible volumetric changes result from particle rearrangement and porosity alterations entailing the expulsion of free water and gases. Irrecoverable deformations are set up by the stresses which exceed a specified limit and interfere with structural bonding. The deformation itself may also bring about structural changes and, therefore, is frequently called 'structural deformation'.

Irrecoverable volumetric changes are commonly at their maximum in those cases when a random initial structure is rearranged into a parallel orientation on being exposed to a load.

On the subject of volumetric changes due to particle rearrangement and porosity alterations, recent studies at the Moscow and Leningrad State Universities have given indications that volumetric deformation is influenced not only by the total amount of porosity but also by pore-size distribution (Larionov, 1970).

Pores are commonly classified into three types: ultra-micropores of a size less than 1 μm, inter-particle pores ranging in size between 1 and 100 μm, and coarse pores exceeding 100 μm.

Experiments have shown that the bulk of a volumetric deformation resulting from a short-term compaction of soil is attributed to the shrinkage of pores measuring 0.02 mm. This is the so-called active porosity. A long-term load application causes shrinkage of the smaller pores and inter-particle pores; this process gives rise to a volumetric creep. The relation between active and passive porosity is helpful in determining the nature of the volumetric deformation suffered by a soil.

Some analogies

Changes in the volume of a soil materially differ in their essence from the volumetric deformation of solids. If the deformation of a solid is attributed to reversible dimensional changes of the lattice, soil deformation results mainly from porosity alterations and is predominantly irrecoverable.

Taking this into account, it seems appropriate to identify the volumetric deformation of soil with the compression of gases. In fact, Boyle's law describing the compression of a perfect gas is expressed by the well-known formula (for θ = const.):

$$pV = C \tag{a}$$

where p is the pressure, V is the volume, and C is a constant.

Soils satisfy a similar equality, though in a more complex form:

$$pf(\bar{V}) = C \tag{b}$$

where $f(\bar{V})$ is a function of a relative volume $\bar{V}$, which is the ratio of the total volume V and the volume of solid phase m, i.e., $\bar{V} = V/m = 1 + e$ (here e is the void ratio).

Viscous deformations

Coming under this category are all time-dependent shear deformations of the soil and volumetric deformations of the soil skeleton. Distinction is to be made between viscoelastic deformations that are recoverable with time and the irrecoverable deformations of viscous flow. The recoverable viscoelastic deformations are sometimes referred to as structural-adsorptive deformations, apparently because they are induced by Van der Waals forces of adsorption. The irrecoverable viscous deformations are commonly referred to as flow.

Until quite recently many authors considered viscous soil deformation to be the effect of the time-dependent motion of a water film. In particular, it was believed that viscous volumetric deformation was the result of a migration of the molecules of adsorbed water, displacing either from the contacts towards the voids, or in the reverse direction. This was apparently the reason for calling such a deformation an 'adsorptive deformation'.

Another, more realistic, viewpoint suggests that the water film is static and functions as a lubricant. When the particles, each surrounded by a film of bound water, displace relative to one another, a water molecule associated with one particle slides over a molecule associated with another particle. The water content increases as extra moisture is sucked into the voids which expand owing to the increased distance between the particles that have been induced to turn.

In the case of volumetric creep, the rearrangement of particles and their movement (sliding over the water film) changes the inter-particle distance, with the result that some water is expelled from the points of contact.

It must be stressed that viscous deformation is inherent not only in soils with water-colloidal bonding. This type of deformation is encountered in soils and rocks with crystallization bonds resulting from cementation (including soft and hard rocks, clay soil in air-dry condition, etc.). This will be the subject of a subsequent discussion.

4.8 THERMODYNAMIC RELATIONS

Some thermodynamic notion

It is known that all physical processes occurring in nature are due to an exchange of energy between a given thermodynamic system and adjacent systems or an external field. A system is said to be in thermodynamic equilibrium when all macroscopic changes, whatever they are, come to an end; to upset the equilibrium, the system must be exposed to some external action

Any change in the thermodynamic state of a system is a thermodynamic process. When the system steadily passes through a series of states of equilibrium, this is said to be an equilibrium process; when the system passes through non-equilibrium states, this is known as a non-equilibrium process. An equilibrium process may also proceed in the reverse direction, then being called a reversible process. An equilibrium process may be either reversible or irreversible; a non-equilibrium process is always irreversible.

The linear elastic deformation of a body will be a reversible equilibrium process. The same applies to the process of non-linear elastic deformation (see Fig. 4-5(a)), whereas a non-linear elastoplastic deformation (see Fig. 4-5(b)) is an equilibrium, but irreversible, process. Viscous deformations come under the heading of irreversible non-equilibrium processes.

Work of deformation

Whenever external forces act on a body, causing its deformation, they do some work. Since this work is always referred to a unit volume of the body, we speak about the work of the stresses which have produced this deformation or, simply, about the unit work (rate of work) of deformation. An increment in the unit work of deformation is given by:

$$\delta A = \sigma_{ij} \delta \varepsilon_{ij} \tag{4-77}$$

where σ_{ij} and ε_{ij} are the components of the stress tensors and strain tensors, respectively ($i, j = x, y, z$).

The elemental work in the above expression is denoted by δA rather than dA, being an inexact differential in the general case. The unit work of deformation will be:

$$A = \int_0^{\varepsilon_{ij}} \sigma_{ij} d\varepsilon_{ij} \tag{4-78}$$

The unit power of the stresses required to produce a deformation is defined by the expression:

$$N = \frac{\delta A}{\delta t} = \sigma_{ij} \dot{\varepsilon}_{ij} \tag{4-79}$$

where $\dot{\varepsilon}_{ij}$ are the components of the strain rate tensors.

The work of deformation A is the sum of the work of the volumetric change A_V and the work of the distortion of shape A_D; the power $N = \dot{A}$ setting up these deformations is the sum of the powers N_V and N_D, respectively:

$$A = A_V + A_D \tag{4-80}$$

$$N = \dot{A}_V + \dot{A}_D = N_V + N_D \tag{4-81}$$

The work of volumetric change and that of the distortion of shape are given, respectively, by:

$$A_V = \int_0^{\varepsilon_m} \sigma_m d\varepsilon_m \quad \text{and} \quad A_D = \int_0^{\gamma_i} \tau_i d\gamma_i \tag{4-82}$$

Hence, we arrive at the expression for the total work of deformation:

$$A = \int_0^{\varepsilon_m} \sigma_m \, d\varepsilon_m + \int_0^{\gamma_i} \tau_i \, d\gamma_i \qquad (4\text{-}83)$$

Work of elastic deformation

The work of deformation of an elastic body sustaining a load is determined by substituting relations 4-3 into expression 4-83. Thus, we obtain:

$$A_V^e = k \int_0^{\varepsilon_m} \varepsilon_m \, d\varepsilon_m = \frac{k\varepsilon_m^2}{2} = \frac{\sigma_m^2}{2k} \qquad (4\text{-}84)$$

$$A_D^e = G \int_0^{\gamma_i} \gamma_i \, d\gamma_i = \frac{G\gamma_i^2}{2} = \frac{\tau_i^2}{2G} \qquad (4\text{-}85)$$

The body, on being relieved of the load, recovers from the deformation. This means that its elemental particles, now unrestrained by external forces, return into their original positions due to free oscillations. The work done in this case is similar to the work of loading, but of opposite sign:

$$A_V^{(-e)} = \int_{\varepsilon_m}^0 \sigma_m \, d\varepsilon_m = -\frac{\sigma_m^2}{2k}$$

$$A_D^{(-e)} = \int_{\gamma_i}^0 \tau_i \, d\gamma_i = -\frac{\tau_i^2}{2G} \qquad (4\text{-}86)$$

Hence, the total work of deformation of an elastic body done in the course of a "loading–unloading" cycle equals zero. Apparently, the same result will be obtained in the case of a non-linear yet totally recoverable deformation (see Fig. 4-5(a)). In fact, the work of the distortion of shape due to loading will then be the same in its absolute value, although of opposite sign:

$$A_D^{(+e)} = \int_0^{\gamma_i} \phi(\gamma_i) \, d\gamma_i$$

$$A_D^{(-e)} = \int_{\gamma_i}^0 \phi(\gamma_i) \, d\gamma_i = -\int_0^{\gamma_i} \phi(\gamma_i) \, d\gamma_i \qquad (4\text{-}87)$$

Thus, the work done by external forces inducing an elastic deformation (linear or non-linear) in a body while this is being loaded is accumulated in the body due to the transformation of kinetic energy into potential energy. This latter energy is totally retrieved during unloading, being transformed into the kinetic energy of moving

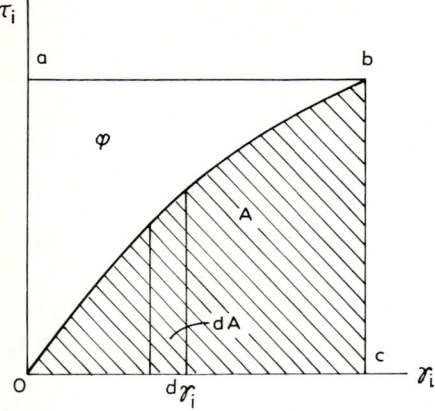

Fig. 4-16. Work of deformation of a body.

particles when the body recovers from deformation. It will be recalled that the kinetic energy possessed by a body is a measure of its mechanical motion, whereas the potential energy is the energy dependent only upon the relative position of the elemental interacting particles of the body.

The work of deformation may be represented on a stress–strain graph (as the area Obc in Fig. 4-16). In the case of non-linear deformation, this work:

$$A = \int_0^{\gamma_i} \tau_i d\gamma_i$$

is not equal to the quantity:

$$\Phi = \int_0^{\tau_i} \gamma_i d\tau_i$$

which is called supplementary work (the area Oab in Fig. 4-16). When the deformation is linear, $A = \Phi$.

The notion of the elastic potential, Π, is helpful in describing the process of elastic deformation. The value of Π may be determined from the relation:

$$\sigma_{ij} = \frac{d\Pi}{d\varepsilon_{ij}} \qquad (4\text{-}88)$$

satisfying the condition that, for an elastic (and only elastic) body, the increment in work is an exact differential.

Work of irrecoverable deformation

The work in question is assessed by taking into account eqs. 4-7 and 4-9, and proceeding from the expressions:

$$A_V^p = \int_0^{\varepsilon_m} \phi^*(\varepsilon_m^p) \, d\varepsilon_m^p \qquad A_D^p = \int_0^{\gamma_i^p} \phi(\gamma_i^p) \, d\gamma_i^p \qquad (4\text{-}89)$$

The above expressions are identical to expressions 4-87 defining the work of non-linear elastic deformation with the exception that they hold for the process of loading only. During unloading, the work will be zero. This implies that the energy required to produce an irrecoverable deformation is not stored in the body but is totally dissipated, being transformed into the internal energy of a heat-induced random motion of particles. Thus, the work of irrecoverable deformation is called the work of dissipation, which is always positive.

In dealing with a process of viscous deformation (in which we consider strain rates rather than strains), it is practical to operate on power instead of work. In such a case, the power is defined as:

$$N = \delta A/\delta t = F(dr/dt) \qquad (4\text{-}90)$$

where F is the force; dr is the displacement, dt is the time interval available for the work, $\delta A = F dr$, to be done.

Because the viscous deformation is irrecoverable, the energy of the viscous deformation dissipates, in the same way as the energy of plastic deformation, without being stored in the body.

Energy of dissipation. In the case of plastic deformation, the energy dissipation is not influenced by strain rate and is due to dry, Coulomb, friction Whenever a viscous deformation is encountered, the energy dissipation is proportional to the flow rate and is associated with the viscous internal friction. In either case, the dissipation energy dispersing throughout the body, (a) is converted into heat which warms the body, and (b) brings about irreversible structural changes in the body. A rearrangement on these lines leads to weakening of the material whose elemental particles tend to occupy the positions corresponding to the minimum free and surface energy.

Work of elastoplastic deformation

The work of soil deformation is the sum of elastic work and dissipated work each of which, in its turn, includes the works of volumetric change and distortion of shape:

$$dA = dA^e + dA^p = (dA_V^e + dA_D^e) + (dA_V^p + dA_D^p) \qquad (4\text{-}91)$$

or:

$$dA = \sigma_m(d\varepsilon_m^e + d\varepsilon_m^p) + \tau_i(d\gamma_i^e + d\gamma_i^p)$$

The work of elastoplastic deformation will be defined, for example, by the expression:

$$A = A_V + A_D^e + A_D^p = \frac{\sigma_m^2}{2k} + \frac{\tau_i^2}{2G} + \int_0^{\gamma_i^p} \phi(\gamma_i^p)\,d\gamma_i^p \qquad (4\text{-}92)$$

in which the first term on the right-hand side represents the work of volumetric (elastic) deformation and the second and third terms define the work of shearing (elastic and plastic) deformation.

The above expression is the strain-energy form of the equation from the deformation theory of plasticity (the theory of small elastoplastic deformations) discussed in Chapter 3.

The equation of the theory of flow may be represented in strain-energy form as:

$$N = \dot{A} = \dot{A}_V + \dot{A}_D^e + \dot{A}_D^p \qquad (4\text{-}93)$$

Description of deformation in terms of thermodynamics

According to the first law of thermodynamics, the sum of the increment in the density of internal energy dE and the increment in the work of deformation, $\delta A = \delta A_V + \delta A_D$, equals the amount of heat δQ supplied to the body during the process of deformation:

$$\delta Q = dE - \delta A \qquad (4\text{-}94)$$

According to the second law of thermodynamics,

$$\delta Q \leqslant \theta\,dS \qquad (4\text{-}95)$$

where θ is the thermodynamic temperature (K), S is the entropy (J/K) indicating the amount of heat Q (in joules) which is supplied to the system due to an infinitesimal quasi-static change in state during a reversible process and is referred to as the thermodynamic temperature (in the case of a reversible process, an increment in entropy includes dissipation). Substituting eq. 4-95 into eq. 4-94, we arrive at the relation:

$$\theta\,dS \geqslant dE - (\delta A_V + \delta A_D) \qquad (4\text{-}96)$$

from which it follows that an increment in the internal energy is a function of volumetric deformation, shearing deformation, and temperature. The sign "equal to" in expression 4-96 is used in conjunction with reversible processes and the sign "greater than" applies to irreversible processes.

Since dE is an exact differential, we can show that:

$$dE = \frac{\partial E}{\partial \varepsilon_m}\delta\varepsilon_m + \frac{\partial E}{\partial \gamma_i}\delta\gamma_i + \frac{\partial E}{\partial \theta}\delta\theta \qquad (4\text{-}97)$$

Substituting this relation into expression 4-96, we obtain:

$$\theta \, dS \geq \left(\frac{\partial E}{\partial \varepsilon_m} - \sigma_m\right) \delta \varepsilon_m + \left(\frac{\partial E}{\partial \gamma_i} - \tau_i\right) \delta \gamma_i + \frac{\partial E}{\partial \theta} \delta \theta \qquad (4\text{-}98)$$

It must be noted that the deformations ε_m and γ_i include recoverable and irrecoverable parts:

$$\varepsilon_m = \varepsilon_m^e + \varepsilon_m^p \quad \text{and} \quad \gamma_i = \gamma_i^e + \gamma_i^p$$

Accordingly, an increment in entropy can be regarded as the sum $dS = dS^e + dS^p$, where dS^e is an external increment in entropy and dS^p is an internal increment occurring in the body due to the conversion of the kinetic energy of an irrecoverable deformation into heat. In other words, an irrecoverable deformation adds to the entropy. If no irrecoverable deformation is present, $dS = dS^e$, i.e. the entropy of the body is not increased. Consequently, inequality 4-98 becomes an equality, as this follows from the spirit of the second principle of thermodynamics.

A point to be noted is that the pattern of deformation may be described only in its most general form in terms of the laws of thermodynamics. Specific relations of the pattern must be established on the basis of micro- and macro-experiments.

Thermodynamic potentials. Thermodynamic processes are studied by the recourse to eigenfunctions called thermodynamic potentials. Coming under this heading are entropy S, free energy F, Gibbs free energy G, and thermal function (enthalpy) H. The quantity S is determined above; other potentials are defined by the following expressions:

$$F = E - \theta S \quad G = E - \theta S - pV \quad H = E + pV \qquad (4\text{-}99)$$

where p is the pressure and V is the volume.

The relations which follow provide an explanation of the physical meaning of the potentials:

$$\sigma_{ij} = \left(\frac{\partial E}{\partial \varepsilon_{ij}}\right)_S \quad \theta = \left(\frac{\partial E}{\partial S}\right)_{\varepsilon_{ij}} \quad \sigma_{ij} = \left(\frac{\partial F}{\partial \varepsilon_{ij}}\right)_\theta$$

$$S = \left(\frac{\partial F}{\partial \theta}\right)_{\varepsilon_{ij}} = -\left(\frac{\partial G}{\partial \theta}\right)_{\sigma_{ij}} \quad \varepsilon_{ij} = -\left(\frac{\partial G}{\partial \sigma_{ij}}\right)_\theta \qquad (4\text{-}100)$$

where the subscripts outside the parentheses indicate which of the quantities are constant.

By way of illustration, the stress tensor components σ_{ij} are defined by the derivatives of the internal energy with respect to the strain tensor components when the entropy S is constant (adiabatic process) and by the derivatives of the free energy with respect to the strain tensor components when the temperature is constant (isothermal process).

The above relations are conducive to obtaining expressions explaining the thermodynamic meaning of the stress tensor invariants I_1 and I_2:

$$I_1 = 3\frac{\partial F}{\partial J_1} + 2J_1\frac{\partial F}{\partial J_2} = 3\frac{\partial E}{\partial J_1} + 2J_1\frac{\partial E}{\partial J_2}$$

$$I_2 = 3\left(\frac{\partial F}{\partial J_1}\right)^2 + 4J_1\left(\frac{\partial F}{\partial J_1}\right)\left(\frac{\partial F}{\partial J_2}\right) + 4J_2\left(\frac{\partial F}{\partial J_2}\right) = 3\left(\frac{\partial E}{\partial J_1}\right)^2$$

$$+ 4J_1\left(\frac{\partial E}{\partial J_1}\right)\left(\frac{\partial E}{\partial J_2}\right) + 4J_2\left(\frac{\partial E}{\partial J_2}\right) \tag{4-101}$$

The relation between the strain tensor invariants J_1 and J_2 and the potentials H and G may be presented in a similar form.

4.9 REFERENCES

Barkan, D.D., 1948. Dynamics of Foundations. Mashstroiizdat Publ., Moscow (in Russian).
Bingham, E.C., 1922. Fluidity and Plasticity. New York, N.Y.
Bingham, E.C. and Green, H., 1919. In: Proc. Am. Soc. Testing Mater., II, 19.
Bolzmann, L., 1876. Zur Theorie der elastischen Nachwirkung. Ann. Phys. Chem., Erg. Bd 7.
Botkin, A.I., 1939. Studies of state of stress in cohesionless and cohesive soils. News Hydraul. Eng. Res. Inst., XXIV, Leningrad (in Russian).
Botkin, A.I., 1940a. On strength of cohesionless brittle materials. News Hydraul. Eng. Res. Inst., XXVI, Leningrad (in Russian).
Botkin, A.I., 1940b. On equilibrium of cohesionless and brittle materials. News Hydraul. Eng. Res. Inst., XXVIII, Leningrad (in Russian).
Denisov, N.Yu., 1956. Engineering Properties of Clay Soils and Their Use in Hydraulic Engineering. Gosenergoizdat Publ., Moscow, Leningrad (in Russian).
Ermolaeva, A.N., Krylova, V.I. and Reltov, B.F., 1968. On the problem of the nature of creep threshold and long-term strength of cohesive soils. In collective volume: Investigations of Rheological Properties of Soils. Energiya Publ., Leningrad (in Russian).
Gersevanov, N.M. and Polshin, D.E., 1948. Theoretical Fundamentals of Soil Mechanics and Their Practical Application. Gosstroiizdat Publ., Moscow (in Russian).
Gor'kova, I.M., 1965. Theoretical Principles of Assessing Sedimentary Rocks for Engineering and Geological Purposes. Nauka Publ., Moscow (in Russian).
Grigoryan, S.S., 1960. On basic concepts of soil dynamics. Appl. Math. Mech., 24(6) (in Russian).
Haefeli, R., 1953. Creep problem in soils, snow and ice. Proc. 3rd ICSMFE, Zürich, Vol. 3.
Il'yushin, A.A., 1963. Plasticity. USSR Acad. Sci., Moscow (in Russian).
Karaulova, Z.M., 1968. Creep threshold and coefficient of viscosity in clay soils. In collective volume: Investigations of Rheological Properties of Soils. Energiya Publ., Leningrad (in Russian).
Lambe, R.W., 1958. The structure of compacted clay. J. Soil Mech. Found Div., Proc. ASCE, V. 84, No. SM4.
Larionov, A.K., 1970. Compacting of clay soils as function of structural particulars. In collective volume: Problems of Engineering Geology. Reports of Soviet Scientists Int. Cong. Assoc. Eng. Geol., All-Union Sci. Tech. Inf. Inst., Moscow (in Russian).
Maslov, N.N., 1968. Fundamentals of Soil Mechanics and Engineering Geology. Vysshaya Shkola, Moscow (in Russian).

Maximov, A.P. and Kiiko, V.I., 1969. Methods and results of studies of relaxation viscosity in Donbass rocks. Proc. All-Union Symposium Prob. Rock Rheology and Relaxation of Solids. Naukova Dumka Publ. Kiev (in Russian).

Meschyan, S.R., 1957. Experimental studies of stress–creep strain relation in cohesive soils. Proc. Acad. Sci. Armenian SSR, v. 24, No. 2 (in Russian).

Mikhailov, N.M. and Rebinder, P.A., 1955. On structural and mechanical properties of dispersed and high-polymeric systems. Colloid. J., V. XVIII, No. 2 (in Russian).

Mogilevskaya, S.E., 1968. Results of studies of creep in loess soil in connection with deformation of water-development works. In collective volume: Studies of Rheological Properties of Soil. Energiya Publ., Leningrad (in Russian).

Ostwald, W., 1925. Über die Geschwindigkeitsfunktion der Viscosität disperser Systeme. Kolloid Z., 36: 99.

Ostwald, W. and Auerbach, R., 1926. Über die Viscosität kolloider Lösungen in Struktur-, Lamimar- und Turbulenzgebiet. Kolloid Z., 38: 261–280.

Reiner, M., 1958. Rheology. Springer, Berlin.

Roza, S.A., 1954. Properties of silt below earth dam. Hydraul. Eng. Proj., No. 3 (in Russian).

Sokolovsky, V.V., 1960. Statics of Cohesionless Medium. Publ. Books on Phys. Math., Moscow (in Russian).

Sorokina, G.V., 1965. Building properties of marine silts. In: Proc. All-Union Conf. Construction in Weak Saturated Clay Soils. Tallin (in Russian).

Thomson, I.I., 1888. Application of Dynamics to Physics and Chemistry. London, New York.

Voitkovsky, K.F., 1960. Mechanical Properties of Ice. USSR Acad. Sci. Publ. House, Moscow (in Russian).

Volterra, V., 1913. Fonctions de liques. Gauthier Villard, Paris.

Volterra, V., 1931. Theory of Functions and of Integro-Differential Equations. London.

Chapter 5

CREEP OF SOILS*[1]

5.1 MECHANISM OF CREEP

Attenuating and non-attenuating creep

As has already been mentioned, the process of creep may proceed at an either decreasing or increasing rate, respectively referred to as attenuating creep and non-attenuating creep (Fig. 5-1). In either case the deformation is the sum of a hypothetically instantaneous deformation, γ_0, set up immediately after the application of load and a deformation developing with time:

$$\gamma = \gamma_0 + \gamma(t) \tag{5-1}$$

The deformation $\gamma(t)$ associated with the process of attenuating creep, represented in Fig. 5-1(a) by segment AB, progresses at a decreasing rate tending to zero, $d\gamma/dt \to 0$. Accordingly, the deformation $\gamma(t)$ itself tends to a certain finite value $\gamma_\infty = $ const. dependent upon load.

Non-attenuating creep, in addition to hypothetically instantaneous deformation, displays three stages: (I) a stage of attenuating, non-steady state creep (segment AB); (II) a stage of steady-state flow (segment BC); and (III) a stage of progressive flow. At stage I, the deformation develops at a decreasing rate; at stage II, the deformation rate reaches a minimum and then stays more or less constant, $\dot{\gamma} = $ const; this stage is sometimes called the stage of viscoplastic flow. At stage III, the deformation develops at an increasing rate and ends in a failure (brittle or viscous), the process being referred to as the failure stage.

Strictly speaking, stage III may be subdivided into two substages. At one (segment CE), the plastic deformation goes on developing without causing failure; at the other (segment ED), numerous microcracks and a catastrophically rapid development of deformation culminate in failure. This differentiation is in principle practical, for the first substage may develop over a rather long period during which the bearing capacity of some soils remains unexhausted. Creep processes are, however, met with quite frequently, when the deformation, although developing at a steadily decreasing rate, fails to stabilize and shows up an unconfined increase. This process, described as $\dot{\gamma} \to 0$, and $\gamma \to \infty$, is referred to as "secular" creep coming under the non-attenuating category although some authors are inclined to call it attenuating creep.

---

*[1] Although the notation used in Chapters 5 and 6 refers to shear strains and shear stresses, all the laws examined also hold for shear strain intensities γ_i, shear strain rates $\dot{\gamma}_i$, and shear stress intensities τ_i.

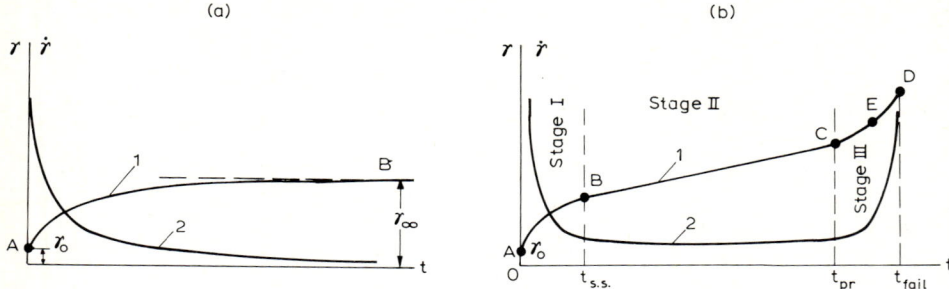

Fig. 5-1. Deformation versus time: (a) attenuating creep; (b) non-attenuating creep.

In general, the singling out of certain creep stages and the subdivision of creep into the attenuating and non-attenuating kinds is an arbitrary classification the success of which depends on the time available for observations and the accuracy of measurements. Deformations which are thought of as stabilized may appear to be developing if observed for another period of time. Another deformation, treated as a constant-rate flow, may in fact either slowly die out or, conversely, develop at an increasing rate. Nevertheless, this classification, widely used in the theory of creep, is practically convenient and reasonable on the whole, for the stages referred to occur in real bodies, soils included, in a more or less explicit way.

The duration and effect of any particular stage of creep vary with the type of soil and load. This is vividly illustrated by Fig. 1-4 depicting a family of creep curves which correspond to various values of a shear load sustained by the soil. The higher the load, the shorter stage II and the sooner stage III, which is the stage of failure, takes shape. If the load is very high, this stage follows stage II almost immediately after the load application with the result that the creep curve acquires an S-like configuration. Moderate loadings distinctly produce all three stages of creep.

A viscoplastic flow resulting from moderate loadings changing insignificantly, and also from tests of certain kind, may steadily develop without passing over into the failure stage III. Finally, low loadings not exceeding a certain limit may fail to give rise to stages II and III. In this case, the process bears an attenuating nature.

Referring to Fig. 5-1, the time of transition from stage to stage will obviously be different: the beginning of steady-state viscoplastic flow is denoted by t_{ss}; the transition to the stage of progressive flow by t_{pr}; and the time of failure by t_{fail}.

Creep curve analysis. The aggregate deformation developing with time may be presented as the sum:

$$\gamma = \gamma_0 + \gamma_I \Big|_0^{t_{ss}} + \gamma_{II} \Big|_{t_{ss}}^{t_{pr}} + \gamma_{III} \Big|_{t_{pr}}^{t_{fail}} \qquad (5\text{-}2)$$

where γ_0 is the hypothetically instantaneous deformation which, strictly speaking, develops at the speed of sound but is measured in practice during a given finite time interval eventually referred to as $t = 0$; γ_I is the attenuating deformation developing over the period $0 < t \leq t_{pr}$; γ_{II} is the deformation of a steady-state flow developing

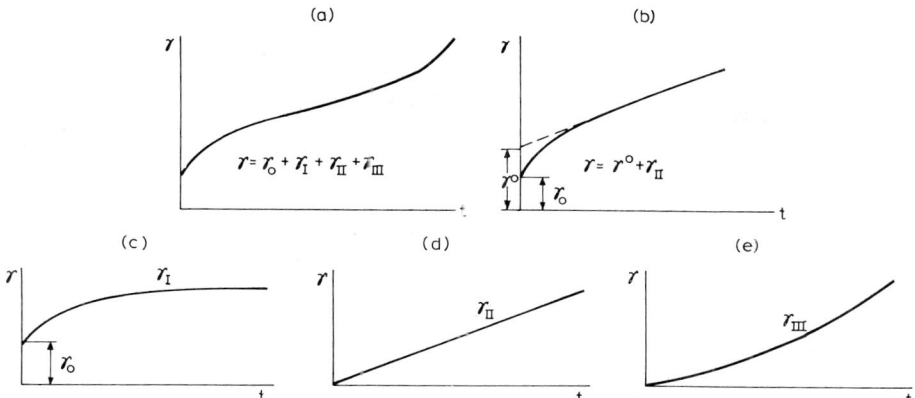

Fig. 5-2. (a) Non-attenuating creep deformation and (b) attenuating creep deformation as the sum of the components; (c) attenuating deformation (γ_I); (d) steady-state deformation (γ_{II}); (e) progressive deformation (γ_{III}).

over the period $t_{ss} < t < t_{pr}$; γ_{III} is the progressive deformation developing over the period $t_{pr} < t \leqslant t_{fail}$.

A statement of the form of eq. 5-2 indicates that each of the deformations mentioned develops within a specified interval of time and is to be summed in succession. However, the variability in the values of t_{ss}, t_{pr}, t_{fail} (which depend upon the level of stress) substantially obscures the prospect of representing the process of deformation as given by eq. 5-2.

It is more convenient to express the aggregate deformation as the sum of the deformations induced at the respective stages, assuming that they all develop simultaneously:

$$\gamma = \gamma_0 + \gamma_I + \gamma_{II} + \gamma_{III} \tag{5-3}$$

Accordingly, the curve of the aggregate deformation is defined by the sum of the ordinates of the three curves (Fig. 5-2). It must be noted that, given the above assumption, the aggregate curve will lack a strictly linear segment. But taking into account that the main factor causing creep during short periods is the attenuating deformation γ_I and that the steady-state deformation γ_{II} and the progressive deformation γ_{III} are the main causes of creep during medium and long intervals, the overall appearance of the aggregate creep curve satisfactorily agrees with reality.

In most cases, it is undesirable to allow soil to operate under the conditions of stage III. Therefore, the progressive deformation γ_{III} is commonly excluded from the analysis and the aggregate deformation of creep is considered (Andrade, 1910, 1962) as the sum of the instantaneous, attenuating and viscoplastic deformations:

$$\gamma = \gamma_0 + \gamma_I + \gamma_{II} \tag{5-3'}$$

Naturally, if the creep attenuates, $\gamma_{II} = 0$ and the deformation is described by the expression $\gamma = \gamma_0 + \gamma_I$.

To simplify the problem, the non-attenuating creep is often represented as the sum of the deformation of steady flow γ_{II} and a hypothetically instantaneous deformation γ^0 which, in its turn, consists of the instantaneous deformation γ_0 proper and the attenuating deformation γ_I projected on the axis of the ordinates, $\gamma = \gamma^0 + \gamma_{II}$ (Fig. 5-2(b)). This approach is possible if the attenuating deformations develop within a short interval and the process is basically confined to stage II. Further simplified, the deformation may be represented as the sum of the instantaneous deformations and the deformation of flow: $\gamma = \gamma_0 + \gamma_{II}$.

Recoverable and residual deformations

When a soil specimen is relieved of load at any time (Fig. 5-3), some of the deformation appears to be of the recoverable kind. The recovery from the initial, hypothetically instantaneous deformation γ_0 is initiated as soon as the load is removed and proceeds until the deformation is gone either totally or partially. Total recovery is the case when the initial deformation is of the purely elastic type, $\gamma_0 = \gamma^e$, so that segment 0–2 of the loading curve (Fig. 5-3(a)) is the same as segment 3–4 of the unloading curve.

Partial recovery takes place when the initial deformation consists of an elastic deformation (segment 0–1) and a plastic deformation (segment 1–2), $\gamma_0 = \gamma_0^e + \gamma_0^p$; consequently, the body recovers from the elastic deformation γ_0^e only.

The recovery from the attenuating deformation γ_I only partly takes place with time (segment 4–5 of the curve), the deformation itself consisting of an elastic deformation γ_I^e (segment 5–6) and a plastic after effect γ_I^p (segment 5–7), $\gamma_I = \gamma_I^e + \gamma_I^p$.

The deformations at stages II and III, the steady-state and progressive flows, respectively, are plastic and totally irrecoverable: $\gamma_{II} = \gamma_{II}^p$ and $\gamma_{III} = \gamma_{III}^p$.

Correspondingly, included into the terms γ_0 and γ_I of expression 5-3 are the recoverable and irrecoverable deformations and into the terms γ_{II} and γ_{III}, the irrecoverable only. In general, the aggregate deformation of creep set up at any time

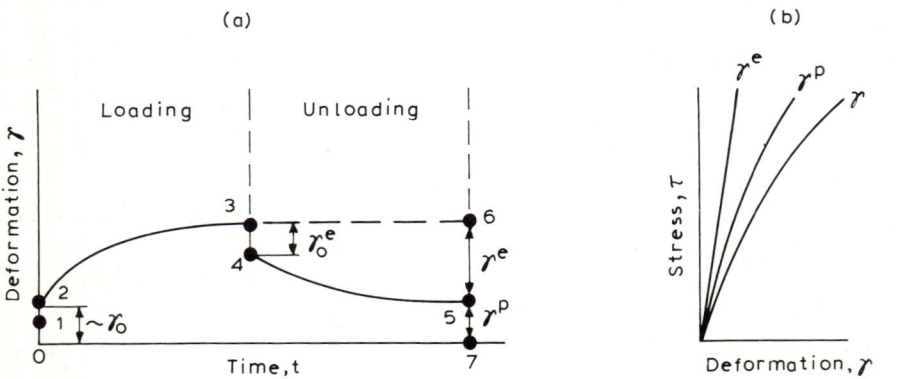

Fig. 5-3. Recoverable, γ^e, and residual, γ^p, components of deformation as presented on (a) the creep curve and (b) the stress-deformation graph.

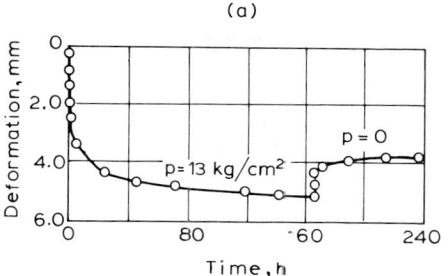

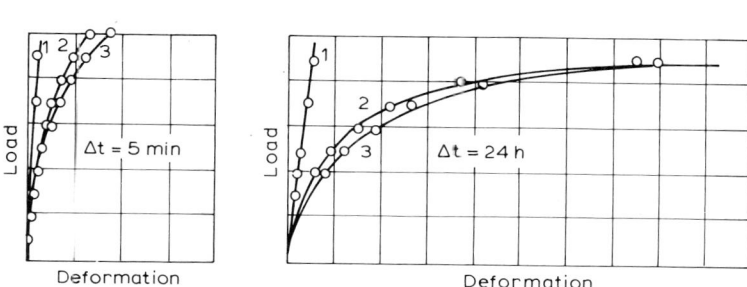

Fig. 5-4. Spherical indenter tests of a body: (a) with unloading; (b) showing recoverable constituent (*1*) and residual constituent (*2*) of aggregate deformation (*3*) for load applications of different duration.

t consists of a recoverable deformation and a residual deformation:

$$\gamma = \gamma^e + \gamma^p \tag{5-4}$$

Illustrated in Fig. 5-4 are the results of the tests with compact banded clay using a settlement plate, conducted by the author of the book. Fig. 5-4(a) depicts a settlement-versus-time curve obtained from a loading–unloading cycle, and Fig. 5-4(b) represents load–deformation curves resulting from stepwise loadings of different duration Δt. As can be seen, the aggregate deformation and its recoverable and irrecoverable components both increase with Δt.

5.2 TIME-DEPENDENT STRESS–STRAIN RELATION

Mechanism of creep

The mechanism of creep may be represented in the form of a dependence of either strain rate $\dot{\gamma}$ or strain γ upon the stress τ and the time t:

$$\dot{\gamma} = f_1(\tau, t) \qquad \gamma = f_2(\tau, t) \tag{5-5}$$

$$\tau = \phi_1(\dot{\gamma}, t) \qquad \tau = \phi_2(\gamma, t) \tag{5-6}$$

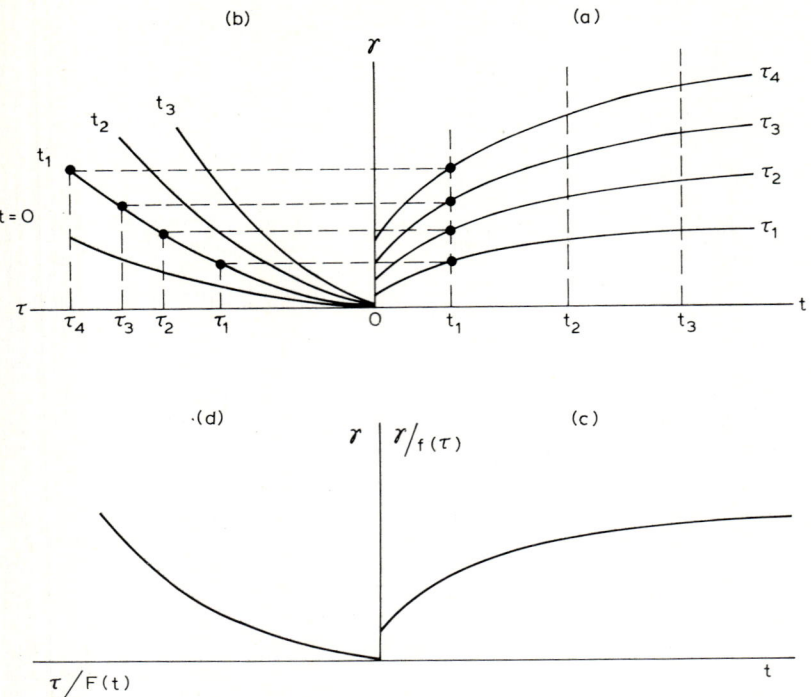

Fig. 5-5. (a) Creep curves. (b) Isochrones. (c) Transformation of the creep curves and isochrones into single curves.

Relations 5-5 are depicted by a family of $\dot{\gamma}$–t curves or γ–t curves corresponding to various values of τ, and relations 5-6 are represented by a family of τ–γ curves for various instants of time t. The γ–t curves for various τ are called creep curves and the τ–γ curves at various t are termed isochrones; the latter are obtained by replotting the creep curves (Fig. 5-5). Note that at $t = 0$ an isochrone becomes the curve of instantaneous deformation and at $t \to \infty$ the isochrone is transformed into the curve of stabilized deformations; this latter curve only exists in the case of attenuating creep.

Stress–strain relation. Isochrones representing the τ–γ relation may occur in three different forms:

(1) Dissimilar curves (Fig. 5-6(a)), each satisfying a function, $\tau = \phi_j(\gamma)$, of its own.

(2) Similar curves, all satisfying the same function $\tau = \phi(\gamma)$ except the curve of instantaneous deformation $t = 0$ satisfying the function $\tau = \phi_0(\gamma)$ (Fig. 5-6(b)).

(3) Curves similar at any instant of time $0 \leqslant t$, all satisfying the same function $\tau = \phi(\gamma)$ (Fig. 5-6(c)).

Similarity between isochrones and creep curves. The condition of similarity between isochrones may be put in the form:

$$\phi(\gamma) = \tau\psi(t) \tag{5-7}$$

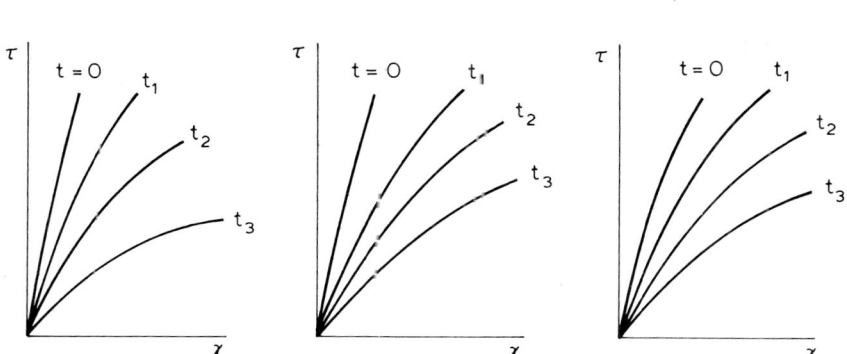

Fig. 5-6. Stress–strain relationships at various times: (a) dissimilar curves; (b) curves similar at all times t_j except $t = 0$; (c) curves similar at all times $t_j \geq 0$.

The condition of similarity between creep curves proper is represented by the relation:

$$\gamma = f(\tau)\Phi(t) \tag{5-8}$$

In the above expressions, $\phi(\gamma)$ and $f(\tau)$ are the functions connecting the stress and strain at any time; $\psi(t)$ and $\Phi(t)$ are the time functions and $\Phi(t)$ is the creep function.

The functions $\Phi(t)$ and $\psi(t)$ should be chosen so that they define the instantaneous deformation at $t = 0$, i.e., $\Phi(0) = \psi(0) = 1$; should this deformation be disregarded, we shall have $\Phi(0) = \psi(0) = 0$.

Suppose, the strain rate rather than the strain proper is to be examined. Then relations 5-7 and 5-8 take the form:

$$\phi(\dot{\gamma}) = \tau K(t) \quad \text{and} \quad \dot{\gamma} = f(\tau)\kappa(t) \tag{5-9}$$

where the functions K and κ are connected with the functions Φ and ψ by the following relations:

$$\psi(t) = 1 + \int_0^t K(t)\,dt \quad \Phi(t) = 1 + \int_0^t \kappa(t)\,dt \tag{5-10}$$

In the case where the creep curves and isochrones are similar, each of the families may be reduced to a single invariant curve. To that end, the creep curves must be plotted on the $\gamma/f(t) - t$ coordinates (see Fig. 5-5(c)) and the isochrones, on the $\tau/F(t) - \gamma$ coordinates, where $F(t) = 1/\psi(t)$ (see Fig. 5-5(d)).

A true similarity between creep curves and isochrones is not always achievable. A significant spread in the values of creep unavoidable in testing various materials, especially soils (for metals, a 20% spread in the test data is not regarded as significant), renders an accurate analytical description of creep curves impractical. Therefore, in most cases it is expedient to use approximations convenient for practical

Deformation function $\phi(\gamma)$

Consider a function of the type $\phi(\gamma)$ at a given instant t_j. Unlike traditional materials that develop plastic deformation only beyond the yield point, soils are likely to manifest elastic and plastic deformations almost as soon as they are loaded. Hence, in accordance with expression 5-4, the stress–strain relation may be described by the binomial formula:

$$\gamma = \tau/G + f(\tau) \tag{5-11}$$

where the first term defines the elastic deformation and the second the plastic deformation.

Since the elastic deformation has a relatively small effect on the configuration of the aggregate $\tau - \gamma$ curve, this may be described by a monomial non-linear relation $\gamma = f(\tau)$ or $\tau = \phi(\gamma)$.

Relations of the above type were considered in Section 4.3. In the case of soils, preference is commonly given to the power function (eq. 4-29) and the linear–fractional function (eq. 4-36).

Time function

Although in the general case the time function is to be adopted, according to eq. 5-3, as the sum of several functions, simplicity calls for the use of a monomial relation holding over a narrow range of stresses. The functions used to describe attenuating and non-attenuating processes are different in this case.

Since the type of function is selected phenomenologically, various authors have suggested numerous expressions. Most often used are power, logarithmic and linear–fractional relations. The present author has suggested a universal time function enabling one to derive in particular cases the relationships referred to above.

Taking into account eq. 5-9, the time function $K(t)$ may be adopted in the form:

$$K(t) = \left(\frac{T_2}{T_1 + t}\right)^n \tag{5-12}$$

The applicability of this function to soils is proved by an apparent rectification of the experimental curves plotted on the $\ln \dot{\gamma} - \ln t$ coordinates. Some of them for various stress intensities τ_i, as plotted in accordance with the data provided by Murayama and Shibata (1961), are exemplified in Fig. 5-7. It will be noted that a departure from eq. 5-12 is evident in one case only, when the stresses are too high. Also to be noted is that the parallelism of the straight lines plotted is an evidence of the constancy of the exponent n (as defined by the slope of the lines).

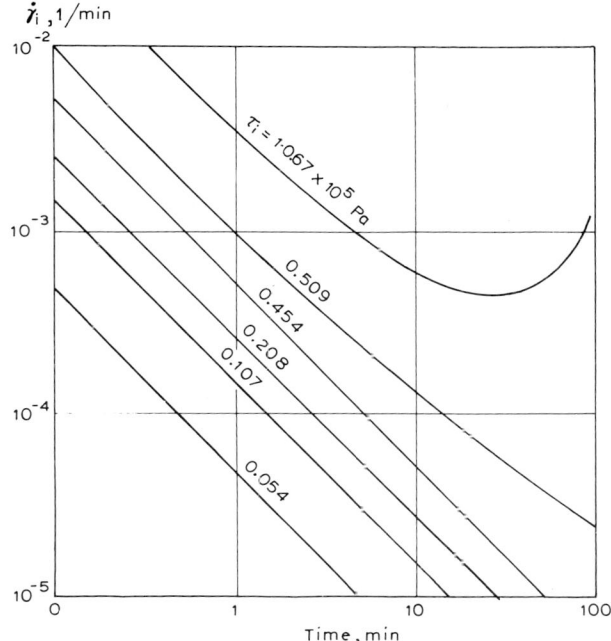

Fig. 5-7. Changes in strain rate with time for various values of tangential stress intensity τ_i, obtained from triaxial compression tests of alluvial clay (after Murayama and Shibata, 1961).

Particular cases of eq. 5-12. This formula provides a basis for deriving, depending on the value of exponent n and taking into account relation 5-10, the following expressions for the creep function.

For $n = 1 - \alpha$ (where $0 < \alpha \leqslant 1, n < 1$) and $T_1 = 0; T_2 = (\alpha\delta/T^\alpha)^{\frac{1}{1-\alpha}}$.

$$\psi(t) = 1 + \delta \left(\frac{t}{T}\right)^\alpha \tag{5-13}$$

for $n = 1$ and $T_1 = T; T_2 = \delta$:

$$\psi(t) = 1 + \delta \ln \frac{t + T}{T} \tag{5-14}$$

for $n = 2$ and $T_1 = T; T_2 = T(\delta - 1)^{1/2}$:

$$\psi(t) = \frac{T + \delta t}{T + t} = 1 + (\delta - 1) \frac{t}{T + t} \tag{5-15}$$

5.3 EQUATIONS OF CREEP

General form of equation

In this section we shall consider an equation describing the development with time of a shear strain induced by a constant stress. This equation is commonly represented in the form given in eq. 5-7 or 5-8.

Let us recall that eq. 5-8 describes similar creep curves and eq. 5-7 defines similar isochrones. These two relations are dissimilar, coinciding only in some cases which are to be examined below.

Thus, to reveal the mechanism of creep, the form of the function $\phi(\gamma)$ or $f(\tau)$ must be established experimentally, using isochrones, and then the form of the function $\psi(t)$ or $\Phi(t)$ is determined, using creep curves. Since the possible forms of these functions are already presented above, we shall now consider some of their combinations.

Power relation. Most commonly in use is a combination of the power functions 4-29 and 5-13. Substituting them into eq. 5-7, we obtain:

$$\gamma^m = \frac{\tau}{A_0}\left[1 + \delta\left(\frac{t}{T}\right)^\alpha\right] \tag{5-16}$$

Proceeding from eq. 5-8, we arrive at:

$$\gamma = \left(\frac{\tau}{A_0}\right)^{1/m}\left[1 + \bar{\delta}\left(\frac{t}{T}\right)^\beta\right] \tag{5-17}$$

where A_0 is the modulus of instantaneous deformation (at $t = 0$) (in Pa); δ, $0 < m \leq 1, 0 < \alpha \leq 1, 0 < \beta \leq 1$ ($\alpha \neq \beta$) are dimensionless quantities (for soils $m = 0.2$–1 and $\beta = 0.5$–1); T is an arbitrary time which may be taken as unity.

The first terms in the relations 5-16 and 5-17 represent an instantaneous deformation $\gamma_0 = (\tau/A_0)^{1/m}$. Ignoring this, the two relations 5-16 and 5-17 adopt the same form:

$$\gamma = \left(\frac{\delta T}{A_{in}}\right)^{1/m}\left(\frac{t}{T}\right)^\beta = \gamma_{in}^* t^\beta \tag{5-18}$$

where $\beta = \alpha/m \leq 1$; $\delta = \bar{\delta}^m$; $\gamma_{in}^* = \gamma_{in}\delta^{1/m}T^{-\beta}$; $\gamma_{in} = (\tau/A_{in})^{1/m}$ and A_{in} are the initial deformation and the initial modulus of deformation corresponding to an instant of time, $t_{in} = T\delta^\alpha$, close to zero.

Relation 5-18 corresponds to the case referred to above when the creep curves and isochrones are similar. Relations 5-7 and 5-8 lead to identical results. Should any of the above curves appear to be dissimilar, the exponents will become variable, $m = m(t)$ or $\beta = \beta(t)$. Strictly speaking, the changes in m with time and in β depending on loading, fit better to the pattern of actual soil behaviour, implying that the condition $m = $ const. and $\beta = $ const., if fulfilled, is an approximation.

Eqs. 5-16 through 5-18 describe a process of creep taking place at a decreasing strain rate $\dot{\gamma}$ tending to zero (at $t \to \infty$) while the deformation proper develops in an unconfined way, $\gamma \to \infty$ (at $t \to \infty$). These equations enjoy wide-spread application in the theory of the creep of metals, soils, rocks and other materials.

The main advantage offered by the power relations of eqs. 5-16–5-18, is their relative simplicity and universality. They have proved to be applicable to soils of diversified kind ranging from fluid and soft varieties to compact clays and from frozen soil and ice to rocks of various hardness.

Logarithmic relation. Taking the functions $\phi(\gamma)$ and $\psi(t)$ according to expressions 4-29 and 5-14 and substituting them into relation 5-7, we obtain the equation of creep:

$$\gamma^m = \frac{\tau}{A_0}\left[1 + \delta \ln \frac{t+T}{T}\right] \tag{5-19}$$

Proceeding, however, from relation 5-8, we arrive at:

$$\gamma = \left(\frac{\tau}{A_0}\right)^{1/m}\left[1 + \bar{\delta} \ln \frac{t+T}{T}\right] \tag{5-20}$$

The values of A_0, δ, $\bar{\delta}$ and T in eqs. 5-19 and 5-20 are the same as in eqs. 5-16 and 5-17.

When the instantaneous deformation $\gamma_0 = (\tau/A_0)^{1/m}$ is disregarded, eqs. 5-19 and 5-20 take the form:

$$\gamma^m = (\gamma^*_{in})^m \ln \frac{t+T}{T} \qquad \gamma = \gamma^*_{in} \ln \frac{t+T}{T} \tag{5-21}$$

In these formulae $\delta = \bar{\delta}^m$; $\gamma^*_{in} = \gamma_{in}\bar{\delta}^{1/m}$, where $\gamma_{in} = (\tau/A_0)^{1/m}$ and A_{in} correspond to an initial instant of time $t_{ir} = (e^{1/\delta} - 1)T$.

As it can be seen, eqs. 5-21 are not identical – a fact which follows from the difference between eqs. 5-7 and 5-8.

Eqs. 5-19–5-21, as well as 5-16–5-18, describe a process of creep in which the strain rate $\dot{\gamma}$ diminishes, tending to zero (at $t \to \infty$), while the deformation γ develops in an unconfined way, tending to infinity (at $t \to \infty$). However, eqs. 5-19–5-21 represent an increase in deformation which is less intensive than is the case in eqs. 5-16–5-18, for the increase according to eqs. 5-19–5-21 obeys a logarithmic law and that corresponding to eqs. 5-16–5-18, a power law.

For small values of α and average values of t eqs. 5-16–5-18 and 5-19–5-21 yield close results, for $t^\alpha = \exp(\alpha \ln t) \approx 1 + \alpha \ln t$ in this case.

The relation of the form given by eqs. 5-19–5-21 was originally employed by Pokrovsky (1933, 1939) and Buisman (1936) to describe a long-term ("secular") process of soil settlement; eventually it was widely used in describing the process of secondary consolidation in soils. (This is the subject of a detailed discussion found in Chapter 8.)

At a later date, it was demonstrated that the logarithmic law of eqs. 5-19–5-21 might be applied to describing the process of shear creep. Colloid systems, peats, plastics and even metals appeared to be other fields of application of the law.

Linear–fractional relation. Taking the function $\phi(\gamma)$ and $\psi(t)$ as given by eqs. 4-36 and 5-15 and substituting into relation 5-7 we obtain, on solving it for γ:

$$\gamma = \frac{\tau(T + \delta t)}{G_0[T(1 - \tau/\tau_s) + t(1 - \delta\tau/\tau_s)]} \tag{5-22}$$

Solving this equation for stress τ, we arrive at:

$$\tau = \frac{\gamma G_0(T + t)}{(T + \delta t)(1 + \gamma G_0/\tau_s)} \tag{5-22'}$$

and in the case where the initial deformation is disregarded:

$$\gamma = \frac{\tau \delta_{in} t}{G_{0(in)}[T + t(1 - \delta_{in}\tau/\tau_{s(in)})]} \tag{5-23}$$

In the above expressions G_0 is in Pa, τ_s is also in Pa, T is in hours; $\delta > 1$ is a dimensionless quantity and $G_{0(in)}$, $\tau_{s(in)}$ and δ_{in} are parameters, the meaning of which will be considered below.

Putting $t = 0$ in eq. 5-22, we obtain the value of the instantaneous deformation:

$$\gamma_0 = \frac{\tau}{G_0(1 - \tau/\tau_s)} \tag{5-24}$$

Assuming that $t = \infty$, we arrive at the value of the finite, stabilized deformation:

$$\gamma_\infty = \frac{\delta\tau}{G_0(1 - \delta\tau/\tau_s)} \tag{5-24'}$$

Hence, taking into account the above expressions, eq. 5-22 may be written as:

$$\gamma = \frac{\gamma_\infty(T + \delta t)}{T\gamma_\infty/\gamma_0 + \delta t} = \gamma_0 + (\gamma_\infty - \gamma_0)\frac{t}{T^* + t} \tag{5-25}$$

where:

$$T^* = \frac{1}{\delta}\frac{\gamma_\infty}{\gamma_0}T = \frac{1}{\delta}\frac{\tau_s - \tau}{\tau_s/\delta - \tau}T$$

Isochrones and creep curves corresponding to eq. 5-22 are depicted in Fig. 5-8. At $t = 0$ an isochrone is described by eq. 5-24 and at $t = \infty$, by eq. 5-24'; for $0 < t < \infty$, the isochrone is described by eq. 5-22.

The quantities G_0 and τ_s are the parameters of the isochrone for instantaneous deformation at $t = 0$; G_0 corresponds to the tangent of the angle made by a tangent line to the $\tau - \gamma_0$ curve at a point $\gamma_0 = 0$, and τ_s is the ordinate of an asymptote to the above curve (for $\gamma_0 \to \infty$); this latter parameter corresponds to the yield point in the case of a rapid load application.

Corresponding to each isochrone (at $t = t_j$) there is a certain value of $G_{0(j)}$ and $\tau_{s(j)}$. So, corresponding to the isochrone at $t \to \infty$, i.e. the curve of finite, stabilized

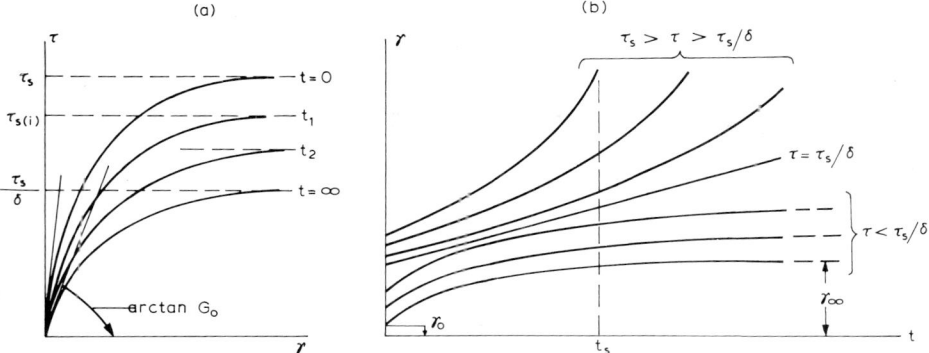

Fig. 5-8. (a) Isochrones and (b) creep curves corresponding to eq. 5-22.

deformation γ_∞, are the relations $G_0/\delta = G_\infty$ and $\tau_s/\delta = \tau_\infty$. Thus, the parameter δ is defined by the relation $\delta = G_0/G_\infty = \tau_s/\tau_\infty$. It will be shown hereinafter that τ_s is the hypothetically instantaneous strength, $\tau_s = \tau_0$, and that τ_s/δ is the ultimate long-term strength.

When the initial deformation is ignored, the parameters $G_{0(in)}$ and $\tau_{s(in)}$ of the isochrone at an instant t_{in} close to the initial time are introduced into eq. 5-23.

The configuration of the creep curves described by eq. 5-22 varies depending on the level of stress τ (Fig. 5-8(b)). At a low stress, $\tau < \tau_\infty$, the curves assume an attenuating shape; at $t \to \infty$, the strain rate $\dot{\gamma} \to 0$ and the deformation proper acquire a finite value defined by eq. 5-24.

A stress $\tau = \tau_\infty$ induces an unconfined flow at a constant rate $\dot{\gamma} = \delta\tau_s/G_0 T$. High stresses, $\tau_\infty < \tau < \tau_s$, produce deformations which develop at a steadily increasing rate and reach an infinitely large value $\gamma \to \infty$ at the instants $t_s = T(\tau_s - \tau)/(\delta\tau - \tau_s)$. A stress $\tau = \tau_s$ brings about a deformation developing in an unconfined way ($\gamma \to \infty$) immediately on the application of load ($t_s = 0$).

Note that a linear stress–strain relation, when the strain–time relation is linear–fractional, will be a particular case of eq. 5-22:

$$\gamma = \frac{\tau(T + \delta t)}{G_0(T + t)} \tag{5-26}$$

However, in some cases the non-linear stress–strain relation in eqs. 5-22 and 5-23 must be strengthened. This is effected by introducing γ^m into the left side of the formula.

The linear–fractional relation as given by eq. 5-22 was first applied to metals by Oding Oding et al. (1959) and to soils, in the form represented by eq. 5-26, by Nichiporovich and Tsybulnik (1961).

Generalized time function (eq. 5-12). As it can be seen, the time function of eq. 5-12 is satisfactorily universal, describing creep processes of various nature ranging from progressive to attenuating. Apart from eq. 5-22, the equation of attenuating creep may be represented in another form if we put that $n > 1$ in eq. 5-12. Taking that in eq. 5-12 $n = 1 + \alpha > 1$, $T_1 = 1$ and $\delta = T_2^{1+\alpha}/\alpha$, and substituting the value of $K(t) = \kappa(t)$ obtained into eq. 5-8 and integrating this equation with respect to the boundary conditions $t = \infty$ and $\gamma = \gamma_0$, we arrive at the following expression:

$$\gamma = \gamma_\infty - (\gamma_\infty - \gamma_0)(1 + t)^{-\alpha} \tag{5-27}$$

where $\gamma_0 = (\tau/A_0)^{1/m}$ and $\gamma_\infty = \gamma_0(1 + \delta)$.

It must be borne in mind that the various values of the exponent n introduced in formula 5-12 each hold for a given range of stresses. In examining a wide range of stresses, a higher degree of coincidence with the experiment may be obtained by representing the creep process as the sum of an attenuating deformation and a steady-state flow, using eq. 5-3. The term $(\gamma_0 + \gamma_1)$ in this relation is then determined by one of the formulae 5-16–5-28 discussed above and the term γ_{II} is adopted as $\gamma_{II} = f(\tau)t$. Hence:

$$\gamma = f_1(\tau)\psi(t) + f_{II}(\tau)t \tag{5-28}$$

Exponential relation. The process of attenuating creep may also be described in terms of an exponential law of the form:

$$\gamma = \gamma_\infty - (\gamma_\infty - \gamma_0)\,e^{-t/T} \tag{5-29}$$

where $\gamma_0 = \tau/G_0$ and $\gamma_\infty = \tau/G_\infty$; in this connection, G_0 and G_∞ are the initial and ultimate long-term shear moduli, respectively.

For $\gamma_0 = 0$, we obtain:

$$\gamma = \gamma_\infty(1 - e^{-t/T}) \tag{5-30}$$

A more complicated law may be employed by introducing an exponent $(t/T)^\alpha$ due to Kohlrausch (1863, 1866, 1877) into eqs. 5-29 and 5-30.

In Chapter 7, it will be shown that eq. 5-29 may be derived on the basis of the theory of viscoelastic deformation (Chapter 1: Reiner, 1949; Eirich, 1956; Rzhanitsin, 1968). Since this theory is one of the keystones of rheology, eq. 5-29 has come into wide-spread application. The exponential time function has also gained recognition in the theory of hereditary creep, particularly in its application to concrete (Arutyunyan, 1952; Maslov, 1955), from where it has been borrowed and applied to the creep of soils by Florin and others (Chapter 1: Florin, 1962; Meschyan, 1978). However, experiments have shown that the exponential relation, as represented by eq. 5-29, lacks adequate agreement with the test data. This will be discussed in Chapter 6 along with evaluation of the formulae examined above.

5.4 EQUATIONS OF FLOW

General form of flow equation

As mentioned above, the mechanism of creep can be established by considering either strain or strain rate. In this latter case, the rheological equation of state is taken as being given by the second relation 5-9. If stage II dominates, i.e. the stage of steady-state creep at constant rate (e.g. in considering the flow of natural or artificial slopes or the deformation of soil having a fluid consistency), only this stage is subjected to investigation (see Fig. 5-1(b)). Hence, assuming that $K(t) = \kappa(t) =$ const. and putting $\dot{\gamma}$ = const., we arrive at the equation of flow (eq. 4-58'):

$$\tau = \varphi(\dot{\gamma}) \quad \text{or} \quad \dot{\gamma} = f(\tau) \tag{5-31}$$

Various forms of the function $f(\tau)$ were considered in Section 4.5.

Mechanism of viscoplastic flow in soils. In deciding whether or not the equation of flow 5-31 is applicable to soils, one has first to find out whether the constant-rate flow is induced by any load or by one exceeding a certain limiting stress, the yield limit τ_y.

Relationships between the steady-state flow rate and stress obtained by various authors in testing various types of soil are depicted in Fig. 5-9. The graphs in (a), (b)

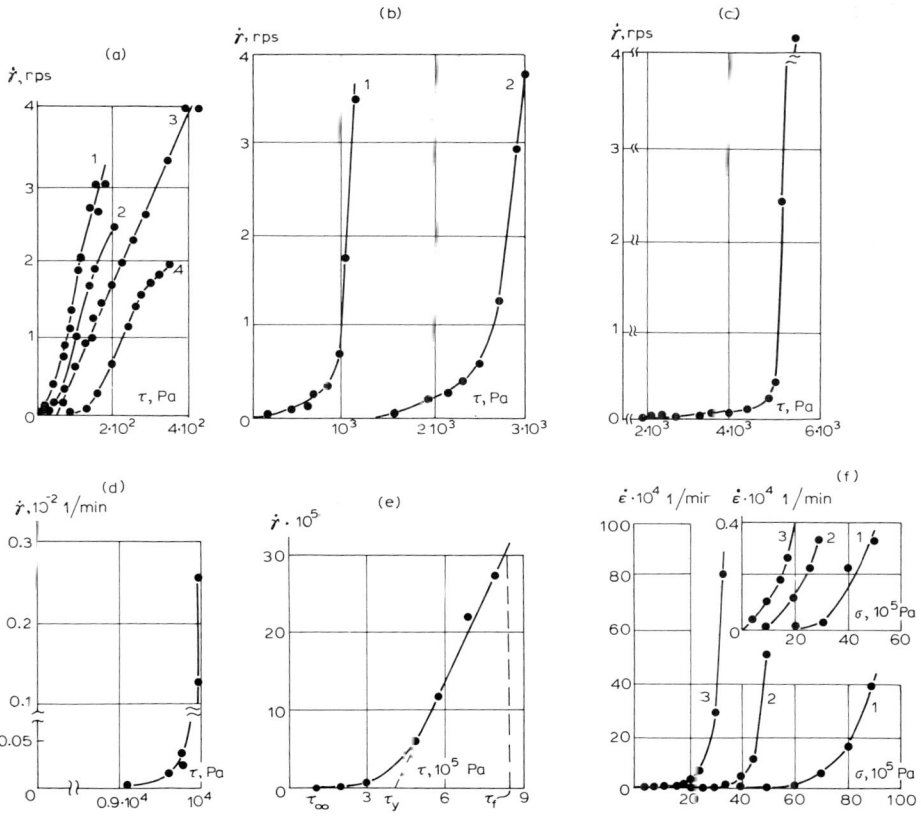

Fig. 5-9. Relationships between the steady-state flow rate and stress for various soils. (a) Quicksands (1 = sands from Salekhard formation, $W = 34\%$; 2 = postglacial clay, $W = 65\%$; 3 = Nizhny Volzhsk sands, $W = 33\%$; 4 = postglacial clay, $W = 54\%$). (b) Silts (1 = Black Sea silt, $W = 126\%$; 2 = ditto, $W = 54\%$). (c) Khvalynsk remoulded clay, $W = 92\%$. (d) Soft Jurassic clay, $W = 32\%$. (e) Stiff clay. (f) Frozen sandy loam, silty, $W = 26\%$: $1 = -20°C$; $2 = -10°C$; $3 = -5°C$.

and (c) refer to the shear tests in torsion of soft soils ranging from quicksands and silts to clays of fluid consistency conducted by Gor'kova (1965). Fig. 5-9(d) is from Pekarskaya and Vyalov (1973) and shows testing of plastic clays for shear in torsion, and (e) presents the findings of Babitskaya and Goldstein who tested stiff clays for uniaxial compression (Goldstein et al., 1962). Fig. 5-9(f) was obtained by E.P. Shusherina experimenting with frozen soil in uniaxial compression at various temperatures (in Vyalov et al., 1965).

Although the soils covered are quite different, varying between fluid and stiff, the flow rate versus stress curves obtained are all of identical configuration analogous with the shape of the rheological curve of a quasi-solid body in Fig. 4-11(b).

An analysis of the curves, as presented in Fig. 5-9(e), reveals three critical values of stress.

The first critical stress is the apparent elastic limit τ_k below which no flow is induced although the deformation itself can be not only elastic but plastic as well as irrecoverable. Generally speaking, one may single out the limit $\tau_e < \tau_k$ below which the deformation will be purely elastic. However, this limit is very low and its introduction is practically useless (as far as static problems are concerned). For unstructuralized soils, mainly those with water-colloidal bonding, the limit τ_k is very low. Assuming that $\tau_k = 0$, such soils can be treated as liquid bodies (see Fig. 4-11(a)).

The second critical stress is the yield limit τ_y. Before this is exceeded, creep deformations are likely to develop. However, their rate is either very low or the deformations die out and the soil structure remains intact. In the case where the yield limit is exceeded, the soil structure breaks down and the flow rate increases by several orders of magnitude.

The third critical stress τ_f brings about total structural failure. The flow rate–stress relation is non-linear in structured and unstructured soils alike, the non-linearity being at its maximum during the transition from high to low flow rates.

In general, the mechanism of viscoplastic flow may be described in soils (Vyalov, 1959) by the relationship:

$$\dot{\gamma} = \frac{1}{\eta_p} \left(\frac{\tau - \tau_\infty}{\tau^*} \right)^{1/m} \tag{5-32}$$

where η_p is the coefficient of plasticity reduced to unit stress: τ^* is a parameter measured in units of stress and taken as unity; $m \leq 1$; τ_∞ is the ultimate long-term strength which may be taken as $\tau_\infty \cong \tau_k$ (see Fig. 4-11(b)).

The segment of the $\dot{\gamma} - \tau$ curve between the limits $\tau_y \leq \tau \leq \tau_f$ may be approximated by a straight line, as shown in Fig. 5-9(e). The process of flow in soils within the above range of stresses may be described by Bingham's law (eq. 4-67):

$$\dot{\gamma} = \frac{\tau - \tau_y}{\eta_p} \tag{5-33}$$

where η_p is the coefficient of plastic viscosity. Note that as far as soils are concerned, Bingham's law is applicable only over the range of time $t_y \leq t \leq t_{pr}$, during which

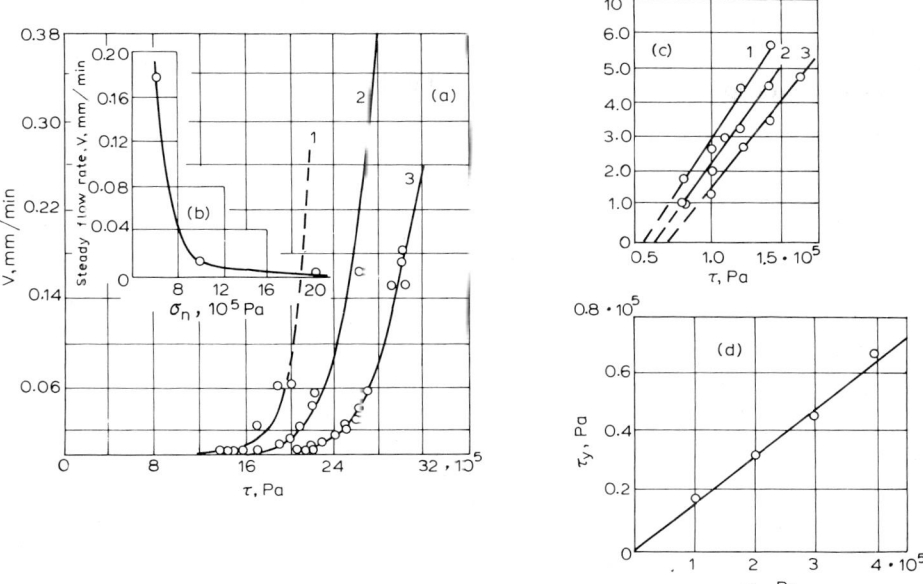

Fig. 5-10. Effect of normal stress σ_n on flow. (a) Results of testing frozen sandy loam, silty, at $-10°C$ represented by rheological curves for various values of σ_n: $1 = 6 \cdot 10^5$ Pa; $2 = 10 \cdot 10^5$ Pa; $3 = 20 \cdot 10^5$ Pa. (b) Relationship between flow rate and σ_n. (c) Results of testing bituminous clay at σ_n as follows: $1 = 2 \cdot 10^5$ Pa; $2 = 3 \cdot 10^5$ Pa; $3 = 5 \cdot 10^5$ Pa. (d) Relationship between yield point τ_y and σ_n for remoulded Cambrian clay.

stage II of the steady-state creep, having a roughly constant flow rate, is developing; and only over the range of stresses $\tau_y \leqslant \tau \leqslant \tau_f$.

The effect of normal stress. Creep in a state of combined stress is a subject of subsequent discussions. Here we must point out that a normal stress σ_n, inducing flow simultaneously with other factors, has a pronounced effect on the intensity of the deformation process.

As first noted by Maslov (1955), this effect can be vividly seen in Fig. 5-10. Depicted are the results of shear tests of frozen soil for various σ_n (Pekarskaya, 1963), plastic clay (Maslov, 1968) and remoulded Cambrian clay (Ermolaeva, see Sect. 4.9). The flow rate versus stress curves materially differ from one another depending on σ_n (Fig. 5-10, (a) and (c)), an increase in σ_n causing a reduction of the flow rate (Fig. 5-10(b)). Also changing in accordance with σ_n are the critical stresses τ_k and τ_y. The $\tau_y - \sigma_n$ relation may be assumed to be approximately linear (Fig. 5-10(d)):

$$\tau_y = \tau_0 + f\sigma_n \qquad (5\text{-}34)$$

where τ_0 is the yield limit in pure shear and f is the friction coefficient of flow.

Investigations carried out by the author (Vyalov, 1959) show that in this case the shear strength of soil in the state of flow, taking into account eq. 5-31 is:

$$\tau = \tau_0 + f\sigma_n + \eta_p \dot{\gamma} \tag{5-35}$$

For $\dot{\gamma} = 0$, this expression is transformed into the Coulomb condition and for $f = 0$ it becomes the Bingham condition.

On creep threshold. According to Maslov's (1968) conception, the shear strrength of a clay soil is defined by the relation:

$$\tau = \sigma_n \tan \phi_w + \Sigma_w + c_s \tag{5-36}$$

where ϕ_w is the 'true' angle of internal friction varying with the moisture content W of the soil; Σ_w is the cohesiveness of water-colloidal nature varying with W (bonding of this kind is characteristic of plastic clay soils); c_s is the structural cohesion inherent in cemented rocks and attributed to rigid crystallization bonds.

Creep is provoked whenever the following inequality is valid:

$$\sigma_n \tan \phi_w + c_s < \tau < \sigma_n \tan \phi_w + \Sigma_w + c_s$$

Hence, the limiting stress giving rise to creep, called by Maslov the creep threshold, is:

$$\tau_{\lim} = \sigma_n \tan \phi_w + c_s \tag{5-37}$$

This expression applies to the soils which have been called by the author crypto-plastic soils. For plastic soils, $\phi_w = 0$, $c_s = 0$ and, consequently, $\tau_{\lim} = 0$, i.e., creep is brought about by any stress no matter how small this may be. This interpretation seems to be quite logical. However, in applying the notion "creep threshold", it is necessary to specify which of the creep deformations, time-dependent or non-attenuating, will be developing when $\tau < \tau_{\lim}$. From our point of view, $\tau_{\lim}$ should be adopted at a level before exceeding which the time-dependent deformations should either stabilize or not exist at all.

Non-linearly viscous flow of soil

Let us go back to the rheological curve of Fig. 5-9(e). As can be seen, the value of the ultimate long-term strength $\tau_\infty = \tau_k$ is defined by the intercept of the $\dot{\gamma} - \tau$ curve on the axis of stress. However, taking into account that the curve often approaches the x-axis in a gentle slope, the value of τ_k may be ignored in an analytical description of the curve. Thus, putting $1/\eta_p = \dot{\gamma}^*$, eq. 5-32 is transformed into the equation of non-linearly viscous flow (eq. 4-61'):

$$\dot{\gamma} = \dot{\gamma}^* (\tau/\tau^*)^{1/m} \tag{5-38}$$

Eq. 5-38 is widely used in describing the steady flow of metals at high temperatures, plastics and a host of other materials. It is frequently employed in dealing with soils, mainly frozen, and is particularly popular in the case of ice. Also widely used is relation 4-64:

$$\gamma = \dot{\gamma}^* \sinh(\tau/\tau^*) \tag{5-39}$$

If we consider the hyperbolic function of eq. 5-39 it has certain distinctions. At low values of the independent variable $x \ll 1$, the function $\sinh x = x$ approximately and at high values $x \gg 1$, $\sinh x = 1/2\, e^x$ approximately. The first approximation is obtained from expanding into the series:

$$\sinh x = 1/2(e^x - e^{-x}) = x + \frac{x^3}{1} + \frac{x^5}{3!} + \ldots$$

and restricting oneself to the first term for $x \ll 1$. The second approximation results if the quantity e^{-x} in the expression $\sinh x = 1/2(e^x - e^{-x})$ is neglected, for $x \gg 1$. Hence, at low stresses, $\tau/\tau^* \ll 1$, eq. 5-39 may be used in the linear form:

$$\dot{\gamma} = \tau \frac{\dot{\gamma}^*}{\tau^*} = \tau/G \tag{5-40}$$

At high stresses, $\tau/\tau^* \gg 1$, the above relation becomes:

$$\dot{\gamma} = \tfrac{1}{2}\dot{\gamma}^* e^{\tau/\tau^*} \tag{5-41}$$

The difference between eqs. 5-38, 5-39, 5-40 and 5-41 can be visualized from plotting the relations they represent on the $\tau/\tau^* - \ln \dot{\gamma}$ coordinates (Fig. 5-11). The only straight line yielded by the plot is line *2* representing the exponential relation 5-41. Its slope defines τ^*, for $\tau^* = d\tau/d \ln \dot{\gamma}$; the value of $\dot{\gamma}^*$ is determined from the expression $\ln \dot{\gamma}^* = \ln 2\dot{\gamma} - \tau/\tau^*$.

The relation $\dot{\gamma}/\dot{\gamma}^* = \sinh(\tau/\tau^*)$ is rectified only for $\tau/\tau^* > 1.3$, for beginning with this value it coincides (curve *1*) with the exponential formula. This implies that for $\tau/\tau^* > 1.3$ (or, practically, for $\tau/\tau^* = 1$ approx.), eq. 5-39 may be replaced by eq. 5-41. Beginning, however, with the values $\tau/\tau^* \leqslant 1$, eq. 5-41 yields data lacking agreement with the experiment, e.g. for $\tau/\tau^* = 0$, $\dot{\gamma} = 0.5\dot{\gamma}^*$, whereas in reality $\dot{\gamma} = 0$. However, taking into account that under real conditions τ/τ^* is always greater than 1, the exponential relation 5-41 appears to be quite suitable for describing the behaviour of soils.

The value $\tau/\tau^* = 1.3$ may be regarded as a kind of critical relative stress. As long as this value is not exceeded, the soil flows linearly, which is evidence that the structural changes are small. For $\tau/\tau^* = 1.3$, the flow becomes non-linear, entailing structural changes in the soil. In other words, the quantity $\tau/\tau^* = 1.3$ may be considered as a conventional yield limit τ_y/τ^*.

The power relation 5-38 was scrutinized in Section 5.3. It is now not irrelevant to mention that for not very high values of m (say, $m = 0.35$, which is the value used in plotting curve *3* of Fig. 5-11) the power curve differs little from hyperbolic curve *1* and exponential curve *2*. At higher values of m (particularly for $m = 1$), curve *4* significantly departs from curve *1*, coinciding only for $0 < \tau/\tau^* < 1$ because over this range of stresses $\sinh(\tau/\tau^*) = \tau/\tau^*$ approx.

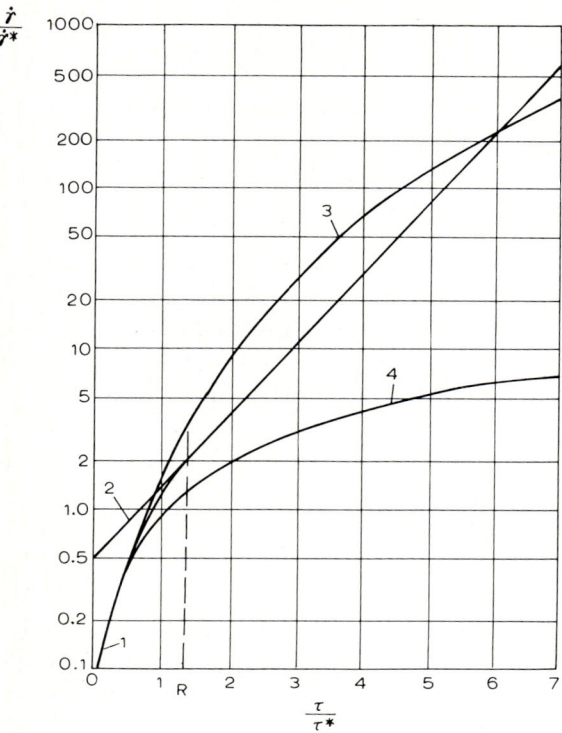

Fig. 5-11. Relationship between strain rate $\dot{\gamma}/\dot{\gamma}^*$ and stress τ/τ^*: $1 = \dot{\gamma}/\dot{\gamma}^* = \sinh(\tau/\tau^*)$; $2 = \dot{\gamma}/\dot{\gamma}^* = 0.5\,e^{\tau/\tau^*}$; $3, 4 = \dot{\gamma}/\dot{\gamma}^* = (\tau/\tau^*)^n$ (3 for $n = 0.5$; 4 for $n = 1$).

Time-dependent changes in strain rate

The equations of flow referred to above describe a steady-state creep having a constant strain rate. If the process of soil creep is considered as a whole, one must take into account the time-dependent changes in strain rate.

Experimental data characterizing such changes are represented in Fig. 5-12; they were acquired by testing bentonite, silt, and kaolinite (Tsytovich et al., 1965), windblown clay loam and clay (Maslov and Kodzhemanov, 1970) and frozen sandy loam (Vyalov, 1959).

Fig. 5-12, (a) and (b) depict a process of attenuating creep at a strain rate steadily decreasing during the experiment. Fig. 5-12(c) demonstrates the development of the process at all stages of deformation. So, under a load slightly less than yield limit ($\tau/\tau_y = 0.9$), the deformation of the attenuating nature and the strain rate asymptotically approaches zero (curve *1*). At a load $\tau/\tau_y = 1.15$, the process has a non-attenuating character with all the three stages of creep – unsteady, steady state, progressive – being markedly shown. Finally, at a load of $\tau/\tau_y = 1.3$, a transition from stage I to stage III, practically by-passing the stage of steady-state flow, is observed (curve *3*).

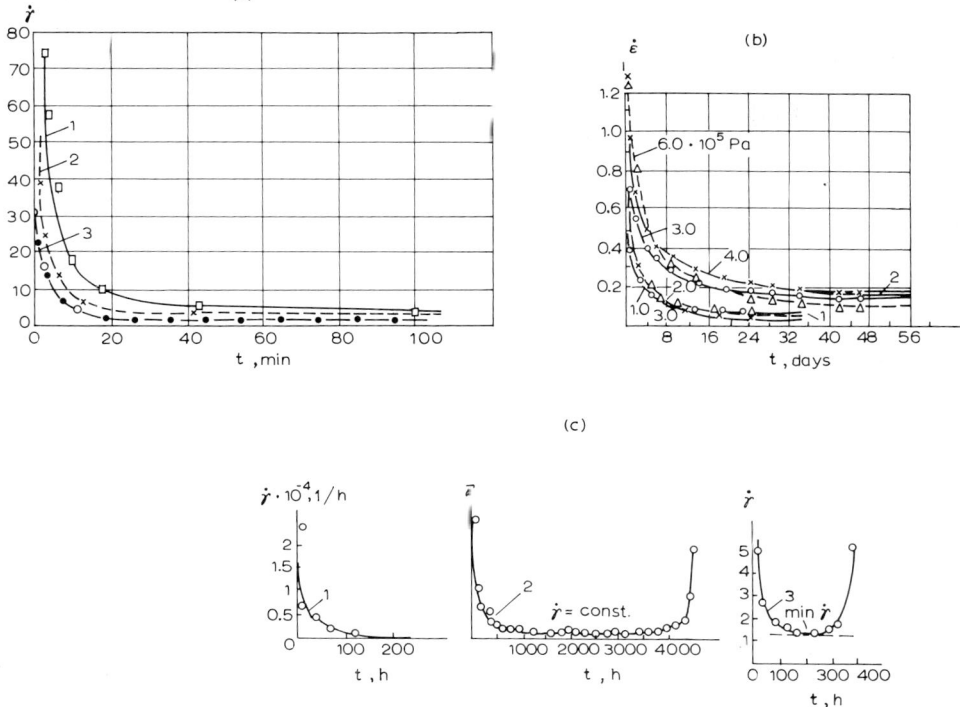

Fig. 5-12. Changes in strain rate with time in response to constant loading. (a) Shear tests (1 = bentonite remoulded clay, $W = 106\%$, $e = 2.6$; 2 = river silt, $W = 53\%$, $e = 1.4$; 3 = remoulded kaolinite, $W = 50\%$, $e = 1.3$). (b) Compression tests under various loads (1 = silty clay loam, $W = 8\%$, $e = 0.58$; 2 = quaternary clay, $W = 12\%$, $e = 0.84$). (c) Shear tests of frozen sandy loam, $W = 28\%$, $\theta = -0.4°C$; loading, τ: $1 = 1.3 \cdot 10^5$ Pa; $2 = 1.5 \cdot 10^5$ Pa; $3 = 1.7 \cdot 10^5$ Pa.

Flow equation allowing for changes in strain rate

Time-dependent changes in strain rate are allowed for by introducing a time function $K(t)$ into the rheological equation of state 5-9. Taking this function as given by eq. 5-12 for $T_1 = 0$ and $T_2 = T$, and proceeding from relation 5-38, we arrive at the following equation of flow:

$$\dot{\gamma} = \dot{\gamma}^* \left(\frac{\tau}{\tau^*}\right)^{1/m} \left(\frac{t}{T}\right)^{-n} \tag{5-42}$$

where $n \leqslant 1$; τ^* and T may be assumed to equal unity.

Eq. 5-42, on being integrated, yields (depending on the value of n) the eqs. 5-17, 5-20 and 5-27 examined above. By analogy with the exponent β in eq. 5-18, the exponent n in formula 5-42 is, strictly speaking, a variable. During shear tests of clay soil (W is 20–21%) by Feda et al. (1973), the value of n varied between 0.75 and 1.25

depending on the level of stress and increased directly with the stress. The authors attributed these changes to an increase in the number of distributed interparticle contacts; these may be estimated by statistical methods, using the Gaussian function of errors.

Changes in viscosity coefficient with time

Changes in strain rate with time can be explained by the changes in the viscous properties of a body which occur in the course of deformation. Accordingly, eq. 5-9 may be written in the form:

$$\dot{\gamma} = \frac{f(\tau)}{\eta(t)} \tag{5-43}$$

where $\eta(t)$ is a time-dependent viscosity coefficient.

In eq. 5-42, this coefficient is $\eta(t) = (t/T)^n$ or, in a more correct form:

$$\eta(t) = \eta_0 \left(\frac{t+T}{T}\right)^n$$

where η_0 is the initial viscosity (at $t = 0$).

The value of $\eta(t)$ has been examined by Kharkhutta and Ievlev (1961), assuming that $n = 1$. The authors have established that $\eta(t)$ may vary, depending on the soil, between $3 \cdot 10^7$ P for loose clay and $140 \cdot 10^7$ P for compact clay, while the value of T is more or less constant for all soils, being 0.5 s.

Another form of time-dependent viscosity coefficient suitable for use is the exponential relation suggested by Persoz (1953) and Maslov (1968):

$$\eta(t) = \eta_\infty - (\eta_\infty - \eta_0) e^{-t/T} \tag{5-44}$$

where η_0 and η_∞ are the initial (at $t = 0$) and final (at $t = \infty$) values of the viscosity coefficient, respectively, where $\eta_\infty \gg \eta_0$, and T is a parameter (measured in hours) which is defined by the relation:

$$T = t/\ln\left(\frac{\eta_\infty - \eta_0}{\eta_\infty - \eta}\right)$$

Both authors have adopted the function $f(\tau)$ in its linear form and Maslov has correlated eq. 5-43 with Bingham's condition, expressing 5-43 in the form:

$$\dot{\gamma} = \frac{\Delta\tau}{\eta_\infty - (\eta_\infty - \eta_0) e^{-t/T}} \tag{5-45}$$

where $\Delta\tau = \tau - \tau_{\lim}$ is the creep threshold.

By integrating expression 5-45, we arrive at the following equation of deformation:

$$\gamma = \gamma_0 + \frac{\Delta\tau}{\eta_\infty}\left[t + T \ln \frac{\eta_\infty - (\eta_\infty - \eta_0) e^{-t/T}}{\eta_0}\right] \tag{5-46}$$

It follows from eq. 5-46 that the strain rate decreases in time from an initial value $\dot{\gamma} = \Delta\tau/\eta_0$ to a finite constant value $\dot{\gamma}_\infty = \Delta\tau/\eta_\infty$, approaching the latter asymptotically. Accordingly, the deformation of creep, as defined by eq. 5-46, increases with time (at a steadily decreasing rate), being transformed for $t = \infty$ into a steady-state flow progressing at a constant rate, $\dot{\gamma}_\infty = $ const.

Note that all the above expressions, including eq. 5-45, are phenomenological equations because the time-dependent changes in the viscosity of soils they are describing have been derived empirically and not by examining the physical essence of the process.

5.5 EXPERIMENTAL DATA

Curves of soil creep

The classical creep curves illustrated in Fig. 5-1 are typical of most soils, as has been proved experimentally by various authors, i.e. curves of the attenuating character at low stresses and of the non-attenuating type at high stresses. All three stages of deformation: unsteady, steady-state, and progressive, are as a rule traceable on the non-attenuating creep curves. Creep curves of this kind are typical for all kinds of clay soils with a consistency varying from fluid to hard, irrespective of the type of test from which the curves have been plotted — shear, uniaxial compression, triaxial compression, etc.

Creep curves of plastic clays

Referring to Fig. 5-13, we can see the results of testing plastic clays for creep.

Fig. 5-13(a), from Vyalov and Pekarskaya (1965), refers to the tests on Jurrasic clay specimens from a Bat-baioss bed — an extremely fine and plastic soil containing 56% of particles less than 0.005 mm; its water content was $W = 32\%$ at $W_L = 48\%$ and $W_P = 26\%$. The specimens, hollow cylinders unsupported laterally, were tested for pure shear due to torsion under a constant shearing load constituting a certain fraction of the hypothetically instantaneous strength τ_s, i.e. the strength ascertained under a short-term load applied over a period of 0.5 to 1 min; $\tau_s = 0.167 \cdot 10^5$ Pa. A load $\tau \leq 0.575\,\tau_s$ brought about an attenuating deformation, the deformation resulting from a load $\tau \geq 0.575\,\tau_s$ was of the non-attenuating type and ended in failure. It is of interest to note that the transition from the attenuating to non-attenuating stage was caused by a load increment $\Delta\tau$ as small as $\Delta\tau = 0.015\,\tau_s = 0.025 \cdot 10^5$ Pa.

Fig. 5-13(b) represents the data acquired by Murayama and Shibata (1961) from testing samples of undisturbed alluvial clay with a high plasticity ($W = 65\%$, $W_L = 63-83\%$, 100% water saturation) from an Osaka bed. The clay samples were subjected to constant compressive loads taken between $0.9\,\tau_s$ and $0.63\,\tau_s$, where $\tau_s = 0.9 \cdot 10^5$ Pa. They all gave rise to non-attenuating creep with three stages of deformation distinctly discernible.

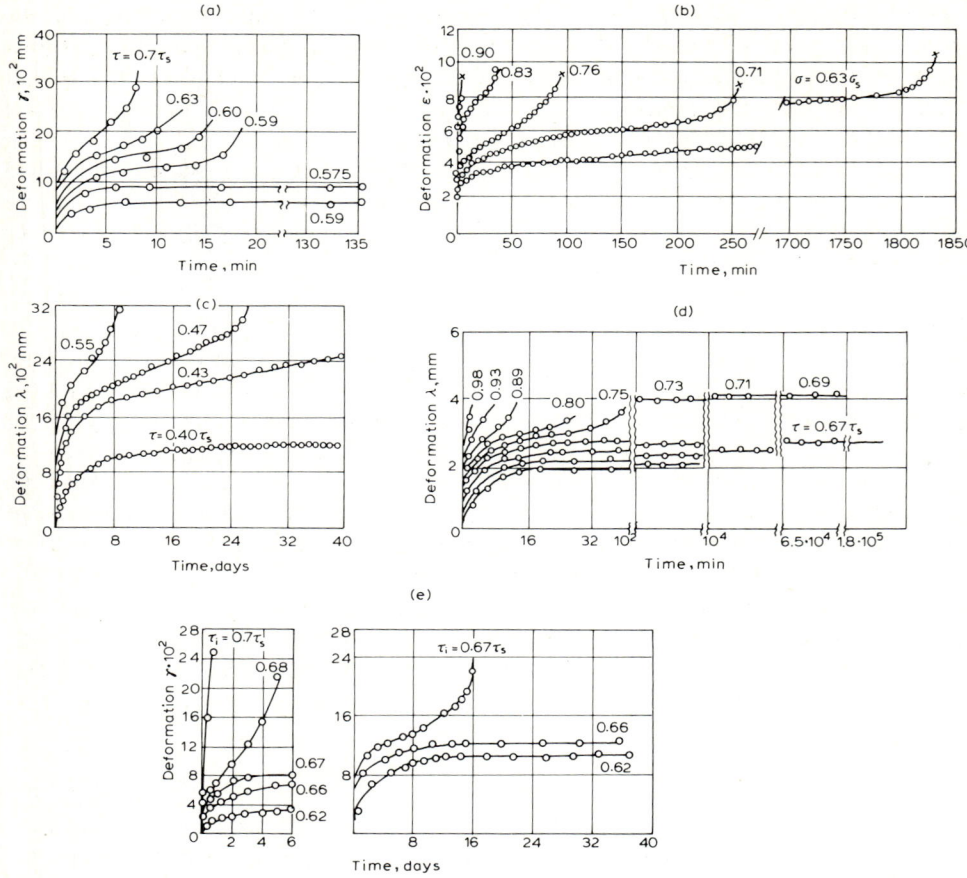

Fig. 5-13. Creep curves of plastic clays. (a) Jurassic clay, shear in torsion test. (b) Alluvial Osaka clay of undisturbed structure, compression test. (c) Clay, tests by means of shear–strain apparatus. (d) Loess soil, tests by means of shear–strain apparatus. (e) Compacted silt, triaxial compression test.

Fig. 5-13(c) depicts the results of the shear tests on clay soil conducted by O.G. Diasamidze (see Maslov, 1968), using a shear–strain apparatus. The shear load applied was between 40 and 55% of τ_s. The deformation induced by a load $\tau = 0.4\tau_s$ was of the attenuating nature; at $\tau > 0.4\tau_s$, a viscoplastic flow, eventually transformed into the failure-culminating progressive stage, was obtained.

The results of the tests of loess soil with an undisturbed structure carried out by Aliev (1972), are shown in Fig. 5-13(d). The loess with a natural moisture content $W = 9.3\%$ at $W_L = 31.2\%$ and $W_P = 20.8\%$ was soaked in water, preparatory to the tests, so as to become totally water-saturated and compacted by a load $\sigma_n = 1.0 \cdot 10^5$ Pa until stabilized. The tests, with the specimens submerged in water, were carried out on a shear–strain apparatus at $\sigma_n = 1.0 \cdot 10^5$ Pa, $2.0 \cdot 10^5$ Pa and

3.0 · 10⁵ Pa under various constant shear loads, τ = const. Fig. 5-13(d) presents the results of the test with σ_n = 1.0 · 10⁵ Pa, τ_s being 0.75 · 10⁵ Pa in this case. As can be seen, a non-attenuating creep was set up at $\tau > 0.73\,\tau_s$, at $\tau < 0.73\,\tau_s$ the derformation had an attenuating nature.

Winding up, Fig. 5-13(e) illustrates the results of triaxial compression tests of compacted silt (structure disturbed) conducted by Sorokina and Stroganov (1971). The silty soil as occurring in nature (W_L = 120% and W_P = 51%) with a humus content of up to 9% was precompacted under a load between 0.25 · 10⁵ and 3.0 · 10⁵ Pa and tested for creep under various stress intensities τ_i but at a constant mean normal stress σ_m. Fig. 5-13(e) represents the results obtained at σ_m = 1.0 · 10⁵ Pa and at a value of τ_i equalling a fraction of the hypothetically instantaneous strength τ_s = 1.16 · 10⁵ Pa. By analogy with the preceding examples, the creep of silty sand due to triaxial compression was — contingent on the value of stress deviator (or, to be exact, upon the relation between the stress deviator and all-around pressure) — of either attenuating or non-attenuating nature. In the latter case, the steady-state flow was transformed into progressive flow.

Creep curves of frozen soils. Fig. 5-14 presents the data acquired from testing frozen soils of various composition by loads of various kinds.

Fig. 5-14(a) depicts the results of testing frozen clay loam (W = 38%, θ = −0.°C) for shear along a rod frozen into each specimen (Vyalov, 1959; modelling the performance of a pile driven into a frozen ground) under constant loads each constituting a certain fraction of the hypothetically instantaneous strength τ_s. This strength, as determined from load applications lasting 3 s, was adopted at τ_s = 4.6 · 10⁵ Pa for the tests.

Fig. 5-14(b) displays the shear test data acquired by Pekarskaya (Vyalov et al., 1962) from experiments with Callovian sandy loam (W = 26%, θ = −10°C) under a normal stress σ_n = 20 · 10⁵ Pa, using a shear–strain apparatus. The hypothetically instantaneous shear strength, as determined from a 2-min load application, was τ_s = 36.2 · 10⁵ Pa.

Fig. 5-14(c) and (b) represent the data due to S.E. Gorodetsky and E.P. Shusherina (Vyalov et al., 1962) testing Callovian sandy loam and compact Bat-baioss clay (W = 22%, θ = −20°C) under uniaxial compression. The hypothetically instantaneous strength, corresponding to a 20-min test, was σ_s = 128 · 10⁵ Pa for the sandy loam and σ_s = 65.5 · 10⁵ Pa for the clay.

Frozen soils clearly display all three stages of creep. The transition from attenuating to non-attenuating creep is also distinct. It occurred under a stress amounting to between 30 and 50% of the hypothetically instantaneous strength in tests (b), (c) and (d) (Fig. 5-14) and at a 24% stress in test (a). Note that the non-attenuating creep of frozen soils subjected to tension took shape under a load amounting only to some 8–10% of the hypothetically instantaneous strength.

Characteristic of frozen soil deformation is a long unsteady stage I, lasting sometimes a good hundred hours or longer. The transition to the progressive stage takes place gradually, the process developing over a period measurable by hundreds, if not thousands, of hours. So, according to the experimental data depicted in

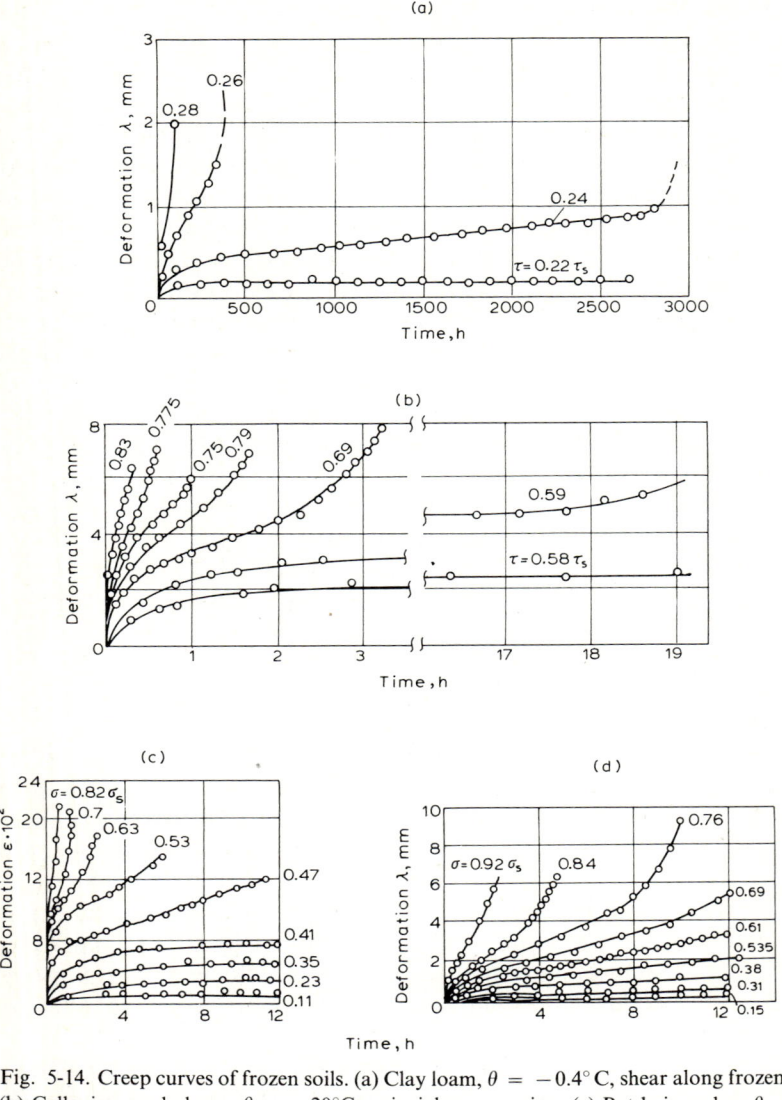

Fig. 5-14. Creep curves of frozen soils. (a) Clay loam, $\theta = -0.4°$ C, shear along frozen-in cylindrical rod. (b) Callovian sandy loam, $\theta = -20°$C, uniaxial compression. (c) Bat-baioss clay, $\theta = -20°$C, uniaxial compression.

Fig. 5-12(c), the 4,735-h total period broke down by stages as follows: stage I, around 1,000 h; stage II, 3,000 h; and stage III (ending at the instant of failure), 735 h. The corresponding deformation at each of the stages, taking the aggregate deformation at the time of progressive flow as 100%, was (according to the data given in Fig. 5-14(b)) as follows: instantaneous deformation, 2–7%; deformation due to attenuating creep (stage I), 49–73%; deformation of steady flow, 25–44%.

Steady and progressive flow in soils

It follows from the data presented above that a non-attenuating creep of soils leads quite frequently to a progressive flow at an increasing rate. However, there are also cases when a constant-rate steady flow develops during a considerable period without changing over into the progressive flow. Such were the results of tests by Geuze and Tan, Meschyan and others.

Experimental studies by Geuze and Tan (see Chapter 1: Geuze and Tan, 1954) were among the earliest attempts to probe into the creep of soils. Specimens in the form of hollow cylinders prepared from illite clay of plastic consistency (particles of a size less than 0.002 mm accounting for 50%; $W = 47.5\%$ at $W_L = 93.5\%$ and $W_P = 27.4\%$) were subjected to torsion. The resulting shear load amounted to between 20 and 80% of a short-term (10 min) strength at failure τ_s. Very soon (in 5–10 h) almost any load application induced a constant-rate steady flow.

The experiments by S.R. Meschyan (see Chapter 1: Meschyan, 1967), exemplified in Fig. 5-15(b), covered a by far longer period of 110 days. Clay loam specimens

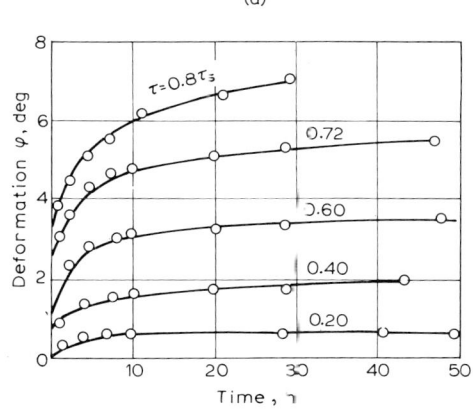

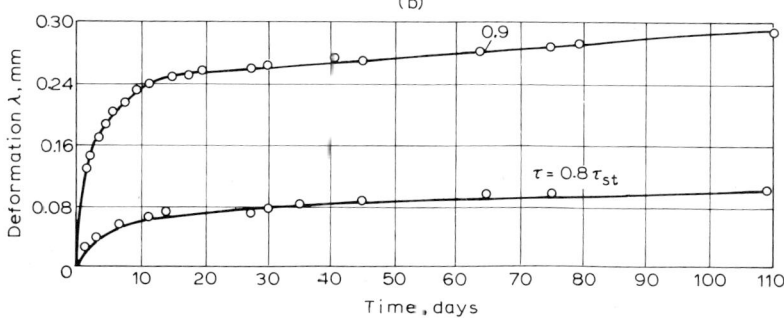

Fig. 5-15. Steady-state flow. (a) Plastic illite clay with disturbed structure, shear in torsion test. (b) Plastic clay with disturbed structure, tests in ring shear apparatus.

($W = 26.5\%$, $W_L = 31.3\%$, $W_P = 18.6\%$) were tested in a ring shear apparatus with a simultaneous application of a torque and an axial pressure ($\sigma_z = 2 \cdot 10^5$ Pa). The shear stress applied was between 80 and 90% of the 'standard' strength τ_{st}. This latter strength is understood as to be determined in accordance with the standard procedure of a slow shear test under the conditions of stabilized deformation at each loading stage as stipulated by USSR Standard GOST 12248-66. The test period is 1–8 days. The strength thus determined is 20% less than the hypothetically instantaneous value.

During the experiments in question it was found that the creep deformation resulting from a load $\tau < \tau_{st}$ was confined to either the attenuating or steady-state stages. This gave reason to be doubtful about the very existence of the progressive stage. It was said that the acceleration of deformation, noted in other experiments, might result from experimental conditions such as reduction in the cross-sectional area of specimen, stress concentrations, etc.

Let us scrutinize this problem. Way back when the results of the first experiments with frozen solids were published (Vyalov, 1955) the present author came up with the suggestion that the development of progressive creep leads to failure — a brittle failure entailing a loss of continuity or viscous failure with the loss of stability — contingent upon the properties of soil and not influenced by the conditions of the experiment.

Speculations about the effect of experimental conditions are groundless even in the light of the fact that the progressive flow is produced during tests of various type. This is proved by Figs. 5-13, 5-14 and the results of other experiments, including those causing no changes in the cross-sectional area of specimen (torsion of cylinders) and even experiments during which this area increases (uniaxial compression).

The physical background of soil deformation at an increasing rate and the failure of soil in creep under constant stress will be explained later by examining the microstructure of soil. Here it is necessary to point out that a soil subjected to creep gets simultaneously stronger and weaker, the loss of strength being attributed to a breakdown of structural bonds, triggering of microcracks, etc.

Steady flow results from a balance established between the broken and re-formed structural bonds. In lightly structured soils, this state may last indefinitely. In structuralized soils with rigid bonds, one of the phenomena will finally prevail with the result that the deformations will either attenuate or develop at an increasing rate. The predominance of the weakening effect over the strengthening effect is the cause of progressive flow and soil failure.

One needs to stress that the setting up of progressive flow may be handicapped in experiments conducted under conditions of restrained deformation, as is the case during tests conducted in the ring shear apparatus (Fig. 5-15(b)). Rigid grips holding fast the specimen in the apparatus or metal rings applied to the inside and outside surfaces of the specimen prevent its sidewise expansion. Since the development of progressive flow is associated with the phenomenon of dilatancy — an increase in the volume of specimen caused by microcracks in response to shear — it is obvious that the creation of conditions eliminating this phenomenon prevents the setting up of progressive flow.

This has been proved by the comparison torsion tests of cylindrical soil specimens conducted by S.R. Meschyan (see Chapter 1: Meschyan, 1967). To prevent the possibility of sidewise expansion, some of the specimens were tested while protected by a set of rings (functioning as grips) while the others were tested unprotected. The former specimens exhibited a steady flow at a constant rate developing over a long period of time and in the latter, the steady flow had transformed into a progressive flow ending in failure.

Time scale is a factor of paramount importance in assessing the configuration of creep curves. Offering a noteworthy example to that end are the results of the tests by Ter-Stepanyan published in 1973 and presented in Fig. 5-16 (Ter-Stepanyan et al., 1973). Subjected to testing in a ring shear apparatus was a sample of diatomite clay with a disturbed structure from a lake bed of Upper Miocene age ($W = 81.6\%$, $W_L = 117\%$, $W_P = 58.5\%$, $\gamma = 1.5 \text{ g/cm}^3$).

The creep curves obtained during the test, which lasted around 350 days, were plotted to different scales. Curve A refers to an initial 90-s interval, curve B covers a 10-min period, curve C depicts a 100-min one and so on, the last curve G reproducing the results of the 350-day period.

Examining the curve for each interval separately, one can single out the segments of attenuating deformation (before 10 to 20 days) and those of steady flow (after 25 to 30 days) on curve F. Assessing, however, curve G as a whole, one comes to the conclusion that starting from $t = 25$ or 30 days and onward there was a continuous stepwise increase in the strain rate attributed to structural changes in the soil.

Hence, one may conclude, as was said before, that an attempt to determine the nature of a deformation in the course of a short-term observation is likely to invite an error.

In a subsequent work Ter-Stepanyan (1977) suggested that the process of creep falls into two stages only: the stage of mobilization (of the resistance force) and the stage of failure. At the first stage, the deformation bears an attenuating nature and at the second stage it is of the progressive kind, being approximated — according to the author — by a logarithmic law ($\gamma \sim \ln t + \Delta t/t$) or a parabolic law ($\gamma \sim t^2$).

The problem arising due to the property of soils to flow at a truly constant rate requires a long lasting and scrupulous study. Below we shall show that stiff clays and rocks may be devoid of this property. In practical computations pertaining to plastic soils one may assume the existence of a range of loadings over which deformations develop slowly at a rate that changes little and can be regarded therefore as constant.

At the same time field observations indicate that steady flow seems to be transformed into progressive flow with time.

Typical in this respect is an incident involving long-term creep and failure of a retaining wall at Kensal Green described by Skempton (see Chapter 1: Skempton and Hutchinson, 1969). Erected in 1912 to hold back the slope of an excavation in stiff-fissured London clay ($W = 33\%$, $W_L = 83\%$, $W_P = 30\%$), the wall collapsed on two occasions, in 1929 and 1941. The rate of slope displacement, as recorded in the course of a few years immediately after the rebuilding of the wall in 1929, was 0.6 cm per year. Eventually, the displacement rate gradually rose and just before the second failure the total displacement amounted to 46 cm (Fig. 5-17).

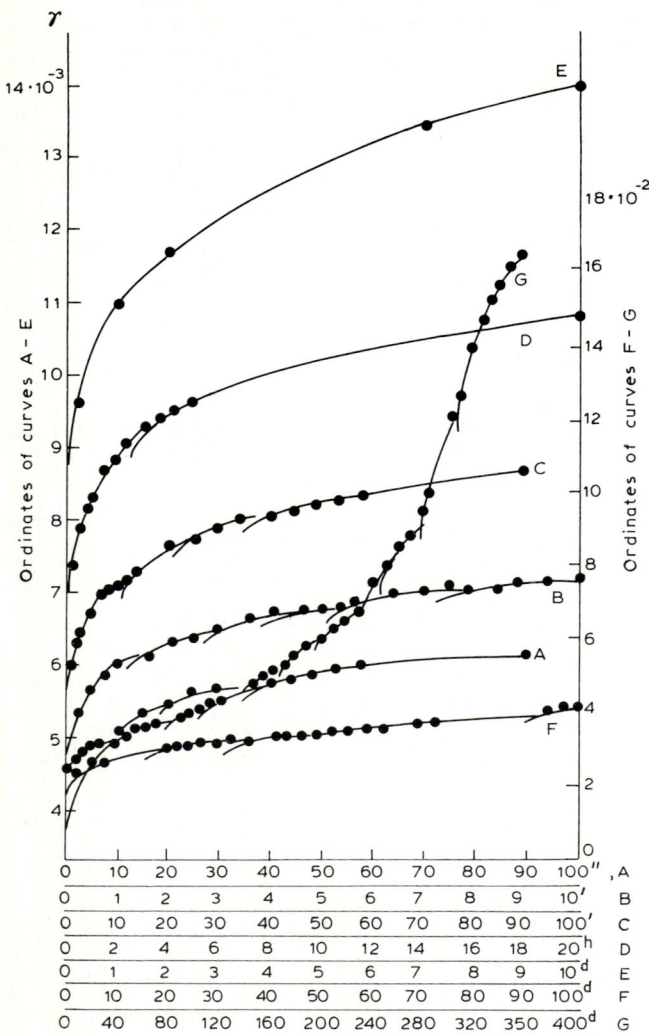

Fig. 5-16. Creep curves of diatomite clay plotted on different time scales. Tests in ring shear apparatus, $\sigma_z = 10^5$ Pa, $\tau = 0.4 \cdot 10^5$ Pa (after G.I. Stepanyan).

Reversing the reasoning, it was established that the shear stress on the plane of failure was $\tau = 0.19 \cdot 10^5$ Pa whereas the peak strength (for the same value of $\sigma_n = 0.39 \cdot 10^5$ Pa) was $0.3 \cdot 10^5$ Pa, and the residuary drag was $0.11 \cdot 10^5$ Pa. In other words, the actual stress was 63% of the peak (short-term) strength.

Creep curves of stiff clays

Consider the mechanism of creep in stiff, structuralized clays. Fig. 5-18(a) depicts

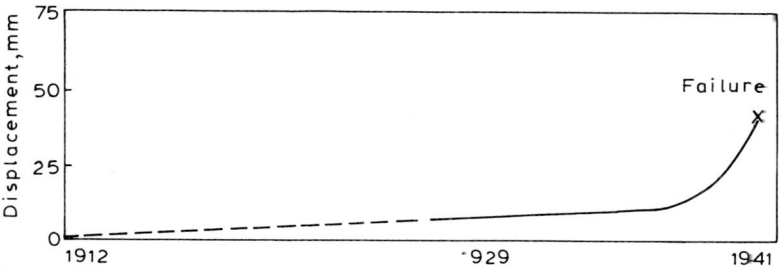

Fig. 5-17. Displacement of a retaining wall at Kersal Green (Great Britain) due to progressive long-term flow of soil.

experimental data, due to A.M. Skibitsky, obtained from Tertiary beidellite–montmorillonite clay of an undisturbed structure extracted from a depth of 50 m and which were published in 1957 (Vyalov and Skibitsky, 1957).

The soil composition was 37% silt and 62% clay, 70% of this latter fraction being colloid particles. The natural moisture content was $W = 33\%$, which corresponded to a plastic limit $W_P = 34.4\%$ and a liquid limit $W_L = 73.5\%$. The experiments were conducted on a shear–strain apparatus at $\sigma_n = 7 \cdot 10^5$ Pa; the shear stress was applied in fractions of the short-term (1 h) strength $\tau_s = 2.75 \cdot 10^5$ Pa.

The deformations set up by the shear loads τ varying between 0.3 τ_s and 0.7 τ_s over the test period of 2.5 months were of attenuating nature, stabilizing towards the end. A stress $\tau = 0.9 \tau_s$ induced a failure in 32 days preceded by an attenuating deformation obeying a logarithmic law. No steady flow was observed before a speedy brittle failure.

Similar experimental results were obtained by Bishop and Lovenbury (1969) (Fig. 5-18(b)). Undisturbed samples of fissured London clay in the form of heavily overcompacted Eocene deposits were subjected to 3.5-year tests (the natural compacting load had been estimated to be as high as $30 \cdot 10^5$ Pa). The clay particle content was 58%, the natural moisture content was $W = 29.3\%$, at $W_P = 29\%$ and $W_L = 76\%$. The test tool was a triaxial compression apparatus (with provision for draining) operating at various values of $\sigma_i = \sigma_1 - \sigma_3$ and a constant value of $\sigma_3 = \sigma_2 = 1.41 \cdot 10^5$ Pa.

The short-term (peak) strength, as determined by a standard 5-day procedure preparatory to the experiments, appeared to be $\sigma_{i(s)} = \max(\sigma_1 - \sigma_3) = 2.27 \cdot 10^5$ Pa. The values of σ_i used during the creep tests were fractions of $\max(\sigma_1 - \sigma_3)$. The graph presents only the maximum and minimum values of σ_i during six repetitive tests.

The creep deformation induced by all loadings was a logarithmically attenuating one. However, at loads varying between 90 and 106% (e.g., 100% on the average) and between 80 and 94% (89% on the average) the deformation ended in a failure of the specimen. This took place 2 days after the load application in the former case and in 1,250 days in the latter case, in either case the rate of deformation being slightly

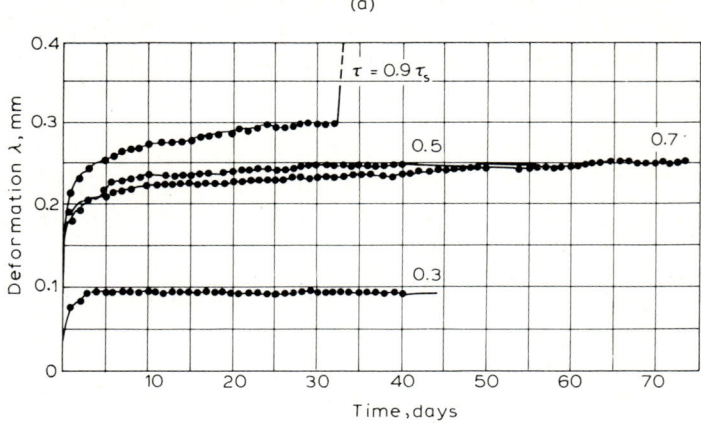

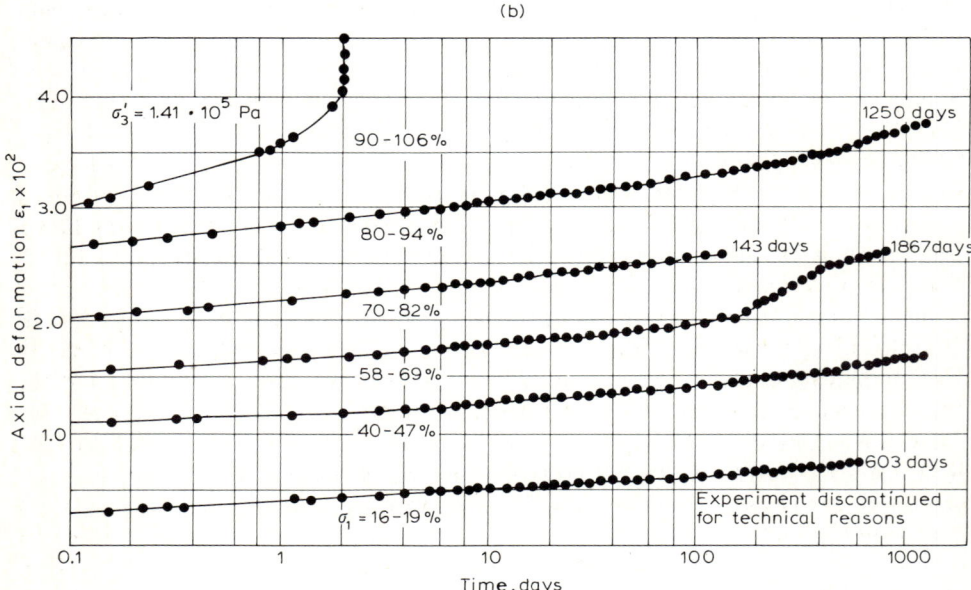

Fig. 5-18. Creep curves of stiff clays. (a) Tertiary montmorillonite clay with undisturbed structure ($W = 33\%$, $W_p = 34.4\%$), tests by means of shear–strain apparatus. (b) Eocene London clay with undisturbed structure ($W = 29.3\%$, $W_p = 29\%$), triaxial compression tests with provision for draining.

variable before failure. A conspicuous stage of progressive flow preceded the failure in both cases. A steady flow was absent as in the experiments represented in Fig. 5-18(a).

A difference in the character of deformation depending on the level of stress was also observed in the experiments by Ecu and Grisolia (1977) who tested over-compacted clay for long-term (up to 100 days) creep in shear. The tests, called for by the necessity to estimate the stability of an excavation varying in depth between 25 and 100 m, revealed an attenuating creep for $\tau/\tau_s < 0.4$, a steady flow for $0.4 < \tau/\tau_s < 0.9$, and a progressive flow ending in failure for $\tau/\tau_s > 0.9$ (where τ_s was the short-term strength).

Creep curves of soft and hard rocks

Deformation without the stage of steady flow is a salient feature of most rocks, soft and hard alike (irrespective of the test duration). Although deforming the attenuating creep style, these rocks are likely to fail eventually under a load close to the instantaneous strength.

A wealth of information pertaining to the creep of rocks of various hardness is available nowadays (Erzhanov, 1970, referenced in Chapter 1). Creep curves from this reference and other works (Problems of Rheology of Rocks, 1970) appear in Fig. 5-19.

Fig. 5-19(a) depicts the data acquired by Kartashov (1970), from testing siltstone marl ($W = 26\%$, $\varrho = 1.89 \text{ g/cm}^3$) which may be treated as either a soft rock or a very stiff clay. The load applied during uniaxial compression tests was 20–70% of the hypothetically-instantaneous strength, $\sigma_s = 30 \cdot 10^5$ Pa. The loads amounting to $0.6\,\sigma_s$ brought about stabilized deformation, $\sigma_z = 0.7\,\sigma_s$ caused rapid failure.

Erzhanov (1970, see Sect. 1.5) tested siltstone ($\varrho = 2.61 \text{ g/cm}^3$, $W = 1\%$) for creep by applying bending loads constituting 20–73% of the hypothetically-instantaneous strength to specimens in the form of $20 \times 20 \times 160$ mm beams. All the tests ended in an attenuating creep deformation (Fig. 5-19(b)) amounting after an accumulation for 48 h to 340% of the hypothetically-instantaneous elastic deformation.

It will be noted that the creep deformation of argillite was 320% of the elastic deformation and that of sandstone up to 40%. According to Glushko et al. (1970, see Sect. 1.5), the deformation of creep in sandstones is 8–44% of the elastic deformation. The recoverable deformation observed on removal of load was never higher than 8% of the aggregate deformation suffered by the specimen.

The above figures demonstrate the significance of creep in the process of rock deformation. In either of the above experiments the stress–strain relation was linear as long as the stress remained within the confines of 50–70% of the stress at failure.

A similar picture was observed during the uniaxial compression tests on siltstone, argillite and fine-grained sandstone cemented with clay (up to 15%) by N.F. Arenzhigalov (1968). The results of testing sandstone in water-saturated and air-dry states are represented in Figs. 5-19, (c) and (d). The deformations were all of the attenuating character, those of the water-saturated specimens being roughly twice as large as the deformations of the air-dry specimens.

Some rocks do not fail directly due to creep, displaying before failure a progressive stage as dispersed soils.

This fact is represented in Fig. 5-19(e) by a creep curve plotted by Shiforenko and Oxenkrug (1973) from the uniaxial compression tests of rock salt having a medium-crystalline structure. All loads between 50 and 80% of the hypothetically-instantaneous

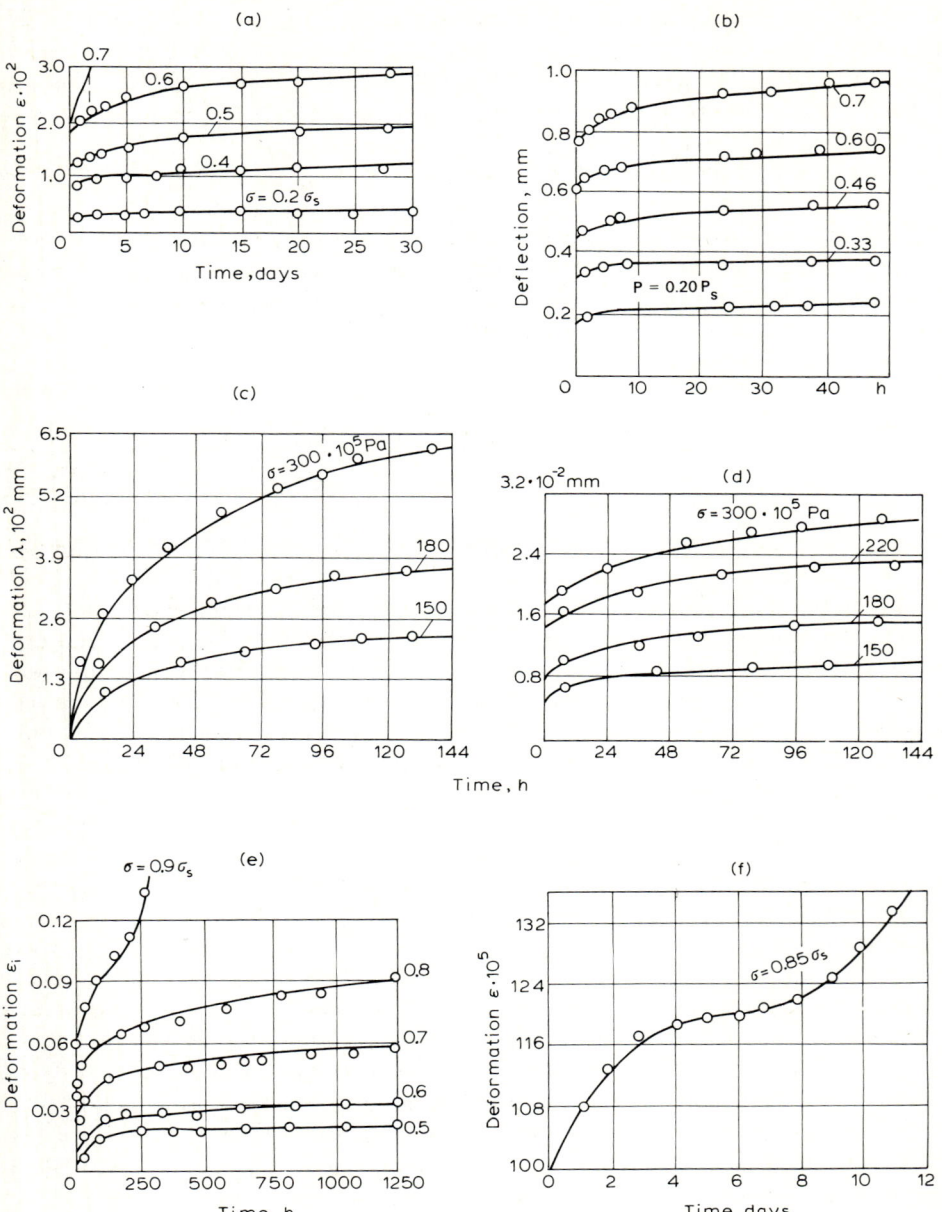

Fig. 5-19. Creep curves of soft and hard rocks. (a) Uniaxial compression tests of siltstone marl. (b) Bending tests of beam-shaped siltstone specimens. (c) Uniaxial compression of fine-grained sandstone in a water-saturated condition. (d) Uniaxial compression test of fine-grained sandstone in air-dry condition. (e) Uniaxial compression of rock salt with medium-crystalline structure. (f) Uniaxial compression of sandstone.

strength caused attenuating deformations, the load $\sigma_z = 0.9\,\sigma_s$ resulted in failure of the specimen in 250 h.

The development of a stage at failure is also illustrated by the creep curve in Fig. 5-19(f) due to V.T. Glushko testing a clay-cemented Donbass sandstone under a load amounting to 85% of the instantaneous load at failure $\sigma_s = 200 \cdot 10^5$ Pa (see Chapter 1: Glushko et al., 1970).

It follows from the above data that the failure of soft and even of some hard rocks is preceded by progressive flow (stage III) which may develop over a period of 10 days or longer. Apparently this is the property of rocks either cemented by means of a viscous material (e.g. sandstone cemented with clay) or of those having a crystalline matrix displaying rheological properties (e.g. rock salt).

5.6 SIMILARITY OF THE PATTERNS OF SOIL DEFORMATION

Dependency of rheological properties of soils on the nature of structural bonds

The data examined above provide evidence that there is inherent in all soils, from soft clay to hard rock, a similar pattern of rheological behaviour.

Contingent on load, type of soil and arrangement of stress, there may develop an attenuating creep ending in a stabilized deformation, a "secular" creep characterized by an unconfined deformation developing, however, at a slowing-down rate without failure, an hypothetically steady creep in the form of a viscoplastic flow, and, finally, a progressive creep leading to failure. Whichever of the creep stages will be the dominating one, the effect of each stage will depend to a considerable extent on the type of soil.

Pekarskaya (1976), credited with carrying out an interesting analysis of the patterns of soil behaviour in the light of structural bonding, has come to the conclusion that the nature of the deformation process is defined before all by the type of structural bonds. So, soft clays with predominantly water-colloidal bonds (coagulation structures) display all stages of deformation of the classical creep curve. The softer the soil, the sooner begins the stage of steady flow and the more significant is the effect of this stage on the process of deformation. In soils of a fluid consistency, the steady flow dominates the process and the soils themselves display properties resembling those of viscous liquids. In soft soils with water-colloidal bonding, the progressive flow is induced by loads amounting to 70–80% of the short-term strength and develops over a long period, ending in failure of the viscous kind as a rule.

In stiff clay soils with both coagulation and crystallization bonds, the stage of attenuating deformation dominates. The progressive stage is caused by loads constituting at least 70–80% of the short-term strength, and the transition from attenuating deformation to progressive flow omits the stage of steady flow in most cases. The progressive stage is commonly comparatively short, and the failure is predominantly brittle.

Similar patterns of behaviour are inherent in soft rocks and some harder rocks with crystallization bonds. However, the behaviour of these materials is decided mainly by the nature of the cementing bonds. A rock with a cementing substance possessing explicitly rheological properties will also behave itself rheologically.

A typical example of such rocks is frozen soil cemented with ice — a classical non-linearly viscous material. The behaviour of a frozen soil is represented by the typical creep curve having all the stages characteristic of classically creeping media and showing a transition into the stage of progressive flow in response to loads of only 10–25% of the instantaneous strength. Note that metals at high temperatures are as a rule also capable of developing a non-attenuating creep in response to a minimum load.

Another representative of rocks cemented with a viscous material is clay-cemented sandstone. Since the cementing clay is by far less viscous than ice, dominating in this case is the stage of attenuating deformation, the stage of steady flow being short, if present at all. The progressive stage in such sandstone is only likely to exist due to a high load.

The behaviour of the rocks cemented with a material displaying poor viscous properties is approaching that of brittle bodies. However, in this case the properties of the crystals the rock is composed of are to be taken into account. It is known that crystals with predominantly covalent bonds are reluctant to deform plastically due to the sliding of atomic layers in single crystals. Such crystals deform only elastically and show brittle failure.

Crystals with ionic bonds behave themselves on brittle lines in response to short-term load application. However, they display viscous properties in sustaining a long-term load. This is exemplified by rock salt whose creeping properties were determined in Weinberg's experiments way back in 1903.

In concluding it must be pointed out that, as far as the soils referred to are concerned, the deformations developing with time should be the chief consideration because they exceed the instantaneous deformations many times.

Creep of soil skeleton

In clay soils, creep is attributed not only to the presence of the water films surrounding solid particles. Of course, these films that display explicitly viscous properties significantly facilitate the development of rheological processes in water-saturated soils with water-colloidal bonding. However, the condensation bonds set up between the particles due to the short-range intermolecular forces during the "dry" contacts also offer a viscous resistance to particle displacements. This implies that soils with condensation bonds also display creep.

Depicted in Fig. 5-20 are the results of the compression tests of clay reduced to powder (62% of silty particles and 37% of particles less than 0.002 mm; $e = 2.83$) carried out by L. Šuklje (see Chapter 1: Šuklje, 1969). Each specimen was tested twice, i.e. in a dry state and on being totally wetted. From the graph, presenting the results obtained due to a load increment from $0.27 \cdot 10^5$ Pa to $0.5 \cdot 10^5$ Pa, we can see that the deformation of both wet and air-dry specimens developed in time. In either case it obeyed a logarithmic law.

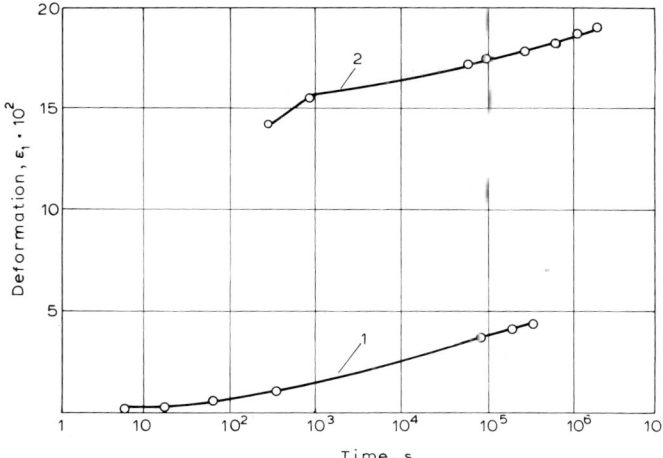

Fig. 5-20. Creep curves obtained from compressing dried clay soil without provision for lateral expansion in an oedometer: 1 = in dry state; 2 = after wetting.

Similar results were obtained in 1968–1969 by Singh and Mitchell (1968, 1969) from the triaxial compression tests of illite clay in both wet ($W = 34\%$) and air-dry conditions. The applied loads $\sigma_1 - \sigma_3$ (σ_3 = const.) were fractions of the long-term strength. In both cases, wet and dry specimens being implied, there was noted an appreciable creep. In particular, a load $\sigma_1 - \sigma_3$ amounting to 60% of the short-term strength failed to attenuate the deformation even after a 24-h test period.

A point to be noted is that in testing clay powder and illite clay, the deformations of creep induced in the wetted and dried specimens have obeyed the same law, a logarithmic law in the former case and a power law in the latter case. Although the deformation suffered by the wetted soil has been by far greater in its absolute value than the deformation of the dried soil, the intensity at which both deformations have been developing is roughly the same for both specimens: the slope of the ε_z versus $\ln t$ straight lines in Fig. 5-20 is almost the same.

Thus, we come to the conclusion that the creep of dispersed soils is attributed to the viscous resistance offered by the long-range forces of the double layer which are set up between the particles surrounded by the film of bound water and by the short-range forces arising at the points of direct contact between the particles. The mechanism of creep in soils with coagulation, condensation and crystallization bonds may be described by the same relationship given by eq. 5-42:

$$\dot{\gamma} = A\left(\frac{t}{T}\right)^{-n}$$

where the parameter n is the same for soils in the wetted and dried states. This is evidence that the rheological processes taking place in soils are all of the same character.

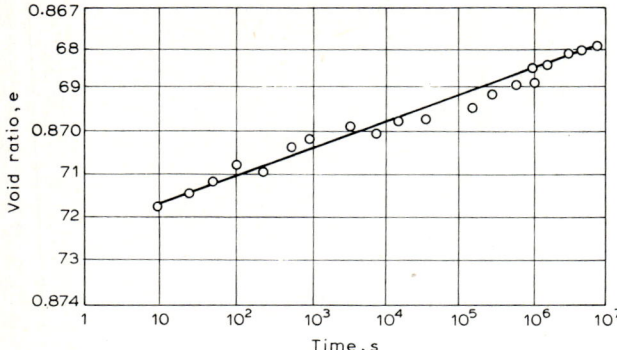

Fig. 5-21. Creep curves obtained from compressing dry fine-grained sand without provision for lateral expansion in an oedometer (see Chapter 1: Šuklje, 1973).

Creep of sand

Fig. 5-21 depicts the results of Šuklje's tests of a dry fine-grained sand (particles of a size less than 0.006 mm, 1%; between 0.006 and 0.2 mm, 50%; 0.2–0.6 mm, 49%; original void ratio $e = 0.8896$) for uniaxial compression under a constant load without provision for lateral expansion (see Chapter 1: Šuklje, 1969). From the curve representing the deformation produced by an increment in load from 10^5 Pa to $2 \cdot 10^5$ Pa, one can see that the property of creep is also inherent in sand — the deformation of compaction developing with time in accordance with a logarithmic law. Identical results were obtained by Nonveiller (1963) in experiments with fine-grained sand soaked with water.

The creep property of sand may be witnessed in everyday life, for example, as the gravity-induced trickling of the sand in an hourglass, the migration of barkhans across a desert, etc. Sand may be treated as if it were a large-scale model of liquid molecules arranged at random and bonded loosely. By analogy with this liquid, soil obeys the statistical law of viscous flow. However, in sand this process is less conspicuous than in dispersed soils.

Creep of soils in compression and tension

It was said in Chapter 2 that soils resist the deformations of compression and tension in different ways. This property of soils is evident also in the case of creep.

Fig. 5-22 illustrates the data acquired by Grechishchev (1963) from the uniaxial compression and tension tests on frozen clay loam (W is 29–33%, $\theta = -3°C$) under constant and stepped loads including a deformation of creep at a constant flow rate. From the graphs of the rate of flow vs. stress one can see the difference between the curves in compression and those in tension. Although both these curves satisfy the

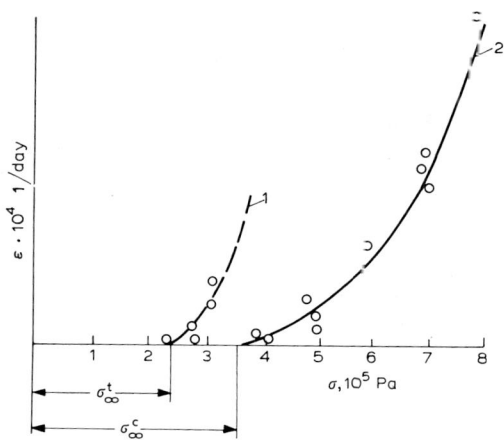

Fig. 5-22. Rheological curves obtained from (1) compressing and (2) tensioning frozen clay loam, $\theta = -3°C$ (S.E. Grechishchev's experiments).

same equation:

$$\dot{\varepsilon}_z = \frac{1}{\lambda}(\sigma_z - \sigma_\infty)^{1/m} \tag{5-47}$$

the coefficient of plastic viscosity λ and the ultimate stress σ_∞ have different values in compression and tension.

Similar results were obtained from testing frozen sand (W is 18–24%, $\theta = -3°C$). Simultaneous pure shear tests of the same sand in twisting, using specimens in the form of hollow cylinders, yielded results almost coinciding with the tension test data but appreciably differing from the compression test data. The values of the parameters of eq. 5-47 are given in Table 5-I (note that σ_∞ is in 10^5 Pa).

The exponent, $1/m$, appeared to be the same in the case of compression, tension and shear, implying that the rheological curves are all identical in the above cases.

Proceeding from eq. 4-22, one can establish a connection between the ultimate strengths in shear τ_∞, compression σ_∞^c, and tension σ_∞^t. To that end, substitute into

TABLE 5-I

Experimental values of parameters in eq. 5-47

Soil	Compression			Tension			Shear		
	σ_∞	λ	$1/m$	σ_∞	λ	$1/m$	σ_∞	λ	$1/m$
Frozen sand	6.5	0.4	1.0	1.8	0.10	1.0	1.7	0.05	1.0
Frozen clay loam	3.6	1.2	1.64	2.5	0.67	1.69	–	–	–

formula 4-22 the values of τ_i and σ_m for the case of compression and tension, as taken from Table 3-I: $\tau_i^c = (1/\sqrt{3})\sigma_z^c$; $\tau_i^t = (1/\sqrt{3})\sigma_z^t$; $\sigma_m^c = (1/3)\sigma_z^c$; $\sigma_m^t = -(1/3)\sigma_z^t$. Thus, we obtain:

$$\left(\frac{1}{\sqrt{3}} + \frac{\tan \Psi}{3}\right)\sigma_\infty^c = \left(\frac{1}{\sqrt{3}} - \frac{\tan \Psi}{3}\right)\sigma_\infty^t = \tau_\infty$$

Hence:

$$\sigma_\infty^t = \frac{3\tau_\infty}{\sqrt{3} + \tan \Psi} \qquad \sigma_\infty^c = \frac{3\tau_\infty}{\sqrt{3} - \tan \Psi}$$

$$\tan \Psi = \frac{\sqrt{3}(\sigma_\infty^c - \sigma_\infty^t)}{\sigma_\infty^c + \sigma_\infty^t} \qquad \tau_\infty = \frac{2\sigma_\infty^c \sigma_\infty^t}{\sqrt{3}(\sigma_\infty^c + \sigma_\infty^t)} \qquad (5\text{-}48)$$

Substituting expressions 5-48 into eq. 5-47, we can establish the connection between the coefficients of plastic viscosity in shear η, compression λ^c and tension λ^t:

$$\lambda^t = \frac{3^{2+m/2m}}{(\sqrt{3} + \tan \Psi)^{1/m}} \qquad \lambda^c = \frac{3^{2+m/2m}}{(\sqrt{3} - \tan \Psi)^{1/m}}$$

$$\eta = \frac{2^{1/m}\lambda^c \lambda^t}{3^{1+m/2m}[(\lambda^c)^m + (\lambda^t)^m]^{1/m}} \qquad (5\text{-}49)$$

This comparatively simple form of connection holds for $m^c = m^t = m^{sh} = m$.

The data presented above refer to the stage of steady flow. The question arising then is what will be the relation between the strain and strain rates in shear, compression and tension in the course of the process as a whole, including the stage of non-steady flow? In other words, will the time functions $\Phi(t)$ and $K(t)$ in eqs. 5-8 and 5-9 retain their similarity?

If we proceed from eq. 5-47, as was done by Grechishchev (1963), assuming, however, that the parameters λ and σ_∞ are time-dependent variables, we obtain:

$$\lambda^{c,t}(t) = \frac{1}{K^{c,t}(t)} = \frac{\lambda_0^{c,t}}{1 + \xi_1^{c,t}(t)}$$

$$\eta(t) = \frac{1}{K^{sh}(t)} = \frac{\eta_0}{1 + \xi_1^{sh}(t)}$$

$$\sigma_\infty^{c,t}(t) = \frac{\sigma_\infty^{c,t}}{1 + \xi_2^{c,t}(t)} \qquad (5\text{-}50)$$

$$\tau_\infty(t) = \frac{\tau_\infty}{1 + \xi_2^{sh}(t)}$$

where $\xi(0) = 0$; $\sigma_\infty(0) = \sigma_\infty$; $\tau_\infty(0) = \tau_\infty$.

If the time functions $\xi(t)$ for loadings of all kinds are the same:

$$\xi_{1,2}^c(t) = \xi_{1,2}^t(t) = \xi_{1,2}^{sh}(t) = \xi(t) \qquad (5\text{-}51)$$

the coefficients of viscosity are connected by the following relation:

$$\lambda^{c,t} = \frac{3^{2+m/2m}\eta_0}{(\sqrt{3} \mp \tan \Psi)^{1/m}[1 + \xi(t)]} \quad (5\text{-}52)$$

provided:

$$\left[\frac{\lambda^c(t)}{\lambda^t(t)}\right]^m = \alpha = \text{const.}$$

Experiments have proved that eq. 5-51 holds for frozen sand ($m = 1$). In the case of frozen sandy loam, the value of α varies, being $\alpha = 1.1$ at $t = 1$ day; $\alpha = 1.25$ at $t = 5$ days, and $\alpha = 1.4$ at $t = \infty$. This is an indication that eq. 5-51 fails to hold for some soils. However, it significantly simplifies the pattern of deformation, rendering the above assumption justifiable in many cases.

Longitudinal and lateral deformations of creep

As has been pointed out in analyzing 4-45, the coefficient of lateral expansion $v = \varepsilon_x/\varepsilon_z$ will be constant under conditions of non-linear deformation only then when the pattern of shearing deformation and that of the deformation due to an all-around pressure are described by the same functions.

It has also been shown that the coefficient of lateral expansion $v = \dot{\varepsilon}_x/\dot{\varepsilon}_z$ will be constant under the conditions of viscous flow when the coefficients of viscosity in shear and that in volumetric strain remain either constant or change with time in accordance with the same law. This condition is satisfied as long as the relation between the rates or longitudinal and lateral strains is constant in the course of flow.

The above is true in determining the coefficient v under the conditions of creep. Apparently, this coefficient may be expressed in terms of eqs. 5-8 and 5-9 as:

$$v = \frac{f(\sigma_x)\Phi_x(t)}{f(\sigma_z)\Phi_z(t)} = \frac{f(\sigma_x)\kappa_x(t)}{f(\sigma_z)\kappa_z(t)} \quad (5\text{-}53)$$

The coefficient v will be constant only when both the stress functions $f(\sigma)$ and the time functions $\Phi(t)$ or $\kappa(t)$ for the longitudinal (z) and the lateral (x) deformation are identical; for example, if both deformations may be described by eq. 5-18 using the same exponents.

Consider the actual conditions with reference to the experiments by Shusherina (1966), devoted to the creep of frozen sandy loam ($W = 26\%$, $\gamma = 1.81 \text{ g/cm}^3$) and stiff Bat-baioss clay ($W = 23\%$, $\gamma = 2.07 \text{ g/cm}^3$) in uniaxial compression at the temperatures of -5, -10, and $-20°C$. The longitudinal and lateral strains, as measured during the tests, were developing obeying different laws. Consequently, the coefficient of lateral strain changed both in time and with the stress applied.

The results of one test are shown in Fig. 5-23. As can be seen, the coefficient v suffers appreciable changes, starting from the initial period, i.e. the stage of unsteady flow. Particularly notable are the changes at the stage of progressive flow — the higher

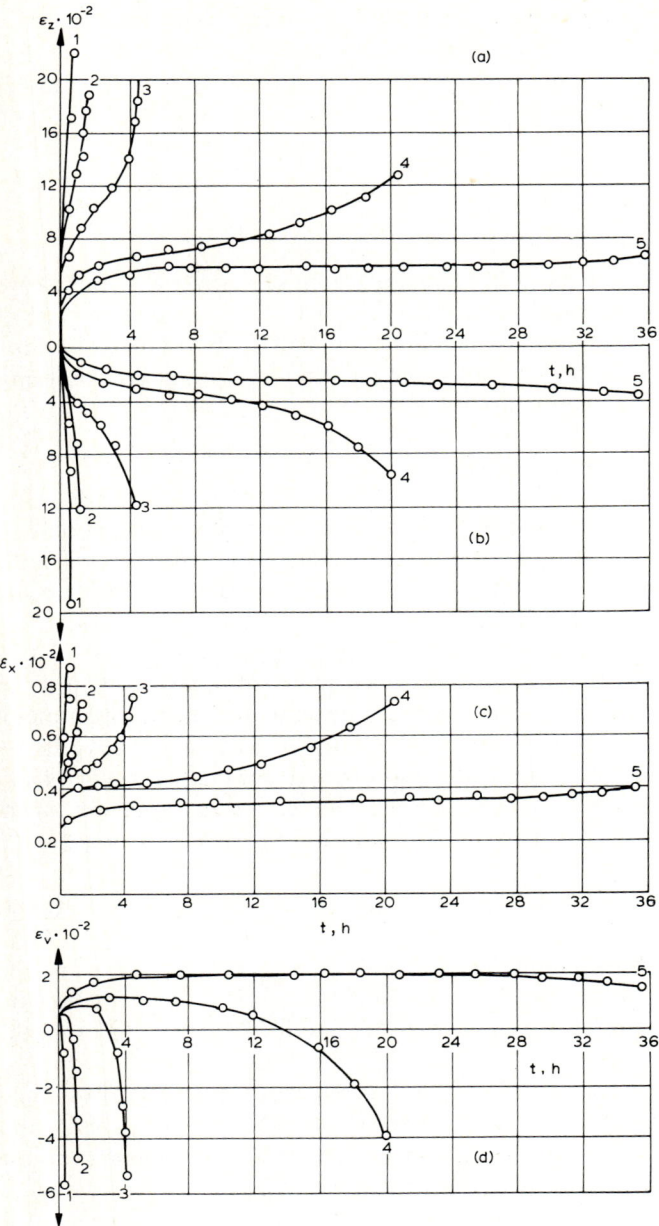

Fig. 5-23. Changes in (a) longitudinal strain, (b) lateral strain, (d) volumetric strain and (c) coefficient of lateral strain during creep (after E.P. Shusherina). Uniaxial compression of frozen sandy loam, $\theta = -5°C$ at the following values of σ_2: $1 = 40 \cdot 10^5$ Pa; $2 = 35 \cdot 10^5$ Pa; $3 = 30 \cdot 10^5$ Pa; $4 = 25 \cdot 10^5$ Pa; $5 = 22.5 \cdot 10^5$ Pa.

the applied stress, the more pronounced being the changes. At the stage of steady flow the high value of v remains more or less constant, being influenced, however, by the magnitude of load. In general, for frozen sandy loam, the coefficient varies between 0.18 at the beginning of the process and 0.77 at the end; in the case of frozen clay, the range of change is 0.14–0.73.

Identical results were obtained by Grechishchev from the experiments with frozen sand and frozen clay loam referred to above. The values of v, as obtained from the measurements of longitudinal and lateral deformations, increased from 0.31 at the first day of the process to 0.53 at the fifth day in the case of clay loam and from 0.37 to 0.55 over the same pereiod of tests with the sand.

Noteworthy is the fact that the value of v may be greater than 0.5. The explanation of this is as follows. Since the volumetric strain is expressed by:

$$\varepsilon_v = \varepsilon_z(1 - 2v) \tag{5-54}$$

the expression $v > 0.5$ implies an increment in the volumetric strain, i.e. in increase in volume. This fact was directly recorded during the experiments considered above (Fig. 5-23). The explanation of the increase in volume by the development of microcracks in soil, offered by Shusherina, seems to be plausible.

Note that eq. 5-54 holds for small strains only. In the case of larger strains, it is transformed into:

$$\varepsilon_v = 1 - (v\varepsilon_z + 1)^2(1 - \varepsilon_z)$$

where the choice of the $+$ or $-$ sign before ε_v is decided by both the value of v and the magnitude of ε_z.

The next chapter will be devoted to examining, in the light of experimental data, the assets and limitations of the empirical formulae considered in the opening sections of this chapter. Recommendations helpful in selecting the right formula will also be found in the next chapter.

5.7 REFERENCES

Aliev, S.K., 1972. On long-term strength of saturated loess soils. In collective volume: Problems of Soil Mechanics. Baku (in Russian).
Andrade, E.N., 1910. On the Viscous Flow of Metals and Allied Phenomena. Proc. R. Soc. Lond., 84, (567).
Andrade, E.N., 1962. The validity of the $t^{1/3}$ law of flow of metals. Phil. Mag., 1 (84).
Arenzhigalov, N.F., 1968. Studies on Rheological Properties of Soils. Proc. Coordinating Conf. Hydraul. Eng., Issue 38. Energiya, Moscow (in Russian).
Arutyunyan, N.Kh., 1952. Some Problems in the Theory of Creep. Gostekhteorizdat, Moscow (in Russian).
Bishop, A.W. and Lovenbury, H.T., 1969. Creep characteristics of two undisturbed clays. Proc. 7th ICSMFE, Mexico City, Vol. 1.
Buisman, K.A.S., 1936. Results of long duration settlement tests. Proc. 1st ICSMFE, Cambridge, Vol. 2.
Ecu, F. and Grisolia, M., 1977. Creep characteristics of an overconsolidated clay. Proc. 9th ICSMFE, Tokyo, Vol. 1.

Feda, J., Kamenov, B. and Klablena, P., 1973. Investigation of creep and structure of clayey materials. Proc. 8th ICSMFE, Moscow, Vol. 1.

Goldstein, M.N., Babitskaya, S.S. and Mizyumsky, V.A., 1962. Methodology of testing soils for creep and long-term strength. In collective volume: Problems of geotechnique, No. 5. Dnepropetrovsk (in Russian).

Gor'kova, I.M., 1965. Structural and Deformation Peculiarities of Sedimentary Rocks. Nauka, Moscow (in Russian).

Grechishchev, S.E., 1963. Creep of frozen soils in the state of combined stress. In collective volume: Strength and creep of frozen soils. USSR Acad. Sci., Moscow (in Russian).

Kartashov, Yu.M., 1970. Accelerated tests of weak rocks for creep. In collective volume: Problems of Rock Rheology. Naukova Dumka, Kiev (in Russian).

Kharkhutta, N.I. and Ievlev, V.M., 1961. Rheological Properties of Soils. Avtotransizdat, Moscow (in Russian).

Kohlrausch, F., 1863, 1866, 1877. Experimentale Untersuchungen über die elastische Nachwirkung bei der Torsion, Ausdehnung and Biegung. Pogg. Ann., Bd. 119, 1863; Bd. 128, 1866; Bd. 160, 1877.

Maslov, N.N., 1955. Condition of Stability of Artificial and Natural Slopes in Hydro Power Projects. Gosenergoizdat, Moscow (in Russian).

Maslov, N.N., 1968. Long-Term Stability and the Displacement Deformation of Retaining Walls. Energiya, Moscow (in Russian).

Maslov, N.N., and Kodzhemanov, K.T., 1970. Some problems in forecasting the settlement of structures built in nonsaturated soils. SMFE, 4 (in Russian).

Murayama, S. and Shibata, T., 1961. Rheological properties of clays. Proc. 5th ICSMFE, Paris, Vol. 1.

Nichiporovich, A.A. and Tsybulnik, T.I., 1961. Forecasting Settlements for Water-Development Projects in Cohesive Soils. Gosstroiizdat, Moscow (in Russian).

Nonveiller, E., 1963. Settlement of a grain silo on fine sand. Proc. Europ. Conf. Soil Mech. Found. Eng., Wiesbaden, Vol. 1.

Oding, I.A., Ivanova, V.S., Burduksky, V.V. and Geminov, V.N., 1959. Theory of Creep and Long-Term Strength of Metals. Metalurgizdat, Moscow (in Russian).

Pekarskaya, N.K., 1963. Shear Strength of Frozen Soils Depending on Texture. USSR Acad. Sci., Moscow (in Russian).

Pekarskaya, N.K., 1976. Creep of soils depending on the character of structural bonds. In collective volume: Proc. 2nd All-Union Symp. Rheology of Soils. Erevan State University (in Russian).

Persoz, B., 1953. Contribution à l'étude du fluage des risines organiques: le fluage du polyisobutylene. Recherche Aéronautique, No. 34.

Pokrovsky, G.I., 1933. Application of Boltzmann's principle to computations of foundation settlements. In collective volume: Foundations, No. 1. Gosstroiizdat, Moscow–Leningrad (in Russian).

Pokrovsky, G.I., 1939. On physical principles of evaluating soil deformation. In collective volume: Proc. Conf. Found. Eng. Gosstroiizdat, Moscow–Leningrad (in Russian).

Problems of Rheology of Rocks, 1970. Naukova Dumka, Kiev (in Russian).

Pekarskaya, N.K. and Vyalov, S.S., 1973. In: Proc. All-Union Symposia on Soil Rheology (collective volume). Erevan State University. Erevan (in Russian).

Shiforenko, E.M. and Oxenkrug, E.S., 1973. Study of rheological properties of rock salt. In collective volume: Proc. 1st All-Union Symposium Rheology of Soils. Erevan State University, Erevan (in Russian).

Shusherina, E.P., 1966. On coefficients of lateral strain and volumetric strain of frozen soils in the course of creep. In collective volume: Permafrost Research. Moscow State University, Moscow, Issue 5 (in Russian).

Singh, A. and Mitchell, J.K., 1968. General stress–strain–time functions for soils. J. Soil. Mech. Found. Div., Proc. ASCE, Vol. 94, No. SMI.

Singh, A. and Mitchell, J.K., 1969. Creep potential and creep rupture of soils. Proc. 7th ICSMFE. Mexico City, Vol. 1.

Sorokina, G.V. and Stroganov, A.S., 1971. The strength of clay soils dependent on time factor. 4th Asian Reg. Conf. Soil Mech. Found. Eng., Bangkok.

Ter-Stepanyan, G.I., Meschyan, S.K. and Galstyar, K.K., 1973. Investigations of creep of clay soils at shear. Proc. 8th ICSMFE, Moscow, Vol. 1.2.
Ter-Stepanyan, G.V., 1977. Behaviour of clays close to failure. Proc. 9th ICSMFE, Tokyo, Vol. 1.
Tsytovich, N., Patvardkan, A.M. and Abelev, M.Yu., 1965. Studies of strength properties of weak saturated clay soils in shear. In collective volume: Proc. All-Union Conf. Constructions in Weak Saturated Clay Coils. Tallin (in Russian).
Vyalov, S.S., 1955. Creep and long-term strength of frozen soils. Proc. USSR Acad. Sci., 104 (6) (in Russian).
Vyalov, S.S. and Skibitsky, A.M., 1957. Rheological processes in frozen soils and dense clays. Proc. 4th ICSMFE.
Vyalov, S.S., 1959. Rheological Properties and Bearing Capacity of Frozen Soils. USSR Acad. Sci., Moscow; translated into English by USA Cold Regions Res. Eng. Lab. Transl. No. 74, Hanover, N.H., 1965.
Vyalov, S.S. and Pekarskaya, N.K., 1965. Survey of long-term strength of clay soils. In collective volume: Proc. All-Union Conf. Constructions in Weak Saturated Clay Soils. Tallin (in Russian).
Vyalov, S.S. et al., 1962. The Strength and Creep of Frozen Soils and Calculations for Ice-Soil Retaining Structures. Translated into English by USA Cold Regions Res. Eng. Lab., Transl. No. 75, Hanover, N.H., 1965.

Chapter 6

METHODS OF EXPERIMENTAL DATA HANDLING

6.1 TESTING SOILS FOR CREEP

Specific conditions for creep tests

The rheological processes occurring in soils under natural conditions develop over decades or even centuries. The limited facilities of the laboratory preclude experiments which last longer than a few hours or days. Only on rare occasions can an experiment be scheduled to last a few months or years.

This implies that the data of short-term soil tests must be extrapolated over long periods of time. In this case, maximum care is needed, otherwise an error may easily be committed, as is exemplified in Fig. 5-16 in which a misinterpretation of the time scale may lead to the attenuating deformation being identified as a stage of steady flow. The methods of creep testing are outlined in works of Meschyan (1974), Vyalov et al. (1966, referenced in Chapter 1) and others. Here we shall discuss the method of handling experimental data.

An assessment of the rheological properties of a soil from the test data begins with selecting an empirical equation of creep and determining its parameters. Once the parameter is known, one may predict from this equation the time-dependent behaviour of the soil, be it the soil below a foundation or the material of a structure. The extrapolation of test data obtained in the laboratory over periods that are longer than the test period by several orders of magnitude, calls for careful evaluation and analysis of these data.

Constant and stepwise increasing loads

In soil creep tests commonly a series of specimens, or, occasionally, only one specimen is used. In the former case, each specimen is tested by a constant load which differs from the load applied to other specimens in the series; in the latter case, the specimen is subjected to a stepwise-increasing load applied over an interval Δt. This interval can be the same at all loading stages, or may induce an arbitrary stabilized deformation of an attenuating nature at each stage. The data acquired during the test is represented in Fig. 5-5 in the form of a family of creep curves and corresponding isochrones.

Constant-load tests of soils require a set of similar specimens. Since complete similarity between a number of specimens cannot be ensured — especially when specimens with an undisturbed structure are required — a significant spread of test data is unavoidable during tests of this kind. On the other hand, constant-load tests eliminate the effect of loading conditions, making the test results more reliable.

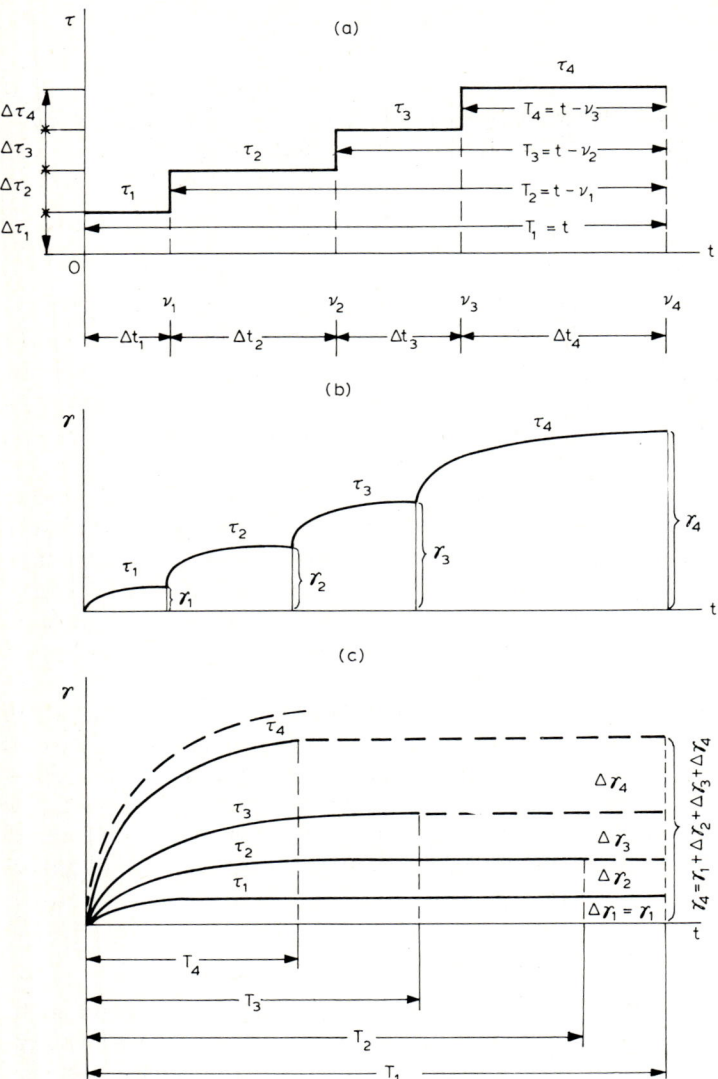

Fig. 6-1. Principle of superposition. (a) Steps of loading. (b) Stepwise-increasing deformation. (c) Summation of deformations.

Stepped-load tests have the advantage that their results spread to a lesser extent because only one specimen is involved. However, these tests are time-consuming, as the sum of Δt for all stages appears to be of a significant duration, particularly if testing is continued until stabilized deformation is obtained. The effect of a previous loading stage on the deformation induced at a succeeding stage (in the form of strain hardening of the soil) is also a limitation of the stepped-load tests.

Principle of superposition

Since Boltzmann's principle of strain superposition is theoretically applicable to a homogeneous isotropic medium, the creep deformation induced by a load varying with time may be determined as the sum of the strains produced by elemental stress increments.

Let a body be loaded in a stepwise manner (Fig. 6-1) so that the loads $\Delta\tau_j = \tau_j - \tau_{j-1}$ are increased at the instants $v_1, v_2, v_3 \ldots$ and the durations of load application at each step are $\Delta t_j = v_j - v_{j-1}$. Thus, a load $\Delta\tau_1 = \tau_1$ will be acting on the body over a period $T_1 = t$, the load increment $\Delta\tau_2 = \tau_2 - \tau_1$ will be exerting its action during the time $T_2 = t - v_1$, and the load increment $\Delta\tau_j = \tau_j - \tau_{j-1}$ will be applied in the course of the interval $T_j = t - v_{j-1}$.

According to the principle of superposition, the strain induced by a stress difference $\tau_m = \tau_1$ equals the sum of the strain increments produced by the stress increments $\Delta\tau_j$:

$$\gamma_m = \gamma_j + \sum_{1}^{m} \Delta\gamma_j(T_j) \qquad (6\text{-}1)$$

For example, a strain due to stress τ_4 is:

$$\gamma_4 = \gamma_1 + \Delta\gamma_2 + \Delta\gamma_3 + \Delta\gamma_4$$

However, in the case of soils, the strain resulting from an increment in stress appears to be less than the strain set up by a constant load of the same magnitude applied over the same period T_j (shown by the dashed lines in Fig. 6-1(c)). Studies by the present author in cooperation with Pekarskaya (Vyalov and Pekarskaya, 1968) indicate that this situation arises from the strain-hardening of soil at some previous loading stage, so that the strain produced at a succeeding stage is less than one determined in accordance with the principle of summation (eq. 6-1).

Thus, constant-load tests are preferred because of the accuracy of their results. On the other hand, when allowance should be made for a gradual application of load, as occurs in natural conditions, it is more practical to conduct stepped-loading tests over a sufficiently long period.

6.2 CHOICE OF EMPIRICAL FORMULA

Experimental data handling by statistical methods

Lack of homogeneity in soils, a disturbance in their natural structure, changes in the soil properties during the taking of samples, and inaccuracy of tests cannot but cause a spread in the soil characteristics obtained from tests. The spread is commonly so large that a 10 to 20% difference between the values obtained is regarded as a satisfactory or even a good closeness of fit, whereas a 5% difference is treated as doubtful. Therefore, experimental data should be handled by statistical methods. By referring to the special literature (e.g., Komarov, 1972) those who wish may acquire a more extensive knowledge of the subject, but here we shall confine ourselves to basic formulae and definitions.

If a series of repetitive determinations carried out with the same precision yields n values of some quantity A, then, assuming that the errors of measurements follow a distribution approximating the Gaussian normal form, the most probable value of A is the arithmetical mean:

$$A_{mean} = \frac{1}{n} \sum_{1}^{n} A_j \qquad (6\text{-}2)$$

If we denote the deviation of a given value A_j from the arithmetical mean as $\varepsilon_j = A_j - A_{mean}$, then the root mean square or standard deviation of an individual test will be:

$$\sigma_A = \sqrt{\frac{\sum_{1}^{n} \varepsilon_j^2}{n-1}} \qquad (6\text{-}3)$$

and a simple error will be:

$$\eta_A = \frac{\sum_{1}^{n} |\varepsilon_j|}{\sqrt{n(n-1)}} = \sigma_A \sqrt{\frac{2}{\pi}} \qquad (6\text{-}4)$$

On computing the arithmetical mean, A_{mean}, we determine the standard deviation σ_A — defining the spread of points — by eq. 6-3.

Also, a simple error η_A may be determined by eq. 6-4 and the values of σ_A, as computed by eqs. 6-3 and 6-4, are then compared. If they coincide, this is an indication that the measured values of A_j obey the normal form of distribution.

The standard deviations σ_j, as determined by eq. 6-3, refer to individual values of A_j. The overall precision of measurements is commonly expressed in terms of the standard error σ of the arithmetical mean, A_{mean}:

$$\sigma = \frac{\sigma_A}{\sqrt{n}} = \sqrt{\frac{\sum_{1}^{n} \varepsilon_j^2}{n(n-1)}} \qquad (6\text{-}5)$$

A relative error is assessed from the relation:

$$v = \frac{\sigma}{A_{mean}} \qquad (6\text{-}6)$$

called the coefficient of variation.

The true value of quantity that is being determined, A_{true}, lies inside an interval called the confidence interval; its confines, termed the confidence limits, depend on the error σ and the confidence level α, indicating that the true value of A is probably inside a given interval:

$$A_{true} = A_{mean} \pm t_\alpha \sigma \qquad (6\text{-}7)$$

where t_α is a factor dependent upon a given value of α and the number of tests n. Applicable values of t_α can be found in any textbook on mathematical statistics.

Proof and permissible values of soil characteristics

According to the "USSR Building Codes and Regulations, BC&R II–15–74", the mechanical properties of soils are expressed in terms of the proof values of strength and deformation. The proof value of a soil property is the mean of repetitive determinations of this property, $A^{proof} = A_{mean}$. However, the soil below a foundation is evaluated in terms of the permissible values of its properties:

$$A = \frac{A^{proof}}{K_s} \tag{6-8}$$

where $K_s = 1/(1 \pm \varrho)$ is a factor of safety; ϱ is the precision factor of the mean, $\varrho = vt_\alpha$, in which v is the coefficient of variation as determined by eq. 6-6; and t_α is a factor explained in eq. 6-7. Substituting the value of ϱ into eq. 6-18 we arrive at eq. 6-7.

In determining the factor of safety K_s, the choice of the $+$ or $-$ sign must provide for a safety margin in the permissible value of a given soil characteristic. The confidence level of the permissible value of strength is specified in the BC&R II–15–74 as $\alpha = 0.95$ when the behaviour of the soil below a foundation is assessed in terms of bearing capacity and $\alpha = 0.85$ when deformation is the criterion.

Method of curve rectification

In selecting a creep equation from the formulae examined in Chapter 5, preference is given to the empirical expression which gives the closest fit to the experimental curve. For the test of fit we use the method of curve rectification, wherein the experimental curve is replotted on coordinates yielding a straight line.

Given a non-linear relationship:

$$y = f(x) \tag{6-9}$$

we transform it into a straight line equation by introducing new variables X and Y, fulfilling the conditions:

$$X = \phi_1(x, y) \quad \text{and} \quad Y = \phi_2(x, y) \tag{6-10}$$

and we then select the functions ϕ_1 and ϕ_2 to connect the values of X and Y by a linear relation. Thus, we arrive at a straight line (Fig. 6-2) plotted on the X–Y coordinates and described by the equation:

$$Y = BX + D \tag{6-11}$$

Apparently, the straight line of eq. 6-11 must be drawn to fit the experimental points, plotted on the coordinates of 6-10, as closely as possible. This requirement will be met if the sum of the squares of vertical deviations of the experimental points from the straight line, $\sum_1^n \varepsilon_j^2$, is at a minimum.

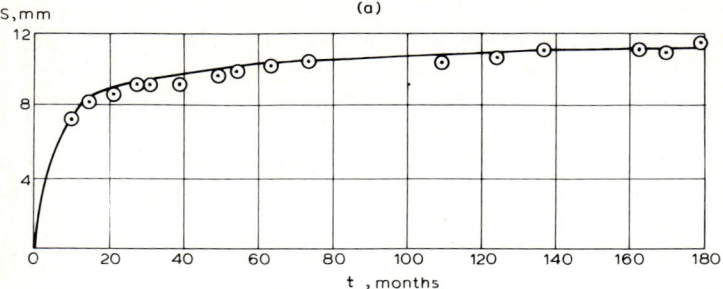
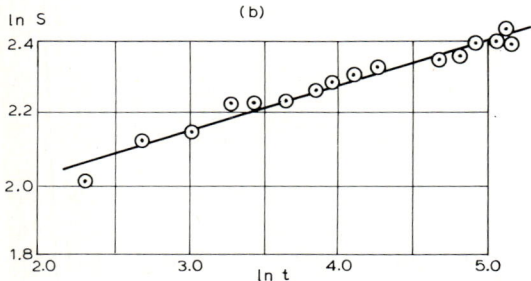

Fig. 6-2. Rectification of curve $y = f(x)$ by replotting it on Y versus X coordinates. Data acquired from observing the settlement of silt below a dam.

Proceeding from the normal Gaussian form of error distribution, the parameters B and D may be determined from two equations, referred to as normal equations:

$$nD + B \sum_1^n X_j = \sum_1^n Y_j \qquad D \sum_1^n X_j + B \sum_1^n X_j^2 = \sum_1^n X_j Y_j$$

where n is the number of experimental points.

Solving the two equations simultaneously, we can easily find the values of B and D:

$$B = \frac{n \sum_1^n X_j Y_j - \sum_1^n Y_j \sum_1^n X_j}{n \sum_1^n X_j^2 - \left(\sum_1^n X_j\right)^2}$$

$$D = \frac{\sum_1^n Y_j \sum_1^n X_j^2 - \sum_1^n X_j \sum_1^n X_j Y_j}{n \sum_1^n X_j^2 - \left(\sum_1^n X_j\right)^2} = \frac{\sum_1^n Y_j - B \sum_1^n X_j}{n}$$

(6-12)

It is essential to have highly accurate computations without rounding-off intermediate values, as the differences of large numbers enter into formula 6-12.

Correlation coefficient. The straight line Y–X obtained in Fig. 6-2(b) corresponds to the mean values Y_{mean} and X_{mean}. It is called the regression line and eq. 6-11 is known as the regression equation. Taking into account the scale of the graph, the tangent of the slope of the line Y–X defines parameter B and its intercept on the axis of ordinates decides parameter D.

On establishing the confidence limits for Y_{mean}, we can determine the so-called confidence region. There is a high probability that the true regression line falls in this region.

The measure of relationship provided by the linear correlation of eq. 6-11 is expressed in terms of the correlation coefficient:

$$r = \frac{n \sum_1^n X_j Y_j - \sum_1^n X_j \sum_1^n Y_j}{\sqrt{\left[n \sum_1^n X_j^2 - \left(\sum_1^n X_j\right)^2\right]\left[n \sum_1^n Y_j^2 - \left(\sum_1^n Y_j\right)^2\right]}} \qquad (6\text{-}13)$$

The relationship is said to be high when $r > 0.8$ and low when $r < 0.8$.

Polynomial function

In addition to those functions that were examined in Chapter 5, the approximating function $y = f(x)$ used in handling experimental data may be taken in the form of a polynomial:

$$y = a + b_1 x + b_2 x^2 + \ldots + b_n x^n \qquad (6\text{-}14)$$

where the parameters $a, b_1, b_2, \ldots, b_n$ are determined from the experimental data.

Usually, the three first terms of the equation will suffice:

$$y = a + b_1 x + b_2 x^2 \qquad (6\text{-}15)$$

This provides for handling the experimental data by the method of rectification examined above, reducing eq. 6-15 to the form:

$$Y = (b_1 + b_2 x_1) + b_2 x \qquad (6\text{-}16)$$

where $Y = (y - y_1)/(x - x_1)$; x_1, y_1 are the coordinates of an arbitrary point on curve of eq. 6-15. On determining the values of b_1 and b_2 from eq. 6-16, we compute the parameter a by the equation:

$$\sum y = b_1 \sum_1^n x^2 + b_2 \sum_1^n x + na$$

where n is the number of experimental points.

Experimental data handling procedure

On reducing the original eq. 6-9 to the linear form of eq. 6-11, compute the parameters B and D by eqs. 6-12. Determine the correlation coefficient r (using formula 6-13), which defines the goodness of fit of the linear approximation of eq. 6-11 adopted.

Reversing the reasoning with respect to eqs. 6-10, determine the parameters of eq. 6-9 from the known values of parameters B and D of eq. 6-11.

Check eq. 6-9 for closeness of fit. Denote the ratio of the experimental value, y_i, to the computed value as $y_{i(\text{exp})}/y_{i(\text{comp})} = y_i^*$. The mean value of this ratio is:

$$y_{\text{mean}}^* = \frac{1}{n}\sum_1^n y_i^*$$

If the experimental and computed points perfectly fit one another, $y_{\text{mean}}^* = 1$. Hence, the deviation of a given value of y_i^* from y_{mean}^* will be defined by the difference:

$$\varepsilon_j = y_i^* - y_{\text{mean}}^* = \frac{y_{i(\text{exp})}}{y_{i(\text{comp})}} - 1 = \frac{y_{i(\text{exp})} - y_{i(\text{comp})}}{y_{i(\text{exp})}} \qquad (6\text{-}17)$$

Compute the standard error (σ_A) of a single determination (y_i^*) by eq. 6-3, and the standard error (σ) of y_{mean}^* by eq. 6-5, putting, at $n - 2$, the denominator of the term under the radical sign in computing σ_A. Compute the coefficient of variation v by eq. 6-6. The smaller the coefficient v and the greater the coefficient r, the closer eq. 6-9 fits the experimental data.

To determine the permissible values of the parameters in eq. 6-9, use formula 6-7 and find the confidence limits which contain the true values of the parameters of eq. 6-11 for a given confidence level α:

$$B_{\text{true}} = B_{\text{comp}} \pm t_\alpha \sigma_B \qquad D_{\text{true}} = D_{\text{comp}} \pm t_\alpha \sigma_D \qquad (6\text{-}18)$$

where B_{comp} and D_{comp} are the values of the parameter computed by formulae 6-12 corresponding to the mean values $B_{\text{comp}} = B_{\text{mean}}$ and $D_{\text{comp}} = D_{\text{mean}}$; t_α is a coefficient which varies with the given values of α and then $n - 2$ degrees of freedom (for applicable values of t_α consult, for example, BC&R II–15–74); and σ_B and σ_D are standard deviations given by:

$$\begin{aligned}\sigma_B &= \sigma_y \sqrt{\frac{n}{n\sum_1^n X_j^2 - \left(\sum_1^n X_j\right)^2}} \\ \sigma_D &= \sigma_y \sqrt{\frac{\sum_1^n X_j^2}{n\sum_1^n X_j^2 - \left(\sum_1^n X_j\right)^2}} \\ \sigma_y &= \sqrt{\frac{\sum_1^n [Y_{j(\text{comp})} - Y_{j(\text{exp})}]^2}{n - 2}}\end{aligned} \qquad (6\text{-}19)$$

in which σ_y is the standard deviation of the experimental data $Y_{j(\text{exp})}$ from the mean (computed) values of $Y_{j(\text{comp})}$, given by $Y_{j(\text{comp})} = D_{\text{comp}} + B_{\text{comp}} X_j$.

6.3 PRACTICAL SELECTION OF FORMULA

Initial data

Consider several cases exemplifying the selection of empirical formulae on the basis of the data represented in Fig. 1-12, which refer to the settlement of the soil below the Kakhovka Dam. We shall approximate experimental curve 4, which corresponds to a loading of $p = 1.95 \cdot 10^5$ Pa. Since the initial settlement is not given in Fig. 1-12, we shall focus our attention on the formulae derived when $S_0 = 0$

Example 1. Consider the approximation of the experimental curve in Fig. 1-12 by eq. 5-8 taking the form:

$$S = S_{in}^* t^\beta \tag{6-20}$$

where $S_{in}^* = S_{in} \bar{\delta}$ (in which $S_{in} = (p/A_{i_1})^{1/m}$) and t, the time, is a dimensionless quantity $t = t/T$ (in which $T = 1$).

Taking the logarithms of 6-20, $\ln S = \ln S_{in}^* + \beta \ln t$, and denoting $Y = \ln S$, $X = \ln t$, $D = \ln S_{in}^*$, and $B = \beta$, we reduce this expression to linear form (see eq. 6-11). The computed data are tabulated in Table 6-I[*1] and are shown graphically in Fig. 6-2.

The initial data of the in-situ observations are given in columns (1), (2), (3) of Table 6-I (the values of settlement S are given in centimetres, and not as percentages of the depth of compacted layers as in Fig. 1-12).

The values of parameters B and D, according to formulae 6-12, are:

$$B = \frac{16 \times 149.446 - 64.839 \times 36.490}{16 \times 275.156 - (64.839)^2} = 0.1268$$

$$D = \frac{36.490 \times 275.156 - 64.839 \times 149.446}{16 \times 275.156 - 64.839^2} = 1.767$$

Comparing the computed value of D with the value determined by the second of formulae 6-12 — i.e., $(36.490 - 0.1268 \times 64.839)/16 = 1.767$ — we can see that the computations seem to be correct.

Compute the correlation coefficient by eq. 6-13:

$$r = \frac{16 \times 149.446 - 64.839 \times 36.490}{\sqrt{(16 \times 275.156 - 64.839^2)(16 \times 183.425 - 36.490^2)}} = 0.986$$

The correlation index appears to be sufficiently high. Now compute the parameters of eq. 6-20 from the known values of B and D:

$$\beta = B = 0.127 \qquad S_{in}^* = e^D = e^{1.767} = 5.85$$

[*1] Computations pertaining to the examples were carried out by S.M. Tikhomirov.

TABLE 6-I

Experimental data reduction by eq. 6-20

Ref. number of points (1)	t (months) (2)	S (cm) (3)	$X = \ln t$ (4)	$Y = \ln S$ (5)	X^2 (6)	Y^2 (7)	XY (8)	S_{comp} (9)	ε_j (10)	ε_j^2 (11)
1	10	7.5	2.303	2.015	5.304	4.060	4.640	7.83	−0.0421	0.00166
2	15	8.4	2.708	2.128	7.333	4.528	5.762	8.25	0.0182	0.00033
3	21	8.5	3.045	2.140	9.272	4.580	6.516	8.61	−0.0128	0.00016
4	28	9.1	3.332	2.208	11.102	4.875	7.357	8.93	0.0190	0.00038
5	32	9.2	3.466	2.219	12.013	4.924	7.691	9.08	0.0132	0.00017
6	40	9.3	3.689	2.230	13.609	4.973	8.226	9.35	−0.0053	0.00003
7	50	9.6	3.912	2.262	15.304	5.117	8.849	9.61	−0.0010	0
8	54	9.9	3.989	2.292	15.212	5.253	9.143	9.71	0.0196	0.00038
9	64	10.2	4.159	2.323	17.297	5.396	9.661	9.92	0.0282	0.00080
10	75	10.4	4.318	2.342	18.645	5.485	10.112	10.13	0.0267	0.00071
11	112	10.4	4.719	2.342	22.269	5.485	11.052	10.64	−0.0446	0.00051
12	125	10.6	4.829	2.361	23.319	5.574	11.401	10.79	−0.0176	0.00031
13	137	10.9	4.921	2.389	24.216	5.707	11.756	10.92	−0.0018	0
14	164	11.0	5.101	2.398	26.020	5.750	12.232	11.18	−0.0161	0.00026
15	173	11.0	5.154	2.398	26.564	5.750	12.359	11.26	−0.0231	0.00053
16	180	11.5	5.194	2.443	26.977	5.968	12.689	11.32	0.0159	0.00025
	$\Sigma =$		64.839	36.490	275.156	83.425	149.446		0.2832	0.00657

Eq. 6-17 thus takes the form

$$S = 5.85 \, t^{0.127} \tag{6-21}$$

where t is given in months and S in centimetres.

The settlement values (S_{comp}) computed by eq. 6-21 are given in column (9) of Table 6-I. If the settlement is computed by this formula for an interval longer than the period of observation, we obtain by extrapolation that $S = 13.2$ cm for $t = 600$ months (50 years) and $S = 14.4$ cm for $t = 1200$ months (100 years). Taking into account that the actual settlement observed during the 180-month period was $S = 11.5$ cm, the computed values of S_{600} and S_{1200} appear to be trustworthy.

Let us assess the closeness of fit provided by the equation in question in approximating the experimental data. Compute the deviation from the mean, $\varepsilon_j = (S_j - S_{comp})/S_{comp}$, and the square of this deviation for every test value S_j; then enter the results in columns (10) and (11) of Table 6-I. Compute the standard deviation by formula 6-3:

$$\sigma_A = \sqrt{\frac{\sum_1^n \varepsilon_j^2}{n-2}} = \sqrt{\frac{0.00657}{14}} = 0.0217$$

Since the deviation ε_j is known to be a dimensionless quantity, σ_A is also dimensionless. Bearing in mind that in the case under consideration $A_{mean} = 1$, find the coefficient of variation from eq. 6-6 which serves as a precision index of the test data handling:

$$v = \frac{\sigma A}{1\sqrt{n}} = \frac{0.0217}{\sqrt{16}} = 0.0054$$

As can be seen, eq. 6-20 fits the experimental data quite satisfactorily.

The values of S_i^* and β entering eq. 6-21 are mean (proof) values. To compute the permissible values, use eq. 6-19 to find the standard deviations σ_B and σ_D of parameters B and D. We also compute:

$$\sigma_y = \sqrt{\frac{0.0067}{16-2}} = 0.0022$$

where:

$$0.0067 = \sum_1^{16} [Y_{j(comp)} - Y_{j(exp)}]^2$$

(The values of $Y_{j(comp)} = D_{comp} + B_{comp} X_j$ have been omitted in Table 6-I.) Hence, we arrive at:

$$\sigma_B = 0.0022 \sqrt{\frac{16}{16 \times 275.156 - 64.839^2}} = 0.0006$$

and:

$$\sigma_D = 0.0022 \sqrt{\frac{275.156}{198.370}} = 0.0024$$

The coefficient t_α in eq. 6-7 is taken from a table in the BC&R II–15–74. Adopting the confidence level at $\alpha = 0.85$, provided that the number of degrees of freedom $(n - 2)$ is 14, we find that $t_\alpha = 1.08$. Then, according to eq. 6-7, we obtain:

$B_{true} = 0.1268 \pm 1.08 \times 0.0006 = 0.1271 \pm 0.001$

$D_{true} = 1.767 \pm 1.08 \times 0.0024 = 1.767 \pm 0.003$

Accordingly, the permissible values of parameters S_{in}^* and β of the approximating eq. 6-20 will be:

$S_{in}^* = e^{1.767 \pm 0.003}$ and $\beta = 0.127 \pm 0.001$.

Example 2. Consider the approximation of the test curve in Fig. 1-12 by eq. 5-21 taking the form:

$$S = S_{in}^* \ln(t + 1) \tag{6-22}$$

where $S_{in}^* = S_{in} \delta^{1/m}$ (in which $S_{in} = (p/A_{in})^{1/m}$) and t is a dimensionless quantity, $t = t/T$ (in which $T = 1$).

Putting $Y = S$, $X = \ln(t + 1)$, $B = S_{in}^*$, $D = 0$, we reduce formula 6-22 to a linear form (see eq. 6-11). Next, we tabulate the results of computations in Table 6-II and find the value of the coefficient B, using eq. 6-12:

$$B = \frac{16 \times 656.093 - 157.5 \times 65.17}{16 \times 277.211 - 4247.129^2} = 1.239$$

From the meaning of eq. 6-22, one may consider the value of D to be zero. However, computations show that:

$$D = \frac{157.5 - 1.239 \times 65.17}{16} = 4.797$$

This implies that eq. 6-22 must be written for the experiment in question as:

$$S = S_{in}^* \ln \frac{t + 1}{T} \tag{6-23}$$

where $S_{in}^* = S_0/\ln(1/T)$; $S_0 = D$. Hence, $T = e^{-S_0/S_{in}^*}$ or, in our case, $t = \exp(-4.797/1.239) = 0.021$ months.

The difference between eqs. 6-22, 6-23 and a similar formula:

$$S = S_{in}^* \ln t \tag{6-24}$$

is as follows. At $t = 0$, we obtain from eq. 6-22 that $S_0 = 0$, whereas eqs. 6-23 and 6-24 yield $S_0 = S_{in} \ln(1/T)$ and $S_0 = -\infty$, respectively. A zero settlement, $S = 0$, is obtained from eqs. 6-23 and 6-24 for $t = -1$ and $t = 1$, respectively.

TABLE 6-II
Experimental data reduction by eq. 6-23

Ref. number of points	t (months)	S (cm)	$X = \ln(t+1)$	$Y = S$	XY	X^2	Y^2	S_{comp}	ε_j	ε_j^2
1	10	7.5	2.395	7.5	17.962	5.736	56.25	7.76	−0.0335	0.00112
2	15	8.4	2.769	8.4	23.260	7.667	70.56	8.23	+0.0207	0.00043
3	21	8.5	3.087	8.5	26.240	9.530	72.25	8.62	−0.0139	0.00019
4	28	9.1	3.363	9.1	30.603	11.310	82.81	8.96	0.0156	0.00024
5	32	9.2	3.492	9.2	32.126	12.194	84.64	9.12	0.0088	0.00008
6	40	9.3	3.710	9.3	34.503	13.764	86.49	9.39	−0.0096	0.00009
7	50	9.6	3.928	9.6	37.709	15.429	92.16	9.66	−0.0062	0.00004
8	54	9.9	4.003	9.9	39.630	16.024	98.01	9.76	0.0143	0.00021
9	64	10.2	4.170	10.2	42.534	17.389	104.04	9.96	0.0241	0.00058
10	75	10.4	4.326	10.4	44.990	18.714	108.16	10.16	0.0236	0.00056
11	112	10.4	4.723	10.4	49.119	22.307	108.16	10.65	−0.0235	0.00055
12	125	10.6	4.830	10.6	51.198	23.329	112.36	10.78	−0.0167	0.00028
13	137	10.9	4.923	10.9	53.661	24.236	118.81	10.90	0	0
14	164	11.0	5.102	11.0	56.122	26.030	121.00	11.20	−0.107	0.00011
15	173	11.0	5.156	11.0	56.716	26.584	121.00	11.19	−0.0170	0.00029
16	180	11.5	5.193	11.5	59.720	26.967	132.25	11.23	0.0240	0.00058
$\Sigma =$			65.170	157.5	656.093	277.211	1568.95		0.2622	0.00535

Eq. 6-22 thus takes the form:

$$S = 4.8 + 1.24 \ln(t + 1) \qquad (6\text{-}25)$$

where t is in months and S in centimetres.

The settlement values (S_{comp}) computed by this formula are found in Table 6-II. An extrapolation to 600 and 1200 months yields $S_{600} = 12.7$ cm and $S_{1200} = 13.6$ cm, which seems to be logical.

Determine the precision indices of the approximation.

According to eq. 6-13 the coefficient of correlation is:

$$r = \frac{14.576}{\sqrt{11.765 \times 18.559}} = 0.986$$

The standard deviation as given by eq. 6-3 is:

$$\sigma_A = \sqrt{\frac{0.00535}{14}} = 0.0195$$

and the coefficient of variation defined by eq. 6-6 appears to be:

$$v = \frac{0.0195}{\sqrt{16}} = 0.0049$$

The permissible values of the parameters of formula 6-25 are determined exactly in the same way as was shown in Example 1.

Example 3. Consider the approximation of the experimental curve in Fig. 1-12 by formula 5-25 taking the form:

$$S = S_\infty \frac{t}{T^* + t} \qquad (6\text{-}26)$$

where $T^* = T/[1 - \delta_{in} p/p_{s(\infty)}]$.

Putting $Y = t/S$, $X = t$, $D = T^*/S_\infty$, and $B = 1/S_\infty$, we transform the above formula into linear form (see eq. 6-11).

On tabulating the computations in Table 6-III, we determine the values of $B = 0.0864$ and $D = 0.713$ by eq. 6-12.

The parameters of eq. 6-26 are as follows:

$$S_\infty = \frac{1}{B} = \frac{1}{0.0864} = 11.57 \text{ cm}$$

$$T^*_\infty = 0.713 \times 11.57 = 8.25 \text{ months}$$

The computed values (S_{comp}) are given in Table 6-III. The corresponding values extrapolated to 600 and 1200 months are 11.4 and 11.55 cm, respectively, i.e. close to the value $S_{180} = 11.5$ cm.

Determine the precision indices of the approximation. The coefficient of correlation, as computed by eq. 6-13, appears to be sufficiently high, $r = 0.999$.

TABLE 6-III

Experimental data reduction by eq. 6-26

Ref. number of points	t (months)	S (cm)	$X = t$	$Y = \dfrac{t}{S}$	Y^2	XY	S_{comp}	ε_j	ε_j^2
1	10	7.5	10	1.333	1.777	13.330	6.34	0.1830	0.03348
2	15	8.4	15	1.786	3.190	26.790	7.50	0.1200	0.01440
3	21	8.5	21	2.471	6.106	51.891	8.31	0.0229	0.00052
4	28	9.1	28	3.077	9.468	86.156	8.94	0.0179	0.00032
5	32	9.2	32	3.478	12.096	111.296	9.20	0	0
6	40	9.3	40	4.301	18.499	172.040	9.59	−0.0302	0.00091
7	50	9.6	50	5.208	27.123	260.400	9.93	−0.0332	0.00110
8	54	9.9	54	5.455	29.757	294.570	10.04	−0.0139	0.00019
9	64	10.2	64	6.275	39.376	401.600	10.25	−0.0049	0.00002
10	75	10.4	75	7.212	52.013	540.900	10.43	−0.0029	0.00001
11	112	10.4	112	10.769	115.971	1206.128	10.78	−0.0353	0.00124
12	125	10.6	125	11.792	139.051	1474.000	10.86	−0.0239	0.00057
13	137	10.9	137	12.569	157.980	1721.953	10.92	−0.0018	0
14	164	11.0	164	14.909	222.278	2445.076	11.02	−0.0018	0
15	173	11.0	173	15.652	247.339	2720.771	11.05	−0.0045	0.00002
16	180	11.5	180	15.652	244.985	2817.360	11.07	0.5350	0.00151
$\Sigma =$			1280.0	122.014	1327.009	14344.261		0.5350	0.05429

The standard deviation, as given by eq. 6-3, is:

$$\sigma_A = \sqrt{\dfrac{0.0543}{14}} = 0.0622$$

hence the coefficient of variation, according to eq. 6-6, will be:

$$v = \dfrac{0.0622}{\sqrt{16}} = 0.156$$

Example 4. Consider the approximation of the test curve in Fig. 1-12 by eq. 5-30 in the form:

$$S = S_\infty(1 - e^{-t/T}) \tag{6-27}$$

The above formula is transformed into linear form (see 6-11) by putting:

$$Y = \ln \dfrac{S_\infty}{S_\infty - S}; \quad X = t; \quad B = 1/T; \quad D = 0$$

Assuming that the last measurement of S is the stabilized settlement $S_\infty = 11.5$ cm, we compute the values of Y_j (see Table 6-IV). On finding, by eq. 6-12, that $B = 0.0119$, we determine that $T = 84$ months.

The coefficient D has been adopted at zero. However, a check-up by eq. 6-12 reveals that $D = 1.189$. This implies that for the test in question eq. 6-27 takes the form:

$$S = S_\infty\left[1 - \exp\left(-\dfrac{t}{T} - D\right)\right] \tag{6-28}$$

TABLE 6-IV

Experimental data reduction by eq. 6-27

Ref. number of points	t (months)	S (cm)	$X = t$	$Y = \ln \dfrac{S_\infty}{S_\infty - S}$	Y^2	XY	S_{comp}	ε_j	ε_j^2
1	10	7.5	10	1.055	1.113	10.550	8.39	−0.1061	0.01125
2	15	8.4	15	1.309	1.713	19.635	8.57	−0.0198	0.00039
3	21	8.5	21	1.342	1.801	28.182	8.77	−0.0308	0.00095
4	28	9.1	28	1.565	2.449	43.820	9.00	0.0111	0.00012
5	32	9.2	32	1.608	2.586	51.456	9.11	0.0099	0.00010
6	40	9.3	40	1.652	2.729	66.080	9.33	−0.0032	0.00001
7	50	9.6	50	1.799	3.236	89.950	9.57	0.0031	0.00001
8	54	9.9	54	1.970	3.881	106.38	9.66	0.0248	0.00062
9	64	10.2	64	2.177	4.739	139.328	9.87	0.0334	0.00112
10	75	10.4	75	2.344	5.494	175.800	10.07	0.0320	0.00107
11	112	10.4	112	2.344	5.434	262.528	10.58	−0.0170	0.00029
12	125	10.6	125	2.545	6.477	318.125	10.71	−0.0103	0.00011
13	137	10.9	137	2.950	8.702	404.150	10.82	0.0074	0.00005
14	164	11.0	164	3.132	9.809	513.648	11.00	0	0
15	173	11.0	173	3.132	9.809	541.836	11.05	−0.0045	0.00002
16	180	11.5	180				11.09		
$\Sigma =$			1100	30.924		70.034	2771.468		0.3142	0.01611

At $t = 0$, we obtain that $S_0 = S_\infty(1 - e^{-D})$. Then:

$$D = \ln \frac{S_\infty}{S_\infty - S_0}$$

The value of S may equal zero only if $t = -TD < 0$. Hence, eq. 6-27 may be rewritten as:

$$S = 11.5 \left[1 - \exp\left(-\frac{t}{84} - 1.19\right)\right]$$

At zero instant, we obtain $S_0 = 11.5(1 - e^{1.19}) = 8.00$ cm. This value is acceptable, for the actual settlement reached 7.5 cm in only 10 months. The values of S corresponding to 600 and 1200 months are $S_{600} = S_{1200} = S = 11.5$ cm approx.

Compute the indices of precision of the approximation. The coefficient of correlation, as given by eq. 6-13, is $r = 0.976$. The standard deviation according to eq. 6-3 amounts to $\sigma_A = \sqrt{0.011611/14} = 0.0352$. Hence, the coefficient of variation computed by eq. 6-6 is:

$$v = \frac{0.0352}{\sqrt{16}} = 0.0088$$

The precision indices show that eq. 6-7 fits the experimental points satisfactorily. However, the approximated values of the initial and final settlements S are doubtful, being overestimated for $t = 0$ and underestimated for $t = \infty$.

However, if we assume that the final settlement, S_∞, is unknown and should be determined, the procedure is as follows. Denoting $z = e^{-\beta t}$, we reduce eq. 6-27 to the form:

$$S = S_\infty(1 - z) \tag{6-29}$$

Taking two arbitrary points t_1 and $t_2 = 2t_1$ on the x-axis we determine their ordinates S_1, S_2 and set up two equations:

$$S_1 = S_\infty(1 - z_1)$$

and:

$$S_2 = S_\infty(1 - z_1^2)$$

Finding from the equations that $z_1 = (S_2 - S_1)/S_1$, we obtain:

$$S_\infty = \frac{S_1}{1 - z_1} \tag{6-30}$$

If, in our example, we take $t_1 = 40$ months and $t_2 = 2t_1 = 80$ months, with corresponding settlement values of $S_1 = 9.3$ cm and $S_2 = 10.3$ cm, we find $z_1 = (S_2 - S_1)/S_1 = 0.107$. Hence:

$$S_\infty = \frac{S_1}{1 - z_1} = \frac{9.3}{0.893} = 10.4 \text{ cm} \tag{6-31}$$

If we take some other values of t_1 and t_2 (e.g., $t_1 = 10$ months, $t_2 = 20$ months, with corresponding settlement values of $S_1 = 7.5$ cm and $S_2 = 8.45$ cm), we obtain $z_1 = 0.126$; hence, $S_\infty = 7.5/0.874 = 8.58$ cm.

It can be seen from the above that the final settlement value is significantly underestimated, being less than the actual settlement after 180 months ($S_{180} = 11.5$ cm). Moreover, there is an S_∞ value corresponding to each of the selected points t_1 and t_2. This indicates a lack of fit between eq. 6-27 and the experimental data in the case under consideration.

6.4 COMPARISON OF THE RESULTS OF COMPUTATIONS

Presentation of data

The results of handling experimental data by eqs. 6-20, 6-23, 6-26 and 6-28 as prescribed by statistical methods, are tabulated in Table 6-V; the settlement-vs-time curves plotted by these formulae are depicted in Fig. 6-3.

Experimental data analysis. The examples presented above enable us to draw some conclusions. First, the results obtained exponentially from the use of formula 6-28 are totally unacceptable. The values of the initial settlement differ appreciably one from another, and those of the final settlement S_∞ fail to be single-valued. Thus, eq. 6-28 is not recommended for use in cases other than those when its applicability is proved experimentally.

TABLE 6-V

Results of experimental data approximation by various equations

Equation number	Equation	Settlement (cm) computed for a given time (months)					Indices of precision	
		0	180	600	1200	∞	v	r
6-20	$S = S_{in}^* t^\beta$	0	11.3	13.2	14.4	∞	0.54	0.986
6-23	$S = S_{in}^* \ln \dfrac{t+1}{T}$	4.8	11.2	12.7	13.6	∞	0.49	0.986
6-26	$S = S_\infty \dfrac{t}{t+T^*}$	0	11.1	11.4	11.55	11.6	1.56	0.999
6-28	$S = S_\infty \left[1 - \exp\left(-\dfrac{t}{T} - D\right)\right]$	8.0	11.1	11.5	11.5	11.5	0.88	0.976

Lack of fit between experimental data and the results of computations by eq. 6-28 has been observed by many other authors experimenting with various soft rocks such as siltstone, argillite, sandstone (see Chapter 1: Erzhanov et al., 1970), frozen soil (see Chapter 1: Vyalov, 1959), and loess. On the other hand, the power formula (eq. 6-20) provided satisfactory agreement with experiments in the same cases.

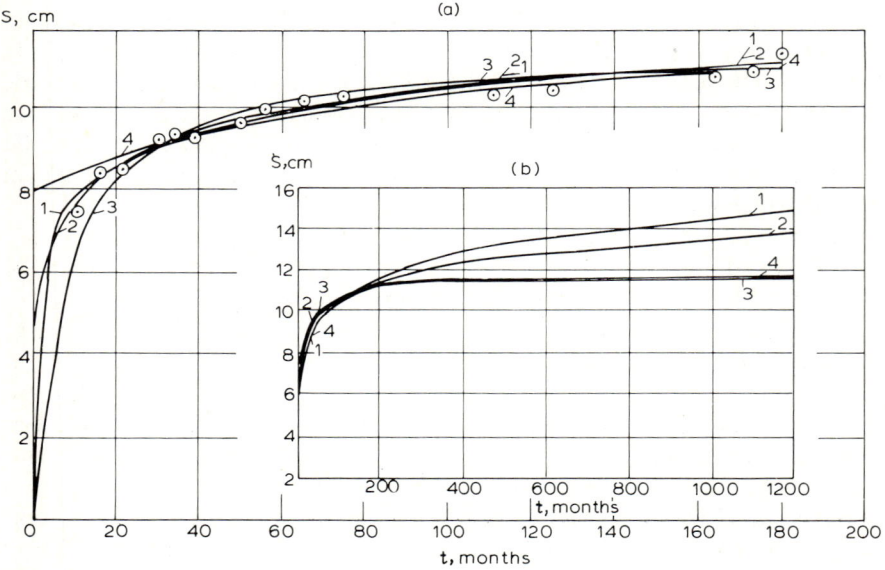

Fig. 6-3. Approximating experimental values of soil settlement below the Kakhovka Dam by various formulae. (a) Approximation of test points. (b) Extrapolation of the analytical curves over a 100-year period. *1*. Eq. 6-20, $S = S_{in}^* t^\beta$; *2*. eq. 6-23, $S = S_{in}^* \ln [(t+1)/T]$; *3*. eq. 6-26, $S = S_\infty(t/t+T)$; *4*. eq. 6-28, $S = S_\infty [1 - e^{-(t/T+D)}]$.

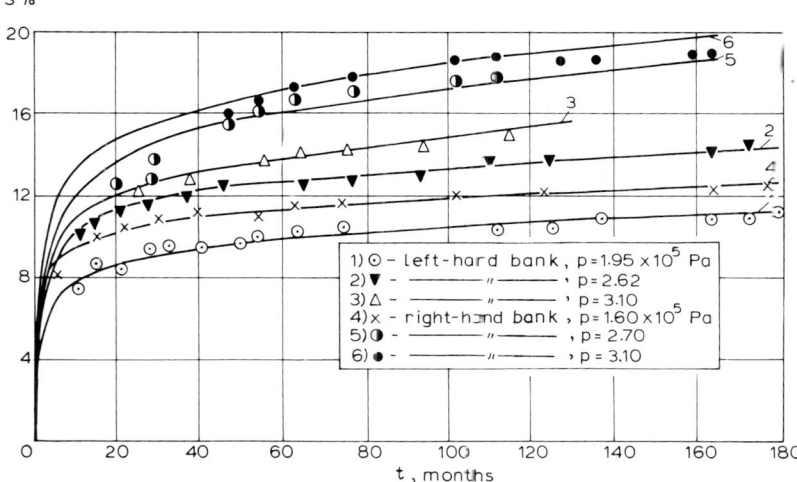

Fig. 6-4. Comparison of the settlement data obtained from tests and computed by eq. 6-20. Curves *1–6* indicate the settlement as observed at the Kakhovka Dam in percent of the depth of the compressed soil (silt) resting on bedrock.

Concerning power (eq. 6-20), logarithmic (eq. 6-23) and linear rational (eq. 6-26) relations, the results satisfactorily fit one another and the experimental data. All precision indices are satisfactory but the coefficients of variation in eqs. 6-20 and 6-23 are better than those in eq. 6-26. The close fit of eqs. 6-20 and 6-23 is particularly evident at an early stage of the process, the best-fitting curves plotted by these formulae being those representing the period between 0 and 30 months (Fig. 6-3(a)).

A noticeable lack of fit is evident in extrapolating the results of computations by the above three formulae beyond the period of observations. This can be seen from Fig. 6-3(b) in which the data are extrapolated over a 100-year period.

The final settlement and one in 180 months, as computed by eq. 6-26, differ but little, being 11.6 and 11.1 cm, respectively. This is an indication that in computing by the linear rational formula 6-26 we obtain a speedy dying-out of the settlement. On the other hand, the power and logarithmic formulae (eqs. 6-20 and 6-23) serve to define an unceasing settlement build-up, however, at a slow rate. which is smaller for eq. 6-23 than for eq. 6-20. Thus, it is good practice to employ eq. 6-26 for an apparently attenuating deformation, reserving eqs. 6-20 and 6-23 for use when "secular" deformation is evident.

However, under the conditions of moderate loadings which give no rise to a conspicuous steady flow, the three formulae referred to provide a satisfactorily close fit even if the data are extrapolated to a distant time. So, the settlement computed by formulae 6-20, 6-23 and 6-26 for $t = 100$ years is 14.4, 13.6 and 11.55 cm, respectively, the maximum discrepancy being 33%. The somewhat higher settlements obtained from computations by formulae 6-20 and 6-23 provide a stability margin for the soil below the dam.

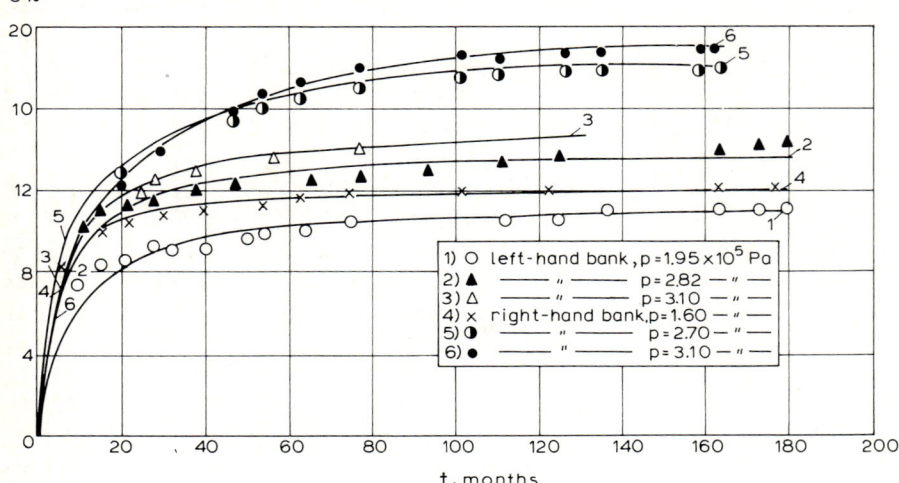

Fig. 6-5. Comparing the settlement obtained from tests with the data computed by formula $S = S_\infty(t/t + T)$ (see Fig. 6-4).

In conclusion, we refer to Figs. 6-4 and 6-5 providing a comparison of the results of computations by formulae 6-20 and 6-26 with the settlement test data. These data have also been obtained from observations made at the Kakhovka Dam (see Fig. 1-12) but at various places and for various loadings. The comparison corroborates the above statement that both formulae satisfactorily fit the test data, but the power relation appears to give the best fit for the initial segment of the experimental curve.

6.5 TESTING FORMULAE FOR CLOSENESS OF FIT

Use of isochrones and creep curves

Practical selection of the formulae mentioned above was based on experimental data acquired from only one constant-load test. However, full-program testing involves the use of various loadings, yielding a family of creep curves and isochrones (see Fig. 5-5). Examine the way these experimental curves are rectified and used in testing equations of the form of eqs. 5-16, 5-20 and 5-22 for closeness of fit.

Testing power formulae. The power relation is defined by eqs. 5-16, 5-17 and 5-18 which, in the case under consideration, respectively take the form:

$$\gamma^m = \frac{\tau}{A_0}\left[1 + \left(\frac{t}{T^*}\right)^\alpha\right] \tag{6-32}$$

$$\gamma = \left(\frac{\tau}{A_0}\right)^{1/m}\left[1 + \left(\frac{t}{T^*}\right)^\beta\right] \tag{6-33}$$

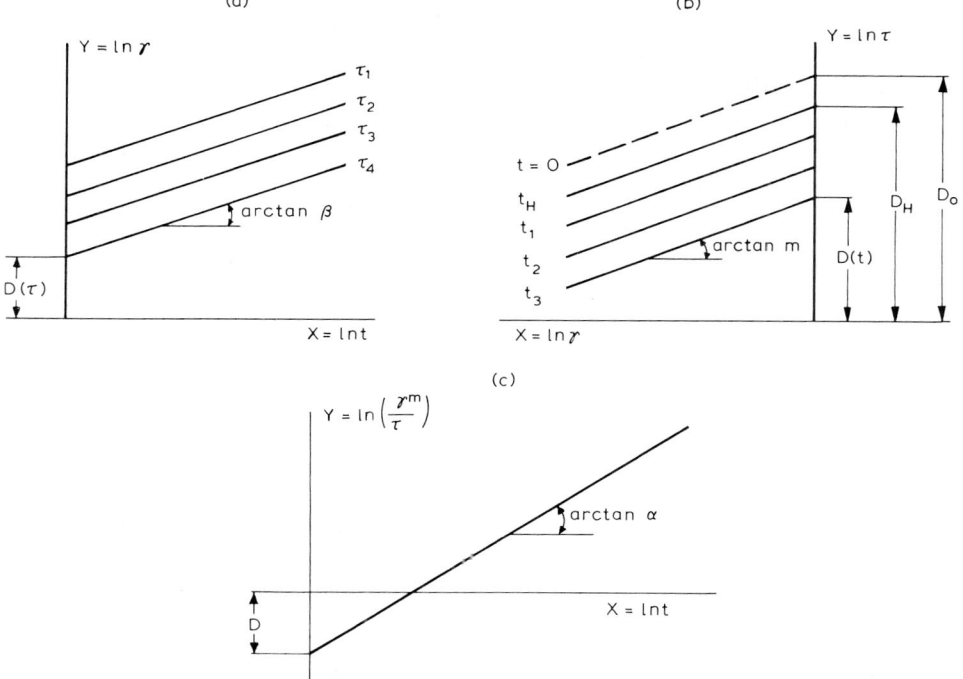

Fig. 6-6. Test of fit of eqs. 6-32–6-34.

$$\gamma = \left(\frac{\tau}{A_{in}}\right)^{1/m} \left(\frac{t}{T^*}\right)^{\beta} \quad \text{(for } \gamma_0 = 0\text{)} \tag{6-34}$$

where, in eq. 6-32, $T^* = \delta^{-1/\alpha} T$; and in eq. 6-33, $T^* = \delta^{-1/\beta} T$; $T = 1$.

We start by testing eq. 6-34. Transforming this equation into:

$$\ln \gamma = \frac{1}{m} \ln \frac{\tau}{A_{in}} + \beta \ln \frac{t}{T^*} \tag{6-35}$$

and putting:

$$Y = \ln \gamma, \quad X = \ln t, \quad B = \beta, \quad \text{and} \quad D(\tau) = \ln \left[\left(\frac{1}{T^*}\right)^{\beta} \left(\frac{\tau}{A_{in}}\right)^{1/m}\right]$$

we have a linear equation of the form given by eq. 6-11:

$$Y = BX + D(\tau)$$

Replotting the γ–t creep curves on the X–Y coordinates, we obtain a family of straight lines (Fig. 6-6(a)) each corresponding to a stress $\tau = $ const. The tangent of the slopes of these lines defines the values of the parameter β; the fact that the slopes of all lines are the same is an evidence of the similarity of the creep curves.

Putting in eq. 6-11:

$$Y = \ln \tau, \quad X = \ln \gamma, \quad B = m, \quad \text{and} \quad D(t) = -\ln\left[\frac{1}{A_{in}}\left(\frac{t}{T^*}\right)^{\beta m}\right]$$

we replot the τ–γ isochrones on the X–Y coordinates. The result is a family of straight lines for various instants t (Fig. 6-6(b)). The tangents of the slopes of these lines define (with due regard for the scale of the graph) the parameter m; the sameness of the slopes heralds the similarity of the isochrones. The intercept made by the straight line for $t = t_{in}$ will define the quantity $D_{in} = \ln A_{in}$, from which we find the parameter $A_{in} = e^{D_{in}}$. Once A_{in} and β are known, we can compute T^* by the expression:

$$D(t) = -\ln\left[\frac{1}{A_{in}}\left(\frac{t}{T^*}\right)^{\beta m}\right]$$

This quantity must be constant for all values of t. To test eq. 6-34 for closeness of fit, we plot a generalized curve (Fig. 6-6(c)) by putting in eq. 6-11:

$$Y = \ln(\gamma^m/\tau), \quad X = \ln t, \quad B = \alpha \quad \text{and} \quad D = -\ln[A_{in}(T^*)^\alpha]$$

If all the experimental points fall on the straight line plotted on the generalized X–Y coordinates, this is evidence that formula 6-34 is applicable. The parameters B and D determined from the generalized straight line are used to check the values of A_{in}, m and $\alpha = m\beta$ which have been obtained from the curves in Figs. 6-6, (a) and (b). Note that only two graphs, Fig. 6-6, (b) and (c), may suffice.

If the straight lines in Fig. 6-6, (a) and (b), have different slopes, this is an indication that the parameters m and α are variables dependent on time, $m = m(t)$, and stress, $\alpha = \alpha(\tau)$.

In case the instantaneous deformation must be taken into account, eqs. 6-32 and 6-33 are tested for fit and preference is given to the formula that shows better agreement with the experimental data.

To test eq. 6-32 for fit, we put:

$$Y = \ln \tau, \quad X = \ln \gamma, \quad B = m, \quad \text{and} \quad D(t) = -\ln\left\{\frac{1}{A_0}\left[1 + \left(\frac{t}{T^*}\right)^\alpha\right]\right\}$$

The experimental points must come down on the straight lines of Fig. 6-6(b), the intercept of the $t = 0$ straight line (shown in a dashed line) on the axis of ordinates defining $D_0 = \ln A_0$. Hence $A_0 = e^{D_0}$; the parameter m is defined, as before, by the tangent of the slope of the straight lines.

To plot a generalized graph, we put:

$$Y = \ln\left(\frac{\gamma^m}{\tau} - \frac{1}{A_0}\right), \quad X = \ln t, \quad B = \alpha, \quad \text{and} \quad D = -\ln[A_0(T^*)^\alpha]$$

In this case, too, the experimental points for all creep curves must fall on a straight line similar to that of Fig. 6-6(c). From its slope we determine the parameter α, and from the intercept on the axis of ordinates we find the value of D. Hence, we obtain the parameter T^*.

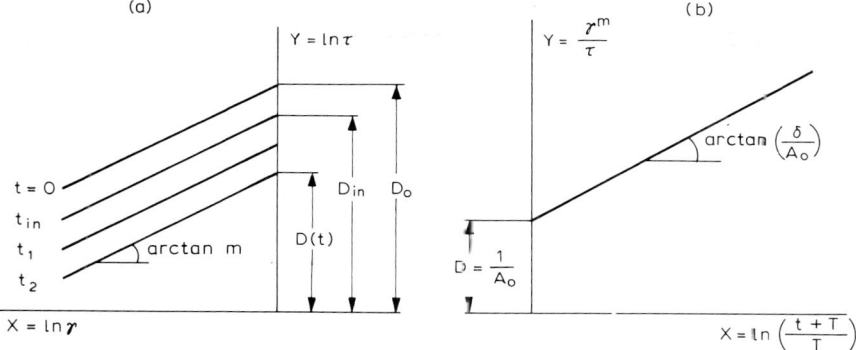

Fig. 6-7. Test of fit of eq. 6-36.

To test eq. 6-33 for fit, we transform it into a linear equation by putting:

$$Y = \ln \tau, \quad X = \ln \gamma, \quad B = m, \quad \text{and} \quad D(t) = m \ln \left\{ \left(\frac{1}{A_0}\right)^{1/m} \left[1 + \left(\frac{t}{T^*}\right)^\beta\right]\right\}$$

The experimental points are plotted in Fig. 6-6(b). A generalized graph is obtained by putting:

$$Y = \ln(\gamma/\tau^{1/m}) - 1/A_0^{1/m}, \quad X = \ln t, \quad B = \beta, \quad \text{and} \quad D = -\ln[A_0^{1/m}(T^*)^\beta]$$

Testing logarithmic formula. The logarithmic relation represented by eqs. 5-19 and 5-20 assumes the form:

$$\gamma^m = \frac{\tau}{A_0}\left(1 + \delta \ln \frac{t+T}{T}\right) \tag{6-36}$$

$$\gamma = \left(\frac{\tau}{A_0}\right)^{1/m}\left(1 + \delta \ln \frac{t+T}{T}\right) \tag{6-37}$$

where T is an arbitrary moment (e.g., $T = 1$).

The isochrones corresponding to formula 6-36 are rectified into straight lines if we put:

$$Y = \ln \tau, \quad X = \ln \gamma, \quad B = m, \quad \text{and} \quad D(t) = -\ln\left[\frac{1}{A_0}\left(1 + \delta \ln \frac{t+T}{T}\right)\right]$$

On plotting the graph of Fig. 6-7(a) on these coordinates, we obtain the parameters m and $A_0 = e^{D_0}$.

Next, putting:

$$Y = \gamma^m/\tau, \quad X = \ln[(t+T)/T], \quad B = A_0^{-1}\delta, \quad \text{and} \quad D = A_0^{-1}$$

we plot a generalized graph (Fig. 6-7(b)) in which the family of creep curves becomes transformed into a single straight line. From this graph we determine the parameter $\delta = BA_0$.

If instantaneous deformation is disregarded, the isochrones lend themselves to rectification in accordance with formula (5-21) by putting, as before, $Y = \ln \tau$, $X = \ln \gamma$, $B = m$, but adopting that

$$D(t) = -\ln\left(\frac{1}{A_0} \ln \frac{t+T}{T}\right)$$

Hence, $A_{in} = e^{D_{in}}$.

The creep curves become a generalized straight line if we adopt:

$$Y = \gamma^m/\tau, \quad X = \ln[(t+T)/T], \quad B = \delta/A_{in}, \quad \text{and} \quad D = 0$$

For the test of fit of eq. 6-37, the isochrones are plotted as before on the coordinates $Y = \ln \tau$, $X = \ln \gamma$ which yield straight lines, but the parameter D is then taken as:

$$D = -m \ln\left[\left(\frac{1}{A_0}\right)^{1/m}\left(1 + \delta \ln \frac{t+T}{T}\right)\right]$$

Hence, $A_0 = e^{D_0}$.

A generalized plot is obtained by putting:

$$Y = \gamma/\tau^{1/m}, \quad X = \ln[(t+T)/T], \quad B = \delta/A_0^{1/m}, \quad \text{and} \quad D = 1/A_0^{1/m}$$

Preference is given to whichever of eqs. 6-36 and 6-37, providing better agreement between the experimental points and the generalized graph. Note that, for $m = 1$ the formulae are identical.

Testing linear–fractional formula. The linear–fractional relation, as represented by eq. 5-22, takes the form:

$$\gamma = \frac{\tau(T + \delta t)}{G_0[T(1 - \tau/\tau_s) + t(1 - \delta\tau/\tau_s)]} = \gamma_0 + (\gamma_\infty - \gamma_0)\frac{t}{T^* + t} \quad (6\text{-}38)$$

where γ_0 and γ_∞ are the instantaneous and finite deformations defined by eqs. 5-24 and 5-24', respectively, and $T^* = \gamma_\infty T/\delta\gamma_0$.

The curves described by eq. 6-38 are schematically represented in Fig. 5-8. The isochrones of the given shape become straight lines if eq. 6-38 is modified into the expression:

$$\frac{\tau}{\gamma} = \frac{G_0(T + t)}{T + \delta t} - \frac{G_0}{\tau_s}\tau \quad (6\text{-}39)$$

Then, putting $Y = \tau/\gamma$, $X = \tau$, $B = -G_0/\tau_s$, $D(t) = [G_0(T+t)]/(T+\delta t)$, and plotting the above-mentioned curves on the X–Y coordinates, we obtain a family of straight lines (Fig. 6-8), all having the same slope, the tangent of which is $G_0/\tau_s = \text{const}$. The intercepts on the axis of ordinates give the values of D for various instants t which vary over the range between $D_0 = G_0$ at $t = 0$ and $D_\infty = G_0/\delta$ at $t = \infty$.

The creep curves lend themselves to rectification if eq. 6-38 is transformed into:

$$\frac{t}{\gamma - \gamma_0} = \frac{T^*}{\gamma_\infty - \gamma_0} + \frac{1}{\gamma_\infty - \gamma_0}t \quad (6\text{-}40)$$

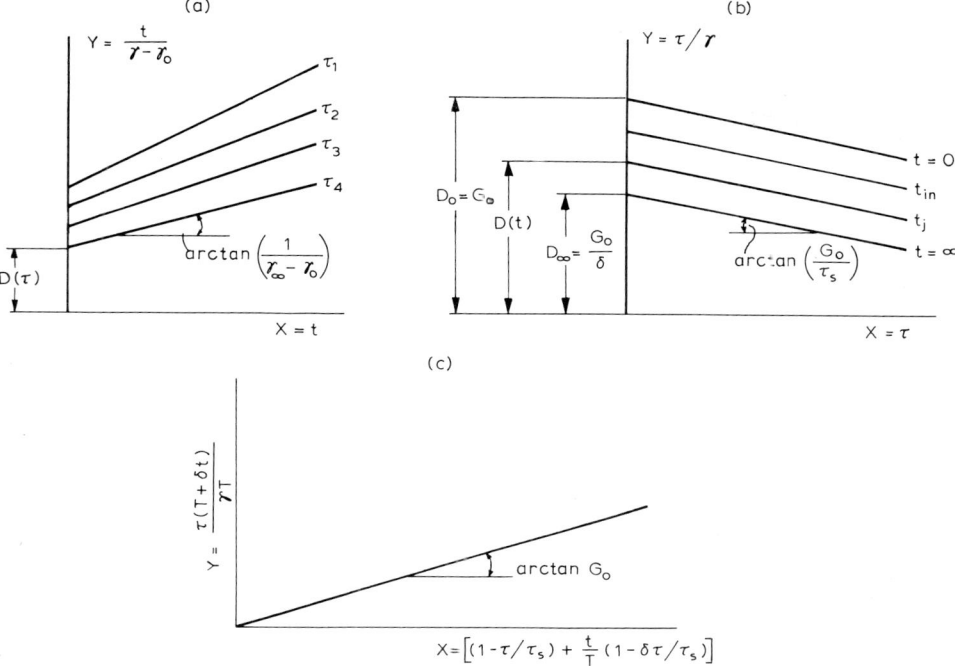

Fig. 6-8. Test of fit of eq. 6-38.

Then, putting $Y = t/(\gamma - \gamma_0)$, $X = t$, $B = 1/(\gamma_\infty - \gamma_0)$, and $D(\tau) = T^*/(\gamma_\infty - \gamma_0)$, we plot the curves on the X–Y coordinates.

The tangents of the slopes of the straight lines will give the value of B, from which we can compute the finite deformation $\gamma_\infty = (1/B) - \gamma_0$. The intercepts on the axis of ordinates will define the values $D(\tau)$, from which we find $T^* = D(\tau)(\gamma_\infty - \gamma_0)$.

Thus, the curves in Fig. 6-8, (a) and (b), enable us to determine the finite, stabilized deformation γ_∞ as well as the parameters of eq. 6-38, i.e. G_0, δ, τ_s and $T = T^*\delta\gamma_0/\gamma_\infty$.

The above parameters can be verified by comparing the computed values $D_j = [G_0(T + t)]/(T + \delta t_j)$ with the values of this quantity as defined by the intercepts at the corresponding instants in Fig. 6-8(b).

For the final test of fit of eq. 6-38, we plot a generalized curve on the coordinates:

$$Y = \frac{\tau(T + \delta t)}{\gamma T}$$
$$X = \left[(1 - \tau/\tau_s) + \frac{t}{T}(1 - \delta\tau/\tau_s)\right]$$
(6-41)

The experimental points (for any values of τ and t) must fall on a straight line $Y = BX$, where $B = G_0$ (Fig. 6-8(c)).

If instantaneous deformation is disregarded, the problem becomes somewhat simpler. In this case we adopt $D = G_{0(in)}(T + t)/(\delta_{in} t)$ and $B = G_{0(in)}/\tau_{s(in)}$, where the subscripts (in) connote that the parameters refer to the isochrones at the instant $t = t_{in} > 0$; for example, $t_{in} = 1$ (Fig. 6-8).

A generalized graph is plotted adopting:

$$Y = \tau t/\gamma, \quad X = t[1 - \delta_{in}\tau/\tau_{s(in)}], \quad D = TG_{0(in)}/\delta_{in}, \quad \text{and} \quad B = G_{0(in)}/\delta_{in}.$$

Evaluating empirical formula. In this chapter we have examined equations describing the pattern of deformation in time, and have demonstrated the technique of checking the formulae for agreement with the experimental data. We have deliberately covered as many formulae as possible from those used in practice. Although overloading the text, this approach seems to be practical, for any phenomenological formula fitting the experimental data can be used, albeit within certain limits. The problem is to select the right formula for a given case.

The above does not imply that in selecting an equation describing soil deformation we need to test many empirical formulae for closeness of fit. The most simple and popular formula — such as the power, logarithmic and linear–fractional forms — will obviously provide a solution to your problem. The criterion helpful in making the right choice is the strain rate in the soil.

Examining strain rate in soil

It was shown in Chapter 5 that the power, logarithmic and linear–fractional equations of creep may be derived from a generalized strain rate function (eq. 5-12). Assuming that $K(t) = \kappa(t)$ and putting:

$$\kappa(t) = \left(\frac{T_2}{T_1 + t}\right)^n \tag{6-42}$$

we may define the pattern of time-dependent deformation from the values of n and T. It will be recalled that for $n < 1$, $T_1 = 0$, $T_2 \neq 0$, we obtain the power relation; for $n = 1$, $T_1 = 0$ and $T_2 \neq 0$, the relation is logarithmic; and for $n = 2$, $T_1 \neq 0$, $T_2 \neq 0$, we arrive at the linear–fractional relation.

Let us put the equation of the γ–t curves into the form:

$$\dot{\gamma} = A\kappa(t) \tag{6-43}$$

where $A = f(\tau) = $ const. and rectify these curves on corresponding coordinates. Putting in eq. 6-42 $T_1 = 0$, $T_2 = T$, we may write eq. 6-43 as:

$$\dot{\gamma} = A\left(\frac{t}{T}\right)^{-n} \tag{6-44}$$

Denoting $Y = \ln \dot{\gamma}$, $X = \ln t$, and $B = -n$, we plot an X–Y graph for various values of τ from which we obtain the value of n. If the plot yields straight lines for $n \neq 1$, this is an indication that the power equation of creep is the best fitting formula; for $n = 1$, the logarithmic relation describes the mechanism of creep.

To test the fit of the linear–fractional equation of creep, we modify eq. 6-43 into:

$$\dot{\gamma} = A \frac{T}{(t + T)^2} \tag{6-45}$$

and then transform it into a linear equation by putting $Y = \dot{\gamma}^{-0.5}$, $X = t$, and $B = (AT)^{-0.5}$.

6.6 REFERENCES

Komarov, I.S., 1972. Data Acquisition and Handling in Engineering and Geological Studies. Nedra, Moscow (in Russian).

Meschyan, S.R., 1974. Mechanical Properties of Soil and Methods of Determining Them in the Laboratory. Nedra, Moscow (in Russian).

Vyalov, S.S. and Pekarskaya, N.K., 1968. Long-term strength of soils. In collective volume: Proc. Coordinating Conf. Hydraul. Eng., Issue 38, Study of Rheological Properties of Soils. Energiya, Leningrad (in Russian).

Chapter 7

THEORIES OF CREEP

7.1 THE THEORY OF LINEAR VISCOELASTIC DEFORMATION

Equation of state

The theory of linear viscoelastic deformation, one of the early fundamentals of rheology, examines a joint manifestation of the elastic and viscous properties of a body. The equation of viscoelasticity was first formulated by Maxwell to describe the phenomenon of relaxation; later Kelvin and Voigt suggested the equation of after-effect.

Equation of linear viscoelasticity were scrutinized in the works of Reiner, Rzhanitsin (see Chapter 1: Reiner, 1958, 1960; Rzhanitsin, 1968), Ishlinsky (1945) and others. The equation in its general form, as postulated by Hohenemser and Prager (1932), is:

$$\alpha_0 + \alpha_1 \tau + \alpha_2 \dot{\tau} = \beta_1 \gamma + \beta_2 \dot{\gamma} \tag{7-1}$$

where $\alpha_0 = -\tau_y$ is the limiting shear stress according to Bingham; $\alpha_1 = 1$; $\alpha_2 = T_r = \eta/G_0$ is the relaxation time; $\beta_2 = \eta$ is the coefficient of viscosity; $\beta_1 = G_\infty$; and G_0 and G_∞ are, respectively, the instantaneous and limiting long-term values of the shear modulus.

Rheological equations of state connect stress, strain, strain rate and time (in their explicit or implicit form). The nature of this connection depends on the accepted hypothesis which gives its name to the theory of creep in question.

Mechanical models

A method of simulating rheological properties of bodies by means of models (Fig. 7-1) is widely used in the theory of linear viscoelasticity. This method is based on the notion that the properties in question can be represented by a combination of elastic, viscous and plastic characteristics.

The elastic properties of a body are simulated by a model in the form of an elastic element — a spring — obeying Hooke's law, $\tau = G\gamma$, and denoted by the symbol H. The model of viscous bodies is a liquid-filled dashpot with a perforated piston moving down the cylinder at a rate obeying Newton's law, $\tau = \eta\dot{\gamma}$, and is denoted by N. The plastic properties are simulated by a dry-friction element obeying Saint-Venant's law $\tau = \tau_y$, where $\tau_y = \tau_s$ is the stress below which no strain is induced; this element is denoted by SV. A series connection of the above elements is symbolized by a dash (e.g., H–N); a parallel connection is indicated by a vertical line (e.g., H|N).

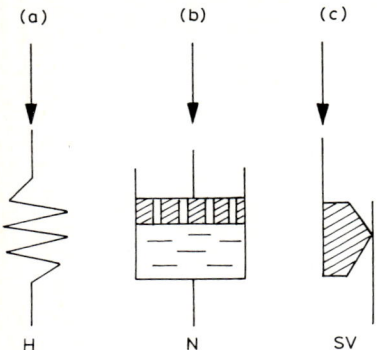

Fig. 7-1. Mechanical models: (a) elastic Hookean body; (b) viscous Newtonian body; (c) plastic Saint-Venant body.

Obviously, the three elements referred to may be combined in a variety of ways so that the combinations obtained will describe various aspects of viscoelastic behaviour.

Models have won general recognition by virtue of their simplicity and their capacity to demonstrate the properties of a body visually. There is hardly a book on rheology which does not include a reference to mechanical models of viscoelastic bodies.

At the same time, the evidence from numerous experiments indicates that model-simulated concepts, and the equations derived from them, lack agreement with the actual behaviour of various materials. This is not surprising as no model — be it a simple device or an intricate combination used by some authors — is capable of representing the real properties of bodies.

However, if models are employed to examine only the qualitative aspect of the problem, the method is an asset owing to its explicitness. Taking this into account and desiring to present a complete picture about rheological theories in their historical development, we are going to make a survey of models.

Kelvin–Voigt body. An elastic element connected in parallel with a viscous element gives the model of a Kelvin body, denoted by the letter K (Fig. 7-2(a)):

$$K = H|N \tag{7-2}$$

The rheological equation of state of a Kelvin body (sometimes referred to as a Kelvin–Voigt body) may be derived from the equation of the model:

$$\tau = \tau^H + \tau^N \tag{7-3}$$

where τ^H and τ^N are the stresses in the elastic (Hookean) and viscous (Newtonian) elements, respectively.

Substituting the values $\tau^H = G_\infty \gamma$ and $\tau^N = \eta \dot{\gamma}$ into this equality, we obtain an equation of deformation in the form of eq. 4-69 examined above:

$$\tau = G_\infty \gamma + \eta \dot{\gamma} \tag{7-4}$$

This expression is a particular case of eq. 7-1 obtained for $\tau_y = T_r = 0$. Relation

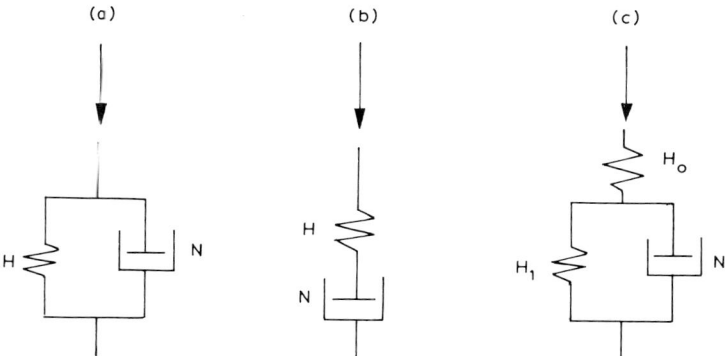

Fig. 7-2. Models of viscoelastic bodies: (a) Kelvin–Voigt body; (b) Maxwell body; (c) generalized viscoelastic body.

7-4, on being integrated at a constant stress (τ = const.), takes the form:

$$\gamma = C e^{-(G_\infty/\eta)t} + \frac{1}{G_\infty} \tau \tag{7-5}$$

An arbitrary constant C is determined from the initial conditions at $t = 0$ when the deformation $\gamma = 0$. Hence, $C = -\tau/G_\infty$, and the law of time-dependent changes in deformation becomes:

$$\gamma = \frac{\tau}{G_\infty}(1 - e^{-(G_\infty/\eta)t}) \tag{7-6}$$

Note that this expression is similar to the equation of elastic aftereffect examined above (see eq. 4-70).

Eq. 7-6 describes an increase in deformation (Fig. 7-3(a)) from zero to a finite, stabilized value $\gamma_\infty = \tau/G_\infty$. In consequence, the parameter G_∞ is treated as the limiting long-term modulus of elasticity (in shear). The quantity $\eta/G_\infty = T_p$ is called the time of aftereffect or the time of deformation lag.

It follows from eq. 7-4 that if the strain is constant (γ = const.), the stress is also constant, being defined by $\tau = G_\infty \gamma$ (Fig. 7-3(a)). Thus, a Kelvin–Voigt body may be referred to as a body possessing the property of aftereffect but devoid of the property of relaxation.

Maxwell body. Linking an elastic element in series with a viscous element, we obtain the model of a Maxwell body, denoted by the letter M (Fig. 7-3(b)):

$$M = H-N \tag{7-7}$$

The rheological equation of state of a Maxwell body may be derived from the equation of the model:

$$\gamma = \gamma^H + \gamma^N \tag{7-8}$$

where γ^H and γ^N are the deformations of the elastic and viscous elements, respectively.

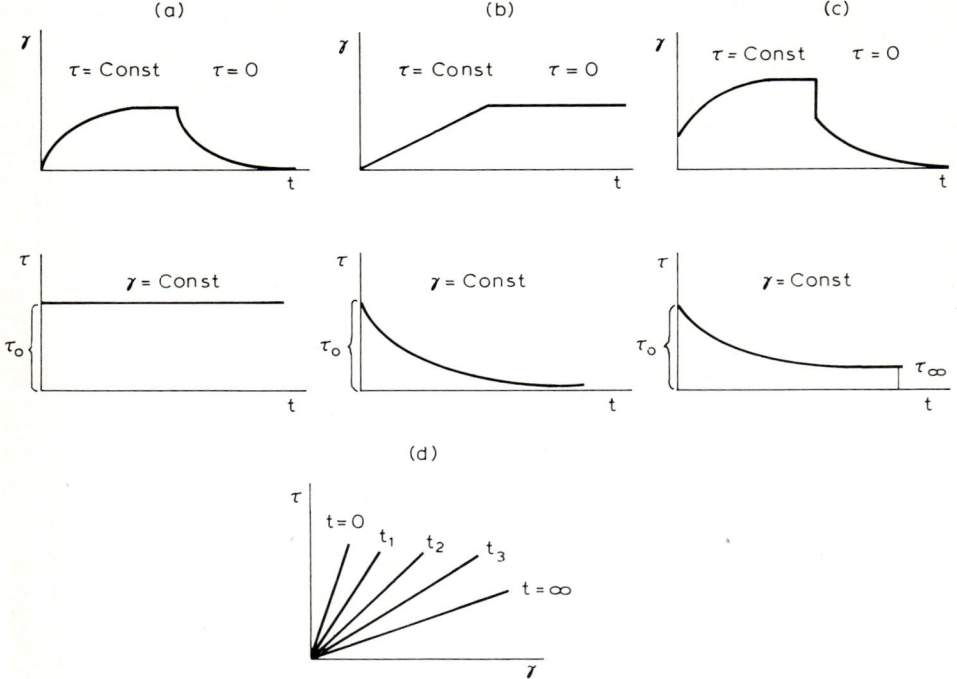

Fig. 7-3. Creep and relaxation of viscoelastic bodies: (a) Kelvin–Voigt body; (b) Maxwell body; (c) generalized viscoelastic body; (d) changes in modulus of deformation G with time.

Taking $\gamma^H = \tau/G_0$ and $\dot{\gamma}^N = \tau/\eta$ and differentiating the above equality, we arrive at eq. 4-73 examined above:

$$\tau + T_r \dot{\tau} = \eta \dot{\gamma} \tag{7-9}$$

where T_r is the relaxation time. This expression is a particular case of eq. 7-1 for $\tau_y = G_\infty = 0$.

When the deformation is constant (γ = const.) and eq. 7-9 is solved for the initial condition $t = 0$ ($\tau = \tau_0$), we arrive at the equation of relaxation 4-74 considered above:

$$\tau = \tau_0 e^{-t/T_r} \tag{7-10}$$

In accordance with eq. 7-10, the stress relaxes in time from an initial value τ_0 to zero (Fig. 7-3(b)). Consequently, the parameter G_0 in this equation is to be considered as the initial hypothetically instantaneous modulus of elasticity (in shear).

Whenever the stress is constant (τ = const.), eq. 7-9 is transformed into Newton's equation, $\dot{\gamma} = \tau/\eta$, corresponding to a continuous increase in strain at a constant rate (Fig. 7-3(b)). This implies that a Maxwell body may be treated as a body possessing the property of relaxation but devoid of the property of aftereffect.

Generalized viscoelastic body. The Maxwell and Kelvin–Voigt models each represent just one aspect of the process of viscoelastic deformation, i.e. either the aftereffect or the relaxation.

The Hohenemser and Prager equation (eq. 7-1) is more general, describing both processes simultaneously. The properties of a body described by this equation may be simulated by a combination of the Maxwell and Kelvin–Voigt models (Fig. 7-2(c)):

$$HP = H_0 - (H_1 | N) = H-K \tag{7-11}$$

Let us denote the elastic moduli of the Hookean elements H_0 and H_1 by G_0 and G_1, respectively, and the coefficient of viscosity of the Newtonian element N by η. The stresses in these elements will be $\tau_0^H = G_0\gamma_0^H$; $\tau_1^H = G_1\gamma_1^H$; $\tau^N = \eta\dot{\gamma}^N$, where the symbol H refers to the Hookean elements and the symbol N to the Newtonian elements.

The equations of the system will take the form:

$$\tau = \tau_0^H = \tau_1^H + \tau^N \qquad \gamma = \gamma_0^H + \gamma_1^H \qquad \gamma_1^H = \gamma^N$$

Substituting the values $\tau_0^H = G_0\gamma_0^H$; $\tau_1^H = G_1\gamma_1^H$; and $\tau^N = \eta\dot{\gamma}^N$ into these equations, we obtain, after manipulation:

$$\frac{G_1 + G_0}{G_0}\tau + \frac{\eta}{G_0}\dot{\tau} = G_1\gamma + \eta\dot{\gamma}$$

Putting in the above equality:

$$G_\infty = \frac{G_1 G_0}{G_1 + G_0} \qquad T_r = \frac{\eta}{G_0 + G_1} \tag{7-12}$$

we arrive at the equation of deformation:

$$\tau + T_r\dot{\tau} = G_\infty\gamma + G_0 T_r\dot{\gamma} \tag{7-13}$$

which, for $\tau_y = 0$, coincides with eq. 7-1.

Solving eq. 7-13 for constant loading (τ = const.) we obtain:

$$\gamma = C\exp\left(-\frac{G_\infty}{G_0}\frac{1}{T_r}t\right) + \frac{1}{G_\infty}\tau \tag{7-14}$$

Hence, under the initial conditions when $\gamma = \gamma_0$ at $t = 0$, the equation of elastic aftereffect takes the form:

$$\gamma = \frac{\tau}{G_\infty} - \tau\left(\frac{1}{G_\infty} - \frac{1}{G_0}\right)e^{-t/T_p} \tag{7-15}$$

where $T_p = (G_0/G_\infty)T_r$.

According to the above equation we have, at $t = 0$, $\gamma_0 = \tau/G_0$ and at $t = \infty$, $\gamma_\infty = \tau/G_\infty$. Expression 7-15 may then be rewritten (Fig. 7-3(c)) as:

$$\gamma = \gamma_\infty - (\gamma_\infty - \gamma_0)e^{-t/T_p} \tag{7-16}$$

This clarifies the meaning of the initial G_0 and finite G_∞ deformation moduli, the former defining the connection between the stress and the instantaneous strain γ_0 and the latter determining the relation between the stress and the finite, stabilized strain γ_∞ (Fig. 7-3(d)).

In the case of unloading ($\tau = 0$), eq. 7-13 yields the following solution:

$$\gamma = \left[\gamma_0 - \frac{\tau_0}{G_0}\right] \exp\left(-\frac{t - t_0}{T_p}\right) \tag{7-17}$$

where t_0 is the instant of unloading and τ_0 and γ_0 are, respectively, the stress and strain at this instant.

When deformation is constant (γ = const.), eq. 7-13 describes the relaxation of stresses (Fig. 7-3(c)):

$$\tau = G_\infty \gamma + (G_0 - G_\infty) e^{-t/T_r} = \tau_\infty + (\tau_0 - \tau_\infty) e^{-t/T_r} \tag{7-18}$$

where τ_0 is the initial value of the stress (at $t = 0$) and τ_∞ is the finite value (at $t = \infty$).

Thus, the generalized eq. 7-13 describes both the mechanism of elastic aftereffect and that of relaxation, taking account — unlike eq. 7-4 — of the instantaneous deformation and — unlike eq. 7-9 — of a reduction in the stress to some finite value rather than to zero.

Bingham and Schwedoff bodies. An elastoplastic Prandtl body (see Fig. 4-1(b)) is simulated by a Saint-Venant element and an elastic Hookean element connected in series. The model thus obtained demonstrates that the stresses $\tau < \tau_y$ give rise to an elastic deformation, $\gamma = \tau/G$, whereas the stress $\tau = \tau_y$ induces an unconfined deformation, $\gamma \to \infty$.

The model of viscoplastic Bingham body consists of an elastic element H, a viscous element N and a Saint-Venant element SV (Fig. 7-4(b)):

$$B = H - (N|SV) \tag{7-19}$$

The pattern of deformation of a Bingham body may be defined by the condition

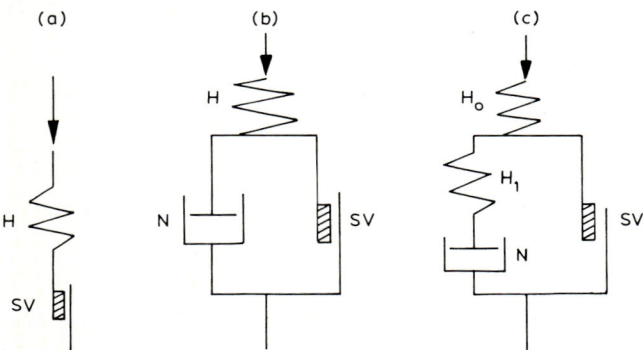

Fig. 7-4. Models of elastoplastic and viscoplastic bodies: (a) Prandtl body; (b) Bingham body; (c) Schwedoff body.

$\tau = \tau^H + \tau_y$. Hence, it follows that:

for $\tau < \tau_y$ $\quad \tau = G\gamma$

for $\tau \geqslant \tau_y$ $\quad \tau - \tau_y = \eta_{pl}\dot{\gamma}$ (7-20)

where G is the elastic modulus of the Hookean element; η is the coefficient of viscosity of the Newtonian element; τ_y is the ultimate shear strength of the Saint-Venant element. These expressions correspond to eq. 4-68.

The model of another viscoplastic body may be made from two elastic elements H_0 and H_1 connected to a viscous element N and a Saint-Venant element SV (Fig. 7-04(c)). This is known as a Schwedoff body:

$$SW = H_0 - [(H_1|N)|SV] = H_0 - (M|SV) \tag{7-21}$$

The equation of the deformation of a Schwedoff body corresponds to eq. 4-76 examined above:

$$\dot{\gamma} = \frac{\tau - \tau_y}{\eta_{pl}} + \frac{\dot{\tau}}{G} \tag{7-22}$$

Combination models

In addition to the models considered above, there are other models in which elastic, viscous and Saint-Venant elements are represented in various combinations. The Poynting–Thomson model, PT, is a Hookean element and a Maxwell element connected in parallel, $PT = H|M$; the Lethersich model, L, is a Newtonian element and a Kelvin element connected in series, $L = N–K$; the Jeffreys model, J, represents a parallel connection of a Newtonian element with a Maxwell element, $J = N|M$; the Burgers model, Bu, consists of series-connected Maxwell and Kelvin elements, $Bu = M–K$.

The Jeffreys model described by the equation:

$$\tau + \tau_{r(1)}\dot{\tau} = \eta(\dot{\gamma} + T_{r(2)}\ddot{\gamma}) \tag{7-23}$$

which, for $\tau =$ const., gives:

$$\gamma = \gamma_0 + A[1 - e^{-t/T_{r(2)}}] + \frac{\tau}{\eta}t \tag{7-24}$$

This model has been devised to describe the behaviour of the earth's crust, assuming that the relaxaton time of rock is $T_r = 10^8$ s and the viscosity is $\eta = 5 \cdot 10^{20}$ P approx.

Multi-element models. A significant departure of the simulated properties of real bodies from their actual behaviour has urged some authors to use more complicated models comprising a large number of elastic and viscous elements. A joint manifestation by the elements of different elastic (G_i) and viscous (η_i) properties is used to define the macro-behaviour of a body. By way of illustration, such a model may be obtained from series-connected Kelvin models or parallel-connected Maxwell models. Some of the models referred to will be examined in Chapter 8 while considering the process of soil consolidation.

In the general case (eq. 7-1), the differential equation of a model comprising a large (yet finite) number of elements will, for $\alpha_0 = 0$, take the form:

$$\alpha_1 \tau + \alpha_2 \frac{d\tau}{dt} + \alpha_3 \frac{d^2\tau}{dt^2} + \ldots + \alpha_{n+1} \frac{d^n\tau}{dt^n}$$

$$= \beta_1 \gamma + \beta_2 \frac{d\gamma}{dt} + \beta_3 \frac{d^2\gamma}{dt^2} + \ldots + \beta_{n+1} \frac{d^n\gamma}{dt^n} \qquad (7\text{-}25)$$

Apparently, when $\alpha_3 = \ldots = \alpha_{n+1} = \beta_3 = \ldots = \beta_{n+1} = 0$, this equation is transformed into eq. 7-13.

Integral form of deformation equations

Returning to the equation of deformation (7-13), whose solutions for $\tau = $ const. and $\gamma = $ const. have already been examined, let us derive a general solution for the stresses and strains changing with time, $\tau(t)$ and $\gamma(t)$, according to any law.

Solving the differential eq. 7-13 for strain, we obtain the expression:

$$\gamma = \gamma_0 e^{-(\beta_1/\beta_2)t} + \frac{\alpha_1}{\alpha_2} \int_0^t \left[\tau(v) + \frac{\alpha_2}{\alpha_1} \dot{\tau}(v) \right]^{-(\beta_1/\beta_2)t-v} \qquad (7\text{-}26)$$

Integrating by parts we arrive at:

$$\gamma = \frac{1}{G_0} \left[\tau(t) + \int_0^t \tau(v) K(t - v) \, dv \right] \qquad (7\text{-}27)$$

where:

$$K(t - v) = \frac{G_0 - G_\infty}{G_\infty T_p} e^{-(t-v)/T_p}$$

is the kernel of the integral relation of eq. 7-27; $G_\infty = \beta_1/\alpha_1$; $G_0 = \beta_2/\alpha_2$; $T_p = \beta_2/\beta_1 = T_r G_0/G_\infty$; and $T_r = \alpha_2/\alpha_1$.

Solving eq. 7-13 for stress, we obtain:

$$\tau = G_0 \left[\gamma(t) - \int_0^t \gamma(v) R(t - v) \, dv \right] \qquad (7\text{-}28)$$

where:

$$R(t - v) = \frac{G_0 - G_\infty}{G_0^2 T_r} e^{-(t-v)/T_r}$$

is the kernel of the integral relation of eq. 7-28.

If we set up a generalized model comprising a finite number n of elements with parameters G_j and η_j each having a value of its own, K and R will be expressed by the sum of the exponential relations given above. In the case of an infinitely large number of elements, the summation will be replaced by integration.

7.2 MECHANICAL MODELS OF SOIL

Models simulating soil consolidation

Numerous attempts to simulate the processes of soil deformation by means of mechanical models have been made in soil mechanics.

The classical Terzaghi–Gersevanov model (see Chapter 1: Gersevanov, 1937), describing the process of consolidation of a water-saturated soil, is represented in its traditional form as a dashpot filled with a viscous liquid sinking into which is a perforated piston connected to an elastic spring (Fig. 7-5(a)). The liquid in the dashpot is the analog of pore water, the piston perforations simulate soil capillaries, and the spring represents the soil skeleton. It is assumed that the load p, transmitted

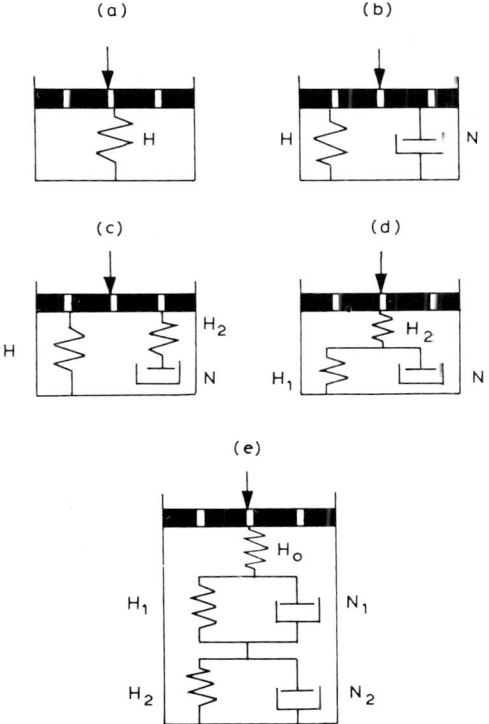

Fig. 7-5. Mechanical models simulating soil consolidation: (a) Terzaghi–Gersevanov model; (b) Taylor model; (c) Tan model; (d) Gibson–Lo model (e) Lo model.

at first to pore water, is gradually taken up by the spring as the piston continues to sink, being thus shared by the liquid in the dashpot (pore pressure u) and the elastic spring (effective pressure σ'). The process terminates when all the load has been transmitted to the spring ($\sigma' = p$) and this ceases compressing. Since the piston and spring are connected in parallel, this model is identical with the Kelvin–Voigt model depicted in Fig. 7-2(a).

Further modifications of the Terzaghi model were all aimed at taking account of secondary consolidation. According to Taylor (1942) the process of soil consolidation may be simulated by an analog consisting of a Kelvin model submerged into a liquid-filled dashpot (Fig. 7-5(b)). Since, in the case under consideration, the Kelvin model simulates the behaviour of the soil skeleton, this is endowed with viscoelastic properties, i.e. it is capable of secondary consolidation.

Anagnosty (1979) made a more complex model by adding another elastic element. He produced, in fact, two models: a Kelvin model simulating the isotropic strain tensor and a Maxwell model representing the deviatory tensor.

Tan (1957), a pioneer of the research into the behaviour of soils by means of rheological models, used a Poynting–Thompson model submerged in a liquid-filled dashpot (this model is illustrated in Fig. 7-5(c)). In the model due to Gibson and Lo (1961), depicted in Fig. 7-5(d), the soil skeleton is represented by a Hohenemser–Prager model simulating V.A. Florin's theory of soil consolidation.

Referring to this model, we will consider the simulation of primary and secondary consolidation. At the initial instant, the effective stress is taken up by elastic element H_2 and viscous element N. As the piston continues to sink down the large cylinder while the pressure is being redistributed between the pore water and the soil skeleton, the effective stress rises with the result that spring H_1 begins compressing along with spring H_2. When the effective stress reaches its full value of $\sigma' = p$, spring H_2 stops compressing, indicating that primary consolidation has ended. A further process goes on in response to the full stress p sustained by the elastic element H_1 and viscous element N; this is the stage of secondary consolidation.

Subsequently, Lo (1961) devised a more complex model in which the model referred to above was supplemented by a Kelvin element connected through a Saint-Venant element arranged in series to represent the structural strength of the soil (Fig. 7-5(e)).

In an effort to improve the simulation of soil consolidation, some researchers employed multicomponent models. Thus, a model designed by Schiffman et al. (1966) consisted of two or three Kelvin elements connected in series by analogy with the model of Fig. 7-5(e)[*1].

It can be seen that all the mechanical models simulating the consolidation of water-saturated soil are based on the notion that pressure is redistributed between the pore water and the soil skeleton. Simulating the pore water in all models is a liquid-filled external dashpot, while the soil skeleton is represented by elastic or viscous elements in various combinations. It is the addition of a viscous element to

---

[*1] Introduced between the first and second Kelvin elements in the Lo model is a Saint-Venant element omitted in Fig. 7-5(e). No such element is incorporated into the Schiffman model.

the model which allows us to simulate the viscous resistance of the soil skeleton producing secondary consolidation.

None of the combinations of elastic and viscous elements in use is arranged in accordance with some principle. They are just a means of representing certain particulars of the process. For example, if all the models we have referred to dispense with the external dashpots, they will simulate the process of consolidation of a non-saturated soil in which the entire load is sustained by the skeleton only.

Models simulating creep in shear

To simulate creep in shear, use is made of models representing both attenuating and non-attenuating deformations.

The author has obtained such a model (see Chapter 1: Vyalov, 1959) by connecting a Hohenemser–Prager model in series with a Bingham model (Fig. 7-6(a)). In this case, the equation of deformation may be derived from the sum $\gamma^{HP} + \gamma^B$, where γ^{HP} is the deformation of the Hohenemser–Prager model and γ^B is the deformation of the Bingham model.

For a constant load ($\tau = $ const.) the value of γ^{HP} is defined by eq. 7-15 and that of γ^B by eq. 7-20. For a stress $\tau < \tau_y$, we obtain:

$$\gamma = \frac{\tau}{G_\infty} - \tau \left(\frac{1}{G_\infty} - \frac{1}{G_0} \right) e^{-t/T_p} \tag{7-29}$$

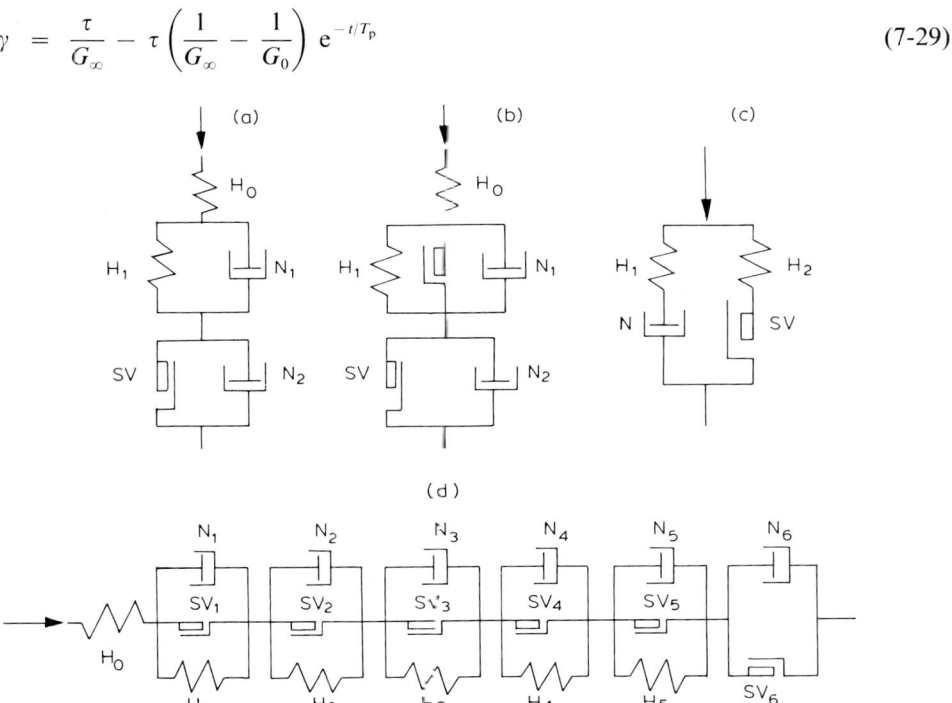

Fig. 7-6. Mechanical models simulating the process of creep in shear: (a) Vyalov model; (b) Goldstein model; (c) Kisiel–Folque model; (d) Fedder–Bredth model.

and for $\tau \geqslant \tau_y$:

$$\gamma = \frac{\tau}{G_\infty} - \tau\left(\frac{1}{G_\infty} - \frac{1}{G_0}\right) e^{-t/T_p} + \frac{\tau - \tau_y}{\eta_{pl}} t \qquad (7\text{-}30)$$

where G_0 and G_∞ are the initial and finite shear moduli; T_p is the time of aftereffect (deformation attenuates); and η_{pl} is the coefficient of plastic (Bingham) viscosity.

Thus, for a stress $\tau < \tau_y$, the model describes attenuating deformation, and for $\tau \geqslant \tau_y$ it simulates non-attenuating deformation including the deformation of elastic aftereffect and the viscoplastic flow of soil. On removing the load, the soil recovers from the instantaneous deformation and the deformation of elastic aftereffect; the deformatin of viscoplastic flow is irrecoverable.

By adding another Saint-Venant element to the model in Fig. 7-6(a) we obtain a model suggested by M.N. Goldstein (Fig. 7-6(b)). This model presumes the existence of two limiting stresses in soil — a proportional limit SV_1 and a yield point SV_2 — which, as long as they are not exceeded, give rise only to elastic deformation and only to viscoplastic flow, respectively (see Chapter 1: Goldstein, 1979).

A model with just one viscous element, similar to the Schwedoff model, is illustrated in Fig. 7-6(c). It has been suggested by Kisiel for use in conjunction with soils as it serves to simulate the elastic aftereffect for $\tau < \tau_y$ and the Maxwell flow for $\tau \geqslant \tau_y$ (see Chapter 1: Kisiel and Lysik, 1966). An identical model was devised by Folque (see Chapter 1: Folque, 1961) to describe the consolidation of unsaturated soil. Adding another spring after the Saint-Venant element, he used it to represent the process of consolidation of an overcompacted soil.

Some soil models in use are composed of several other models as, for example, the model according to Fedder and Bredth (1973) depicted in Fig. 7-6(d).

Mechanical models of soils having variable viscosity

In all the models examined above it was assumed that the viscosity of soil, represented by a Newtonian element N, is constant. There are, however, models which take into account changes in viscosity with load. This applies to the model (Fig. 7-7(a)) devised by Murayama and Shibata (1966). Proceeding from Eyring's physical theory of rates process, the authors came to the conclusion that the viscosity of bound water present in soil might be expressed by the non-linear relation:

$$\eta = \frac{\tau_N}{A(\tau - \tau_y) \sinh\left(\dfrac{B\tau_N}{\tau - \tau_y}\right)} \qquad (7\text{-}31)$$

The variable viscosity η thus obtained is inherent in the Newtonian element $N(\tau)$.

The equation of the model may be obtained from the conditions $\gamma = \gamma_1 + \gamma_2$ and $\tau = \tau_1 = \tau_2$, where the subscript 1 refers to the elastic element H_1 and the subscript 2 refers to the elastic element H_2 and the non-linearly viscous element $N(\tau)$ connected

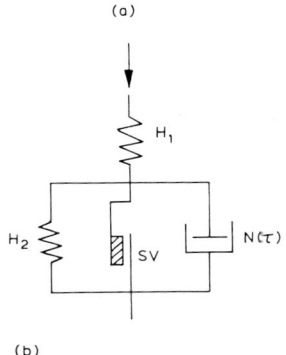

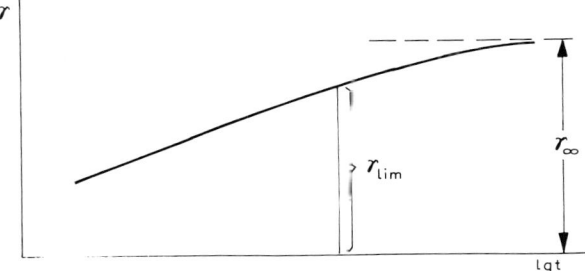

Fig. 7-7. Murayama and Shibata model allowing for variable viscosity of soil: (a) model; (b) creep curve.

thereto. Hence, we arrive at the following equation of deformation:

$$\gamma = \frac{\tau}{G_1} + \frac{\tau - \tau_y}{G_2} + \frac{\tau - \tau_y}{2G_2} \lg\left(\frac{AB}{2} G_2 t\right) \qquad (7\text{-}32)$$

which is a logarithmic equation of a time-dependent deformation within the confines $\gamma \leqslant \gamma_{\lim} = \gamma_0 + (\tau - \tau_y)(2B - 1)/2BG_2$, where $\gamma_0 = \gamma_1 = \tau/G_1$.

For $\gamma \geqslant \gamma_{\lim}$, the deformation develops with time in accordance with a more complex law, acquiring at $t \to \infty$ a finite value:

$$\gamma_\infty = \frac{\tau}{G_1} + \frac{\tau - \tau_y}{G_2} \qquad (7\text{-}32')$$

Unlike the traditional models, indicating that the deformation obeys an exponential law, the Murayama model provides for a logarithmic law. The corresponding creep curve is depicted in Fig. 7-7(b).

Soil models based on the assumption that the elastic elements obey Hooke's law and that the viscous elements are non-linear, being defined by Eyring's theory, were suggested by Christensen (1964), Abdel-Hardi (1966) and others. They differ from Murayama's model by the arrangement of the elastic and viscous elements.

An attempt to devise a model simulating changes in soil structure was made by Budin (see Chapter 1: Budin, 1974). In his model the load is transmitted through an

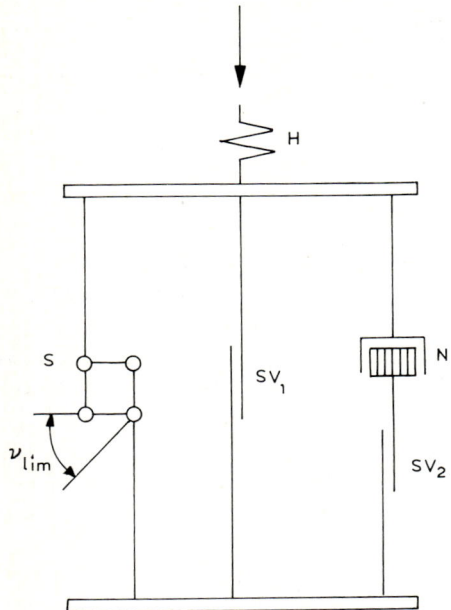

Fig. 7-8. Budin model allowing for structural changes in soil.

elastic element H connected in series with a viscous element N and an element S representing non-linear structural deformation (Fig. 7-8). This last element is capable of responding to the changing orientation of soil particles due to deformation; the changes in orientation which alter the strain rate are simulated by a skewing arrangement incorporated into the structural deformation element.

The viscous element obeys Newton's law, $\tau_N = \eta \gamma_N$. The deformation of the structural element is described by the non-linear relation:

$$\gamma_s = \gamma^* \tanh(\tau_s t / \eta_s)$$

where $\gamma^* = \gamma^p / \sin v_s$ (here γ^p is the relative shear at the end of structural rearrangement); and v_s is the ultimate skewing angle of the structural element S. The model also includes two Saint-Venant (or, to be precise, Coulomb) elements SV_1 and SV_2 with yield points $\tau_{y(1)}$ and $\tau_{y(2)}$, respectively.

The equation of the model is obtained from the condition $\gamma = \gamma_s + \gamma^N$ and $\tau = \tau_s = \tau^N$. For a load $\tau < \tau_{y(1)}$, no creep is present; a load $\tau > \tau_{y(2)}$ gives rise to an attenuating creep associated with the rearrangement of soil particles:

$$\gamma = \gamma^* \tan\left\{[(\tau - \tau_{y(1)})t]\frac{1}{\eta_s}\right\} \tag{7-32''}$$

The attenuating creep described by this expression terminates when the particles cease to rearrange themselves ($v = v_{\lim}$) and further deformation may take place in the form of steady flow:

$$\tau = \tau_{y(2)} + \tau_{y(1)} + \eta\dot{\gamma} \tag{7-33}$$

provided that:

$$\tau_{y(1)} < \tau < [\tau_{y(1)} + \tau_{y(2)}]$$

Conclusion. In concluding, we stress once more that mechanical models are capable of providing only schematic and qualitative representations of the real rheological processes taking place in soils. They fail, as a rule, to give a quantitative fit with the experimental data. One of the reasons of this limitation is that most equations that are derived with the aid of models yield exponential creep curves only — a fact explained by our original differential eq. 7-13.

7.3 THE THEORY OF HEREDITARY CREEP

Linear hereditary creep

It was shown in the preceding section that the differential equation of deformation (eq. 7-1) may be replaced by integral relations 7-27 and 7-28, having kernels $K(t - v)$ and $R(t - v)$ in the form of exponential functions or as the sum of these functions. However, by specifying in advance the form of the deformation equation (exponential), we drastically reduce the field of its application. On the other hand, the integral equation of deformation may be derived in a more general way and its kernels, $K(t - v)$ and $R(t - v)$, may have any form fitting the experimental data of any given material.

The pattern of deformation presented in such a form is called the law of hereditary creep, due to Boltzmann (see Chapter 4: Boltzmann, 1876) and Volterra (see Chapter 4: Volterra, 1913, 1931). It is based on the principle of superposition according to which the strain at any instant t caused by loads applied during the preceding instants v equals the sum of the strain which would develop by the time t if each of the loads acted independently. In other words, the strain at a given time depends not only on the magnitude of the stress at that time but also on the history of the preceding deformation, as if inheriting the effect of the preceding stress. Hence the term hereditary creep.

Let a load $\tau(v)$ applied to a body at an instant v for a short time interval Δv induce a strain $\gamma(v)$. The strain at any arbitrary time $t > v$ will be proportional to the stress $\tau(v)$, the period of stress application Δv and a function $K(t - v)$ dependent upon the body properties and the time interval $t - v$ elapsed between the instant v of load application and the instant t at which the strain $\gamma(t)$ is to be determined. Moreover, if at the instant t under consideration a time-dependent stress $\tau(t)$ is applied to the body, it will induce an instantaneous strain $\gamma = \tau(t)/G_0$, where G_0 is the instantaneous modulus of elasticity. Thus, the aggregate strain at an instant $t > v$ will be:

$$\gamma = \frac{\tau(t)}{G_0} + \frac{1}{G_0}\tau(v)K(t - v)\Delta v$$

If a body sustains various stresses $\tau(v_j)$ set up at various instants v_j and acting during various time intervals Δv_j, its aggregate strain at an instant t will be defined — in accordance with the principle of superposition — by the sum of the strains due to each of the stresses $\tau(v_j)$. In the case of continuous loading, the summation is replaced by an integration from which we obtain a creep equation similar to eq. 7-27:

$$\gamma = \frac{1}{G_0}\left[\tau(t) - \int_0^t K(t - v)\tau(v)\,\mathrm{d}(v)\right] \tag{7-34}$$

The first term on the right-hand side represents the instantaneous strain γ_0 produced by the time-dependent stress $\tau(t)$ at the instant t, and the second term expresses the time-dependent strain caused by the time-dependent stress $\tau(v)$.

Solving integral eq. 7-34 with respect to stress, we obtain an equation of relaxation similar to eq. 7-28:

$$\tau = G_0\left[\gamma(t) - \int_0^t R(t - v)\gamma(v)\,\mathrm{d}v\right] \tag{7-35}$$

The first term on the right-hand side of the equation represents the initial stress at the instant t arising due to the time-dependent strain $\gamma(t)$, and the second term reflects the attenuation of the stress in time with the changes in the strain $\gamma(v)$.

Eqs. 7-34 and 7-35 are referred to as quadratic Volterra equations. Eq. 7-35 is the solution of eq. 7-34 for τ and, similarly, eq. 7-34 is the solution of eq. 7-35 for γ.

The kernels $K(t - v)$ and $R(t - v)$ of the integral equations are functions of the two variables, t and v. In such a case, the function $R(t - v)$ — the relaxation kernel — is the resolvent of the kernel $K(t - v)$ and the function $K(t - v)$ — the creep kernel — is the resolvent of the kernel $R(t - v)$. Hence, it is sufficient to determine just one of the functions. The measurement unit of K and R is 1/h.

The lower limit of integration of eqs. 7-34 and 7-35, corresponding to the beginning of the period of load-application, is taken as zero on the assumption that neither stresses nor strains are present in the body at this instant. If, however, the body is already in a state of stress and strain at the instant of load application (as this is, generally speaking, the case in rocks), the lower limits of integration of the above relations are taken at $-\infty$.

For a constant load, $\tau = \mathrm{const.}$, eq. 7-34 takes the form:

$$\gamma = \frac{\tau}{G_0}\left[1 + \int_0^t K(t)\,\mathrm{d}t\right] \tag{7-36}$$

and for a constant strain, $\gamma = $ const., eq. 7-35 becomes:

$$\tau = G_0\gamma\left[1 - \int_0^t R(t)\,dt\right] \tag{7-37}$$

Kernels of integral equation

Differentiating relations 7-36 and 7-37, we obtain:

$$\begin{aligned}\frac{1}{G_0}K(t) = \bar{K}(t) &= \frac{1}{\tau}\frac{d\gamma}{dt} \\ G_0 R(t) = \bar{R}(t) &= -\frac{1}{\gamma}\frac{d\tau}{dt}\end{aligned} \tag{7-38}$$

Hence, the kernel $\bar{K}(t)$ is a function defining the time-dependent changes in strain rate for a unit stress $\tau = 1$, and the kernel $\bar{R}$ is a function of defining the time-dependent changes in stress required to maintain a constant strain $\gamma = 1$.

This provides for a simple method of determining the functions $\bar{K}(t)$ and $\bar{R}(t)$ from experimental data. The function $\bar{K}(t)$ defines the time-dependent alteration of creep rate under a unit load, and the function $\bar{R}(t)$ describes the time-dependent changes in relaxation rate due to a unit strain.

A feature of the functions $K(t)$ and $R(t)$ is that, in the case of an attenuating creep, $K(0) = R(0) = \infty$ and $K(\infty) = R(\infty) = 0$. Accordingly, for $t = \infty$, the integrals of eqs. 7-36 and 7-37 have finite values. Thus, at $t = \infty$, we obtain from eq. 7-36:

$$\int_0^\infty K(t)\,dt = (G_0 - G_\infty)/G_\infty$$

where $G_\infty = \tau/\gamma_\infty$ is the modulus of finite strain.

In the case of a non-attenuating creep transformed into a constant rate flow at $t = \infty$:

$$K(\infty) = \text{const.} \quad \text{and} \quad \int_0^\infty K(t)\,dt = \infty$$

Integral equations are versatile and conducive to all the above relationships of viscoelastic deformation, provided that the correct type of functions $K(t)$ and $R(t)$ are chosen.

Non-linear hereditary creep

We have examined the equations of hereditary creep for a linear stress–strain relation. The theory of non-linear hereditary creep, which has been developed by Rabotnov (see Chapter 1: Rabotnov, 1966), enables us to take account of a non-linear

$\tau-\gamma$ relation. To that end, we introduce the functions of stress and strain $f(\tau)$ and $\phi(\gamma)$, respectively, defining the $\tau-\gamma$ relation, into integral eqs. 7-14 and 7-15 instead of the stresses and strains proper. Three different kinds of integral relation can be obtained in this case, depending on the configuration of isochrones mentioned above (see Fig. 5-6). According to Rozovsky (1951, 1955) these relations are as follows.

(1) A $\tau-\gamma$ relation in which isochrones are dissimilar at various times and each is described by a law $\phi_j(\gamma)$ of its own:

$$\gamma = f_0[\tau(t)] + \int_0^t Q(\tau, t - v) \, dv$$

$$\tau = \phi_0[\gamma(t)] - \int_0^t R(\gamma, t - v) \, dv \qquad (7\text{-}39)$$

where $f_0(\tau) = \gamma_0$ and $\phi_0(\gamma) = \tau_0$ are the initial strains and stresses, respectively; $Q(\tau, t - v)$ and $R(\gamma, t - v)$ are the kernels of integral equations which are functions of the time $t - v$ and the stress τ or the strain γ.

(2) A $\tau-\gamma$ relation in which isochrones are similar at all times except at the initial instant; the functions $\phi(\gamma)$ and $f(\tau)$ have in this case two values, (i) for $t = 0$, $\phi_0(\gamma)$ and $f_0(\tau)$, and (ii) for $t > 0$, $\phi(\gamma)$ and $f(\tau)$:

$$\gamma = f_0[\tau(t)] + \int_0^t Q(t - v) f[\tau(v)] \, dv$$

$$\tau = \phi_0[\gamma(t)] - \int_0^t R(t - v) \phi[\gamma(v)] \, dv \qquad (7\text{-}40)$$

(3) A $\tau-\gamma$ relation in which isochrones are similar at all times, including the initial instant $0 \leq t \leq \infty$, and are described by the same function $\phi(\gamma)$ and $f(\tau)$. In this case, according to Yu.N. Rabotnov, the equation of relaxation takes the form:

$$\tau = \phi[\gamma(t)] - \int_0^t R(t - v) \phi[\gamma(v)] \, dv \qquad (7\text{-}41)$$

The equation of creep will be:

$$\phi(\gamma) = \tau(t) + \int_0^t K(t - \tau) \tau(v) \, dv \qquad (7\text{-}42)$$

which, when the creep curves are similar, becomes:

$$\gamma = f[\tau(t)] + \int_0^t Q(t - v) f[\tau(v)] \, dv \qquad (7\text{-}43)$$

For τ = const. and γ = const., the above relations are:

$$\phi(\gamma) = \tau\left[1 + \int_0^t K(t)\,dt\right]$$

$$\gamma = f(\tau)\left[1 + \int_0^t Q(t)\,dt\right] \qquad (7\text{-}44)$$

$$\tau = \phi(\gamma)\left[1 - \int_0^t R(t)\,dt\right] \qquad (7\text{-}45)$$

Relation 7-43 may be presented as $\gamma = \gamma_0 \bar{Q}(t)$, where $\bar{Q} = 1 + \int_0^t Q(t)\,d\tau$. Hence, the problem of creep (linear and non-linear) is amenable to solution (with some reservations) in terms of the theory of linear or non-linear elasticity for a hypothetically instantaneous state; the time factor is allowed for by multiplying the strain by the time function $\bar{Q}(t)$.

$R(t)$ and $K(t)$ — the kernels of integral equations — in eqs. 7-41 through 7-45 are reciprocal resolvents:

$$K(t) = \frac{1}{\tau}\frac{d}{dt}\phi(\gamma) \qquad R(t) = -\frac{1}{\phi(\gamma)}\frac{d\tau}{dt} \qquad (7\text{-}46)$$

The kernel $Q(t)$ is not the resolvent of $R(t)$; it is defined by the formula:

$$Q(t) = \frac{1}{f(\tau)}\frac{d\gamma}{dt} = K(t)\frac{\tau}{d(\tau)}\frac{1}{\phi'(\gamma)} \qquad (7\text{-}47)$$

For example, for the power form of the functions $\phi(\gamma) = A_0\gamma^m$ and $f(\tau) = (\tau/A_0)^{1/m}$, the kernels are:

$$K(t) = \frac{A_0 m \gamma^{m-1}}{\tau}\frac{d\gamma}{dt}$$

$$Q(t) = \left(\frac{A_0}{\tau}\right)^{1/m}\frac{d\gamma}{dt}$$

$$R(t) = -\frac{1}{A_0\gamma^m}\frac{d\tau}{dt}$$

Although the function $K(t)$ is of a more complex form than the function $Q(t)$, eq. 7-42 appears to be more convenient in some cases.

Note that the functions K, R and Q are readily replaceable by $\bar{K}$, $\bar{R}$ and $\bar{Q}$. For the power τ–γ relation, the values of $\bar{K}$, $\bar{R}$ and $\bar{Q}$ are given by:

$$\bar{K}(t) = \frac{K(t)}{A_0} = \frac{m\gamma^{m-1}}{\tau} \frac{d\gamma}{dt}$$

$$\bar{R}(t) = A_0 R(t) = -\frac{1}{\gamma^m} \frac{d\tau}{dt}$$

$$\bar{Q}(t) = \frac{Q(t)}{A_0^{1/m}} = \left(\frac{1}{\tau}\right)^m \frac{d\gamma}{dt}$$

The use of creep equations in the form given by eqs. 7-41 through 7-43 simplifies the solution. It is good practice to give them preference in all cases in which the condition of similarity of isochrones may be adopted, even with a significant degree of approximation.

Connection between kernels $K(t, v)$ and $R(t, v)$

Consider the connection between the kernels $K(t, v)$ and $R(t, v)$ as expressed by Rzhanitsin (see Chapter 1: Rzhanitsin, 1968). Substituting eq. 7-35 into eq. 7-34, we obtain an integral equation of the type:

$$K(t) - R(t) = \int_0^t R(v) K(t-v) \, dv \qquad (7\text{-}48)$$

where:

$$\int_0^t R(v) K(t-v) \, dv = \int_0^t R(t-v) K(v) \, dv$$

To solve eq. 7-48, we shall apply the Laplace transformation:

$$K^*(\lambda) = \int_0^\infty K(\varrho) e^{-\lambda \varrho} \, d\varrho \qquad R^*(\lambda) = \int_0^\infty R(\varrho) e^{-\lambda \varrho} \, d\varrho \qquad (7\text{-}49)$$

where $\varrho = t - v$.

It will be recalled that $K^*(\lambda)$ and $R^*(\lambda)$ are called the unilateral transforms of the functions $K(\varrho)$ and $R(\varrho)$ and these latter functions are termed the originals of the transforms $K^*(\lambda)$ and $R^*(\lambda)$. In this case, the functions proper are determined by inverse transformation:

$$f(\varrho) = \frac{2}{2\pi i} \int_{\delta - i\infty}^{\delta + i\infty} f^*(\lambda) e^{\lambda t} \, d\lambda \qquad (7\text{-}50)$$

Eq. 7-48 may be written, with due regard for eq. 7-49, as:

$$K^*(\lambda) - R^*(\lambda) = R^*(\lambda) K^*(\lambda)$$

Hence:

$$R^*(\lambda) = \frac{K^*(\lambda)}{1 + K^*(\lambda)} \qquad (7\text{-}51)$$

Thus, to find the form of the kernel $R(t, v)$ from a known value of the kernel $K(t)$ (or vice versa), we determine the transform $K^*(\lambda)$ by eq. 7-49, compute the transform $R^*(\lambda)$ by eq. 7-51, and find the original $R(\varrho)$ from eq. 7-50.

Suppose the transform of the kernel $K(\varrho)$ is of the form:

$$K^*(\lambda) = \frac{M\Gamma(\alpha)}{(T\lambda)^\alpha}$$

or:

$$K^*(\lambda) = \frac{a}{\lambda + \beta}$$

where $\Gamma(\alpha)$ is the gamma function of an independent variable α, then the original will be, respectively:

$$K(t - v) = \frac{M}{T}\left(\frac{t - v}{T}\right)^{\alpha - 1}$$

or:

$$K(t - v) = a\, e^{-\beta(t - v)}$$

7.4 KERNELS OF THE INTEGRAL EQUATIONS OF HEREDITARY CREEP

Since the form of the kernel $K(t, v)$ is established experimentally, it stands to reason that the creep function examined in Chapter 5 can be used as the kernel. However, these functions are amenable to generalization in the form of any of the two combined kernels considered below.

Combined power kernel

This kernel, which has been suggested by the present author, corresponds to the time function in eq. 5-12, i.e its value is defined by the expression:

$$K(t - v) = \left[\frac{T_2}{T_1 + (t - v)}\right]^n \qquad (7\text{-}52)$$

Depending on the exponent n, eq. 7-52 may be reduced to a number of special cases which we shall consider below.

Power kernel of the integral equation. For $n = 1 - \alpha$, where $\alpha < 1$ and $T_1 = 0$,

$T_2 = (M/T^\alpha)^{1/1-\alpha}$, relation 7-52 takes the form of the Düffing-Abel power kernel:

$$K(t - v) = \frac{M}{T}\left(\frac{t - v}{T}\right)^{\alpha-1} \tag{7-53}$$

If we express this kernel in the form of eq. 7-42, then, on putting $\phi(\gamma) = A_0\gamma^m$, we obtain:

$$\gamma^m = \frac{1}{A_0}\left[\tau(t) + \frac{M}{T^\alpha}\int_0^t \frac{\tau(v)\,dv}{(t-v)^{1-\alpha}}\right] \tag{7-54}$$

For $\tau = $ const. and $M = \alpha\delta$, eq. 7-54 is transformed into eq. 5-16.
The resolvent of the kernel in eq. 7-53 is of the form:

$$R(t - v) = \frac{1}{t-v}\sum_{j=1}^{\infty}(-1)^{j+1}\frac{[M\Gamma(\alpha)]^j}{\Gamma(j\alpha)}\left(\frac{t-v}{T}\right)^{j\alpha} \tag{7-55}$$

where $\Gamma(\alpha)$ is the gamma function.
One can approximately adopt (as shown by M.I. Rozovsky) that:

$$\int_0^t R(t-v)\,dv \simeq 1 - \exp\left[\frac{t^\alpha M}{T^\alpha}\Gamma(\alpha)\right] \tag{7-56}$$

Hyperbolic kernel. For $n = 1$, $T_1 = T$, $T_2 = \delta$, relation 7-52 is transformed into the hyperbolic Boltzmann kernel:

$$K(t - v) = \delta[T + (t - v)]^{-1} \tag{7-57}$$

Substituting this kernel into eq. 7-42, and putting $\phi(\gamma) = A_0\gamma^m$, we obtain:

$$\gamma^m = \frac{1}{A_0}\tau(t) + \delta\int_0^t \frac{\tau(v)\,dv}{T + (t-v)} \tag{7-58}$$

which, for $\tau = $ const., becomes eq. 5-19.
Linear–fractional kernel. For $n = 2$, $T_1 = T$, $T_2 = [T(\delta - 1)]^{1/2}$, eq. 7-52 is transformed into the linear–fractional kernel suggested by Zaretsky (1972):

$$K(t - v) = \frac{T(\delta - 1)}{[T + (t - v)]^2} \tag{7-59}$$

Taking the function $\phi(\gamma)$ in the form of the linear–fractional relation of eq. 4-37 and substituting the above kernel into eq. 7-42, we arrive at:

$$\frac{G_0\tau_s}{\tau_s + G_0}\gamma = \tau(t) + T(\delta - 1)\int_0^t \frac{\tau(v)\,dv}{[T + (t-v)]^2} \tag{7-60}$$

Eq. 7-60, if solved for γ and $\tau = $ const., is transformed into eq. 5-22.
It can be seen that the kernels of eqs. 7-53 and 7-57 represent the process of non-attenuating creep (or, to be exact, a process characterized by an unconfined

deformation developing at a decreasing rate), and kernel 7-59 defines the process of an attenuating, stabilizing creep. Putting $n = 0$, $T_2 = (1/\eta)^{1/n}$ in eq. 7-52 we obtain a kernel in the form:

$$K(t - v) = \frac{1}{\eta} = \text{const.} \tag{7-61}$$

which describes the Newtonian constant-rate flow.

Combined exponential power kernel

The kernel of this form, derived by Rzhanitsin and applied to frozen soils by Vyalov (see Chapter 1: Rzhanitsin, 1968, and Vyalov, 1959), is a combination of the exponential and power functions:

$$K(t - v) = (t - v)^{\alpha - 1} e^{-\beta(t - v)} N \tag{7-62}$$

For $\beta > 0$ the kernel describes an attenuating (stabilized) creep; for $\beta < 0$ it represents a non-attenuating creep; and for $\beta = 0$ the kernel defines a steady flow.

In using the kernel of 7-62, it is expedient to adopt $Q = K$ and $f(\tau) = \tau^{1/m}$. Hence, for $\tau = \text{const.}$, we obtain from eq. 7-44:

$$\gamma = \gamma_0 + N \frac{\tau^{1/m}}{\beta^\alpha} \Gamma(\beta t, \alpha) \tag{7-63}$$

where:

$$\Gamma(\beta t, \alpha) = \int_0^{\beta t} e^{-z} z^{\alpha - 1} \, dz$$

is the bounded gamma function. In the case of a finite, stabilized deformation, $\Gamma(\beta t, \alpha)$ is transformed into the entire gamma function:

$$\Gamma(\alpha) = \int_0^\infty e^{-z} z^{\alpha - 1} \, dz$$

The rate of settlement data, as computed by eq. 7-62, is in good agreement with the data acquired from frozen soil experiments (sandy loam, $\theta = -0.4°C$) using a spherical indenter (see Chapter 1: Vyalov, 1959), as can be seen from Fig. 7-9.

The resolvent of kernel 7-62 is represented by:

$$R(t - v) = \frac{e^{-\beta}(t - v)}{t - v} \sum_{j=1}^{\infty} (-1)^{j+1} \frac{[cM\Gamma(\alpha)]^j}{\Gamma(j\alpha)} \left(\frac{t - v}{T}\right)^{j\alpha} \tag{7-64}$$

where $M = (1/\alpha) N T^\alpha$.

Particular case (exponential kernel). For $\alpha = 1$, the kernel in eq. 7-62 may be expressed in the form of an exponential function:

$$K(t - v) = N e^{-\beta(t - v)} \tag{7-65}$$

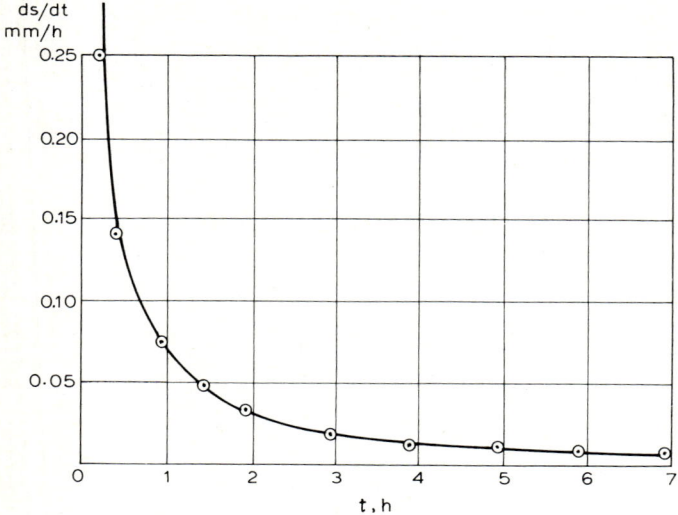

Fig. 7-9. Test of fit of a kernel by eq. 7-62. Dots represent rates of settlement as obtained from testing frozen sandy loam by spherical indenter at $-0.4°C$. Solid curve was obtained from computations by eq. 7-62.

and the resolvent may be represented by the function:

$$R(t - v) = N e^{-(N+\beta)(t-v)} \tag{7-66}$$

The above kernels coincide with eqs. 7-27 and 7-28 derived from differential eq. 7-13 to describe the process of attenuating deformation.

The kernel K and the resolvent R, represented in exponential form, significantly simplify the mathematical treatment of problems. Therefore, the kernels in eqs. 7-65 and 7-66 are widely used in practical computations. However, as shown above, the kernel in the form of an exponential function fails to show good agreement with the experiment.

A better coincidence is obtained if the kernel is represented as the sum of exponential functions:

$$K(t - v) = \sum_{j=1}^{n} N_j e^{-\beta_j(t-v)} \tag{7-67}$$

in which two terms only appear to be sufficient:

$$K(t - v) = N_1 e^{-\beta_1(t-v)} + N_2 e^{-\beta_2(t-v)} \tag{7-68}$$

Numerous parameters in eq. 7-67, and even in eq. 7-68, invite computation difficulties, which limit the application of these kernels.

Binomial kernel

Considering the process as the sum of an attenuating creep γ_I and a steady flow γ_{II} (see Fig. 5-2), the kernel of the integral equation may be represented in the form of a binomial expression, in which the kernel Q is commonly used for convenience:

$$Q(t - v) = Q_I(t - v) + Q_{II}$$

where Q_{II} = const.

Adopting, for example:

$$Q_I = \frac{\alpha\delta}{T}\left(\frac{t}{T}\right)^{\alpha-1}$$

$$Q_{II} = \left(\frac{\tau}{A_0}\right)^{-1/m_1}\left(\frac{\tau - \tau_y}{\eta_{pl}}\right)^{1/m_2} t$$

and substituting into eq. 7-44, we obtain:

$$\gamma = \left(\frac{\tau}{A_0}\right)^{1/m_1}\left[1 + \delta\left(\frac{t}{T}\right)^{\alpha}\right] + \frac{(\tau - \tau_y)^{m_2}}{\eta_{pl}} t \tag{7-69}$$

where, in a particular case, $m_1 = 1$, $m_2 = 1$. Likewise, we may obtain a formula for γ in the form:

$$\gamma = \left(\frac{\tau}{A_0}\right)^{1/m_1}\left[1 + \delta \ln\frac{t + T}{T}\right] + \frac{(\tau - \tau_y)^{1/m_2}}{\eta_{pl}} t \tag{7-70}$$

The above expressions compare favourably with monomials from the point of versatility. However, the difficulties encountered in determining an increased number of parameters, and in obtaining a solution, prevent a wide-spread application of kernels in the form of binomial functions.

Kernel with variable viscosity

Representing a kernel of the integral equation by a function of time t means nothing other than taking implicitly an account of the changes in viscosity with time. A changing viscosity may also be explicitly allowed for by introducing into the kernel a quantity $\eta(t)$, for example, in the form of the function:

$$Q(t - v) = \frac{1}{\eta_\infty - (\eta_\infty - \eta_0) e^{-(\tau - v)/T}} \tag{7-71}$$

Substituting the function into eq. 7-44 for $f(\tau) = \tau$, we obtain eq. 5-46.

Variable-parameter kernels. The parameters entering into the kernels $K(t, v)$ examined above are constants. There are, however, materials whose properties change with time. Typical in this respect is concrete, the strength and rheological properties of which are dependent upon age. In soils, too, the changes in mechanical and physical properties with time are sometimes significant factors. This applies to thixotropic strain hardening, alterations of moisture content and density due to compaction, changes in the temperature of frozen soils with time, etc.

Taking into account the changes in material properties with time, the theory of creep is also referred to as the creep theory of ageing materials or the theory of elastically creeping bodies. It has been formulated, for concrete, by G.N. Maslov and N.Kh. Arutyunyan (in Arutyunyan, 1952) and applied to soils by A.V. Florin (see Section 1.5) and S.R. Meschyan (see Section 6.6). The linear behaviour of an ageing material may be represented in an integral form by:

$$\gamma = \frac{1}{G(t)}\left[\tau(t) + \int_{v_1}^{t} \tau(v)K(t, v)\,dv\right] \tag{7-72}$$

where v_1 is the age of the material at the time of load application, and $G(t)$ is the shear modulus, which changes with the age of the material, hence $\tau(t)/G(t)$ is the instantaneous elastic strain at time t. The kernel $K(t, \tau)$ is adopted as:

$$K(t, \tau) = -\frac{1}{G(v)}\frac{d}{dv}\delta(t, v) \tag{7-73}$$

$$\delta(t, v) = \frac{1}{G(v)} + C(t, v) \tag{7-74}$$

where $\delta(t, v)$ is the total strain due to the unit load $\tau = 1$ and $C(t, v)$ is a measure of creep, i.e. the strain of creep at time t due to the unit load τ applied at time v; apparently, $C(v, v) = 0$. According to Arutyunyan (1952), the kernel of eq. 7-72 is of the form:

$$K(t, v) = G_\infty(1 - b\,e^{-\alpha t})\left[\frac{\alpha b\,e^{-\alpha v}}{G_\infty(1 - b\,e^{-\alpha v})^2} + \frac{T}{v^2}\right]$$

$$+ \left(C_\infty \beta + \frac{T\beta}{v} - \frac{T}{v^2}\right)e^{-\beta(t-v)}\right] \tag{7-75}$$

As we can see, the expression defining the kernel of an ageing material is complex and its use in engineering practice invites difficulties. Therefore, it is advisable to reserve the theory of ageing for use in those cases when ageing has a pronounced effect on the process of deformation, simplifying the relationships referred to above as much as possible. For example, we may neglect the changes in the modulus of instantaneous strain by taking $b = 0$, or disregard instantaneous strain entirely. We may also simplify the form of the function $C(t, v)$.

Consider, by way of illustration, the most simple method of allowing for changes in the temperature of frozen soil. Proceeding from the integral eq. 7-42 for $\gamma_0 = 0$, $\phi(\gamma) = A_{in}\gamma^m$, and $K(t, v) = \alpha\delta(1 - v)^{\alpha-1}$, we have:

$$\gamma^m = \int_0^t \frac{\alpha\delta}{A_{in}}\tau(v)(t - v)^{\alpha-1}\,dv \tag{7-76}$$

According to the experimental data of S.E. Gorodetsky (see Chapter 1: Vyalov et al., 1962), the parameters m and α are not influenced by the temperature of frozen soil but the quantity A/δ is a function of temperature:

$$A/\delta = \bar{a} + \bar{b}\theta^n \tag{7-77}$$

where θ is the temperature in degrees centigrade (omitting the "minus" sign).

Assuming that the temperature is a function of time $\theta(v)$, and substituting eq. 7-77 into eq. 7-76, we obtain:

$$\gamma^m = \int_0^t \frac{\alpha\tau(v)}{\bar{a} + \bar{b}[\theta(v)]^n} (t - v)^{\alpha-1} \, dv \tag{7-78}$$

Application of the theory of hereditary creep to soils

The complexity of the equation of hereditary creep (eq. 7-74) has apparently been the reason for the scepticism of some authors about its practical use in soil mechanics in general. However, equations of hereditary creep are not always as complex.

The theory is intended to take into account several factors (inherited preceding loading, load alternations, unloading, ageing of material, etc.). It goes without saying that the larger the number of factors taken into consideration, the more parameters are introduced, and the more complex is the initial equation. If extra factors are disregarded, the equations are reduced to a very simple form. For example, for a constant loading, they coincide with the simplest empirical equations examined in Chapter 5.

The possibility of complicating the equations when there may be a need to take extra factors into consideration is an asset of hereditary theory. At the same time, a word of caution should be given to avoid the introduction of numerous factors, enabling us to set forth a comprehensive equation of deformation. The resulting formulae become very complicated with many parameters, the determination of which appears to be outside the realms of reality.

7.5 ENGINEERING THEORIES OF CREEP

Equations of engineering theories of creep

If, in the theories of creep examined above, the rheological equations of state were presented in differential or integral form, then equations similar to those used in engineering theories could establish relations between strain, strain rate and time in a direct, explicit form. Accordingly, the rheological equations of state are presented as relations between strain, stress and time:

$$\gamma = f(\tau, t) \tag{7-79}$$

or between strain rate, stress and time:

$$\dot{\gamma} = f(\tau, t) \tag{7-80}$$

or between strain rate, stress and strain proper:

$$\dot{\gamma} = f(\tau, \gamma) \tag{7-81}$$

The first equation of state corresponds to the so-called theory of ageing; the second represents the theory of flow; and the third defines the theory of strain hardening. Although in eqs. 7-79 and 7-80 time enters in its explicit form, which is a limitation of the theories of ageing and flow, in eq. 7-81 it is represented in its implicit form.

Theory of ageing

Formulated by Soderberg in 1936 this theory treats the aggregate strain as the sum of elastic strain γ^e and creep strain γ^c. Consequently, eq. 7-79 takes the form:

$$\gamma = \gamma^e + \gamma^c = \frac{\tau}{G} + f(\tau)\Phi'(t) \tag{7-82}$$

The equation of relaxation is arrived at by solving eq. 7-82 for τ when $\gamma = \gamma_0 = \tau_0/G = \text{const.}$:

$$\tau = \tau_0 - Gf(\tau)\Phi'(t) \tag{7-83}$$

The function $\Phi'(t)$, defining the time-dependent retardation of deformation, represents in fact the changes in material properties with time — the "ageing" of the material. This peculiarity of the material's behaviour is reflected in the name of the theory. However, it should not be confused with the phenomenon of ageing in the physics of solids or with the theory of hereditary ageing considered above.

A convenient version of the theory of ageing in computations has been suggested by Yu.N. Rabotnov (see Section 1.5) taking isochrones into consideration (Fig. 5-6). For similar isochrones, the equation of state is:

$$\phi(\gamma) = \tau\psi(t) \tag{7-84}$$

The creep function $\psi(t)$ is to be adopted so that $\psi(0) = 1$; hence, the function $\phi(\gamma)$ describes the isochrone at $t = 0$. The relaxation of stresses is described by the relation:

$$\tau = \phi(\gamma)F(t) \tag{7-85}$$

where $F(t) = 1/\psi(t)$.

It can be seen that eq. 7-84 (i.e. the equation of state according to the theory of ageing) is identical with eq. 5-7. However, unlike eq. 5-7 representing the mere fact of similarity of isochrones obtained during a given experiment, eq. 7-84 establishes a connection between strain, time and load in their various combinations, describing the simultaneous processes of creep and relaxation. The equations of the theory of ageing have been scrutinized and developed by Malinin (1968).

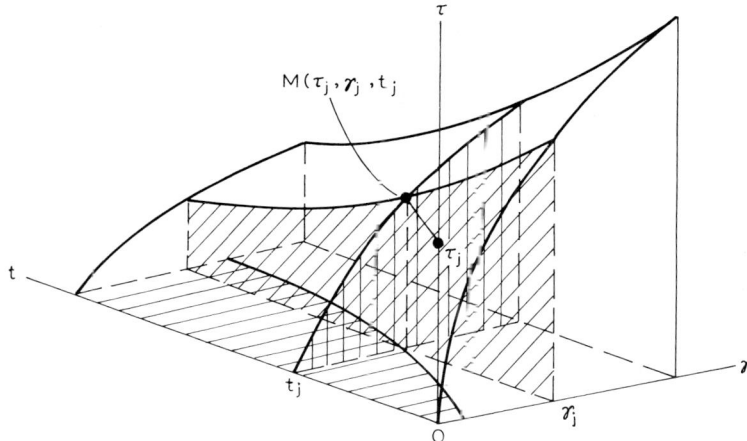

Fig. 7-10. (τ–γ–t) surface according to ageing theory.

(τ–γ–t) *surface.* The equation of state 7-79 of the theory of ageing may be set forth as $F(\tau, \gamma, t) = 0$ and represented graphically as a three-dimensional (τ–γ–t) surface. Cross-sections of the surface projected on the τ–γ, γ–t and τ–t planes give the isochrones, creep curves and relaxation curves, respectively (Fig. 7-10).

The creep and relaxation curves may be plotted from the isochrones. The points where the isochrones intersect the straight lines τ_j = const. and γ_j = const. give the points of the creep curves and relaxation curves, respectively.

The relaxation curves may also be plotted directly from the creep curves. If we intersect the creep curves by a straight line γ_j = const., the points of intersection of the line and the γ–t curves for various τ will define the points of a τ–t relaxation curve. On the other hand, if we cross the relaxation curves by a straight line τ_j = const., we can plot a γ–t creep curve from the points of intersection with the τ–t curves for various γ.

The functions $\phi(\gamma)$ and $\psi(t)$ entering into the equation of state 7-84 may be of various forms. The most widely used formulae by virtue of their good agreement with experimental data, are eqs. 4-29 and 4-37 for the function $\phi(\gamma)$ and eqs. 5-13, 5-14 and 5-15 for the function $\psi(t)$. Hence, the equation of creep takes the form of expressions 5-16, 5-19 and 5-22. The equation of relaxation is obtained by solving for τ, according to eq. 7-85, expressions 5-16, 5-19 and 5-22.

The values of τ and γ entering into these equations may be treated as variables, though with the reservations considered below.

Equation of ageing theory in parametric form. Whenever the function $\phi(\gamma)$ is used in the power form, the equation of deformation may be set forth in the following convenient form (Chapter 1: Vyalov, 1959):

$$\tau = A(\gamma)\gamma^m \tag{7-86}$$

where $A(t)$ is a coefficient of strain treated as a time-dependent quantity: at $t = 0$, $A(0) = A_0$.

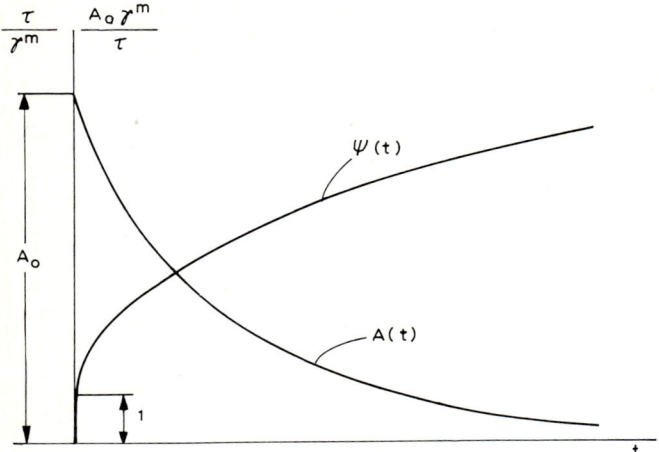

Fig. 7-11. Creep function $\psi(t)$ and coefficient of deformation $A(t)$.

The advantage of eq. 7-86 is that the function of creep is represented in the form of a strain coefficient. Consequently, in solving creep problems, the time factor may be taken account of in the functional and the parametric form. In the latter case, the value of the parameter $A(t_j)$ may be determined directly from the isochrone at a given time t_j.

This procedure is convenient when we must determine deformation at a given instant rather than over the period of service of the soil. Moreover, statements in the form of eq. 7-86 simplify the handling of experimental data, enabling the determination of the time function from the isochrones directly. In such a case, eq. 7-86 is reduced to the linear equation:

$$\ln \tau = m \ln \gamma + \ln [A(t)]$$

Accordingly, on plotting the experimental points on a graph of $\ln \gamma$ versus $\ln \tau$ (as is shown in Fig. 6-6(b)), we obtain a family of straight lines. Their slope defines the parameter, m, and the intercepts on the axis of ordinates determine the value of $A(t)$. By plotting the curve on the (τ/γ^m)–t coordinates (Fig. 7-11) we obtain a graphical representation of the function. Next, we select an empirical formula to describe the $A(t)$ curve obtained.

When the deformation increases in time according to the power law, the parameter $A(t)$ takes the form:

$$A(t) = \frac{1}{\dfrac{1}{A_0} + \dfrac{1}{\xi}\left(\dfrac{t}{T}\right)^\alpha} \tag{7-87}$$

where $T = 1$ and $\xi = A_0/\delta$.

The graph of this function lends itself to rectification if eq. 7-87 is represented in the form:

$$\ln\left[\frac{1}{A(t)} - \frac{1}{A_0}\right] = \ln\frac{1}{\xi} + \alpha \ln t$$

Comparing eq. 7-87 with eq. 5-13, we can see the connection between the coefficient of strain $A(t)$ and the creep function $\psi(t)$:

$$\psi(t) = \frac{A_0}{A(t)} \qquad (7\text{-}88)$$

The scope of ageing theory applicability. Speaking about an allowance for changes in load made by the equation of ageing theory, we must bear in mind that this is an assumption of a significant degree. The theory states in fact that a strain induced at an instant t_j is not affected by the load applied at a time preceding t_j. However, in the case of a decreasing load, an assumption of this kind leads to untrustworthy results. Suppose, for example, a load reduced by an amount $\Delta\tau$ causes the strain to decrease by $\Delta f(\gamma)$, then, if the stress becomes zero, the strain will also be zero. This is absolutely wrong. Thus, the theory of ageing is applicable only in those cases when the load is constant or is increasing slowly and monotonously.

Theory of flow

Put forward by Davenport (1938) and developed by L.M. Kachanov (see Section 1.5), this theory assumes that the total strain rate is the sum of the elastic strain rate and the creep strain rate:

$$\dot{\gamma} = \dot{\gamma}^e + \dot{\gamma}^c \qquad (7\text{-}89)$$

Adopting $\dot{\gamma}^e = (1/G)(d\tau/dt)$ and $\dot{\gamma}^c = f(\tau)\kappa(t)$, we obtain:

$$\dot{\gamma} = f(\tau)\kappa(t) + \frac{1}{G}\frac{d\tau}{dt} \qquad (7\text{-}90)$$

For an attenuating creep, $\kappa(\infty) = 0$; in the case of a non-attenuating creep, $\kappa(\infty) = \infty$; and for a constant-rate steady flow, $\kappa(\infty) = \infty = \text{const}$.

Eq. 7-90 is the general form of the Maxwell eq. 7-9 which may be obtained for $f(\tau) = \tau$ and $\kappa(t) = 1/\eta = \text{const}$.

The name "theory of flow" originates from an analogy of the equation describing this theory with the equation of viscous flow, although, unlike this latter flow, the theory of flow deals with a time-dependent strain rate. The name of this theory should not be confused with a similar name adopted for one of the theories of plasticity (see Chapter 3); the two theories coincide only in so far as they both deal with strain rates, not strains.

The equation of creep is obtained by integrating eq. 7-90, and the equation of relaxation is derived by solving eq. 7-90 for τ when $\gamma = \gamma_0 = \text{const}$. For $\tau = \text{const}$, eq. 7-90 takes a form coincident with eq. 5-9. The difference consists in that eq. 5-9 describes only the configuration of the creep rate curves whereas the equation of flow

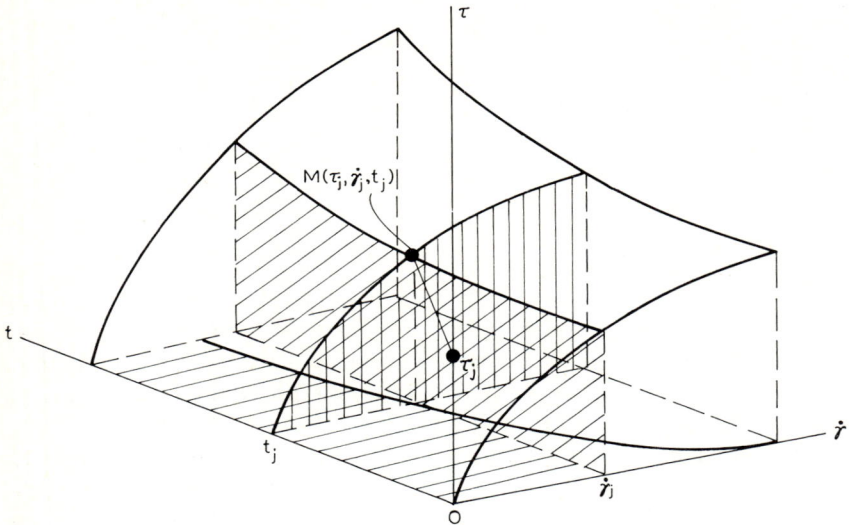

Fig. 7-12. (τ–$\dot{\gamma}$–t) surface according to the theory of flow.

theory 7-90 establishes the connection between the flow rate, time and time-dependent load, describing both the process of creep and that of relaxation.

(τ–$\dot{\gamma}$–t) *surface*. The equation of state 7-80 in terms of the theory of flow, which is set forth as $F(\tau, \dot{\gamma}, t) = 0$, may be represented graphically by a three-dimensional surface (τ–$\dot{\gamma}$–t). Cross-sections of the surface projected on the τ–$\dot{\gamma}$, $\dot{\gamma}$–t and τ–t planes will yield "stress versus strain rate" (rheological) curves for various τ = const., curves depicting strain rate alterations at various t, and relaxation curves, respectively (Fig. 7-12).

The form of function $f(\tau)$ adopted depends on the form of eqs. 4-61, 4-64 and 4-65 examined above; the function $\kappa(t)$ may be adopted by analogy with function 5-12.

If we adopt that $f(\tau) = (\tau/A_0)^{1/m}$, eq. 7-90 takes the form:

$$\dot{\gamma} = \left(\frac{\tau}{A_0}\right)^{1/m} \kappa(t) + \frac{1}{G}\frac{d\tau}{dt} \tag{7-91}$$

whereas the equation of relaxation will be set forth as:

$$\tau = \tau_0 \left[1 + \frac{1-m}{m} G \frac{\tau_0^{m/(m-1)}}{A_0^{1/m}} \Phi(t)\right]^{-m/(1-m)} \tag{7-92}$$

where $\Phi(t) = \int_0^t \kappa(t)\, dt$.

The scope of flow theory applicability. The flow theory equation and the ageing theory equation both incorporate time in the explicit form. Consequently, the flow theory is sometimes referred to as the second theory of ageing. Therefore, the equation of this theory is not invariant to an alteration of the initial time and is recommended for use only in the case of a gradually changing load.

Theory of strain hardening

According to this theory, due to Nadai (see Section 1.5) and Davenport (1938), the connection between creep strain rate, stress and strain proper is represented in the form:

$$\dot{\gamma} = \frac{f(\tau)}{\phi(\gamma)} \tag{7-93}$$

It follows from this relation that the strain rate decreases as the strain increases, i.e. as if the body strain hardens; hence the name of the theory.

The equation of state of the strain-hardening theory, as given by eq. 7-81, may be represented in the form $\Phi(\tau, \dot{\gamma}, \gamma) = 0$, which is equivalent to the concept of a three-dimensional $(\tau-\dot{\gamma}-\gamma)$ surface. It must be stressed once more that time does not enter into eq. 7-81 in its explicit form.

The functions entering into eq. 7-39 may be of various kinds. Yu.N. Rabotnov (see Section 1.5) and F.S. Churikov have suggested the adoption of $\phi(\gamma) = \gamma^\alpha$ and $f(\tau) = a\, e^{\tau/b}$.

Hence, for $(\dot{\gamma}\gamma^\alpha) > a$:

$$\tau = b \ln \frac{\dot{\gamma}\gamma^\alpha}{a} \tag{7-94}$$

and for $(\dot{\gamma}\gamma^\alpha) \leqslant a$:

$$\tau = 0 \tag{7-95}$$

For τ = const. we obtain:

$$\gamma = [a(\alpha + 1)]^{1/(\alpha+1)} \exp\left[\frac{\tau}{b(\alpha + 1)}\right] t^{1/(\alpha+1)} \tag{7-96}$$

According to Davis (1943), $\phi(\gamma) = \gamma^\alpha$, $f(\tau) = a\tau^\beta$, with eq. 7-93 consequently taking the form:

$$\dot{\gamma}\gamma^\alpha = a\tau^\beta \tag{7-97}$$

The equation of creep is arrived at by integrating the above equality:

$$\int_0^\gamma \gamma^\alpha\, d\gamma = \int_0^t a\tau^\beta\, dt \tag{7-98}$$

For τ = const. and provided that $\gamma^c = 0$ at $t = 0$, we obtain:

$$\gamma = [a(\alpha + 1)\tau^\beta t]^{1/(\alpha+1)} \tag{7-99}$$

The equation of relaxation, corresponding to the creep eq. 7-93 for $\phi(\gamma) = \gamma^\alpha$, is obtained by substituting into eq. 7-93 the quantity $\gamma = \tau_0/G - \tau/G$. On integrating the expression obtained with respect to the initial condition $\tau = \tau_0$ at $t = 0$, we arrive at the solution:

$$t = \frac{1}{G^{\alpha+1}} \int_\tau^{\tau_0} [\tau_0 - \tau]^\alpha \frac{d\tau}{f(\tau)} \tag{7-100}$$

Connection between time functions according to various theories

A point to be noted is that all the above theories yield similar results for constant loads. (The solutions obtained by means of the theory of viscoelastic deformation will coincide with those arrived at by other theories only if the time functions entering into the equations are in an exponential form.)

Accordingly, the time functions in the equations of the theories referred to are connected by certain relationships:

$$\Phi(t) = 1 + \int_0^t \kappa(t)\,dt = 1 + \int_0^t Q(t)\,dt$$

$$\psi(t) = 1 + \int_0^t K(t)\,dt \tag{7-101}$$

The function $A(t)$ from eq. 7-86 is connected with the function $K(t)$ by the relation:

$$A(t) = \frac{A_0}{\psi(t)} = \frac{A_0}{1 + \int_0^t K(t)\,dt} \tag{7-102}$$

Allowance for unloading

The mechanism of unloading can be described in terms of the theory of hereditary creep.

Let a body be exposed to a load over a period between $t = 0$ and $t = t_1$ during which a strain $\gamma(t_1)$ accumulates in the body. Determine the recoverable strain $\gamma^e(t_2)$ the body is relieved of during a period between t_1 and t_2 following the removal of load. Apparently, the strain sought is:

$$\gamma^e(t_2) = \gamma(t_1) - \gamma^e - \gamma^p(t_2)$$

where γ^p is the irrecoverable strain at t_2.

Since it may be assumed that the relation between the stress and recoverable strain is linear during the unloading, this strain may be determined from:

$$\gamma^e(t_2) = \frac{\tau}{G}[\tilde{K}(t_1) + \tilde{K}(t_2) - \tilde{K}(t_1 + t_2)] \tag{7-103}$$

where $\tilde{K}(t) = \int_0^t K(t)\,dt$.

Experimental verification of creep theory

To prove the theory, two lines of enquiry may be followed. Firstly, we may carry out comparative tests for creep and relaxation and, on establishing the mechanism of one process, determine the mechanism of the other process analytically, using relevant equations. The results are to be checked against the experimental data. For most materials, the relaxation curves plotted in accordance with the ageing theory appear above the test curves, and those drawn as prescribed by the flow theory appear below the test curves. The best agreement with the test data is provided by the theory of strain hardening.

Secondly, we may perform testing by a stepwise increasing load and, on reducing the data acquired in accordance with any of the theories, compare the results of computations with the test curves. The technique of experimental data reduction is outlined below (Malinin, 1968).

Let a soil be tested by stepwise increasing stresses $\tau_1, \tau_2 \ldots$ (Fig. 7-13) applied at instants $t_1, t_2 \ldots$ for arbitrary intervals $\Delta t_1, \Delta t_2 \ldots$. At the first stage of loading, τ_1, all the theories (ageing, flow, strain hardening, and hereditary creep) obviously yield the same creep curve $0A$.

After the second-stage loading, τ_2, the strain must instantaneously increase according to the ageing theory by an amount AB and continue to increase in accordance with the same law it would obey if the stress τ_2 was applied beginning with the instant $t = 0$. The creep curve at the second stage of loading is represented in Fig. 7-13 by curve BC.

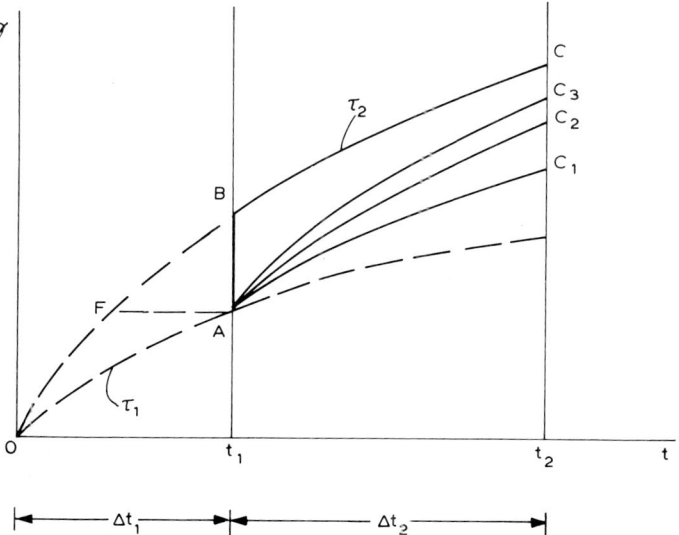

Fig. 7-13. Creep curves for stepped loading plotted in accordance with (C) ageing theory, (C_1) theory of flow, (C_2) strain hardening theory, and (C_3) theory of hereditary creep.

According to the theory of flow, the rise in stress from τ_1 and τ_2 increases the strain rate so that at the instant t_1 it is equal to the value $\dot{y}_2$ which would exist at that instant if the stress τ_2 was applied over the interval between $t = 0$ and $t = t_1$; in other words, the strain rate $\dot{y}_2$ at the instant t_1 is defined by the slope of a tangent to the creep curve OB at point B. Hence, the creep curve at the second stage of loading may be obtained (segment AC_1) in accordance with the flow theory if curve BC is lowered parallel to its initial position until point B coincides with point A.

According to the theory of strain hardening, the strain rate experienced when the stress increases from τ_1 to τ_2 is influenced by both the stress τ_2 and the strain accumulated at the instant t_1. The strain rate is defined by the slope of a tangent to the creep curve for τ_2 at the point F where a straight horizontal line beginning at A meets curve OB. Corespondingly, the creep curve AC_2 at the second stage of loading is a segment of the creep curve for τ_2 shifted parallel to its original position until point F coincides with point A.

According to the theory of hereditary creep, the stress increase from τ_1 to τ_2 requires us to take account of the changing strain rate and the strain accumulated due to the preceding stress τ_1. Correspondingly, the creep curve AC_3 at the second stage of loading may be plotted from the ordinates set up, beginning from curve AC_1 and equal to the difference between the ordinates of curve OB and those of curve OA. In other words, segments $C_1 C_3$ and AB must be the same.

Thus, we have obtained creep curves of the following types: curve $OABC$, according to the ageing theory; curve OAC_1, according to the flow theory; curve OAC_2, according to the strain hardening theory; curve OAC_3, according to the hereditary creep theory. Similar plots result from other stages of loading. Referring to the soil test by a stepped load, we must point out once more that in most cases their results lack coincidence with the constant-load tests, the creep curves for a stepwise increasing load appearing, as a rule, below the $\tau = $ const. curve. As mentioned before (see Fig. 6-1), this is attributed to an augmented resistance of the soil strain-hardening during the process of deformation.

S.R. Meschyan's experiments (see Chapter 1: Meschyan, 1967) have shown that the curves plotted according to the ageing and hereditary creep theories are located above the experimental curves, and those plotted as prescribed by the strain hardening theory are below but close to the experimental curve. In general, it appears that the curves plotted according to the various theories differ but little from the experimental data. This implies that the theories examined above may all be applied when analyzing soil deformations.

Note that there is another way of reducing the data acquired from stepped-load tests. Firstly, the parameters of the deformation equation are determined from the constant-load test (using, for example, the first stage of loading); these parameters must all be the same, whatever the theory. Secondly, assuming that the stepped loading increases uniformly ($\tau(t) = vt$, where $v = \Delta\tau/\Delta t$), substitute the value of $\tau(t)$ into the deformation equations of the various theories and determine analytically the pattern of deformation increase. The best agreement between the computed and experimental data indicates which of the theories should be used.

The scope of applicability of creep theories

Since all the creep theories examined above (except the theory of viscoelastic deformation) provide for a reasonably satisfactory agreement with the test data, the choice of the most suitable theory is contingent in soil mechanics on the condition of the problem. If given a constant load, or one changing gradually and over a narrow range, it is good practice to use the simplest theory — the ageing or flow theory. The flow theory is given preference when it is expedient to deal with strain rates — for example, in examining the flow of artificial and natural slopes.

Both theories are simple, particularly the ageing theory. They also allow the use of initial creep equations as the basis for setting up a deformation equation, reflecting in their final form all the intricate microprocesses taking place during soil deformation.

However, significant changes in loading with time, combined loadings (after Ilyushin) with allowance for unloading, and changing soil properties require the use of the theory of hereditary creep.

As far as the theory of strain hardening is concerned, it holds out a special promise, though being more complex than the other theories. However, at present we lack adequate experimental data on the patterns of soil deformation in accordance with this theory.

The equation of state corresponding to any of the creep theories is regarded as a physical equation entering into the set of equations from the plasticity theory: eqs. 3-78, 3-80, 3-82, and 3-82. If we apply either the ageing theory or the theory of hereditary creep, use is made of the equation of the theory of plastic deformation (eq. 3-90); the equation of plastic flow 3-102 is employed in connection with the theory of flow.

7.6 THE MOLECULAR THEORY OF FLOW

On physical theories of deformation

The theory of the flow of viscous media presented below examines the pattern of motion of elemental particles (molecules and atoms) in a force field. It has been formulated to describe the flow of an ideal, Newtonian liquid by Frenkel (1934, 1945). Eyring (1936) has developed the so-called theory of rate process based on similar physical concepts and has set up the equation of flow of a non-Newtonian medium.

Heat motion in solids and liquids. It is known that crystalline solids are characterized by an orderly arrangement of closely-packed atoms and by a regular repetition of the pattern of this arrangement by atoms spaced more widely apart. In other words, crystalline solids display a short- and long-range atomic order. In liquids, only the closely located molecules are arranged in an orderly way; consequently, they are said to have a short-range order.

However, an orderly arrangement of solids and liquids does not mean it is perfect. A solid lattice may have imperfections and defects such as vacant lattice sites, or vice versa, or extra atoms wedged between the atoms at a point of the lattice. A vacant lattice site is called "vacancy" or (according to Frenkel) a "hole" and an extra atom occupying a position at a point is referred to as an "interstitial" atom. The defects attributed to the above discontinuities of the lattice or an irregular alternation of atomic planes are termed "dislocations". A plastic deformation of a solid is attributed to dislocation displacements, and creep represents an accumulation of these displacements in time.

In liquids, structural defects are represented by voids of a size commensurate with the size of molecules but which are not occupied by molecules. Capable of expanding and contracting spontaneously, these voids may disappear at one point and reappear at another as if migrating, by analogy with the migration of holes through a solid lattice. In other words, the molecules of a liquid are unceasingly displaced, being in a state of thermal motion. Oscillating for some period $t \gg t_0$ about one equilibrium position at a frequency of $1/t_0$, they migrate to another position of equilibrium, continuing to oscillate there. Every such displacement occurs in hops. In the absence of a force, the hops are omnidirectional and the motion of the particles lacks orientation. A stress applied for a period of $t > t_0$ causes the particles to assume an oriented displacement. This motion of molecules is referred to as flow.

Thus, the flow of liquids and the creep of solids are of the same nature, both being the manifestation of an oriented motion of elemental particles (molecules, atoms) induced by structural defects. However, since the molecular bonds in liquids are weaker than the atomic bonds in solid lattices, the mobility of liquids is much more than that of solids. Consequently, the flow of liquids is induced by very small loads and has a rate that is higher than the rate of solid flow.

Activation energy. Consider the motion of an individual molecule in a force field. The field may be imagined as an assembly of potential wells separated by ridges, the spacing δ of the wells being commensurate with the separation of adjacent particles (Fig. 7-14). A particle in a stable position has always a minimum of potential energy. This energy, U_1, will be possessed by the molecule located in a potential well. The molecule topping a ridge will have a maximum energy U_2. Consequently, the molecule is capable of hopping from well to well on acquiring a kinetic energy $U = U_2 - U_1$ during a time interval t_0.

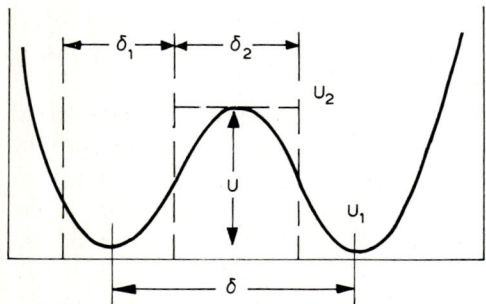

Fig. 7-14. Schematic representation of a force field.

Once the molecule has hopped into a new well, covering a distance δ during an interval t_0, the energy U it has acquired is transferred to the surrounding particles and the molecule itself becomes stuck in the well for some time t. Its energy — in this case, U — equals the height of the potential barrier it has to escape. This energy, called the activation energy, is attributed to the bonding of a molecule with adjacent particles and is defined as the work to be done by a molecule in order to force its way through surrounding molecules into a neighbouring position.

Boltzmann distribution law. The pattern of the behaviour of a medium composed of a collection of particles (atoms, molecules) may be established by recourse to statistical physics. This paves the way to describing macro-processes in terms of the data characterizing the pattern of motion of individual particles comprising the microstructure. Thus, the pattern of distribution of elemental particles which depend on the energy they possess in a force field at a heat equilibrium is described by the Boltzmann distribution law:

$$N = N_0 \, e^{-U/k\Theta} \tag{7-104}$$

where U is the activation energy (J); Θ is the thermodynamic temperature (K); N is the average number of activated particles per unit volume; N_0 is the number of activated particles at $\Theta \to 0$; k is the Boltzmann constant, $k = 1.38 \cdot 10^{-23}$ J/K $= 3.29 \cdot 10^{-24}$ cal/K. This constant establishes a connection between the entropy S of a system and the thermodynamic probability p of a given macro-state of the system, $S = -k \ln p$.

The Boltzmann law may be regarded as a probability law defining the relative number of particles whose potential U_j is between the limits $U \leqslant U_j \leqslant (U + \Delta U)$.

Kinetic theory of viscosity

Assuming that the average time t_s during which a particle is in a transient state of equilibrium, varies inversely with the number of activated particles N, we obtain, from eq. 7-104:

$$t_s = t_0 \, e^{U/k\Theta} \tag{7-105}$$

where t_0 is the period of oscillations of the molecule at a transient equilibrium position. It can be shown that the time t_s corresponds to the relaxation time T_r, which is the "settled life" period of molecules.

The reciprocal of t_s is called particle mobility, $q = \delta^2/6k\Theta t_s$, defining the random thermal (Brownian) motion of molecules; this motion is connected with diffusion by the relation $q = D/k\Theta$, where D is the diffusion coefficient.

Considering the Newtonian pattern of the plane-parallel motion of two liquid layers (Fig. 4-10) and assuming that the molecules escape the energy barrier due to the action of a shearing force $F = \tau\delta^2$ (where τ is the force applied to a particle with an area δ^2), the velocity of the liquid may be expressed by the relation $v = qF$.

Since, according to eq. 4-50, $dv/dy = v/\delta = \tau/\eta$, we obtain the following formula for the coefficient of viscosity:

$$\eta = A\,e^{U/k\Theta} \tag{7-106}$$

where $A = 6k\Theta t_0/V$; $V = \delta^3$ is the molecular volume. Since A is influenced by Θ to a much smaller extent than the factor $e^{U/k\Theta}$, it is commonly assumed that $A = $ const.

Eq. 7-106, obtained by Ya.I. Frenkel[*1] (1934), reveals the physical meaning of the coefficient of viscosity which, as can be seen, varies with activation energy, temperature, and particle size. It is assumed that the particle sizes δ', δ'', δ''' are approximately commensurate with the distance between adjacent particles $\delta' = \delta'' = \delta$.

Theory of flow of a non-Newtonian medium

This theory, developed by Eyring (1936), is also called the theory of rate process, being based on the examining of absolute rates of chemical reactions described by an equation according to Avenarius, which is similar in form to eq. 7-105.

Eq. 7-105 reflects the random motion of molecules having an equal opportunity to move in any direction; eq. 7-106 has been deduced on the assumption that the molecules move in an orderly manner, being acted upon by an external force. However, the application of this force does not change the activation energy, and, consequently, the coefficient of viscosity of a perfect Newtonian liquid is not affected by an external force.

In a structuralized non-Newtonian medium, the activation energy will vary with the force applied. In fact, the molecules of such a medium, moving due to a force F from a potential well to the top of a barrier (Fig. 7-15), do some work, $\delta F/2$. This means that they acquire extra energy $U^* = \delta F/2$ while the height of the energy barrier

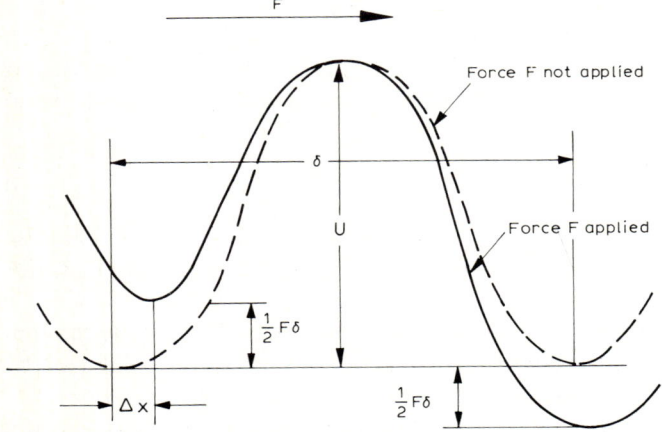

Fig. 7-15. Effect of shearing force on energy barrier.

[*1] Note that Ya.I. Frenkel has derived another formula of viscosity differing from eq. 7-106 by the value of a constant factor.

decreases by the same amount. Thus, the activation energy required to set the molecules into motion is $U = U_0 - U^* = U_0 - \delta F/2$. When the direction of motion is reversed, the height of the energy barrier apparently increases by the amount $U = U_0 + \delta F/2$.

If the number of times a molecule escapes the energy barrier each second, in either direction, is denoted by $\vec{j}_s$ and $\overleftarrow{j}_s$, the flow rate of the liquid will be equal to the difference: $\dot{\gamma} = \vec{j}_s - \overleftarrow{j}_s$. The number of escapes j_s may, in turn, be regarded as a unit rate of particle motion varying inversely with the period t_s of the equilibrium of the particles:

$$j_s = j_0 \, e^{-U/k\Theta} \qquad (7\text{-}107)$$

where $j_0 = X/t_0 = Xk\Theta/h$, in which $h = 16.625 \cdot 10^{-34}$ J s is Planck's constant and X is the coefficient of particle concentration defining the number of moving molecules.

Substituting $U = U_0 \pm \delta F/2$ into eq. 7-107, we obtain:

$$\dot{\gamma} = \vec{j}_s - \overleftarrow{j}_s = j_0 \left[\exp\left(-\frac{U_0 - \frac{1}{2}\delta F}{k\Theta} \right) - \exp\left(-\frac{U_0 - \frac{1}{2}\delta F}{k\Theta} \right) \right]$$

$$= j_0 \, e^{-U_0/k\Theta} (e^{\delta F/k\Theta} - e^{-\delta F/k\Theta})$$

Since $e^x - e^{-x} = 2 \sinh x$, we finally arrive at:

$$\dot{\gamma} = 2 j_0 \, e^{-U_0/k\Theta} \sinh \frac{\tau V}{2k\Theta} \qquad (7\text{-}108)$$

where $j_0 = Xk\Theta/h$ is the unit rate of oscillations of a molecule at a transient equilibrium position (1/s); X is the function of molecule concentration; k is Boltzmann's constant (J/k); h is Plank's constant (J s); Θ is the thermodynamic temperature (K); U_0 is the activation energy acquired by a single molecule (1/J); V is the molecular volume (cm³); δ is the average distance between successive equilibrium positions of molecules (cm); and $\tau = F/\delta^2$ is the force acting per unit area (Pa).

Denoting $\tau^* = 2k\Theta/V$ and $\dot{\gamma}^* = 2 j_0 \, e^{-U_0/k\Theta}$, we obtain eq. 7-108 in a form similar to eq. 4-64:

$$\dot{\gamma} = \dot{\gamma}^* \sinh\left(\frac{\tau}{\tau^*} \right) \qquad (7\text{-}109)$$

Eq. 7-108 describes a non-linear flow (inherent in structuralized non-Newtonian liquids) whose shearing resistance increases with stress. The viscosity of such a liquid is variable and may be determined by eq. 4-66. (In case the activation energy is referred to a mole rather than a molecule, the quantity $U_0/k\Theta$ is transformed into $\bar{U}_c/R\Theta$, where $R = k/N = 8.31 \cdot 10^{-3}$ J/(kmol K) is the characteristic gas constant, in which $N = 6.02 \cdot 10^{26}$ kmol^{-1} is the Avogadro constant; $\bar{U}_0 = UN$ J/mol.)

Particular cases of eq. 7-108. If the stress applied is so low that $\tau V \leqslant 2k\Theta$, then, taking into account that for $x \ll 1$ the equality $\sinh x \cong x$ is correct, eq. 7-108 takes a form which corresponds to a Newtonian pattern of flow:

$$\dot{\gamma} = \tau j_0 \frac{V}{k\Theta} e^{-U_0/k\Theta} \qquad (7\text{-}110)$$

The value of the coefficient of viscosity in this expression is:

$$\eta = \frac{k\Theta}{j_0 V} e^{U_0/k\Theta} \qquad (7\text{-}111)$$

which coincides with eq. 7-106 due to Frenkel, with the exception of the value of the constant factor A.

For media in which the flow is produced by quite high stresses (i.e. in the case $\tau V/2k\Theta \gg 1$) and provided that, for $x \gg 1$, the approximation $\sinh x = 1/2\, e^x$ is correct, eq. 7-108 takes the form:

$$\dot{\gamma} = j_0\, e^{-U_0/k\Theta}\, e^{\tau v/2k\Theta} = 1/2 \dot{\gamma}^*\, e^{\tau/\tau^*} \qquad (7\text{-}112)$$

An examination of eqs. 7-109 and 7-112 in Section 5.4 (see Fig. 5-11) has shown that approximation 7-112 holds for $\tau/\tau^* \geqslant 1.3$, and even for $\tau/\tau^* = 1.0$ it gives only small discrepancies.

Eq. 7-112 and its modifications are widely used to describe the process of steady flow of various materials ranging from viscous liquids and dispersed systems to plastics and metals. Considering flow to be a manifestation of the combined effect of the various deformation patterns, Kauzmann (1941) suggested that eq. 7-112 should be expressed as the sum of nth physical characteristics.

Kinetic theory of failure

The kinetic concept has gained also recognition in describing long-term failure of solids (Regel et al., 1974). Accordingly, the failure of a solid is attributed to the breakage of interatomic bonds due to the action of an external force. The force itself causes no breakage, it only intensifies the process of thermal fluctuation leading to the breaking of bonds between adjacent atoms.

Let us recall that fluctuation is understood to be a deviation of the energy possessed by an atom from an average value. The probability of fluctuation obeys Boltzmann's distribution law. If the energy of an atom at the instant of fluctuation is denoted by E_{fl}, the average interval of time t_{fl} between two successive fluctuations will be expressed by a relationship similar to eq. 7-105, namely, $t_{fl} = t_0 \exp(E_{fl}/k\Theta)$, where t_0 is the period of heat-induced oscillations of the atom, $t_0 = 10^{-13}$ s.

Thus, according to the kinetic theory of strength, the failure of solids is attributed to a breakage of interatomic bonds succeeded by a gradual loss of continuity. Accordingly, it is assumed that the time before failure t_f (durability of material) is defined by the same expression as the fluctuation period, i.e.:

$$t_f = t_0\, e^{U_b(\tau)/k\Theta} \qquad (7\text{-}113)$$

where $U_b(\tau)$ is the breakage activation energy of a stressed bond between two atoms varying with stress τ; here $U_b \geqslant E_{\text{fl}}$.

Flow of dispersed media

N.M. Mikhailov and P.A. Rebinder (see Section 4.9) have applied the findings of Frenkel and Eyring in the kinetic theory to dispersed media. They have assumed that the behaviour of the colloid particles forming the spatial structural framework is similar to the behaviour of molecules of a liquid. However, the pattern of behaviour of a dispersed medium differs from eq. 7-108.

According to eq. 7-108, the relation between the flow rate $\dot{\gamma}$ and the stress τ is represented by the smooth curve of Fig. 4-9(b), whereas the curves defining real structuralized media display inflections (see Fig. 4-11).

Assuming that the effective viscosity $\eta(\tau)$ is an outcome of structure-destroying and structure-reconstructing processes, Mikhailov and Rebinder have derived the following expression:

$$\eta(\tau) = \eta_m + (\eta_0 - \eta_m) \frac{\tau/\tau^*}{\sinh(\tau/\tau^*)} \tag{7-114}$$

where η_0 and $\eta_m = \eta_\infty$ are the maximum (at $t \to 0$) and minimum (at $t \to \infty$) values of viscosity. For $\tau/\tau^* \ll 1$ we obtain $\eta(\tau) \cong \eta_0$, where η_0 is determined from eq. 7-111.

Applicability of rate process theory to soils

It was stated in Section 7.2 that Murayama and Shibata (referenced in Section 5.7), Soderberg (1936), Christensen (1964) and other researchers had employed Eyring's theory of rate process in devising mechanical models of soils. They explained the viscous behaviour of soils by water molecules "hopping" in the bonds between solid particles. Consequently, these authors applied an equation similar to eq. 7-108 to the viscous elements of models, while using Hooke's law for the elastic elements as before. However, Murayama and Shibata departed from Eyring's equation, suggesting that the number of activated molecules depends on whether or not the stress applied exceeds the yield point τ_y.

Noteworthy was the attempt by Goldstein (1961) to apply Boltzmann's distribution law to describe soil particle displacement and to set up an equation of soil deformation in terms of that displacement.

Andersland and Akkili (1967), Mitchell et al. (1968) and Singh and Mitchell (1969) applied eq. 7-108 to soils directly, being of the opinion that soil deformation is attributed to the "hopping" of molecules in the interparticle bonds of the soil at the points of contact between the mineral particles. Let us scrutinize this aspect.

Soil creep as a heat-activated process

The conceptions referred to provide for treating the creep of soil as a heat-activated process. Applying Eyring's flow equation (eq. 7-108) to soils, Mitchell and his fellow workers examined in its light such problems as the activation energy of soils, the effect of temperature on soil deformation, strength and number of interparticle bonds, etc. To interpret the results of soil creep tests in a triaxial compression apparatus, they used eq. 7-112, defining the relation between the flow rate and soil temperature, and also the expression:

$$\ln \frac{\dot{\gamma}}{\Theta} = \ln \frac{kX}{h} - \frac{U_0 - 0.5V\tau}{k} \frac{1}{\Theta} \qquad (7\text{-}115)$$

obtained by taking the logarithm of eq. 7-112. The authors showed that the test points fell quite reasonably on a straight line plotted on the $\ln(\dot{\gamma}/\Theta)$ versus $1/\Theta$ coordinates. With due regard for the scale of the graph, the slope of this line defined the amount of activation energy equal to:

$$\frac{1}{k}(U_0 - 0.5V\tau) = \frac{d \ln (\dot{\gamma}/\Theta)}{d (1/\Theta)}$$

The activation energy of the soils tested (water-saturated illite with a disturbed structure, W is 30–43%; air-dry illite; marine silt with an undisturbed structure, W is 69.5–74%; and dry sand) was between 25 and 40 kcal./mol ($105 \cdot 10^3$–$168 \cdot 10^3$ J/mol). Approximately the same values of U (between 23 and 32 kcal./mol) were obtained by other authors testing undisturbed and disturbed lake clay (Christensen and Wu, 1964) and alluvial clay having a moisture content of 52–92% (Murayama and Shibata, referenced in Section 1.5). The activation energy of a frozen soil according to Andersland and Akkili (1967) was $U = 94$ kcal./mol. These results compare with the data available in the literature for various materials, U kcal./mol: water, 4–5; plastics, 7–14; ice, 20–28; soils, 23–40; metal, 50; concrete, 54; frozen soil, 90–95.

A point to note is that the activation energy, when referred to a mole, appeared to be unaffected by the moisture content of the soil. Moreover, the activation energy of a water-saturated soil (illite, W is 30–43%) was roughly the same as in the air-dry soil (W is 1%), being $U = 23$–40 kcal./mol in the former case and $U = 37$ kcal./mol in the latter. The value of U obtained for dry sand was approximately of the same order (25 kcal./mol).

This fact, in conjunction with the rather high values of U obtained, was the reason for Mitchell's assumption that the interparticle bonds which are formed in soils are due to contact forces acting between the particles. The number of contacts depends on the thickness of the film of bound water (this is the way bound water influences the strength of bonds), and the number of bonds per contact depends on the effective compacting pressure.

However, there is another point of view. Pusch (1976) pointed out a marked non-uniformity in the state of stress of the soil skeleton, which always contained weakened spots — places of eventual structural failure. Consequently, the crushing energy appeared to have one of the lowest values of those obtained by Mitchell who assumed a uniform energy distribution. The energy computed by Pusch was approximately 10 kcal./mol, i.e. more or less corresponding with the energy of hydrogen bonds. This paved the way to the conclusion that the main factor counteracting shear is the water film rather than the dry contacts.

At the same time, the role of hydrogen bonds should not be overestimated, nor should the process of soil deformation be reduced to the movement of the water film. In fact, this process results from the displacement of individual particles and aggregates along the film of bound water separating them. The forces of interaction between the particles separated by a water film counteract this displacement. Despite the benefits of assessing the behaviour of soils by Eyring's equation (7-108), its use as a means of describing the process of soil creep as a whole is a question requiring further clarification. The equation is undoubtedly applicable only to the stage of steady flow, whereas creep is developing at a changing rate.

This was noted by Singh and Mitchell (1969) who found it necessary to introduce an empirical time function into eq. 7-108. The reason why Eyring's equation is not applicable in describing soil creep is obvious; it has been derived for a viscous liquid flowing continuously and steadily without structural changes caused by displacing elemental particles. The deformation of soils results primarily from the displacement of mineral particles, incurring significant structural changes in the soil. Therefore, the kinetic theory may be used in assessing the behaviour of soil only if this feature of its deformation is duly taken into account. A relevant theory is discussed in Chapter 10.

7.7 REFERENCES

Abdel-Hardy, M. and Herrin, M., 1966. Characteristics of soil-asphalt as a rate process. J. Highway Div., Proc. ASCE, v. 92, NOHWI, Proc. Pap. 4719.

Anagnosty, P.V., 1979. Creep properties of overconsolidated clay and rigid foundation design. In: Design Parameters in Geotechnical Engineering, 7th Europ. CSMFE, Brighton, v. 1.

Andersland, O.B. and Akkili, W., 1967. Stress effect of creep rates of a frozen clay soil. Géotechnique, 47(1).

Arutyunyan, N.Kh., 1952. Some Problems in the Theory of Creep. Gostekhteorizdat, Moscow (in Russian).

Christensen, R.W. and Wu, T.H., 1964. Analysis of clay deformation as a rate process. J. Soil Mech. Found. Div., Proc. ASCE, v. 90, No. SM6, Proc. Pap. 4147.

Davenport, C.C., 1938. Correlation of creep and relaxation properties of copper. J. Appl. Mech., 60: 2–56.

Davis, E.A., 1943. Creep and relaxation of oxygen-free copper. J. Appl. Mech., 10(2).

Eyring, H., 1936. Viscosity, plasticity and diffusion as examples of absolute reaction rates. J. Chem. Phys., 4(4). (Also in: The Theory of the Rate Process. McGraw-Hill, New York, 1941.)

Fedder, D. and Bredth, H., 1973. Rheological investigation by a new apparatus. Proc. 8th ICSMFE, v. 1.1, Moscow.

Frenkel, Ya.I., 1934. The Theory of Liquids and Solids. OHTN, Moscow (in Russian).

Frenkel, Ya.I., 1945. The Kinetic Theory of Liquids. USSR Acad. Sci., Moscow-Leningrad (in Russian).

Gibson, R.E. and Lo, K.Y., 1961. A Theory of Consolidation for Soils Exhibiting Secondary Compression. Norway Geotechn. Inst., Publ. No. 41.

Goldstein, M.N., 1961. On strength of clay soils. SMFE, No. 3 (in Russian).

Hohenemser, K. and Prager, W., 1932. Fundamental equations and definitions concerning the mechanics of isotropic continua. J. Rheol., No. 3.

Ishlinsky, A.Yu., 1945. Deformation Equation for Not Totally Elastic and Visco-Plastic Bodies. News USSR Acad. Sci., Eng. Sci. Sect., No. 1–2 (in Russian).

Kauzmann, W., 1941. Flow of solid metals from the standpoint of the chemical-rate theory. Trans. Am. Inst. Min. Metall. Eng., 143.

Lo, K.Y., 1961. Stress–strain relationship and pore water pressure characteristics of a normally consolidated clay. Proc. 5th ICSMFE, v. I, Paris.

Malinin, N.N., 1968. Applied Theory of Plastics and Creep. Mashinostroenie, Moscow (in Russian).

Mitchell, J.K., Campanella, K.G. and Singh, A., 1968. Soil creep as a rate process. J. Soil. Mech. Found. Div., Proc. ASCE, v. 94, No. SM1.

Murayama, S. and Shibata, T., 1966. Flow and stress relaxation of clay. Rheology and Soil Mechanics, IUTAM Symp., Grenoble, 1964, Springer Verlag, Berlin.

Push, R., 1976. Shear deformation of clay microstructure. Proc. 7th Int. Congr. Rheology, Gotenbur

Regel, V.R., Slutsker, A.I. and Tomashevsky, E.K., 1974. Kinetic Nature of Strength in Solid Bodies. Nauka, Moscow (in Russian).

Rozovsky, M.I., 1951. Creep and long-term rupture of metals. J. Eng. Phys., v. 21, no. 2 (1951) (in Russian).

Rozovsky, M.I., 1955. On linear equations of creep. J. Eng. Phys., no. 13 (in Russian).

Schiffman, R.L., Ladd, Ch.C and Chen, A., 1964. The secondary consolidation of clay. Rheology and Soil Mechanics, IUTAM Symp., Grenoble, 1964.

Singh, A. and Mitchell, J.K., 1969. Creep potential and creep rupture of soils. Proc. 7th ICSMFE, v. 1, Mexico City.

Soderberg, C.R., 1936. The interpretation of creep tests for machine design. Trans. ASME, v. 58, no. 8.

Tan, Tjong-Kie, 1957. Three dimensional theory of the consolidation and flow of clay layers. Scientia Sinica, v. 6, no. 1.

Taylor, D.W., 1942. Research on consolidation of clays. Mass. Inst. Technol., Dept. Civ. Sanitary Eng., Publ. Ser. 82.

Zaretsky, Yu.K., 1972. On rheological properties of plastic frozen soils and assessment of time-dependent settlements of plates. SMFE, 2 (in Russian).

Chapter 8

THE THEORY OF SOIL CONSOLIDATION

8.1 VOLUMETRIC CREEP

Effective and neutral stresses

As already stated, the pressure sustained by a water-saturated soil is distributed between the water contained in voids and the soil skeleton, i.e. the mineral particles surrounded by water films and interlinked into a single framework. The stress set up in the pore water is a hydrostatic pressure; it neither causes an appreciable compression of the soil skeleton nor influences the capacity of the soil to resist shear.

A stress in excess of the atmospheric pore-water pressure is called neutral pressure and is denoted by u or p_w. It may be positive or negative. A positive neutral pressure is termed, after Terzaghi, the pore-water pressure.

The compressive stress set up in the soil skeleton is called effective pressure, denoted by σ' or p'. The values of σ' and u change with time because consolidation causes a redistribution of the pressure between the pore water and soil skeleton; however, their sum is always constant for a given constant external load. For a completely water-saturated soil (provided the problem is a single-valued one), the sum is:

$$\sigma_z = \sigma'_z + u = \text{const.} \tag{8-1}$$

In the case of a partially saturated soil we have, according to Bishop (1959):

$$\sigma'_z = \sigma_z - [u_w + (1 - \kappa)(u_a - u_w)] \tag{8-1'}$$

where u_w is the pore-water pressure; u_a is the air pressure in soil voids; κ is a coefficient determined experimentally. In the case of complete saturation with water (practically, when $G > 90\%$), $\kappa = 1$ and eq. 8-1' is transformed into eq. 8-1.

For the state of combined stress, eq. 8-1 takes the form:

$$T = T_{\sigma'} + T_u \tag{8-2}$$

where $T_{\sigma'}$ is an effective stress tensor and T_u is a neutral stress tensor.

The tensor $T_{\sigma'}$, in its turn, may be resolved into a spherical tensor of effective stress $T^o_{\sigma'}$, causing a volumetric deformation of the soil skeleton, and an effective stress deviator $D_{\sigma'}$, inducing shearing deformation:

$$T_{\sigma'} = T^o_{\sigma'} + D_{\sigma'} \tag{8-3}$$

The neutral stress tensor T_u builds up a pore-water head without causing shear, the spherical tensor thus being $T_u = T^o_u$.

Volumetric deformations of creep

The volumetric deformations of soil skeleton produced by the spherical tensor of effective stresses T_σ°, i.e. by the mean stress $\sigma'_m = (\sigma'_1 + \sigma'_2 + \sigma'_3)/3$, are time dependent in the general case. They include a recoverable deformation, a residual deformation and are non-linear.

The volumetric deformations of the soil skeleton develop with time owing to the viscous resistance of interparticle bonds. Since these deformations are not associated with the squeezing out of free pore water, they manifest themselves in the most refined form in heavily compacted soils the liquid phase of which is represented by bound water.

Moreover, as shown above, the creep deformations of a soil skeleton are likely to develop even in air-dry soils (see Fig. 5-20). Volumetric deformation entailing no water percolation may also be observed in frozen soils containing water in either solid form (ice) or firmly-bound state (unfrozen water). So, the volumetric deformation of frozen soils, as obtained by the present author (see Chapter 1: Vyalov, 1959) from a 3000-h confined compression test by a load $\sigma_z = 20 \cdot 10^5$ Pa, was 5% at $\theta = -1.4°C$ and 11% at $\theta = -0.3°C$; the residual and recoverable parts each accounted for one half of the total in either case.

Shedding more light on the volumetric deformation of creep in frozen soils are the creep curves and isochrones (Fig. 8-1) obtained by Gorodetsky (1969) from triaxial compression tests under the conditions of all-around pressure, σ_m = const. It can be seen that the relation between the volumetric strain $\varepsilon_m = (\varepsilon_1 + \varepsilon_2 + \varepsilon_3)/3$ and the all-around pressure $\sigma_m = \sigma'_m$ is non-linear.

The volumetric creep of heavily compacted soils is exemplified in Fig. 8-2 depicting data acquired by R. Schiffman (see Section 7.7) in testing blue Boston clay ($W_L = 33\%$, $W_P = 18\%$) in an oedometer and a triaxial compression apparatus; the test load used in this latter case was an all-around pressure $\sigma_1 = \sigma_2 = \sigma_3$ = const. The results are presented in terms of void ratio versus time curves rectified on a semi-log scale. It will be noted that when the creep induced by the triaxial compression was transformed into a progressive flow (at $t > 90$ days), a reversal of the sign of the volumetric deformation was observed — a fact coinciding with the data considered above (see Fig. 5-23).

Generally speaking, the volumetric deformation resulting from testing heavily compacted clays containing bound water may be treated as volumetric creep. In normally compacted soils and, even more, in undercompacted ones, volumetric deformations result from both the creep of the soil skeleton and the percolation of moisture. The process of percolation is observed mainly at an early stage of the test and the skeleton creep is evident, eventually, on expelling the bulk of moisture; dissipation of the pore pressure becoming zero is an indication of the creep.

Equation of volumetric creep. By analogy with eq. 5-8, this equation may be written in the form:

$$\varepsilon_m = f^*(\sigma'_m) \Phi^*(t) \tag{8-4}$$

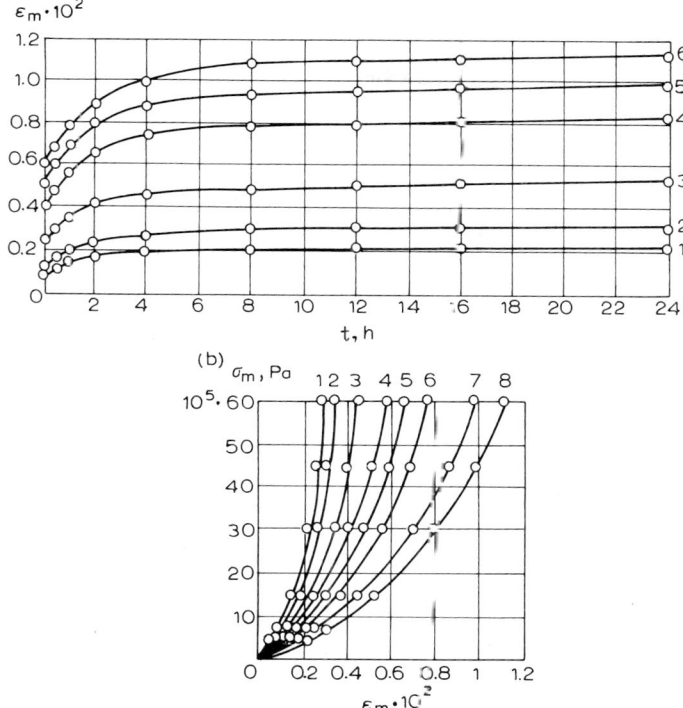

Fig. 8-1. Curves of volumetric creep, frozen soil (sandy loam, $W = 26\%$, $\theta = -10°C$) subjected to all-around pressure.
(a) Curves of volumetric creep $\varepsilon_m = (\varepsilon_1 + \varepsilon_2 + \varepsilon_3)/3$ induced by the stress $\sigma_m = (\sigma_1 + \sigma_2 + \sigma_3)/3$ of the following values: 1. $5 \cdot 10^5$ Pa; 2. $7.5 \cdot 10^5$ Pa; 3. $15 \cdot 10^5$ Pa; 4. $30 \cdot 10^5$ Pa; 5. $45 \cdot 10^5$ Pa; 6. $60 \cdot 10^5$ Pa.
(b) Curves of ε_m versus σ_m at the following times: 1. 30 s; 2. 1 min; 3. 5 min; 4. 15 min; 5. 30 min; 6. 1 h; 7. 4 h; 8. 24 h.

where $f^*(\sigma'_m)$ is a function defining the connection between the mean normal effective stress σ'_m and the mean linear strain ε_m; $\Phi^*(t)$ is a function of volumetric creep.

Unlike the curve of shear deformation, the curve of the $\sigma'_m - \varepsilon_m$ relation (see Fig. 4-8) is concave towards the axis of stress, for with an increase in σ'_m the volumetric deformation tends to a certain limit ε_s corresponding to the maximum compaction. This limit is attained, however, under a sufficiently high all-around pressure only.

The functions $f(\sigma'_m)$ take in accordance with eqs. 4-41, 4-42 and 4-44 the following form:

$$\varepsilon_m = \left(\frac{\sigma'_m}{D}\right)^{1/\kappa} \quad \text{where} \quad \kappa \geqslant 1 \tag{8-5}$$

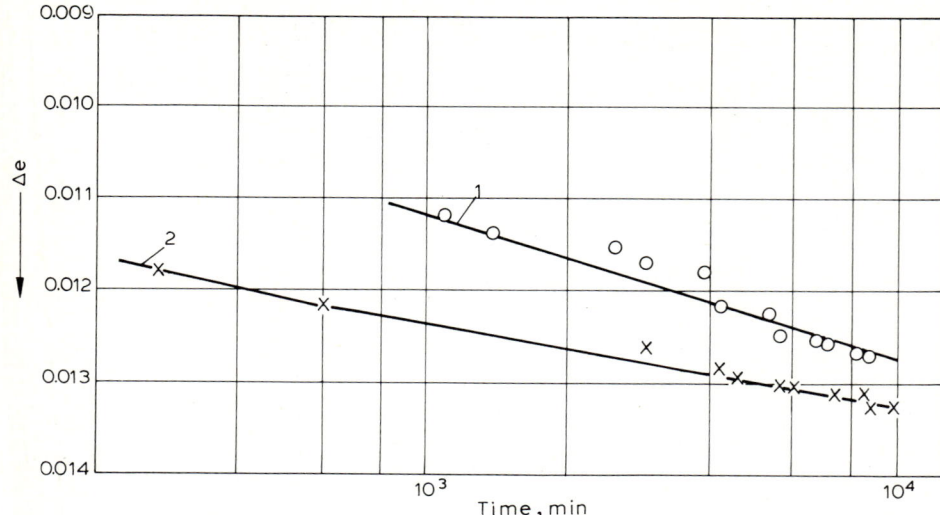

Fig. 8-2. Curves of creep for overcompacted clay: 1 = test in oedometer at σ_1 = const.; 2 = triaxial compression test at σ_m = const.

$$\varepsilon_m = \frac{\sigma'_m \varepsilon_s}{k_0 \varepsilon_s + \sigma'_m} \quad (8\text{-}6)$$

$$\varepsilon_m = \varepsilon_s (1 - e^{-b\sigma'_m}) \quad (8\text{-}7)$$

The curves corresponding to eqs. 8-6 and 8-7 have asymptotes $\varepsilon_m \to \varepsilon_s$ for $\sigma'_m \to \infty$. Curve 8-5 has no asymptote, but coincides well with the experimental data, provided the stress is not too high.

Since the non-linearity of volumetric deformation at low stresses manifests itself little, the $\sigma'_m - \varepsilon_m$ relation may be assumed to be linear. In any case, linear approximation appears to be better applicable to volumetric deformations than to shearing ones. Taking this into account, the practice of treating shearing deformations as non-linear and volumetric deformation as linear, which is followed for the sake of simplicity, seems to be justifiable.

A time-dependent volumetric deformation is an attenuating one, tending to a finite, stabilized value with the progress of time. Possible forms of the volumetric creep function $\Phi^*(t)$ correspond to eqs. 5-13, 5-14 and 5-15.

Volumetric creep can be described in terms of any of the creep theories examined in Chapter 7 provided all pertinent considerations are observed with respect to volumetric deformations.

Relation between volumetric and shearing deformations

Figure 8-3 presents the experimental findings of Koizumi and Ito (1964) testing water-saturated clay ($W = 120\%$, $W_L = 136\%$, $W_P = 81\%$) for triaxial com-

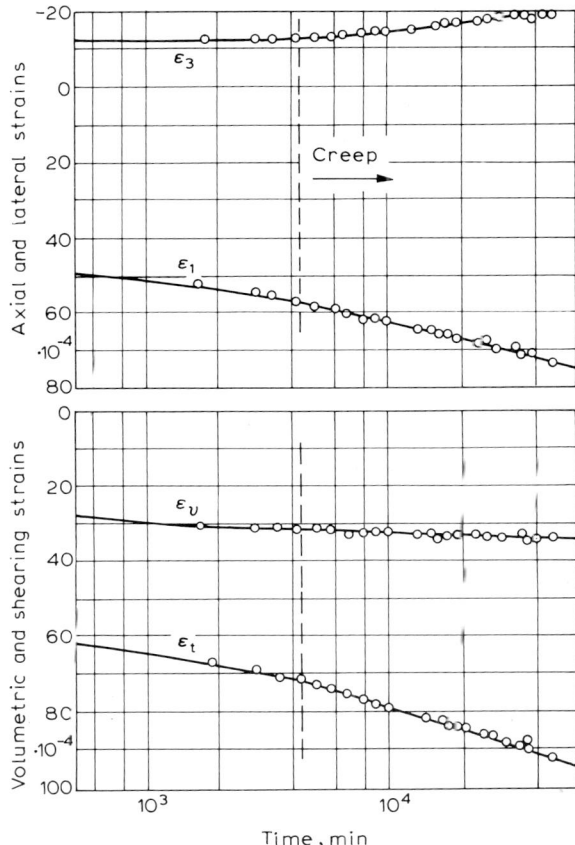

Fig. 8-3. Curves of volumetric creep and creep in shear, water-saturated clay sustaining triaxial compression ($\sigma_m = 0.27 \cdot 10^5$ Pa; $\tau_i = 1.85 \cdot 10^5$ Pa).

pression under a constant mean normal stress σ_m and a constant shearing stress τ_i. The manner in which the axial, ε_1, and lateral, ε_3, strains were developing in the specimen are shown in the upper graph while the lower graph depicts the volumetric and shearing strains $\varepsilon_v = (\varepsilon_1 + 2\varepsilon_3)/3$ and $\varepsilon_t = \gamma_i = \varepsilon_1 - \varepsilon_3$, respectively. It can be seen that percolation stopped $4 \cdot 10^3$ min into the test as soon as the creep of soil skeleton began to take shape.

The curves of volumetric and shearing deformations are easily amenable to rectification on semi-log coordinates. Thus the two curves may be described by logarithmic functions of the type:

$$\Phi(t) = a \ln (t/T) \tag{8-8}$$

with different values of a and T. This means that the curves of volumetric and shearing creep are dissimilar in the case under consideration.

The question of whether or not the creep curves and isochrones for volumetric and shearing deformation are similar is important, for it decides whether or not the coefficient of lateral expansion v entering into the initial equation of any problem other than a single-dimensional one, is constant.

Coefficient of lateral expansion. It was stated in Chapter 5 while analysing eq. 5-53, that the coefficient v is constant provided the functions $f(\sigma)$ and $\Phi(t)$ of the longitudinal and lateral deformations are identical. The same applies to the functions f and Φ of the volumetric and shearing deformations.

Thus, the coefficient v is constant if the condition

$$\frac{f(\tau_i)\Phi(t)}{f^*(\sigma'_m)\Phi^*(t)} = \text{const.} \tag{8-9}$$

is satisfied; where the functions f and Φ refer to the shearing deformations and the functions f^* and Φ^* describe the volumetric deformations.

In general, soils cannot satisfy condition 8-9. In fact, the functions $\phi(\gamma_i)$ and $\phi(\varepsilon_m)$ cannot be of the same form (provided they are non-linear) because, in the case of a stress increase, $\gamma_i \to \infty$ whereas $\varepsilon_m = \text{const.}$ (see Fig. 1-2).

The functions $\Phi(t)$ and $\Phi^*(t)$ will also be different because the shearing creep may be of both the attenuating and non-attenuating nature whereas the volumetric creep always attenuates. Nevertheless, since a variable parameter v complicates computations, one must try to fulfil condition 8-9, at least approximately. As far as the stress function f is concerned, this is quite possible, provided that the loading changes over a narrow range, in this case the τ_i–γ_i and σ'_m–ε_m relations may be approximately described by a function of the same type.

The creep functions $\Phi(t)$ and $\Phi^*(t)$ may also be assumed to be identical for a not too high stress that induces shearing creep of the attenuating character. Taking into account the significant spread of soil characteristics in the natural conditions, assumptions of this kind appear to be practical.

Finally, in some cases one can make the assumption that the volumetric deformation of soil skeleton is not present at all (condition of incompressibility) or that it is only an elastic one.

8.2 PERCOLATION CONSOLIDATION OF SOILS

Equation of consolidation

The theory of percolation consolidation of soils, formulated by Terzaghi and developed by Gersevanov (see Chapter 1: Terzaghi, 1926, 1944; Gersevanov, 1937, 1948), was based on the assumption that the spherical tensor of effective stress T_δ^o induced only an elastic volumetric strain of the soil skeleton and that the process of consolidation was attributed to the squeezing out of pore moisture by the pore pressure tensor T_u in accordance with Darcy's law of percolation. This theory treats the soil as being completely saturated (soil mass), disregards the compression of the liquid phase and assumes that the volumetric strains of the soil skeleton are increasing directly with the effective pressure.

Under the above assumption, the equation of consolidation for a single-dimensional problem takes the form:

$$\frac{\partial \sigma'_z}{\partial t} = C_v \frac{\partial^2 \sigma'_z}{\partial z^2}$$

or: (8-10)

$$\frac{\partial e}{\partial t} = -aC_v \frac{\partial^2 u}{\partial z^2}$$

where $C_v = k_{per}/a_0 \gamma_w$ is the coefficient of soil consolidation in cm²/h; $a_0 = a/(1 + e_m)$ is the modulus of volume change in Pa⁻¹; $a = \Delta e/\Delta \sigma$ is the coefficient of compressibility of the soil in Pa⁻¹; k_{per} is the coefficient of pore moisture percolation in cm/h; e and e_m are the time-dependent and mean (regarded as being constant) values or the void ratio, respectively; σ'_z and u are the effective and neutral (pore) pressure in Pa, respectively, and t is the time.

Deriving eq. 8-10 by means of a mechanical model. The way eq. 8-10 is derived is well known (cf., for example, Chapter 1: Tsytovich, 1976). We are going to demonstrate that it may be obtained from considering mechanical models, in particular a multi-element model comprising a great number of elements of the Kelvin type (see Chapter 1: Rzhanitsin, 1968).

Consider a multi-element model (Fig. 8-4) made of elastic elements connected in series so as to form a single elastic rod of a length h which simulates the soil skeleton. Let us assume that the elastic elements obey the law:

$$\varepsilon_z = a_0 \sigma'_z \qquad (8\text{-}11)$$

where ε_z is the strain of an element.

Connect liquid-filled dashpots with perforated pistons in parallel with the elastic elements. The liquid contained in each dashpot will simulate pore water of the soil and the perforations in the piston will represent soil capillaries. The sinking of the piston down the dashpot, simulating pore water percolation, obeys the law:

$$v = k_{per} i \qquad (8\text{-}12)$$

where v is the velocity of percolation, $i = -\partial H/\partial z = -1/\gamma_w = \partial u/\partial z$ is the hydraulic gradient, and $H = u/\gamma_w$ is the head.

Thus, the element simulating pore water will be similar to the Newtonian element of rheological models (see Fig. 7-1) except that the piston displacement obeys Darcy's law of percolation rather than the Newtonian law of viscous flow.

Let us assume that the parameters a_0 and k_{per} of eqs. 8-11 and 8-12 are the same for each of the elements of the model, representing a soil mass of uniform depth. If allowance must be made for non-uniform depth, the parameters should be different for each of the elements.

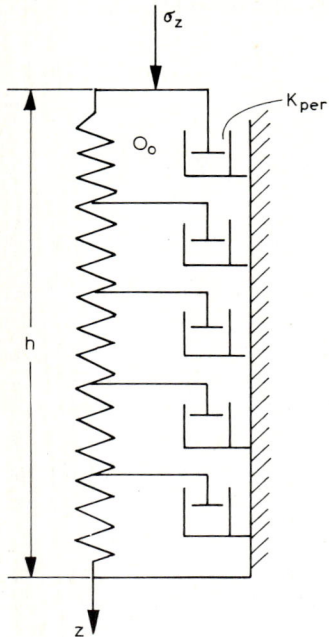

Fig. 8-4. Multi-element model simulating primary consolidation of soil.

According to the layout of the model, the total load σ_z is first taken by the viscous elements, being then redistributed (with the compression of the springs) between the elastic elements supporting a pressure σ'_z and the viscous elements sustaining a pressure u. If the transmission of the load to the soil skeleton is allowed for from the beginning, an elastic element must be added at the top of the model to support the load σ'_z.

In the model of the given arrangement, each of the Kelvin elements is subjected to the same stress σ_z. This corresponds to the assumption that the value of σ_z is constant down the depth of the soil mass which was adopted in deriving eq. 8-10 under the condition of single-dimensional consolidation. The strain of the elastic rod due to compression is:

$$\varepsilon_z = \frac{\partial w}{\partial z} = a_0 \sigma'_z \qquad (a)$$

where w is the displacement of the points of the elastic rod along the z-axis.

On the other hand, taking into account eq. 8-12, the rate of displacement of the perforated piston in each of the viscous elements is:

$$v_{per} = \frac{\partial w}{\partial t} = -k_{per} \frac{u^*}{\gamma_w} \qquad (b)$$

where $u^* = \Delta u / \Delta z$ is the pressure per unit length of the model.

The condition of equilibrium of a length Δz of the elastic rod may be represented in the differential form as:

$$\frac{\partial \sigma'_z}{\partial z} + u^* = 0 \tag{c}$$

Substituting expressions a and b, we obtain:

$$\frac{\partial w}{\partial t} = \frac{k_{per}}{a_0 \gamma_w} \frac{\partial^2 w}{\partial z^2} \tag{d}$$

On differentiating with respect to z and substituting expression a, we arrive at eq. 8-10.

Degree of consolidation. The settlement of soil due to compaction reaches its finite, stabilized value $S = S_\infty$ when $t = \infty$ and $\sigma'_z = \sigma_z$. The ratio of the settlement S at any time to the stabilized settlement S_∞ is called the degree of consolidation:

$$U = \frac{S}{S_\infty} \tag{8-13}$$

In the case of single-dimensional consolidation (compression of a soil layer with a depth of $2h$ under the conditions of two-way percolation), the value of U, as obtained from solving eq. 8-10, is known (see Chapter 1: Tsytovich, 1976) to be:

$$U = 1 - \frac{8}{\pi^2}(e^{-Nt} + \tfrac{1}{9}e^{-9Nt} + \tfrac{1}{25}e^{-25Nt} + \ldots) \tag{8-14}$$

or approximately:

$$U \cong 1 - \frac{8}{\pi^2} e^{-Nt} \tag{8-15}$$

where $N = \pi^2 C_v / 4h^2$.

Equation of consolidation and aftereffect

Compare the equation of consolidation 8-15 as set forth in the form:

$$S = S_\infty \left(1 - \frac{8}{\pi^2} e^{-Nt}\right) \tag{8-16}$$

with the equation of viscoelastic deformation (aftereffect) (eq. 7-6) represented by:

$$S = S_\infty (1 - e^{-t/T_p}) \tag{8-17}$$

It can be seen that both equations are of identical form, both describing the attenuation of deformation in accordance with an exponential law. This is quite natural, because they are both derived from a model of the same kind, i.e. the Kelvin–Voigt model.

At the same time, eqs. 8-16 and 8-17 materially differ from one another. The exponent t/T_p in the equation of aftereffect 8-17 is not influenced by the depth h of the layer compressed and, consequently, the value of the relative strain S/S_∞ is not affected by h. The exponent Nt in the equation of consolidation 8-16 depends upon the length of percolation path, being thus a function of h. Accordingly, the degree of consolidation S/S_∞ appears to be dependent upon the depth of the soil compressed, h.

This serves the purpose of telling a percolation from a creep process during laboratory experiments. If the data acquired from testing specimens of various height are not affected by h, this means that the process is creep, otherwise it is consolidation (or a combination of the two processes).

Development of percolation theory

The percolation theory has been improved by introducing into eq. 8-10 additional factors influencing consolidation and providing solutions to problems with various boundary conditions, including three-dimensional problems.

It has been shown that at the initial moment only a fraction of the external load may be transmitted to the pore water, the rest being sustained by the soil skeleton. Tsytovich explains this by the structural strength possessed by soils, $\sigma_{str} = \sigma(1 - \beta_0)$, where $\beta_0 = u_0/\sigma$ is the coefficient of initial pore-water pressure; u_0 is the initial pore-water pressure. (See Chapter 1: Tsytovich, 1976.)

The coefficient β_0 varies between the limits $0 \leqslant \beta_0 \leqslant 1$. For $\beta_0 = 1$, all the pressure comes on the water at $t = 0$, $u = u_0 = \sigma$; for $\beta_0 = 0$, $u = u_0 = 0$ and the pressure is sustained by the soil skeleton alone.

Another factor influencing consolidation, which also arises due to the interparticle binding of soil, is the initial head gradient i_0 bringing into balance the surface tension forces of water films. Pore water is capable of migrating only when the hydraulic gradient exceeds the value of i_0. Accordingly, the velocity of percolation will be equal to $v_m = k_{per}(i - i_0)$, where:

$$i_0 = -\frac{1}{\gamma_w}\frac{\partial u_0}{\partial z}$$

In contrast to the assumption that pore water is incompressible, it has been demonstrated that water containing air bubbles and dissolved gases is as a rule capable of changing its volume. This property is defined by the coefficient of pore water compressibility.

Moreover, Florin (1959), Šuklje (1969) (both referenced in Chapter 1) and other authors have suggested taking into account the non-linear stress–strain relation and the changes in percolation coefficient and initial head gradient during consolidation.

Method of body forces

Biot (1941, referenced in Chapter 1) and Florin (1959, referenced in Chapter 1) have proposed treating the process of consolidation in terms of the so-called method of 'body forces'. It takes into consideration the interaction between the solid and liquid constituents of soil and presents the forces of the interaction in the form of body forces resulting from a decrease in the weight of the soil skeleton as this becomes suspended owing to the pore-water pressure. The stress–strain relation is assumed to be linear but allowance for the compressibility of pore water is made. The way the external pressure is redistributed between pore water and soil skeleton is assumed to be in accordance with the Terzaghi percolation theory.

The theory of 'body forces' has gained wide-spread recognition, being developed in the works of Tan (see Section 7.7), Mandel (1957) and others.

8.3 PRIMARY AND SECONDARY CONSOLIDATION OF SOIL

'Secular' soil consolidation

It has already been pointed out that in 1956 K. Buisman (referenced in Chapter 1) had employed a logarithmic function similar to eq. 5-14 to describe the process of secondary consolidation:

$$S = ph(\alpha' + \alpha'' \lg t) \tag{8-18}$$

where α' is a unit settlement (for $p = 1$ and $h = 1$) caused by primary consolidation; α'' is a unit settlement brought about due to the development of secondary consolidation during a time interval $t > 1$; p and h are the load and depth of the layer compressed, respectively, and t is time ($t > 1$).

When primary consolidation is to be separated from secondary consolidation, eq. 8-18 is written as:

$$S = ph\left(\alpha' + \alpha'' \lg \frac{t}{t'}\right) \tag{8-18'}$$

where t' is the time of the end of primary consolidation.

Reporting at the 2nd ICSMFE, Croce (1948) showed that the parameters α' and α'' were connected by the relation $\alpha'' = 10\alpha'$ and varied non-linearly with the load: for 10^5 Pa, $\alpha' = 0.081$; for $2 \cdot 10^5$ Pa, $\alpha' = 0.105$; for $4 \cdot 10^5$ Pa, $\alpha' = 0.128$.

Eq. 8-18 is based on a linear settlement–load (S–p) relation and describes, as is inherent in the logarithmic law, a slow and unconfined increase in the settlement, hence the name of the formula "the law of secular settlement".

This formula has been derived on the assumption that secondary consolidation begins only after the primary consolidation has ended. This does not fully agree with the nature of the process, for primary and secondary consolidations take place, in fact, simultaneously. However, taking into consideration that the percolating mechanism of compaction dominates the early stage of the process, being succeeded by the viscous stage, and that a subdivision of the process on such lines is convenient in handling experimental data, the above assumption appears to be justified. Taken commonly as the onset of the stage of secondary consolidation is the instant when the settlement versus curve plotted on the logarithmic scale begins to rectify (Fig. 8-3).

Several researchers (Casagrande, Taylor, Brinch Hansen) have suggested a special technique of pinpointing the moment of the transfer from primary to secondary consolidation (see Chapter 1: Šuklje, 1969; Tsytovich, 1976). Particulary, it has been found that the law of primary consolidation (eq. 8-15) may be replaced within the range of up to $U = 60\%$ approx. by the power relationship:

$$U = \left(\frac{4C_v}{\pi h^2} t\right)^{0.5} \tag{8-19}$$

Accordingly, the instant of the transfer from primary to secondary consolidation may be determined from a settlement curve plotted on the S versus $t^{1/2}$ and $\lg t$ combined coordinates (see Fig. 1-5).

The power relation t^n with the exponent $n \neq 0.5$ has also been subsequently employed to describe secondary consolidation. In a paper presented at the 7th ICSMFE, Hansen and Inan (1969) disclosed a formula taking account of both the changes in void ratio with time according to a power law and the power relation between void ratio and load:

$$e_j = e_0 \left\{ 1 + \frac{\sigma'(t_j)}{\sigma_0'} \left[\int_0^{t_j} \left(\frac{\sigma'(t)}{\sigma'(t_j)}\right)^b \frac{dt}{t_0} \right]^c \right\}^{-a} \tag{8-20}$$

where e_0 and e_j are the void ratios at the initial ($t = 0$) instant and at a given ($t = t_j$) instant, respectively; $\sigma'(t)$ is the effective pressure varying over the range between 0 and $\sigma'(t_j)$ for $0 \leq t \leq t_j$; σ_0' is a constant in Pa connected with e_0 by the relation $\sigma_0 = ne_0^4$; t_0 is an arbitrary time, e.g. $t_0 = 1$ day; and $c < 0.05$, $a < 0.3$, $b > 10$ are parameters.

Secondary consolidation as hereditary creep

V.A. Florin (see Section 1.5) has suggested describing secondary soil consolidation by the equation of hereditary creep (eq. 7-72), as interpreted by G.N. Maslov and N.Kh. Arutyunan (see Section 5.7: Arutyunyan, 1952) and using an exponential kernel. The volumetric strain of the soil skeleton will be expressed by:

$$e_0 - e_j = \sigma_z' \{a_{inst} + a_{sec}[1 - \exp(-\beta(t - v_1))]\} + \int_{v_1}^{t} \frac{d\sigma_z'}{dv}$$
$$\times \{a_{inst} + a_{sec}[1 - \exp(-\beta(t - v))]\} dv \tag{8-21}$$

where e_0 and e_j are the initial ($t = 0$) void ratio and the void ratio at a given instant, respectively; and a_{inst} and a_{sec} are the coefficients of the instantaneous and secondary compaction of the soil skeleton.

Considering the equation of percolation consolidation 8-10 and the equation of soil skeleton creep 8-21 simultaneously, we can solve the problem of linear soil compaction with due regard for primary and secondary consolidations. Taken into account in this case are the compressibility of pore water and that of mineral particles, initial head gradient (if present), the effect of structural strength and that of the ageing of material, variability of the coefficient of percolation, the effect of entrapped air, etc.

Mechanical model taking account of primary and secondary consolidation

A mechanical model designed to take account of secondary soil consolidation must also incorporate, in addition to the percolation elements, viscous, Newtonian elements connected to the elastic elements simulating the soil skeleton. As a result, the soil skeleton possesses not only elastic properties but viscosity as well. Such a model (Fig. 8-5), as devised by the author of this book, is a combination of the model depicted in Fig. 8-4 and extra viscous elements.

All viscoelastic elements, each representing a Kelvin–Voigt model, are connected in series so as to form the setup of the model illustrated in Fig. 7-2(c). This model represents a soil skeleton. Assuming that the parameters of the repeating elements in the model are all the same, and the soil mass is of uniform depth, the equation of

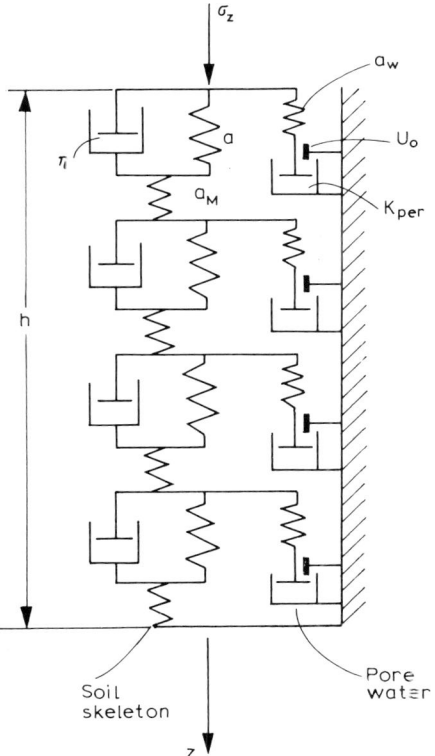

Fig. 8-5. Multi-element model simulating primary and secondary consolidation of soil.

deformation of the soil skeleton will take a form similar to eq. 7-27:

$$\varepsilon = a_0' \left[\sigma'(t) + \int_0^t \sigma'(v) K(t-v) dv \right] \tag{8-22}$$

where $\sigma'(t)$ is the effective pressure changing with time and

$$K(t-v) = \bar{a}_0 e^{-(1/T_p)(t-v)} \tag{8-23}$$

in which $\bar{a} = (a_0'' - a_0')/(a_0' T_p)$; a' and a'' are the instantaneous (at $t=0$) and the finite (at $t=\infty$) values of the modulus of volume change, respectively, and T_p is the period of aftereffect.

A pore-water model consists of percolation elements coupled, by analogy with the Maxwell model, with elastic elements which have a coefficient of compressibility a_w and simulate the compressibility of pore water. A dry friction element connected to the piston rod simulates initial pore-water pressure u_0, defining the initial head gradient:

$$i_0 = -\frac{1}{\gamma_w} \frac{\partial u_0}{\partial z}$$

The piston is capable of displacing only if the pore-water pressure exceeds the value of u_0. Thus, the mechanism of deformation of the elastic and percolation elements of the model will be defined by the relations:

$$\varepsilon = a_w u \qquad v_m = -\frac{k_{per}}{\gamma_w} \frac{\partial \bar{u}}{\partial z} \qquad \bar{u} = u - u_0$$

Examining the displacements of points in the model and setting up, as this was done with respect to the model in Fig. 8-4, a differential equation of equilibrium, we obtain the equation of deformation:

$$\frac{\partial e}{\partial t} + a_w e_m \frac{\partial u}{\partial t} = \frac{1 + e_m}{\gamma_w} k_{per} \frac{\partial^2}{\partial z^2} (u - u_0) \tag{8-24}$$

where the symbols are the same as in eq. 8-10.

Substituting the equation of soil skeleton deformation (for $\varepsilon = \Delta e$) into eq. 8-24, we obtain a law governing soil compaction with regard to primary and secondary consolidation, the compressibility of the liquid phase and the presence of initial pore-water pressure.

Eq. 8-24, derived with the aid of the model, coincides with the analytical solution obtained earlier by Florin (referenced in Section 1.5).

Unlike the various mechanical models representing certain aspects of the consolidation process, which were examined in Section 7.2, the multi-element models of the type presented in Fig. 8-5 pave the way to setting up the equation of soil skeleton deformation in the integral form according to the theory of hereditary creep.

The theory of consolidation has been scrutinized by Ŝuklje (see Section 1.5). Studies of the theory were reviewed by Poorooshasb (1969) and Gorbunov-Posadov and Davydov (1973).

Recent theories of soil consolidation

Tsytovich and Ter-Martirosyan (Tsytovich et al., 1967) have come up with an equation describing the mechanism of settlement due to compaction in the following form:

$$S = ph(a_0^f U^f + a_0^c U^c) \qquad (8\text{-}25)$$

where a_0^f and a_0^c are the moduli of volume change during primary and secondary consolidation and U^f and U^c are the degrees of primary and secondary consolidation, respectively.

Thus, in eq. 8-25 the settlement due to compaction is separated into the settlement resulting from percolation and the settlement due to creep. The former form of settlement is determined by solving eq. 8-14 and the latter is found by solving the equation of soil skeleton deformation.

Note that resolving the settlement into primary and secondary constituents is possible only in the case of a single-dimensional problem. In a soil mass subjected to three-dimensional stress and strain the percolation process ends at different times from point to point so that no boundary can be drawn between the primary and secondary consolidation.

To take care of the state of three-dimensional stress in soil, Zaretsky (see Section 1.5) has suggested a general form of the Biot–Florin 'body force' model allowing for a simultaneous (linear) development of primary and secondary consolidation, time-dependent stress variations at any point of the soil mass and partial transmission of external pressure to water at the initial moment. The solution boils down to the following set of simultaneous equations:

$$\left. \begin{array}{l} \bar{G}(\Delta w_j) + \tfrac{1}{3}(\bar{G} + \bar{k})(w_{k,kj}) = -\dfrac{1}{\bar{\beta}_0}(u, j) \\[4pt] (k, j = 1, 2, 3) \\[4pt] \dfrac{\partial}{\partial t}(\sigma_{k,k}) + \dfrac{3}{\bar{\beta}_0}\dfrac{\partial u}{\partial t} = \dfrac{3}{\bar{\beta}_0}\bar{C}_v \Delta u \end{array} \right\} \qquad (8\text{-}26)$$

where $\bar{G}$, $\bar{k}$, $\bar{\beta}_0$ and $\bar{C}_v$ are the shear modulus, bulk modulus, coefficient of initial pore-water pressure and coefficient of consolidation, respectively, all changeable in time; $\sigma_{kk} = 3\sigma_m$ is the sum of the principal stresses and w represents the displacements. The comma before the symbol j denotes differentiation with respect to the variable j:

$$A_{k,j} = \frac{\partial A_k}{\partial x_j}$$

the symbol $A_{k,jj}$ denotes the sum:

$$A_{k,11} + A_{k,22} + A_{k,33} = \Delta A_k$$

where

$$\Delta = \frac{\partial^2}{\partial x_1^2} + \frac{\partial^2}{\partial x_2^2} + \frac{\partial^2}{\partial x_3^2}$$

is the Laplacian operator.

The first relation in eq. 8-26 represents the condition of equilibrium of the displacements w. The second relation describes the migration of liquid in the deformed medium. In the case of a single-dimensional problem, both are solved to obtain:

$$S = ha'_0 p \left[1 + \int_0^t K(t-v)\,dv - \frac{8}{\pi^2} \sum_{m=1,2,\ldots}^{\infty} \frac{1}{m^2} \bar{\Psi}_m(t) \right] \tag{8-27}$$

where the first term on the right-hand side represents the instantaneous settlement, the second term expresses the settlement due to the creep of soil skeleton, and the third indicates the settlement brought about by moisture percolation with regard to the interaction between the liquid and solid constituents of the soil medium. This function is defined by the expression:

$$\bar{\Psi}_m(t) = \Psi_m(t) + \int_0^t K(t-v)\,\Psi_m(v)\,dv \tag{8-28}$$

in which

$$\Psi_m(t) = \exp\left(-\frac{m\pi}{h} C_v t\right)$$

is the function defining the settlement due to percolation; here $\Psi(0) = 1$, $\Psi(\infty) = 0$.
The kernel $K(t, v)$ is of the exponential form according to eq. 7-62.

Time of consolidation

Since the time of percolation consolidation varies with the depth of the soil layer compressed, the times of consolidation t_1 and t_2 are connected with the depths of layers h_1 and h_2 by the relation:

$$\frac{t_1}{t_2} = \left(\frac{h_1}{h_2}\right)^n \tag{8-29}$$

If the process is of a purely percolating nature, then $n = 2$ in accordance with eq. 8-15. When soil creep alone is experienced, $n = 0$. Since in real soils the two processes take place, as a rule, simultaneously, the exponent n is commonly taken as $0 < n < 2$. This point was noted by N.N. Maslov way back in 1949 (see Chapter 1: Maslov, 1968a). In a later work (1972) he found that n is a function of such physical characteristics of soil as plasticity index and consistency:

$$n = \varrho \ln I_L + v \qquad n = aI_p + b$$

where I_L is the liquidity index and ϱ, v, a, and b are parameters. So, for soils of a weak consistency, $n \to 2$; for soils of a firm consistency, $n = 1.5$ approx. and for soils of a stiff and hard consistency, $n \to 0$.

The fact that the exponent n may vary ($0 \leqslant n \leqslant 2$, defining the relation between primary and secondary consolidation) is also evident from an analytical solution of the problem (Tsytovich and Zaretsky et al., 1967).

8.4 REFERENCES

Bishop, A.W., 1959. The principle of effective stress. Teknisk Ukeblad, 106(39).
Croce, A., 1948. Secondary time effect in the compression of unconsolidated sediments of volcanic origin. Proc. 2nd ICSMFE, Rotterdam, v. 1.
Gorbunov-Posadov, M.I. and Davydov, S.S., 1973. Interaction of soil bases and structures (Prediction of settlements, design of massive foundations based on the limiting state, design of flexible foundation beams and slabs). Proc. 8th ICSMFE, v. 3.
Gorodetsky, S.E., 1969. Effect of shearing stresses on the volumetric deformation of soils. In collective volume: Foundations and Underground Structures. Stroiizdat, Moscow, no. 58 (in Russian).
Hansen Brinch, I. and Ivar, S., 1969. Tests and formulas concerning secondary consolidation. Proc. 7th ICSMFE, Mexico City, v. 1.
Koizumi, Y. and Ito, K., 1964. Compressibility of a certain volcanic clay. Rheology and Soil Mechanics, IUTAM Symp., Grenoble.
Mandel, J., 1957. Consolidation des couches d'argiles. Proc. 4th ICSMFE, London, v. 1.
Maslov, N.N. and Le Ba Liong, 1972. To the problem of increasing the strength and bearing capacity of clay soils with time. SMFE, no. 1 (in Russian).
Poorooshasb, H.B., 1969. Advances in consolidation theories for clays. Proc. 7th ICSMFE, Spec. Session, Mexico City, v. 3.
Tsytovich, N.A., Zaretsky, Yu.K., et al., 1967. Forecasting Settlement Rates of Soil Below Foundation (Consolidation and Creep of Multiphase Soils). Stroiizdat, Moscow (in Russian).

Chapter 9

LONG-TERM STRENGTH OF SOILS

9.1 CREEP AND LONG-TERM STRENGTH

Durability and long-term strength

The phenomenon of creep is associated with a body property known as durability. This is understood to be the resistance of a body to failure in response to a long-term load application or, in other words, the long-term strength.

It has already been pointed out more than once that the process of non-attenuating creep eventually becomes a progressive flow proceeding at an increasing rate and culminating in brittle or viscous failure. Thus, a long-term failure of soil results from a stress of a magnitude which may be lower than the strength displayed under the conditions of short-term loading; the lower the stress applied, the longer the period before failure.

If a test specimen prepared from a material displaying creep is loaded at a rapid rate to the point of failure, the load applied will indicate the so-called hypothetically instantaneous strength. This concept approaches the concept of 'ultimate strength'.

If an identical specimen is loaded to a point somewhat lower than the hypothetically instantaneous load, it will also fail, not immediately however, but after some time. The response to other, smaller loadings will be the same, until the deformation resulting from a load will begin to attenuate. A decrease in the stress at failure with an increase of the period before failure is an evidence of strength deterioration.

Long-term strength curve. A curve of strength deterioration may be obtained by replotting creep curves. Let us take a family of non-attenuating creep curves plotted from testing specimens under the loadings $\tau_1 > \tau_2 > \tau_3 \ldots$ causing failure during the time intervals $t_1 < t_2 < t_3$, respectively (Fig. 9-1).

Projecting the values of the times before failure $t_1, t_2, t_3, \ldots$ on the x-axis and laying off as ordinates the corresponding values of the stresses at failure $\tau_1, \tau_2, \tau_3, \ldots$, we obtain a curve representing the relation between the stress at failure and the time before failure. It is called the curve of long-term strength.

Accordingly, distinction is to be made between the following values of strength.

(a) Hypothetically instantaneous strength τ_0 (or R_0), i.e. the maximum strength defining the resistance of material to rapid failure; it is determined by the initial (at $t = 0$) ordinate of the curve of long-term strength.

(b) Long-term strength $\tau(t)$ or $R(t)$ defined by the stress causing the failure of material within a given interval of time; this value is represented by a current position on the curve of long-term strength.

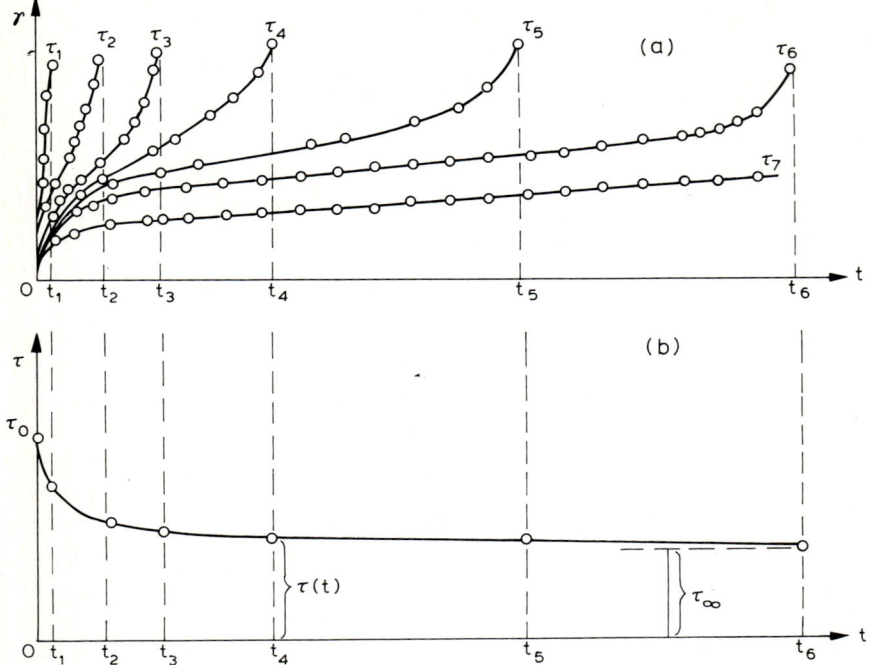

Fig. 9-1. (a) Creep curves and (b) long-term strength curves of soil.

(c) Ultimate long-term strength τ_∞ (or R_∞) corresponding to a stress at and below which the deformation is of the attenuating nature and no failure takes place within any practically observable period of load application; a stress in excess of τ_∞ brings about a non-attenuating creep resulting in failure with the progress of time. This ultimate value is represented by an asymptote of the curve of long-term strength.

In some materials, a non-attenuating creep leading to failure is induced by any stress other than zero. Hence, for such materials $\tau_\infty = 0$ and the curve of long-term strength asymptotically approaches the x-axis.

On criteria of soil failure

Since a progressing non-attenuating creep results in failure, the very fact that an attenuating creep becomes a steady flow (point B on the curve in Fig. 9-2(a)) and, moreover, is eventually transformed into a progressive flow (point C), heralds a potential failure. Accordingly, the instants t_y and t_{pr} of the transition from stage I of creep into stage II and from that stage into stage III, respectively, are critical points. The deformations γ_y and γ_{pr} corresponding to these points and the deformation γ_f at the instant of failure t_f may be regarded as the criteria of long-term failure. Later we shall consider a general criterion based on the analysis of micro-structural changes in soil but at the moment we will proceed on the above assumptions.

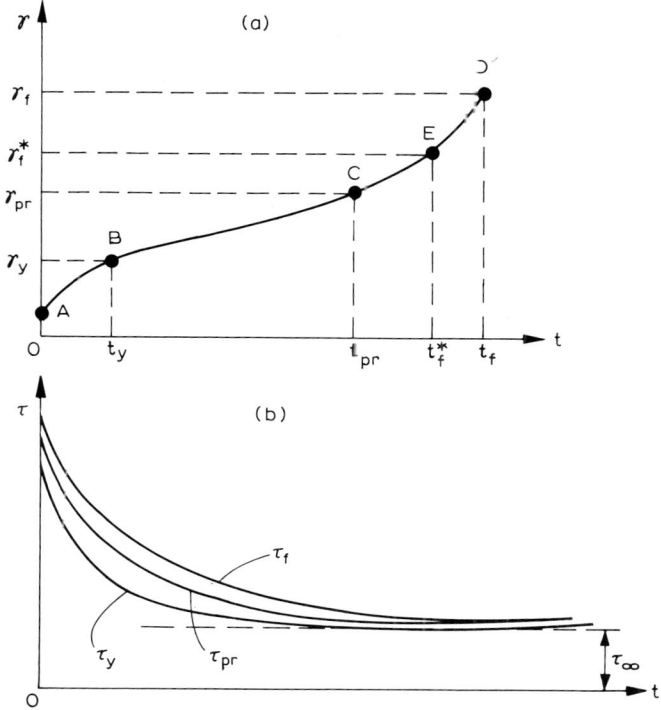

Fig. 9-2. (a) Critical points on the creep curve of a soil, and (b) curves of long-term strength corresponding to the critical points.

Thus, we note three critical states during the process of non-attenuating creep. The first state, defined by a deformation $\gamma = \gamma_y$, corresponds to the onset of steady flow and indicates the possibility of a subsequent failure. The second critical state, characterized by a progressing deformation $\gamma = \gamma_{pr}$, corresponds to the transition into the stage of progressive flow and forecasts an imminent failure. The third stage, described by a deformation $\gamma = \gamma_f$, corresponds to the instant of failure.

The three critical states may be produced by any stress τ exceeding the ultimate long-term strength τ_∞. However, the times t_y, t_{pr} and t_f when the deformations γ_y, γ_{pr} and γ_f take shape vary with the stress, increasing as the latter decreases. The relation between the stress τ and the times t_y, t_{pr} and t_f may be represented by curves as is shown in Fig. 9-2(b). All three curves have a common asymptote τ_∞.

Brittle and viscous failure

Soil failure may be either brittle or viscous depending on the type of soil, the condition and the type of loading. Brittle failure, manifesting itself in the form of cleavage, results from a deformation developing to a comparatively small extent.

Viscous failure is preceded by a large deformation entailing a necking-down of the specimen during a tensile test or its distortion to a barrel-shape without the loss of continuity in the course of a compression test. Brittle failure is inherent in rocks, compact clays, and frozen soil at low temperatures. Viscous failure is observed in soft soils including frozen soil at a temperature close to zero.

Unlike brittle failure, the onset of which is clear-cut, viscous failure lacks explicitness. Therefore, the onset of this failure is regarded as the instant corresponding to an unconfined increase in the deformation rate $\dot{\gamma} \to \infty$. Should difficulty be experienced in determining this instant, a conventional criterion of failure is a deformation $\gamma_f^* = k\gamma_{pr}$, where k may be assumed to be equal to 1.5; the time when the deformation reaches this level is denoted by t_f^*. Also the second critical state, i.e. the beginning of the stage of progressive flow $\gamma = \gamma_{pr}$, may be adopted as the criterion of failure which will have a certain margin in this case.

Irrespective of the failure criterion adopted, the soil below a foundation should sustain a stress not exceeding the ultimate long-term strength τ_∞, otherwise deformations of non-attenuating creep may be set up. However, in some cases the deformation of flow is tolerable, provided the aggregate deformation induced during a given period (e.g. service life of the structure) remains within the allowable limits, $\gamma \leqslant \gamma_{allow}$.

A stress τ_{crp} producing a given deformation $\gamma = \gamma_{allow}$ during a specified period t_{crp} is called the limit of creep (not to be confused with the term "threshold of creep"). The τ_{crp}–t_{crp} relation is represented by a curve similar to the curve of long-term strength.

The notion of long-term strength is obviously useless in speaking about the materials in which creep develops in the form of a continuous flow without transition into the stage of failure. Coming under this category are, in particular, fluidized soils, suspensions, etc. The notion of "limit of creep" is applicable to these materials.

Deterioration of soil strength during creep. The deterioration of strength due to creep observed in a wide range of materials — metals, wood, concrete, plastics, etc. — has been the subject of numerous studies.

At an early stage of the investigations, this fact was regarded as doubtful. The onset of progressive flow and specimen failure under a load less than the ultimate strength were ascribed to the experimental conditions such as the reduction in the cross-sectional area of the specimen due to necking during a tensile test. Later it was shown that long-term failure is a natural process inherent in all materials capable of displaying creep. Being of kinetic nature, this process is associated with intracrystalline flow (during a viscous failure), intercrystalline flow (during a brittle failure), and with the triggering of micro-cracks.

Probing into the long-term strength of soils, some researchers have found that the progressive stage is an outcome of the experimental conditions, and should their effect be eliminated the deformation of creep would develop infinitely without causing failure. However, numerous studies have shown that the long-term failure of soils is not influenced by the conditions of the experiment and is observed in soils of the most diverse kind during tests of the most diverse nature, including compression tests when the cross-sectional area of the specimen increases rather than contracts (see Section 6.6: Vyalov and Pekarskaya, 1968).

Also in dispute was the question of the quantity which should be adopted as the reference value in assessing the loss of strength. Some authors compared the long-term strength not with the hypothetically instantaneous stress, as is commonly practised, but with the so-called standard strength which is the strength of soil as determined by a standard technique of low-rate loading.

This technique, as stipulated by a USSR State Standard (GOST 12248-66), specifies a stepwise loading of the specimen during a test period between 1 and 8 h with a pause at each stage until the deformation becomes stabilized. Since, as will be shown later, this is exactly the period of maximum strength deterioration, some authors have failed to detect an appreciable difference between the long-term strength and standard strength. Consequently, they have come to a wrong conclusion that the strength of soil does not deteriorate due to creep. The relation between the standard and long-term strength will be examined in Section 9.8. For now, we shall only state that in ascertaining a loss of strength with time the reference to be used is the hypothetically instantaneous strength determining the maximum resistance offered by the soil before failure.

9.2 EXPERIMENTAL DATA

Long-term strength of frozen soils

Investigations of the long-term strength of soils were pioneered by M.N. Goldstein in the 1940s, who studied the forces causing soils to freeze together. In the 1950s, the author of this book succeeded in establishing the main factors governing long-term strength of frozen soils (see Section 1.5: Vyalov, 1959) during the integrated studies of their long-term resistance to various loadings conducted in Igarka. Subsequently, the problem of long-term strength of frozen soils has been tackled at the former Institute of Permafrostology, USSR Academy of Sciences; Foundation Research Institute (see Section 1.5: Vyalov et al., 1962, 1966); Department of Permafrostology, Moscow State University; Irkutsk Institute of Permafrostology, USSR Academy of Sciences; Research and Design Institute of Industrial Construction, Krasnoyarsk; No. 1 All-Union Research Institute, USSR; U.S. Cold Region Research and Engineering Laboratory; and Division of Building Research, National Research Council, Canada.

The way frozen soil mechanics is concerned with long-term strength is readily explicable. Ice, a constituent of frozen soils, possesses a sufficiently high strength in resisting rapid failure but tends to flow whatever the load and is devoid of ultimate long-term strength. Correspondingly, the strength of frozen soils deteriorates with time by a considerable amount, their long-term strength appearing to be between 1/2 and 1/15 of the instantaneous strength.

Curves of the loss of strength by frozen soils

Consider the experimental curves of long-term strength of frozen soils obtained by the author and his colleagues (see Section 1.5; Vyalov et al., 1962, 1966) which are depicted in Figs. 9-3 and 9-4.

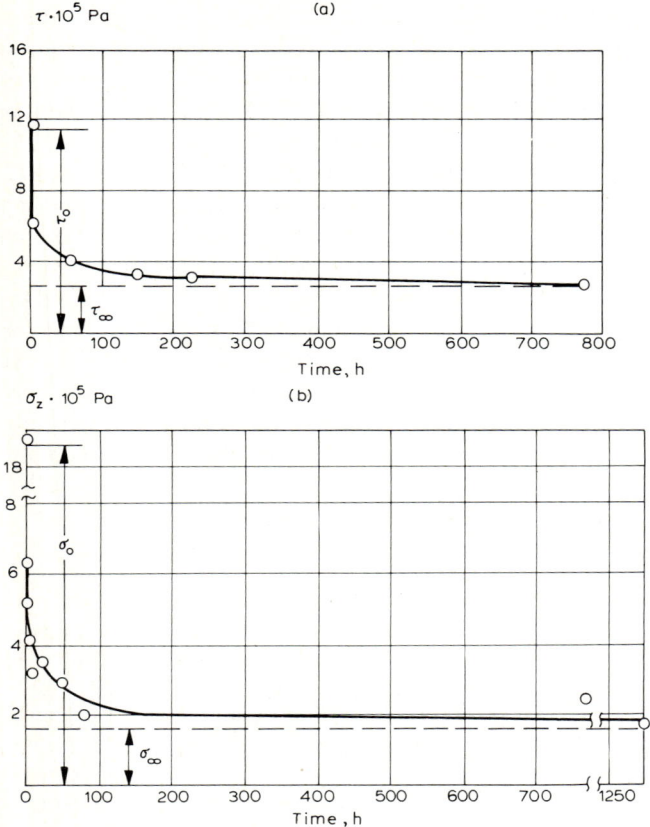

Fig. 9-3. Curves of long-term strength, frozen soil. (a) Shear test of sandy loam, $\theta = -3.6°C$; (b) break test of sandy loam, $\theta = -4.2°C$ (Vyalov's experiments).

Fig. 9-3(a) represents the data acquired from shear-at-side surface tests of a soil-below-foundation model (the freezing-together strength), Fig. 9-3(b) illustrates the results of break tests of dumb-bell pieces, and Fig. 9-4 presents uniaxial compression test data. The test load used was constant, and the test data were assessed in accordance with the technique represented in Fig. 9-1.

The first thing to be noted is the fact that the long-term strength is significantly smaller than the instantaneous strength — a salient feature of frozen soils as stated above. The loss of strength varies with the type of soil, ice-and-moisture content, and temperature; but the key factor is the arrangement of stress.

Minimum loss of strength is associated with compression, although in frozen soil the amount lost is significant (the ratio τ_∞/τ_0 varies between 0.6 and 0.3). More is lost of the freezing-together strength ($\tau_\infty/\tau_0 = 0.4$–0.15). The maximum loss of strength is evident when frozen soils are subjected to tension, the ratio τ_∞/τ_0 being between 0.08 and 0.06 which indicates that the strength decreases by a factor of 12 to 15.

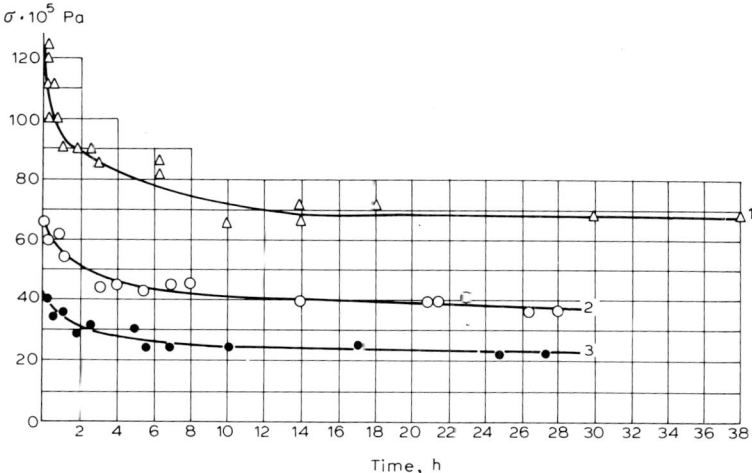

Fig. 9-4. Curve of long-term compression strength, frozen Callovian sandy loam at the temperatures: 1. $-20°C$; 2. $-10°C$; 3. $-5°C$ (experiments by E.P. Shusherina and S.S. Vyalov).

An extended period of time-dependent strength deterioration is another feature of frozen soils: a number of specimens failed in 250 to 350 days after the load application. But most of the loss was observed at an early stage of the process; after that the strength deteriorated at a sharply reduced rate. So, taking the ratio τ_∞/τ_0 as 100%, the loss of shearing stress τ/τ_0 amounted to 60% during the first 30 min, to 70% in the course of the first hour and to 80% at the end of the first 8-h period.

Long-term strength of clay soils

Among the early experimental studies of the long-term strength of unfrozen clay soils were the uniaxial compression tests of clay and clay shale conducted by A. Casagrande and S. Wilson in 1950 (reference in Chapter 1) under various constant loads.

The tests showed that creep was the cause of strength deterioration. Thus, specimens failed under a load between 80 and 40% (at times even 25%) of the quick-test (1 min) strength within a period of up to 30 days. Particularly significant was the loss of strength by the clay shales from a landslide in the Panama Canal zone. Analysing the situation, Casagrande came to the conclusion that the cause of a sudden bank collapse, only a few years after the completion of the project, was the loss of strength by the soil. Unfortunately, this factor was not allowed for in the computation of the stability of the slopes.

Other examples of similar soil failures were examined in Chapter 1. One may add the report by E. Geuze and Tan Tjong-Kie at the 3rd ICSMFE on the collapse of bridge piers after two years of service. Miscalculations of the pressure sustained by the piers, which was computed on the basis of quick tests without allowance for loss of the strength, was the cause (see Chapter 1: Geuze, 1953). In Geuze's words, from that time and onward he tackled all problems in the light of rheology. Similar conclusions were arrived at by Haefeli when presenting a paper at the same conference (see Chapter 1: Haefeli et al., 1953). The problem of long-term soil strength was discussed by Hvorslev (1960) at a conference devoted to the shearing resistance of cohesive soils (Colorado, 1960). Dwelling on the rheological components of the resistance, he identified it in his paper with the components of effective cohesion decreasing with time.

Detailed investigations of the long-term strength of soils carried out by Goldstein and his colleagues at the Dnepropetrovsk Institute of Railway Engineers (Goldstein and Babitskaya, 1959, 1964; Goldstein, 1964) have proved the applicability to soils of the classical mechanism of the loss of strength due to creep (see Fig. 9-1).

Creep tests of various clays under the conditions of uniaxial and triaxial compression showed that long-term failure of specimens was caused by loads amounting to between 90 and 45% of the hypothetically instantaneous strength σ_0, i.e. of the resistance to failure as determined from a 50- to 60-sec test by a stepped loading applied in increments each of a 5-sec duration. Illite clay suffered failure from a load equal to 90, 80, 70 and 60% of the instantaneous strength in 23 min, 151 min, 3.5 days and 77 days, respectively. Smaller loads caused attenuating deformation without failure of the soil. The ultimate long-term compression strength of polymineral lean clay, montmorillonite clay, illite clay and dense Kinel clay was $\sigma_\infty = 0.45\,\sigma_0$ to $0.8\,\sigma_0$.

Identical results were obtained by other investigators from shearing tests of various clay soils (Fisenko, 1964) and loesses (Mogilevskaya, 1960). Specimens failed in 20 to 90 days under a load amounting to 70 and 80% of the quick-test strength, and the ratio τ_∞/τ_0 was between 0.5 and 0.85.

Dense clays. The strength of dense soils also deteriorates although to a lesser degree than is experienced in plastic clays. Experimenting with stiff and hard clays of the kaolinite, montmorillonite and mica hydrate groups with undisturbed structure, Stepanenko (1968) found that the ultimate long-term strength in shear and uniaxial compression averaged 90% of the quick-test strength for brown clay ($W = 20$–30%) and 70% for variegated clay ($W = 16$–28%). Some of the tests yielded a strength as low as 50% of the quick-test value (this was determined from tests by a stepped loading applied in increments each of 1-min duration).

Experimental studies of A.M. Skibitsky in 1957 (referenced in Section 5.7) showed that the ultimate long-term shear strength of stiff Lower Cretaceous clays was 70–75% of the quick-test strength obtained due to a 1-h load. For stiff Kinel clay and hard beidelite–montmorillonite Tertiary clay ($W = 34\%$ at $W_P = 34.5\%$) the relevant figures were 60–70% and 90%, respectively.

Curves of the loss of strength by clay soils. Depicted in Fig. 9-5 is a curve of the long-term strength of a plastic clay plotted from the creep tests under uniaxial compression illustrated in Fig. 5-13(b). For an overall ratio $\tau_\infty/\tau_0 = 0.6$ in these soils, the strength rapidly deteriorates within a comparatively short interval. By analogy with frozen soils, two stages can be seen: an early stage of a steep decrease in strength and a later stage characterized by a gradual loss of strength.

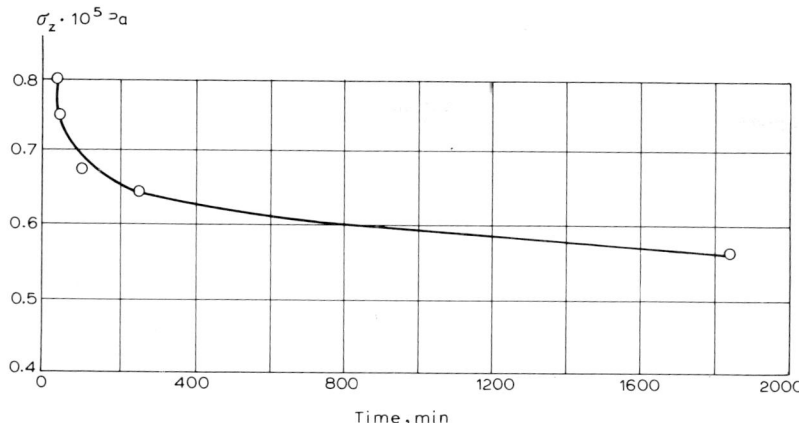

Fig. 9-5. Curve of long-term strength, plastic clay, $W = 65\%$. Uniaxial compression tests (by Murayama and Shibata).

These stages are clearly discernible on the curves of long-term strength plotted on the log–log coordinates, the inflection of the lg τ–lg t straight line indicating the boundary.

A curve of long-term strength may be conveniently presented in the form of a 'relative strength τ/τ_0 versus logarithm of time lg t' graph. In addition to the logarithmic time scale which is an asset in itself, the graph enables the estimation of the rate of strength loss as a percent of strength reduction per unit of the logarithm of time, i.e. $(\Delta\tau/\tau_0):(\Delta \log t)$. Summarized data on the uniaxial compression tests by Casagrande and Wilson referred to above, and plotted on these coordinates, are depicted in Fig. 9-6. For a general loss of strength σ_∞/σ_0 between 0.8 and 0.55, where σ_0 was the strength determined from a 1-min test and σ_∞ was the strength obtained from a 30-days test, the value of $(\Delta\sigma/\sigma_0):(\Delta \lg t)$ was 4.5–9%.

Long-term strength of fluid clays and clay mortars. The above examples all refer to tests with plastic and dense clays. Let us now consider the long-term strength of clays having a fluid consistency. Although, in principle, quick soils of a vague structure may exhibit unconfined deformations of flow at a constant rate, experimental studies indicate that even in clay suspensions, provided these have some sort of structure, the flow assumes a progressive pattern leading in time to viscous failure.

The loss of strength by such soils was examined by N.K. Pekarskaya (see Section 6.6: Vyalov and Pekarskaya, 1968) on the basis of experimental data. Good examples of this were the results obtained by Kabakhadze et al. (1957) with a 4% suspension of bentonite clay.

Shear tests of the suspension revealed that loads between 5.9 and 16 Pa (5.6–16 dyne/cm^2) produced a flow at a constant rate of about 10^{-8}–10^{-9} I/s. An increase in the load up to 26.8 Pa, transformed the steady flow into progressive flow causing structural failure shown by an infinite increase in the deformation rate, $\dot\gamma \to \infty$.

The corresponding curve of the long-term strength of the suspension is given in

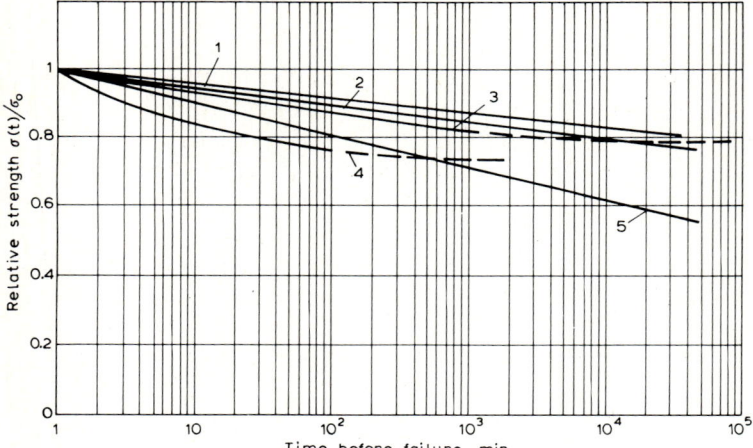

Fig. 9-6. Curve of long-term strength obtained from uniaxial compression tests of various soils: *1* = Mexico City clay; *2* = Cambridge clay; *3* = Berno clay shale; *4* = Mississippi gumbo; *5* = Oahu bentonite.

Fig. 9-7. The ratio τ_∞/τ_0 was 0.6, i.e. within the confines characteristic of plastic clays. However, mortars deteriorate in strength much quicker, in 2 or 3 h.

In the light of the data presented, one must disagree with the existing point of view that a loss of strength is inherent in rigidly bonded soils only. As can be seen, a deterioration of strength is observed in artificially remoulded clays with predominantly water-colloidal bonding, not the cementing one, and even in suspensions devoid of rigid bonds at all.

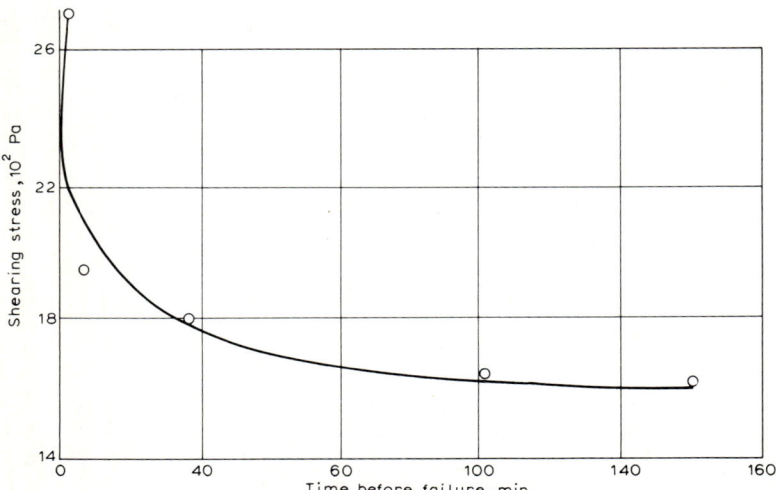

Fig. 9-7. Curve of long-term strength, bentonite clay suspension.

Drained and undrained tests

Since the loss of strength by a soil is ascribed to creep, tests aimed at determining long-term strength are commonly conducted under conditions facilitating the development of creep but preventing compaction due to percolation and ensuring a constant density and moisture content of the soil throughout the test. Pure shear tests or triaxial compression tests of the undrained type (employing the "confined" scheme) meet this requirement.

A comparative study of the results of drained and undrained tests of various clays for long-term strength based on the data presented by A. Skempton (see Section 1.5: Skempton and Hutchinson, 1969) at the 7th ICSMFE, Mexico City, 1969, is depicted in Fig. 9-8. The data refer to the triaxial compression tests of soft Fornebu clay with a normal density (by Bjerrum et al., 1958), brown London clay (by Skempton and La-Rochelle, see Sect. 1.5), Cambridge clay (by Casagrande and Wilson, see Sect. 1.5), and remoulded Weald clay (by Bishop and Henkel, 1957). All the above tests were conducted employing the "confined" scheme except the last-named test which used the "unconfined" scheme.

As can be seen from curve *4*, the loss of strength was a minimum, amounting to 4.5% per time interval, during the drained tests. The explanation is that the soil, being compacted during a drained test, strain-hardens at the same time to a point reducing the effect of strength loss. The undrained tests reduced the strength by 5% (curve *3*) to 14% (curve *1*) per unit of the logarithm of t. During the last-named test, the total loss of strength was $\sigma/\sigma_0 = 0.2$ in 1 month — a value not often met with in testing clays. The significant loss of strength resulted from the building-up of pore-water pressure and a corresponding reduction of the effective stresses.

The effect of pore-water pressure on the process of undrained creep is a problem that is creating interest, for the loss of strength depends to a considerable extent on whether this pressure is constant or changing with time. The relevant data so far available are contradictory.

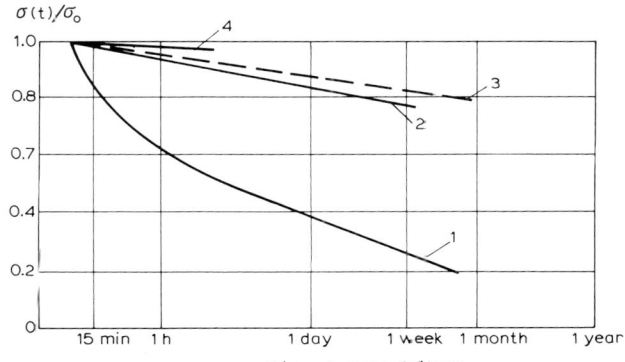

Fig. 9-8. Curves of long-term strength obtained from triaxial (undrained) compression tests of the following clays: *1* = Forneby clay, Oslo; *2* = London clay; *3* = Cambridge (Mass., U.S.A.) clay, drained test; *4* = Weald clay.

Experimenting with Lida clays, Coats and McRostie (1963) found that soils failed due to the deformation of creep without an appreciable change in pore-water pressure. On the other hand, Shibata and Karube (1969) obtained a 25–30% increase in pore pressure during creep ending in failure of the clay specimens subjected to triaxial compression under the "confined" conditions. An increase in pore pressure during "confined" tests was proved experimentally by Bishop and Henkel (1957).

The creep tests of soil by Goldstein (Goldstein and Babitskaya, 1959, 1964; Goldstein, 1964) of between 1 and 3 months duration resulted in a decrease in strength by a factor of 1.5 to 3, as in the examples referred to above, when the confined scheme was used and in no loss of strength at all when the unconfined scheme was used.

These data corroborate the point of view already stated that triaxial compression tests for long-term strength should be conducted under the conditions of the "confined" scheme over an adequately long period of time; quick tests may give a strength which is considerably overestimated. If the long-term strength of a drained soil is to be assessed, the relevant test must be undertaken when the process of soil compaction comes to its end.

Long-term strength of soft and hard rocks

Investigations have shown that the deterioration of strength due to creep is also inherent in soft and hard rocks. This can be seen, for example, from Fig. 5-19 presenting data on the failure of marl, sandstone and rock salt under a load below the quick-test strength. Uniaxial compression tests of sylvinite (a potassium salt) specimens for creep carried out by Stavrogin (1968) at the Mine Surveying Research Institute in 1966 and 1967 resulted in failure under a load amounting to 85%, 70% and 60% in 1.4 days, 4.5 days and 34 days, respectively, whereas a lower load (50% and 30%) produced only an attenuating deformation. Accordingly, for sylvinite, $\sigma_\infty = 0.55\sigma_0$. A similar behaviour, i.e. failure due to a load constituting 78% and 55% of the quick-test load in 49 days and 174 days, respectively, was also observed in rock salt by Avershin et al. (1970).

A deteriorating rock strength has been noted in mine workings under natural conditions. As pointed out at the All-Union Symposium on Problems of the Rheology of Rocks (Kiev, 1969), pillars lost their stability due to a loss of 30–60% of the initial strength in 12–24 months after finishing with mining in the adjacent rooms.

Fig. 9-9 depicts curves of the long-term strength of limestone (in compression), sandstone, shale, and coal (in bending) set forth in the article by Artemov and Vodop'yanov (1970). These curves vividly corroborate the fact that the strength of rocks deteriorates due to creep.

Some conclusions

Loss of strength is invariably observed during the process of non-attenuating creep (its physical meaning will be examined later) in all soils ranging from those of fluid

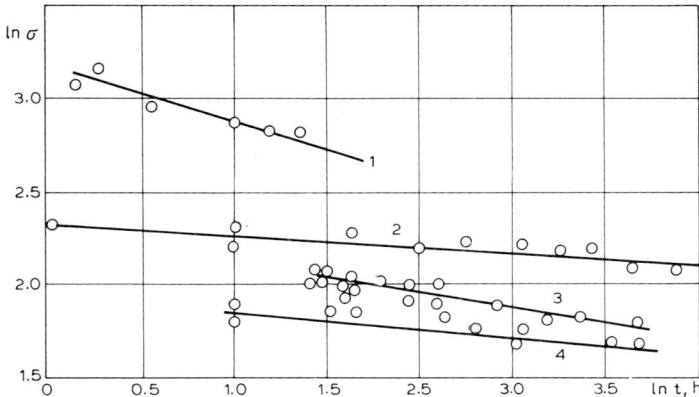

Fig. 9-9. Curves of long-term rock strength: 1 = uniaxial compression of sandstone; 2 = bending of sandstone; 3 = bending of shale; 4 = bending of coal.

consistency to rocks. However, the rate of loss varies appreciably with the type of soil and soil density. Although the values τ_∞/τ_0 given above are difficult to compare, because τ_0 was determined by different authors at different rates of load application, we shall try to establish the limits of the changes in the ratio.

The strength of ice decreases slowly but unconfined up to the limit $\tau_\infty/\tau_0 = 0$.

The strength of frozen soils also deteriorates slowly, although not as slow as in ice, the ratio τ_∞/τ_0 being 0.15–0.5.

Plastic clay soils lose their strength relatively quickly, the ratio τ_∞/τ_0 varying between 0.2 and 0.6. For dense clays, the ratio is 0.5–0.8 or even 0.9.

For soft and hard rocks, the ratio τ_∞/τ_0 varies over the range between 0.6 and 0.8 depending on the viscous properties of cementing bonding and crystals proper.

The rate of strength deterioration is also influenced by the arrangement of stress. Compression affects strength less than shear, let alone tension.

In the case of combined stress, the loss of strength varies with the ratio of the deviatory and spherical components of the stress tensor: the higher the mean normal stress σ_m, the less the reduction in strength. However, under the conditions of all-around pressure, the problem of long-term strength is senseless.

Field data

The loss of strength by soils due to creep is observed also under natural conditions. The slope failures in stiff-fissured London clays of morainic origin mentioned in Chapter 1 resulted from loss of strength by the soils. It was stressed that they occurred after 13–54 years of service.

In a paper presented at the 4th ICSMFE (London, 1957), Henkel showed that for $c' = 0.125 \cdot 10^5$ Pa and $\phi' = 20°$, as determined from triaxial compression tests, the coefficient of stability of the retaining walls was 1.18–1.35 (see Section 1.5: Henkel, 1957). Assuming that at the instant of failure the coefficient was 1 and the reduction

TABLE 9-I

Loss of soil strength with time for different sites

Site	Time before failure (years)	Cohesion	
		$\times 10^5$ Pa	%
Laboratory (test data)	several hours	0.125	100
Wembley Hill	13	0.083	66.5
Uxbridge	29	0.067	46
Northolt	35	0.043	35
Wood Green	54	0.057	55

in shear strength was attributed to cohesion alone, while friction was constant, Henkel reversed the reasoning and determined the shear strength at the moment of the loss of stability and assessed the reduction in cohesion of the soil at the instant of each failure; these data, illustrating a typical loss of strength by the soil with time, are given in Table 9-I.

Other slope failures also in the same London clays (at Northolt, Kensal Green and Sudbury Hill) were investigated by A. Skempton (see Section 1.5: Skempton and Hutchinson, 1969); a curve of the long-term displacement of a Kensal Green retaining wall is depicted in Fig. 5-17 as an illustration of the way the stage of progressive flow develops under natural conditions.

Reversing the reasoning, Skempton determined the normal and tangential components of the effective stress coming on the area of slip at the moment of failure. Utilizing these data, L. Šuklje computed the shear strength at failure for a normal stress $\sigma = 0.35 \cdot 10^5$ Pa which was the same in all cases (see Section 1.5: Šuklje, 1969).

The results of these computations were identical with Henkel's findings, except that the angle of internal friction was not assumed to be constant. The computed shearing strength at failure was compared with the data acquired from quick laboratory tests.

TABLE 9-II

Loss of shear strength with time for different sites

Site	Time before failure (years)	Shearing strength	
		$\times 10^5$ Pa	%
Laboratory (test data)	hours	0.283	100
Northolt	19	0.179	63
Kensal Green	29	0.167	59
Sudbury Hill	49	0.134	47
Natural slopes	∞	0.100	35

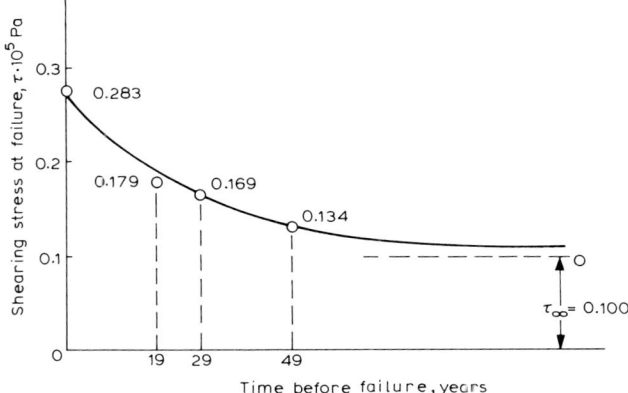

Fig. 9-10. Curve of loss of shearing strength in the natural conditions (after L. Šuklje).

Field surveys showed that the inclination of natural slopes averaged 16°. Adopting this angle as the ultimate value ensuring long-term stability of slopes and reversing the reasoning, Šuklje found the ultimate shear strength. The results of his computations are tabulated in Table 9-II.

A curve of long-term strength plotted on the basis of Šuklje's data is represented in Fig. 9-10. As can be seen, its configuration is similar to that of the curves of the loss of strength obtained from the laboratory tests and discussed above. It will be shown later that the former curve is described by the same law as the latter curves. It will also be noted that the ratio $\tau_\infty/\tau_0 = 0.35$ obtained from the field surveys coincides with the laboratory data.

The example referred to above leaves no doubt that the loss of strength due to the creep of a soil mass should be allowed for whenever the stability of the mass is to be ascertained. To assess the actual factor of safety, the analysis should be based on the long-term strength rather than on the quick-test data.

9.3 LOSS OF SHEAR STRENGTH BY SOILS

Effect of normal stress on long-term shear strength

The long-term strength under the conditions of a combined stress will be the subject of subsequent discussions. Here we are going to consider the most simple problem — the relation between the long-term strength and the effective normal stress σ'_n. In the general case, we are interested in the relation between the long-term strength and the total stress σ_n, for we are considering creep processes when percolation does not exist (or has come to an end) and the pressure is sustained by soil skeleton.

The relationship in question is illustrated by the graphs in Fig. 9-11: an increase in σ_n retards the development of creep with the result that soil failure occurs at a later time, provided the shearing stress τ remains the same (Fig. 9-11(a)). Thus, if three

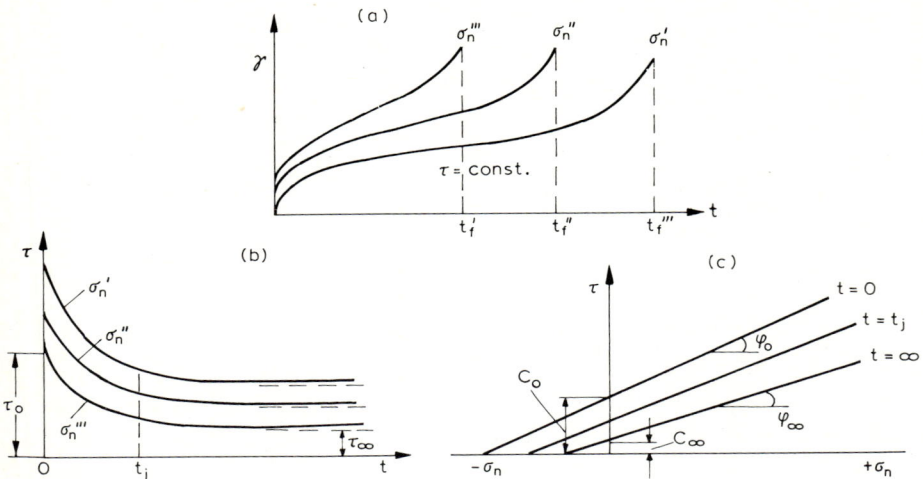

Fig. 9-11. Long-term shearing strength of clays: (a) creep curves for τ = const. and various values of normal stress $\sigma'_n > \sigma''_n > \sigma'''_n$; (b) curves showing the deterioration of shearing strength for various values of σ_n; (c) graph of the shears at various times.

specimen batches each sustaining a different normal stress $\sigma'_n > \sigma''_n > \sigma'''_n$ are tested for creep, we shall obtain three families of creep curves and three corresponding curves of long-term strength (Fig. 9-11(b)). The long-term strength curve for the highest stress σ'_n will be located above the rest of the curves and, accordingly, the values of τ_0 and τ_∞ will be at a maximum in this case.

If the curves of long-term strength are replotted on the τ–σ_n coordinates and the relationship thus obtained is approximated by a straight line, the creep graph will show a family of τ–σ_n straight lines for various t values (Fig. 9-11(c)). These lines will define the shear strength of soil at various times ranging from $t = 0$ (instantaneous strength) to $t = \infty$ (ultimate long-term strength).

Condition of long-term strength of soil. To allow for a loss of strength by soil with time, the Mohr–Coulomb failure condition (eq. 4-8) takes the form (Avershin et al., 1970):

$$\tau = c(t) + \sigma_n \tan \phi(t) \tag{9-1}$$

where $c(t)$ and $\phi(t)$ are the cohesion and angle of internal friction, respectively, varying with time (Fig. 9-11).

The values of $c(t)$ and $\phi(t)$ vary over the range between the hypothetically instantaneous values c_0, ϕ_0 and the ultimate long-term values c_∞, ϕ_∞ (Fig. 9-12). Consequently, the shear strength will vary between the maximum hypothetically instantaneous value and the minimum ultimate long-term value:

$$\begin{aligned} \tau_0 &= c_0 + \sigma_n \tan \phi_0 \\ \tau_\infty &= c_\infty + \sigma_n \tan \phi_\infty \end{aligned} \tag{9-2}$$

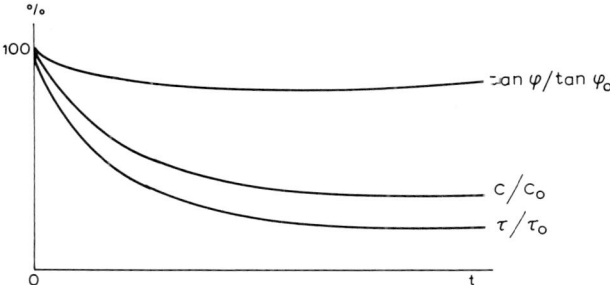

Fig. 9-12. Time-dependent deteriorations of shearing strength $\tau(t)$, cohesion $c(t)$ and internal friction $\phi(t)$ in clay (percent of hypothetically instantaneous values).

Experimental evidence indicates that soil cohesion significantly changes with time whereas the angle of friction alters little, being treated as a constant in some cases.

Experimental data. Fig. 9-13 presents the shear curves of a plastic frozen soil (clay loam, $\theta = -0.3°C$) obtained by the author way back in 1950 (see Section 1.5: Vyalov, 1959) from tests under the conditions of a hypothetically instantaneous resistance τ_0 and an ultimate long-term resistance τ_∞, using a shear–strain apparatus. It can be seen that the soil strength changes with time mainly due to the weakening of the forces of cohesion while the angle of friction suffers less alteration; thus, in the case under consideration, $c_\infty/c_0 = 0.145$ whereas $\tan \phi_\infty/\tan \phi_0 = 0.7$.

Consider the envelopes of the circles of stress at failure constructed by N.K. Pekarskaya (see Section 1.5: Vyalov et al., 1962) on the basis of the data acquired from integrated tests for compression, tension and pure shear of frozen sandy loam at $\theta = -10°C$ (Fig. 9-14). In this case too, a family of envelopes for various instants of time has been obtained, all being of the curvilinear shape. Consequently, in the

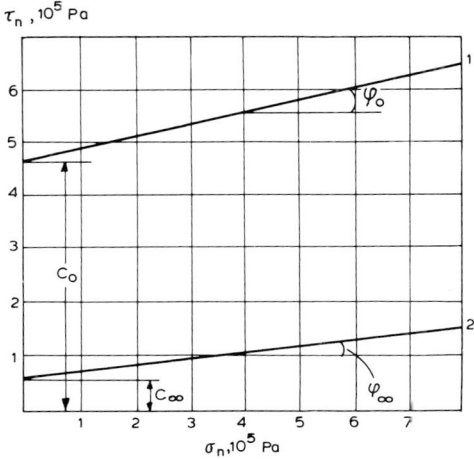

Fig. 9-13. Curves of (*1*) instantaneous, and (*2*) ultimate long-term shearing strength, frozen clay loam, $\theta = -0.3°C$ (Vyalov's experiments).

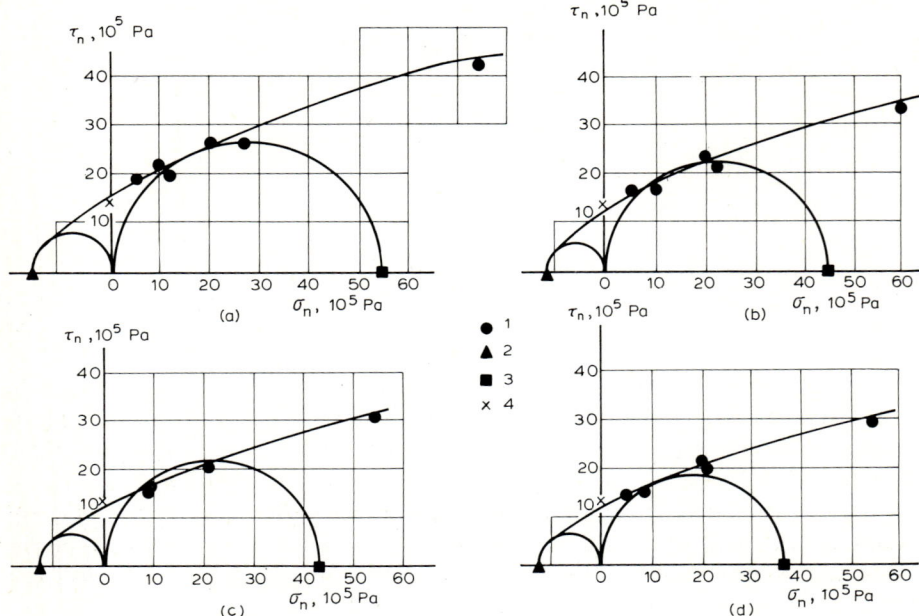

Fig. 9-14. Envelopes of the circles of stress at failure for frozen Callovian sandy loam, at various times t, $\theta = -10°C$: (a) 1 h; (b) 4 h; (c) 12 h; (d) 24 h. Key to test designations: 1 = shear tests; 2 = break tests; 3 = compression tests; 4 = spherical indenter tests. (N.K. Pekarskaya's experiments.)

general case, the equation of long-term strength is written as:

$$\tau = f(\sigma_n, t) \qquad (9\text{-}3)$$

9.4 "PEAK" AND RESIDUAL SOIL STRENGTH

Constant deformation rate tests

Widely practised as a means of assessing mechanical properties of materials in the laboratory is the method of testing at a constant deformation rate. Specifying a constant rate of displacement for the load-applying unit of the test apparatus, we measure the stress increasing directly with the specimen strain. The data acquired are represented in the form of either a stress–strain curve or a stress–time curve (the constant-rate strain increases directly with time).

The above test method was first applied to soils by Tiedemann (1937), M. Hvorslev and R. Haefeli in 1937–1938 (see Section 1.5: Hvorslev, 1937; Haefeli et al., 1953). It was examined by A. Skempton in 1964–1969 (see Section 1.5: Skempton, 1964; Skempton and Hutchinson, 1969) and also by Goldstein and Babitskaya (1964). Having gained ground in soil mechanics since then, the method is employed in investigating land-

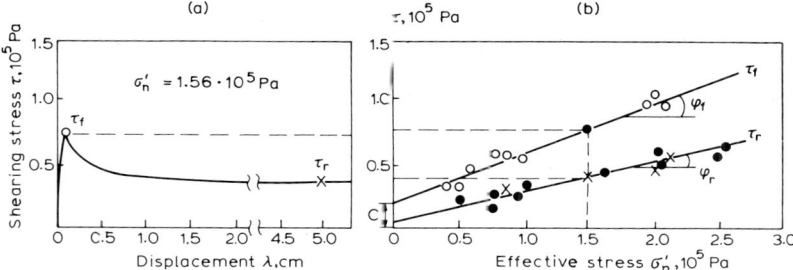

Fig. 9-15. Shear tests at constant deformation rate. (a) Changes in shearing strength τ with increase in displacement λ (for effective pressure σ'_n = const.). (b) Curves of peak (τ_f) and residual (τ_r) resistance to shear.

slides on the assumption that the resistance to shear at a constant rate of displacement simulates, to a certain extent, the conditions under which a moving landslide is performing its work.

Maximum soil displacement is the main requirement to be met during a constant deformation rate test. Convenient to that end are ring shear apparatus, although triaxial compression apparatus may also be used. Soils are tested by various effective stresses but the stress applied during a test is constant.

Peak and residual soil resistance. Fig. 9-15 depicts the results of a test carried out by A. Skempton (1964, referenced in Section 1.5), using an undisturbed sample of the clay from the Waltons Wood landslide. The graph is typical for the test under consideration. Fig. 9-15(a) represents the change in the shear stress measured, τ, depending on the displacement, λ, for the same effective normal stress, σ'_n. A salient feature of the graph is an extremum on the τ–λ curve. At an early stage, the stress increases with the displacement, reaching a maximum called the peak resistance, τ_f. After the maximum, the stress sharply decreases and, as the displacement increases, the stress goes on decreasing gradually and insignificantly until a minimum value, called by Haefeli the residual (recqual) resistance, τ_r, is attained. Further displacement causes no changes in the stress.

Thus, with a progress of deformation, the resistance to shear decreases or, in Skempton's words, the soil weakens due to deformation. Apparently, the physical essence of the process is the same as the deterioration of strength due to creep which was discussed above.

On carrying out a series of tests at the same rate of displacement but for various values of $\sigma'_n = \sigma_n$ = const. the results can be represented in the form of a creep graph for the peak resistance τ_f and residual resistance τ_r (Fig. 9-15(b)). These values will be given by:

$$\tau_f = c_f + \sigma_n \tan \phi_f \quad \text{and} \quad \tau_r = c_r + \sigma_n \tan \phi_r \tag{9-4}$$

where c_f, c_r and ϕ_f, ϕ_r are the cohesion and angle of friction for the "peak" and residual resistance, respectively. In the case under consideration, $c_f = 0.16 \cdot 10^5$ Pa and $c_r = 0$; and $\phi_f = 21°$ and $\phi_r = 13°$.

Relationship 9-4 and the graphs of peak and residual resistances to shear (Fig. 9-15) are identical with relationship 9-2 and the corresponding graphs of instantaneous and long-term strength (see Fig. 9-13). In both cases the soil strength has deteriorated mainly due to the weakening of the cohesive forces whereas friction has changed to a lesser degree.

Relation between the maximum and minimum strength of soil

Referring to Fig. 9-15, the ratio τ_r/τ_f is 0.475 in the case under consideration. The "peak" strength corresponds to the maximum resistance of the soil to shear at the moment preceding the breaking of bonds when the forces of resistance — the viscous component of cohesion, crystallization bonds, friction — actively come into play.

The residual strength, on the other hand, corresponds to a total or almost total breakage of stiff crystallization bonds. They inevitably break in response to significant displacements of particles, and the shearing between the plate-shaped particles takes place along the slip area formed due to the face-to-face contact.

Accordingly, the bulk resistance to shear is ascribed to the friction between plate-shaped particles and the viscous component of cohesion. Quite naturally, the role of the normal stress σ_n increases in this case. All researchers are unanimous in their conclusion that, with a slowing-down of the rate of displacement, the "peak" strength of soil significantly deteriorates. So, according to the data acquired at the Dnepropetrovsk Institute of Railway Engineers (Goldstein and Babitskaya, 1964), a decrease in the deformation rate by a factor of 10^4 resulted in a reduction of τ_f by 26% for clay loam and by 40% for clay. Accordingly, the "peak" of the τ–λ curve (Fig. 9-16) decreased or, according to Rebinder (1966), vanished completely, if the deformation was very slow.

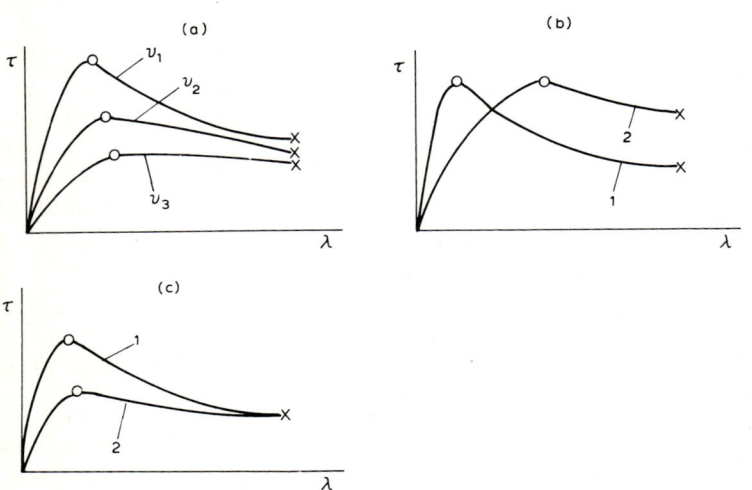

Fig. 9-16. Stress-displacement curves: (a) for different displacement rates $v_1 > v_2 > v_3$; (b) for brittle (*1*) and plastic (*2*) soils; (c) for overcompacted (*1*) and normally compacted (*2*) soils.

The residual soil strength depends upon the rate of displacement to a lesser degree in this case. Skempton's experiments with London clays and Edale shales showed that a slowing down of the deformation rate by a factor of 10^3 (from 2 cm/year to 20 cm/day) caused only a 4% reduction in the value of τ_f. In Goldstein and Babitskaya's experiments referred to above, the effect of deformation rate on the value of τ_f was not even noted.

Thus, if the stability of a soil mass is assessed in terms of "peak" strength, a coefficient taking account of the relation between the laboratory rate of soil displacement and the field rate should be introduced. A. Skempton (1964, referenced in Section 1.5) suggests that the decrease may be assumed to be equal to 3.5 per unit log of time. However, should the residual strength (as determined, for example, at a displacement rate amounting to 1 cm/day) be used as the criterion of stability, the possible error will not exceed 2–5% of the field data even if the rate of soil displacement (during the sliding of an artificial or natural slope) is adopted anywhere over as wide a range as 1 cm/year and 100 cm/day.

Effect of density and moisture content on soil strength

The configuration of the τ–λ curve and, accordingly, the relation between "peak" and residual strengths may appreciably vary with the density and moisture content of the soil. So, if a brittle clay shows a characteristic curve with an explicit "peak" strength by far exceeding the residual strength, the curve representing a plastic clay has almost no "peak", i.e. the peak and residual strengths of this clay are almost the same (Fig. 9-16(b)). Bishop (1967) suggested a special criterion called brittleness index:

$$I_B = \frac{\tau_f - \tau_r}{\tau_f} \tag{9-5}$$

For $I_B = 1$, the soil is brittle, for $I_B = 0$ it is plastic. Curve *1* in Fig. 9-16(b) corresponds to $I_B = 0.6$; curve *2*, to $I_B = 0.1$.

The curves for an over-compacted and a normally compacted clay are depicted in Fig. 9-16(c). Curve *1* for an over-compacted clay has a distinct "peak", the difference between τ_f and τ_r being substantial in this clay. For a normally compacted clay (curve *2*), the "peak" of the τ–λ curve is less explicit and the difference between τ_f and τ_r is smaller. Here, the value of τ_r for the over-compacted clay and the normally compacted one is the same.

At the same time, Kanji and Wolle (1977) showed that residual strength varied with soil consistency. So, in particular, the dependence of the angle of friction (in shear along a hard surface) was expressed as $\phi_r = A/I_p^a$.

Relation between instantaneous, long-term, "peak" and residual strengths

Let us consider the relation between a hypothetically instantaneous strength τ_0 and an ultimate long-term strength τ_∞ on the one hand (see Fig. 9-1) and the relation between a "peak" strength τ_f and a residual strength τ_r on the other hand (see Fig. 9-15).

It was mentioned above that the "peak" resistance τ_f increases with an increase in the deformation rate. Apparently, when the deformation develops at a practically possible maximum rate, the "peak" strength will correspond to the hypothetically instantaneous strength $\tau_f \rightarrow \tau_0$.

According to the experiments carried out by Turovskaya at the Dnepropetrovsk Institute of Railway Engineers, the residual strength τ_r of plastic soils with water-colloidal bonds is of a value close to the ultimate long-term strength τ_∞ (Babitskaya and Turovskaya, 1975). For soils with rigid bonds (dense clay, soft rocks), the value of τ_r is less than τ_∞. In fact, the ultimate long-term strength is the resistance of a soil with an undisturbed structure which manifests itself in the form of both cohesion and friction. The residual strength is the resistance of a soil with broken bonds which is attributed mainly to friction only.

Thus, in assessing stability, when neither non-attenuating deformation nor significant shearing are tolerable, the analysis should be based on the ultimate long-term strength. However, in investigating a process of soil transportation (e.g., a landslide or a long-term flow of a slope) and determining the shearing resistance to the displacement, the residual strength of the soil is the criterion.

9.5 CRITERIA OF LONG-TERM FAILURE

Condition of long-term strength

The mechanism of the long-term strength of a soil may be deduced from the equation of creep if a criterion of long-term failure is introduced into the equation.

Vyalov (1956) and Goldstein (1957) came up with the idea that soil failure results from plastic deformation (deformation of creep), γ^c, building up to some constant limit, γ_f^c, and that failure is imminent when that limit is exceeded. This implies that the criterion of failure adopted is a point when the deformation of creep reaches a constant value:

$$\gamma^c = \gamma_f^c = \text{const.} \tag{9-6}$$

This is a soil constant affected by neither the stress nor the time before failure.

Stability of limiting strain

Strictly speaking, the magnitude of the strain γ_f^c (or ε_f^c in the case of compression) at the moment of failure changes with the time before failure t_f or, which is the same, depending on the level of the stress τ: increasing to a certain maximum at an early stage and decreasing eventually as the time t_f increases (Fig. 9-17). The changes in γ_f^c γ_f^c during a short time before failure can be distinctly seen, e.g. in Fig. 9-17(a).

If short time intervals t_f before failure are omitted from consideration, γ_f^c will

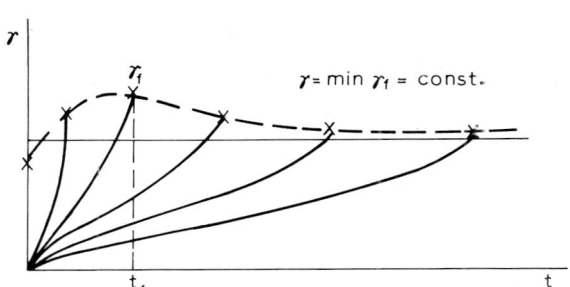

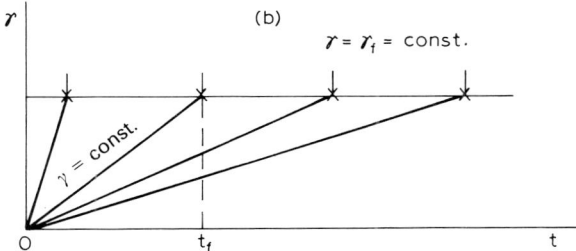

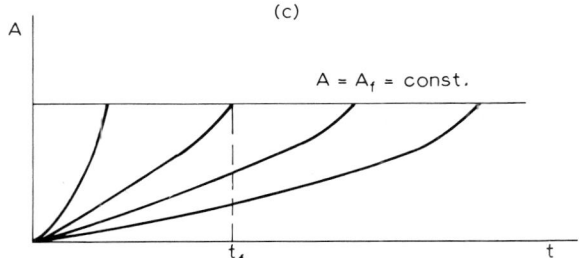

Fig. 9-17. Conditions of long-term failure of soils: (a) stability of limiting strain $\gamma = \gamma_f = $ const.; (b) constancy of the product of strain rate and time before failure $\dot{\gamma} t_f = $ const.; (c) constancy of the work of deformation $A t_f = $ const.

eventually change insignificantly so that this value may be assumed to be a constant in many cases. The fact that the amount of plastic strain γ_f^c corresponding to the moment of failure eventually remains stable is proved by the configuration of the creep curves (for uniaxial compression) depicted in Fig. 5-13(b). For all these curves, the value of ε_f^c varies between 0.09 and 0.1, i.e. remains practically the same.

Goldstein and his colleagues reported similar results: the value of ε_f^c, as observed over the period between 1 min and 10^5 min before failure, varied within the usual spread of the experimental points (Goldstein, 1957).

A further evidence of small changes in the value of ε_f^c is provided by the data acquired from testing frozen soils, particularly those illustrated in Figs. 5-14 and 5-23. The curves in Fig. 5-14, (c) and (d) indicate that the ultimate strain, ε_f^c, averages 0.16 ± 0.03 for sandy loam and 0.06 ± 0.015 for Bat-baioss dense clay.

The value of γ_f^c (or ε_f^c) varies with the type of soil, being higher in plastic soils failing viscously. This value seems also to be affected by the arrangement of stress and is different, for example, in the case of compression, tension or shear.

Returning to eq. 9-6, it is not, strictly speaking, a physical value. Such a value and its connection with eq. 9-6 will be discussed eventually. Here it will be sufficient to state that, whether or not γ_f^c is a soil constant, the adoption of the above conditions means that soil performance is confined by the value of limiting strain. In other words, the condition referred to combines the assessment of soil behaviour in terms of ultimate long-term strength with that in terms of creep. Should the actual strain at failure exceed the adopted value of γ_f^c = const. during a certain period, the limitation adopted will add to the factor of safety.

Substituting condition 9-6 into the equation of strain 5-8, $\gamma_f^c = f(\tau)\Phi(t_f)$, and solving it with respect to τ, we obtain the equation of long-term strength:

$$\tau = \frac{\text{const.}}{\psi(t_f)} \tag{9-7}$$

connecting the stress at failure τ and the time before failure t_f. The connection is of the hyperbolic type, the stress being inversely proportional to the time function.

Constancy of the product of strain rate and time before failure in soils

The deformation of creep γ^c at any moment of time t during the stage of steady flow may be expressed in terms of the rate of flow, $\gamma^c = \dot{\gamma}t$. At the moment of failure $t = t_f$, the deformation will be $\gamma_f^c = \dot{\gamma}t_f$; hence eq. 9-6 may be written in the form:

$$\dot{\gamma}t_f = \text{const.} \tag{9-8}$$

Condition 9-8, well known in the theory of creep, thus follows from condition 9-6. The applicability of condition 9-8 to soils was shown by M. Saito and H. Uezawa in the papers presented at the 5th, 6th and 7th ICSMFE (Saito and Uezawa, 1961; Saito, 1965, 1969) and by Liam Finn and Shead (1973) reporting at the 8th ICSMFE.

Depicted in Fig. 9-18 is a combined graph of the relationship between the flow rate and the time elapsed before failure plotted on the basis of the data referred to above and data supplied by other authors. The graph covers the results of laboratory experiments and field surveys. The laboratory data refer to uniaxial and triaxial compression tests and the field data include observations of the displacement rates of artificial and natural slopes.

Although the data presented have been obtained for various soils (including those with disturbed and undisturbed structure, varying density and moisture content), the experimental points come down fairly close together next to a common straight line. The dashed lines in Fig. 9-18 delineate the area of spread, covering 95% of the

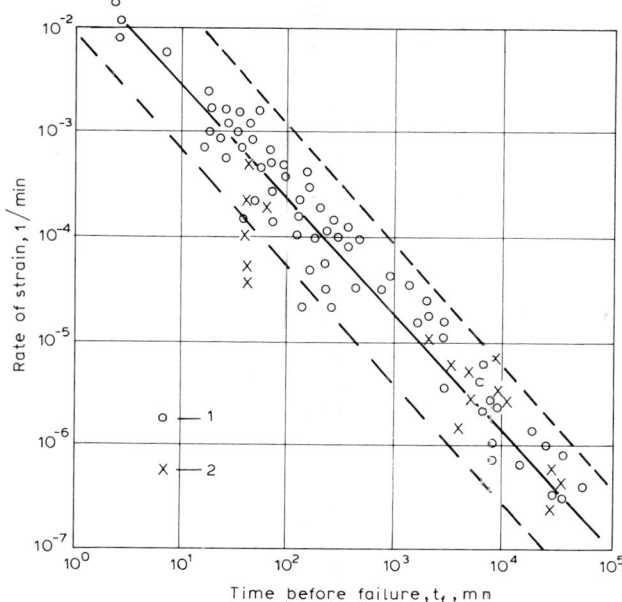

Fig. 9-18. Relationship between strain rate and time before failure: *1* = data acquired by various investigators from laboratory experiments; *2* = data acquired from field observations and large-scale experiments.

experimental points. Thus, the relation between the time before failure and the strain rate $\dot{\varepsilon}$ (or, in the general case, $\dot{\gamma}$) may be expressed as $\lg t_f = \lg C + b \lg \dot{\gamma}$, hence:

$$t_f = C/\dot{\gamma}^b \qquad (9\text{-}9)$$

where C = const.

Since the parameter b varies between 0.92 and 1.08, it may be adopted that $b = 1$. Then, eq. 9-9 coincides with eq. 9-8:

$$t_f = C/\dot{\gamma} \qquad (9\text{-}10)$$

The dimensionless constant C equals 0.023 according to Saito and 0.017 after Finn, i.e., varies over a narrow range despite the diversity of experimental conditions, soil types, etc. The question whether the constant C is the same for all soils under all conditions of deformation or varies with these factors requires special investigation. In our opinion the second assumption is more probable. However, small deviations of the experimental values of C (falling inside the 95% area as shown in Fig. 9-18) characteristic for such a diversity of soil types and experimental conditions give good reason to believe that the value of C depends on the kind of soil and arrangement of stress to an insignificant amount.

Forecasting the period elapsing before slope failure

Condition 9-8 is a specific case of the more general condition 9-6; the point is that condition 9-6 holds for a curve of non-attenuating creep of any configuration whereas condition 9-8 is applicable, strictly speaking, to a steady Newtonian flow only, whose rate is constant from the moment of load application $t = 0$ and to the moment of failure t_f.

At the same time, insignificant changes in the value of C entering into condition 9-10 render this applicable to a creep process with an altering rate of deformation. In such a case, we may substitute into condition 9-10 the value of $\dot{\gamma}$ corresponding to the deformation rate at the stage of steady flow, $\dot{\gamma} = \dot{\gamma}_y = $ const., counting off the time before failure from the moment this stage takes shape, i.e. adopting that $t_f - t_y = C/\dot{\gamma}_y$. If the stage of flow is inconspicuous one may use the minimum rate of deformation as reference, counting off t_f from the moment t_0, corresponding to this rate.

Finally, condition 9-8 may be applied (as was shown by Saito (1969) at the 7th ICSMFE) to the progressive stage of creep with an increasing rate of deformation. In this case, condition 9-10 is replaced by the relationship:

$$\gamma = c \lg \frac{t_f - t_0}{t_f - t} \tag{9-11}$$

where γ is the deformation of creep at any moment t; t_0 is the time adopted as reference; and t_f is the time before failure counted off from the moment t_0.

Eq. 9-11 can be solved graphically for the value of t_f we are interested in, using three points on the curve of creep (inside the segment of progressive flow). Given arbitrary moments t_1, t_2, t_3 which meet the requirement that the difference between the deformations corresponding to each of these moments is the same, $\gamma_2 - \gamma_1 = \gamma_3 - \gamma_2$, one may transform eq. 9-11 into:

$$t_f - t_1 = \frac{0.5(t_2 - t_1)^2}{(t_2 - t_1) - 0.5(t_3 - t_1)} \tag{9-12}$$

The value of t_f is conveniently determined graphically.

Relationships 9-10 and 9-11 can be used to forecast possible failures of slopes, natural and man-made, from the observations of displacements.

Constancy of the work of deformation. Comparing with eq. 9-8, a more general condition of long-term failure will be one stating that the product of the rate of the work of plastic deformation and the time before failure is constant:

$$\dot{A} t_f = \text{const.} \tag{9-13}$$

where $\dot{A}$ is the unit power of stress defined by eq. 4-79. If $\dot{A}$ changes with time, condition 9-13 is set forth in the form:

$$\int_0^{t_f} \dot{A}(t) \, dt = \text{const.} \tag{9-14}$$

It follows from conditions 9-13 and 9-14 that soil failure occurs at the moment the work of deformation reaches a certain ultimate value (see Fig. 9-17(c)).

Since we are considering visco-plastic deformations, condition 9-15 refers to dissipated work:

$$A = \sigma_{ij} \dot{\varepsilon}_{ij}^p$$

where σ_{ij} and $\dot{\varepsilon}_{ij}^p$ are the components of the stress tensors and the plastic strain rate, respectively.

However, the most general approach is one in terms of thermodynamics when the change in entropy due to deformation is being examined. Accordingly, the criterion of soil failure is a point at which an increment, ΔS (influenced, in its turn, by external forces, temperature and internal thermodynamic changes), in the density of the entropy accumulated during the process of creep reaches some critical value, ΔS_{crit}, varying with the degree of the structural changes occurring in the body depending on a given form of failure:

$$\Delta S(t_f) = \Delta S_{crit} \qquad (9\text{-}15)$$

This criterion has been formulated by I.I. Goldenblat (Bolotin et al., 1973) and, independently, by Chudnovsky (1973).

9.6 EQUATIONS OF LONG-TERM STRENGTH

Initial equation

The physical essence of the process of long-term soil failure will be discussed in Chapter 10. Here we are going to derive an equation of long-term soil strength, proceeding from the condition of long-term failure (eq. 9-6) and the equation of creep based on macroexperiments. Taking this into account and using eq. 5-7, we may put down:

$$\phi(\gamma_f) = \tau \psi(t_f) = \text{const.} \qquad (9\text{-}16)$$

where $\gamma_f = $ const. is the strain of creep at the moment of failure $t = t_f$; $\phi(\gamma_f)$ is a function defining the relation between the strain γ_f and the stress τ; $\psi(t_f)$ is a time function defining the time t during which the strain of creep reaches the critical value $\gamma = \gamma_f = $ const.

Solving eq. 9-16 for stress, we obtain relationship 9-7 examined above:

$$\tau = \frac{\text{const.}}{\psi(t_f)} \qquad (9\text{-}17)$$

It is more expedient to obtain relationship 9-17 from an integral equation of creep of the form of eq. 7-44, adopting that:

$$\phi(\gamma_i^c) = \int_0^{t_f} K(t)\, dt \qquad (9\text{-}18)$$

If the kernel of the equation of creep is adopted in the form given by eq. 7-52:

$$K(t) = \left(\frac{T_2}{T_1 + t}\right)^n \tag{9-19}$$

the equation of long-term strength derived from eq. 9-18 takes the form:

$$\tau = \frac{N}{\int_0^{t_f} \left(\frac{T_2}{T_1 + t}\right)^n dt} \tag{9-20}$$

Logarithmic equation of long-term soil strength

Putting $n = 1$, $T_1 = T$ and $T_2 = \delta$ in relationship 9-19, we obtain from eq. 9-20:

$$\tau = \frac{\beta}{\ln \frac{t_f + T}{T}} \tag{9-21}$$

where $\beta = N/\delta$ (in Pa); $N = A_0(\gamma_f^c)^m$ (in Pa); A_0 is the coefficient of deformation, in Pa; and δ and m are dimensionless parameters.

Eq. 9-21 lacks agreement with the initial condition, for at $t = 0$ we have $\tau_0 = \infty$. More accurate is the statement:

$$\tau = \frac{\beta}{\ln \frac{t_f + t^*}{T}} \tag{9-22}$$

where t^* is an arbitrary short interval (e.g. $t^* = 1$ s) measured in units (s, min, h) shorter than the total period of observations by several orders of magnitude.

Disregarding t^* when the interval is $t \gg t^*$, we obtain:

$$\tau \cong \frac{\beta}{\ln (t_f/T)} \tag{9-22'}$$

holding for $t_f > T$.

Formula 9-22 was applied to soils by the author of this book (see Section 1.5: Vyalov, 1959; Vyalov et al., 1962) and has been proved experimentally in the works of many other researchers; particularly it is used to assess the strength of frozen soils.

It follows from expression 9-22 that the hypothetically instantaneous strength (at $t = 0$) is:

$$\tau_0 = \frac{\beta}{\ln (t^*/T)} = \frac{\beta}{\ln (1/T)}$$

and, accordingly, formula 9-22 may be written for $t^* = 1$ as:

$$v\frac{\tau_0}{\tau} = \ln \frac{t_f + 1}{T}$$

or:

$$v \frac{\tau_0 - \tau}{\tau} = \ln(t_f + 1) \qquad (9\text{-}23)$$

where $v = \ln(1/T)$; $(\tau_0 - \tau)/\tau$ is the level of stress.

As far as metals, plastics and some other materials are concerned, an equation of long-term strength somewhat differing from eq. 9-22 is in use. It may also be applied to soils:

$$\tau = \beta \ln \frac{T}{t_f + t^*} = \beta \ln \frac{T}{t_f} \text{ approx.} \qquad (9\text{-}24)$$

where $\beta \ln(T/t^*) = \tau_0$.

Eq. 9-24 may be arrived at from considering expressions 9-17 and 4-65 simultaneously.

The results of computations by eqs. 9-24 and 9-22, if compared with each other and experimental data, are satisfactorily correlated within a certain interval of time. However, for long intervals, preference is given to eq. 9-22; because, unlike eq. 9-24 according to which the strength τ decreases to zero for $t_f = T$, eq. 9-22 provides for the decrease only at $t_f \to \infty$.

Consider the problem of loss of strength in response to an infinitely long period of load application. According to eq. 9-22, the strength decreases infinitely and the curve of long-term strength asymptotically approaches the x-axis, i.e. for $t \to \infty$ the strength $\tau \to 0$. Therefore, the strength corresponding to some ultimate time before failure t_∞ is adopted as a conventional ultimate long-term strength:

$$\tau_\infty = \frac{\beta}{\ln \dfrac{t_\infty}{T}} \qquad (9\text{-}25)$$

The value of t_∞ is either decided by the service life or is computed by the expression:

$$t_\infty = (100\, T^{0.03})^{1/1.03} \qquad (9.26)$$

where t_∞ and T are in years. The value of t_∞ in eq. 9-26 is an interval after which the loss in strength will be less than 3% in 100 years, i.e. t_∞ is the time satisfying the condition:

$$\frac{\tau_\infty - \tau_{100}}{\tau_{100}} = 0.03$$

where τ_{100} is a theoretical value of strength for $t = 100$ years; τ_∞ is the computed value of the conventional ultimate long-term strength.

Power equation of long-term strength

Putting $n = 1 - \alpha$, $T_1 = 0$, $T_2 = (\alpha\delta/T^\alpha)^{1/1-\alpha}$ in eq. 9-19, we obtain from eq. 9-20 an equation of long-term strength in the form also widely used in the theory of creep:

$$\tau = \left(\frac{t_f}{T^*}\right)^{-\alpha} \tag{9-27}$$

where $T^* = T(N/\alpha)^{1/\alpha}$; $N = A_0(\gamma_f^c)^m$.

Since, for $t = 0$, we obtain from eq. 9-27 that $\tau_0 = \infty$, we may put it forth in a form corresponding to the actual conditions. Taking the upper limit of the integral in eq. 9-20 at $t + t^*$, we arrive at:

$$\tau = \left(\frac{t_f + t^*}{T^*}\right)^{-\alpha} \tag{9-28}$$

where t^* is an arbitrary short interval of time, e.g. $t^* = 1$ s (or min or h). For $t = 0$, we obtain $\tau_0 = (T^*)^\alpha$, and for $t^* = 1$, formula 9-28 takes the form:

$$\frac{\tau}{\tau_0} = (t_f + 1)^{-\alpha} \tag{9-29}$$

By analogy with the case considered above, the strength of soil τ according to eq. 9-29 deteriorates infinitely (for $t_f \to \infty$, $\tau \to 0$).

Linear–fractional equation of long-term strength

Putting $n = 2$, $T_1 = T$, $T_2 = [T(\delta - 1)]^{1/2}$ in eq. 9-19, we obtain from eq. 9-20 a linear–fractional equation of long-term soil strength:

$$\tau = \frac{N}{\delta - 1} \frac{T + t_f}{t_f} \tag{9-30}$$

where $N = \phi(\gamma_f) - \phi(\gamma_0)$, provided:

$$\phi(\gamma_f) = \frac{G_0 \tau_s}{\tau_s + G_0 \gamma_f} \gamma_f$$

When use is made of the above relationship, the condition at failure is produced by an infinitely large deformation $\gamma_f = \infty$ (see Fig. 5-8), hence $\lim\limits_{\gamma \to \infty} \phi(\gamma_f) = \tau_s$.

On the other hand, it follows from eq. 5-24 that:

$$\phi(\gamma_0) = \frac{G_0 \tau_s}{\tau_s + G_0 \gamma_0} \gamma_0$$

Putting $N = \tau_s - \tau$, we obtain eq. 9-30 in the form:

$$\tau_s - \tau = \tau(\delta - 1) \frac{t_f}{T + t_f}$$

For $t_f = 0$, we shall have $\tau_0 = \tau_s$; for $t = \infty$, we obtain $\tau_0 - \tau_\infty = \tau_\infty(\delta - 1)$, hence $\tau_\infty = \tau_0/\delta$ as this was postulated in Chapter 5 while analysing eq. 5-22. Substituting the value $N = \tau_0 - \tau$ into eq. 9-30, we obtain, after simple manipulations, the equation of long-term soil strength:

$$t_f = \frac{T}{\delta} \frac{\tau_0 - \tau}{\tau - \tau_\infty} \tag{9-31}$$

where $\delta = \tau_0/\tau_\infty$; $(\tau_0 - \tau)/(\tau - \tau_\infty)$ is the level of stress in the soil.

Eq. 9-31, obtained by Zaretsky and Vyalov (1971) in another way, describes the mechanism of strength deterioration down to a limiting value which is the ultimate long-term strength τ_∞; accordingly, the curve of long-term strength has an asymptote $\tau \to \tau_\infty$ for $t \to \infty$.

It will be shown later that the linear–fractional formula 9-31 lacks satisfactory agreement with experimental data and that the situation may be improved by representing the relation between the time before failure and the level of stress in the form of a power equation:

$$t_f^\alpha = \left(\frac{T}{\delta}\right)^\alpha \frac{\tau_0 - \tau}{\tau - \tau_\infty} \tag{9-32}$$

This formula may be derived from eq. 5-22 if the time t is entered into it with an exponent α. Moreover, expressing the deformation γ as a power function γ^m, we may represent the equation of creep 5-22 in a more universal form:

$$\gamma^m = \frac{\tau(1 + \delta t^\alpha)}{G_0[T(1 - \tau/\tau_s) + t(1 - \delta\tau/\tau_s)]} \tag{9-33}$$

9.7 METHODOLOGY OF EXPERIMENTAL DATA HANDLING

Testing formulae for closeness of fit[*1]

To test the equations of long-term strength 9-22, 9-28, 9-31 and 9-32 for closeness of fit, the curves of long-term strength are rectified by transforming the above equations into the linear form $Y = BX + D$ as was shown in Sections 6.2 and 6.3.

Formula 9-22 may be reduced to linear form on being transformed into:

$$\frac{1}{\tau} = \frac{1}{\beta} \ln (t_f + 1) - \frac{1}{\beta} \ln T \tag{a}$$

Formula 9-28 is also amenable to linearization:

$$\ln \tau = \alpha \ln (t_f + 1) - \alpha \ln T^* \tag{b}$$

Formula 9-31, in its turn, may be reduced to the following linear form:

$$\frac{t_f}{\tau_0 - \tau} = \frac{1}{\tau_0 - \tau_\infty} t_f + \frac{T}{\delta} \frac{1}{\tau_0 - \tau_\infty} \tag{c}$$

where τ_0 is obtained from experiments.

[*1] The closeness-of-fit tests of the formulae in Sections 9.7, 10.8, 11.6 and 13.5 were carried out by M.E. Slepak.

TABLE 9-III

Results of testing frozen soil for long-term strength

	Ref. number of points:						
	1	2	3	4	5	6	7
Stress ($\times 10^5$ Pa)	4.8	2.3	2.1	1.9	1.7	1.5	1.3
Time before failure (in s):	14	$19.1 \cdot 10^3$	$11.9 \cdot 10^4$	$15.45 \cdot 10^5$	$2.84 \cdot 10^6$	$17.05 \cdot 10^6$	Without failure
(in h):	$3.9 \cdot 10^{-3}$	5.3	33	429	788	4,736	

Finally, formula 9-32 may be transformed into:

$$\ln \frac{\tau_0 - \tau}{\tau - \tau_\infty} = \alpha \ln t_f - \alpha \ln (T/\delta) \tag{d}$$

Preparatory to solving eq. d, one needs to assess τ_∞. If the value of τ_0 is known from experiments, the value of τ_∞ may be computed from three points on the experimental curve of long-term strength the abscissas of which are connected by the relationship $(t_1/t_2)^2 = t_1/t_3$. Substituting this relationship into eq. d and eliminating its right-hand side, we find that the ordinates τ_1, τ_2 and τ_3 of the points in question are connected by the relationship:

$$\frac{(\tau_3 - \tau_\infty)(\tau_1 - \tau_\infty)}{(\tau_2 - \tau_\infty)^2} = \frac{(\tau_0 - \tau_1)(\tau_0 - \tau_3)}{(\tau_0 - \tau_2)^2} \tag{e}$$

from which we find the value of τ_∞.

The rest of the parameters entering into eqs. a–d are determined by formulae given in Sections 6.2 and 6.3. Relevant formulae from the same sections are used in evaluating the experimental data (computing r.m.s. error ε_j, standard deviation σ, correlation coefficient r and coefficient v).

Experimental data evaluation

Let us evaluate by way of illustration the data acquired from testing frozen soil (silty sandy loam at $\theta = -0.4°$) for long-term strength in shear along the side surface of piling (see Chapter 1: Vyalov, 1979). The times before failure t_f obtained under relevant constant loads are given in Table 9-III.

The experimental value of ultimate long-term strength was adopted at $1.30 \cdot 10^5$ Pa to $1.35 \cdot 10^5$ Pa approximately.

Testing formula 9-22. On reducing formula 9-22 to the linear form $Y = BX + D$ according to eq. a, we adopt $X = \ln (t_f - 1)$ (in seconds) and $Y = 1/\tau$.

Plotting experimental points on the graph of Fig. 9-19(a) in given coordinates, we can see that they fall on the straight line. This is an indication of the goodness of fit of formula 9-22. The angle the line makes with the x-axis defines the parameter $B = 1/\beta$, and the intercept on the y-axis determines the value of $D = -(1/\beta) \ln T$.

The values of the above parameters may be determined analytically by eq. 6-12 on tabulating the computed values of X, Y, X^2, Y^2 and XY in Table 9-IV. Relevant computations yield:

$$B = \frac{70.01 \times 2.90 - 6 \times 37.79}{70.01^2 - 6 \times 948} = 0.0304$$

$$D = \frac{70.01 \times 37.79 - 2.90 \times 948}{70.01^2 - 6 \times 948} = 0.128$$

Hence we find the parameters β and T entering into the initial eq. 9-22: $\beta = 1/0.0304 = 32.9 \cdot 10^5$ Pa; $T = e^{-\beta D} = \exp(-32.9 \times 0.128) = 0.015$ s.

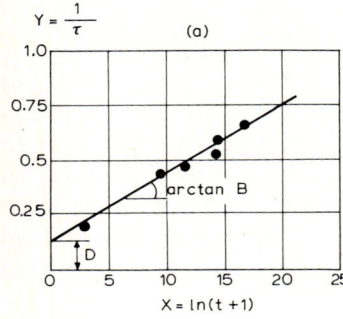
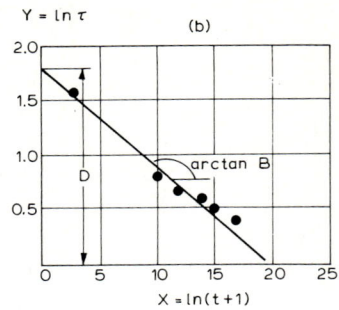
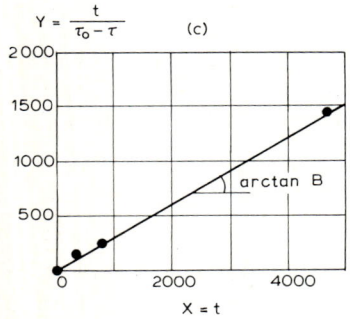
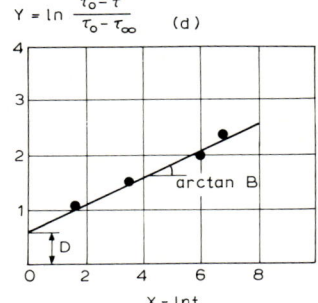

Fig. 9-19. Rectification of long-term strength curves by: (a) eq. 9-22; (b) eq. 9-28; (c) eq. 9-31; (d) eq. 9-32.

Thus in the case under consideration, the equation of long-term strength will take the form:

$$\tau = \frac{32.9}{\ln[1.5 \cdot 10^{-2}(t_f - 1)]} \qquad (f)$$

where t_f is the time in seconds; τ is the strength in 10^5 Pa.

The values of strength τ_{comp} computed by this formula are given in Table 9-IV, column 9. The theoretical value of instantaneous strength computed by eq. f for $t_f = 0$ is:

$$\tau_0 = \frac{32.9}{\ln(1.5 \cdot 10^{-2})} \cong 7.8 \cdot 10^5 \text{ Pa}$$

The ultimate time t_∞ and the value of τ_∞ corresponding to that time are computed by eqs. 9-25 and 9-26. From eq. 9-26 we obtain:

$$t_\infty = 100\left[\left(\frac{0.015}{60 \times 60 \times 24 \times 365}\right)^{0.03}\right]^{1/1.03} = 47 \text{ years}$$

TABLE 9-IV

Handling long-term strength test data for soil by eq. 9-22

Ref. number of points	t_f (s)	τ (10^5 Pa)	X	Y	X^2	Y^2	XY	τ_{comp}	ε_j	ε_j^2
(1)	(2)	(3)	(4)	(5)	(6)	(7)	(8)	(9)	(10)	(11)
1	14.0	4.8	2.708	0.208	7.4	0.043	0.564	4.75	0.01	10^{-4}
2	$19.1 \cdot 10^3$	2.3	9.852	0.435	97.3	0.191	4.29	2.34	−0.02	$4 \cdot 10^{-4}$
3	$11.9 \cdot 10^4$	2.1	11.687	0.475	138	0.227	5.57	2.07	0.01	10^{-4}
4	$15.45 \cdot 10^5$	1.9	14.254	0.526	204	0.278	7.52	1.78	0.06	$36 \cdot 10^{-4}$
5	$2.84 \cdot 10^6$	1.7	14.857	0.588	221	0.348	8.75	1.73	−0.02	$4 \cdot 10^{-4}$
6	$17.05 \cdot 10^6$	1.5	16.655	0.666	280	0.447	11.1	1.58	−0.05	$25 \cdot 10^{-4}$
$\Sigma =$	—	—	70.01	2.90	948	1.534	37.79	—	—	$7 \cdot 10^{-3}$

Hence, according to eq. 9-25:

$$\tau_\infty = \tau_{47} \text{ approx.} = \frac{32.9}{\ln\left(\dfrac{47 \times 365 \times 24 \times 60 \times 60}{0.015}\right)} = 1.3 \cdot 10^5 \text{ Pa}$$

The value obtained is close to the value $\tau_\infty = 1.35 \cdot 10^5$ Pa determined directly from the experiments.

Note that if t_∞ is adopted at 100 years, which is the service life of the structure, the value of τ_∞ remains almost unchanged: $\tau_\infty = \tau_{100}$ approx. $= 1.28 \cdot 10^5$ Pa.

The above computations thus provide evidence that the ultimate long-term strength of soil may be assessed by eq. 9-25 with an accuracy acceptable in engineering practice.

The closeness of fit of the equation of long-term strength (eq. 9-22) is determined by analogy with the procedure outlined in Chapter 6. Compute the deviation of each experimental value ε_j from the mean:

$$\varepsilon_j = \frac{\tau_j - \tau_{\text{comp}}}{\tau_{\text{comp}}}$$

and compute the square of this value ε_j^2, entering the results into columns 10 and 11 of Table 9-IV, respectively.

Compute from these values the standard deviation σ, the coefficient of variation v and the correlation coefficient r, using formulae 6-3, 6-6 and 6-13, respectively. Thus, we obtain:

$$\sigma = \sqrt{\frac{7 \cdot 10^{-3}}{6-2}} \times 100 = 4.2\% \qquad v = \frac{4.2}{1\sqrt{6}} = 1.7\%$$

$$r = \frac{6 \times 37.79 - 70.01 \times 2.90}{\sqrt{(6 \times 948 - 70.01^2)(6 \times 1.534 - 2.90^2)}} = 0.949$$

As can be seen, the indices of the precision of approximation of the experimental data by eq. 9-22 are satisfactory. Theoretical curve *1* plotted in accordance with this formula is depicted in Fig. 9-20. The curve, as is evident from Table 9-IV, does not depart from the experimental points by more than 6%.

Testing eq. 9-28. Taking into account eq. b, we adopt $X = \ln(t_f + 1)$ and $Y = \ln \tau$. On plotting a graph on these coordinates, we can see that the experimental points fall on the straight line (see Fig. 9-19(b)); the slope of the line defines the parameter $B = \alpha$ and the intercept on the y-axis yields the value of $D = -\alpha \ln T$.

To compute the values of B and D analytically, we use eq. 6-12 tabulating the results of preliminary computations in Table 9-V. We obtain:

$$B = \frac{70.01 \times 4.72 - 6 \times 44.93}{70.01^2 - 6 \times 948} = -0.078$$

and

$$D = \frac{70.01 \times 44.93 - 4.72 \times 948}{70.01^2 - 6 \times 948} = 1.7$$

TABLE 9-V

Handling long term strength test data for soil by eq. 9-28

Ref. number of points	t_r (s)	τ (10^5 Pa)	X	Y	X^2	Y^2	XY	τ_{comp}	ε_j	ε_j^2
(1)	(2)	(3)	(4)	(5)	(6)	(7)	(8)	(9)	(10)	(11)
1	14.0	4.8	2.708	1.569	7.4	4.25	2.460	4.43	0.08	$64 \cdot 10^{-4}$
2	$19.1 \cdot 10^3$	2.3	9.852	0.833	97.3	8.21	0.700	2.54	−0.09	$81 \cdot 10^{-4}$
3	$11.9 \cdot 10^4$	2.1	11.687	0.742	138	8.67	0.552	2.20	−0.04	$16 \cdot 10^{-4}$
4	$15.45 \cdot 10^5$	1.9	14.254	0.642	204	9.15	0.413	1.80	0.05	$25 \cdot 10^{-4}$
5	$2.84 \cdot 10^6$	1.7	14.857	0.531	221	7.89	0.285	1.72	−0.01	10^{-4}
6	$17.05 \cdot 10^6$	1.5	16.655	0.406	280	6.76	0.165	1.49	0.01	10^{-4}
$\Sigma =$			70.01	4.72	948	44.93	4.58	—	—	$1.9 \cdot 10^{-2}$

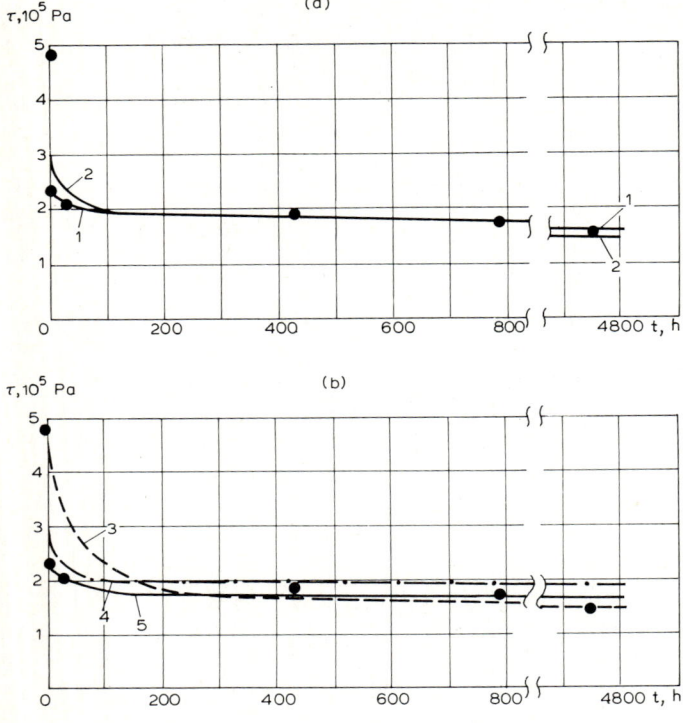

Fig. 9-20. Curves of long-term strength. Test data are marked by dots and the results of analytical computations are presented by curves. The formulae used in computations are as follows: *1* = eq. 9-22; *2* = eq. 9-28; *3* = eq. 9-31; *4* = eq. 9-31 in conjunction with the data of quick tests; *5* = eq. 9-32.

Hence, the parameters of the initial eq. 9-28 are $\alpha = -0.078$ and $T = \exp(1.7/0.078) = 29.5 \cdot 10^8$ s.

Thus, the equation of long-term strength 9-28 takes, in the case under consideration, the form:

$$\tau = \left(\frac{t_f + 1}{29.5 \cdot 10^8}\right)^{-0.078} \tag{g}$$

where t_f is the time in seconds and τ is the strength in 10^5 Pa.

The values of strength τ_{comp} computed by this formula are given in column 9 of Table 9-V. The theoretical value of instantaneous strength will be $\tau_0 = (29.5 \cdot 10^{-8})^{-0.078} = 5.47 \cdot 10^5$ Pa.

The long-term strength, as assessed for a period $t_f = 47 = 50$ years approx. and $t_f = 100$ years, will be $\tau_{50} = 1.05 \cdot 10^5$ Pa and $\tau_{100} = 1.0 \cdot 10^5$ Pa, respectively.

The deviations ε_j of the computed values τ_{comp} from the experimental values τ are tabulated in columns 10 and 11 of Table 9-V; the indices of precision appear to be $\sigma = 6.9\%$, $v = 2.8\%$ and $r = -0.96$.

Theoretical curve 2 plotted in accordance with the data computed by eq. 9-28 is depicted in Fig. 9-20. It departs, according to Table 9-V, from the experimental points by not more than 9%.

Testing formula 9.31. Taking eq. c, we adopt that $X = t_f$ and $Y = t_f/(\tau_0 - \tau)$, where τ_0 is assumed to be known from the experiments; in the case under consideration, $\tau_0 = 4.8 \cdot 10^5$ Pa.

A rectified curve plotted on these coordinates is represented in Fig. 9-19(c). Since the time is laid off on the graph to a scale which is not transformed, the initial points coincide with the origin (although accurate computation proves that $D \neq 0$) and are omitted from consideration; this is a limitation of formula 9-31.

The analytical parameters B and D are determined by formula 6-12 as in previous cases, by tabulating the results of preliminary computations in Table 9-VI. Thus, we obtain $B = 0.302$ and $D = 8.89$. Hence, the parameters of the initial eq. 9-28 are: $\tau_\infty = \tau_0 - 1/B = 1.5 \cdot 10^5$ Pa; $\delta = \tau_0/\tau_\infty = 3.2$; $T = D\delta(\tau_0 - \tau_\infty) = 94.3$ h.

The equation of long-term strength (eq. 9-31) takes, in the case under consideration, the form:

$$t_f = 29.43 \frac{4.8 - \tau}{\tau - 1.5} \tag{h}$$

where t_f is the time in hours and τ is the strength in 10^5 Pa.

The values of τ_{comp} computed by this formula and the deviations of τ_{comp} from the experimental data, ε_j, are tabulated in columns 8, 9, 10 of Table 9-VI. The indices of precision, as determined by eqs. 6-3, 6-6 and 6-13, are $\sigma = 29\%$, $v = 12\%$ and $r = 1.006$, respectively, they cannot be regarded as satisfactory.

Consequently, the computed values of strength appreciably depart (up to 45%) from the experimental values (see Table 9-VI) and the theoretical curve (curve 3 in Fig. 9-20) lacks fit with the experimental points. The lack-of-fit is particularly pronounced within the initial segment of the curve. If the first four experimental points are used for the test of fit, then, conversely, the theoretical curve will coincide with the experimental points over the initial segment (see curve 4 in Fig. 9-20) but will depart from them at distant times. Especially significant is the lack-of-fit of the theoretical value of τ_∞ which appears to equal $1.94 \cdot 10^5$ Pa in this case.

Testing formula 9-32. A better closeness of fit, compared with formula 9-31, is obtained by using the power–fractional eq. 9-32. For the test of fit of the equation, we adopt in accordance with eq. d that $X = \ln t_f$ and $Y = \ln[(\tau_0 - \tau)/(\tau - \tau_\infty)]$.

Given three points on the experimental curve of long-term strength, we determine τ_∞ by eq. e, taking $\tau_0 = 4.8 \cdot 10^5$ Pa as obtained from the experiments. For better results the above points must be separated by long intervals (e.g. $t_3 = 10t_2 = 100t_1$).

Thus, adopting $t_1 = 5.3$ h, $t_2 = 53$ h and $t_3 = 530$ h, the corresponding values of strength will be $\tau_1 = 2.3 \cdot 10^5$ Pa, $\tau_2 = 2.0 \cdot 10^5$ Pa and $\tau_3 = 1.8 \cdot 10^5$ Pa, respectively. Substituting these into eq. e, we find that $\tau_\infty = 1.5 \cdot 10^5$ Pa.

The values of X and Y for the adopted values of τ_0 and τ_∞ are tabulated in Table 9-VII, and a curve on these coordinates, yielding a straight line, is represented in Fig. 9-19(d). The slope of this line defines $B = \alpha$ and the intercept on the y-axis

TABLE 9-VI

Handling long-term strength test data for soil by eq. 9-31

Ref. number of points (1)	$X = t_f$ (h) (2)	τ (10^5 Pa) (3)	Y (4)	Y^2 (5)	XY (6)	X^2 (7)	τ_{comp} (8)	ε_j (9)	ε_j^2 (10)
1	$3.9 \cdot 10^{-3}$	4.8	–	–	–	–	4.799	10^{-4}	10^{-8}
2	5.3	2.3	2.1	4.5	11.2	28.1	4.25	-0.45	0.2
3	33	2.1	12.1	150	403	1089	3.1	-0.32	0.1
4	429	1.9	148	22100	63500	$184 \cdot 10^3$	1.71	0.12	0.014
5	788	1.7	254	64800	$2 \cdot 10^5$	$621 \cdot 10^3$	1.60	0.11	0.012
6	4735	1.5	1435	$206 \cdot 10^4$	$67.9 \cdot 10^5$	$224 \cdot 10^5$	1.50	0	0
$\Sigma =$	5990	–	1851	$214.7 \cdot 10^4$	$70.6 \cdot 10^5$	$232 \cdot 10^5$	–	–	0.33

TABLE 9-VII

Handling long-term strength test data for soil by eq. 9-32

Ref. number of points (1)	t_f (h) (2)	τ (10^5 Pa) (3)	X (4)	Y (5)	XY (6)	X^2 (7)	Y^2 (8)	τ_{comp} (9)	ε_j (10)	ε_j^2 (11)
1	$3.9 \cdot 10^{-3}$	4.8	—	—	—	—	—	3.88	0.231	0.0491
2	5.3	2.3	1.67	1.14	1.904	2.789	1.300	2.35	−0.021	0.0004
3	33	2.1	3.50	1.50	5.250	12.250	2.250	2.07	0.014	0.0002
4	429	1.9	6.06	1.98	11.999	36.724	3.920	1.81	0.050	0.0025
5	788	1.7	6.67	2.74	18.276	44.489	7.508	1.76	−0.034	0.0012
6	4735	1.5	—	—	—	—	—	1.71	−0.123	0.0151
$\Sigma =$	—	—	17.9	7.36	37.429	96.252	14.978	—	—	0.0685

TABLE 9-VIII

Summary of the results of testing various equations for closeness of fit

Equation number	Equation	Strength, τ, changing with time, t, (10^5 Pa)				Indices of precision:			
		10 years	20 years	50 years	100 years	max. error of the average, ε_j (%)	standard deviation, σ (%)	coefficient of variation, v (%)	correlation coefficient, r
9-22	$\tau = \beta \left[\ln \dfrac{t+1}{T} \right]^{-1}$	1.38	1.34	1.30	1.28	6.0	4.2	1.7	0.92
9-28	$\tau = [(t+1)/T]^\alpha$	1.19 1.5	1.13 1.5	1.05 1.5	1.0 1.5	9.0	6.9	2.8	0.96
9-31	$t = T(\tau_0 - \tau)/\delta(\tau - \tau_\infty)$	1.94	1.94	1.94	1.94	45.0	29.0	12.0	1.01
9-32	$t^\alpha = T^\alpha(\tau_0 - \tau)/\delta^\alpha(\tau - \tau_\infty)$	1.58	1.56	1.55	1.54	12.3	13.1	5.3	0.93

gives the value of $D = -\alpha \ln(T/\delta)$. Computing the values of $B = 0.278$ and $D = 0.595$ by formula 6-12, we find the parameters of the initial eq. 9-32:

$$\alpha = B = 0.278 \quad \text{and} \quad (T,\delta)^\alpha = \exp(-0.595) = 0.55$$

The equation of long-term strength, eq. 9-32, takes in the case under consideration the form:

$$t_f^{0.278} = 0.55 \frac{4.8 - \tau}{\tau - 1.5} \tag{i}$$

where t_f is the time in hours and τ is the strength in 10^5 Pa.

The values of τ_{comp} computed by this formula and their deviations from experimental points are tabulated in columns 9, 10, 11 of Table 9-VII. The indices of precision appear to be $\sigma = 13.1\%$, $v = 5.3\%$ and $r = 0.931$.

Theoretical curve 1 plotted on the basis of eq. d is depicted in Fig. 9-20(b). It departs from the experimental points, as can be seen from Table 9-VII, by not more than 12.3%.

Comparison of formulae. A summary of the results of testing various formulae for closeness of fit is tabulated in Table 9-VIII and depicted in Fig. 9-20. It will be recalled that the adopted experimental value of ultimate long-term strength was $\tau_\infty = 1.3 \cdot 10^5 - 1.35 \cdot 10^5$ Pa.

From Table 9-VIII it can be seen that the best fitting formula in point of the indices of precision and in agreement with the experimental value of τ_∞ is eq. 9-22. Moreover, the way the strength is decreasing according to the formula — slowly, over an extended period — corresponds to reality.

The results computed by formula 9-28 provide for closeness of fit over comparatively short intervals of time. The values of τ_∞ computed at long-term t's appear to be underestimated, for formula 9-28 is of a type that represents the mechanism of loss of strength at a more higher rate.

The results of computation by formula 9-31 are given in two versions: above the line are shown the strength data obtained from using all the experimental points in each case under consideration and below the line are indicated the results of employing the first four points only. In both cases a lack-of-fit is evident.

Therefore, formula 9-31 must be used with great care despite all its advantages that are attractive to some researchers (including the author of this book). If used in a more complex form, as represented by eq. 9-32, this formula provides for a closer fit but the indices of precision will be worse than those obtained for eqs. 9-22 and 9-28. Entering into eq. 9-32 are three parameters (or even four, including τ_0) against two parameters in each of the eqs. 9-22 and 9-28. This is a limitation of eq. 9-32.

In general, eq. 9-22 is recommended for use as the best fitting one. However, when one must take account of the condition $\tau = \tau_\infty$ for $t = \infty$, formula 9-32 may be employed.

Concluding, we refer to Fig. 9-21 which depicts the curves of long-term strength obtained by various authors in testing various soils by various loadings which are rectified on the $1/\tau$ versus $\ln t$ coordinates. As can be seen, the computed data fit well the actual results.

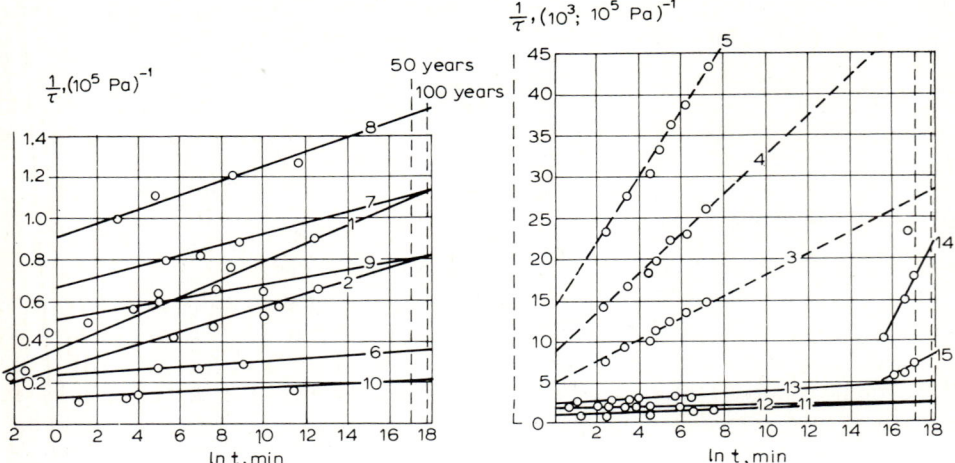

Fig. 9-21. Curves of long-term strength rectified by eq. 9-22. Plotted from test data acquired by various investigators. *1, 2* = frozen sandy loam, $\theta = -0.4°C$, shear along a concrete surface (Vyalov's experiments); *3, 4, 5* = frozen sandy loam, $\theta = -20°C, -10°C, -5°C$, respectively; $W = 25\%$; uniaxial compression; curves plotted on the scale of $1/\varepsilon \cdot 10^{-3}$ (Gorodetsky's experiments); *6, 7* = stiff clay, $W = 23\%$, uniaxial compression (Goldstein's experiments); *8, 9, 10* = illite, montmorillonite and polymineral clays (experiments by Goldstein and Babitskaya); *11, 12, 13* = alluvial clay, $W = 65\%$, compression (experiments by Murayama and Shibata); *14, 15* = London fissured clays, compact, failure of slopes. (Data acquired from field observations by Henkel, Skepton, and Šuklje.)

9.8 EFFECT OF LOADING CONDITIONS

Augmentation of soil strength due to creep

Reporting at the 3rd ICSMFE in 1953, R. Haefeli stressed the fact that the strength of the soil tested for long-term creep had augmented after the test (see Section 1.5: Haefeli et al., 1953). Testing clay soils in a ring shear apparatus under a constant load for two months, he then destroyed some of the specimens by a stepped loading. Comparing the ultimate load at failure with the strength of the specimens not subjected to the stepped loading, he detected that the specimens tested for creep showed a 20–30% augmentation in strength.

Similar results were obtained by Trollope and Chan (1960) from testing silt clay and bentonite for uniaxial compression by a load amounting to 60% of the hypothetically instantaneous strength applied for 10 days and then causing failure of the specimens by a load. The load at failure appeared to be higher than the hypothetically instantaneous strength by 12–15% for the silt clay and by 25% for the bentonite (obtained from 10- to 15-min load applications). Seed et al. (1958), observing an augmentation in the strength of the soil tested for creep, provided convincing evidence that this phenomenon cannot be attributed to an increase in the soil density during the test, for the void ratio and moisture content of the soil remained the same throughout the experiment.

An increase in the strength of soil due to creep was noted by Meschyan (1965) and Zhikhovich (1963) not only in comparison with the hypothetically instantaneous strength but also with respect to the so-called standard strength, i.e. one determined according to the standard procedure of stepped loading with a pause at each step to obtain a conditional stabilization of deformation. So, the strength of a clay soil after a shear test by a load equalling 80 and 90% of the standard strength increased by 30 and 12.5%, respectively. Loess soil showed a 10 and 20% strength increase in response to the loads amounting to 70 and 86% of the standard strength applied for 25 and 45 days, respectively.

Strengthening and weakening of soil

In an effort to establish the mechanism of soil strengthening on the one hand and that of its weakening on the other hand, a series of parallel tests of shear due to torsion were undertaken in 1968 by the present author and N.K. Pekarskaya (see Section 6.6: Vyalov and Pekarskaya, 1968). Involved were two batches of clay soil specimens (kaoline of disturbed structure, $W = 40\%$, $W_P = 38\%$ and $W_L = 58\%$), one batch being tested as usual, and the specimens of the other batch being deformed by a 20-h constant shearing load preparatory to testing. The load applied ($\tau_1 = 66 \cdot 10^2$ Pa) was selected so as to confine the deformation to the stage of attenuating creep (curve 3, Fig. 9-22(a)). The two batches were then tested for creep under various loads greater than τ_1. Depicted in Fig. 9-22(a) are some of the creep curves thus obtained (curves 1 and 2). It can be seen that the preliminary loading contributed to a sharp increase in the resistance of specimens to deformation: the rate of steady flow slowed down by a factor of 4.5 compared with the non-deformed specimens and, correspondingly, the period before failure increased 4.5 times, being 97 min against $t_f = 21$ min for the non-deformed specimens.

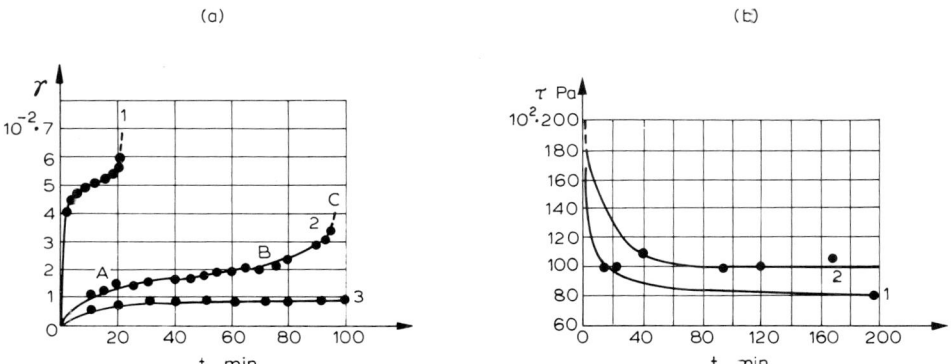

Fig. 9-22. Strengthening of kaolin soil, $W = 40\%$, after creep resulting from shear due to torsion (experiments by N.K. Pekarskaya and S.S. Vyalov): (a) creep curves; (b) curves of long-term strength.

Note, that the relations between flow rates and the times before failure obtained in the example referred to ($\dot{\gamma}_1/\dot{\gamma}_2 = t_{f(2)}/t_{f(1)}$) prove that condition 9-8 is fulfilled.

Two tests, one carried out on the lines described above and the other being a routine long-term strength test, yielded two curves of long-term strength — for the hardened soil and the non-hardened one (Fig. 9-22(b), curves 2 and 1, respectively). The former curve will be located above the latter and, consequently, all kinds of strength, from the instantaneous to the ultimate long-term one, of the hardened soil will be higher than those of the non-hardened one.

The hardening of soils has been observed also during stepped load testing. Non-predeformed specimens failed at the 8th step of load-application at $\tau = 75 \cdot 10^2$ Pa whereas predeformed specimens failed at the 10th step at $\tau = 91 \cdot 10^2$ Pa. The moisture content of the specimens before and after the test changed by not more than $\pm 0.5\%$, i.e. was practically constant.

Thus, a soil in the state of creep (in any case, one at the stage of attenuating creep) displays an increase in strength. This phenomenon is similar to the wear strain hardening of metals i.e. similar to an increase in resistance to deformation due to a repeated loading. However, in a strain-hardened soil, the strength will decrease compared with the instantaneous value after the application of another constant load of sufficient magnitude.

It follows that the strengthening and weakening of soil are not two mutually excluding phenomena and that the process of creep is accompanied by the weakening of internal bonding on the one hand and by the strengthening thereof on the other hand. However, the prevailing development of either phenomenon depends on the stage of creep. At the stage of attenuating creep, strain hardening will prevail; during non-attenuating creep, especially at the stage of progressive flow, weakening of interparticle bonding will dominate, inviting a deterioration of soil strength.

Effect of the rate of load application on the strength of soil

An immediate consequence of the strengthening and weakening of soils is the dependence of soil strength upon the conditions of loading. The strength of soil specimens subjected to a stepped test loading of various durations, Δt, decreases with an increase in Δt, i.e. with the decrease in the rate of load application.

This phenomenon, well known in rheology, was first detected in soils by Casagrande and Wilson in 1951 and observed by other researchers later. Reporting at a conference on earth pressure (Brussels, 1958), Bjerrum et al. (1958) indicated that the strength of soil decreased to 86% of the instantaneous value in response to a reduction in the rate of load application from $10 \cdot 10^{-5}$ 1/h to $3.6 \cdot 10^{-5}$ 1/h during triaxial compression tests under confined conditions.

The triaxial compression tests of Mexico City clay under consolidated undrained conditions, carried out by Alberro and Santoyo (1973) at a constant rate of deformation, showed that the angle of internal (effective) friction decreased from 41° to 34° in response to a change in the rate of 94 to 0.0045% per hour. A similar picture, i.e. a decrease in shearing resistance with a reduction of the rate of deformation, was

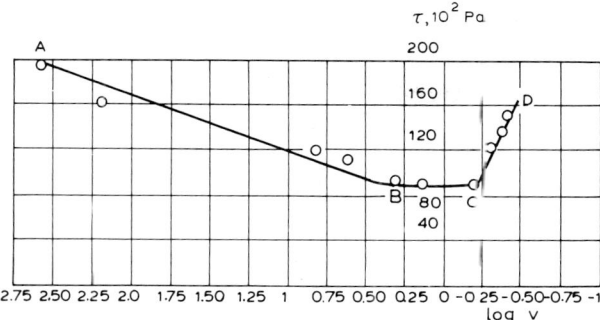

Fig. 9-23. Variations in the strength ($\tau \cdot 10^2$ Pa) of kaolin soil, $W = 40\%$, depending on the rate of load application ($v \cdot 10^2$ Pa/min); shear due to torsion (Pekarskaya's experiments).

observed by Berre and Bjerrum (1973) during triaxial compression tests of a normally compacted clay at a constant deformation rate.

It will be noted that drained tests of a duration in excess of 1 day caused no changes in soil strength, apparently due to the strain-hardening effect of secondary consolidation. Goldstein and Babitskaya (1964) found that the compression strength of a lean polymineral clay decreased by 16.5%, that of montmorillonite clay by 20% and of illite clay by 34% due to an increase in the period of each load application Δt from 5 s to 24 h. Meschyan (1978), changing Δt from 5 s to 12 h, i.e. extending the duration of the test from 1 min to 7920 min, obtained a reduction in the strength of clay by 20–30%.

Thus, the fact that the strength of soil changes with the rate of load application has been proved experimentally.

To establish the nature of the change, Vyalov and Pekarskaya (1968) carried out a study which has shown that the curve depicting the relation between soil strength and the rate of load application is of a concave configuration. It has an extremum which is particularly pronounced when the curve is plotted to a semi-log scale (Fig. 9-23). At an early stage (segment AB of the curve) corresponding to a high rate of load application, the slowing-down of the rate leads to a decrease in the strength which may be assumed to vary directly with the logarithm of loading rate. Upon reaching a certain minimum, the strength becomes constant (segment BC of the curve) and then increases, as the rate of loading goes on decreasing (segment CD).

Thus, the zone of high loading rates, where new interparticle bonds are unable to form to compensate for the broken bonds, corresponds to a loss of strength; in the zone of low loading rates, new bonding can form, providing for an increase in strength. At loading rates of average duration, the strength is a minimum corresponding to the ultimate long-term strength τ_∞.

Instantaneous, standard and long-term strength of soil. Instantaneous strength τ_0 is obtained from a test carried out at a maximum rate of load application and, as pointed out above, is the highest strength. Ultimate long-term strength τ_∞ is the lowest strength. Standard strength varies with the technique of its determination.

It has already been said that standard strength is understood to be the strength of soil obtained from a standard test by a stepped loading. However, the period of load application Δt at each step may vary from a few minutes to several hours or even days. The total duration of the test may thus vary from a few tens to several thousands of minutes. However, as can be seen from Fig. 9-23, the strength will appreciably change in these cases, acquiring a constant value corresponding to τ_∞ only within a certain narrow range of loading rates.

According to Meschyan (1978), the shear tests of soil to the USSR Standard GOST 12248-66 by a stepped loading with a pause at each step until a conditional attenuating deformation (0.01 or 0.005 mm/day) is obtained, yield standard strength values approaching the ultimate long-term strength if the total duration of the test is 1–8 h. However, one must note that the coincidence of the two values is approximate and may vary depending on soil type. In other words, the point is that the strength of soil determined by the standard tests is not invariant.

9.9 TECHNIQUES OF TESTING SOIL FOR LONG-TERM STRENGTH

Testing by constant loading

The most reliable values, τ_∞, of long-term strength are acquired from creep tests of a series of identical specimens. Loading them until failure, we note the elapsed time. The first specimen is tested by a quickly applied load to determine the hypothetically instantaneous strength τ_0. The rest of specimens, totalling at least five or six, are loaded each by a constant load given by:

$$\tau_j = \tau_0 \left(1 - \frac{n}{m}\right)$$

where n is the consecutive number of the specimen and m is a coefficient varying with rheological properties of the soil; $m = 10$ for soils losing much of the strength with time, $m = 20$ to 30 for soils whose strength deteriorates but insignificantly.

From the test data, we plot a curve of long-term strength (see Fig. 9-1) and determine the value of τ_∞ from the curve as exemplified in Section 9.7.

Testing by stepped loading

Required for the test is just one specimen loaded in the same increments $\Delta \tau_1 = \Delta \tau_2 = \Delta \tau_3 = \ldots$ equal to a fraction of the hypothetically instantaneous load τ_0, $\Delta \tau = \tau_0/m$, where m is between 10 and 20. Each load increment is of a duration ensuring a conditional stabilization of the deformation; the index of stabilization is an increase in the deformation not greater than 0.01% in 12 or 24 h.

The loading is increased until at some step a non-attenuating creep is induced, having an either constant or increasing rate of development. This loading and the loadings of two succeeding steps are maintained for at least 3 days in order to be sure that the process of deformation has entered the steady-state stage.

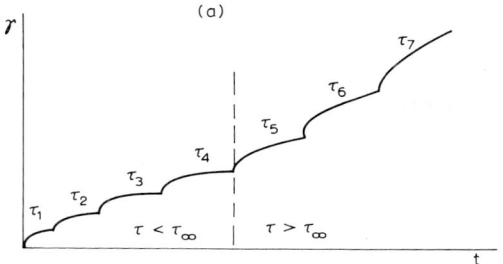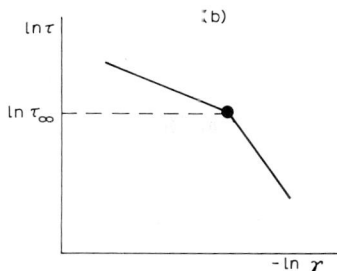

Fig. 9-24. Stepped test load. (a) Curve depicting the progress of deformation. (b) Relationship between stress (ln τ) and strain (ln γ).

On plotting a strain–time curve (Fig. 9-24(a)) from the test data, we determine τ_∞ as the maximum stress causing the strain stabilization. For a check, it is good practice to plot stress–strain curves on the conventional and logarithmic coordinates (Fig. 9-24(b)). The point of inflection of the curve resulting from a sharp increase in the strain at the stage of non-attenuating creep defines the value of τ_∞. The inflection on the log plot is particularly conspicuous.

The strength determined by the above technique is either within the confines of segment BC (Fig. 9-23) or, more probably, outside it, i.e. in segment CD. In this latter case we obtain an overestimated value of τ_∞. However, we must take into account that under real conditions the load is applied to the soil not at once but gradually, in proportion to the rise of the structure. The rate of load application is, in any case, not less than the rate of the stepped loading used during the test. Hence, the strength assessed from a stepped load application with an extended pause at each step will define the ultimate long-term strength corresponding to the condition of a gradual increase in the load applied to the soil during the period of construction.

Shortening the test period. The process of testing soil by stepped loadings may be speeded up if each load increment is applied during a short interval $\Delta t_1 = \Delta t_2 = \ldots$ which is the same at all steps (e.g. from 1 to 24 h). Then, we plot a ln τ–ln γ graph (where γ is the deformation at the end of each step) and determine the long-term strength τ_t for a given Δt from the inflection point of the straight line. The ultimate long-term strength in this case will be $\tau_\infty = k\tau_t$, where k varies between 0.7 and 0.9 depending on the duration of Δt and soil type.

Determining ultimate long-term strength of soil from a flow rate graph. Whatever the test load, a constant or a stepped one, the value of τ_∞ may be determined from the data defining the rate of steady flow. It is known that the ultimate long-term strength is the stress at the boundary between the attenuating creep and steady flow. Hence, if we determine from tests the rates of steady flow corresponding to each value of the stress and plot a τ–$\dot{\gamma}$ curve (see Fig. 5-9), the point of intersection of the curve with the τ-axis defines that stress at which the rate of steady flow $\dot{\gamma} = 0$, i.e. the value of τ_∞.

Testing soil by simultaneous application of shearing and normal stresses

Tests of this kind are carried out on the same lines as those described above except that each series of tests (by either a constant or stepped loading) is conducted under a constant normal loading σ_n, differing, however, from series to series. Consequently we obtain a family of long-term strength curves for various values of σ_n. These are replotted into shear graphs for various values of the time before failure t_f (see Fig. 9-11), which are described by condition 9-1:

$$\tau = c(t) + \sigma_n \tan \phi(t) \qquad (9\text{-}34)$$

The mechanism of changes in cohesion $c(t)$ and in the angle of internal friction $\phi(t)$ with time is determined from the graph in Fig. 9-12. Alternatively, in accordance with eq. 9-22 it may be defined applicable for $c(t)$ and $\phi(t)$ with the parameters β_c, β_ϕ and T_c, T_ϕ. In a special case, $T_c = T_\phi = T$, hence eq. 9-22 takes the form:

$$\tau = \frac{\beta_c + \sigma_n \beta_\phi}{\ln \dfrac{t_f + t^*}{T}} \qquad (9\text{-}35)$$

This case refers to the condition when the parameter $H = c(t) : \tan \phi(t) = \text{const.}$ is independent of time. However, more realistic is the case when cohesion significantly changes with time, $c = c(t)$, whereas the angle of friction changes but little and may be adopted as $\phi = \text{const.}$ In this latter case, eq. 9-22 takes the form:

$$\tau - \sigma_n \tan \phi = \frac{\beta_c}{\ln \dfrac{t_f + t^*}{T}} \qquad (9\text{-}36)$$

Accelerated soil tests

The author of this book has invented an accelerated method of testing soils for long-term strength, using a "dynamometric" apparatus (Vyalov, 1966) shortening the test period.

Schematically, the apparatus is depicted in Fig. 9-25(a). Specimen *1* is loaded by way of a load gauge (*4*) set in a fixed position once an initial stress σ_0 is applied, i.e. the distance $l_0 = l' + l''$ is constant (where l' is the height of the load gauge and l'' is the height of the specimen).

The stress applied to the specimen induces a creep strain, $\varepsilon = \Delta l'/l''$, enabling the load gauge to expand, the stress set up therein consequently decreases. In other words, the soil specimen is simultaneously tested for creep under a changing stress and for relaxation under a changing strain, the stress and strain alterations being interrelated.

During the test, the time-dependent reduction of the arbitrary initial stress σ_0 applied through the load gauge and the time-dependent development of the specimen strain are recorded. The curves thus obtained (Fig. 9-25(b)) define the type of creep equation and the values of the parameters entering into it. For example, adopting the

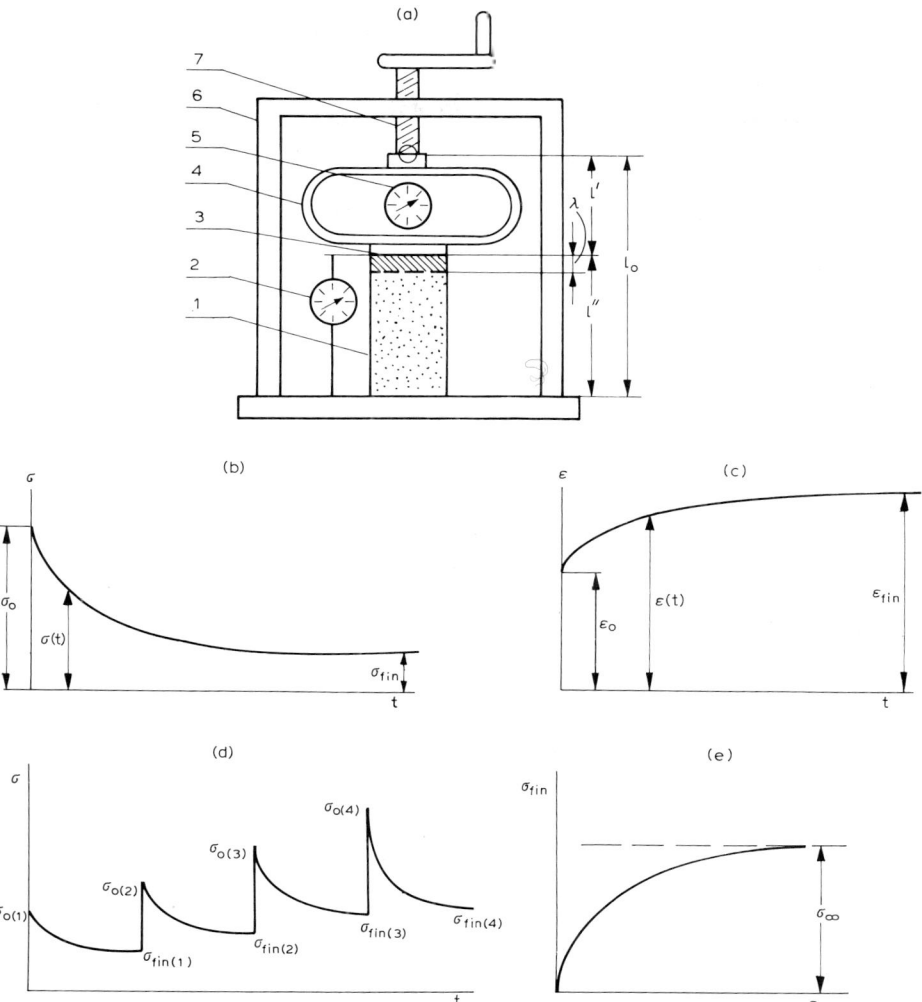

Fig. 9-25. "Dynamometric" apparatus used in long-term strength and creep tests. (a) Schematic diagram of the apparatus: 1 = specimen; 2 = dial indicating specimen strain; 3 = plate; 4 = load gauge; 5 = dial indicating load gauge deformation; 6 = frame; 7 = loading screw. (b, c) Test data obtained from a single loading. (d, e) Test data obtained from stepped loading.

equation of creep in the form given by eq. 5-7:

$$\phi(\varepsilon) = \sigma\psi(t) \tag{9-37}$$

and assuming that $\phi(\varepsilon) = A_0 \varepsilon^m$, we determine the creep function $\psi(t)$ as:

$$\psi(t) = A_0 \frac{[\varepsilon(t)]^m}{\sigma(t)} \tag{9-38}$$

where $\varepsilon(t)$ is the specimen strain as determined from Fig. 9-25(c); $\sigma(t)$ is the changing stress reading of the load gauge as determined from Fig. 9-25(b). The test of fit is carried out in accordance with Sections 6.2 and 6.3.

If an initial stress σ_0 is given which is close to the hypothetically instantaneous strength (as determined in advance), the finite value of stress σ_{fin}, corresponding to the stabilized strain ε_{fin}, will approach the ultimate long-term strength, as strain stabilization may be regarded as the striking of a balance between the applied load and the internal forces of soil resistance:

$$\sigma_\infty = \xi \sigma_{\text{fin}} \quad \text{where} \quad \xi = \left(\frac{\varepsilon_0}{\varepsilon_{\text{fin}}}\right)^m \tag{9-39}$$

In a modified testing technique the stepped initial stress σ_0 is applied in increments each of a duration enabling the finite value of stress σ_{fin} to be attained (Fig. 9-25(d)) and, at the same time, the specimen strain $\varepsilon(t)$ is being recorded. These data provide for plotting a σ_{fin}–ε_{fin} graph (Fig. 9-25(e)) from which we determine the type of function, $\phi(\varepsilon)$, and the value of σ_∞.

Since the process of relaxation progresses at a substantially greater pace than creep (the period of relaxation $T_r \ll$ the period of aftereffect T_p), the testing by means of the "dynamometric" apparatus takes much less time than the conventional tests described above.

Spherical indenter test. The long-term cohesion of soil may be effectively determined by the spherical indenter test under a constant load P suggested by N.A. Tsytovich in 1947 (see Section 1.5: Tsytovich, 1975, 1976). The value of cohesion is computed from the test by the formula:

$$c(t) = 0.18 \frac{P}{\pi d S(t)} \tag{9-40}$$

where d is the diameter of the spherical indenter; $S(t)$ is the depth of indentation (settlement) varying with time from an initial value S_0 to the finite, stabilized value corresponding to which are the hypothetically instantaneous cohesion c_0 and the ultimate long-term cohesion c_∞, respectively.

Eq. 9-40 was obtained from a strict solution (Ishlinsky, 1944) of the problem with respect to an ideally plastic St. Venant body, appearing, consequently, to be applicable to soils devoid of internal friction. However, later it was shown (Vyalov and Tsytovich, 1956) that the value of c, as determined by the above formula, may be regarded as an equivalent cohesion, c_{equiv}, taking account of both cohesion and internal friction. In particular, the ultimate load sustained by a foundation (circular or square in plan) may be determined for cohesive soils by the formula:

$$p_{\text{ult}} = 5.55 \, c_{\text{equiv}} = \frac{P}{\pi d S_\infty}$$

9.10 REFERENCES

Alberro, I. and Santoyo, E., 1973. Long-term behaviour of Mexico City clays. Proc. 8th ICSMFE, Moscow, v. 1.1.
Artemov, V.G. and Vodop'yanov, V.L., 1970. On the prospect of using the law of summation of fatigue damages in studying long-term strength of rocks. In collective volume: Problems of Rock Rheology. Naukova Dumka, Kiev (in Russian).
Avershin, S.G., Yalymov, N.G. and Stepanov, V.Ya , 1970. Pillar calculations with allowance for rheological properties of soils. In collective volume: Problems of Rock Rheology. Naukova Dumka, Kiev (in Russian).
Babitskaya, S.S. and Turovskaya, A.Ya., 1975. Stages of deformation and assessment of rock strength. In: Problems of Geotechnics, no. 24. Dnepropetrovsk Railway Eng. Inst. (in Russian).
Berre, T. and Bjerrum, L., 1973. Shear strength of normally consolidated clays. Proc. 8th ICSMFE, Moscow, v. 1.1.
Bishop, A.W. and Henkel, J., 1957. The Measurement of Soil Properties in the Triaxial Tests. Arnold, London.
Bishop, A.W. and Little, A.L., 1967. The influence of the size and orientation of the sample on the apparent strength of the London Clay at Meldon. Proc. Geotech. Conf., Oslo, 1967.
Bjerrum, J., Simons, N. and Torbloa, I., 1958. The effect of time on the shear strength of a soft marine clay. Proc. Conf. Earth Pressure, Brussels, 1958, v. 1.
Bolotin, V.V., Goldenblat, I.I., et al., 1973. Structural Mechanics, The State of the Art and Outlook. Stroiizdat, Moscow (in Russian).
Chudnovsky, A.I., 1973. On failure of macro-bodies. In collective volume: Research into Elasticity and Plasticity. Leningrad State Univ., no. 9 (in Russian).
Coates, D.F. and McRostie, G.C., 1963. Some deficiencies in testing Leda Clay. Proc. ASTM, STP, no. 36.
Fisenko, T.L., 1964. The Role of Strength and Creep of Rock in Shaping Landslides. Kiev State Univ., Kiev (in Russian).
Goldstein, M.N., 1957. Creep and long-term strength of clay soils. In collective volume: Proc. Conf. Eng. Geol. Prop. Rocks and Methods Study, Moscow (in Russian).
Goldstein, M.N., 1964. To the problem of soil strength. SMFE, no. 4 (in Russian).
Goldstein, M.N. and Babitskaya, S.S., 1959. Methodology of determining long-term strength of soils. SMFE, no. 4 (in Russian).
Goldstein, M.N. and Babitskaya, S.S., 1964. On long-term strength of cohesive soils. In collective volume: Problems of Geotechnics, no. 7. Transport Publ. (in Russian).
Hvorslev, M.J., 1960. Physical components of the shear strength of saturated clays. ASCE Res. Conf. on Shear Strength of Cohesive Soils, Univ. Colorado, 1960.
Ishlinsky, A.Yu., 1944. Axisymmetrical plasticity problem and Brinell test. Appl. Math. Mech., v. 8 (in Russian).
Kabakhadze, E.I., Shishniashvili, N.N. and Serb-Serbina, M.E., 1957. Tixotropic and structural-mechanical properties of suspensions and ascagel depending on composition of exchange complex. Colloid J., 9(1) (in Russian).
Kanji, M.A. and Wolle, C.M., 1977. Residual strength — new testing and microstructure. Proc. 9th ICSMFE, Tokyo, v. 1.
Liam Finn, W.D. and Shead, D., 1973. Deformation and creep rupture of an undisturbed sensitive clay. Proc. 8th ICSMFE, Moscow, v. 1.
Meschyan, S.R., 1965. On long-term resistance of clay soils to shear. News Acad. Sci. Armenian SSR, Phys. Math. Ser., XXVIII, no. 3 (in Russian).
Meschyan, S.R., 1978. Initial and Long-Term Strength of Soils. Nedra, Moscow (in Russian).
Mogilevskaya, S.E., 1960. Problems of long-term strength and deformability of loess soils used as foundation for water development works. News, All-Union Hydraul. Eng. Res. Inst., Leningrad, no. 64 (in Russian).
Rebinder, P.A., 1966. Structure-shaping processes in dispersed systems. In collective volume: Physico-Chemical Mechanics of Soils and Structural Mechanics. FAK Publ., Tashkent (in Russian).

Saito, M., 1965. Forecasting the time of occurrence of a slope failure. Proc. 6th ICSMFE, Montreal, v. 2.
Saito, M., 1969. Forecasting time of slip failure by tertiary creep. Proc. 7th ICSMFE, Mexico City, v. 2.
Saito, M. and Uezawa, H., 1961. Failure of soil due to creep. Proc. 5th ICSMFE, Paris, v. 1.
Seed, H.B., McNeil, R. and De Guenin, J., 1958. Increased resistance to deformation of clay caused by repeated loading. J. Soil Mech. Found. Div., Proc. ASCE, v. 84, SM2.
Shibata, A. and Karube, K., 1969. Creep rate and creep strength of clays. Proc. 7th ICSMFE, Mexico City, v. 1.
Stavrogin, A.N., 1968. Experimental studies of creep and durability of rocks. Proc. Coordinating Conf. Hydraul. Eng., Issue 38: Study of Rheological Properties. Energia, Leningrad (in Russian).
Stepanenko, G.P., 1968. Rheological properties of brown and variegated clays. Proc. Coordinating Conf. Hydraul. Eng., Issue 38: Study of Rheological Properties. Energia, Leningrad (in Russian).
Tiedemann, B., 1937. Über die Schubfestigkeit bindiger Böden. Bautechnik, v. 45.
Trollope, D. and Chan, C.K., 1960. Soil structure and step strain phenomenon. J. Soil Mech. Found. Div. Proc. ASCE, v. 86, no. SM2.
Vyalov, S.S., 1956. Stress-strain relationship in frozen soils with allowance for time factor. Trans. USSR Acad. Sci., v. 108, no. 6 (in Russian).
Vyalov, S.S. and Pekarskaya, N.K., 1968. Long-term strength of soils. SMFE, no. 3 (in Russian).
Vyalov, S.S. and Tsytovich, N.A., 1956. Assessing bearing capacity of cohesive soils from spherical indenter tests. Trans. USSR Acad. Sci., v. 111, no. 6 (in Russian).
Zaretsky, Yu.K. and Vyalov, S.S., 1971. Problems of structural mechanics in clay soils. SMFE, no. 3 (in Russian).
Zhikhovich, V.V., 1963. On creep, "standard" and long-term strength of dense meotic clays. SMFE, no. 4 (in Russian).

Chapter 10

THE KINETIC THEORY OF SOIL STRENGTH AND SOIL CREEP

10.1 SOIL DEFORMATION AS A THERMO-ACTIVATED PROCESS

Applicability of kinetic theory to soils

The preceding chapters were devoted to discussing phenomenological theories of creep based on the data of macro-experiments. Outlined in this chapter is the physical theory of creep and long-term strength of soils taking into account the kinetic concept of deformation and long-term failure and based on data acquired from the study of micro-processes occurring in soils.[*1]

In accordance with the concepts of kinetic theory (see Section 7.6), the processes of deformation and failure are thermo-activated processes, in the course of which elemental particles activated by an external force escape the energy barrier and assume new positions of equilibrium.

Such a process bears a fluctuating nature. It arises from a difference between the heat energy of individual particles and the average amount of energy because the heat energy is non-uniformly distributed between the particles due to their random thermal motion. Under such circumstances, the pattern of distribution of the energy between the elemental particles may be described in terms of the Boltzmann distribution law.

However, before applying this law to a soil system, one must decide what should be regarded as its elemental particle.

Apparently, thermo-fluctuating processes take place both in the lattice of mineral particles and in the molecular structure of the liquid film interlinking the particles. Since the loadings commonly sustained by soils are only a small fraction of the intracrystalline forces, they cannot activate the atoms of the lattice to a point causing their oriented displacement. Such processes are likely to occur only in massive crystalline rocks exposed to stresses of considerable magnitude. In loose rocks, deformation is induced by low loads which are incapable of activating the atoms in mineral particles, but are quite adequate to activate the molecules of bound water (and, in some cases, even the cementitious material bonding the mineral particles).

However, it is incorrect to treat soil deformation as an outcome of the migration of bound water at the molecular level, for taking part in deformation are not only the water films but the mineral particles as well. Therefore, it is appropriate to regard soil

---

[*1] The source materials used in writing this chapter were author's contributions published in the Proceedings of the 7th ICSMFE (1969), 2nd International Conference on Permafrostology (1973), All-Union Symposium on Soil Rheology (1975) and various articles written in co-operation with Yu.K. Zaretsky, R.V. Maximyak, N.K. Pekarskaya and published between 1968 and 1973.

deformation as the displacement of the structural elements, i.e. mineral particles and their micro-aggregates comprising the soil system.

Kinetic nature of soil deformation and failure

Schematically, the micro-structure of soil may be regarded as a random assemblage of bound water-enshrouded solid particles, interlinked by the forces of mutual interaction (orientation of particles is not excluded), and voids filled with water and air.

Interparticle bonds, the nature of which was discussed in Chapter 2, impart positional stability to micro-structural elements whose state of equilibrium is defined by the interparticle distance and corresponds to the minimum of potential energy. From this point of view, the curve of the potential binding energy of a soil particle is analogous with the curve of the potential binding energy of molecules in a liquid medium (see Fig. 7-14).

To displace particles, their equilibrium must apparently be disturbed. The displacement is possible only because in a soil structure there are voids whose dimensions are commensurate with the dimensions of particles and their aggregates. This displacement is analogous with the movement of molecules in a liquid medium, which is possible due to the "holes" in the structure. However, to disturb particle equilibrium and induce particle motion, the particle must be activated so that the energy it acquires is higher than the binding energy of the neighbouring particles.

By analogy with a thermo-activated process, particle displacement may be regarded as the escaping of an energy barrier by the particle, and the energy it acquires may be treated as an "activation energy". In what follows we use this term with reference to soil particles without quotation marks in spite of certain ambiguity, for it is applied to the energy possessed by atoms and molecules.

The binding energy of individual elements in a soil system varies over a wide range, depending on shape, size and location of the particles. Random particle arrangement and the size, which is smaller than any volume of soil under consideration, suggest the employment of the statistical approach. Consequently, we may apply the Boltzmann distribution law on the assumption that the number of activated particles whose energy U is equal to, or greater than, the average amount of the binding energy between the particles is defined by exponential law (eq. 7-104). Assuming further, that the average time, t, of particle equilibrium varies inversely with the number of activated particles and that their displacement is brought about by an external force and is time-dependent, we may write by analogy with eq. 7-105:

$$t = t_0 \, e^{U(\tau,t)/k\theta} \tag{10-1}$$

where the notation is the same as in eq. 7-105: $t_0 = h/k\theta$ is the period of the thermal oscillations of elemental particles (s); θ is the temperature (K); h is the Planck constant (J s); k is the Boltzmann constant (J/K); and $U(\tau, t)$ is the activation energy treated as a quantity varying with the stress applied τ and the period of stress application t. However, a point to be noted is that eq. 7-105 has been derived from studying the behaviour of elemental particles (atoms, molecules) in the force and temperature fields whereas eq. 10-1 describes the behaviour of soil particles interlinked by a bonding of a more complex type.

Taking this into account and proceeding from the analogy with eq. 7-105, the activation energy of soil particles in eq. 10-1 is to be treated as an averaged value of energy $\langle U \rangle$ referred to a characteristic volume of soil, as has been done in eq. 2-10. As far as the constants k and h are concerned, one must bear in mind that using their values in eq. 10-1 at $k = 1.38 \cdot 10^{-23}$ J/K and $h = 16.625 \cdot 10^{-31}$ J s is an assumption conditional to a certain extent.

Kinetics of deformation and structural changes

Consider the dependence of the activation energy U and, consequently, of the activation process as a whole upon the applied stress τ and time t. The initial activation energy, which we denote by U_c, corresponds to the original binding energy of the soil structure elements which is determined by their initial disposition relative to each other, shape, size, orientation, chemical composition of the solid and liquid phases of soil system, interparticle distance, etc., i.e. by all the factors giving rise to the force field which were considered in Sections 2.4 and 2.5. In other words, the amount of the initial activation energy U_0 corresponds to the binding energy of soil particles and is determined by the physical properties of the soil in its state before deformation.

External forces applied to soil break and dislocate interparticle bonds with the result that the binding energy of particles changes. This means that, before every successive displacement, the particles must acquire an activation energy differing from the preceding energy. This new energy may be higher than U_0 if the deformation is accompanied by strain hardening (e.g. the soil is compacted) or lower than U_0 if the soil becomes looser (for example, a non-attenuating creep is induced). In any case, because of the displacement developing with time, U_0 changes with time as well.

Note, that eq. 10-1 takes care of pure shear. In the general case, the activation energy is the function of all stress tensor components, as is shown in Chapter 11.

An alteration of the activation energy, due to external force, is associated primarily with changes in the micro-structure of soil. Experimental evidence indicates that these changes consist of aggregate disintegration, particle dislocation, rearrangement and reorientation, etc., and also include the development of structural defects in soil.

Structural defects in soil

The term "defect" is understood to define, in this case, any point of weakness, discontinuity or disturbance of the bonding in the soil skeleton. No matter how random the arrangement of mineral particles in a soil system may be, each particle is linked to the neighbouring particles by at least two bonds. If only one bond is present, however, this is the so-called discontinuity in the soil skeleton, i.e. a disturbance of the soil structure.

Such disturbances are likely to occur when the interparticle distance is greater than the range of the intermolecular forces. According to A.K. Larionov, 1968 (reference in Section 2.9) this distance is determined by the condition $\delta_{\lim} \geqslant x_0$, where x_0 is the range of Van der Waals forces (4–10 Å). Defects of comparatively dense clay soils manifest themselves in the form of micro-cavities, voids, micro-cracks, cleavage, etc. It will be recalled that, in accordance with the definition adopted above, voids containing free water and gases do not come under the category of defects, being inseparable constituents of the three-component soil system.

Apparently, the term "defect of soil structure", as defined above, differs from the term "defect of solid lattice", the analogy between the two terms being conditional.

The defects arising in soil are the main factors provoking long-term failure; ranking first among the defects in point of effect, as is shown later, are micro-cracks. Thus, in order to determine the mechanism of changes in the activation energy due to deformation and long-term failure, and establish the kinetic nature of these processes, it is necessary, firstly, to find the changes in micro-structure brought about by deformation, and define the pattern of these changes.

10.2 CHANGES IN THE MICRO-STRUCTURE OF SOIL DUE TO DEFORMATION

The changing micro-structure of soils was examined by R.V. Maximyak and N.K. Pekarskaya under the author's guidance, at the Research Institute for Foundations and Underground Structures, between 1967 and 1976. In testing identical specimens of clay soil for creep, the investigators also focused their attention on the changes in micro-structure at various stages of deformation.

The specimens were prepared from hard, stiff and firm monomineral clays (Glukhovka kaolins) and from a polymineral (Jurassic) clay. The kaolin had a moisture content $W = 38-40\%$, a liquid limit $W_L = 58\%$ and a plastic limit $W_P = 38\%$. The moisture content of the Jurassic clay was $W = 38\%$ at $W_L = 50\%$ and $W_P = 26\%$.

The creep tests were conducted under the conditions of pure shear in torsion, using hollow cylindrical specimens. The constant shearing loads varied from test to test, from those producing instantaneous failure, to the loads inducing only attenuating deformation. The methodology of the tests is set forth in an article by R.V. Maximyak (1968).

The curves of the shearing deformation γ plotted against time on the log scale and the curves of the deformation rates $\dot{\gamma}$ plotted against time, as obtained from a series of tests, are presented in Fig. 10-1. It can be seen that low stresses induced an attenuating creep, and high stresses gave rise to a non-attenuating process with the three stages of flow: non-steady, steady, and progressive, ending in failure.

The creep tests were interrupted from time to time, to take soil samples for microscopic investigations of the structural changes at various stages of deformation, and under different loads.

Soil structure was investigated under optical and electron microscopes. The

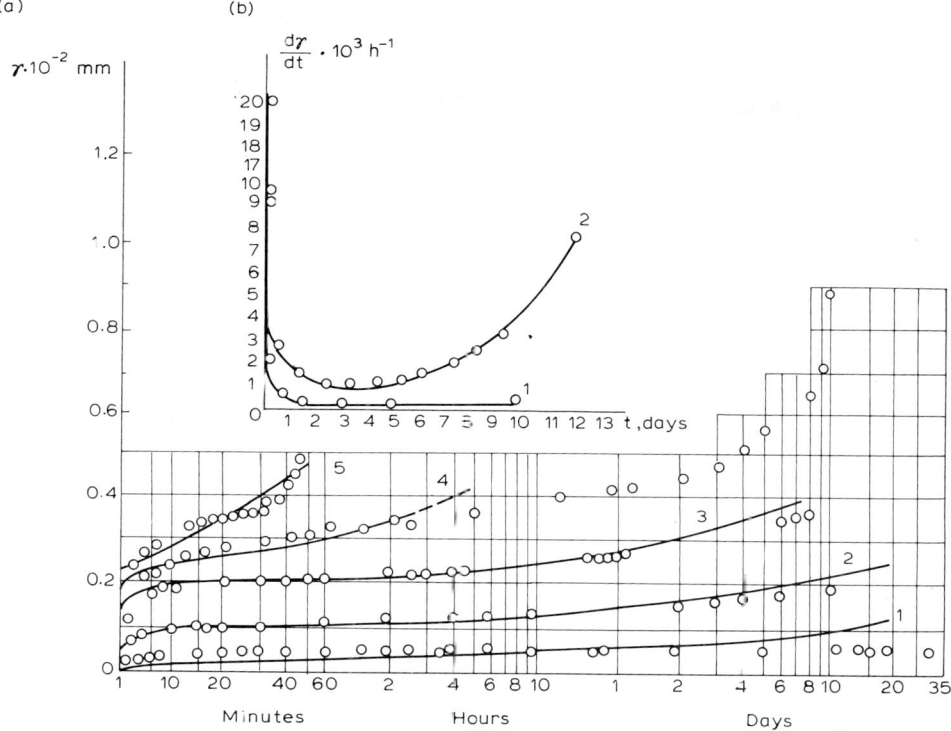

Fig. 10-1. Creep curves for kaolin soil in creep. $W = 38\%$ (Maximyak's experiments). (a) Progress of deformation γ with time for the following loadings τ: $1 = 83 \cdot 10^2$ Pa; $2 = 90 \cdot 10^2$ Pa; $3 = 100 \cdot 10^2$ Pa; $4 = 135 \cdot 10^2$ Pa; $5 = 165 \cdot 10^2$ Pa. (b) Strain rate versus time curves plotted by eq. 10-60 for the following values of τ: $1 = 60 \cdot 10^2$ Pa; $2 = 180 \cdot 10^2$ Pa.

changes in the structure, such as particle disintegration and reorientation, were examined by means of conventional petrographic techniques, using sections. To trace the development of defects and other structural disturbances, the soil was impregnated with a liquid polymer (in vacuum) stabilizing the defects after polymerization. The structural changes were evaluated in terms of the degree of structural damage (defect density) ω and the degree of orientation Ω:

$$\omega = \frac{s_{\text{def}}}{s} 100\% \qquad \Omega = \frac{s_{\text{orient}}}{s} 100\% \tag{10-2}$$

where s_{def} is the area of section under the structural defects; s_{orient} is the area of section occupied by the particles oriented in the direction of shear; s is the aggregate area of the section at right angles to the direction of shear.

The quantities mentioned were measured by means of an optical microscope with an integrating stage. Strictly speaking, the values of ω and Ω were to be referred to a unit volume of soil but, taking into account that the pinpointing of structural changes in three dimensions was a formidable problem, the changes had been referred to the area of the section.

Soil cracks may be classified into sub-microscopic cracks (in the Å-size range), microscopic cracks (in the μ-size range), and macroscopic cracks (in the mm-size range).

This classification stems from the faculty of perceiving the defects. So, sub-microscopic cracks can only be perceived under an electron microscope. In the case of microscopic cracks an optical microscope is required, and macroscopic cracks are visible to the naked eye. On the other hand, such classification is not devoid of a physical meaning. Sub-microscopic cracks are initial, incipient cracks; microscopic cracks are propagating cracks, and the macroscopic ones are the cracks at failure (main cracks).

In evaluating the degree of structural damage during the experiments under consideration, only the micro-cracks were taken into account. Consequently, the values quoted must be treated as having a relative meaning, but the increments in these values are conducive to solving the main problem — assessing the mechanism of crack formation.

The initial soil structures were of both perfect and random arrangement, 60–70% of the volume being occupied by the aggregates composed of discretely oriented parts and 40–30% by the randomly oriented clay particles filling the space between the aggregates. At the contacts between the aggregates there were observed voids and cavities — the weakest points of the structure. The defects were readily perceptable under an optical microscope giving a magnification of $\times 40$ and upwards. The degree of damage of the initial structure of the koalin was $\omega_0 = 22-25\%$ and that of the Jurassic clay, $\omega_0 = 20-22\%$.

The first stage of the microscopic studies carried out by Pekarskaya and Sheveleva (1971) has revealed that structural changes are significantly influenced by the type of deformation, being different not only in the case of attenuating and non-attenuating deformation, but varying also with the stages of the non-attenuating process. Further investigations have enabled Maximyak (Vyalov et al., 1970) to determine the quantitative mechanism of structural changes and establish the criteria of failure.

Structural changes in soil due to creep

A summary of the structural changes noted in a deformed soil, and expressed in terms of the degree of damage (defect density) and the degree of reorientation at various stages of creep, will be found in Table 10-I; these changes are illustrated in Fig. 10-2 (after R.V. Maximyak).

It can be seen from Table 10-I that the attenuating creep reduces before all the number of structural defects. Micro-cracks tend to close, and cavities and voids contract and expand in the direction of shear. The soil compacts, and new inter-particle bonds are formed instead of the broken ones; consequently, the deformation attenuates in time. The bulk of structural defects diminished during an early period of the deformation when this was at the peak of the development. So, during one of

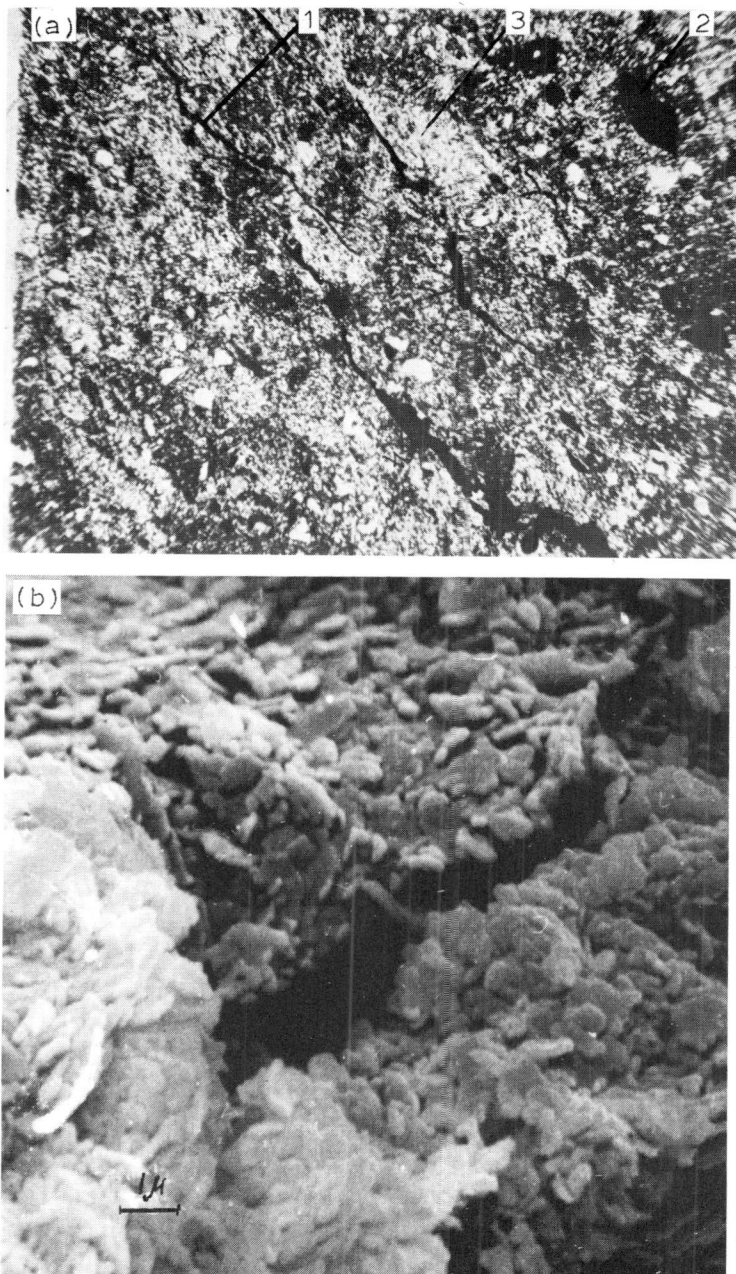

Fig. 10-2. Structural changes in Jurassic clay ($W = 32\%$) due to creep (at the stage of failure) (Maximyak's experiments): (a) magnification × 40; (b) magnification × 500; *1* = micro-cracks; *2* = voids; *3* = cavities.

TABLE 10-I

Structural changes in kaolin ($W = 40\%$) resulting from deformation

Stage of deformation	Stress (τ, 10^2 Pa)[*1]	Test period (t, h)	Degree of damage (ω, %)	Degree of orientation (Ω, %)
Initial condition	0	0	24.1	Dominating orientation absent
Attenuating creep	91	144	21.3	Dominating orientation absent
	83	408	23.1	
	83.6	768	21.1	
	75	528	20.3	
Steady flow	100	6	28.5	21.9
	100	28	27	—
	100	55	33.8	22.3
	100	72	34.8	25
	100	144	35	41.1
	133	1.2	—	18.4
	133	6.3	—	24.3
	125	21	27.6	—
	133	72	34.2	28.3
	117	168	34.6	—
	125	192	34.2	—
Progressive flow	166	0.03	36.8	Dominating orientation absent
	160	0.15	—	14.3
	138	1.2	36.3	18.4
	135	3.8	37	—
	135	4.2	37	—
	133	984	—	65.8
	100	840	37.2	50

[*1] 10^2 Pa = 1 g/cm^2

the experiments, the initial degree of structural damage $\omega_0 = 25\%$ decreased to 22% in 2 days and to 20% in 6 days.

Attenuating creep caused no noticeable changes in the degree of particle reorientation, Ω, at least over a period of 770 h. However, not excluded is the possibility that the value of Ω may change in the course of a longer period of deformation.

Non-attenuating soil deformation sharply changed the nature of structural transformations, significantly varying with the period of deformation which was influenced, in its turn, by the load.

In the case of a rapid soil failure, no particle reorientation took place, due to lack of time, and the structure remained practically unchanged (see Table 10-I, $t = 0.03$ h). Nevertheless, micro- and macro-cracks were triggered, and the degree of

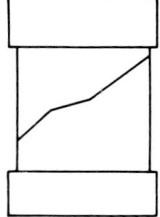

Fig. 10-3. Brittle failure of specimen in stiff Jurassic clay after shear test (in torsion).

structural damage increased, for example, from 24.1 to 36.8% for the time before failure $t = 0.03$ h.

During a longer period of soil deformation, the intensity of structural changes varied with the stage of creep. Thus, at the beginning, when the creep was confined to the unsteady stage I, the structure suffered insignificant changes: slight shrinking of cavities, and early particle reorientation at the weakest points between the aggregates. Following the onset of stage II (steady flow), more conspicuous structural changes came to notice (see Fig. 10-6): incipient aggregate disintegration and reorientation of particles tending to turn with their basal surfaces in the direction of shear; the oriented areas noticeably increasing in number. Along with the continuing "healing" of defects, fresh structural disturbances in the form of micro-cracks came into being. Triggering at the weakest points, next to cavities, they frequently propagated from one cavity to another.

Further development of the deformation, and its transition to the stage of progressive flow, induced a more pronounced aggregate disintegration and particle reorientation. However, a salient feature of the stage of progressive flow was an intensive propagation of micro-cracks (see Figs. 10-6 and 10-2). Eventually, these merged into a main crack, causing failure of the specimen. Note that the failures involving the formation of main cracks (Fig. 10-3) were observed in clay soils of the hard, stiff and firm type (kaolin and Jurassic clay).

The structural changes resulting from non-attenuating creep described above are vividly illustrated in Table 10-I.

The pattern of changes in the degrees of structural damage and orientation under the conditions of non-attenuating deformation induced by a constant loading $\tau = 100 \cdot 10^2$ Pa is presented in Fig. 10-4.

Regularity of structural changes in soil

First it will be pointed out that, according to the experimental data, the degrees of orientation observed during the deformation caused by different loads of the same duration have appeared to be roughly the same. Hence, it may be assumed in the first approximation that the process of orientation is not influenced directly by the stress applied and is governed, initially, by the duration of deformation (non-attenuating).

However, the process of crack triggering is a function of both the period of

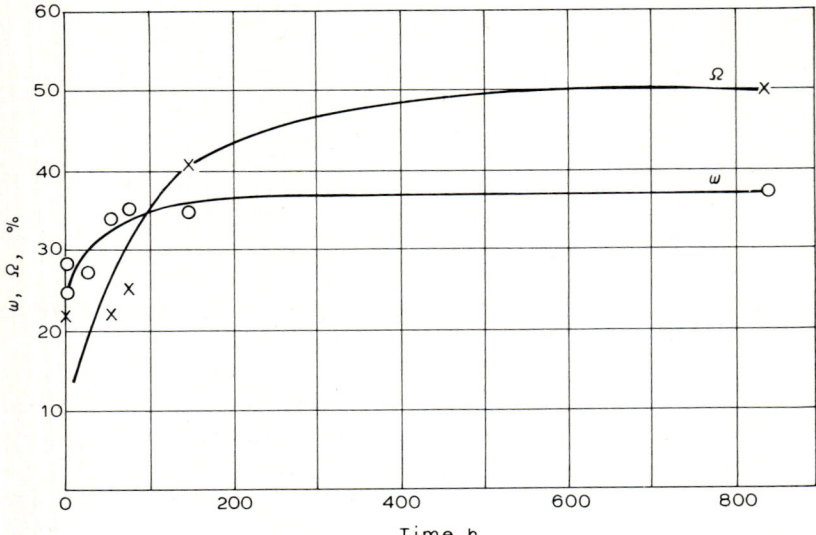

Fig. 10-4. Curves depicting changes in the degree of orientation Ω and the degree of damage ω in stiff kaolin ($W = 38\%$) due to non-attenuating creep under constant load $\tau = 100 \cdot 10^2$ Pa.

deformation and the level of stress. In other words, $\Omega = \Omega(t)$ and $\omega = \omega(\tau, t)$. The type of these functions defining the pattern of structural changes is determined by the expressions:

$$1 - \omega = (1 - \omega_0)(t + 1)^{-\kappa_1 \bar{\tau}} \tag{10-3}$$

$$1 - \Omega = (1 - \Omega_0)(t + 1)^{-\kappa_2} \tag{10-4}$$

where κ_1 and κ_2 are parameters and $\bar{\tau}$ is a dimensionless stress function, the meaning of which will be considered below.

Here and in what follows t is understood to be a dimensionless quantity equalling t/t^*, where t^* is a parameter measured in units of time; one may adopt that $t^* = 1$.

The quantities $1 - \omega$ and $1 - \Omega$ define the undamaged and non-oriented unit areas of soil, respectively, ω and Ω being expressed here in fractions of unity; correspondingly, $1 - \omega_0$ and $1 - \Omega_0$ are the same quantities for the initial condition. Note that $\Omega > 0$, because the orientation of some of the particles will coincide with the direction of the would-be shear even if their arrangement is a random one.

Eqs. 10-3 and 10-4 are presented graphically in Fig. 10-5 which has been produced by replotting the curves of Fig. 10-4 in an appropriate way. The appropriate relativity of eqs. 10-3 and 10-4 is proved by the fact that on transforming these formulae into:

$$\ln(1 - \omega) = \ln(1 - \omega_0) - \kappa_1 \bar{\tau} \ln(t + 1) \tag{10-5}$$

$$\ln(1 - \Omega) = \ln(1 - \Omega_0) - \kappa_2 \ln(t + 1) \tag{10-6}$$

the log plots of the above curves yield straight lines (the lower graphs in Fig. 10-5).

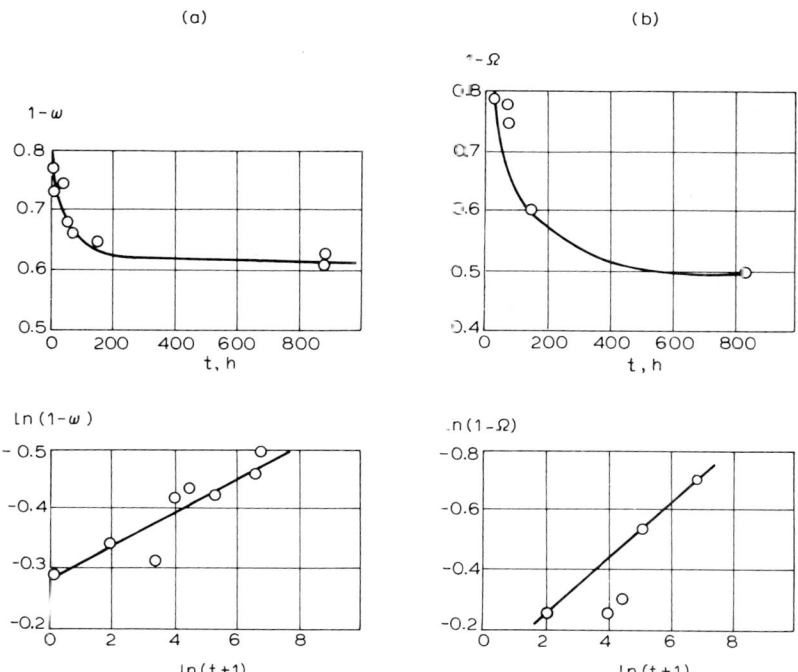

Fig. 10-5. Ordinary and log plots of structural changes in stiff kaolin soil ($W = 38\%$) due to creep ($\tau = 100 \cdot 10^2$ Pa). (a) Changes in damage. (b) Changes in orientation.

Criterion of soil failure

The results of micro-structural investigations have been helpful in revealing the effect of structural changes on the process of soil deformation and long-term failure. In establishing the direct connection between the degree of structural damage and long-term soil failure, R.V. Maximyak (in Vyalov et al., 1970) has found that failure is always associated with the same degree of structural damage ω_f (Table 10-II).

This finding is of particular significance, indicating that the degree of structural damage, ω_f, at the instant of failure may be regarded as a critical soil damage. The value of ω_f is a constant for the soil of a given type. So, in the case of the kaolin examined above, $\omega_f = 37.5 \pm 2\%$; for the Jurassic clay, $\omega_f = 40.9 \pm 2\%$.

Hence, the degree of damage resulting from an accumulation of critical damages, ω_f, may be regarded as the criterion of long-term failure. This value is constant for the soil of a given type, $\omega_f = $ const.

We point out once more that the data presented above referred to clay soils varying in consistency from hard to firm. However, all considerations may apparently be applied to soft soils. The criterion of failure of the soils having fluid consistency is a problem requiring special investigation. One may assume that the failure of these soils, obviously of viscous nature, is caused by the displacement of particles over a

TABLE 10-II

Degree of structural damage (ω_f, %) suffered by soil at the instant of failure

Kaolin Stress applied (τ, 10^2 Pa):						Jurassic clay Stress applied (τ, 10^2 Pa):			
100	133	160	166	180	200	416	425	460	550
36.9	39.1	36.8	37	38.4	38	40.3	43	39.9	41
37.3	37.1	36.8	37.2	—	36	41.1	41.6	41.4	—
39.7	36.3	37.4	38.5	—	37.5	—	39.6	—	—
40.5	37	35.9	—	—	37.3	—	—	—	—
Average values:						Average values:			
38.6	37.3	36.7	37.6	38.4	37.2	40.7	41.4	40.6	41
$\omega_{f(av)} = 37.5$						$\omega_{f(av)} = 40.9$			

distance greater than the range of interparticle forces. The displacement may be treated as the breaking of bonds on the lines analogous with the breaking of bonds in structuralized soils described above.

In concluding it will be noted that structural defects play an important part in the phenomenon of dilatancy of clay soils (see Section 12.1) and that, as shown by E.P. Shusherina, 1966 (reference in Section 5.7), the nature of volumetric deformation is associated with the propagation of micro-cracks.

Mechanism of deformation and long-term failure in soils

In summary, we may try to imagine the mechanism of deformation and long-term failure in soils. As already shown by the present author (Vyalov, 1963), at the heart of soil creep are two mutually excluding phenomena — the hardening and the softening of soil. If the former (hardening) is dominating, the deformation dies out, causing no failure. Should softening prevail, a non-attenuating creep leading to failure is induced in the soil.

The phenomena of soil hardening and softening result from structural changes which, in turn, are caused by the load applied. The changes can be represented graphically by a diagram depicted in Fig. 10-6. When an external load is applied to a soil, the stress concentrations produced here and there overstrain and break some of the bonds. The breakage commonly occurs at the weakest points of the structure; as a result, the particles with damaged bonds tend to displace into another, more stable position.

At an early stage, the displacing particles form a more closely packed arrangement; simultaneously new bonds are formed, while the structural defects — cavities, voids, micro-cracks — shrink and decrease in number. The "healing" of defects is accompanied by the shrinking of soil volume, i.e. by the compaction of soil. Consequently,

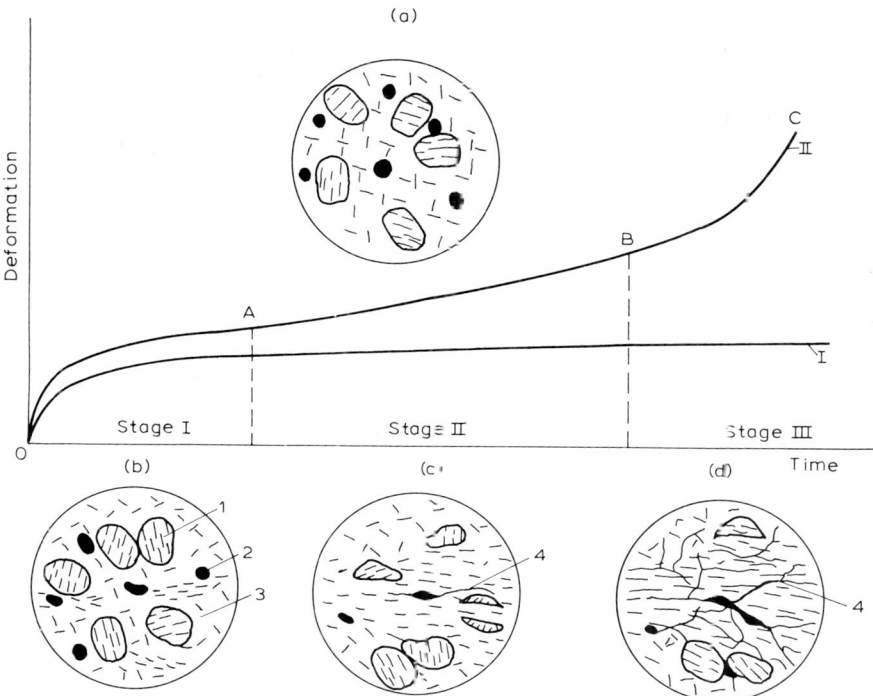

Fig. 10-6. Schematic representation of changes in microstructure of soil due to creep: (a) initial structure; (b) structure at stage I; (c) structure at stage II; (d) structure at stage III of creep; 1 = micro-aggregates of clay particles; 2 = cavities and voids; 3 = cementing clay; 4 = micro- and macro-cracks.

the soil hardens. All this may occur even under the conditions of pure shear, but an application of an all-around compressive stress will obviously increase the intensity of the process.

The "healing" of structural defects is accompanied by the springing up of new disturbances in the soil which weaken the structural bonds. However, if the load is low, the re-formed bonds and "healed" defects will outnumber the broken bonds and fresh defects. As a result, the process of hardening will prevail, providing an explanation of the attenuating character of soil deformation. The hardening of soil structure was the exact cause of the increase in the resistance of the soil after its deformation which was discussed in Section 9.8.

Under a sufficiently high load, the phenomenon of soil hardening takes place at the stage of non-steady flow only (i.e. at stage I of the creep process) when the weakening of bonds is offset by the strengthening of the structure. As the creep progresses, the breaking of bonds in ever-increasing numbers, aggregate disintegration, particle reorientation and further structural damage cause a more rapid deterioration of the resistance of the soil to the loading — its softening. However, at some stage, the softening is compensated for by hardening so that the creep takes the form of a quasi-steady flow at a roughly constant rate (stage II).

The changes in the structure enhanced by the deformation (under a high load) and the accumulation of structural damages lead to the prevailing of the softening over the hardening. As a consequence, the deformation rate increases and the process of soil deformation enters stage III, the stage of progressive flow. At this stage, the defects develop at the maximum of intensity and, as soon as the structural damage (defect density) attains a critical value ω_f = const., the soil suffers failure.

10.3 KINETIC NATURE OF LONG-TERM STRENGTH

Regularity of long-term soil failure

In presenting the theory of long-term soil failure, the basic principles of which were discussed by the author of the book in a paper presented at the 2nd International Conference on Permafrostology, USSR, 1973 (Vyalov, 1973), we shall proceed from the following four conditions mentioned in Section 10.2.

(1) Soil failure (at least the failure of hard, stiff and firm soils) results from the triggering and propagation of structural defects — micro- and macro-cracks.

(2) The development of the structural defects are indicated by the changes in the defect density with time, or in other words, the changes in the degree of structural damage, ω, varying over the range:

$$\omega_0 \leq \omega \leq \omega_f \quad \text{and} \quad (1 - \omega_0) \geq (1 - \omega) \geq (1 - \omega_f) \tag{10-7}$$

where ω_0 and ω_f are the degrees of damage at the initial moment ($t = 0$) and the moment of failure ($t = t_f$), respectively; $1 - \omega_0$ and $1 - \omega_f$ are the undisturbed unit areas at the moments $t = 0$ and $t = t_f$, respectively.

(3) The degree of structural damage at its critical value:

$$\omega_f = \text{const.} \tag{10-8}$$

is regarded as a criterion of soil failure.

(4) The intensity of the process of long-term soil failure varies with the rate of defect development and soil structure, the rate, in its turn, being a function of the stress applied and time:

$$\frac{d\omega}{dt} = f(\tau, t) \tag{10-9}$$

The type of the function $f(\tau, t)$ may be determined by eq. 10-5 deduced from the data of micro-structural investigations. On differentiating this relationship, we obtain:

$$\frac{d\omega}{1 - \omega} = \kappa_1 \bar{\tau} \frac{dt}{t + 1} \tag{10-10}$$

This equation explicitly indicates that the mechanism of crack triggering is a stochastic process; it may be deduced also in terms of the theory of probability.

Suppose that in a volume of soil consisting of N elemental volumes a set of circumstances causes the triggering of cracks which, in propagating, bring about failure of these volumes. Denote, by N_t, the number of elemental volumes remaining intact at the moment t, by dN, the number of elemental volumes which have cracked and failed during an interval of time between t and $t + dt$ and, by $m(t)dt$, the probability of crack triggering during a time interval dt. Denoting now, by $p = N_t/N_0$, the probability of the fact that no failure will occur before the moment t and, by $dp = dN_t/N_0$, the probability of the fact that the failure of elemental volumes will take place over the time interval between t and $t + dt$, we may write:

$$- dp = p(t)m(t)dt \qquad (a)$$

Hence:

$$- dN = N_t m(t) dt \qquad (b)$$

This relationship proves the postulate reading that the number of elemental volumes failing over an infinitesimal interval of time dt varies directly with the number of elemental volumes (samples) staying intact at this time. In our case, $N_0 = 1 - \omega_0$; $N_t = 1 - \omega$; $dN_t = -d\omega$. Putting $m = \kappa_1 \bar{\tau}/(t + 1)$, we arrive at eq. 10-10.

Deducing the equation of long-term soil strength

From eq. 10-5, taking into account eq. 10-7, or by integrating eq. 10-10 directly, we obtain:

$$\int_0^{t_f} \bar{\tau} \frac{dt}{t + 1} = \frac{1}{\kappa_1} \ln \frac{1 - \omega_0}{1 - \omega_f} = \text{const.} \qquad (10\text{-}11)$$

in which $\bar{\tau}$ is the level of stress taken as:

$$\bar{\tau} = \frac{\tau}{\tau_0 - \tau} \qquad (10\text{-}12)$$

where τ_0 is the hypothetically instantaneous strength and τ is the applied stress which, in the general case, varies with time, $\tau = \tau(t)$.

Eq. 10-11 represents the condition of long-term failure deduced from condition 10-8. It establishes a connection between the time before failure and the load applied. The integral form of the expression allows for any loading conditions, load variations, time-dependent soil properties, etc.

Whatever the case, soil failure will occur when the ratio of the intact area of a unit cross-section in the initial condition and at some moment t_f reaches the critical value $(1 - \omega_0)/(1 - \omega_f) = \text{const.}$ The lower the load, the longer the period elapsed before this critical state is attained — evidence of the phenomenon of long-term strength in soil.

For τ = const., we shall have:

$$\frac{\tau_0 - \tau}{\tau} = \frac{1}{v} \ln (t_f + 1)$$

$$v = \frac{1}{\kappa_1} \ln \frac{1 - \omega_0}{1 - \omega_f} = \text{const.} \quad (10\text{-}13)$$

It will be noted that eq. 10-13 coincides with empirical formula 9-23 or, if we denote $v = \ln (1/T)$ and $\tau_0 = \beta/v = \beta/\ln (1/T)$, with formula 9-22:

$$\tau = \frac{\beta}{\ln \left(\dfrac{t_f + 1}{T}\right)} = \frac{\beta}{\ln (t_f/T)} \text{ approx.} \quad (10\text{-}14)$$

However, unlike the empirical formula, the parameters of eq. 10-13 have a quite definite physical meaning based on the concept of soil failure as the process of accumulation of structural defects. For example, the parameter $T = [(1 - \omega_f)/(1 - \omega_0)]^{1/\kappa_1}$ of this expression defines the relation between the intact area of soil cross-section at the initial moment and at the moment of failure. Obviously, for $t = 0$, $T = 1$ and $v = 0$.

Accordingly, the quantity τ_0 is the failure force F coming on a unit intact area in the initial condition, $\tau_0 = F/(1 - \omega_0) \times 1 \times 1$. However, the strength τ is the force coming on a unit intact area at the moment of failure, $\tau = F/(1 - \omega_f) \times 1 \times 1$.

Since the goodness of fit of formula 9-22 was proved in Chapter 9, we have reasons to believe that the physical formula 10-13 is appropriate as well.

Incidentally, the equation of long-term failure 10-11 may be obtained from the thermodynamic criterion of strength (eq. 9-15). Substituting the rate of increment in entropy density for the increment in entropy density in this equation, we obtain according to A.I. Chudnovsky, 1973 (reference in Section 9.10):

$$\int_0^{t_f} \dot{S}(t) \, dt = \Delta S_{\text{crit}} \quad (10\text{-}15)$$

The critical value of entropy density ΔS_{crit} is decided by the critical degree of structural damage ω_f and may be assumed, in accordance with the data considered above, to equal the value of v as determined from eq. 10-13. The increment in entropy density, in its turn, may be presented as the sum of an increment resulting from the interchange of energy with the surroundings, dS_e, and an increment arising due to the irreversible processes inside the system, dS_i, particularly, due to the dissipation of mechanical energy. Then,

$$\dot{S}_e = \frac{dQ}{dt} \frac{1}{\theta}$$

$$\dot{S}_i = \Phi(T_\sigma, T_\varepsilon^p, \theta, f) \quad (10\text{-}16)$$

where dQ is the influx of non-mechanical energy; θ is the temperature (K); T_σ and T_ε^p are the tensors of stress and irrecoverable strain, respectively; f is the density of free energy.

Eq. 9-15 is instrumental in treating long-term failure as a result of a thermal and mechanical impact on the one hand and as an outcome of physical and chemical processes in the soil (diffusion, osmosis, sorption) on the other hand.

If we assume that $S_e = 0$ and neglect the effect of internal physical and chemical processes, putting $f = 0$ because of a comparatively small influence of the relevant factors, and, furthermore, by using the functions S_i and ΔS_{crit} in the form according to eq. 10-10, eq. 10-15 will coincide with eq. 10-11.

Kinetic nature of long-term failure

According to the concepts of kinetic theory, the failure of bodies is regarded as a nonequilibrium process of breaking and re-forming of bonds which is activated by the action of external forces and bears a thermofluctuating character (see Section 7.6).

Assuming that the time before failure t_f corresponds to the interval between the moment of load application and the moment a particle gets loose, i.e. to the period of "settled life" of the particle, the kinetic theory of strength describes the relationship between the stress applied τ and the time before soil failure, t_f, by means of eq. 10-1 taking in this case the form:

$$t_f = t_0 \, e^{U_f/k\theta} \tag{10-17}$$

The quantity U_f in this expression is regarded as the activation energy of failure which is to be imparted to the elemental particle in order to enable it to separate from the neighbouring particles.

Consider the kinetic essence of eq. 10-14 in the light of the concepts referred to above. Presenting it as:

$$t_f = T \, e^{\beta/\tau} \tag{10-18}$$

and comparing with eq. 10-17, we can conclude that eq. 10-14 will acquire the sense of a kinetic equation if the parameter β is a function of the activation energy U_f, stress τ and temperature θ, i.e. if $\beta/\tau = U_f/k\theta$.

To see if this assumption holds true, one needs to find out firstly whether the expression $\beta \sim 1/\theta$ is correct. Checking frozen soils in this way is good practice because of the marked effect of temperature on their behaviour.

The check-up was carried out on the basis of the data acquired from the spherical indenter tests of sandy loam and clay for long-term strength in compression and long-term cohesion at the temperatures of $-20°$, $-15°$, $-10°$, $-5°$ and $20°C$ (Vyalov et al., 1965, referenced in Section 1.5).

The test points plotted on the $1/\tau$–$\ln t_f$ coordinates in accordance with formula

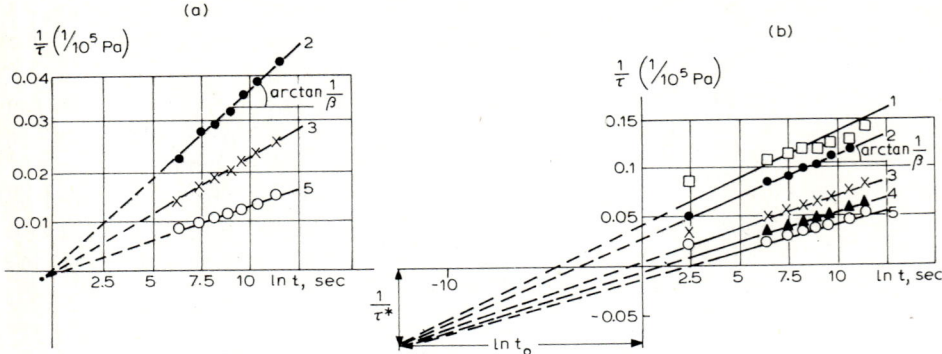

Fig. 10-7. Curves of long-term strength of soil on $1/\tau$ versus ln (t, c) coordinates: (a) uniaxial compression test of sandy loam; (b) tests of soil for cohesion at the temperatures; $1 = -20°C$; $2 = -5°C$; $3 = -10°C$; $4 = -15°C$; $5 = -20°C$.

10-14 represented as:

$$\frac{1}{\tau} = \frac{1}{\beta}(\ln t_f - \ln T) \tag{10-19}$$

yielded a family of straight lines (Fig. 10-7) the slopes of which defined the values of β at various θ. As can be seen from the graph, all $1/\tau - \ln t_f$ straight lines converged at the same pole with the coordinates $1/\tau^*$, ln t_0. Hence, eq. 10-19 may be represented in the form:

$$\ln\left(\frac{t_f}{t_0}\right) = \beta\left(\frac{1}{\tau} + \frac{1}{\tau^*}\right) \tag{10-20}$$

Now, by plotting the test points on the $1/\theta$–ln t_f, coordinates corresponding to eq. 10-17 transformed into:

$$\ln t_f = \ln t_0 + \frac{U_f}{k}\frac{1}{\theta} \tag{10-21}$$

we obtain a family of straight lines (Fig. 10-8) the slopes of which define the values of U_f/k for various τ. The lines converge at the same pole with the coordinates $1/\theta^*$, ln t_0 so that, consequently, eq. 10-21 may be presented in the form:

$$\ln\left(\frac{t_f}{t_0}\right) = \alpha\left(\frac{1}{\theta} - \frac{1}{\theta^*}\right) \tag{10-22}$$

Note that the point corresponding to 20°C ($1/\theta = 3.41 \cdot 10^{-3}$ 1/K) significantly departs from the straight lines obtained (Fig. 10-8(b)). The cause will be discussed below.

The values of t_0 from eqs. 10-20 and 10-22 appeared to be totally coincident in both cases (see Figs. 10-7 and 10-8), equalling $5.49 \cdot 10^{-1}$ s for compression (sandy loam) and $2.25 \cdot 10^{-6}$ s for cohesion (clay).

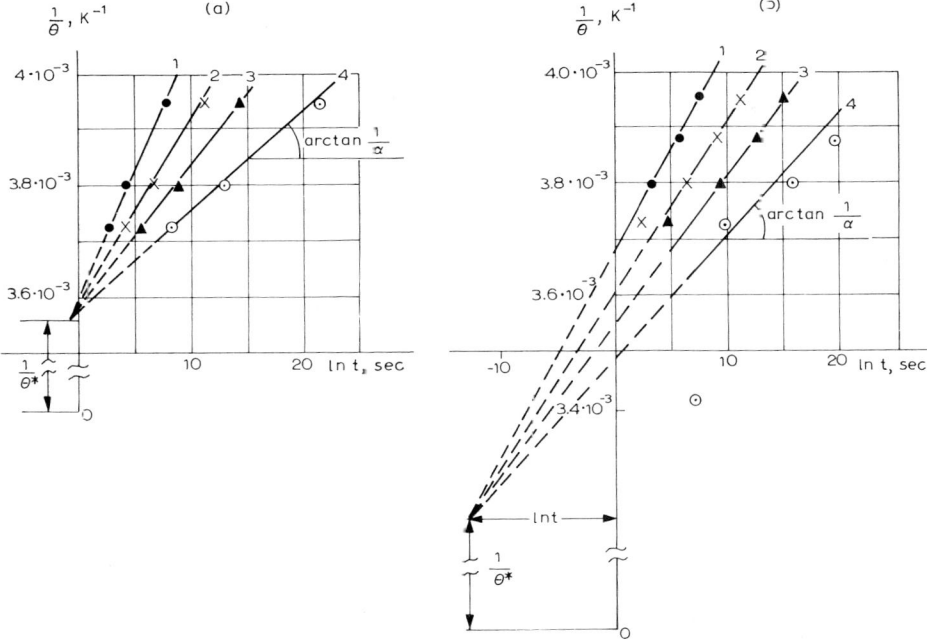

Fig. 10-8. Relationship between time before failure t_f and temperature θ, K, for various values of stress τ. (a) Uniaxial compression test of sandy loam, with τ-values: $1 = 100 \cdot 10^5$ Pa; $2 = 67 \cdot 10^5$ Pa; $3 = 50 \cdot 10^5$ Pa; $4 = 33 \cdot 10^5$ Pa. (b) Tests of clay for long-term cohesion with τ-values: $1 = 67 \cdot 10^5$ Pa; $2 = 33 \cdot 10^5$ Pa; $3 = 14 \cdot 10^5$ Pa; $4 = 9 \cdot 10^5$ Pa.

According to the kinetic theory of strength, the value of t_0 in eq. 10-17 is taken (Regel et al., 1974, referenced in Section 7.7) equal to the period of the thermal oscillations of atoms, $t_0 = h/k\theta$. It is assumed that inside a narrow range of temperature changes ($\theta = 300$ K approx.), $t_0 = 10^{-13}$ s approx. and is a constant.

For soils (see Fig. 10-7), the time $t_0 \gg 10^{-13}$ s and is not a constant, varying with the type of soil and type of test. Apparently, the explanation is that the value of t_0 has a somewhat different physical meaning, for we study the process of soil failure at the micro-structural rather than molecular level. Comparing eqs. 10-20 and 10-22, we obtain:

$$\beta = m\left(\frac{1}{\theta} - \frac{1}{\theta^*}\right) \qquad \alpha = m\left(\frac{1}{\tau} + \frac{1}{\tau^*}\right) \tag{10-23}$$

The goodness of fit of these relations is proved by the graphs of Fig. 10-9: the test points fall on the straight lines. The values of m, as defined by the slope of the straight lines, appear to be the same.

It follows from eqs. 10-20, 10-22 and 10-23 that $\ln t_f$ varies with the parameters β and α which, in their turn, are functions of $1/\theta$ and $1/\tau$. Since this form of the dependence of t_f upon θ and τ corresponds to the kinetic eq. 10-17, we can conclude

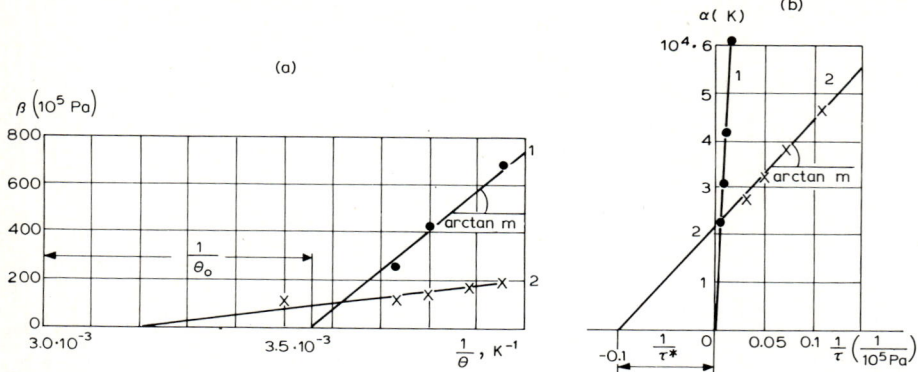

Fig. 10-9. Changes in the parameters of long-term strength according to eqs. 10-23 with (a) temperature and (b) force: *1* = compression tests of sandy loam; *2* = cohesion tests of clay.

that the process of soil failure, and eq. 10-14 describing it, are both of the kinetic nature.

Analysing eq. 10-20 simultaneously with eq. 10-22 and comparing the two expressions with due regard for eq. 10-23, we may express the activation energy of the process of failure by the following relationship:

$$U_f = U^* \frac{\tau^* + \tau}{\tau} \qquad (10\text{-}24)$$

where:

$$U^* = \frac{km}{\tau^*} \frac{(\theta^* - \theta)}{\theta^*}$$

It follows from eq. 10-24 that the activation energy of the process of failure is influenced by the temperature θ. Consider this relation in detail. It is known from thermodynamics that the internal energy is:

$$U = H - \theta S \qquad (10\text{-}25)$$

where S is the entropy of the system and H is the heat of system activation (enthalpy).

It will be recalled that enthalpy, or heat content, is the point function of a thermodynamic system $H = \int_0^\theta c_p \, d\theta + H_0$, where c_p is the heat capacity of the medium at constant pressure and H_0 is the enthalpy of the medium at $\theta = 0\,\text{K}$.

From eqs. 10-25 and 10-24 we obtain:

$$U_f = \frac{\tau^* + \tau}{\tau} (H^* - \theta S^*) \qquad (10\text{-}26)$$

Hence, the time before failure may be expressed in terms of a kinetic relationship:

$$t_f = t_0 \exp\left(\frac{U^*}{k\theta}\frac{1}{\bar{\tau}}\right) = t_0 \exp\left[\left(\frac{H^*}{k\theta} - \frac{S^*}{k}\right)\frac{1}{\bar{\tau}}\right] \quad (10\text{-}27)$$

where $\bar{\tau} = \tau/(\tau^* + \tau)$.

10.4 PHYSICAL MEANING OF LONG-TERM STRENGTH PARAMETERS

Physical meaning of the parameters β and T

Let us go back to the equation of long-term strength 10-14, rewriting it in the form given by eq. 10-18, $t_f = T e^{\beta/\tau}$.

In accordance with eqs. 10-23 and 10-24, the parameter β in the above formula may be expressed as the function:

$$\beta = \frac{\tau^*}{k\theta} U^* \quad (10\text{-}28)$$

From eqs. 10-19 and 10-20 we may express the parameter T as:

$$T = t_0 e^{\beta/\tau^*} = t_0 e^{U^*/k\theta} \quad (10\text{-}29)$$

Thus, the parameters β and T of eq. 10-16 are functions of the activation energy which, in its turn, varies with soil properties and temperature. The values of the above parameters, as obtained from the experiments referred to above (see Figs. 10-7 and 10-8), will be found in Table 10-III.

TABLE 10-III

Experimental values of the parameters β and T entering into equation of long-term strength (eq. 10-14)

Soil	Temperature (θ, °C)	Spherical indenter test		Uniaxial compression test	
		β (10^5 Pa)	T (s)	β (10^5 Pa)	T (s)
Clay	+20	108	0.013	—	—
	−5	120	0.016	329	0.0039
	−10	154	0.497	400	0.0075
	−15	169	1.649	—	—
	−20	186	6.686	596	0.044
Sandy loam	+20	64	$18 \cdot 10^{-8}$	—	—
	−5	—	—	263	0.930
	−10	222	0.0037	429	1.297
	−15	250	0.018	—	—
	−20	258	0.030	692	2.014

Effect of temperature

The way the parameters β and T are influenced by temperature, as expressed by relationships 10-28 and 10-29, is illustrated by the data given in Table 10-III. It can be seen that θ appreciably affects these parameters. To assess the temperature effect on the process of long-term failure, let us return to eq. 10-18. Substituting the values of β and T from eqs. 10-28 and 10-29, we obtain the following relationship between the time before failure and temperature:

$$t_f = t_0 \exp\left(\frac{U^*}{k\bar{\tau}} \frac{1}{\theta}\right)$$

$$\ln \frac{t_f}{t_0} = \frac{U^*}{k\bar{\tau}} \frac{1}{\theta} \tag{10-30}$$

The relationship between the long-term strength and temperature may be determined from the expression:

$$\tau = \frac{\tau^* U^*}{k\theta \ln\left(\frac{t_p}{t_0}\right) - U^*} \tag{10-31}$$

It will be noted that sub-zero temperatures have a different effect on the process of failure than the temperatures above zero, because inherent in frozen soils are ice-cementation bonds whose strength is significantly dependent upon θ. Consequently, the activation energy U^* of a frozen soil is more influenced by θ than is the case in unfrozen soils.

Whenever a soil freezes, the changes in U^* attributed to the ice-cementation bonds formed (other causes such as closer spacing of mineral particles, increased coherence of water films, etc. may also be involved) take place in steps. Apart from that, some of the activation energy is wasted in absorbing the latent heat of water crystallization during rapid changes of phase, i.e. over the range of temperatures close to zero.

Thus, to obtain a complete statement, a factor taking account of the energy absorbed by the changes in the phase of ground water must be introduced into eq. 10-26. Therefore, the time before failure will sharply change on passing the freezing point, as is shown in Fig. 10-10. Segment Oa of the graph corresponds to the range of temperatures above zero, segment ab represents the sub-zero temperatures close to zero at which rapid changes of phase are observed, and segment bc corresponds to the range of sub-zero temperatures devoid of significant changes of phase. The slopes of the $1/\theta$–$\ln(t_f/t_0)$ straight lines within the above segments will be different, demonstrating the fact that the value of U^* changes within the above intervals. This also provides the explanation of the departure of experimental points from the straight lines of the graphs in Fig. 10-8 at the temperatures above zero.

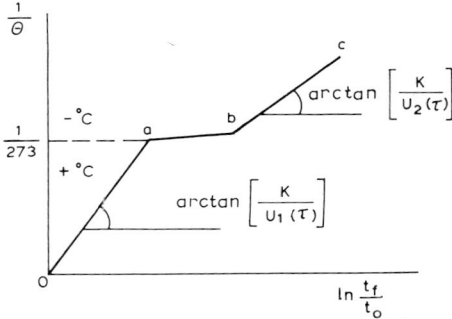

Fig. 10-10. Relationship between soil temperature and period of soil failure with allowance for changes in the phase of ground water.

Activation energy and the number of interparticle bonds

In eq. 10-1, the activation energy is given in joules (or ergs) and is referred to an elemental particle (in accordance with the meaning of the Boltzmann constant). Dividing U by the Avogadro constant $N_A = 6.02 \cdot 10^{23}$ mole^{-1}, we obtain $U/N_A k\theta = U/R\theta$, where $R = kN_A = 1.99 \cdot 10^{-3}$ kcal./mole $= 8.31 \cdot 10^3$ kJ/mole (the gas constant). Thus, the activation energy U is referred to 1 mole of the substance and expressed in J/mole (or cal./mole). Mole, it will be recalled, is the unit of amount of substance.

Since it has been agreed that we are going to examine the process at the microstructural rather than the atomic or molecular level, it is expedient to refer the activation energy to an elemental volume of soil, using the symbol $\langle U \rangle$ by analogy with formula 2-10. Let us assess the number of interparticle bonds per such a volume.

The stress τ controlling the value of U is defined as $\tau = sF$, where F is the force coming on an elemental particle (N); s is the number of particles displaced per unit area. Assuming that F is the force required to break an interparticle bond, the value of s defines the number of single bonds per unit area. Assuming that, to activate the breaking of a particle, it is necessary to break n bonds at a time, we obtain $s = 1/n$.

A clay particle may be treated as a modified sphere with an equivalent diameter averaging 0.0005 mm; then, 1 cm^3 of soil will contain approximately $8 \cdot 10^{12}$ particles. Assuming that in a normal packing each spherical particle has six contacts, the number of contacts in 1 cm^3 of the volume will total $48 \cdot 10^{12}$ and the number of contacts per 1 cm^2 of the area will amount to $48 \cdot 10^{12(2/3)} = 13 \cdot 10^8$. According to Mitchell et al., 1968 (referenced in Section 7.7), the number of bonds per unit area of illite clay is $5 \cdot 10^{10}$ in a wet sample ($W = 40\%$) and $5 \cdot 10^{12}$ in a dry one. Thus, the number of bonds per contact varies between 40 and 4,000.

Consider which of the two factors — the number of the contacts between the particles or the number of bonds — determines soil strength. Interparticle bonds are formed due to the contacting of the particles and the appearance of interparticle forces. These can be of physico-chemical origin or can be set up by an effective pressure σ' imposed by an external load. When $\sigma' = 0$, cohesion is of a purely physico-chemical nature. On the other hand, if cohesion is low, the interparticle forces owe their origin to the normal pressure only. The breaking of bonding (whatever its nature) indicates the onset of soil failure, and, according to Mitchell et al., it may be assumed that the soil strength varies with the number of the interparticle bonds per unit volume rather than with the number of contacts.

On the stress function $\bar{\tau}$

In considering the mechanism of long-term strength in the preceding sections, we proceeded on the assumption that the loss of strength goes on infinitely and that at $t_f \to \infty$ the strength $\tau \to 0$. However, if we assume that the strength decreases to a certain constant value at $t_f \to \infty$, which is the true ultimate long-term strength τ_∞, this value should be introduced into eq. 10-12 defining the form of the stress function:

$$\bar{\tau} = \frac{\tau - \tau_\infty}{\tau_0 - \tau}$$

Accordingly, eq. 10-13 takes the form:

$$\frac{\tau_0 - \tau}{\tau - \tau_\infty} = \frac{1}{\nu} \ln (t_f + 1) \tag{10-32}$$

Taking the $\bar{\tau}$ function in the form:

$$\bar{\tau} = \frac{\tau - \tau_\infty}{\tau_0 - \tau} (t_f + 1)$$

we arrive at an expression coinciding with eq. 9-31:

$$\frac{\tau_0 - \tau}{\tau - \tau_\infty} = \frac{1}{\nu} t_f \tag{10-33}$$

As mentioned in Chapter 9, the question of whether or not a true ultimate long-term strength does exist in reality remains obscure at present. Clarity is lacking even when considering strength in terms of the kinetic theory. According to Regel et al., 1974 (reference in Section 7.7) a sharp attenuation of the curve representing the long-term strength of solids (metals, plastics, etc.) at room temperature has created the illusion of a threshold failure. Hence, the notion of ultimate strength. However, it is further stated that the expansion of the range of test temperatures and that of durability changes has led to denying the existence of ultimate strength as a physical characteristic of a body.

In fact, the kinetic expression, as represented by eq. 10-1 in its general form, indicates that the process of breaking of interparticle bonds is activated by any force applied and that the durability of a material depends, consequently, on the period of application of the force.

10.5 TAKING ACCOUNT OF VARIABLE LOADING

Formulation of the problem

The curve of long-term strength of the type depicted in Fig. 9-1 and eq. 9-18 corresponding to this curve, refers to the case when a constant load is applied and the soil properties and soil condition (density–moisture relation, temperature, etc.) are unchanged with time. At the same time, an assessment of long-term strength under the conditions of a variable loading and time-dependent soil properties appears to be very important.

In fact, the case when the loading sustained by soil is constant beginning from the moment of its application seems to be a hypothetical one; it is used only for simplification. In most cases, the loading gradually increases as the structure is being erected, becoming constant only at the end of the construction period. Load alterations are likely to occur also in the course of service when, for example, the water level in a reservoir or the service load in a warehouse, grain elevator, tower silo and the like are subjected to changes.

Not less important is the necessity to take account of variable loadings in assessing the long-term strength of the soils sustaining a short-term or intermittent load.

The changes in soil properties must also be taken into consideration, for example, in assessing the long-term strength of the consolidating soils in which the density–moisture relation changes with time. In this case, soil compaction may take place both in the course of erecting the structure, i.e. simultaneously with the load increase, and after the construction, i.e. under the constant loading. Finally, a variable temperature is a factor of paramount importance in determining the long-term strength of frozen soils. When used as the foundation of a structure or the medium in which a structure is being erected, frozen soils change their temperature depending on the variations of the air temperature or due to the heat release of the structure. A method of allowing for these factors was presented by the present author at the 2nd International Conference on Permafrostology, USSR, 1973 (Vyalov, 1973).

Principle of linear summation of damage. To take account of the changing soil properties and a variable loading in the equation of long-term strength, use is made of Robinson's (1952) principle of damage summation.

A stress τ_j applied to a specimen during a period t_j incurs damage of a degree equalling the ratio of t_j and $t_f(\tau_j)$, which is the time before failure under the stress applied. Then, in accordance with eq. 10-7, we may write:

$$\frac{\omega_j}{\omega_f - \omega_0} = \frac{t_j}{t_f(\tau_j)}$$

In the case of a variable loading τ and variable soil properties (temperature, density–moisture content relation), the soil damages are summed up:

$$\int_{\omega_0}^{\omega_f} \frac{d\omega}{\omega_f - \omega_0} = \int_{t^*}^{t_f} \frac{dt}{t_f(\tau, N)} = 1 \qquad (10\text{-}34)$$

Substitute the value of t_f from eq. 10-18 in the above equation, assuming that the stress is a given function of time, $\tau(t)$, and that the parameters β and T are functions of the time-dependent soil properties $N(t)$, i.e. $\beta = \beta[N(t)] = \beta(t)$ and $T = T[N(t)] = T(t)$. Eq. 10-34 takes then the form:

$$\int_{t^*}^{t} \frac{1}{T(t)} \exp\left[-\frac{\beta(t)}{\tau(t)}\right] dt = 1 \qquad (10\text{-}35)$$

The expression obtained provides for determining the long-term strength with due allowance for the load varying with time in accordance with a given law and also for the soil properties changing with time as specified by the varying long-term strength parameters $\beta(t)$ and $\tau(t)$. Since $\beta(t)$ and $T(t)$ enter into the integral as exponent and factor, respectively, it is sometimes assumed, for simplicity, that the parameter T is a constant. Consequently, the changing properties and condition of soil are taken account of by using the variable parameter $\beta(t)$.

Variable loading

Consider a few simple cases of taking into account a variable loading.

Loading pattern of Fig. 10-11(a). Loading increases in a stepwise manner so that a loading τ_1 sustained by the soil during a period t_1 is increased to $\tau_2 = k\tau_1 (k > 1)$. Determine the moment t_f of soil failure.

Substituting the values $\tau_1 = $ const. and $\tau_2 = k\tau_1$ into eq. 10-35 and separating the integral into two integrals, we obtain:

$$\int_0^{t_1} e^{-\beta/\tau_1} dt + \int_{t_1}^{t_f} e^{-\beta/\tau_2} dt = T \qquad (10\text{-}36)$$

Hence:

$$t_f = t_1 + (t_m - t_1) \exp\left[\frac{\beta}{\tau_1}\left(\frac{1}{k} - 1\right)\right] \qquad (10\text{-}37)$$

where $t_m = T e^{\beta/\tau_1}$ is the time before the failure due to the action of the loading τ_1 alone.

Obviously, the above solution is applicable to the cases when $t_1 < t_f < t_m$. Put simply the lower limit of integration is $t^* = 0$, although it would be more correct to take $t^* = -1$; this limit satisfies the initial condition in a better way. This correction is applicable to the calculations which follow.

Let us determine the loading, $\tau_2 = k\tau_1$, causing soil failure immediately after its application. Substituting $t_f = t_1$ into eq. 10-37, we obtain — after some manipulations — that the failure will take place at the moment t_1 if the load ratio $\tau_1/\tau_2 = 1/k$

is:

$$\frac{1}{k} = \frac{\tau_1 \ln\left(\frac{1}{t_m - 1}\right)}{\beta \ln (1/T)} + 1$$

Example. Let a soil mass with the long-term strength characteristics $T = 6.5 \cdot 10^{-6}$ 1/h and $\beta = 31.9 \cdot 10^{-5}$ Pa sustain a load $\tau_1 = 1.55 \cdot 10^5$ Pa increasing to $\tau_2 = 1.2 \tau_1 = 1.85 \cdot 10^5$ Pa after an interval $t_1 = 1000$ h.

An instantaneous failure of the mass could occur only under a load of $\tau_0 = \beta/\ln 1/T = 31.9/\ln 10^6/6.5 = 4.27 \cdot 10^5$ Pa. A failure due to the continuous load $\tau_1 = 1.55 \cdot 10^5$ Pa could take place after an interval $t_m = 6.5 \cdot 10^6 \times \exp(31.9/1.55) = 4735$ h. However, if the load τ_1 acts for a period $t_1 = 1000$ h and then increases to $\tau_2 = 1.2\tau_1$, as stipulated by the conditions of the example, failure will occur in $t_f = 1000 + (4735 - 1000) \exp \frac{31.9}{1.2} (\frac{1}{1.2} - 1) = 1109$ h. If the load τ_2 is to be treated as a load at failure causing the disintegration of the soil mass immediately after its application, i.e. at the moment $t_1 = 1000$ h, the magnitude of this load may be determined from the relation:

$$\frac{\tau_1}{\tau_2} = \frac{1}{k} = \frac{1.55}{31.9} \ln \frac{1}{4735 - 1000} + 1 = 0.712$$

Hence, $\tau_2 = 1.55/0.712 = 2.8 \cdot 10^5$ Pa.

Loading pattern of Fig. 10-11(b). Loading decreases in a stepwise manner, i.e. $\tau_1 < \tau_2 = k\tau_1$, where $k < 1$. Eq. 10-36 holds in this case, but $t_1 < t_m < t_f$.

Loading pattern of Fig. 10-11(c). The soil is loaded continuously and stepwise, $\tau_1 < \tau_2 = k_1\tau_1 < \tau_3 = k_2\tau_1 < \ldots < \tau_n = k_{n-1}\tau_1$, with a pause at each step $\Delta t = t_1 = t_2 - t_1 = t_3 - t_2 \ldots$.

In this case, the integral in eq. 10-35 is separated into n integrals, taken respectively between the limits 0 and t_1, t_1 and t_2, ..., t_n and t_f. The solution is obtained in the form:

$$\frac{t_m}{\Delta t} = 1 - \exp\left[\frac{\beta}{\tau_1}\left(1 - \frac{1}{k_1}\right)\right] + \exp\left[\frac{\beta}{\tau_1}\left(1 - \frac{1}{k_2}\right)\right] + \ldots + \left[\frac{t_f}{\Delta t} - n\right]$$

$$\times \exp\left[\frac{\beta}{\tau_1}\left(1 - \frac{1}{k_{n-1}}\right)\right] \tag{10-38}$$

where $t_m > t_f > t_n$.

From this expression we determine the number, n, of the stage at which failure will take place and also the time before failure, t_f.

Loading pattern of Fig. 10-11(d). The loading sustained by the soil increases continuously, obeying a linear law $\tau = mt$. Eq. 10-35 takes in this case the following form conducive to computing t_f:

$$T = \int_0^{t_f} e^{-\beta/mt}\, dt = t_f\, e^{-\beta/mt_f} + \frac{\beta}{m} E_i\left(-\frac{\beta}{mt_f}\right) \tag{10-39}$$

where E_i is an integral exponential function.

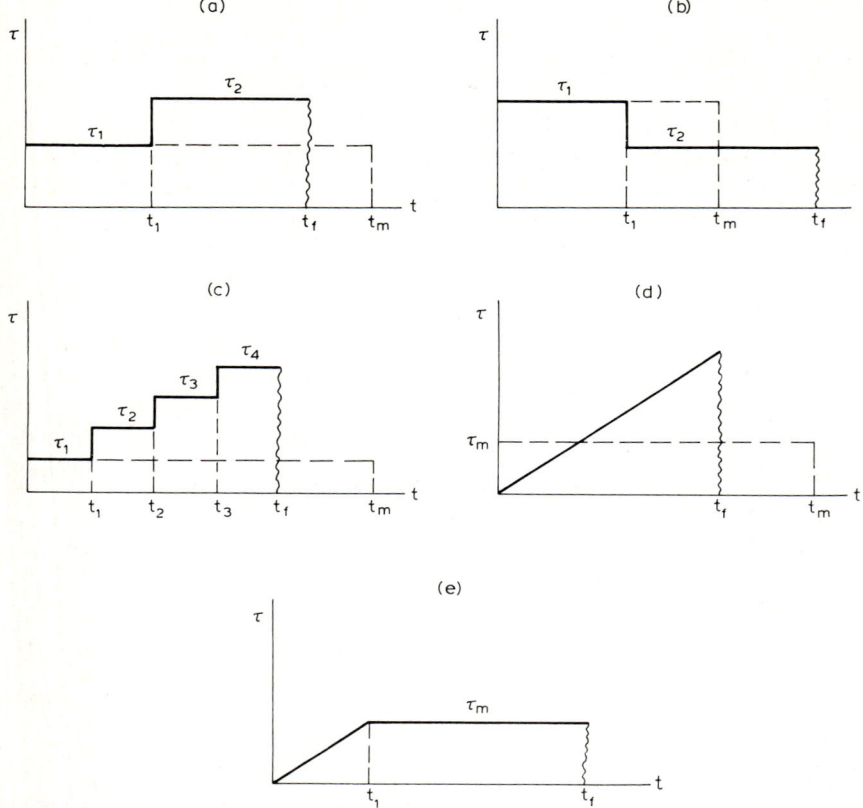

Fig. 10-11. Loading patterns.

Loading pattern of Fig. 10-11(e). The loading increases linearly, $\tau = mt$, during an interval t_1, then becoming constant, $\tau = \tau_m = $ const. This case is of particular practical significance, for the interval between 0 and t_1 can be considered as the period of construction and the time $t > t_1$ as the period of service. Separating the integral in eq. 10-35 into two integrals:

$$\int_0^{t_1} e^{-\beta/mt}\,dt + \int_{t_1}^{t_f} e^{-\beta/\tau_m}\,dt = T$$

we obtain:

$$T = \frac{\beta}{m}\left[\frac{1}{\beta}mt_1\exp\left(-\frac{\beta}{mt_1}\right) + E_i\left(-\frac{\beta}{mt_1}\right)\right] + (t_f - t_1)\exp\left(-\frac{\beta}{\tau_m}\right) \quad (10\text{-}40)$$

From the above equation one can determine either t_f for a given τ_m or, vice versa, find the value of τ_m which may bring about the failure of soil mass during a given time t_f, e.g. the service life.

Taking account of changing soil properties

Consider the mechanism of the loss of strength by frozen soil due to changes in its temperature.

As far as construction in permafrost is concerned, of great interest are the temperature variations in the region of phase changes. As pointed out above, the dependence upon temperature in the form given by eq. 10-31 ceases to be effective in this region and the recourse is to an empirical formula.

Experimental evidence shows that the suitable relationship is:

$$\beta = \frac{a}{(1 + \theta_c)^k}$$

or:

$$\beta = a + b\theta_c^n \tag{10-41}$$

where θ_c is the temperature of frozen soil (omitting the minus sign), in °C. The parameter T can be treated as a constant.

Adopting the second temperature-defining empirical relationship and assuming that the temperature changes with time in a given manner, $\theta_c(t)$, we obtain, on substituting eq. 10-41 into eq. 10-35, that:

$$T = \int_{t^*}^{t_f} \exp\left\{-\frac{a + b[\theta_c(t)]^n}{\tau}\right\} dt \tag{10-42}$$

The way the temperature changes with time, $\theta_c(t)$, is either specified in terms of the boundary condition of the problem or, in the general case, is determined by solving Fourier's equation of heat:

$$\frac{\partial \theta_c}{\partial t} = a_{th} \Delta \theta_c \tag{10-43}$$

where $a_{th} = \lambda/c$ is the thermal diffusivity of the soil; λ is the thermal conductivity; c is the heat capacity per unit volume; and:

$$\Delta = \frac{\partial^2}{\partial x^2} + \frac{\partial^2}{\partial y^2} + \frac{\partial^2}{\partial z^2}$$

is the Laplacian operator.

Capitalizing on the fact that the concept of effective heat capacity includes both the heat capacity proper and the heat of phase changes, Kolesnikov and Martynov (1952, 1953) transformed eq. 10-43 into:

$$\frac{\partial \theta}{\partial t} = a_{eff} \Delta \theta_c \tag{10-44}$$

where:

$$a_{\text{eff}} = \frac{\lambda}{c + \varrho Q \dfrac{dW_{\text{n.f.}}}{dt}}$$

ϱ is the density of the dry soil; Q is the crystallization heat; and $W_{\text{n.f.}}$ is the amount of nonfrozen water in the soil.

Thus, the problem of determining the long-term strength of soil with due allowance for time-dependent temperature is reduced to solving simultaneously the equation of long-term strength (eq. 10-42) and the equation of thermal conductivity (eq. 10-44).

Intermittent temperature changes. Consider, for example, periodic changes in soil temperature occurring due to the fluctuating temperature of outdoor air. The problem posed is of particular interest, being frequently encountered in practice. It is known, that the change in soil temperature at a depth of z from the surface may be determined from the relationship:

$$\theta_c = \theta_{c(0)}\left[1 - e^{-D}\cos\left(\frac{2\pi t}{t_1} - D\right)\right]$$

where $\theta_{c(0)}$ is the average annual temperature of permafrost (°C); $t_1 = 1$ year (7860 h) is the period of temperature fluctuations; $D = z\sqrt{\pi/a_{\text{th}} t_1}$ is the decrement of attenuation; and a_{th} is the thermal conductivity or permafrost (m²/h).

The problem is solved by substituting this expression into eq. 10-43:

$$T = \int_0^{t_f} \exp\left\{-\frac{1}{\tau}\left[a + b\theta_{c(0)}^n\left(1 - e^{-D}\cos\left(\frac{2\pi t}{t_1} - D\right)\right)^n\right]\right\} dt \qquad (10\text{-}45)$$

From this equation, solved numerically, one may find either the time before failure, t_f, for a given load τ (constant or variable) or, assuming that t_f equals the service life, assess the long-term strength τ_∞.

Other formulae of long-term strength

Statement 10-34, taking account of variable loading and changing soil properties, may be applied not only to eq. 10-18 but to the phenomenological formulae of long-term strength as well. Thus, if the term $t_f(\tau, N)$ of the integral in eq. 10-34 is adopted in accordance with eq. 9-27 as:

$$t_f = T^*\tau^{1/\alpha} \qquad (10\text{-}46)$$

eq. 10-34 takes the form:

$$\int_0^{t_f} \frac{1}{T^*(t)} \frac{dt}{[\tau(t)]^{1/\alpha}} = 1 \qquad (10\text{-}47)$$

where $T^* = T^*[N(t)\theta(t)] = T^*(t)$ is the parameter of strength and $\tau = \tau(t)$ is the loading.

Consider the solutions of eq. 10-47 for a variable loading in accordance with the patterns in Fig. 10-11 (for $T^* =$ const.).

In the case of loading patterns (a) and (b) (loading in two steps), the solution is:

$$t_f = \frac{1}{k^\alpha}[t_m + t_1(k^\alpha - 1)] \tag{a}$$

where $t_m = T^*\tau_1^{1/\alpha} = T^*\tau_m^{1/\alpha}$.

For loading pattern (c) of Fig. 10-11 (continuous stepped loading) we obtain:

$$t_m = \Delta t\left[1 + k_1^\alpha + k_2^\alpha + \ldots + k_{n-1}^\alpha\left(\frac{t_f}{\Delta t} - n\right)\right] \tag{b}$$

In case of a continuous loading according to a linear law — loading pattern (d) — the solution is:

$$t_f = [t_m \tau_m^\alpha m^{-\alpha}(\alpha + 1)]^{1/(\alpha+1)} = [T^* m^{-\alpha}(\alpha + 1)]^{1/(\alpha+1)} \tag{c}$$

Loading pattern (e) (linearly increasing load maintained at $\tau =$ const. eventually) yields:

$$t_f = \frac{1}{\tau_m^\alpha}\left[T^* + \tau_m^\alpha t_1 - \frac{m^\alpha t_1^{\alpha+1}}{\alpha+1}\right] \tag{d}$$

from which we can find the time before failure t_f for a given τ_m or the long-term strength $\tau_m = \tau_\infty$ at a given t_f.

10.6 KINETIC NATURE OF CREEP IN SOILS

Kinetic nature of viscosity

In accordance with kinetic theory, the coefficient of viscosity is determined by eq. 7-106:

$$\eta = A\, e^{U/k\theta} \tag{10-48}$$

which has been derived for an ideal Newtonian liquid characterized by a constant flow rate $\dot{\gamma}$ and a linear relation between the flow rate and shear stress, $\eta = \tau/\dot{\gamma} =$ const.

However, soils, as most other real bodies, deform in an unsteady manner, at a rate varying with time, and this should be taken into consideration.

Changes in activation energy during deformation

Owing to the changes in viscosity depending on the stress τ and time t, the activation energy in eq. 10-48 must be treated as a variable subjected to changes during deformation.

These changes result from the alterations in soil micro-structure considered above. Since the alterations, in their turn, are decided by the stress applied and duration of the process, the activation energy appears to be a function of τ and t: $U = U(\tau, t)$. Accordingly, eq. 10-48 takes the form:

$$\eta = A\, e^{U(\tau,\, t)/(k\theta)} \tag{10-49}$$

It was shown above that creep is an outcome of the phenomena of soil hardening and soil softening developing under a load. Consequently, the activation energy may either decrease if the soil softens or increase if the hardening occurs, the former case being equivalent to a lowering of the energy barrier (see Fig. 7-14) and the latter, to the rising thereof. Denoting the relative activation energy as $\bar{U} = U/k\theta$, we may write:

$$\bar{U}(\tau, t) = \bar{U}_0 - \bar{U}_{\text{soft}} + \bar{U}_{\text{hard}}$$

where $\bar{U}_0$ is the initial activation energy (at $\tau = 0$, $t = 0$) varying with the physical and chemical properties of the soil; $\bar{U}_{\text{soft}}$ and $\bar{U}_{\text{hard}}$ is the time-dependent energy wasted in softening and hardening the structure, respectively.

Substituting the above equation into eq. 10-49, we obtain:

$$\eta(\tau, t) = A\, \exp(\bar{U}_0 - \bar{U}_{\text{soft}} + \bar{U}_{\text{hard}}) \tag{10-50}$$

where $\eta(\tau, t)$ is the variable (effective) viscosity.

Eq. 10-50 implies that to effect a further displacement of a particle already on the move, the required energy is different from the initial activation energy $\bar{U}_0$ and is decided by the structure at the moment t under consideration. This new energy $\bar{U}(\tau, t)$, will be less than $\bar{U}_0$ if the preceding particle displacement has softened the structure or greater than $\bar{U}_0$ if soil hardening has taken place.

Assess the activation energy wasted in softening the soil, $\bar{U}_{\text{soft}}$, on the basis of micro-structural investigations.

Since the bulk of structural softening is brought about by the accumulated defects, we may assume that:

$$d\bar{U}_{\text{soft}} = \varrho_1 \frac{d\omega}{1-\omega} \tag{10-51}$$

where ω is the defect density (the degree of damage as fraction of unity); $\varrho_1 = \varrho_1^*/k\theta$ is a dimensionless parameter; and ϱ_1^* is a coefficient (J).

On the other hand, according to eq. 10-10:

$$\frac{d\omega}{1-\omega} = \kappa_1 \bar{\tau}\, \frac{dt}{t+1} \tag{10-52}$$

where:

$$\bar{\tau} = \frac{\tau}{\tau_0 - \tau} \quad \text{and} \quad \kappa_1 = \frac{1}{v}\ln\left(\frac{1-\omega_0}{1-\omega_f}\right)$$

Substituting eq. 10-52 into eq. 10-51, we obtain:

$$d\bar{U}_{\text{soft}} = \kappa_1 \varrho_1 \bar{\tau} \frac{dt}{t+1}$$

Hence:

$$d\bar{U}_{\text{soft}} = \lambda_1 \int_0^t \bar{\tau} \frac{dt}{t+1} \qquad (10\text{-}53)$$

where:

$$\lambda_1 = \varrho_1 \kappa_1 = \frac{\varrho_1}{v} \ln\left(\frac{1-\omega_0}{1-\omega_f}\right)$$

From eqs. 10-53 and 10-11 we can see that eq. 10-11 is equivalent to the statement $\bar{U}_{\text{soft}} = \text{const.}$, allowing that the energy wasted due to soil failure is constant. Since this energy is dissipated, eq. 10-11 corresponds to the condition of long-term failure 9-14.

For $\tau = \text{const.}$, we obtain:

$$\bar{U}_{\text{soft}} = \lambda_1 \bar{\tau} \ln(t+1) \qquad (10\text{-}54)$$

Now, assess the activation energy wasted in hardening soil structure, $\bar{U}_{\text{hard}}$. By analogy with eq. 10-51 we have:

$$d\bar{U}_{\text{hard}} = \varrho_2 \frac{d\bar{\Omega}}{1 - \bar{\Omega}} \qquad (10\text{-}55)$$

where $\bar{\Omega}$ is the index of structure hardening (in fraction of unity); $\varrho_2 = \varrho_2^*/k\theta$; and ϱ_2^* is a parameter (J).

Since the hardening of structure results from the displacement of particles with time, the time-dependent pattern of these displacements may be adopted in accordance with eq. 10-4. Then, taking into account eq. 10-6, we obtain:

$$\frac{d\bar{\Omega}}{1 - \bar{\Omega}} = \bar{\kappa}_2 \frac{dt}{t+1} \qquad (10\text{-}56)$$

Substituting this expression into eq. 10-55, we have:

$$\bar{U}_{\text{hard}} = \varrho_2 \int_{\bar{\Omega}_0}^{\bar{\Omega}} \frac{d\bar{\Omega}}{1 - \bar{\Omega}} = \varrho_2 \bar{\kappa}_2 \int_0^t \frac{dt}{t+1} = \varrho_2 \bar{\kappa}_2 \ln(t+1)$$

or:

$$\bar{U}_{\text{hard}} = \lambda_2 \ln(t+1) \qquad (10\text{-}57)$$

where $\lambda_2 = \varrho_2 \bar{\kappa}_2$.

Taking into account eqs. 10-54 and 10-57, eq. 10-50 takes the form:

$$\bar{U}(\tau, t) = \bar{U}_0 + [\lambda_2 - \lambda_1 \bar{\tau}] \ln(t+1) \qquad (10\text{-}58)$$

Kinetic equation of strain rate

Proceeding from the formula for the coefficient of viscosity, $\eta = \tau/\dot{\gamma}$, the strain rate defined by eq. 10-50 may be presented as the relationship:

$$\dot{\gamma} = \frac{\tau}{A} \exp\left[-(\bar{U}_0 - \bar{U}_{\text{soft}} + \bar{U}_{\text{hard}})\right] \tag{10-59}$$

or, taking into account eq. 10-58, as:

$$\dot{\gamma} = \frac{\tau}{A} e^{-\bar{U}_0} e^{-[\lambda_2 - \lambda_1 \bar{\tau}]\ln(t+1)} \tag{10-59'}$$

Hence:

$$\dot{\gamma} = \frac{\tau}{\eta_0}\left(\frac{1}{t+1}\right)^{n(\tau)} \tag{10-60}$$

In deriving eq. 10-6, t is understood to be a dimensionless time quantity t/t^*, where t^* is a parameter measured in units of time. Accordingly, the above formula can be set forth more accurately as:

$$\dot{\gamma} = \frac{\tau}{\eta_0}\left(\frac{t}{t^*}+1\right)^{-n(\tau)} \tag{10-60'}$$

or, assuming that $t^* \ll 1$, in a simplified form:

$$\dot{\gamma} = \frac{\tau}{\eta_0^*} t^{-n} \tag{10-60''}$$

where $\eta_0^* = \eta_0(t^*)^{-n}$.

In what follows, we are going to use both eq. 10-60 and eq. 10-60″, in which:

$$n(\tau) = \lambda_2 - \lambda_1 \bar{\tau} \qquad \bar{\tau} = \frac{\tau}{\tau_0 - \tau}$$

τ_0 is the hypothetically instantaneous strength; τ is the stress applied; $\eta_0 = A\,e^{U_0/k\theta}$ is the initial viscosity (P); k is the Boltzmann constant (J/K); θ is the thermodynamic temperature (K); U_0 is the initial activation energy required to effect an initial displacement of soil particles (J); $A = 6k\theta t_0/V$; V is the molar volume; t_0 is the period of thermal oscillations of an elemental particle;

$$\lambda_2 = \frac{\varrho_2^* \bar{\kappa}_2}{k\theta} = \frac{\bar{\kappa}_2}{k\theta}(1-\bar{\Omega})\frac{d\bar{U}_{\text{hard}}}{d\bar{\Omega}}$$

is a dimensionless structural parameter characterizing the hardening of structural bonds in soil; $\bar{\kappa}_2$ is the dimensionless coefficient entering into eq. 10-56; ϱ_2^* is the parameter entering into eq. 10-55 (J); $\bar{\Omega}$ is the degree of structural change (due to the displacement, rearrangement and reorientation of particles) relative to the initial condition in a fraction of unity;

$$\lambda_1 = \frac{\varrho_1^* \kappa_1}{k\theta} = \frac{1}{v_{st} k\theta} \ln \frac{1 - \omega_0}{1 - \omega_f}$$

is a structural parameter characterizing the weakening and breaking of interparticle bonds; ϱ_1^* is the parameter entering into relationship eq. 10-51 (J);

$$\kappa_1 = \frac{1}{v_{st} \varrho_1^*} \ln \frac{1 - \omega_0}{1 - \omega_f}$$

is a soil constant defining the conditions under which the failure of structure takes place; ω_0 and ω_f are the degrees of damage in the initial condition and at failure, respectively, in fractions of unity; and v_{st} is the dimensionless coefficient entering into the equation of long-term strength (eq. 10-13).

The activation energy U and energy changes U_{soft} and U_{hard} are expressed in joules per interparticle bond. Dividing by the Avogadro constant, $N_A = 6.02 \cdot 10^{23}$ mole^{-1}, the energy U will be referred to 1 mole, as was mentioned in Section 10.4.

Thus, the equation of soil deformation 10-60 derived by the author in 1976 (Vyalov, 1976) originated from the study of the pattern of deformation at the microstructural level and was based on statements from the kinetic theory. The parameters entering into the equation all have a definite physical meaning and cover both the deformation and strength characteristics of soil which are thus correlated. The exponent $n(\tau)$ in eq. 10-60 has a definite physical meaning too. Moreover, it does not interfere with the dimensions of η_0, thus turning the limitations of the power function referred to in Section 4.3 into an asset.

Eq. 10-60 has been derived as proceeding from the theory of flow. If eq. 10-60 is set forth in the integral form, it may be applied, with assumptions, to solving variable-load problems:

$$\gamma = \frac{1}{\eta_0} \left[\tau(t) + \int_0^t \tau(v) Q(\tau, t - v) dv \right] \qquad (10\text{-}61)$$

where $Q(\tau, t) = (t + 1)^{-n(\tau)}$.

10.7 EQUATION OF DEFORMATION

Pattern of creep

The equation of soil deformation is readily arrived at from eq. 10-60:

$$\gamma = \gamma_0 + \int_0^t \frac{\tau}{\eta_0 (t + 1)^{n(\tau)}} dt$$

or, for τ = const. (provided $n(\tau) \neq 1$):

$$\gamma = \gamma_0 + \frac{\tau}{\eta_0[1 - n(\tau)]}[(t + 1)^{1-n(\tau)} - 1] \tag{10-62}$$

where γ_0 is the hypothetically instantaneous deformation of soil.

It will be noted that the process of deformation described by eqs. 10-60 and 10-62 may vary in its nature depending on the value of the exponent:

$$n(\tau) = \lambda_2 - \lambda_1 \bar{\tau} = \lambda_2 - \lambda_1 \frac{\tau}{\tau_0 - \tau} \tag{10-63}$$

or, in other words, depending on the magnitude of the stress:

$$\tau = \tau_0 \frac{\lambda_2 - n}{\lambda_1 + \lambda_2 - n}$$

The turning point is at $n = 0$: for $n < 0$ the strain rate increases, for $n > 0$ it decreases; for $n = 0$ the flow progresses at a constant rate. Consider these cases in detail (Fig. 10-12).

The case $n < 0$. Observed for the stresses:

$$\tau > \tau_0 \frac{\lambda_2}{\lambda_1 + \lambda_2}$$

this situation corresponds to an unconfined increase in both the strain and strain rate: at $t \to \infty$, the strain rate $\dot{\gamma} \to \infty$ and $\gamma \to \infty$ (curves *1, 2, 3* in Fig. 10-12). However, the way the strain rate is increasing varies, depending on the value of n. Thus, if $n < -1$, i.e. $\tau > \tau_0[(\lambda_2 + 1):(\lambda_1 + \lambda_2 + 1)]$, the strain rate $\dot{\gamma}$ increases at an acceleration which increases as well (curve *1*); if $n > -1$, i.e. $\tau < \tau_0[(\lambda_2 + 1):(\lambda_1 + \lambda_2 + 1)]$, the strain rate increases at a decreasing acceleration (curve *3*); and for $n = -1$, i.e. $\tau = \tau_0[(\lambda_2 + 1):(\lambda_1 + \lambda_2 + 1)]$, the acceleration is constant, $d\gamma/dt$ = const. (curve *2*).

The case $n = 0$. Taking place for the stress:

$$\tau = \tau_0 \frac{\lambda_2}{\lambda_1 + \lambda_2}$$

this situation corresponds to the Newtonian flow at a constant rate, $\dot{\gamma} = \tau/\eta_0$ = const. (curves *4* in Fig. 10-12).

The case $n > 0$ met with for the stresses:

$$\tau < \tau_0 \frac{\lambda_2}{\lambda_1 + \lambda_2}$$

corresponds to the straining at a decreasing rate; at $t \to \infty$, the strain rate $\dot{\gamma} \to 0$. The strain proper may be either a non-attenuating or an attenuating one contingent on the value of n (curves *5, 6* and *7* in Fig. 10-12).

Thus, if $n < 1$, i.e. $\tau > \tau_0[(\lambda_2 - 1):(\lambda_1 + \lambda_2 - 1)]$, the strain increases in an

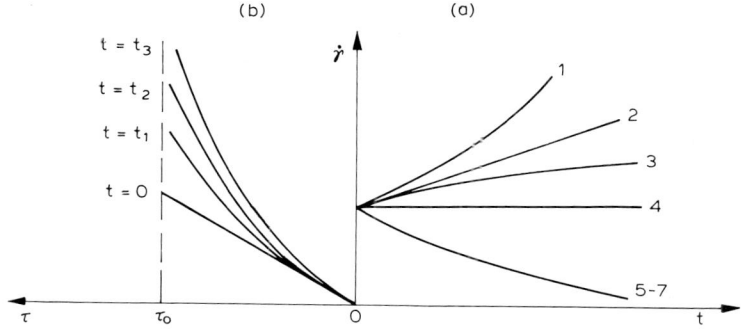

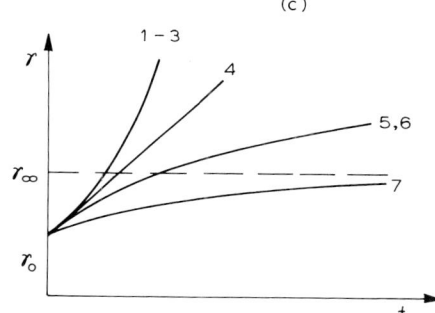

Fig. 10-12. Deformation curves corresponding to eq. 10-60. (a) Changes in strain rates $\dot{\gamma}$ with time for various values of stress τ. (b) Changes in strain rates $\dot{\gamma}$ with stress for various moments of time t_j. (c) Progress of deformation with time for the values of τ as follows: 1. $n < -1$; 2. $n = -1$; 3. $-1 < n < 0$; 4. $n = 0$; 5. $0 < n < 1$; 6. $n = 1$; 7. $n > 1$, where $n = \lambda_1 - \lambda_2[\tau/(\tau_0 - \tau)]$.

unconfined way in accordance with the power law 10-62 and at $t \to \infty$, the strain $\gamma \to \infty$, although the strain rate $\dot{\gamma} \to 0$ (curve 5).

For $n = 1$, i.e. $\tau = \tau_0[(\lambda_2 - 1):(\lambda_1 + \lambda_2 - 1)]$, the soil strain increases in an unconfined manner and at $t \to \infty$, $\gamma \to \infty$; however, by analogy with the preceding case, $\dot{\gamma} \to 0$ (curve 6). Also, the strain increases at a slower pace than this is the case when $n < 1$, obeying the logarithmic law:

$$\gamma = \gamma_0 + \frac{\tau}{\eta_0} \ln(t+1) \tag{10-64}$$

Finally, for $n > 1$, i.e. $\tau < \tau_0[(\lambda_2 - 1):(\lambda_1 + \lambda_2 - 1)]$, an attenuating strain is observed and at $t \to \infty$, the strain rate $\dot{\gamma} \to 0$ and the strain proper $\gamma \to \gamma_\infty$. The equation of soil deformation 10-62 takes the form:

$$\gamma = \gamma_\infty - \frac{\tau}{\eta_0(n-1)}(t+1)^{1-n} \tag{10-65}$$

where:

$$\gamma_\infty = \gamma_0 + \frac{\tau}{\eta_0(n-1)} = \frac{\tau(\tau_0 - \tau)}{[\tau_0(\lambda_2 - 1) - \tau(\lambda_1 + \lambda_2 - 1)]\eta_0} \tag{10-66}$$

If $n \gg 1$, the deformation of creep may be neglected owing to its insignificance and because $\gamma_\infty = \gamma_0$ approx.

Critical stress values

The strain rate–stress relation determined by eq. 10-60 is represented graphically in Fig. 10-12(b). It can be seen that the $\tau - \dot{\gamma}$ curves are dissimilar at various times t. Their non-linearity diminishes with a decrease in time, and at $t = 0$ the curve is transformed into a straight line $\dot{\gamma} = \tau/\eta_0$.

Thus, the kinetic eq. 10-60 describes all deformation processes in real soil depending on the stress applied: attenuating creep ending in deformation stabilization; non-attenuating, secular creep at a decreasing rate when the deformation increases unconfined in accordance with the power or logarithmic law; steady flow at a constant rate; and progressive flow at an increasing rate (obeying various laws of acceleration). The transition from one state of deformation to another is contingent on the stress applied.

Two critical stress values may be singled out, being called the first and second yield limits:

$$\begin{aligned}\tau_{s(1)} &= \tau_0 \frac{\lambda_2 - 1}{\lambda_1 + \lambda_2 - 1} \\ \\ \tau_{s(2)} &= \tau_0 \frac{\lambda_2}{\lambda_2 + \lambda_1}\end{aligned} \tag{10-67}$$

The value of $\tau_{s(1)}$ corresponds to $n = 1$ and that of $\tau_{s(2)}$, to $n = 0$.

For the stresses $\tau < \tau_{s(1)}$, the creep of soil is of the attenuating type (curve 7 in Fig. 10-12). The stresses $\tau_{s(1)} \leqslant \tau < \tau_{s(2)}$ give rise to secular creep at a decreasing rate (curves 5 and 6 in Fig. 10-12). When $\tau = \tau_{s(2)}$, a steady flow comes into existence (curve 4), and at $\tau > \tau_{s(2)}$ a progressive flow is induced, developing at an increasing rate (curves 1–3). It will be appropriate to point out that, in the light of what was said above, the limit $\tau_{s(2)}$ may be compared with the ultimate long-term strength τ_∞, however, the conditional character of this notion should not be overlooked.

The difference between the above limits and the critical values of stress τ_k shown in Fig. 4-11 is that the values of τ_k define the way in which the connection between the stress and the rate of steady flow is changing whereas the yield limits $\tau_{s(1,2)}$ define the connection between the stress and a variable rate of strain with allowance for the time factor.

Note, that the yield limits $\tau_{s(1,2)}$ are expressed in terms of physical (structural) parameters λ_1 and λ_2 having a clearly defined physical meaning, for, we may recall, that these parameters define the capacity of soil for softening (λ_1) or hardening (λ_2).

Accordingly, the values of $\tau_{s(1,2)}$, as determined from eq. 10-67, establish the confines of predominance for each of the processes (softening or hardening), thus determining the nature of deformation.

Approximate formula of soil deformation

Eq. 10-60 describes a family of dissimilar creep curves whose configuration is decided by the stress (Fig. 10-12). This is an asset of the formula which thus creates the prospect of describing real creep processes of various kinds.

However, if the stresses vary insignificantly, the creep curves may be treated as having a similar configuration, adopting a constant exponent n in eq. 10-60 to that end. In this case, the effect of stress must be represented by a factor $f(\tau)$. The relationship which is consequently obtained is simpler than eq. 5-9 and incorporates the time function in the form given by eq. 5-12:

$$\dot{\gamma} = \left(\frac{\tau}{\tau^*}\right)^{1/m}\left(\frac{t + t^*}{t^*}\right)^{-n} \tag{10-68}$$

Equation of deformation in another form

The equation of deformation 10-60 has been derived on the basis of microstructural studies. Quite naturally, some assumptions have been made in analysing the data. In particular, this applies to deducing the stress function $\bar{\tau}$ entering into eq. 10-53. With some approximation, this function can also be adopted in the form given by eq. 10-32. Then, the equation of deformation 10-59 will take the form considered by the author and Yu.K. Zaretsky (Zaretsky and Vyalov, 1971, referenced in Section 9.10):

$$\dot{\gamma} = \frac{\tau \exp\left(\frac{\lambda_1}{T}\frac{\tau - \tau_\infty}{\tau_0 - \tau}t\right)}{\eta_0(1 + t)^{\lambda_2}} \tag{10-69}$$

where λ_1, λ_2, η_0, and T are parameters; and τ_0 and τ_∞ are the instantaneous and ultimate long-term strength of the soil, respectively.

Depending on the relation between τ and τ_∞, eq. 10-69 describes either the process of non-attenuating creep (for $\tau > \tau_\infty$) or the process of attenuating creep (for $\tau < \tau_\infty$), coinciding in this latter case with eq. 7-62, providing the factor τ is introduced therein.

However, if we assume that the value of $\bar{U}_{\text{soft}}$, as defined by eq. 10-53, does not change with time and equals $\bar{U}_{\text{soft}} = \lambda_1\tau/\tau_0$, which is a simple approximation, then the equation of deformation 10-69 may be reduced to the form, provided other assumptions are made, arrived at empirically by Singh and Mitchell, 1969 (referenced in Section 7.7):

$$\dot{\gamma} = \frac{1}{\eta_0}\frac{\exp\left(\frac{\lambda_1\tau}{\tau_0}\right)}{(1 + t)^{\lambda_2}} \tag{10-69'}$$

Finally, putting $\lambda_2 = 0$ in eq. 10-60, we arrive at eq. 7-112 of the kinetic theory of flow of a non-linearly viscous medium; assuming that $\lambda_1 = \lambda_2 = 0$ in eq. 10-60, we shall obtain the equation of flow of an ideally viscous Newtonian liquid in Frenkel's interpretation.

Thus, it can be seen that eq. 10-60 is adequately versatile.

10.8 EXPERIMENTAL DATA HANDLING

Determining the parameters of eq. 10-60

To determine the parameters of eq. 10-60, it is practical to use the simplified version (eq. 10-60″), on transforming it into the linear form:

$$Y = BX + D \tag{10-70}$$

by putting $Y = \ln(\dot{\gamma}/\tau)$, $X = \ln t$, $B = -n$, $D = \ln 1/\eta_0^*$. On plotting the experimental data on the $\ln(\dot{\gamma}/\tau)$–$\ln t$ coordinates, we obtain a family of straight lines (Fig. 10-13(a)) for various values of τ making a $\ln(1/\eta_0^*)$ intercept the y-axis; the numerical values of the tangents of the slopes of the lines determine the values of n. The dotted straight lines on the graph correspond to those values of τ for which the exponents n acquire critical values: $n = 1$ for $\tau = \tau_2 = \tau_{s(1)}$, $n = 0$ for $\tau = \tau_4 = \tau_{s(2)}$, and $n = -1$ for $\tau = \tau_6$.

Taking into account that:

$$n = \lambda_2 - \lambda_1 \frac{\tau}{\tau_0 - \tau} \tag{10-71}$$

we plot an n versus $\tau/(\tau_0 - \tau)$ graph (Fig. 10-13(b)); the intercept on the y-axis made by the straight line obtained gives the parameter λ_2, and the numerical value of the

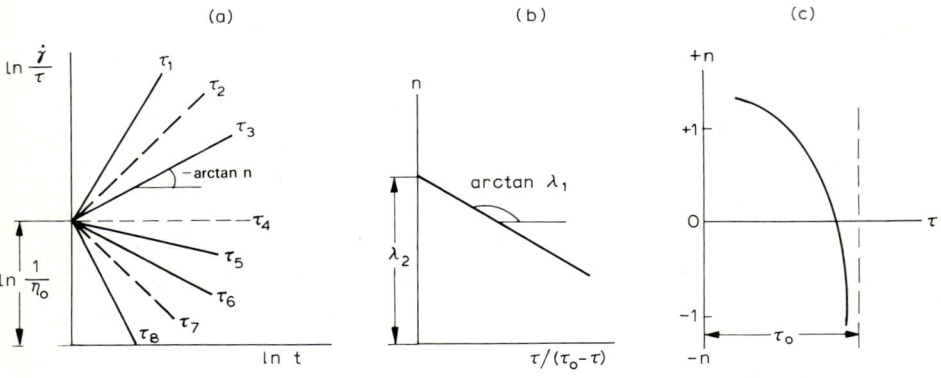

Fig. 10-13. Test data reduction in accordance with eq. 10-60. (a) Graph for determining the parameters η_0 and n. (b) Graph for determining the parameters λ_1 and λ_2. (c) Graph for determining the value of τ_0.

tangent of the slope of this line determines the parameter λ_1 (with due regard for the scale of the graph).

The instantaneous strength τ_0 entering into eq. 10-71 may be determined directly from quick failure tests. In the absence of such data, one may plot an n–τ graph (Fig. 10-13(c)) and determine τ_0 from three points (n_1, τ_1; n_2, τ_2; n_3, τ_3) of the graph, using the formula:

$$\tau_0 = \frac{\tau_3 - k\tau_2}{1 - k}$$

where:

$$k = \frac{(n_1 - n_2)(\tau_3 - \tau_1)}{(n_1 - n_3)(\tau_2 - \tau_1)} \tag{10-72}$$

The parameters of eq. 10-60 may also be determined from the creep curves directly, using the simplified version of eqs. 10-62, 10-64 and 10-65 obtained by integrating eq. 10-60":

for $n < 1$, $\quad \gamma = \gamma_0 + \dfrac{\tau}{\eta_0^*(1-n)} t^{1-n} \tag{10-73}$

for $n = 1$, $\quad \gamma = \gamma_0 + \dfrac{\tau}{\eta_0^*} \ln t \tag{10-74}$

for $n > 1$, $\quad \gamma = \gamma_\infty - \dfrac{\tau}{\eta_0^*(n-1)} t^{1-n} \tag{10-75}$

where γ_0 and γ_∞ are the initial and finite (stabilized) deformations, respectively.

The above equations are reduced to the linear form (eq. 10-70) by means of the following substitutions.

In eq. 10-73:

$Y = \ln(\gamma - \gamma_0), \quad X = \ln t, \quad B = 1 - n, \quad D = \ln\left[\dfrac{\tau}{\eta_0^*(1-n)}\right]$

In eq. 10-74:

$Y = \gamma, \quad X = \ln t, \quad B = \tau/\eta_0^*, \quad D = \gamma_0$

In eq. 10-75:

$Y = \ln(\gamma_\infty - \gamma), \quad X = \ln t, \quad B = -(n-1), \quad D = \ln\left[\dfrac{\tau}{\eta_0^*(1-n)}\right]$

Further data handling and the determination of parameters are carried out as indicated above (Figs. 10-13(b) and (c)). The value of γ_0 in the formulae is adopted from experimental data, and the value of γ_∞ in formula 10-75 is either adopted from experiments or computed by the formula:

$$\gamma_\infty = \frac{\gamma_1 \gamma_3 - \gamma_2^2}{\gamma_1 + 2\gamma_2 + \gamma_3} \tag{10-76}$$

where γ_1, γ_2 and γ_3 are the deformations at the arbitrary moments t_1, t_2 and t_3 correlated by the equation $t_3/t_1 = (t_2/t_1)^2$.

Generally speaking, experimental data handling by means of the strain rate curves, i.e. by using eq. 10-60 directly, is simpler than is the case with creep curves. However, the strain rate values obtained from experimental data appear to be spread more widely than the strain values, for the former are obtained by differentiating the latter. Therefore, it is good practice to handle data by employing both the strain rate curves and creep curves, cross-checking the results.

Test of fit. Let us test the results of analytical computations by eqs. 10-73, 10-74 and 10-75 for closeness of fit with the data of pure shear test (due to torsion) given in Section 10.2 (see Fig. 10-1). Tabulated below are the values of the exponents n of the experimental curves in Fig. 10-1, as obtained from reducing the test data by means of eq. 10-60:

No. of curve	1	2	3	4	5
τ, 10^{-2} Pa	83	90	100	135	165
Experimental value of n	0.92	0.7	0.64	0.49	-0.05
Computed value of n	0.74	0.7	0.64	0.35	-0.05

Tabulated as the computed values of n are the values computed by eq. 10-71 on the basis of the values of λ_1 and λ_2 as determined from the graph in Fig. 10-13(b).

Determining rheological characteristics

To find an experimental value of the instantaneous strength τ_0, plot the curve of the type given in Fig. 10-13(b) and compute τ_0 by eq. 10-72; $\tau_0 = 262 \cdot 10^2$ Pa.

Once the value of τ_0 is known, plot the curve of the type given in Fig. 10-13(b), and determine the values of $\lambda_2 = 1.04$ and $\lambda_1 = 0.65$. Finally, compute the critical stresses $\tau_{s(1)} = 0.058\tau_0 = 15 \cdot 10^2$ Pa and $\tau_{s(2)} = 0.62\tau_0 = 162 \cdot 10^2$ Pa, using eq. 10-67. In fact, for $\tau_{s(1)} < \tau < \tau_{s(2)}$, the strain rate has decreased (curves *1, 2, 3, 4*), and for $\tau > \tau_{s(2)}$, the deformation has acquired a non-attenuating character at an increasing rate (curve *5*).

The analytical creep curves plotted by eqs. 10-73, 10-74 and 10-75 in accordance with the above values of the parameters are shown in Fig. 10-1(a) while the experimental data are marked by dots. Fig. 10-1(b) enables us to compare the experimental values of strain rates with the results of computations by the original eq. 10-60.

Some conclusions. In examining various phenomenological equations of creep in Chapter 5 it was shown that the choice of a formula depends on the type of the creep curve. However, some of the formulae fit the same curve to a more or less acceptable degree. This renders the selection of a phenomenological equation difficult and influenced mainly by objective factors.

Eq. 10-60 deduced in this chapter is free from the limitations mentioned. Eqs. 10-73, 10-74, 10-75 derived from it provide for describing, in terms of the stress applied, the creep curves of any kind that is met in the field. This is evident from the essence of eq. 10-60 obtained from examining creep as the interaction of the processes of loosening and hardening of the soil structure. At the same time, eq. 10-60 may be simplified under certain assumptions and, consequently, coincide with empirical formulae.

From the above we come to the conclusion that a practical way of determining deformation is by the versatile formula 10-60 and its approximate version eq. 10-68 rather than by numerous empirical formulae.

10.9 REFERENCES

Kolesnikov, A.G. and Martynov, G.A., 1952. On changes in mathematical formulation of freezing problems. Trans. USSR Acad. Sci., v. 82, no. 6 (in Russian).

Kolesnikov, A.G. and Martynov, G.A., 1953. On calculating the depth of freezing and thawing in soils. In: Data about Laboratory Investigations of Frozen Soils. Issue 1, USSR Acad. Sci. (in Russian).

Maximyak, R.V., 1968. Structural changes in clay soil due to deformation. In collective volume: Foundations and Underground Structures. Foundations and Underground Structures Res. Inst., Moscow (in Russian).

Pekarskaya, N.K. and Sheveleva, V.S., 1971. Structural studies of soil in creep due to shear. In collective volume: Permafrost Research and Problems of Construction. Issue 4, Komi Publ. House, Syktyvkar (in Russian).

Robinson, E.L., 1952. Effect of temperature variation on the long time rupture strength of steel. Trans. ASME, v. 74, no. 5.

Vyalov, S.S., 1963. Rheology of frozen soils. In collective volume: Strength and Creep of Frozen Soils. USSR Acad. Sci., Moscow (in Russian).

Vyalov, S.S., 1973. Long-term failure of frozen soil as a thermoactivated process. Proc. 2nd Int. Conf. Permafrostology. Issue 4, Yakuts Publ. House (in Russian).

Vyalov, S.S., 1976. Kinetic theory of soil deformation. Proc. 2nd All-Union Symposium Soil Rheology. Erevan State Univ., Erevan (in Russian).

Vyalov, S.S., Pekarskaya, N.K. and Maximyak, R.V., 1970. On physical essence of deformation and failure processes in clay soils. SMFE, no. 1 (in Russian).

Chapter 11

THE THEORY OF COHESIVE SOIL DEFORMATION

11.1 PARTICULARS OF SOIL DEFORMATION IN COMBINED STRESS STATE

Mutual effect of stress tensor invariants

It was mentioned in Section 3.7 that the classical deformation theory of plasticity relies on three basic principles: (1) a distortion of shape is attributed to the stress deviator and is not influenced by the spherical stress tensor; (2) a change in volume is brought about by the spherical stress tensor and is not influenced by the stress deviator; (3) the state of stress and that of strain are similar. Generally, however, these principles, particularly the first two, are inapplicable to soils.

The first principle cannot be fulfilled because of the different resistance soil is offering to deformation in compression and tension. Consequently, the internal friction is evident in both the limiting and sublimiting states. As a result, a shearing deformation is affected not only by the shearing stress intensity but by the mean normal stress as well. This can easily be proved by a simple test in a shear-strain apparatus: not only the ultimate strength but also the shearing strain appear to be dependent upon the normal stress (see, for example, Fig. 5-10).

It will be recalled that the shearing stress intensity is an invariant of the stress deviator and that the mean normal stress (all-around pressure) is an invariant of the spherical stress tensor.

The second principle is inapplicable because volumetric strains are induced in soil, not only by the effect of the mean normal stress, but also as a result of shear (the phenomenon of dilatancy) causing a repacking of soil particles. It was shown in Chapter 2 that a randomly arranged structure assumes a perfect, closely-packed arrangement if subjected to shear (see Fig. 2-12) and changes its volume. Thus, the volumetric strain is influenced by both the mean normal stress and the shearing stress intensity.

Referring to Fig. 2-10 illustrating the effect of stress arrangement on structural changes in soil, one must take into consideration that this factor also affects deformation (both shearing and volumetric). Consequently, the deformation will be dependent on the third invariant of stress tensor defining the arrangement of stress.

The third principle cannot be used owing to anisotropic deformation inherent in soils as was shown in Chapter 2.

It follows that both the volumetric strains J_1 and shearing strains J_2 are functions of all three stress tensor invariants I_1, I_2, and I_3. Introducing the time factor t, we may formulate a generalized rheological equation of state for soils in the form:

$$J_1 = J_1(I_1, I_2, I_3, t) \qquad J_2 = J_2(I_2, I_1, I_3, t) \qquad (11\text{-}1)$$

A similar statement may be set forth with respect to the strain rates:

$$\dot{J}_1 = \dot{J}_1(\dot{I}_1, \dot{I}_2, \dot{I}_3, t) \qquad \dot{J}_2 = \dot{J}_2(\dot{I}_2, \dot{I}_1, \dot{I}_3, t) \tag{11-2}$$

In the limiting condition, the above equations take the form:

$$\Phi(I_1, I_2, I_3, t) = 0 \tag{11-3}$$

Relationships 11-1 to 11-3 define the mutual effect of the three stress tensor invariants on the behaviour of soil in both the sublimiting and the limiting condition. The likelihood of such an effect has been pointed out in some works on the theory of plasticity. However, as far as most of the traditional materials are concerned, the mutual effect of I_1, I_2 and I_3 has never been taken into consideration, especially with the allowance for the time factor, because experiments show that its significance is small in most cases.

Review of studies

It seems that Botkin (see Section 4.9: Botkin, 1939, 1940a) was the first to note the effect of the first stress tensor invariant I_1 on the shearing strains in soil. He described it in the form of a dependence of the shearing strain intensity γ_i upon the shearing stress intensity τ_i and mean normal stress σ_m, assuming that the volumetric strain was elastic:

$$\frac{\tau_i}{H + \sigma_m} = \gamma_i G(\gamma_i) \qquad \sigma_m = k\varepsilon_m \tag{11-4}$$

where H is a parameter of soil cohesion defining the resistance to all-around tension; k is the bulk modulus; and $G(\gamma_i)$ is the variable shear modulus adopted at $G(\gamma_i) = A/(B + \gamma_i)$.

The condition of limiting state was determined by relationship 4-22:

$$\tau_i = (H + \sigma_m) \tan \psi \tag{11-5}$$

where ψ is the friction angle at the octahedral plane. Note, that $H \tan \psi = \tau_s^0$ is the resistance to pure shear.

Relationships 11-4 and 11-5 have been corroborated and developed by many other investigators. Barshevsky (1956), Stroganov (1956), Murayama (1964) and others have shown that not only the shearing strain γ_i but also the volumetric strain ε_m is a function of the ratio τ_i/σ_m.

The effect of mean normal stress also tells on the flow rate of soils. Experimental proofs of the dependence of $\dot{\gamma}_i$ on σ_m have been furnished by Stroganov (1961) and Grechishchev (1961).

Mathematical models capable of taking into account the mutual effect of stress tensor invariants on soil deformation have been suggested in some works. In 1966 Geniev developed a model of an incompressible, rigid and strain-hardening body, as if representing a Voigt model, in which the viscous friction element was replaced by a dry friction Coulomb element:

$$\gamma_i = \frac{1}{G_i^0}(\tau_i - \sigma_m \tan \psi)$$

where the angle ψ may be variable.

The condition of limiting equilibrium is achieved by the model when $\tau_i = \tau_s^0 + \sigma_m \tan \psi$, where $\tau_s^0 = \gamma_s G^0$.

In his later works, Geniev (1970) examined the possibility of taking into account non-linearly elastic phenomena and the phenomenon of dilatancy by means of the model.

In a study undertaken by Ioselevich (1967), the mutual effect of stress tensor invariants was taken into account by assuming that the strain modulus E and Poisson's ratio v were influenced by τ and σ_m and by replacing in the limiting condition the connection between the stress and shearing strain by a stress–strain increment relation.

Nelson and Baron (1971) expressed the bulk modulus as a parabolic function of σ_m and the shear modulus, as a linear function of τ_i and σ_m.

Roscoe and Poorooshasb (1963) suggested a generalized equation of plastic flow in which the increment in principal strains $\Delta\varepsilon_{1,2,3}$ was the sum of the increments resulting from the change in the ratio τ_i/σ_m on the one hand and the change in volumetric strain on the other hand:

$$\Delta\varepsilon_{1,2,3} = \left(\frac{\partial \varepsilon_{1,2,3}}{\partial \eta}\right)_V \Delta\eta + \left(\frac{\partial \varepsilon_{1,2,3}}{\partial V}\right)_\eta \Delta V$$

In this equation, $\Delta\eta$ is the increment in the ratio τ_i/σ_m for constant volume $V = $ const. and ΔV is the volumetric change at $\eta = $ const. The data acquired from triaxial compression tests of sand have subsequently enabled H. Poorooshasb to formulate the notion that the mean normal stress should be taken into account in both the subliminiting and the limiting condition of a soil.

All the experimental work referred to above corroborates relationship 11-1.

Generalized equation of soil deformation

A generalized equation of soil deformation (for a given state of stress) was suggested by the author (see Section 1.5: Vyalov, 1962) in the following form:

$$\gamma_i = f(\tau_i, \sigma_m, t) \qquad \varepsilon_m = f^*(\sigma_m, \tau_i, t) \tag{11-6}$$

It establishes a connection between the shearing and volumetric strains on the one hand and the shearing stress intensity, mean normal stress and time on the other hand. It was shown that the mutual effect of τ_i and σ_m on soil deformation resulted from the difference in the resistance of the soil to deformation in compression and tension.

The relations referred to above were studied during the triaxial compression tests of soils for creep in the Frozen Soil Mechanics Laboratory, Research Institute for Foundations and Underground Structures.

For a number of years, deformation patterns in sandy and clay soils were the subject of experimental studies by G.I. Lomize, A.L. Kryzhanovsky, I.N. Ivastchenko and others (Lomize et al., 1966, 1968) at the Moscow Civil Engineering Institute. Using triaxial compression apparatus ($\sigma_1 > \sigma_2 = \sigma_3$) and apparatus with independent settings of the three principal stresses ($\sigma_1 \neq \sigma_2 \neq \sigma_3$), the investigators have established that shearing and volumetric strains are influenced not only by the first and second invariants of the stress tensor but also by the third stress tensor invariant defining the arrangement of stress and by the loading path.

Methods of testing soils in the state of combined stress

Outlined below are the testing methods commonly employed in studying the state of combined stress of soils. Most widely used is the triaxi-symmetrical compression test (Fig. 11-1(a)) envisaging the application of a vertical pressure σ_z and a radial pressure $\sigma_x = \sigma_y = \sigma_r$ to a solid cylindrical specimen. The usual practice is to apply an all-around hydrostatic pressure p_0 and an additional vertical pressure p_z to the specimen; thus, $\sigma_2 = \sigma_3 = p_0$ and $\sigma_1 = p_z + p_0$.

During testing on such lines, the shearing stress intensity is $\tau_i = p_z/\sqrt{3} = (1/\sqrt{3})(\sigma_1 - \sigma_3)$ (see Table 3-I) and the mean normal stress is $\sigma_m = p_z/3 + p_0 = (\sigma_1 + 2\sigma_3)/3$. Accordingly, the shear strain intensity and mean linear strain are $\gamma_i = (2/\sqrt{3})(\varepsilon_1 - \varepsilon_3)$ and $\varepsilon_m = (\varepsilon_1 + \varepsilon_3)/3$, respectively, where $\varepsilon_1 = \Delta h/h$; $\varepsilon_3 = \Delta d/d < 0$ (h is the height of the cylindrical specimen and d is the specimen diameter).

The test load is commonly applied in a way that causes crushing of the specimen: $\sigma_1 > \sigma_2 = \sigma_3$. In this case, the Lode parameter is $\mu_\sigma = (2\sigma_2 - \sigma_1 - \sigma_3)$: $(\sigma_1 - \sigma_3) = -1$. Alternatively, radial compression may be employed, $\sigma_x = \sigma_y > \sigma_z$. In such a case, $\sigma_z = \sigma_3$, $\sigma_x = \sigma_y = \sigma_1 = \sigma_2$ and the Lode parameter $\mu_\sigma = +1$.

Triaxial compression tests are frequently carried out under a variable vertical stress σ_z but at a constant radial stress, $\sigma_x = \sigma_y = $ const. However, the values of τ_i and σ_m will both vary in this case, rendering the evaluation of their mutual effect more difficult.

It is advantageous to test soils either at a constant value of σ_m and an increasing value of τ_i or at a constant value of the ratio τ_i/σ_m. Accordingly, in either case one must change both σ_z and $\sigma_x = \sigma_y$. Creep tests are carried out with a series of soil samples, each series being tested under a variable τ_i and a constant σ_m.

A complex test method depicted in Fig. 11-1(b) represents a combination of triaxial compression with a torque M_t applied to one of the cylindrical specimen's end faces (the opposite end face is held fixed). Apart from the value of $\mu_\sigma = \pm 1$, this test provides for obtaining $\mu_\sigma = 0$ corresponding to pure shear. To that end, either the torque is applied when $\sigma_x = \sigma_y = \sigma_z = 0$ or the torque is combined with a hydrostatic pressure $\sigma_x = \sigma_y = \sigma_z = p$; in the former case we shall have $\sigma_{1,3} = \pm \tau$ and $\sigma_2 = 0$ and in the latter, $\sigma_{1,3} = p \pm \tau$ and $\sigma_2 = p$.

Fig. 11-1(c) represents the testing of a hollow cylindrical specimen by a vertical load and a torque, and Fig. 11-1(d) depicts the testing of the same specimen by a vertical load, a torque, an internal and an external all-around pressure. In the former case, the Lode parameter may have two values, -1 and 0; in the latter case it may vary between -1 and 1.

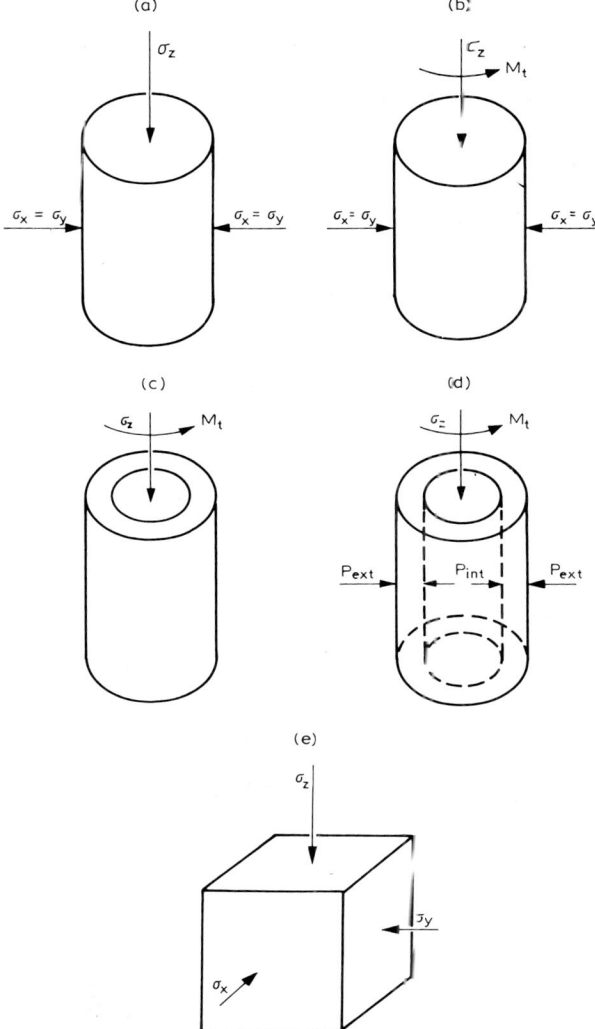

Fig. 11-1. Methods of testing soil in the state of combined stress: (a) triaxi-symmetrical compression ($\sigma_z \lessgtr \sigma_x = \sigma_y$); (b) axi-symmetrical compression ($\sigma_z \lessgtr \sigma_x = \sigma_y$) and torsion M_t; (c) axial compression (σ_z) and torsion of a hollow cylinder; (d) axial compression σ_z, radial internal pressure P_{int}, radial external pressure P_{ext} and torsion M_t of a hollow cylinder; (e) compression of a cube by stresses of arbitrary value ($\sigma_x \neq \sigma_y + \sigma_z$).

Finally, the test method depicted in Fig. 11-1(e) consists of applying arbitrary and mutually independent stresses $\sigma_1 = \sigma_z$, $\sigma_2 = \sigma_x$, and $\sigma_3 = \sigma_y$ to a cubic specimen Correspondingly, the parameter μ_σ may acquire any value from $\mu_\sigma = -1$ for $\sigma_z > \sigma_x = \sigma_y$ and $\mu_\sigma = 0$ for $\sigma_x = (\sigma_z + \sigma_y)/2$ to $\mu_\sigma = +1$ for $\sigma_y > \sigma_z = \sigma_x$. The test methods illustrated in Figs. 11-1(d) and (e) provide for examining the state of combined stress more comprehensively than the rest of methods.

11.2 GENERALIZED RHEOLOGICAL EQUATION OF STATE

Deformation theory

Taking into consideration the particulars of soil deformation considered above, we may formulate the basic principles of deformation theory, as expressed by relationships 3-84 and 3-89, in the following way.

(1) A volumetric change ε_m of a soil is the sum of the strain produced by the mean normal stress σ_m and the strain resulting from the shearing stress intensity τ_i. The volumetric strain develops with time and its constituents are recoverable and irrecoverable strains; the strain is also influenced by the arrangement of stress as defined by the Lode parameter μ_σ:

$$\varepsilon_i = f^*(\sigma_m, \tau_i, \mu_\sigma, t) \tag{11-7}$$

(2) A distortion in the shape γ_i of a soil is brought about by the action of the shearing stress intensity τ_i and mean normal stress σ_m. The strain produced develops with time and consists of recoverable and irrecoverable strains; it is also influenced by the arrangement of stress as defined by the Lode parameter μ_σ:

$$\gamma_i = f(\tau_i, \sigma_m, \mu_\sigma, t) \tag{11-8}$$

The transition into the limiting state of stress occurs when the soil strain reaches a certain critical value $\gamma_i = \gamma_s$; hence, relationship 11-8 takes the form:

$$f(\tau_i, \sigma_m, \mu_\sigma, t) = \text{const.}$$

or:

$$\Phi(\tau_i, \sigma_m, \mu_\sigma, t) = 0 \tag{11-9}$$

(3) The strain tensor is a function of the stress tensor:

$$T_\varepsilon = f(T_\sigma) \tag{11-10}$$

In the specific cases examined below, it may be assumed that the strain deviator D_ε varies with the stress deviator D_σ:

$$D_\varepsilon = \kappa D_\sigma \tag{11-11}$$

This is equivalent to assuming that the state of stress and that of strain are similar:

$$\mu_\sigma = \mu_\varepsilon \tag{11-12}$$

Generalization of Hencky's equations. Hencky's equations (3-90) establishing a connection between the strain and stress components:

$$\begin{aligned}
\varepsilon_x &= \chi(\sigma_x - \sigma_m) + \chi^*\sigma_m & \gamma_{xy} &= 2\chi\tau_{xy} \\
\varepsilon_y &= \chi(\sigma_y - \sigma_m) + \chi^*\sigma_m & \gamma_{yz} &= 2\chi\tau_{yz} \\
\varepsilon_z &= \chi(\sigma_z - \sigma_m) + \chi^*\sigma_m & \gamma_{zx} &= 2\chi\tau_{zx}
\end{aligned} \tag{11-13}$$

hold also for the principles formulated above with the only exception that the functions χ and χ^* entering into the equations are used in a more general form:

$$\chi = \frac{\gamma_i}{2\tau_i} = \frac{f(\tau_i, \sigma_m, \mu_\sigma, t)}{2\tau_i}$$

$$\chi^* = \frac{\varepsilon_m}{\sigma_m} = \frac{f^*(\sigma_m, \tau_i, \mu_\sigma, t)}{\sigma_m}$$

(11-14)

Eqs. 11-13 may be written in a more popular form:

$$\varepsilon_x = \frac{1}{2\bar{G}}(\sigma_x - \sigma_m) + \frac{1}{\bar{k}}\sigma_m$$

$$\gamma_{xy} = \frac{1}{\bar{G}}\tau_{xy}\ldots$$

(11-13')

where $\bar{G} = \tau_i/\gamma_i = \tau_i/f(\tau_i, \sigma_m, \mu_\sigma, t)$; $\bar{k} = \sigma_m/\varepsilon_m = \sigma_m/f(\sigma_m, \tau_i, \mu_\sigma, t)$ are the modified shear and bulk modulus, respectively.

The prospect of describing the general connection between the stress and strain components in soil in terms of Hencky's equations with allowance for the mutual effect of the three stress tensor invariants is created by condition 11-11 adopted above. Evaluating it with respect to the strain components and substituting the values given by eqs. 11-8 and 11-7 into the expressions obtained, we arrive at eqs. 11-13.

These equations hold for the cases of active loading only; to describe a process of unloading, use is made of equations of the type given in eq. 3-92 with due allowance for the modified values of $\bar{G}$ and $\bar{k}$.

Theory of plastic flow

By analogy with the relationships of deformation theory examined above, eqs. 3-98 to 3-100 of the theory of plastic flow may be presented in the generalized form allowing for the fact that the volumetric strain rate $\dot{\varepsilon}_m$ and the intensity of shearing strain rate $\dot{\gamma}_i$ are functions of the shearing stress intensity and mean normal stress. These rates are time-dependent and influenced by the arrangement of stress:

$$\dot{\varepsilon}_m = f^*(\sigma_m, \tau_i, \mu_\sigma, t) \qquad (11\text{-}15)$$

$$\dot{\gamma}_i = f(\tau_i, \sigma_m, \mu_\sigma, t) \qquad (11\text{-}16)$$

We shall assume that the tensor of plastic strain rate is a function of the stress tensor:

$$T_{\dot{\varepsilon}}^p = f(T_\sigma) \qquad (11\text{-}17)$$

It may be assumed that, in a specific case, $D_{\dot{\varepsilon}}^p = \dot{\kappa} D_\sigma$, i.e. the stress field and strain rate field are similar:

$$\mu_\sigma = \mu_{\dot{\varepsilon}} \qquad (11\text{-}18)$$

Expressing eq. 11-17 in terms of the stress components $\sigma_x, \ldots, \sigma_{xy}, \ldots$ and strain increment components $d\varepsilon_x, \ldots, d\varepsilon_{xy}, \ldots$ we obtain, on changing over to the strain rate components:

$$\frac{d\varepsilon_x}{dt} = \dot{\varepsilon}_x \ldots, \frac{d\gamma_{xy}}{dt} = \dot{\gamma}_{xy}, \ldots$$

the following equations:

$$\begin{aligned}
\dot{\varepsilon}_x &= \dot{\chi}(\sigma_x - \sigma_m) + \dot{\chi}^* \sigma_m & \dot{\gamma}_{xy} &= 2\dot{\chi}\tau_{xy} \\
\dot{\varepsilon}_y &= \dot{\chi}(\sigma_y - \sigma_m) + \dot{\chi}^* \sigma_m & \dot{\gamma}_{yz} &= 2\dot{\chi}\tau_{yz} \\
\dot{\varepsilon}_z &= \dot{\chi}(\sigma_z - \sigma_m) + \dot{\chi}^* \sigma_m & \dot{\gamma}_{zx} &= 2\dot{\chi}\tau_{zx}
\end{aligned} \quad (11\text{-}19)$$

where:

$$\begin{aligned}
\dot{\chi} &= \frac{\dot{\gamma}_i}{2\tau_i} = \frac{f(\tau_i, \sigma_m, \mu_\sigma, t)}{2\tau_i} \\
\dot{\chi}^* &= \frac{f^*(\sigma_m, \tau_i, \mu_\sigma, t)}{\sigma_m}
\end{aligned} \quad (11\text{-}20)$$

Eqs. 11-19 differ from the St. Venant–von Mises equations 3-102, firstly, by the function of $\dot{\chi}$ which is of a more general meaning and, secondly, by taking into account the volumetric strain rate. According to eq. 11-15, volumetric changes occur in soils also during a plastic flow whereas eq. 3-97 implies that these changes are purely elastic. Eqs. 11-19 apparently describe the process of active loading only.

Considerations given in Chapter 3 are applicable in analysing generalized eqs. 11-13 of deformation theory and generalized eqs. 11-19 of the theory of plastic flow. Since the conditions of similarity are satisfied better if the strain rate field rather than the strain field is considered, eqs. 11-19 are more suitable for soils than eqs. 11-13. However, as was said above, in the case of simple loading or under conditions close to it, the equations of deformation theory yield results identical with the solutions obtained by the equations of the theory of plastic flow. In other words, any of the above equations may be used. In particular, equations of deformation theory may be employed if time-dependent phenomena are described in terms of ageing theory or the theory of hereditary creep.

11.3 EFFECT OF MEAN NORMAL STRESS

Soil deformation curves; experimental data

Relationship 3-84 used to describe the pattern of deformation in traditional materials, corresponds to an invariant stress–strain graph. This means that the experimental points obtained from tests of any kind will fall on a single curve plotted on the "tangential stress intensity τ_i versus shearing strain intensity γ_i" coordinates.

Metals and similar materials behave as described on the above lines during tests (Fig. 11-2(a)) but quite a different picture is observed in the case of most soils. Results of triaxial compression tests of various soils — sand, sandy loam, clay, dense morainic clay, frozen soil — as obtained by various researchers are presented in Figs. 11-2(b) through (g). In some tests, the value of $\sigma_m = (\sigma_1 + 2\sigma_2)/3$ was constant but τ_i was variable, in others $\sigma_2 = \sigma_3 = $ const while σ_1 was increasing. As can be seen, the $\gamma_i - \tau_i$ graphs appeared to be non-invariant during the tests and were represented by families of $\gamma_i - \tau_i$ curves, each corresponding to the value of its mean normal stress σ_m, i.e.:

$$\gamma_i = f(\tau_i, \sigma_m) \tag{11-21}$$

Surface $(\gamma_i - \tau_i - \sigma_m)$

Let us scrutinize the connection between the tangential stress intensity τ_i, mean normal stress σ_m and shearing strain intensity γ_i at some fixed moment of time t_j (Geniev, 1970).

This connection may be represented by a surface plotted on the $\gamma_i - \tau_i - \sigma_m$ coordinates as depicted in Fig. 11-3(a). The traces obtained at the intersections of the surface with the $(\tau_i - \gamma_i)$, $(\tau_i - \sigma_m)$ and $(\sigma_m - \gamma_i)$ planes yield several families of corresponding curves (Fig. 11-3(b)).

On the $\tau_i - \gamma_i$ plane (quadrant I), there appears a family of curves representing the $\tau_i - \gamma_i$ relationship for various fixed values of the mean normal stress $\sigma_m = $ const. The lower curve of this graph corresponds to pure shear for $\sigma_m = 0$.

The traces on the $\tau_i - \sigma_m$ plane (quadrant II) give a family of curves representing the $\tau_i - \sigma_m$ relationship for various fixed values of strain $\gamma_i = $ const. The τ^0 intercepts on the y-axis correspond to the pure shear stresses ($\sigma_m = 0$) producing a given strain γ_i. The upper curve is the limiting curve corresponding to the limiting condition. On the $\sigma_m - \gamma_i$ plane (quadrant III), we obtain a family of curves showing the effect of σ_m on the shearing strain γ_i for various fixed values of the tangential stress intensity $\tau_i = $ const. The γ^0 intercepts on the y-axis correspond to the strains of pure shear.

It follows from these graphs that relationship 11-21 may be represented in a different form:

$$\gamma_i = f_1(\tau_i) - f_2(\tau_i, \sigma_m)$$

or: $\tag{11-22}$

$$\tau_i = \phi_1(\gamma_i) + \phi_2(\gamma_i, \sigma_m)$$

The first terms in both expressions define the resistance to deformation in pure shear and the second terms characterize the effect of the normal stress σ_m: an increase in the stress causes an increase in the resistance to shear and a slowing down of the deformation progress.

The graphs depicted in Fig. 11-3(b) can be plotted in another way described below. Let us carry out a test for, say, triaxial compression at an increasing τ_i and a constant

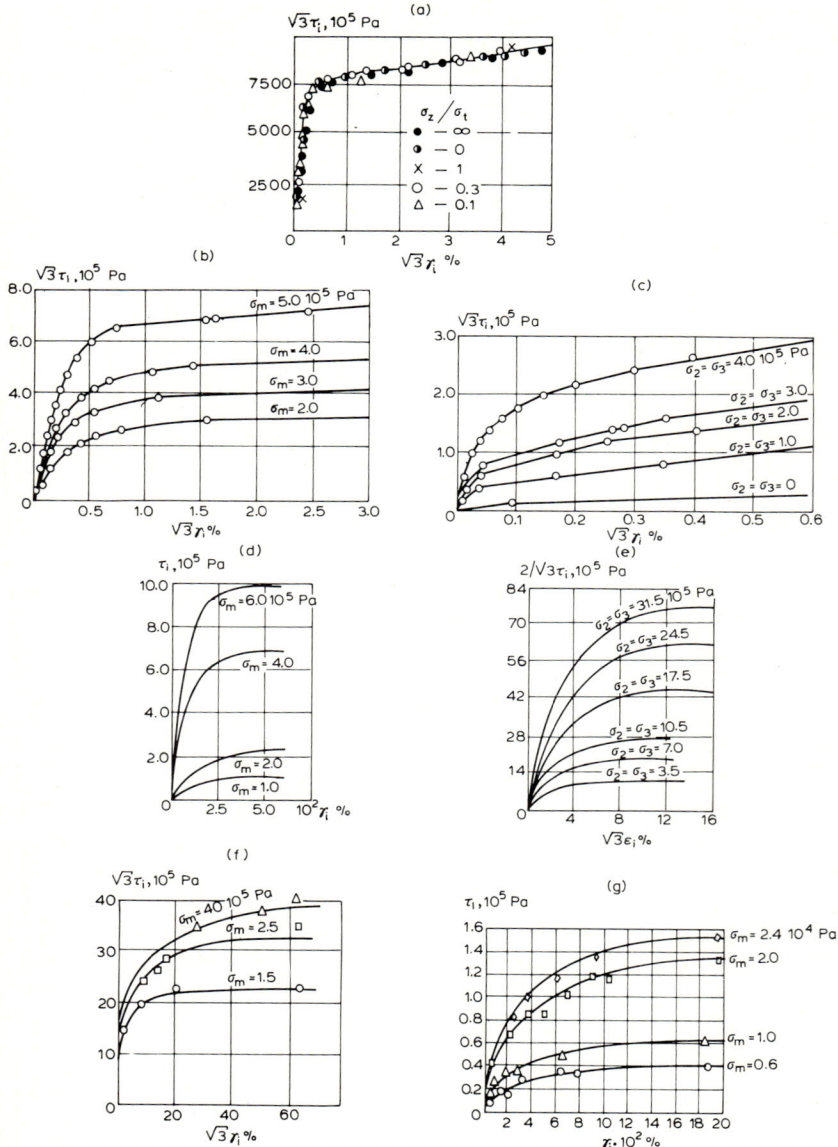

Fig. 11-2. Stress–strain curves indicating the effect of the mean normal stress σ_m in the state of combined stress. (a) Steel, tension and radial pressure for various values of $\sigma_z \sigma_t$ (Zhukov's experiments, 1954). (b) Sand, triaxial compression under various values of σ_m (Botkin's experiments). (c) Clay, triaxial compression under various values of $\sigma_2 = \sigma_3$ (Botkin's experiments). (d) Sand, triaxial compression under various values of σ_m (Fedorov's experiments, 1957). (e) Dense boulder clay, triaxial compression under various values of σ_m (experiments by Insley and Hillis, 1965). (f) Frozen sandy loam at $-10°C$, triaxial compression under various values of σ_m (experiments by Vyalov and Shusherina, 1964). (g) Kaolin clay, triaxial compression under various values of σ_m (tests by Vyalov and Mindich, 1974).

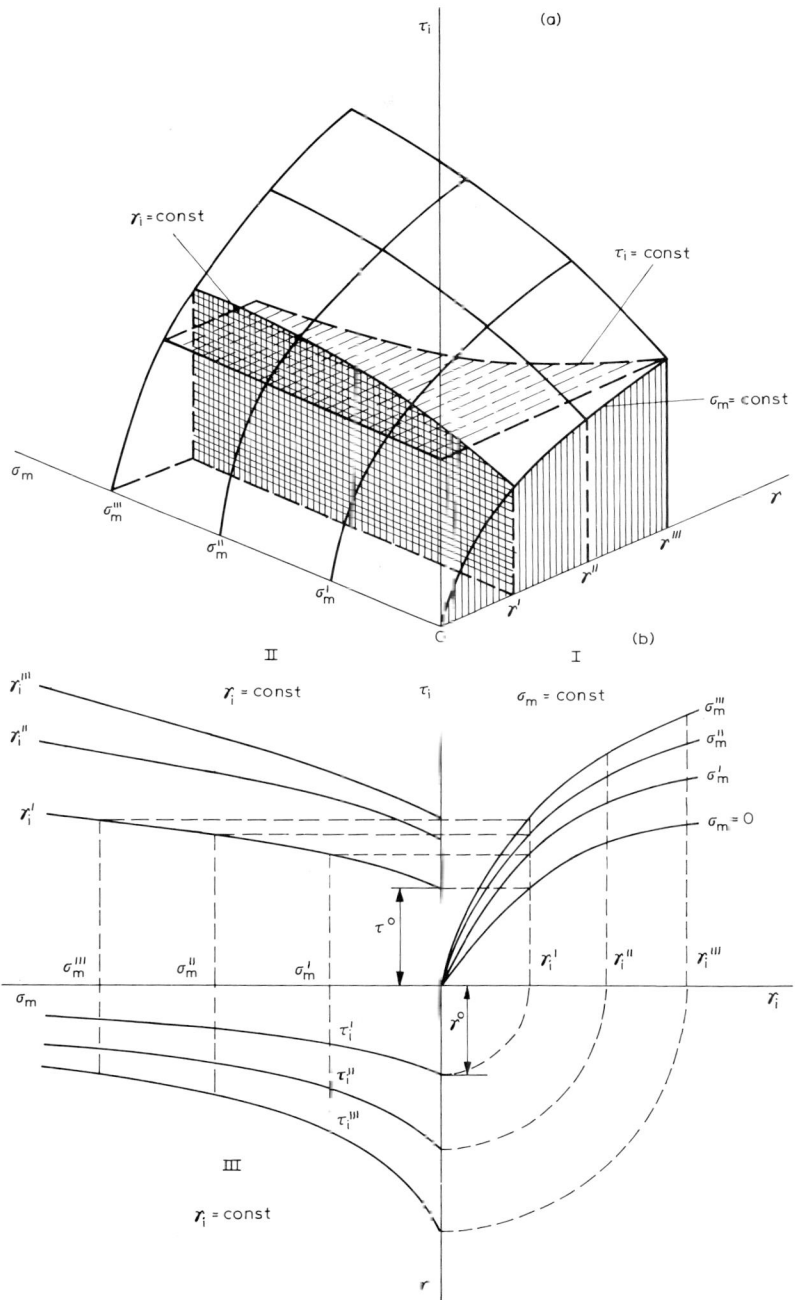

Fig. 11-3. (a) $(\gamma_i-\tau_i-\sigma_m)$ surface, and (b) contours of surface cross-sections on the planes $(\tau_i-\gamma_i)$, $(\tau_i-\sigma_m)$, $(\sigma_m-\gamma_i)$.

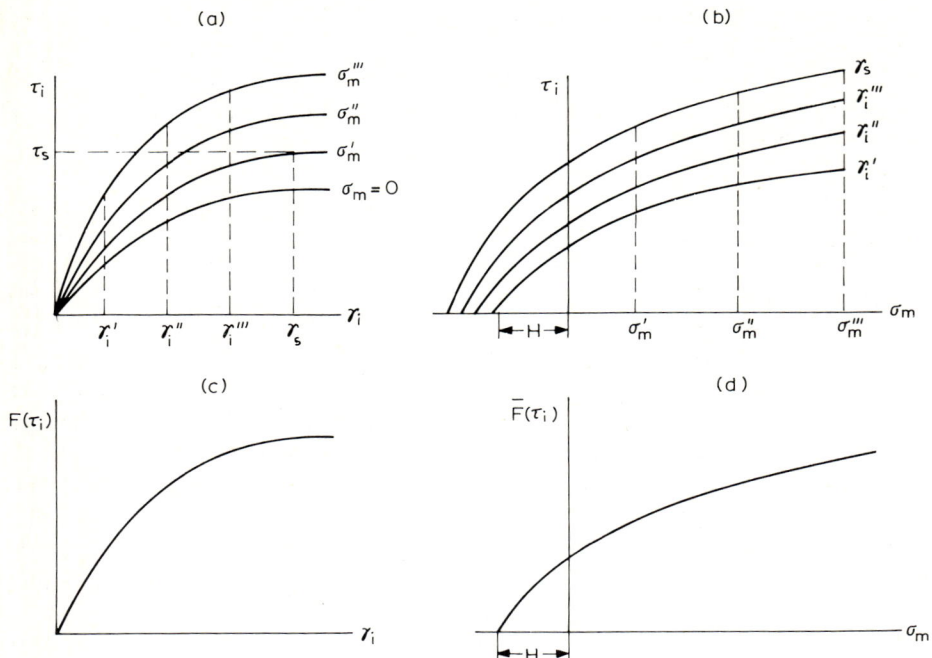

Fig. 11-4. (a) Curves τ_i versus γ_i, and (b) curves τ_i versus σ_m. (c) and (d) Curves reduced to the invariant form.

σ_m, changing it, however, from test to test. The result will be a graph of the τ_i–γ_i relation in the form of a family of τ_i–γ_i curves for various values of σ_m (Fig. 11-4(a)). To convert this graph into a τ_i–σ_m graph for various γ_i, intersect the τ_i–γ_i curves by verticals corresponding to arbitrary values γ'_i, γ''_i,... etc. On transferring the points of intersections to a τ_i–σ_m graph, we obtain a family of curves for various values of γ_i (Fig. 11-4(b)).

The transition into the limiting condition is decided by the strain attaining a certain limiting value $\gamma_i = \gamma_s$. Corresponding to this condition is, as a rule, the transformation of τ_i–γ_i curves into straight horizontal lines. In some instances such a transformation may not be the case, then the limiting state (conditional) is determined by either a sharp inflection of the τ_i–γ_i curve or by the strain γ_i reaching a given limiting value. The stress causing the transition into the limiting condition is $\tau_s = f(\sigma_m)$.

Generally, the limiting strain γ_s is also influenced by the mean normal stress, $\gamma_s = \phi(\sigma_m)$, but in analytical plots it is sometimes practical to assume that $\gamma_s = $ const. The limiting τ_i–σ_m envelope constructed for the value of $\gamma_i = \gamma_s$ (the upper curve of Fig. 11-4(b)) will represent the limiting condition of the soil.

The stress–strain curves depicted in Figs. 11-4(a) and (b) may be presented in an invariant form by reducing the families of τ_i–γ_i and τ_i–σ_m curves each to a single curve. To that end, we replot the curves on the $F(\tau_i)$–γ_i and $\bar{F}(\tau_i)$–σ_m coordinates

(Figs. 11-4(c) and (d)) where, according to eq. 11-22, $F(\tau_i) = \tau_i - \phi_2(\gamma_i, \sigma_m)$ and $\bar{F}(\tau_i) = [\tau_i - \phi_1(\gamma_i)]/\phi_3(\gamma_i)$.

Difference between the resistance of soil to tension and compression and the effect of mean normal stress

It has already been pointed out that variations in the shearing strain with the mean normal stress are attributed to the difference in the resistance of soil to the deformations in tension and compression. Let us prove this.

Consider a graph (Fig. 11-5(a)) of elastoplastic soil deformation under uniaxial compression and tension in the case when the modulus of elasticity and yield limit in compression are greater than in tension, $|+E| > |-E|, |+\sigma_s| > |-\sigma_s|$. Lay off as positive abscissa (compression) and as negative abscissa (tension) the same values of strain, $|+\varepsilon'_z| = |-\varepsilon'_z|$. Obviously, the compressive and tensile stresses corresponding to these values of ε'_z will have different moduli, $|+\sigma'_z| > |-\sigma'_z|$.

Next, construct stress circles for the values of $\pm \sigma'_z$ and draw a tangent $A'B'$ thereto (Fig. 11-5(b)). The slope ψ' of the tangent gives the angle of maximum obliquity, $\tan \psi' = \max [\tau'_n/(\sigma'_n + H')]$, where H' is a parameter of soil cohesion. The values of ψ' and H' will be the parameters of the tangent to the stress circles corresponding to a certain strain $\pm \varepsilon'_z = $ const.

If another value of strain $=\varepsilon''_z$ is now laid off in Fig. 11-5(a), we shall have other corresponding values of the stress $|+\sigma''_z| > |-\sigma''_z|$ and another envelope of the stress circles in Fig. 11-5(b) with the parameters ψ'' and H''. Consequently, the parameters ψ and H appear to be the functions of the strain, $\psi = \psi(\varepsilon_z)$ and $H = H(\varepsilon_z)$.

Setting forth the equation of the envelope in the known form, we obtain:

$$\frac{\sigma_1 - \sigma_2}{2} = \left[H(\varepsilon_1) + \frac{\sigma_1 + \sigma_2}{2}\right] \sin \psi(\varepsilon_1)$$

where $H = [\tau_n(\varepsilon_1)]/[\tan \psi(\varepsilon_1)]$; here $H \sin \psi = \tau_n \cos \psi = \tau^0$ is the pure shear stress.

The terms $(\sigma_1 - \sigma_2)/2$ and $(\sigma_1 + \sigma_2)/2$ are nothing other than the tangential stress intensity τ_i and the mean normal stress σ_m, respectively, in the two-dimensional state of strain. In its turn, the strain due to uniaxial compression/tension determines the shearing strain intensity, $\gamma_i = (2/\sqrt{3})(1 + \nu)\varepsilon_1$. Thus, the relationship obtained may be represented in the form:

$$\tau_i = [H(\gamma_i) + \sigma_m] \sin \psi(\gamma_i) \tag{11-23}$$

Taking $H(\gamma_i) \sin \psi(\gamma_i) = \phi(\gamma_i)$ and $\sin \psi(\gamma_i) = \phi_2(\gamma_i)$, we can see from eqs. 11-22 and 11-23 that the former expression is a specific case of the latter and that, in consequence, the difference in the resistances offered to tension and compression leads to the dependence of the shearing strain upon the value of mean normal stress σ_m.

Allowing for internal friction of soil in limiting and sublimiting conditions. It follows that allowing for the effect of the mean normal stress σ_m is equivalent to taking into consideration the internal friction not only in the limiting but in the sublimiting state as well. Let us explain this by means of a stress–strain graph and corresponding stress circles (Fig. 11-6).

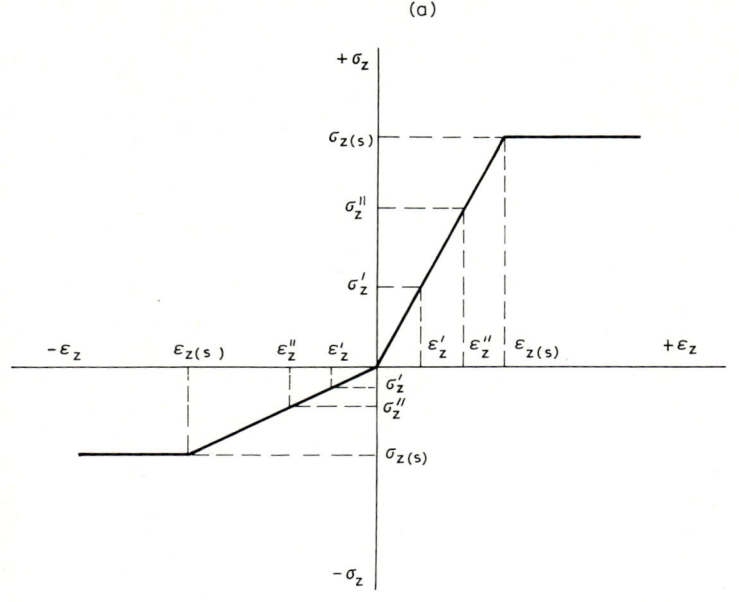
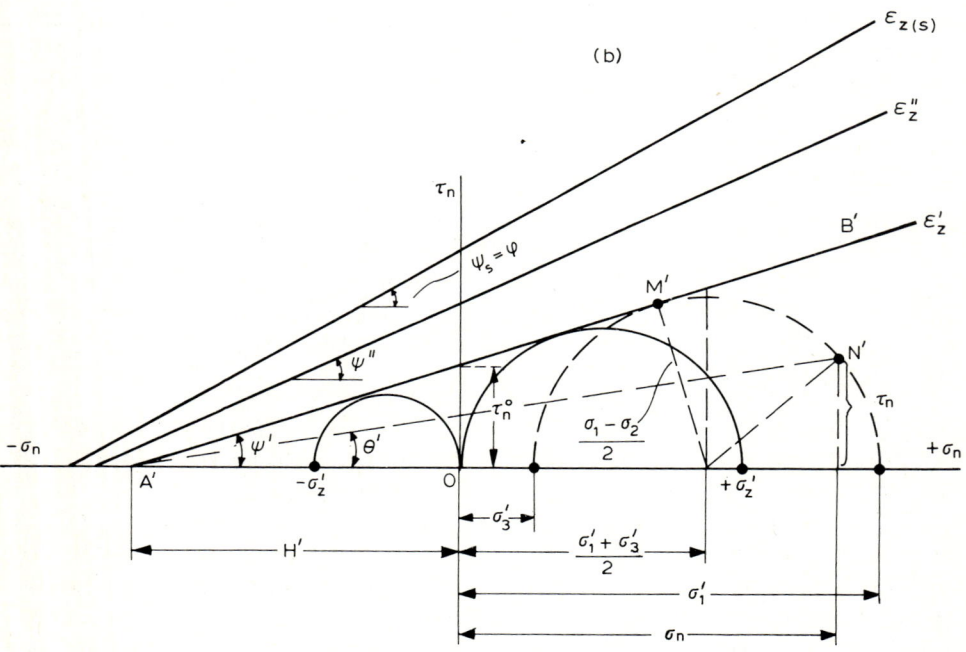

Fig. 11-5. Stress–strain graphs of materials displaying different resistance to compression ($+\sigma$) and tension ($-\sigma$).

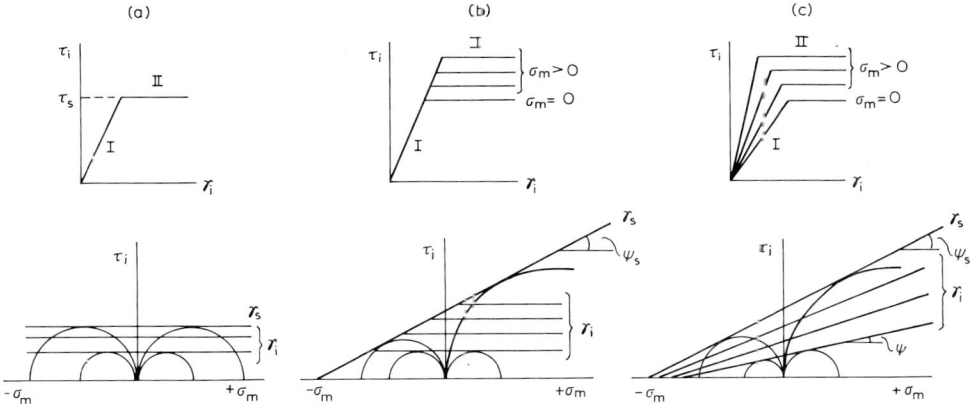

Fig. 11-6. Stress–strain graphs and stress circles for the elastoplastic state.

In graph (a), plotted on the τ_i–γ_i coordinates, the behaviour of elastoplastic materials devoid of internal friction obeys Hooke's law in the subliminating state, $\tau_i = G\gamma_i$, and is described by the St. Venant condition in the limiting state, $\tau_i = \tau_s = $ const. The τ_i–σ_m graph is, consequently, represented by the same stress circles for all values of γ_i, including the limiting value γ_s.

The behaviour of materials having internal friction is also assumed, in the general interpretation, to obey Hooke's law in the subliminating state, $\tau_i = G\gamma_i$, and the von Mises–Botkin (or Mohr–Coulomb) law in the limiting condition, $\tau_s = \tau^0 + \sigma_m \tan \psi$. Accordingly, the subliminating state is represented on the τ_i–γ_i graph of Fig. 11-6(b) by a single straight line and the limiting condition is shown by a family of straight lines $\tau_i = \tau_s$ dependent upon the mean normal stress σ_m. Plotting a τ_i–σ_m graph, we can see that the above assumption lacks logic. In fact, the tangents to the stress circles for all values of strain $\gamma_i < \gamma_s$ (i.e. in the subliminating state) appear to be horizontal lines. The only exclusion is the tangent for $\gamma_i = \gamma_s$ (limiting state) which displays a slope.

This incongruity may be eliminated by making an allowance for friction in both the limiting and subliminating condition. The τ_i–γ_i graph (Fig. 11-6(c)) will then show a family of broken lines each corresponding to a value of σ_m and the τ_i–σ_m graph will be represented by a family of straight lines having a slope ψ. Alterations of the slope over the range $0 < \psi < \psi_s$ represent a gradual transition from the subliminating into the limiting state.

Mechanical model of soil allowing for the effect of σ_m. The effect of the mean normal stress σ_m on the process of deformation may be examined by means of a mechanical model depicted in Fig. 11-7 (whenever the time factor is to be taken into account, this model is combined with the model of Fig. 7-6(a)).

Assuming, for simplicity, that the stress–strain relationship is linear, $\tau_{i(1)} = G\gamma_{i(1)}$, we can simulate the resistance of soil to shear by an elastic element 1. The effect of the mean normal stress σ_m will be represented by a friction, Coulomb element 2,

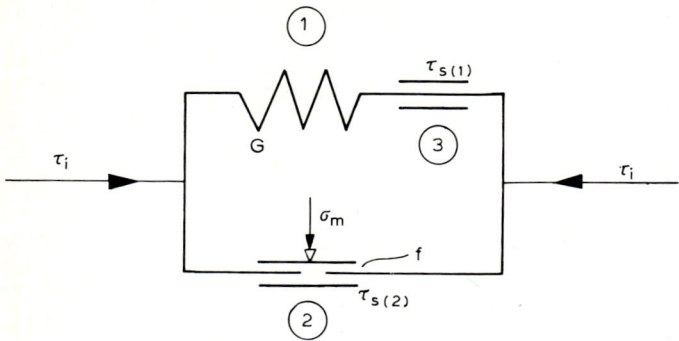

Fig. 11-7. Mechanical model simulating the effect of the mean normal stress σ_m.

assuming that $\gamma_{i(2)} = \tau_{i(2)}/(B\sigma_m)$, where $B = f/\gamma_{i(2)}$ is the friction coefficient referred to as unit strain.

Since the elements are parallel-connected, $\tau_i = \tau_{i(1)} + \tau_{i(2)}$ and $\gamma_i = \gamma_{i(1)} = \gamma_{i(2)}$. Hence:

$$\tau_i = \gamma_i(G + B\sigma_m) = \tau^0\left(1 + \frac{\sigma_m}{H}\right) = \tan\psi(H + \sigma_m) \tag{11-24}$$

where $\tan\psi = B\gamma_i$ is the tangent of the slope decided by the amount of deformation, i.e. $\psi = \psi(\gamma_i)$; $H = G/B = $ const. is a parameter of soil cohesion.

The transition into the limiting state may be represented by connecting a St. Venant element (Geniev, 1966) to the model. It is assumed that spring 1 can compress until $\gamma_i = \gamma_s$ and then the model becomes capable of displacing in an unconfined way. To effect such a displacement, the force required is $\tau_i = \tau_{s(1)} + \tau_{s(2)}$, where $\tau_{s(1)} = G\gamma_s$ and $\tau_{s(2)} = \sigma_m \tan\psi_s$. Consequently, eq. 11-24 is transformed into the von Mises–Schleicher–Botkin limiting condition:

$$\tau_i = \tau_s^0\left(1 + \frac{\sigma_m}{H_s}\right) = \tau_s^0 + \sigma_m \tan\psi_s \tag{11-25}$$

Connection between the parameters of eq. 11-24 and deformation characteristics during compression and tension

The parameters of eq. 11-24 may be expressed in terms of the soil deformation characteristics E_{comp}, γ_{comp}, E_{tens}, v_{tens} corresponding to compression and tension (see Fig. 11-5).

For the state of uniaxial stress, we have from Table 3-I: $\tau_i = \sigma_1/\sqrt{3}$, $\gamma_i = 2\varepsilon_1(1 + v)/\sqrt{3}$, and $\sigma_m = \sigma_1/3$.

Accordingly, for compression we have:

$$\tau_i = \frac{E_{comp}\varepsilon_1}{3} = \frac{E_{comp}\gamma_i}{2(1 + v_{comp})}$$

$$\sigma_m = \frac{E_{comp}\varepsilon_1}{3} = \frac{E_{comp}\gamma_i}{2(1+v_{comp})\sqrt{3}}$$

and for tension:

$$\tau_i = \frac{E_{tens}\gamma_i}{2(1+v_{tens})} \qquad \sigma_m = -\frac{E_{tens}\gamma_i}{2(1+v_{tens})\sqrt{3}}$$

Substituting the above expressions into eq. 11-24, we can determine the parameters entering into it:

$$\left. \begin{aligned} \tan\psi &= \sqrt{3}\frac{(1+v_{tens})E_{comp}-(1+v_{comp})E_{tens}}{(1+v_{tens})E_{comp}+(1+v_{comp})E_{tens}} \\ H &= \frac{E_{comp}E_{tens}\gamma_i}{\sqrt{3}[(1+v_{tens})E_{comp}-(1+v_{comp})E_{tens}]} \\ G &= \frac{E_{comp}E_{tens}}{(1+v_{tens})E_{comp}+(1+v_{comp})E_{tens}} \end{aligned} \right| \qquad (11\text{-}26)$$

In those cases when Poisson's ratios are the same in compression and tension, $v_{comp} = v_{tens} = v$, we obtain the above expressions in a simplified form:

$$\left. \begin{aligned} \tan\psi &= \frac{E_{comp}-E_{tens}}{E_{comp}+E_{tens}} \\ H &= \frac{E_{comp}E_{tens}\gamma_i}{\sqrt{3}(1+v)(E_{comp}-E_{tens})} \\ G &= \frac{E_{comp}E_{tens}}{(1+v)(E_{comp}+E_{tens})} \end{aligned} \right| \qquad (11\text{-}27)$$

A salient feature of the relationships obtained is, firstly, a constant angle ψ. This means that under any given condition the angle of obliquity at any stage of deformation equals the angle of internal friction (this will be examined below). Secondly, deformation can develop only under the stresses $\tau_i > \sigma_m \tan\psi$ — a fact apparent from substituting eqs. 11-26 or 11-27 into eq. 11-24. This latter limitation corresponds to the behaviour of a rigid-hardening body and follows from the particulars of the graphs in Fig. 11-5(a) on which the σ_z-ε_z straight line has an inflection at the origin which, to a certain extent, is conditional.

11.4 EQUATION OF SOIL CREEP ALLOWING FOR THE EFFECT OF MEAN NORMAL STRESS

$(\tau_i$-γ_i-$\sigma_m)$ and $(\tau_i$-γ_i-$t)$ surfaces

The connection between τ_i, γ_i and σ_m was considered in the preceding section, disregarding the time factor. To take this factor into account, one needs to examine the $(\tau_i-\gamma_i-\sigma_m)$ surface of Fig. 11-3 simultaneously with the $(\tau_i-\gamma_i-t)$ surface of Fig. 7-10. Accordingly, the equation of deformation 11-22 takes the form:

$$\left. \begin{array}{l} \gamma_i = f_1(\tau_i, t) - f_2(\tau_i, \sigma_m, t) \\ \tau_i = \phi_1(\gamma_i, t) + \phi_2(\gamma_i, \sigma_m, t) \end{array} \right| \tag{11-28}$$

If the stress–strain curves appear to be similar, we can write:

$$\gamma_i = f_1(\tau_i)\Phi_1(t) - f_2(\tau_i)\bar{\psi}(\sigma_m)\Phi_2(t) \tag{11-29}$$

or:

$$\tau_i = \phi_1(\gamma_i)F_1(t) + \phi_2(\gamma_i)\bar{\Omega}(\sigma_m)F_2(t)$$

where f_1 and ϕ_1 are defined, in accordance with Fig. 11-3, by the configuration of the τ_i–γ_i curves for $\sigma_m = 0$ (quadrant I). Similarly, f_2 and ϕ_2 are defined by the configuration of the same curves for $\sigma_m > 0$ (quadrant I); $\bar{\Omega}$, by the configuration of the τ_i–σ_m curves in quadrant II and ψ, by the same curves in quadrant III.

In a specific case, the functions $\phi_1 = \phi_2 = \phi$ and $F_1 = F_2 = F$. Then:

$$\gamma_i = f(\tau_i)\psi(\sigma_m)\Phi(t) \qquad \tau_i = \phi(\gamma_i)\Omega(\sigma_m)F(t) \tag{11-30}$$

where $\psi(\sigma_m) = 1 + \bar{\psi}(\sigma_m)$ and $\Omega(\sigma_m) = 1 - \bar{\Omega}(\sigma_m)$.

The above relations correspond to the theory of ageing and hold for stresses which are either constant or changing slowly and monotonously.

Integral form of deformation equation

When the pattern of load changes is arbitrary (under the conditions of active loading), the connection between the strain γ_i and stress τ_i with allowance for the mean normal stress σ_m is expressed in terms of the integral Boltzmann–Volterra relationship, as is commonly the case in the theory of hereditary creep. Then, instead of eq. 11-22 the relations 11-31 and 11-32 hold:

$$\gamma_i = f_1[\tau_i(\sigma_m)] - \int_0^{\sigma_m} P(\sigma_m - \xi)f_2[\sigma_m(\xi)]\,d\xi \tag{11-31}$$

$$\tau_i = \phi_1[\gamma_i(\sigma_m)] + \int_0^{\sigma_m} S(\sigma_m - \xi)\phi_2[\gamma_i(\xi)]\,d\xi \tag{11-32}$$

By analogy with eq. 11-22, the first terms in these expressions represent the strain and stress of pure shear and the second terms reflect the retarding effect of mean normal stress.

To take account of the time factor, relations 11-31 and 11-32 are examined simultaneously with the relationships of hereditary creep 7-41 and 7-43. Thus, we obtain:

$$\gamma_i = f_1[\tau_i(\sigma_m, t)] + \int_0^t Q_1(t - v)f_1[\tau_i(v)]\,dv - \int_0^{\sigma_m(t)} P(\sigma_m - \xi)f_2[\tau_i(\xi)]\,d\xi$$

$$- \int_0^t \int_0^{\sigma_m(t)} Q_2(t - v)P(\sigma_m - \xi)f_2[\tau_i(\xi)]\,d\xi\,dv \tag{11-33}$$

$$\tau_i = \phi_1[\gamma_i(\sigma_m, t)] - \int_0^t R_1(t - v)\phi_1[\gamma_i(v)]\,dv + \int_0^{\sigma_m(t)} S(\sigma_m - \xi)\phi_2[\gamma_i(\xi)]\,d\xi$$

$$- \int_0^t \int_0^{\sigma_m(t)} R_2(t - v)S(\sigma_m - \xi)\phi_2[\gamma_i(\xi)]\,d\xi\,dv \tag{11-34}$$

In the above expressions, the notation used is as follows: $f_1(\tau_i) = \gamma^0$ and $\phi_1(\gamma_i) = \tau^0$ are the functions defining the stress–strain relationships for pure shear ($\sigma_m = 0$) at the moment $t = 0$; $f_2(\tau_i)$ and $\phi_2(\gamma_i)$ are the functions defining the stress–strain relationships for $\sigma_m > 0$ at the moment $t = 0$; for $\Phi(t) = 1$, these functions are defined according to eq. 11-29 by the expressions:

$$f_2(\tau_i) = -(\gamma_i - \gamma^0)/\bar{\Psi}(\sigma_m) \quad \text{and} \quad \phi_2(\gamma_i) = (\tau_i - \tau^0)/\bar{\Omega}(\sigma_m)$$

$$Q_1(t) = \frac{1}{f_1(\tau_i)}\frac{d\gamma_i}{dt} \quad \text{and} \quad R_1(t) = -\frac{1}{\phi_1(\gamma_i)}\frac{d\tau_i}{dt}$$

are the kernels of the equations of creep and relaxation for pure creep ($\sigma_m = 0$) which are similar to the kernels of eqs. 7-41 and 7-43:

$$Q_2(t) = -\frac{1}{f_2(\tau_i)}\frac{d}{dt}(\gamma_i - \gamma^0) \quad \text{and} \quad R_2(t) = \frac{1}{\phi_2(\gamma_i)}\frac{d}{dt}(\tau_i - \tau^0)$$

are the kernels of the equations of creep and relaxation for $\sigma_m > 0$ defining the time-dependent changes in the strains and stresses in excess of those due to pure shear:

$$P(\sigma_m) = -\frac{1}{f_2(\tau_i)}\frac{d\gamma_i}{d\sigma_m} \quad \text{and} \quad S(\sigma_m) = \frac{1}{\phi_2(\gamma_i)}\frac{d\tau_i}{d\sigma_m}$$

are the kernels of the integral equations of deformation allowing for the effect of σ_m. These functions define the changes in stresses and strains resulting from an increase in the mean normal stress σ_m and are formally analogous with the creep kernels Q and relaxation kernels R of the equations of hereditary creep 7-41 and 7-43; the only difference is that Q and R define the changes in γ_i and τ_i with t whereas P and S characterize the changes with σ_m.

The first two terms in expressions 11-33 and 11-34 correspond to the strain or stress of pure shear, the first terms representing an instantaneous condition and the second terms taking care of time-dependent conditions. The third and fourth terms of eqs. 11-33 and 11-34 reflect the influence of the mean normal stress σ_m, the third terms characterizing this influence at $t = 0$ and the fourth terms presenting the changes in the influence with time. If we assume that $f_1 = f_2$ and $Q_1 = Q_2$ (or $\phi_1 = \phi_2$ and $R_1 = R_2$), eqs. 11-33 and 11-34 are significantly simplified.

From the values of P, S and eq. 11-29 it follows that:

$$P = \frac{d\bar{\psi}}{d\sigma_m} \qquad S = \frac{d\bar{\Omega}}{d\sigma_m}$$

or:

$$\bar{\psi} = \int_0^{\sigma_m} P \, d\sigma_m \qquad \bar{\Omega} = \int_0^{\sigma_m} S \, d\sigma_m$$

The values of P and S may be substituted into eqs. 11-33 and 11-34 only in those cases when the loading is simple or close to simple.

If the load is constant, $\tau_i = $ const., and so is the deformation, $\gamma_i = $ const., we obtain:

$$\gamma_i = f_1(\tau_i)\left[1 + \int_0^t Q(t)\,dt\right] - f_2(\tau_i)\bar{\psi}(\sigma_m)\left[1 + \int_0^t Q_2(t)\,dt\right] \tag{11-35}$$

$$\tau_i = \phi_1(\gamma_i)\left[1 - \int_0^t R_1(t)\,dt\right] + \phi_2(\gamma_i)\bar{\Omega}(\sigma_m)\left[1 - \int_0^t R_2(t)\,dt\right] \tag{11-36}$$

These expressions are identical with eq. 11-29.

For $f_1 = f_2 = f$, $Q_1 = Q_2 = Q$ and $\phi_1 = \phi_2 = \phi$, $R_1 = R_2 = R$, eqs. 11-35 and 11-36 take the form:

$$\left. \begin{aligned} \gamma_i &= f(\tau_i)\psi(\sigma_m)\left[1 + \int_0^t Q(t)\,dt\right] \\ \tau_i &= \phi(\gamma_i)\Omega(\sigma_m)\left[1 - \int_0^t R(t)\,dt\right] \end{aligned} \right| \tag{11-37}$$

where $\psi(\sigma_m) = 1 + \bar{\psi}(\sigma_m)$ and $\Omega(\sigma_m) = 1 - \bar{\Omega}(\sigma_m)$.

Apparently, these expressions are identical with eq. 11-30.

Functions entering into eqs. 11-33 and 11-34. The functions f, ϕ, Q and R in eqs. 11-33 and 11-34 were examined in Chapter 7. Accordingly, the connection between the stress and strain may be adopted in the form:

$$\phi_1(\gamma_i) = A_0\gamma_i^{m_1} \qquad \phi_2(\gamma_i) = B_0\gamma_i^{m_2} \tag{11-38}$$

As far as the type of the functions Ω and S is concerned, we may accept that on approximating the τ_i–γ_m curves by straight lines — a practice common in soil mechanics — that:

$$\bar{\Omega}(\sigma_m) = \sigma_m \qquad S(\sigma_m) = \frac{d\bar{\Omega}}{dt} = 1 \tag{11-39}$$

On substituting eqs. 11-38 and 11-39 into eqs. 11-36 and 11-29, the equations of deformation, taking account of the effect of σ_m, which have been derived from the theory of hereditary creep in the former case and from the ageing theory in the latter case, will have the following form:

$$\tau_i = A_0 \left\{ [\gamma_i(t)]^{m_1} - \int_0^t R_1(t - v)[\gamma_i(v)]^{m_1} dv \right\}$$

$$+ B_0 \left\{ \sigma_m(t)[\gamma_i(t)]^{m_2} - \int_0^t R_2(t - v)\sigma_m(v)[\gamma_i(v)]^{m_2} dv \right\} \quad (11\text{-}40)$$

and:

$$\tau_i = A_0 \gamma_i^{m_1} F_1(t) + B_0 \sigma_m \gamma_i^{m_2} F_2(t) \quad (11\text{-}41)$$

Denoting $F_1 = A(t)/A_0$ and $F_2(t) = B(t)/B_0$, we obtain from eq. 11-41:

$$\tau_i = A(t)\gamma_i^{m_1} + B(t)\sigma_m \gamma_i^{m_2} \quad (11\text{-}42)$$

The equation of deformation of the above type allows for taking account of time in both functional and parametric forms

The function $A(t)$ in all its possible forms was examined in Chapter 7; the function $B(t)$ may be represented by analogy with the function $A(t)$. A power relation of the type given by eq. 7-87 is one of the simplest relationships fitting the experimental data well:

$$A(t) = \frac{A_0}{1 + \delta_1 t^{\alpha_1}} \qquad B(t) = \frac{B_0}{1 + \delta_2 t^{\alpha_2}} \quad (11\text{-}43)$$

Analysis of deformation equation 11-42

The meaning of the terms in this equation is as follows:

$$A_0 \gamma_i^{m_1} = \tau^0(\gamma_i) \qquad B_0 \gamma_i^{m_2} = \tan \psi(\gamma_i)$$

$$\frac{A_0}{B_0} \gamma_i^{m_1 - m_2} = \frac{\tau^0(\gamma_i)}{\tan \psi(\gamma_i)} = H(\gamma_i) \quad (11\text{-}44)$$

where $\tau^0(\gamma_i)$ are the intercepts on the y-axis of the graph in Fig. 11-8(a) which are formed by the $\tau_i\text{-}\sigma_m$ straight lines and give the resistance to pure shear; $H(\gamma_i)$ are the intercepts on the x-axis which are formed by the $\tau_i\text{-}\sigma_m$ straight lines and determine the resistance to all-around tension; and $\psi(\gamma_i)$ are the slopes of the $\tau_i\text{-}\sigma_m$ straight lines corresponding to the angles of obliquity at the octahedral plane for a given value of γ_i.

All the parameters mentioned are dependent upon deformation and, consequently, upon time, for $\gamma_i = \gamma_i(t)$. However, the values of τ^0, ψ and H may occur in certain combinations so that these give rise to various special cases of eq. 11-42.

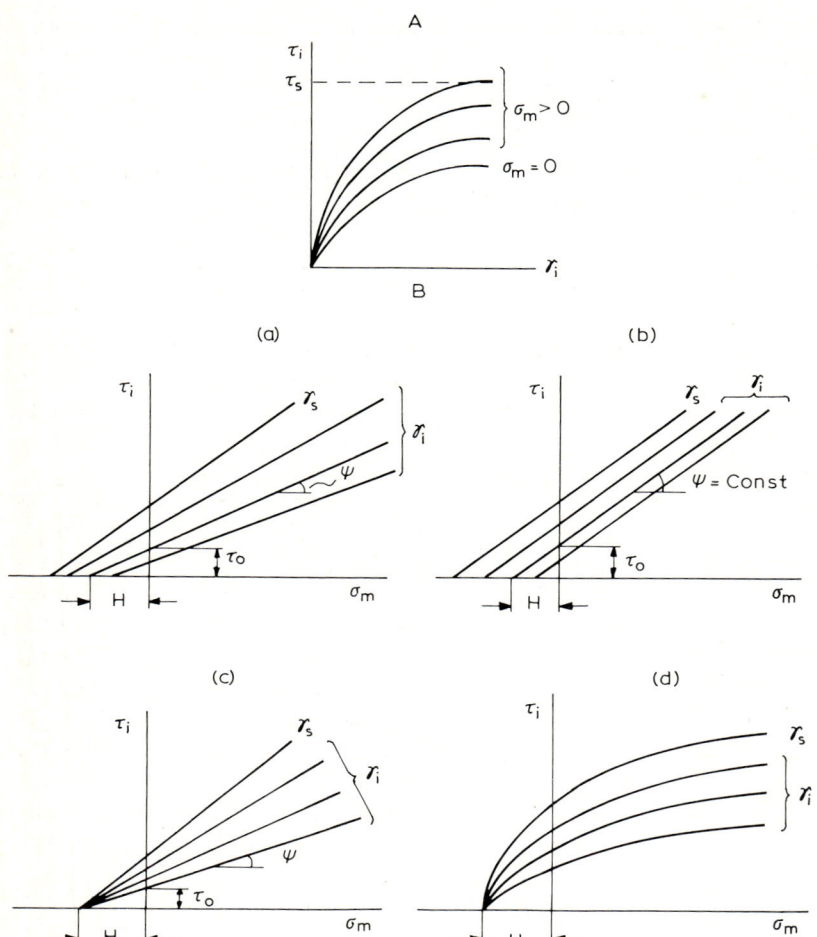

Fig. 11-8. (A) Stress–strain graph, and (B) relationship between τ_i and σ_m for the following cases: (a) $H = H(\gamma_i)$, $\psi = \psi(\gamma_i)$; (b) $H = H(\gamma_i)$, $\psi = $ const.; (c) $H = $ const., $\psi = \psi(\gamma_i)$; (d) $H = $ const., $\psi = \psi(\gamma_i; \sigma_m)$.

Fig. 11-8(a). This is one of the most general cases corresponding to eq. 11-42 for $m_1 \neq m_2$. The equation may be rewritten, taking into account eq. 11-44, as:

$$\tau_i = A(t)\gamma_i^{m_1}\left[1 + \frac{\sigma_m}{H(\gamma_i, t)}\right] \tag{11-45}$$

The parameters of this equation τ, ψ and $H = \tau^0/\tan \psi$ change depending on the amount of deformation and with time.

Fig. 11-8(b). If the exponents in eq. 11-42 are $m_1 \neq 0$ and $m_2 = 0$, the parameter $H = H(\gamma_i, t)$ is a function of γ_i and t whereas the angle ψ is affected neither by deformation nor by time, $\tan \psi = \tan \psi_s = $ const.

The τ_i–σ_m graph is represented by a family of parallel straight lines, and eq. 11-42 may be rewritten in the form:

$$\tau_i = A(t)\gamma_i^{m_1}\left[1 + \frac{\sigma_m}{H(\gamma_i, t)}\right] = A(t)\gamma_i^{m_1} + \sigma_m \tan \psi_s \qquad (11\text{-}46)$$

Note that this expression represents the deformation of a rigid-hardening body which is likely to deform only when $\tau_i > \sigma_m \tan \psi_s$.

Fig. 11-8(c). When the exponents in eq. 11-42 are $m_1 = m_2 = m$, the angle $\psi = \psi(\gamma_i, t)$ is a function of γ_i and t whereas the parameter H is not influenced by γ_i. Moreover, for $A(t)/B(t) = $ const., H is influenced neither by γ_i nor by t, i.e. $H = H_s = $ const. The τ_i–σ_m graph is represented by a family of straight lines emanating from the same pole, and eq. 11-42 takes the form:

$$\tau_i = A(t)\gamma_i^m\left(1 + \frac{\sigma_m}{H_s}\right) \qquad (11\text{-}47)$$

A specific case of the above expression is when $A = 0$ and the τ_i–σ_m graph is represented by a family of straight lines emanating from the origin ($H = 0$):

$$\tau_i = \sigma_m \tan \psi(\gamma_i, t) \qquad (11\text{-}48)$$

Non-linear τ_i–σ_m relation in Fig. 11-8(d). Consider the case when the τ_i–σ_m relation is given in the power form instead of being linear as in eq. 11-42.

This relation has been obtained by Gorodetsky (1975) from experimental data. Consequently, the τ_i–σ_m graph will be represented by a family of curves emanating from the same pole $H = H_s = $ const., and the equation of deformation 11-29 takes the following form, providing we put $\phi_1(\gamma_i) = \phi_2(\gamma_i) = A(t)\gamma_i^m$:

$$\tau_i = A(t)\gamma_i^m\left(1 + \frac{\sigma_m}{H_s}\right)^\lambda \qquad (11\text{-}49)$$

where $A(t)\gamma_i^m = \tau^0(t)$.

For the test of fit of eq. 11-49 and to determine its parameters, we find the value of $\tau^0(\gamma_i)$ from the τ^0–γ_i graph or in some other way. To determine H, we reduce eq. 11-49 to the form $y = q\,e^{kx} + c$, where $y = \sigma_m$, $x = [\ln \tau_i - \ln \tau^0(t)]$, $k = 1/\lambda$, $q = -c = H$.

On plotting a curve on the (y, x) coordinates, we take three arbitrary points on this curve with the abscissas x_1, x_2, $x_3 = (x_1 + x_2)/2$ and find the value of H from the expression $H = (y_3^2 - y_1 y_2)/(y_1 + y_2 - 2y_3)$, where y_1, y_2, y_3 are the ordinates of the points selected.

Once τ^0 and H are known, eq. 11-49 is transformed into a linear expression:

$$\ln \tau_i = \ln \tau^0(t) + \lambda \ln\left(1 + \frac{\sigma_m}{H_s}\right)$$

which is used in the test of fit.

Linear–fractional relation. The non-linear τ_i–σ_m relation may be expressed in terms of a linear–fractional expression. In accordance with eq. 5-22' we have:

$$\tau_i = \frac{G_0 \tau_s \gamma_i (T + t)}{(\tau_s + G_0 \gamma_i)(T + \delta t)} \tag{11-50}$$

where $\tau_s = \lim \tau_i = \Phi(\sigma_m)$ for $\gamma_i \to \infty$, $t \to 0$.

Supposing the function $\Phi(\sigma_m)$ is also of the linear–fractional type, τ_s in eq. 11-50 will take the form:

$$\tau_s = \frac{\tau_s^*(H_s + \sigma_m)\tan\bar{\psi}}{\tau_s^* + (H_s + \sigma_m)\tan\bar{\psi}} \tag{11-50'}$$

where $\tau_s^* = \lim \tau_s$ for $\sigma_m \to \infty$, $\tan\bar{\psi} = \lim [\tau_s/(H_s + \sigma_m)]$ for $(H_s + \sigma_m) \to 0$, and $H_s = \mathrm{const.}$

However, if the function $\Phi(\sigma_m)$ is assumed to be linear, τ_s in eq. 11-50 takes the form:

$$\tau_s = (H_s + \sigma_m)\tan\psi_s = \tau^0\left(1 + \frac{\sigma_m}{H_s}\right) \tag{11-51}$$

where τ_s^0 is the yield point in pure shear.

The value of G_0 in eq. 11-50 is also influenced by σ_m, the relevant relationship being expressed by analogy with eq. 11-50' or 11-51. In this latter case we shall have:

$$G_0 = \tilde{G}(H_s + \sigma_m) = G_0^0\left(1 + \frac{\sigma_m}{H_s}\right) \tag{11-51'}$$

where G_0^0 is the value of G_0 in pure shear; $\tilde{G} = G_0^0/H_s$. Substituting eqs. 11-51 and 11-50' into eq. 11-50, we obtain:

$$\tau_i = (H_s + \sigma_m)\frac{\tilde{G}\tan\psi_s(T + t)\gamma_i}{(\tan\psi_s + \tilde{G}\gamma_i)(T + \delta t)}$$

$$= \left(1 + \frac{\sigma_m}{H_s}\right)\frac{G_0^0 \tau_s^0 \gamma_0(T + t)\gamma_i}{(\tau_s^0 + G_0^0 \gamma_i)(T + \delta t)} \tag{11-52}$$

Invariant graphs

The equations of deformation examined above may be expressed by means of invariant graphs on which the families of τ_i–γ_i and τ_i–σ_m curves are reduced to single curves plotted on the generalized coordinates $F(\tau_i)/\Phi(t) - \gamma_i$ and $\bar{F}(\tau_i)/\Phi(t) - \sigma_m/H$, as is depicted in Fig. 11-9.

The values of the functions mentioned for eq. 11-45 and for eqs. 11-46 and 11-47 derived from this expression will be as follows.

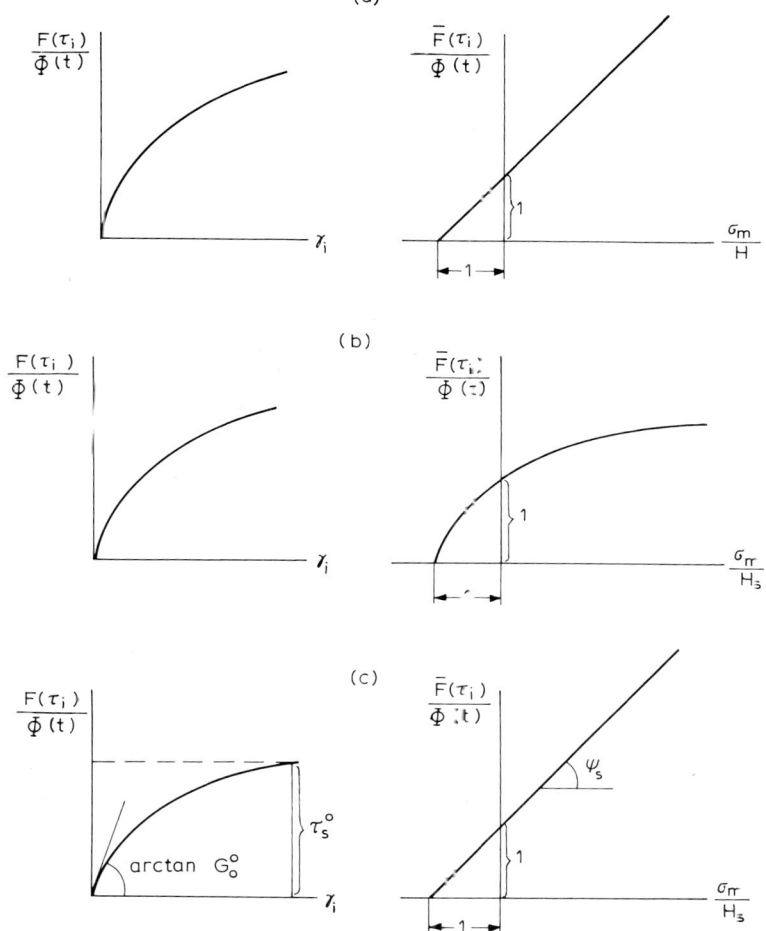

Fig. 11-9. Invariant stress–strain graphs corresponding to: (a) eq. 11-45; (b) eq. 11-49; (c) eq. 11-50.

For eq. 11-45:

$$F(\tau_i) = \frac{\tau_i}{1 + \dfrac{\sigma_m}{H(\gamma_i, t)}} \qquad \bar{F}(\tau_i) = \frac{\tau_i}{\gamma_i^m} \qquad \Phi(t) = A(t)$$

For eq. 11-46:

$$F(\tau_i) = \frac{\tau_i}{1 + \dfrac{\sigma_m}{H(\gamma_i, t)}} \qquad \bar{F}(\tau_i) = \frac{\tau_i}{1 + \dfrac{\sigma_m}{H(\gamma_i, t)}} = \tau_i - \sigma_m \tan \psi_s \qquad \Phi(t) = A(t)$$

For eq. 11-47:

$$F(\tau_i) = \frac{\tau_i}{1 + \dfrac{\sigma_m}{H_s}} \qquad \bar{F}(\tau_i) = \frac{\tau_i}{\gamma_i^m} \qquad \Phi(t) = A(t)$$

For eq. 11-49, the values of the functions will be the same:

$$F(\tau_i) = \frac{\tau_i}{\left(1 + \dfrac{\sigma_m}{H_s}\right)^\lambda} \qquad \bar{F}(\tau_i) = \frac{\tau_i}{\gamma_i^m} \qquad \Phi(t) = A(t)$$

For eq. 11-52, the functions will be as follows:

$$F(\tau_i) = \frac{\tau_i}{1 + \dfrac{\sigma_m}{H_s}} \qquad \bar{F}(\tau_i) = \tau_i \frac{\tau_s^0 + G_0^0 \gamma_i}{\tau_s^0 G_0^0 \gamma_i} \qquad \Phi(t) = \frac{T + t}{T + \delta t}$$

So far insufficient investigation of the problem prevents us from giving positive recommendations about the assets of any of the formulae examined above. Nevertheless, experimental evidence is in favour of eq. 11-45 with the simplifying assumptions that $m_1 = m_2$ or $m_2 = 0$. The assumption that $m_2 = 0$ appears to be in good agreement with the experimental data, indicating that the angle of internal friction is not influenced by the amount of the deformation developed.

Elastoplastic condition

Putting $m_1 = m_2 = 1$ in eq. 11-45, we obtain an equation of linear deformation allowing for the effect of σ_m. All special cases represented in the graphs of Fig. 11-8(b) are taken care of by the equation.

The equations of elastoplastic deformation represented by Fig. 11-6(c) take the form:

$$\frac{\tau_i}{1 + \dfrac{\sigma_m}{H}} = G\gamma_i \quad \text{for } \tau_i < \tau_s$$

$$\frac{\tau_i}{1 + \dfrac{\sigma_m}{H_s}} = \tau_s^0 \quad \text{for } \tau_i = \tau_s$$

Note that the parameters H_s, ψ_s, and τ_s^0 of the limiting condition of a soil may be expressed in terms of the strength characteristics of the soil in uniaxial compression R_{comp} and tension R_{tens}:

$$\tan \psi_s = \sqrt{3}\,\frac{R_{comp} - R_{tens}}{R_{comp} + R_{tens}} \qquad H_s = \frac{2}{3}\frac{R_{comp} R_{tens}}{R_{comp} - R_{tens}} \qquad \tau_s^0 = \frac{2}{\sqrt{3}}\frac{R_{comp} R_{tens}}{R_{comp} + R_{tens}}$$

11.5 EQUATION OF VISCOPLASTIC FLOW OF SOIL ALLOWING FOR THE EFFECT OF MEAN NORMAL STRESS

Connection between $\dot{\gamma}_i$, τ_i, σ_m, and t

We considered above the connection between the shearing strain intensity γ_i, the tangential stress intensity τ_i, the mean normal stress σ_m and time t. Similar connections between the shearing strain rate intensity $\dot{\gamma}_i$ and τ_i, σ_m, and t may be established from eq. 11-16 of the theory of flow.

Since time t enters this relationship in the explicit form, the connection between $\dot{\gamma}_i$, τ_i, σ_m, and t is determined by considering the $(\dot{\gamma}_i - \tau_i - \sigma_m)$ surface simultaneously with the $(\dot{\gamma}_i - \tau_i - t)$ surface. The former is analogous with the surface of Fig. 11-3 and the latter is represented in Fig. 7-12.

Assuming similarity of curves, relationships 11-16 may be presented for $\mu_\sigma = $ const. in the following form:

$$\begin{aligned} \dot{\gamma}_i &= f_1(\tau_i)\kappa_1(t) - f_2(\tau_i)\bar{\psi}(\sigma_m)\kappa_2(t) \\ \tau_i &= \phi_1(\dot{\gamma}_i)\eta_1(t) + \phi_2(\dot{\gamma}_i)\bar{\Omega}(\sigma_m)\eta_2(t) \end{aligned} \qquad (11\text{-}53)$$

In a specific case for $f_1 = f_2 = f$ and $\kappa_1 = \kappa_2 = \kappa$ or $\phi_1 = \phi_2 = \phi$ and $\eta_1 = \eta_2 = \eta$, expressions 11-53 become:

$$\begin{aligned} \dot{\gamma}_i &= f(\tau_i)\psi(\sigma_m)\kappa(t) \\ \tau_i &= \phi(\dot{\gamma}_i)\Omega(\sigma_m)\eta(t) \end{aligned} \qquad (11\text{-}54)$$

where:

$$\psi(\sigma_m) = 1 + \bar{\psi}(\sigma_m); \qquad \Omega(\sigma_m) = 1 - \bar{\Omega}(\sigma_m)$$

Assuming that in eq. 11-54 the functions of ϕ_1 and ϕ_2 are of the power type and the $\tau_i - \sigma_m$ relation is linear, we obtain the equation of non-linearly viscous flow:

$$\tau_i = \eta_1(t)\dot{\gamma}_i^{m_1} + \eta_2(t)\sigma_m\dot{\gamma}_i^{m_2} = \eta_1(t)\dot{\gamma}_i^{m_1}\left[1 + \frac{\sigma_m}{H(\dot{\gamma}_i, t)}\right] \qquad (11\text{-}55)$$

where $H(\dot{\gamma}_i, t) = [\eta_1(t)/\eta_2(t)]\dot{\gamma}_i^{m_1 - m_2}$.

Analysis of equation of non-linear viscous flow, eq. 11-55

The term $\eta_1(t)\dot{\gamma}_i^{m_1} = \tau^0(\dot{\gamma}_i)$ entering the equation defines the resistance to pure shear and the term $\eta_2(t)\dot{\gamma}_i^{m_2}\sigma_m$ indicates the manner in which this resistance is changing due to σ_m; note, that $\eta_2(t)\dot{\gamma}_i^{m_2} = \tan\psi(\dot{\gamma}_i, t)$. The time functions $\eta_1(t)$ and $\eta_2(t)$ in eq. 11-55 represent the changes in viscosity due to deformation.

The functions τ^0, $\tan\psi$, and H may exist in various combinations as was shown in Fig. 11-8.

Case (a). When the exponents of eq. 11-55 are $m_1 \neq m_2$, the $\tau_i - \sigma_m$ graph is analogous to that of Fig. 11.8(a).

Case (b). For $m_1 \neq 0$ and $m_2 = 0$, eq. 11-55 takes the form (Fig. 11-8(b)):

$$\tau_i = \eta_1(t)\dot{\gamma}_i^{m_1}\left[1 + \frac{\sigma_m}{H(\dot{\gamma}_i, t)}\right] = \eta_1(t)\dot{\gamma}_i^{m_1} + \sigma_m \tan\psi_s \qquad (11\text{-}56)$$

Case (c). For $m_1 = m_2$, eq. 11-55 takes the form (Fig. 11-8(c)):

$$\tau_i = \eta_1(t)\dot{\gamma}_i^m\left(1 + \frac{\sigma_m}{H_s}\right) \tag{11-57}$$

where $H_s = \eta_1(t)/\eta_2(t) = \text{const.}$

Steady flow

If only the stage of constant-rate steady flow is considered, the coefficients of friction η_1 and η_2 entering eqs. 11-55 to 11-57 will be constant and $\eta_1/\eta_2 = H = \text{const.}$

Eq. 11-57 will then describe a viscous Newtonian flow with allowance for σ_m:

$$\tau_i = \eta_1\dot{\gamma}_i\left(1 + \frac{\sigma_m}{H_s}\right) \tag{11-58}$$

To obtain the equation of plastic–viscous flow of the Bingham type from eq. 11-58, substitute the difference $\tau_i - \tau_y$ for τ_i in this equation, where $\tau_y = \tau_s^0 + \sigma_m \tan\psi_s$; τ_s^0 is the ultimate shearing strength; ψ_s is the angle of friction at the octahedral plane.

Non-linear viscoplastic flow will be described by eq. 11-55 if we assume that $\eta_1 = \text{const.}$, $\eta_2 = \text{const.}$ and substitute the difference $\tau_i - \tau_y$ for τ_i in the equation. Then, for $m_1 \neq m_2$, we obtain:

$$\tau_i = \tau_s^0\left(1 + \frac{\sigma_m}{H}\right) + \eta_1\dot{\gamma}_i^{m_1} + \eta_2\sigma_m\dot{\gamma}_i^{m_2} \tag{11-59}$$

For $m_1 = m_2 = m$, we have:

$$\tau_i = (\tau_s^0 + \eta_1\dot{\gamma}_i^m)\left(1 + \frac{\sigma_m}{H_s}\right) \tag{11-60}$$

Finally, for $\eta_2 = 0$, we arrive at:

$$\tau_i = \tau_s^0\left(1 + \frac{\sigma_m}{H}\right) + \eta_1\dot{\gamma}_i^{m_1} \tag{11-61}$$

In eq. 11-61 the mean normal stress σ_m influences only the limiting shearing resistance (the first term) and does not affect the viscous resistance (the second term). However, according to eq. 11-60 examined by Stroganov and Grechishchev at an earlier date, σ_m produces an effect on both the limiting and viscous resistance.

Invariant graphs of the non-linear viscous and viscoplastic flows described by eqs. 11-55 and 11-59 as well as by eqs. 11-56, 11-57, 11-60, and 11-61 derived from these equations are depicted in Fig. 11-10. The functions $F(\tau_i)$, $\bar{F}(\tau_i)$, and $\Phi(t)$ on the graphs have the following values.

For non-linear viscous flow (Fig. 11-10(a)), in eq. 11-55:

$$F(\tau_i) = \frac{\tau_i}{1 + \dfrac{\sigma_m}{H(\dot{\gamma}_i, t)}} \qquad \bar{F}(\tau_i) = \frac{\tau_i}{\dot{\gamma}_i^m} \qquad \Phi(t) = \eta_1(t)$$

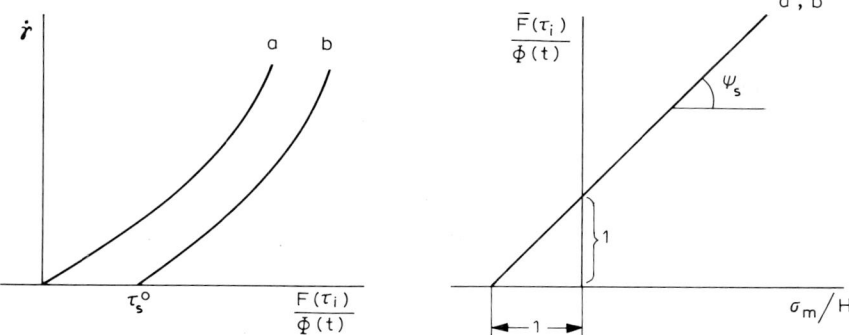

Fig. 11-10. Invariant graphs of flow corresponding to: (a) eq. 11-55; (b) eq. 11-59.

in eq. 11-56:

$$F(\tau_i) = \frac{\tau_i}{1 + \dfrac{\sigma_m}{H(\dot\gamma_i, t)}} = \tau_i - \sigma_m \tan \psi_s \qquad \bar F(\tau_i) = \frac{\tau_i}{\dot\gamma_i^m} \qquad \Phi(t) = \eta_1(t)$$

in eq. 11-57:

$$F(\tau_i) = \frac{\tau_i}{1 + \dfrac{\sigma_m}{H_s}} \qquad \bar F(\tau_i) = \frac{\tau_i}{\dot\gamma_i^m} \qquad \Phi(t) = \eta_1(t)$$

For non-linear viscoplastic flow (Fig. 11-10(b)), in eq. 11-59:

$$F(\tau_i) = \tau_i - \sigma_m \left(\frac{\tau_s^0}{H} + \eta_2 \dot\gamma_i^{m_1}\right) \qquad \bar F(\tau) = \frac{\tau_i - \eta_1 \dot\gamma_i^{m_1} - \eta_2 \sigma_m \dot\gamma_i^{m_2}}{\tau_s^0} \qquad \Phi(t) = 1$$

in eq. 11-60:

$$F(\tau_i) = \frac{\tau_i}{1 + \sigma_m/H_s} \qquad \bar F(\tau_i) = \frac{\tau_i}{\tau_s^0 + \eta_1 \dot\gamma_i^m} \qquad \Phi(t) = 1$$

in eq. 11-61:

$$F(\tau_i) = \tau_i - \tau_s\left(\frac{\sigma_m}{H_s}\right) \qquad \bar F(\tau_i) = \frac{\tau_i - \eta_1 \dot\gamma_i^{m_1}}{\tau_s^0} \qquad \Phi(t) = 1$$

11.6 KINETIC EQUATION OF SOIL DEFORMATION ALLOWING FOR THE EFFECT OF MEAN NORMAL STRESS

General form of the equation

The equations of soil deformation taking into account the effect of the mean normal stress σ_m, which were discussed above, referred to phenomenological theories. Now, we show the way the effect of σ_m is allowed for in the equation of kinetic theory given by eq. 10-60.

In the case of pure shear, this equation takes the form:

$$\dot{\gamma} = \frac{\tau}{\eta_0}(t + 1)^{-n} \qquad (11\text{-}62)$$

where $n = \lambda_2 - \lambda_1 \tau/(\tau_0 - \tau)$ and $t = t/t^*(t^* = 1)$ is the time in a dimensionless form.

In the state of combined stress, the above formula may be presented as:

$$\dot{\gamma}_i = \frac{\tau_i}{\eta_0}(t + 1)^{-n} \qquad (11\text{-}63)$$

or, taking into account the dimensions of t, as:

$$\dot{\gamma}_i = \frac{\tau_i}{\eta_0}\left(\frac{t}{t^*} + 1\right)^{-n} \qquad (11\text{-}63')$$

where $n = \lambda_2 - \lambda_1 \tau_i/(\tau_{s(0)} - \tau_i)$.

In its turn:

$$\tau_{s(0)} = \tau_{s(0)}^0 \left(1 + \frac{\sigma_m}{H_{s(0)}}\right)^{\lambda_3} \qquad (11\text{-}64)$$

where $\tau_{s(0)}$ (or just τ_0) is the hypothetically instantaneous yield limit of the soil in the state of combined stress (the shearing strength at the octahedral plane); $\tau_{s(0)}^0$ (or τ_0^0) is the hypothetically instantaneous yield limit in pure shear; and $H_{s(0)}$ (or H_0) and λ_3 are the parameters of cohesion and hardening for hypothetically instantaneous shear, respectively.

Quite frequently, $\lambda_3 = 1$; then $H_{s(0)} = \tau_{s(0)}^0/\tan \psi_{s(0)}$, where $\psi_{s(0)}$ (or ψ_0) is the angle of internal friction at the octahedral plane. These parameters will be discussed in Section 11.7.

Thus, entering the deformation formula 11-63 are cohesion and the angle of internal friction — fundamental soil characteristics.

Experimental data

For the test of fit of formula 11-63, we shall compare computations by this formula with experimental data. To this end we use the data acquired by Meschyan and Badakhyan (1976) from creep tests of stiff clay soils (structure disturbed, $W = 32.9\%$).

The tests were conducted in an apparatus enabling the application of a torque simultaneously with a vertical pressure σ_z to cylindrical specimens ($r = 50.5$ mm, $h = 24$ mm). During the test, the specimen was encased in a series of rings to prevent its sidewise expansion without interfering with the torsion (shear) (see Section 6.6: Meschyan, 1974). The verticakl test loads σ_z used were $3.0 \cdot 10^5$, $5.0 \cdot 10^5$, and $8.0 \cdot 10^5$ Pa and a series of three specimens was subjected to each loading. The constant torsional loading applied to each of the specimens in a series equalled a certain fraction (0.3, 0.6, 0.8, respectively) of the standard strength, i.e. the ratio of the torque applied and the torque causing failure under a standard loading, M/M_{st}.

The tangential stress τ in the sublimiting condition and the standard shearing strength τ_{st} were determined by the formulae:

$$\tau = \frac{1}{2\pi r^3}\left(3M + \gamma \frac{dM}{d\gamma}\right) \qquad \tau_{st} = \frac{3M_{st}}{2\pi r^3} \qquad (11\text{-}65)$$

where γ is the angular strain.

The tangential stress intensity and mean normal stress were computed by the formulae:

$$\tau_i = \frac{1}{\sqrt{6}}\sqrt{2(1-\xi)^2\sigma_z^2 + 6\tau^2} \qquad \sigma_m = \frac{1+2\xi}{3}\sigma_z \qquad (11\text{-}66)$$

where $\xi = v/(1-v)$.

Supposing that for the soil under consideration $v = 0.3$, we obtained:

$$\tau_i = \sqrt{0.11\sigma_z^2 + \tau^2} \qquad \sigma_m = 0.62\sigma_z \qquad (11\text{-}67)$$

The values of the ultimate long-term shearing strength $\tau_{s(\infty)}$ and those of the hypothetically instantaneous shearing strength $\tau_{s(0)}$ were adopted for the tests under consideration at $\tau_{s(\infty)} = \tau_{st}$ and $\tau_{s(0)} = 2\tau_{s(\infty)}$. For example, at $\sigma_z = 5 \cdot 10^5$ Pa and $M/M_{st} = 0.3; 0.6;$ and 0.8; we had $\tau = 0.7 \cdot 10^5$ Pa; $1.19 \cdot 10^5$ Pa; and $1.93 \cdot 10^5$ Pa; $\tau_i = 1.8 \cdot 10^5$ Pa; $2.04 \cdot 10^5$ Pa, and $2.35 \cdot 10^5$ Pa; $\tau_{s(0)} = 4.82 \cdot 10^5$ Pa. The results of these tests are represented in Fig. 11-11.

Test data handling. For the reduction of test data in accordance with eq. 11-63 and determining the parameters entering this formula it is expedient to follow the procedure outlined in Section 10.8.

Assuming that $t/t^* \gg 1$, we simplify eq. 11-63:

$$\dot{\gamma}_i = \frac{\tau_i}{\eta_0^*} t^{-n} \qquad (11\text{-}68)$$

where $\eta_0^* = \eta_0/(t^*)^{-n}$.

Note, that from the conditions of tests $\dot{\gamma}_i = \dot{\gamma}$, although $\gamma_i \neq \gamma$.

The parameters n and η_0 in eq. 11-68 can be determined by linearizing the equation into:

$$\ln(\dot{\gamma}_i/\tau_i) = \ln(1/\eta_0^*) - n \ln t \qquad (11\text{-}69)$$

and plotting the test points on the $\ln(\dot{\gamma}_i/\tau_i) - \ln t$ coordinates as shown in Fig. 11-12(a). (Note that this graph is a modification of Fig. 10-13(a).)

We obtain a family of straight lines corresponding to various values of τ_i; the slopes of these lines define the value of the exponent $n(\tau_i)$ and the intercept on the axis of ordinates gives the value of η_0^*. The values of n obtained experimentally for the case under consideration are given in Table 11-I.

It can be seen that exponent n is appreciably influenced by stress — a fact stemming

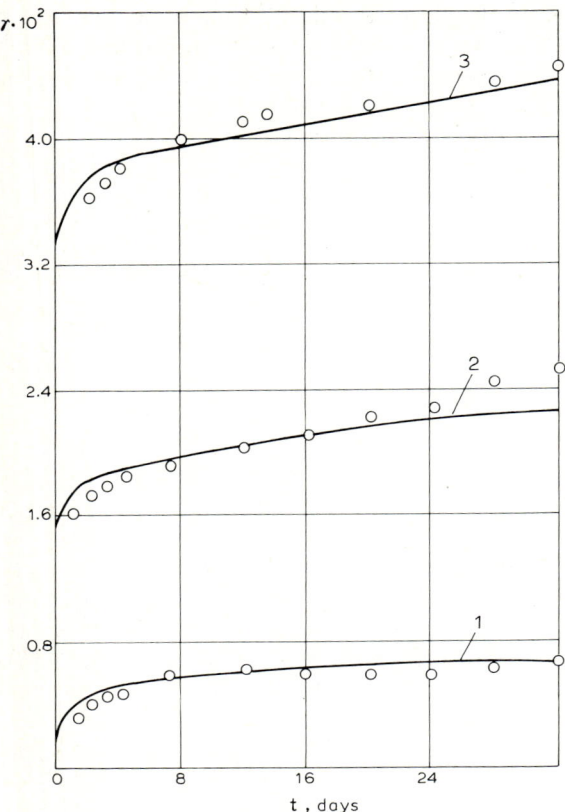

Fig. 11-11. Testing clay soil for creep by applying various values of torque simultaneously with a vertical pressure σ_z. 1: $M/M_{st} = 0.3$. 2: $M/M_{st} = 0.6$. 3: $M/M_{st} = 0.8$. Dots represent test data and curves are plotted from analytical computations by eq. 11-63.

TABLE 11-I

Values of parameter n in eq. 11-68

M/M_{st}	σ_z (Pa): $3.0 \cdot 10^5$		$5.0 \cdot 10^5$		$8.0 \cdot 10^5$	
	experimental	computed	experimental	computed	experimental	computed
0.3	0.71	0.85	0.87	0.78	1.23	0.77
0.6	0.36	0.70	0.41	0.64	0.64	0.65
0.8	0.5	0.44	0.53	0.40	0.38	0.44

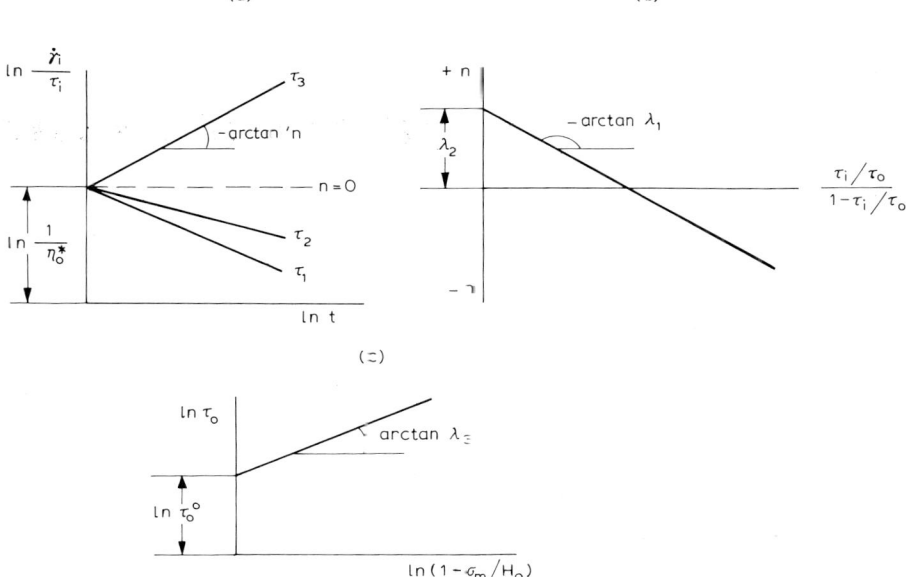

Fig. 11-12. Handling of test data by eq. 11-68: (a) graph for determining the parameters n and η_0^* in eq. 11-68; (b) graph for determining the parameters λ_1 and λ_2 in eq. 11-63; (c) graph for determining the parameters τ_0^0, H and λ_3 in eq. 11-63.

from the physical meaning of the parameter:

$$n = \lambda_2 - \lambda_1 \frac{\tau_i}{\tau_{s(0)} - \tau_i} \tag{11-70}$$

At the same time, as first noted by Meschyan et al. (1976), exponent n is not influenced by σ_m if the values of $\tau_i/\tau_{s(0)}$ are the same. In fact, taking eq. 11-64 into account, we have:

$$\tau_i/\tau_{s(0)} = \frac{\tau_i}{\tau_{s(0)}^0}\left[1 + \frac{\sigma_m}{H_{s(0)}}\right]^{-\lambda_3}$$

and, for $\tau_i/\tau_{s(0)} = $ const. we obtain:

$$n = \lambda_2 - \lambda_1 \frac{\tau_i/\tau_{s(0)}}{1 - \tau_i/\tau_{s(0)}} = \text{const.}$$

Determining the parameters of eqs. 11-70 and 11-64. The obtained values of n may be checked by means of eq. 11-70. To that end, we plot n against $(\tau_i/\tau_{s(0)})/(1 - \tau_i/\tau_{s(0)})$ (Fig. 11-12(b)) and determine from the graph the parameters λ_1 and λ_2 entering eq. 11-70, proceeding as indicated in Fig. 10-14(b). Then, we compute the values of n for any level of stress (these values are given in Table 11-I).

In the case under consideration, the instantaneous strength $\tau_{s(0)}$ is determined experimentally. In the absence of such data, $\tau_{s(0)}$ may be computed by eq. 11-72.

The dependence of $\tau_{s(0)}$ upon σ_m is determined in accordance with eq. 11-64 by plotting a graph on the $\ln \tau_{s(0)}$–$\ln (1 + \sigma_m/H_{s(0)})$ coordinates (Fig. 11-12(c)); the way of estimating the tentative value of $H_{s(0)}$ was considered in analysing eq. 11-49. For $\lambda_3 = 1$, the graph of Fig. 11-12(c) is reduced to an ordinary shear curve on the $\tau_{s(0)}$–σ_m coordinates from which the values of $H_{s(0)}$ and $\psi_{s(0)}$ can be directly determined.

Comparing analytical and experimental data

The final value of n for use in computations by eq. 11-63 is the checked value.

The equation of deformation obtained by integrating eq. 11-63 will have a form analogous with eqs. 10-73, 10-74 and 10-75,

for $n < 1$:

$$\gamma_i = \gamma_{i(0)} + \frac{\tau_i}{\eta_0^*(1-n)} t^{1-n} \tag{11-71}$$

for $n = 1$:

$$\gamma_i = \gamma_{i(0)} + \frac{\tau_i}{\eta_0^*} \ln t \tag{11-72}$$

for $n > 1$:

$$\gamma_i = \gamma_{i(\infty)} - \frac{\tau_i}{\eta_0^*(n-1)} t^{1-n} \tag{11-73}$$

Fig. 11-11 illustrates analytical curves plotted with the aid of these formulae for $\sigma_z = 5 \cdot 10^5$ Pa used during one of the tests referred to above. The curves are compared with the test data. Similar curves were obtained for the rest of σ_z-values.

For the final test of fit of eq. 11-63, the data on γ_i obtained from tests are compared with the analytical data computed by eqs. 11-71 to 11-73 and reduced to a single generalized graph for all values of τ_i (at any value of σ_m).

Such a graph resulting from Meschyan's data (Meschyan et al., 1976) acquired during the testing of clay soil for torsion by a relative torque M/M_{st} under a vertical load σ_z is represented in Fig. 11-13. Points *1, 2, 3*... correspond to the loadings σ_z given in Table 11-II.

TABLE 11-II

Reference numbers of test points (cf. Fig. 11-13)

M/M_{st}	σ_z (Pa):		
	$3 \cdot 10^5$	$5 \cdot 10^5$	$8 \cdot 10^5$
0.3	1	4	7
0.6	2	5	8
0.8	3	6	9

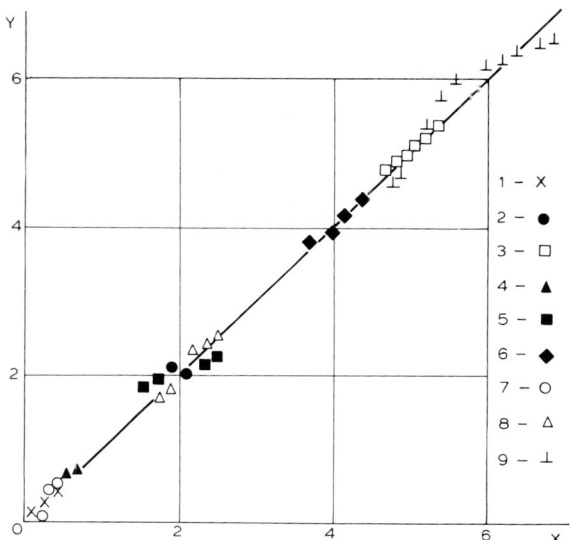

Fig. 11-13. Rectification of creep curves on generalized coordinates. Legend: for explanation of *1–9* see Table 11-II.

Plotting the computed values, $Y = \gamma_{i(comp)} \times 10^2$, on the *y*-axis and the test data, $X = \gamma_{i(exp)} \times 10^2$, on the *x*-axis, we can see from the graph that the experimental points fall on the bisector of the angle $X = Y$ in close groups. This is an indication that the computed values are in agreement with the experimental data.

11.7 EQUATION OF LONG-TERM SOIL STRENGTH ALLOWING FOR THE EFFECT OF MEAN NORMAL STRESS

Equation of limiting stress state

This equation in its general form, as represented by eq. 11-9, has been obtained on the assumption that the transition of soil into the state of limiting stress occurs when the shearing strain intensity reaches a certain critical value $\gamma_i = \gamma_s$. If creep is involved, this value will apparently be reached with time — a fact taken into account by introducing *t* into eq. 11-9.

Graphically, the state of limiting stress is represented by a τ_i–σ_m limiting curve for γ_s (see Fig. 11-4(b)). Apparently, this curve will correspond to a given moment of time t_j. If a series of moments $t_1, t_2, t_3, \ldots$ is subjected to consideration we obtain a family of τ_i–σ_m limiting curves (Fig. 11-14(b)). The changes in the values of τ_s over the range $0 \leq t < \infty$ will be defined by a family of long-term strength curves each corresponding to a certain value of σ_m (Fig. 11-14(a)).

The curves of long-term strength depicted in Fig. 11-14(a) may be adopted from testing at least three series of identical specimens for creep in the state of combined

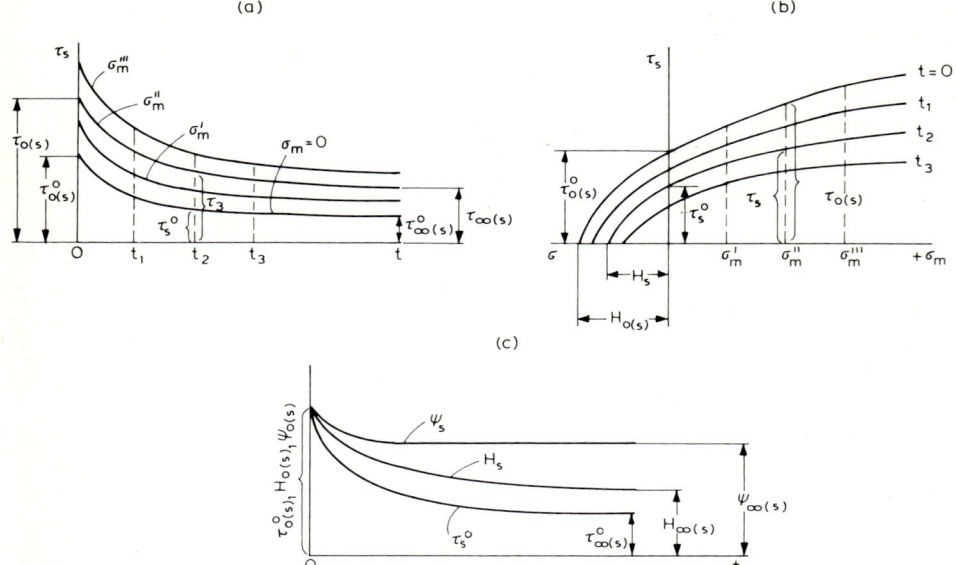

Fig. 11-14. Limiting condition with allowance for the time factor: (a) curves of long-term strength for various values of σ_m; (b) envelopes $\tau_s - \sigma_m$ for various t; (c) curves illustrating the changes in strength parameters with time.

stress, the specimens in each series being loaded up to failure by the same σ_m but different τ_i. The test data corresponding to each value of σ_m are reduced as indicated in Fig. 9-1. The case when $\sigma_m = 0$ corresponds to the resistance in pure shear.

Replotting the τ_s–t curves on the τ_s–σ_m coordinates, we obtain a family of envelopes for various t, as is shown in Fig. 11-14(b).

Equation of long-term soil strength and equation of limiting state

The equation of long-term strength allowing for the effect of σ_m may be obtained from the creep equations into which σ_m is entered, if we change over to the limiting deformation $\gamma_i = \gamma_s = $ const. (see Section 11.4).

The relationship obtained in this way also appears to be the equation of limiting state accounting for the time factor. Thus, putting $\gamma_i = \gamma_s = $ const. in eq. 11-45, we obtain:

$$\tau_s = [H_s(t) + \sigma_m] \tan \psi_s(t) \tag{11-74}$$

This expression corresponds to the von Mises–Botkin equation of limiting state (eq. 4-22) except that the cohesion $H_s(t)$, the angle of friction at the octahedral plane $\psi_s(t)$ and the yield limit in pure shear $\tau_s^0(t)$ are time-dependent parameters. They vary between the hypothetically instantaneous values (at $t = 0$) $\tau_{s(0)}^0$, $H_{s(0)}$, $\psi_{s(0)}$ and ultimate long-term values (at $t \to \infty$) $\tau_{s(\infty)}^0$, $H_{s(\infty)}$, and $\psi_{s(\infty)}$ as is shown in Fig. 11-14(c).

The manner in which the parameters τ_s^0 and $\tan \psi_s$ alter with time may be assumed to be the same as the pattern of changes in long-term strength considered in Chapter 9. In particular, according to eq. 9-22' we obtain:

$$\tau_s^0 = \frac{\beta_1}{\ln(t/T_1)} \qquad \tan \psi_s = \frac{\beta_2}{\ln(t/T_1)} \qquad H_s = \frac{\beta_1 \ln(t/T_2)}{\beta_2 \ln(t/T_1)} \qquad (11\text{-}75)$$

Consider specific cases of eq. 11-74 represented by the graphs in Fig. 11-8. If $\psi_s = $ const., then:

$$\tau_s = [H_s(t) + \sigma_m] \tan \psi_s = \tau_s^0(t) + \sigma_m \tan \psi_s \qquad (11\text{-}76)$$

where:

$$\tau_s^0(t) = \frac{\beta_1}{\ln(t/T_1)} \qquad H_s(t) = \frac{\beta_1}{\tan \psi_s} \frac{1}{\ln(t/T_1)}$$

For $H_s = $ const.:

$$\tau_s = [H_s + \sigma_m] \tan \psi_s(t) = \tau_s^0(t) \left[1 + \frac{\sigma_m}{H_s}\right] \qquad (11\text{-}77)$$

where:

$$\tau_s^0(t) = \frac{\beta_1}{\ln(t/T)} \qquad \tan \psi_s = \frac{\beta_2}{\ln(t/T_2)} \qquad H_s = \beta_1/\beta_2 = \text{const.}$$

An equation of the type similar to eq. 11-77, yet in a more general non-linear form, may be obtained from eq. 11-49 which, for $\gamma_i = \gamma_s$, yields, according to Gorodetsky (1975):

$$\tau_s = \tau_s^0(t) \left[1 + \frac{\sigma_m}{H_s}\right]^{\lambda_3} \qquad (11\text{-}78)$$

where $\tau_s^0 = \beta_1/\ln(t/T_1)$ and $H_s = $ const. as before.

For $\gamma_i \to \infty$, the linear–fractional relation 11-50 becomes a limiting relationship. Then:

$$\tau_s = \frac{\tau_{s(0)}(T + t)}{T + \frac{\tau_{s(0)}}{\tau_{s(\infty)}} t} \qquad (11\text{-}79)$$

where, in accordance with eq. 11-52, $\tau_{s(0)} = (H_s + \sigma_m) \tan \psi_0$ and $\tau_{s(\infty)} = (H_\infty + \sigma_m) \tan \psi_\infty$.

Among the formulae considered above, we give preference to eq. 11-76, because of its simplicity and because its independence of the friction angle from the period of load application is proved experimentally. In those cases when the graph of limiting state is clearly non-linear, it is good practice to use eq. 11-78.

11.8 REFERENCES

Barshevsky, B.N., 1956. On the strength hypothesis of cohesionless soil. News, USSR Acad. Sci., Eng. Sci. Sect., no. 4 (in Russian).
Fedorov, I.V., 1957. On Some Mechanisms of Strength and Deformation in a Cohesionless Medium. VODGEO Publ., Moscow (in Russian).
Geniev, G.A., 1966. On the formulation of the combined problem of elasticity theory and statics of a continuum. SMFE, no. 5 (in Russian).
Geniev, G.A., 1970. Problems of strength and deformation of soil media. In collective volume: Structures. Issue 4 (in Russian).
Gorodetsky, S.E., 1975. Creep and strength of frozen soils in the state of continuous stress. SMFE, no. 3 (in Russian).
Grechishchev, S.E., 1961. On creep rate of frozen soils in the state of combined stress. News, Siberian Branch USSR Acad. Sci., no. 5 (in Russian).
Insley, A.E. and Hillis, S.F., 1965. Triaxial shear characteristics of a compacted glacial till under unusually high confining pressure. Proc. 6th ICSMFE, Tokyo, v. 1.
Ioselevich, V.A., 1967. On mechanisms of deformation in cohesionless soils. SMFE, no. 4 (in Russian).
Lomize, G.M. (Editor), 1966. Problems of strength and deformation of soils. Proc. Sci. Seminar. Aizerbaidzhan State Publ. House, Baku (in Russian).
Lomize, G.M. (Editor), 1968. Engineering Properties of Soils and the Building of Houses on Moist Loess Soils. Checheno-Ingush State Publ. House, Grozny (in Russian).
Meschyan, S.R. and Badakhyan, R.G., 1976. On important mechanisms of creep in clay soils due to shear. SMFE, no. 1 (in Russian).
Meschyan, S.R., Badakhyan, R.G. and Malanyan, R.P., 1976. On the problem of the influence of soil state on deformation creep in shear. In collective volume: Proc. 2nd All-Union Symp. Soil Rheology. Erevan State Univ. Publ. House, Erevan (in Russian).
Murayama, S., 1964. A theoretical consideration on a behaviour of sand. Rheology and Soil Mechanics, IUTAM Symp., Grenoble, 1964.
Nelson, J. and Baron, M.L., 1971. Application of variable moduli; models to soil behaviour. Int. J. Solids and Structures, v. 7, no. 4.
Roscoe, K.H. and Poorooshasb, H.B., 1963. A theoretical and experimental study of strains in triaxial compression tests on normally consolidated clay. Géotechnique, v. 13, no. 1.
Stroganov, A.S., 1956. A method of predicting final settlements of structures. Trans. Moscow Energy Inst., Issue 19. Gosenergoizdat, Moscow (in Russian).
Stroganov, A.S., 1961. Plastic-viscous flow of soil. In collective volume: Reports 5th ICSMFE. Stroiiudat, Moscow (in Russian).
Vyalov, S.S. and Mindich, A.L., 1974. Settlement and limiting equilibrium of a layer of weak soil resting on rigid foundation. SMFE, no. 6 (in Russian).
Vyalov, S.S. and Shusherina, E.P., 1964. Resistance of frozen soils against triaxial compression. In collective volume: Permafrostology Research. Issue 4, Moscow State Univ., Moscow (in Russian).
Zhukov, A.M. and Rabotnov, Yu.N., 1954. Studies of Plastic Deformation of Steel Under a Continuous Loading. Eng. Volume, USSR Acad. Sci., v. 18 (in Russian).

Chapter 12

SOME PECULIARITIES OF SOIL DEFORMATION IN COMBINED STRESS STATE

It was shown in Chapter 11 that the process of soil deformation results from the effect of the three stress tensor invariants; the influence of the first and second invariants on shearing deformation was also examined. Below we investigate the mutual effect of these two invariants on the development of volumetric strains. Moreover, the influence of the third invariant on the process of soil deformation as a whole is also discussed along with other factors.

12.1 DILATANCY

Reynolds' experiments

Experimenting with sand way back in 1885, O. Reynolds discovered a change in volume during simple shearing. His explanation of the phenomenon which he termed dilatancy*[1] was that sand particles repack due to shear, by analogy with the balls changing their arrangement in a known model he used as the example.

When the packing arrangement is loose (each ball is in contact with just four neighbouring spheres), shearing brings the balls into a denser arrangement (each ball contacting six neighbours) so that the body made up of the balls shrinks in volume. Vice versa, a dense initial packing arrangement is changed by shear into a loose one with an increase in volume. Similarly, the shear of a cohesionless soil having a dense packing arrangement leads to the softening of the soil whereas in the case of a loose packing the outcome is a denser arrangement.

Dilatancy is evident during elastic, plastic and viscous deformations, being either positive (hardening) or negative (softening) in any of the cases. An elastic negative dilatancy is described according to Reiner (see Section 1.5: Reiner, 1958, 1960) by the equation:

$$\varepsilon_v = \frac{\sigma_m}{3k} - \frac{\tau_i^2}{\delta} \tag{12-1}$$

where $\varepsilon_v = \varepsilon_1 + \varepsilon_2 + \varepsilon_3$ is the volumetric strain; k is the bulk modulus; and δ is the modulus of volumetric dilatancy.

---
*[1] Also in use is the term "dilatation".

Dilatancy of clay soils

Positive dilatancy of soft (fresh after precipitation) clay soils having an original structure of the card house or "book house" type may be explained by structural changes. The randomly oriented particles are repacked by shear, tending to assume a parallel arrangement which is more compact and conducive to a reduction in volume.

Structural changes are also the cause of dilatancy in compact clays because shearing deformation alters the pattern of particle arrangement. The key factors in this case are the changes in the defect density due to shear.

Experimental studies carried out at Moscow State University have enabled Shusherina (1966, referenced in Section 5.7) to show that the volume decrease of dense soil is attributed to a reduction in the number of defects, microcracks, etc., whereas a volume increase results from the development of defects. The former phenomenon mentioned takes place during an attenuating creep and the second is associated with a non-attenuating one. At the stage of steady flow, the developing defects are at balance with the "healed" defects so that no volumetric change is experienced. These phenomena were discussed in Chapter 10.

Since the accumulation of defects leads to soil failure, we may assert that this process is directly linked with the phenomenon of negative dilatancy. In fact, if experiments are conducted under the conditions preventing the development of structural defects with the associated bulk softening, the failure of soil is impeded, if not hindered at all.

Therefore, the question whether or not failure will occur depends to a considerable extent on the arrangement of stress, the ratio between the deviatory and spherical constituents of the stress tensor being the key factor. This may be exemplified by a triaxial compression: an increase in the ratio of σ_m/τ_i renders failure of a specimen difficult and, under the conditions of hydrostatic pressure, ($\sigma_m = p$, $\tau_i = 0$), even impossible.

Equation of volumetric soil strain in a general form

This equation is represented by eq. 11-1 determining volumetric strain as a function of the three stress tensor invariants and time. Postponing for a while examination of the role of the third invariant, we focus our attention on the mutual effect of the first and second invariants expressed in terms of σ_m and τ_i by analogy with eq. 11-6.

According to Reiner (1960, referenced in Section 1.5), a volumetric soil strain may be expressed as the sum:

$$\varepsilon_v = \varepsilon_v^0 \pm \varepsilon_v^D \quad \text{or} \quad \varepsilon_m = \varepsilon_m^0 \pm \varepsilon_m^D \qquad (12\text{-}2)$$

where $\varepsilon_v^0 = 3\varepsilon_m^0$ is the volumetric strain induced by the spherical stress tensor (all-around pressure σ_m); $\varepsilon_v^D = 3\varepsilon_m^D$ is the volumetric strain caused by the stress deviator (tangential stress intensity τ_i); and the $+$ and $-$ signs in the expression indicate the possibility of either hardening or softening.

The quantity ε_v^0 is the function of σ_m only:

$$\varepsilon_v^0 = f_1^*(\sigma_m) \tag{12-3}$$

The quantity ε_m^D is influenced, as is shown below, by the shearing strain:

$$\varepsilon_v^D = \Lambda \gamma_i \tag{12-4}$$

where Λ is the dilatancy coefficient. Since the shearing strain γ_i, in its turn, is affected by the tangential stress intensity τ_i and mean normal stress σ_m, we have:

$$\varepsilon_v^D = f_2^*(\tau_i, \sigma_m) \tag{12-5}$$

Substituting eqs. 12-3 and 12-5 into eq. 12-2, we obtain the equation of volumetric strain in its general form:

$$\varepsilon_m = f_1^*(\sigma_m, t) \pm f_2^*(\sigma_m, \tau_i, t) \tag{12-6}$$

Note that eq. 12-1 is a particular case of eq. 12-6.

Apparently, the effect of shearing stress on volumetric strain is decided by the relation between the deviatory and spherical constituents of the stress tensor. If the spherical part prevails significantly, the effect of shearing stress on volumetric strain may be disregarded. In such a case we may assume, as was suggested by Grigoryan (1964; see also Section 4.9: Grigoryan, 1960), that the shearing stress–strain relation is linear and that the relation between the soil density and all-around pressure is non-linear.

Dilatancy coefficient. Entering eq. 12-4, the coefficient Λ is a proportionality factor between the increment in volumetric strain, due to the stress deviator, and shearing strain: $\Lambda = \varepsilon_m^D/\gamma_i$.

The dilatancy coefficient is used in equations of the deformation theory of plasticity. In the theory of flow, preference is given to the dilatancy rate as determined by the proportionality factor between plastic volumetric strain rate and shearing strain rate $\dot{\Lambda} = \dot{\varepsilon}_m^D/\dot{\gamma}_i$.

Experiments at the Civil Engineering Institute, Moscow, enabled Lomize and Sukhanov (1973, 1974) to show that the dilatancy rate is influenced by the loading path; when the soil is transformed into the limiting condition, the dilatancy rate becomes constant, $\dot{\Lambda} = \dot{\Lambda}_s = \text{const.}$

The dilatancy angle v rather than Λ, $\dot{\Lambda} = \sin v$, is quite frequently used. Attempts were made to establish a connection between the dilatancy angle and the angle of internal friction, for, according to the associated law, the equality $v = \phi$ must hold. However, experiments have sometimes failed to prove this.

The phenomenon of dilatancy was convincingly illustrated by means of a model by Rowe (1962) — a modification of the Reynolds model. Examining the deformation of a cubic and rhombic arrangement of packed spheres, Rowe obtained the following expressions determining the relation between the principal stresses and the rates of principal strains:

$$\sigma_1/\sigma_3 = \tan\alpha \tan(\beta + \phi) \qquad \dot{\varepsilon}_3/\dot{\varepsilon}_1 = A \tan\alpha \tan\beta \qquad (12\text{-}7)$$

where α is the angle determining the geometry of the packing of the spheres; β is the angle of obliquity the sliding plane of a sphere makes with one of the principal axes; and A is a coefficient varying with the type of packing.

In determining the critical value of the angle of the shear between the particles, $\beta = (\pi/4 - \phi/2)$, we obtain the relationship between the angle of internal friction and the dilatancy angle v (Roscoe, 1970):

$$\sin\phi = \frac{(k-1) + (k+1)\sin v}{(k+1) + (k-1)\sin v} \qquad (12\text{-}8)$$

where $k = \tan^2(\pi/4 + \phi/2)$.

Rowe's dilatant theory of deformation has been experimentally verified by its founder himself, by Barden and Khayatt (1966), by Lee (1966) and other researchers. Applying this model to cohesionless materials with randomly packed spherical particles, Horne (1965) considered the effects of anisotropy. At the same time, Rowe's theory was criticized by some authors (Morgenstern, Roscoe, Trollope) who were doubtful about the mechanism of deformation suggested.

A model of soil structure describing the stress–strain (shearing and volumetric) relation with allowance for particle orientation was suggested by Murayama and Matusoka (1973) from studies of the changes in the microstructure of soil due to creep. The authors also presented their own interpretation of the phenomenon of dilatancy from the standpoint of microstructure.

The problem of dilatancy and that of plasticity potential associated with dilatancy were scrutinized by Nikolaevsky (1972), Roscoe (1970) and also by Scott et al. (1969, reference in Chapter 1). Note that the condition of limiting state allowing for dilatancy presented by Nikolaevsky was arrived at from the considerations given below.

Expressing the increment in plastic deformation in terms of plasticity potential:

$$d\varepsilon_{ij} = d\lambda \frac{df}{d\sigma_{ij}}$$

where $i, j = 1, 2, 3$, and introducing the von Mises–Botkin condition of plasticity and the condition of dilatancy in the form $\dot{\varepsilon}_m - \dot{\Lambda}\dot{\gamma}_i = \dot{\varepsilon}_m - \dot{\varepsilon}_m^D = 0$ into the above expression, Nikolaevsky (1972) arrived at the equation of plastic equipotential line:

$$\tau_i^2 - \dot{\Lambda}(\sigma_m + H)^2 \tan\psi = \tau_i(\sigma_m + H)(\tan\psi - \dot{\Lambda}) \qquad (12\text{-}9)$$

For $\dot{\Lambda} = 0$, eq. 12-9 becomes the von Mises–Botkin condition. The connection between the dilatancy angle v and the angle of internal friction ψ is expressed in soil by the equation $\phi = 2\phi_\mu + v$, where ϕ_μ is a constant defined as the angle of friction between individual soil particles, and ϕ is the angle of the friction of the soil medium as a whole determined by conventional methods.

Dilatancy as consequence of the difference between the resistance of soil to tension and compression

A simple analysis permits the demonstration of the fact that volumetric strains in shear result from the difference between the resistance of soil in tension and that in compression.

In fact, consider the case when an elemental cubic volume is subjected to the action of the same stresses in compression and tension $\sigma_{comp} = \sigma_{tens} = \sigma$ (Fig. 12-1). This stress application will result in pure shear and, provided the modulus of compression E_{comp} and that of tension E_{tens} are the same, the cube will change in shape without changing volume. However, if the moduli are different (e.g., $E_{comp} > E_{tens}$), the strains in tension and compression will also be different, bringing about a volumetric change:

$$\varepsilon_{comp} - \varepsilon_{tens} = \frac{\sigma_{comp}}{E_{comp}} - \frac{\sigma_{tens}}{E_{tens}} = -\frac{\Delta V}{V}$$

Experimental data

The phenomenon of dilatancy may be proved by numerous experimental data such as those depicted in Fig. 12-2. Graph (a) sheds light on the connection between the axial ε_1 and volumetric strains $\varepsilon_m = \varepsilon_v/3$ during the triaxial compression of a sand with various void ratios e under the conditions of a constant stress $\sigma_2 = \sigma_3$.

It can be seen that an increase in the axial strain ε_1 at an early stage causes compaction of the soil, but a further increase in ε_1 results in an intensive softening, increasing inversely with the void ratio.

A similar picture may be observed with respect to the shearing stress τ_i and volumetric strain, as represented in Fig. 12-2(b) plotted from experiments with frozen soils. An early increase in τ_i brings about soil hardening but as soon as the volumetric

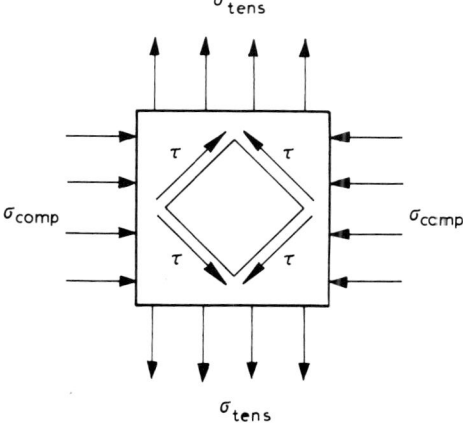

Fig. 12-1. Schematic representation of pure shear.

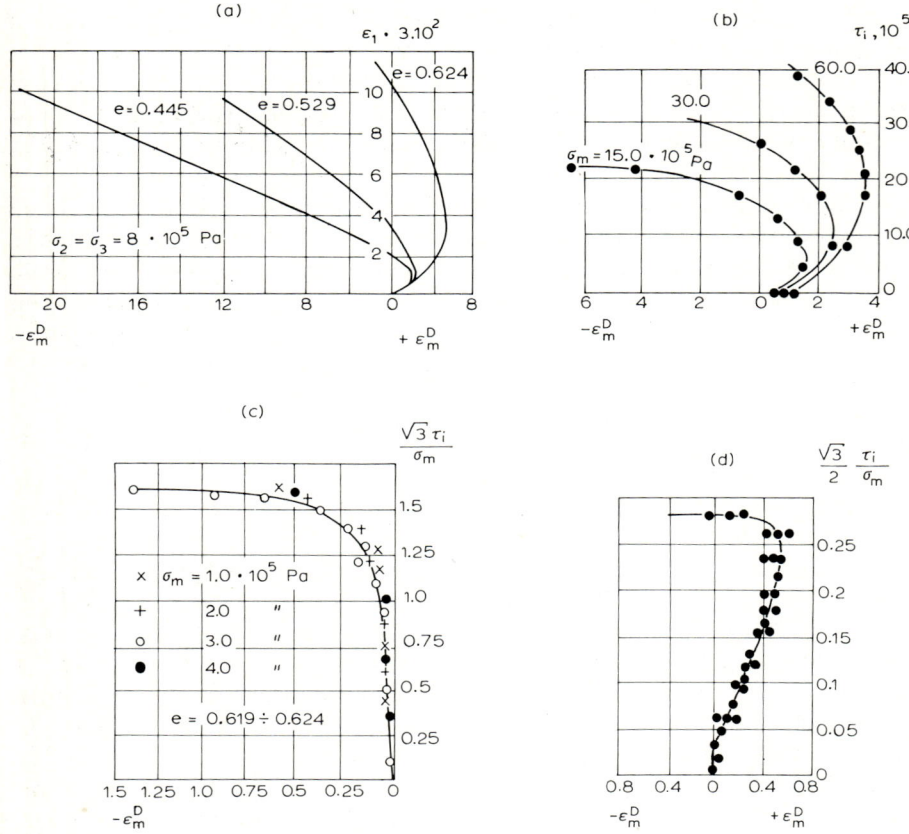

Fig. 12-2. Relationship between volumetric strain ε_m and tangential stress intensity τ_i: (a) relation between ε_m and ε_i in sand of various compactness subjected to triaxial compression (Stroganov's experiments); (b) relation between ε_m and τ_i in frozen sandy loam subjected to triaxial compression at different values of σ_m = const. (Gorodetsky's experiments); (c) relation between ε_m^D and τ_i/σ_m in sand during triaxial compression (Murayama's experiments); (d) relation between ε_m^D and τ_i/σ_m in sand during triaxial compression (experiments by Lomize and Kryzhanovsky). Strains are scaled up times $3 \cdot 10^2$.

strain reaches a certain maximum, a further stress increase leads to softening of the soil.

These experimental data corroborate the notion that the volumetric strain is the sum given by eq. 12-2. According to eq. 12-5, the ε_m^D member of the sum, i.e. the dilatant constituent of the volumetric strain, is a function of both τ_i and σ_m. For sand, this function may be in the form:

$$\varepsilon_m^D = f^*\left(\frac{\tau_i}{\sigma_m}\right) \tag{12-10}$$

which is proved by the graph of Fig. 12-2(c) obtained by Murayama (1964, referenced in Section 11.8) and by the graph of Fig. 12-2(d) obtained by G.I. Lomize and A.L. Kryzhanovsky (see Section 11.8: Lomize et al., 1968). It follows from the graphs that the experimental points for various values of τ_i and σ_m fall on a single curve plotted on the $\varepsilon_m - \tau_i/\sigma_m$ coordinates. Note that the softening of soil may be experienced both at the very beginning of loading (graph (c)) or on the application of a high τ_i, but in this latter case softening is preceded by hardening (graph (d)). The difference in behaviour is explained by the different initial void ratios of the soil.

For cohesive soils, the dependence of volumetric strain on τ_i and σ_m will be of a somewhat more complex nature. In the most simple case, we may assume that:

$$\varepsilon_m^D = f^* \left(\frac{\tau_i}{\sigma_m + H} \right) \tag{12-11}$$

Then, at a fixed moment t, eq. 12-6 takes the form:

$$\varepsilon_m = f_1^*(\sigma_m) \pm f_2^* \left(\frac{\tau_i}{\sigma_m + H} \right) \tag{12-12}$$

Gorodetsky (1969; see also Section 11.8: Gorodetsky, 1975) suggested a more general relationship:

$$\varepsilon_m = f_1^*(\sigma_m) \pm f_2^*(\sigma_m) \tau_i - f_3^*(\sigma_m) \tau_i^2 \tag{12-13}$$

Expression 12-13 is explicitly illustrated by the curves in Fig. 12-2(b). For $\tau_i = 0$, only the strain $\varepsilon_m^0 = f_1^*(\sigma_m)$ caused by the all-around pressure will be experienced; the intercept on the x-axis of Fig. 12-2(b) gives this strain. The strain ε_m^D described by the second and third terms in the right-hand side of eq. 12-13 is represented graphically by the $\varepsilon_m - \tau_i$ curves shifted towards the axis of ordinates by an amount ε_m^0. In this case, the term $f_3^*(\sigma_m) \tau_i^2$ represents softening and the term $f_2^*(\sigma_m) \tau_i$ characterizes either softening (with the minus sign) or additional hardening (with the plus sign).

Deciding the plus or minus sign of the dilatancy is the initial void ratio (density) of the soil. Casagrande (1936) is credited with introducing the notion of critical density ϱ_{cr} of soil which appears to be important. If the initial density ϱ_0 equals the critical value, no volumetric change due to shear will occur in soil. For $\varrho_0 > \varrho_{cr}$, the soil is overcompacted and will soften due to shear. For $\varrho_0 < \varrho_{cr}$, the soil is regarded as being soft and will harden under shear. However, hardening will be observed as long as the density is below the critical value, being succeeded by softening.

The above mechanism may be represented schematically in the form of the graph in Fig. 12-3. The envelope for $\tau_i = 0$ corresponds to the volumetric strain ε_m^0 from an all-around pressure. The curves for $\tau_i > 0$ represent the aggregate volumetric strain caused by both σ_m and τ_i. Accordingly, the difference $\varepsilon_m^0 - \varepsilon_m$ controls the volumetric strain ε_m^D induced by the shearing stress τ_i.

Dilatancy during creep

Dilatancy of clay soils developing in the course of volumetric creep is a time function.

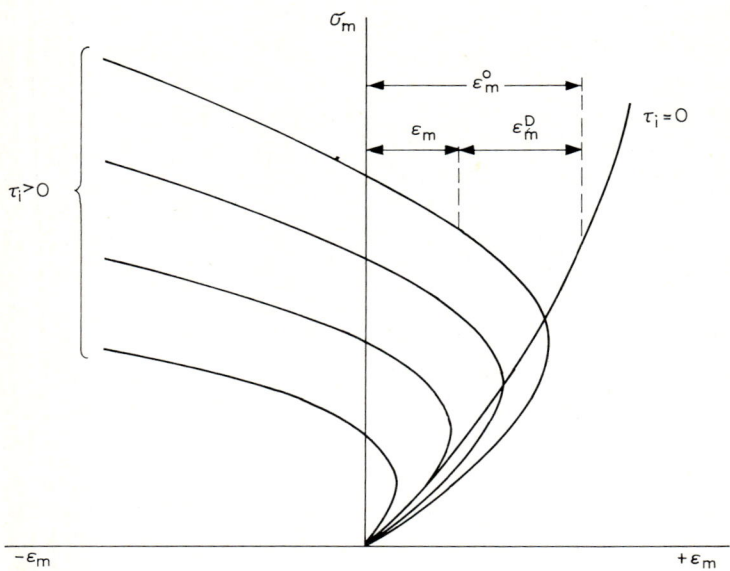

Fig. 12-3. Graph of volumetric strain.

In Fig. 5-23 the curves indicate that dilatancy is inherent in uniaxial compression. The configuration of the curves of volumetric creep due to triaxial compression is similar, as is illustrated by the experimental data acquired by Gorodetsky at the Foundation Research Institute and depicted in Fig. 12-4 (see Section 11.8: Gorodetsky, 1975).

From Fig. 12-4(a) we can see that low shearing stresses τ_i (curves *1–8*) cause only a shrinking of soil volume with time in accordance with the attenuating pattern. In the case of higher stresses τ_i, soil compacts but only at an early stage, then becoming loose to an extent increasing directly with the value of τ_i. Accordingly, the τ_i–σ_m curves change their configuration with time (Fig. 12-4(b)).

Thus, eq. 12-6 allowing for the time factor may be represented in the form:

$$\varepsilon_m = f_1^*(\sigma_m)\Phi_1^*(t) \mp f_2^*(\sigma_m, \tau_i)\Phi_2^*(t) \tag{12-14}$$

where $\Phi_1^*(t)$ is a time function of the attenuating character in all cases and $\Phi_2^*(t)$ is a time function which may be either of the attenuating nature (when additional hardening is described) or of the non-attenuating type (in describing loosening).

In a contribution by Zaretsky and Gorodetsky (1975) it was shown that the dilatant constituent of volumetric strain, ε_m^D, develops synchronously with the shearing strain in creep. Consequently, eq. 12-14 also holds for the difference between the aggregate creep deformation and the deformation of non-steady creep, should this difference be considered. Then, the dilatancy coefficient in the sublimiting state is the function of

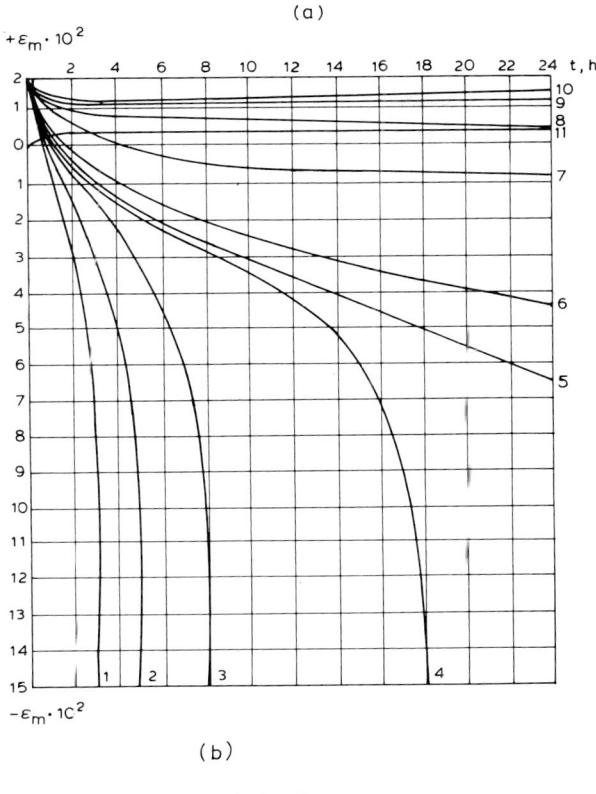

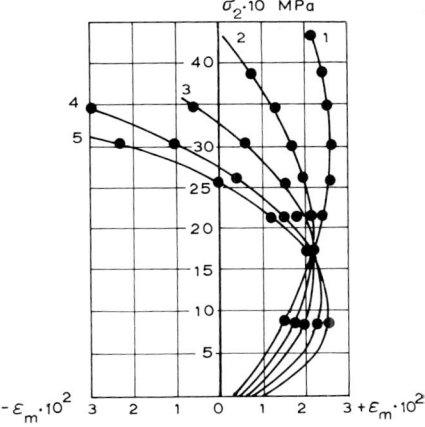

Fig. 12-4. Dilatancy in the course of soil creep. (a) Volumetric creep curves for $\sigma_m = 15 \cdot 10^5$ Pa and τ_j as follows: $1 = 26 \cdot 10^5$ Pa; $2 = 25.1 \cdot 10^5$ Pa; $3 = 24.2 \cdot 10^5$ Pa; $4 = 22.9 \cdot 10^5$ Pa; $5 = 22.5 \cdot 10^5$ Pa; $6 = 21.7 \cdot 10^5$ Pa; $7 = 17.3 \cdot 10^5$ Pa; $8 = 13.0 \cdot 10^5$ Pa; $9 = 8.66 \cdot 10^5$ Pa; $10 = 4.33 \cdot 10^5$ Pa; $11 = 0$. (b) Relation between τ and ε for various times t as follows: $1 = 15$ min; $2 = 1$ h; $3 = 4$ h; $4 = 12$ h; $5 = 24$ h. Triaxial compression test of frozen sandy loam at $\theta = -1°C$ (Gorodetsky's experiments).

both τ_i and σ_m:

$$\Lambda = \Lambda_0 \frac{\tau_{s(\infty)}^{1-\alpha}}{\tau_i} \qquad \text{for } \tau_i < \tau_{s(\infty)} \qquad (12\text{-}15)$$

In the limiting condition, the dilatancy coefficient is the function of σ_m only:

$$\Lambda = -\Lambda_0 \tau_{s(\infty)}^{-\alpha} \qquad \text{for } \tau_i > \tau_{s(\infty)} \qquad (12\text{-}15')$$

where:

$$\tau_{s(\infty)} = \tau_{s(\infty)}^0 \left[1 + \frac{\sigma_m}{H}\right]$$

is the ultimate long-term strength in the state of combined stress (see Section 11.6) and $\tau_{s(\infty)}^0$ is the ultimate long-term strength of the soil in pure shear.

The above statement reflects the fact, proved experimentally, that in the case of attenuating creep ($\tau_i < \tau_{s(\infty)}$) dilatancy is positive (extra soil hardening), for $\tau_i = \tau_{s(\infty)}$ no dilatancy is observed ($\lambda = 0$), and a progressive flow ($\tau_i > \tau_{s(\infty)}$) gives rise to negative dilatancy (softening). The physical background of these phenomena has already been explained: they are associated with the kinetics of the structural changes in soil.

12.2 EFFECT OF THE ARRANGEMENT OF STRESS

Lode parameter as definer of the state of stress

It has already been pointed out that the Lode stress parameter $\mu_\sigma = (2\sigma_2 - \sigma_1 - \sigma_3):(\sigma_1 - \sigma_3)$, determining the relation between the three principal normal stresses, serves to define the arrangement of the stress in a body.

The fact that the μ_σ parameter influences the process of soil deformation, may be established by comparative experiments with various states of stress, using a certain value of μ_σ for each state.

If the experimental points obtained from testing soil at various values of μ_σ fall on the same curve of the τ_i–γ_i graph, this is an indication that the τ_i–γ_i relation is not influenced by the arrangement of stress. If different values of μ_σ yield different τ_i–γ_i curves, this means that the arrangement of stress has a bearing on the process of deformation. For the experiments, it is advantageous to use apparatus capable of producing the loading patterns indicated in Fig. 11-1, (b) through (e), which enable the conduction of tests at various values of μ_σ.

Experimental data

Fig. 12-5 illustrates the results of tests undertaken by Lomize et al. (1969) at the Civil Engineering Institute, Moscow, with sandy loam of natural structure, using the loading pattern of Fig. 11-1(e) and various values of the Lode parameter.

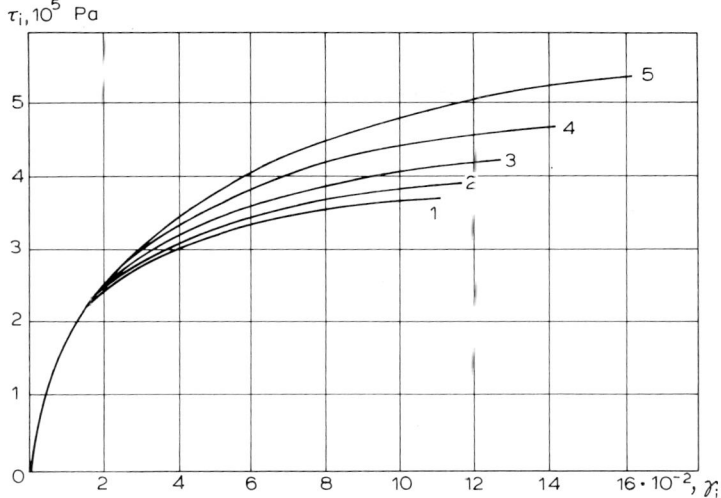

Fig. 12-5. Effect of stress arrangement on the $\tau_i-\gamma$ relation in clay loam: 1. $\mu_\sigma = +1$; 2. $\mu_\sigma = +0.5$; 3. $\mu_\sigma = 0$; 4. $\mu_\sigma = -0.5$; 5. $\mu_\sigma = -1$ (experiments by Lomize and others).

It can be seen that the curves for various values of μ_σ are confluent when the strains are comparatively small (not over 2%). For larger strains, the curves begin to diverge, the rate of divergence increasing with strain. This means that the process of soil deformation is dependent to a certain extent on the arrangement of stress. The influence can be allowed for by introducing a function of the third stress tensor invariant, $Z(I_3)$ or, which is the same, the Lode parameter μ_σ or the angle ω_σ into the equation of state.

Lomize et al. (1967) and Kryzhanovsky et al. (1975) suggested equations of deformation of the following types:

$$\gamma_i^m = \frac{[\tau_s^0 + \sigma_m^\lambda Z(I_3)]^\alpha \tau_i}{G_0[\tau_s^0 + \sigma_m^\lambda Z(I_3) - \tau_i]} \qquad \gamma_i^m = \frac{1}{A}[\tau_i - \sigma_m Z(I_3)] \tag{12-16}$$

A modified form of the equation was derived by Geniev (1968):

$$\tau_i = A\gamma_i^m(1 + \sigma_m/H_s)Z(I_3) \tag{12-16'}$$

Assuming that the density of deformation energy is a function of three strain tensor invariants $W = W(J_1, J_2, J_3)$, T. Chang and H. Ko (see Section 1.5: Scott and Ko, 1969) obtained the equation of deformation in the form:

$$\sigma_{ij} = (2B_1 J_1 + 2B_2 J_1^2 + B_3 J_2)\delta_{ij} + (B_4 + B_3 J_1)\varepsilon_{ij} + B_5 \varepsilon_{il}\varepsilon_{lj} \tag{12-17}$$

where δ_{ij} is the Kronecker delta, $i, j, l = 1, 2, 3$; and B_{1-5} are parameters.

The effect of the arrangement of stress on soil deformation arises from the difference in the resistance offered by the soil to tension and compression. In fact, the pattern of deformation $\varepsilon_z = \sigma_z/E$ will be identical in the case of uniaxial compression

($\mu_\sigma = -1$) and uniaxial tension ($\mu_\sigma = +1$) only then when $E_{comp} = E_{tens}$. However, for $E_{comp} \neq E_{tens}$, we obtain two different diagrams of deformation.

Limiting soil condition

Returning to Fig. 12-5, we point out once more that the effect of the arrangement of stress becomes more pronounced with an increase in deformation. Naturally, the influence will be especially felt after the transition of soil into the limiting condition. Let us discuss the problem, starting with a comparison of the Mohr–Coulomb condition of limiting state with that due to von Mises, Schleicher, and Botkin. The Mohr–Coulomb condition is described by eq. 4-12:

$$\sigma_1 - \sigma_3 = (\sigma_1 + \sigma_3 + 2H) \sin \phi \tag{12-18}$$

where $H = c/\tan \phi$.

In the case of the von Mises–Botkin condition we use eq. 4-23:

$$\sqrt{(\sigma_1 - \sigma_2)^2 + (\sigma_2 - \sigma_3)^2 + (\sigma_3 - \sigma_1)^2} = \sqrt{(2/3)}(\sigma_1 + \sigma_2 + \sigma_3 + 3H) \tan \psi \tag{12-19}$$

where $H = \tau_s^0/\tan \phi$.

It is known that the parameters ϕ and ψ have different meanings. The parameter ϕ is the angle of internal friction after Mohr, which is determined by the slope of a straight line on the graph presenting the relationship between the tangential stress τ_n and normal stress σ_n on the slip plane. However, the value of ψ is decided by the slope of a straight line on the graph presenting the relation between the tangential stress intensity τ_i and mean normal stress σ_m, i.e. the relation between the stresses on the octahedral plane. Therefore, the parameter ψ is frequently treated as the angle of internal friction of soil at the octahedral plane. The parameter ϕ is commonly employed in considering two-dimensional problems and the parameter ψ is used in tackling three-dimensional ones.

The connection between the strength characteristics c and ϕ due to Mohr and H and ψ after von Mises and Botkin varies with the type of soil test. For example, in the case of triaxial compression $\sigma_1 > \sigma_2 = \sigma_3$, on substituting this condition into eqs. 12-18, 12-19 and eliminating σ_1 and σ_3 therefrom, we obtain the following known relations:

$$\tan \psi = \frac{2\sqrt{3} \sin \phi}{3 - \sin \phi} \qquad H = \frac{c}{\tan \phi} \tag{12-20}$$

Using the data on soil tests given in Fig. 11-2(b), Botkin (1940, referenced in Section 4.9) has computed that Mohr's friction angle is $\phi = 30°30'$ whereas $\psi = 34°30'$. However, according to the same author, the difference between ϕ and ψ for clay soils is considerably smaller.

For an ideally cohesive soil $\phi = \psi = 0$. In this case, conditions 11-8 and 11-9 become the St. Venant condition and the von Mises condition differing from one another by the values of c and τ_s only. As was shown in Chapter 3, according to St.

Venant $c = \tau_s = \sigma_s/2$ whereas, according to von Mises $\tau_s = \sigma_s/\sqrt{3}$, where σ_s is the ultimate strength (yield limit) in shear.

Limiting state for various values of parameter μ_σ

The above difference between the strength parameters according to the Mohr–Coulomb theory and the von Mises–Botkin condition originates from the fact that the former theory allows for the effect of the intermediate principal stress σ_2 whereas in the latter case the three principal stresses σ_1, σ_2, and σ_3 are subjected to consideration.

The effect of σ_2 was investigated in a number of experimental studies which have proved that the condition of limiting equilibrium is significantly dependent upon the value of this stress. The exact degree of influence may be determined from comparative studies for various values of μ_σ by using, for example, triaxial compression for $\sigma_2 = \sigma_3$ and $\mu_\sigma = -1$, triaxial tension for $\sigma_2 = \sigma_3$ and $\mu_\sigma = 1$, and pure shear for $\mu_\sigma = 0$.

These studies have revealed a considerable discrepancy between the values of the strength parameters c and ϕ. Fig. 12-6 depicts Cornforth's data (1964) on the triaxial compression and shear tests of sand of various density at $\mu_\sigma = \pm 1$ and $\mu_\sigma = 0$, respectively, along with the results of tests of sand at various values of μ_σ obtained by other researchers (Malyshev et al., 1968; Malyshev, 1969, 1980). The values of the angle of internal friction ϕ after Mohr, as computed from the tests, are plotted against the values of μ_σ. If ϕ was not affected by the type of test, its values would be the same whatever the value of μ_σ. However, we can see that ϕ appreciably changes with μ_σ.

Although taking into account the effect of all three principal stresses σ_1, σ_2, and σ_3,

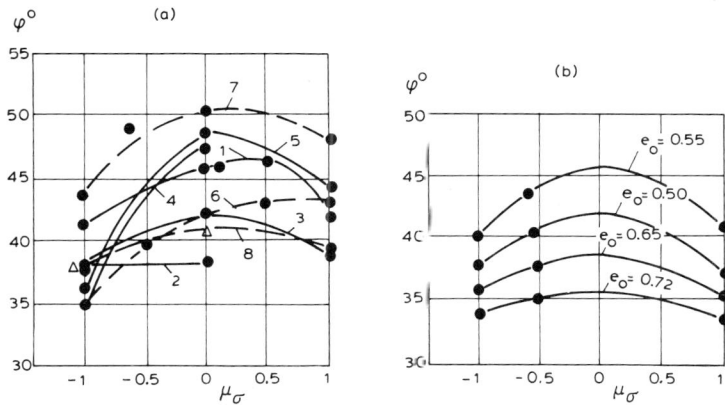

Fig. 12-6. Relationship between the angle ϕ and stress arrangement in sand of various compactness. (a) Results of tests by various authors as presented by Malyshev (1968, 1969, 1980): 1 = Malyshev–Fradis, $e = 0.495$; 2 = Barshevsky, $e = 0.64$; 3 = Kirkpatrick, $e = 0.55$; 4 = Stroganov, $e = 0.56$; 5 = Lomize–Kryzhanovsky, $e = 0.73$; 6 = Hon-Yim Ko and Scott, $e = 0.52$; $7, 8$ = Fradis, $e = 0.61$. (b) Experiments by Cornforth for various values of e.

the von Mises–Botkin condition also yields results influenced by the arrangement of stress. The tests carried out by G.M. Lomize and colleagues at the Civil Engineering Institute, Moscow, between 1966 and 1975 indicated that for sands the value of ψ was influenced by μ_σ in a manner represented by a graph similar to that of Fig. 12-5 (Lomize et al., 1969; Lomize, 1973, 1974; see also Section 11.8: Lomize et al., 1966, 1968). Data on clay soils are, so far, scarce. However, there are reasons to believe that the type of test is telling to a lesser extent in this case.

Thus, the conditions of limiting state according to Mohr and Coulomb, fulfilled by eq. 12-18 as well as to von Mises, Schleicher and Botkin, given by eq. 12-19, provide generally solutions influenced by the arrangement of stress; correspondingly, the values of the strength parameters ϕ and ψ entering these conditions are affected by the type of test. This means that on the shear diagrams τ_n–σ_n (Mohr–Coulomb) and τ_i–σ_m (von Mises–Botkin) we obtain no single straight lines but families of such lines, each corresponding to a given value of μ_σ and having a slope of its own defining ϕ and ψ, respectively. Consequently, these conditions appear to be inadequate as a means of describing the limiting condition of soil in the general form.

Generalized condition of limiting state

To obtain an invariant condition of limiting equilibrium, it is necessary to introduce into this condition either the third stress tensor invariant I_3 or, as equivalent, the Lode parameter μ_σ allowing for the arrangement of stress. In other words, the condition of limiting state is to be adopted in the general form given by eq. 11-3. The exact forms of the relationship may be different.

Geniev (1968) suggested the following modification of the von Mises–Botkin condition arrived at from eq. 12-16' for $A\gamma_i^m = \tau_s = H \tan \psi$:

$$\tau_i = \tan \psi (H + \sigma_m)(1 - k \cos 3\omega_\sigma) \quad (12\text{-}21)$$

where k is a constant and $(1 - k \cos 3\omega_\sigma) = Z(I_3)$.

Eq. 12-21 yields a limiting surface in the form of a non-circular cone. Assuming that the cone is described about the Mohr hexahedron (Fig. 12-7), the strength parameters of the Geniev condition and those of the Mohr condition are correlated by the following expressions:

$$\tan \psi = 6\sqrt{3} \sin \phi / (9 - \sin^2 \phi)$$

$$H = c/\tan \phi \qquad k = \sin \phi / 3$$

The condition of strength, as represented by the shape of cone 3 in Fig. 12-7, was adopted by other researchers. A similar condition suggested by L. Barden (1969) was presented in a report at the 7th ICSMFE (see Sect. 1.5: Scott and Ko, 1969). Shibata and Karube (1965) outlined their version of the condition at the 6th ICSMFE.

The condition of limiting state by G.M. Lomize and A.L. Kryzhanovsky (see Section 11.8: Lomize et al., 1966, 1968) may be given in the form arrived at from

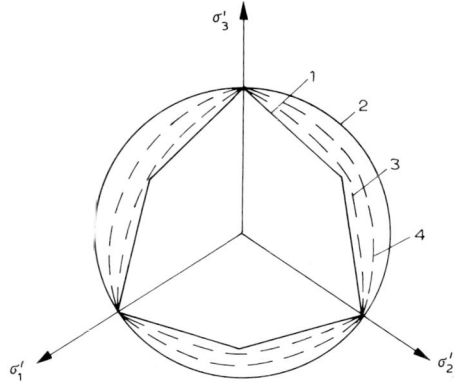

Fig. 12-7. Contours of the $(\sigma_1 - \sigma_2 - \sigma_3)$ limiting surface on the deviatoric plane. *1*. For eq. 12-18. *2*. For eq. 12-19. *3*. For eq. 12-21. *4*. For eq. 12-23.

eq. 12-16 for $\gamma_i^m \to \infty$ and $A\gamma_i^m = \tau_{i(s)}$, respectively:

$$\tau_i = \tau_s^0 + \sigma_m^\lambda Z(I_3) \tag{12-22}$$

$$\tau_i = \tau_s^0 + \sigma_m Z(I_3) \tag{12-23}$$

The limiting surface described by condition 12-23 also has the shape of a cone. However, this cone is inscribed between two von Mises–Botkin cones corresponding to the extreme values of $\mu_\sigma = \pm 1$ (see Fig. 12-7). A limiting surface of the same shape was adopted by Scott and Hon-Yim Ko (1969, referenced in Section 1.5).

Naturally, the fact that the arrangement of stress should be taken into account invites difficulties not only in estimating soil behaviour from tests but in solving the problem itself. Therefore, it is practical to allow for this factor only in projects of paramount importance, keeping in mind that the influence has a greater bearing on sands than on clays.

To simplify computations, the effect of stress arrangement may be disregarded in solving conventional problems. For a good closeness of fit of the data on soil performance acquired from tests, it is advisable to use a technique producing in the specimen a state of stress approaching the state of stress under the natural condition. For example, axi-symmetrical problems (foundations of circular, square and other outlines similar in plane, cylindrical underground structures) should be solved by using the strength parameters H and ψ as obtained from triaxial tests. It is also advisable to continue research into the effect of the arrangement of stress.

12.3 EFFECT OF LOADING CONDITIONS

Loading path

Let a body be loaded so that the stress at point M increases from 0 to some value $M(\sigma_1, \sigma_2, \sigma_3)$. Apparently, this is feasible if each of the components σ_1, σ_2 and σ_3 is increased in a certain manner.

If the process of loading is represented in the $(\sigma_1-\sigma_2-\sigma_3)$ space, we obtain a curve OM (Fig. 12-8) referred to as the loading path. For example, the loading path in the case of all-around pressure ($\sigma_1 = \sigma_2 = \sigma_3$) will be represented by a straight line making the same angles with all three axes.

The path of the triaxi-symmetrical compression ($\sigma_1 > \sigma_2 = \sigma_3$) is a straight line making the same angles with the σ_2- and σ_3-axes. It will be recalled that the arrangement of stress remains constant if the Lode parameter μ_σ is constant, but is variable if the parameter changes in the course of loading.

The loading path may also be presented in the $(\tau_i-\sigma_m-\mu_\sigma)$ space of the invariant characteristics defining the state of stress, or on the $\tau_i-\sigma_m$ plane for $\mu_\sigma = $ const. Consider, for example, various ways of loading over the range between zero and the load at failure.

The condition of limiting equilibrium will be represented on the $\tau_i-\sigma_m$ plane by a curve AB (Fig. 12-9), each point on which determines a $\tau_i-\sigma_m$ relation which fulfills this condition.

Let us apply a load in pure shear. Since, in this case, $\tau_i = \tau$ and $\sigma_m = 0$, the loading path is a segment OA coinciding with the τ_i-axis. Now, testing soil for triaxial compression ($\sigma_1 > \sigma_2 = \sigma_3$), we destroy the specimen at some $\tau_i-\sigma_m$ relation represented by point M on the limiting curve AB. Obviously, this point may be reached by proceeding along various paths. If the stress is increased so that the ratio $\tau_i/\sigma_m = k$ is constant, the loading path will be represented by a straight line OM, the

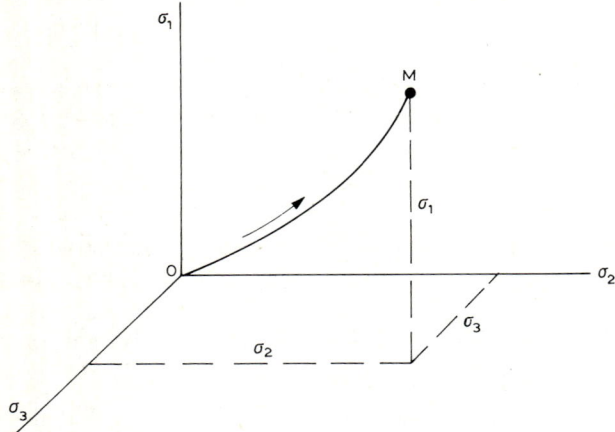

Fig. 12-8. Loading path in stress space $\sigma_1, \sigma_2, \sigma_3$.

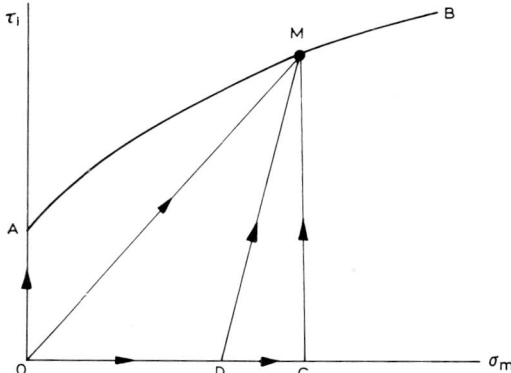

Fig. 12-9. Various loading paths as represented on the τ_i–σ_m plane.

tangent of the slope of which equals k. Since $\tau_i = (\sigma_1 - \sigma_3)/\sqrt{3}$ and $\sigma_m = (\sigma_1 + 2\sigma_3)/3$, the above condition will be fulfilled if the stress components change as the ratio $\sigma_1/\sigma_3 = (\sqrt{3} + 2k) : (\sqrt{3} - k)$.

Let us assume that at the beginning of the test the specimen has been subjected to a hydrostatic pressure (for $\tau_i = 0$) which gradually increases to a value $\sigma_m = p_1$ corresponding to point D in Fig. 12-9. Next, we increase the components σ_1 and $\sigma_2 = \sigma_3$ so that the increment in τ_i is proportional to the increment in σ_m, i.e. $\Delta\tau_i/\Delta\sigma_m = k$. The path of such a loading will be represented by a broken line ODM. Inside the DM segment of the line, the stress components must change so as to satisfy the relation $\sigma_1 = [(\sqrt{3} + 2k)\sigma_3 - 3kp_1] : (\sqrt{3} - k)$. If we increase τ_i so as to maintain the mean normal stress σ_m constant on applying the hydrostatic pressure $\sigma_m = p_2$ to the specimen (point C on the graph), the loading path will be represented by a broken line OCM. In this case, the components $\sigma_2 = \sigma_3$ must decrease so as to satisfy the equality $\sigma_1 + 2\sigma_3 = 3p_2$ with an increase in σ_1.

The Lode parameter during the above tests remained constant ($\mu_\sigma = -1$) whatever the loading path; the loading, however, was a combined one, for the stress components σ_1, $\sigma_2 = \sigma_3$ changed independently.

Similarity between the states of stress and strain in soils

As stated in Section 3.7, the classical theory of plasticity postulates that the state of stress and that of strain are similar. This assumption follows from eq. 3-87 according to which the stress deviator components are proportional to the strain deviator components (or strain rate components) and coincide with one another, i.e. are in alignment.

The alignment condition may be verified by comparing the angles of the principal axes of stresses, α_σ, and those made by the principal axes of strains, α_ε, in various states of stress. To that end, we use a test method allowing the rotation of the axes of the principal stresses in the course of loading.

Suitable tests in this case are by means of a settlement plate pressed into the soil mass so that measurements of the stress and strain components can be taken. Alternatively, apparatus capable of applying loadings of the type shown in Figs. 11-1(b), (c), and (d) can be used. In this latter case, the principal planes of stress change their orientation and the axes of principal stresses turn due to the combined effect of the torque applied to the specimen and the tangential stress this torque sets up. The condition of alignment will be fulfilled if the equality $\alpha_\sigma = \alpha_\varepsilon$ is satisfied in any state of stress.

Experimental data

The effect the loading path and the similarity between the states of stress and strain have on soils has been put to test by experiments only since the 1960s. Among the first studies were the experiments of Ahlvin and Brown (1960) aimed at determining the effect of a loading path in clay loam. Probing with a settlement plate in a tray, they discovered identity in the patterns of the deformations in the tray and in a specimen subjected to triaxial compression only when the loading paths in both cases were the same; otherwise no identity was observed. A significant influence of the loading path was noted by Roscoe and Poorooshasb (1963, referenced in Section 11.8) and also by Lambe (1963).

Among early evidence of failures to fulfill the condition of alignment $\alpha_\sigma = \alpha_\varepsilon$, were the results of tests with kaolin clay carried out by Broms and Casbarian (1965) in an apparatus enabling the rotation of the axis of principal stress. The same evidence was obtained by Gerrard (1967) who analysed the experiments of P. Rowe with sand.

The similarity between the states of stress and strain and the effect of the loading path received exhaustive treatment in a paper by Scott and Ko (1969, referenced in Section 1.5) presented at the 7th ICSMFE. Extensive research into the same problems was carried out at the Civil Engineering Institute, Moscow, by G.M. Lomize, A.L. Kryzhanovsky and others (see Section 11.8: Lomize et al., 1966, 1968). During the studies, apparatus capable of applying a simple or combined loading in the manner illustrated in Figs. 11-1(d) and (e) were used; tests with a square settlement plate in a tray were also employed.

Experiments have shown that the condition $\alpha_\sigma = \alpha_\varepsilon$ is fulfilled quite well when the axes of stresses are turned (so as to change the angle α_σ) simultaneously with an increase in τ_i for $\mu_\sigma = $ const., i.e. when the arrangement of stress remains unchanged. Should the stress arrangement change (changing both α_σ and μ_σ), the principal axes of stresses will fail to coincide directionally with the principal axes of strains.

Similar results were obtained by comparing the experimental values of the Lode parameter obtained in the stressed state, μ_σ, and in the strained state μ_ε (Fig. 12-10).

When test loading was effected in a manner causing no change in the value of μ_σ (equaling 1, 0, −1), the condition $\mu_\sigma = \mu_\varepsilon$ was fulfilled quite satisfactorily; it can be seen that the experimental points fell on the bisector of the angle $\mu_\sigma = \mu_\varepsilon$ in Fig. 12-10. In the case of a combined loading, when the value of μ_σ changed from 1 to −1 for $\tau_i = $ const. and $\sigma_m = $ const. (curve 2) or when, simultaneously with μ_σ, τ_i

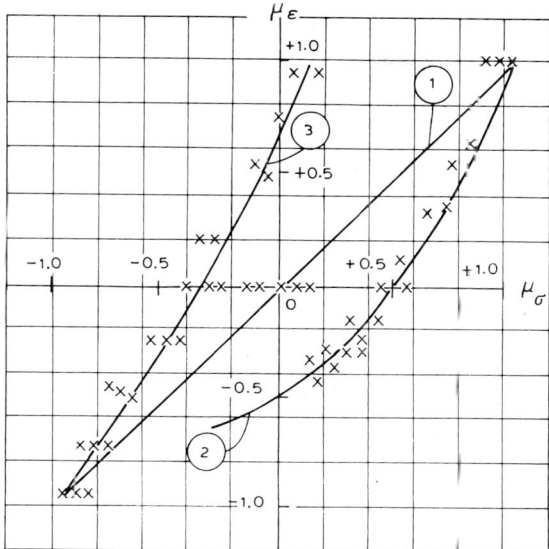

Fig. 12-10. Verifying the condition of similarity between the state of stress and that of strain: 1 = loading for μ_σ = const.; $2, 3$ = loading for $\mu_\sigma \neq$ const. (Triaxial and compression tests of sand at $\sigma_1 \neq \sigma_2 \neq \sigma_3$ by Lomize and Kryzhanovsky.)

and σ_m (curve 3) also changed, the experimental points came down on the curves $\mu_\sigma \neq \mu_\varepsilon$ departing appreciably from the straight line 1.

Thus, the condition of alignment and similarity between the states of stress and strain is fulfilled in soils only when the arrangement of stress is unchanged. In the case of a combined loading, the greater the change in the value of μ_σ the quicker this condition ceases to be fulfilled.

This means that, for a given state of stress, the deformations may vary depending on the conditions of loading. However, to work out a theory which will allow for the lack of similarity between the states of stress and strain is a formidable task. In spite of some efforts made, the formulation of such a theory helpful in solving engineering problems is a question for the future.

Since soils below foundations are loaded practically in a manner differing little from simple loading, the condition of similarity between the state of stress and that of strain, $\mu_\sigma = \mu_\varepsilon$, can be fulfilled with some degree of approximation. The possibility of such an approach can be proved experimentally by taking simultaneous measurements of the stress and displacement components during a test of a layer of clay soil by means of a settlement plate. Relevant data are discussed in Section 13.6.

12.4 REFERENCES

Alvin, R.G. and Brown, D.N., 1960. Duplication of prototype stress–strain relations in soil masses by laboratory tests. Proc. ASTM, 61, 60.

Barden, L. and Khayatt, A.J., 1966. Incremental strain ratios and strength of sand in triaxial tests. Géotechnique, 16.

Barden, L., Khayatt, A.J., et al., 1969. General Deformation Behaviour of a Range of Particulate Material Under a Variety of Test Conditions. Univ. of Manchester Report.

Broms, B.B. and Casbarian, A.O., 1965. Effects of rotation on the principal axes and the intermediate principle stress on the shear strength. Proc. 6th ICSMFE, Montreal, v. 2.

Casagrande, A., 1936. Characteristics of cohesionless soils affecting the stability of slopes and earth fills. In: Contribution to Soil Mechanics. Boston Soc. Civ. Eng.

Cornforth, D.H., 1964. Some experiments on the influence of strain conditions on the strength of sands. Géotechnique, v. 14, no. 2.

Geniev, G.A., 1968. On the problem of generalization of the condition of limiting equilibrium in a cohesionless medium. SMFE, no. 2 (in Russian).

Gerrard, C.M., 1967. Some aspects of the stress–strain behaviour of sand. Aust. Road. Res. Board, v. 3, no. 4.

Gorodetsky, S.E., 1969. The effect of shear stresses on volumetric strains of frozen soils. In collective volume: Foundations and Underground Structures. No. 58, Stroiizdat, Moscow (in Russian).

Grigoryan, S.S., 1964. To the solution of the problem of a subterrain explosion in soft soils. Appl. Math. Mech., v. 28, no. 6 (in Russian).

Horne, M.R., 1965. The behaviour of an assembly of rotund, rigid, cohesionless particles, I and II. Proc. R. Soc., A286.

Kryzhanovsky, A.L., Chevikin, A.S. and Kulikov, O.V., 1975. Effectiveness of foundation calculations with allowance for non-linear properties of soils. SMFE, no. 5 (in Russian).

Lambe, T.W., 1963. In: J. Soil Mech. Found. Div., ASCE, 90.

Lee, I.K., 1966. Stress-dilatancy performance of feldspar. J. Soil Mech. Found. Div., Proc. ASCE, v. 92, no. SM2.

Lomize, G.M. and Sukhanov, E.I., 1973. On limiting state of stress and failure of clay soils. Hydraul. Eng. Projects, no. 8 (in Russian).

Lomize, G.M. and Sukhanov, E.I., 1974. Mechanism of flow in soil at failure. Hydraul. Eng. Projects, no. 6 (in Russian).

Lomize, G.M., Kryzhanovsky, A.L., et al., 1967. Strength of soils. Hydraul. Eng. Projects, no. 3 (in Russian).

Lomize, G.M., Kryzhanovsky, A.L. and Vorontsov, E.I., 1969. Investigation of soil deformability and strength laws in a spatial stressed state. Proc. 7th ICSMFE, Mexico City, v. 1.

Malyshev, M.V., et al., 1968. Strength condition of sandy soils. Acta Tech. Acad. Sci. Hung., T. 63 (1–4).

Malyshev, M.V., 1969. On the applicability of the Huber–V. Mises–Botkin stress condition to cohesionless soils. SMFE, no. 5 (in Russian).

Malyshev, M.V., 1980. Strength of Soils and Stability of Foundations. Stroiizdat, Moscow (in Russian).

Murayama, S. and Matsuoka, H., 1973. A microscopic study on shearing mechanism of soils. Proc. 8th ICSMFE, Moscow, v. 1.2.

Nikolaevsky, V.N., 1972. Engineering properties of soil and plasticity theory. All-Union Inst. Sci. Eng. Inform., Moscow (in Russian).

Roscoe, K.H., 1970. The influence of strains in soil mechanics. Géotechnique, v. 20, no. 2.

Rowe, P.W., 1962. The stress-dilatancy relation for static equilibrium of an assembly of particles in contact. Proc. R. Soc., A.269.

Shibata, T. and Karube, D., 1965. Influence of the variation of the intermediate principal stress on the mechanical properties of normally consolidated clays. Proc. 6th ICSMFE, v. 2.

Zaretsky, Yu.K. and Gorodetsky, S.E., 1975. Dilatancy of frozen soil and formulation of the deformation theory of plasticity. Hydraul. Eng. Projects, no. 2 (in Russian).

Chapter 13

THE THEORY OF NON-LINEAR CREEP: PROBLEMS AND SOLUTIONS

13.1 GENERALIZED EQUATION OF SOIL DEFORMATION

Possible simplifications of the equation

To understand the behaviour of soil in response to loads, one must take account of all the factors influencing the soil's state of stress and strain which were considered earlier — non-linear stress–strain relation, creep, mutual effect of the three stress tensor invariants, the arrangement of stress, loading conditions, etc. However, because of the modern state of the art and the lack of adequate experimental data, we cannot allow for all the factors mentioned in engineering analyses, because this will not only significantly complicate computations but also introduce difficulties in acquiring requisite soil characteristics. Therefore, it is advisable to simplify the equations used in solving practical problems as much as possible. The first assumption to be made is that the condition of similarity of the states of stress and strain is applicable to soils. This and some other assumptions considered below appear to be reasonable because soils operate, as a rule, under the conditions close to those of simple loading. The adoption of the condition of similarity paves the way to applying Hencky's generalized equations of deformation (eq. 11-13) discussed in Section 11.2.

Assuming further that the estimated soil performance is determined under conditions close to field conditions, we may disregard the arrangement of stress and eliminate the Lode parameter μ_σ from eq. 11-13.

Hencky's equations

Let us recall Hencky's equations in their general form (eq. 11-13) which establish a connection between the components of stress and strain:

$$\varepsilon_x = \chi(\sigma_x - \sigma_m) + \chi^*\sigma_m \qquad \gamma_{xy} = 2\chi\tau_{xy}, \ldots \qquad (13\text{-}1)$$

where:

$$\chi = \frac{f(\tau_i, \sigma_m, t)}{2\tau_i(t)} = \frac{\gamma_i(t)}{2\phi(\gamma_i, \sigma_m, t)}$$

$$\chi^* = \frac{f^*(\sigma_m, \tau_i, t)}{\sigma_m(t)} = \frac{\varepsilon_m(t)}{\phi^*(\varepsilon_m, \tau_i, t)}$$

The function χ^* in the second expression reflects the effect of dilatancy. If dilatancy is to be introduced in the explicit form into eq. 13-1, the term $\chi^* \sigma_m$ of this equation may be represented as:

$$\chi^* \sigma_m = \varepsilon_m^0 + \varepsilon_m^D = \varepsilon_m^0 \mp \lambda \gamma_i$$

where $\lambda = \varepsilon_m^D / \gamma_i$.

Substituting into relationships 13-1, we obtain:

$$\varepsilon_x - \varepsilon_m^0 = \chi(\sigma_x - \sigma_m) \mp 2\lambda\tau_i \qquad \gamma_{xy} = 2\chi\tau_{xy} \ldots \qquad (13\text{-}2)$$

It will be recollected that the "plus" sign in the above expressions indicates a positive dilatancy (extra hardening) and the "minus" sign means that the dilatancy is negative (softening); if no volumetric strain is observed, $\lambda = 0$.

Relationships 13-1 are written in the form corresponding to the theory of ageing and are applicable in the case of a constant or monotonically increasing loading. Generally speaking, such a case is typical for soils below foundations. However, if a variable loading is to be taken into account, integral equations 11-33 from hereditary theory are to be introduced into relationships 13-1.

Consider the case of simple loading when the stress components all change in proportion to the same factor. This enables us to introduce

$$\int_0^{\sigma_m} P(\sigma_m - \xi) f_2[\tau_i(\xi)] \, d\xi = f_2(\tau_i) \bar{\psi}(\sigma_m)$$

into relationships 11-33.

Let us introduce the notion of the Volterra integral operator (see Section 1.5: Rabotnov, 1966):

$$\tilde{L}\{y(t)\} = \int_0^t L(t - v) y(v) \, dv \qquad (13\text{-}3)$$

Finally, relationships 13-1 take the form:

$$\varepsilon_x = \chi \left\{ [\sigma_x(t) - \sigma_m(t)] + \int_0^t [\sigma_x(v) K(t - v) - \sigma_m(v) K^*(t - v)] dv \right\}$$

$$+ \chi^* \left\{ \sigma_m(t) + \int_0^t \sigma_m(v) K^*(t - v) \, dv \right\} \ldots$$

$$\gamma_{xy} = 2\chi \left\{ \tau_{xy}(t) + \int_0^t \tau_{xy}(v) K(t - v) \, dv \right\} \ldots \qquad (13\text{-}4)$$

where:

$$\chi = \frac{f_1(\tau_i) - f_2(\tau_i) \bar{\psi}(\sigma_m)}{2\tau_i}$$

$$\chi^* = \frac{f_1^*(\sigma_m) \pm f_2^*(\tau_i, \sigma_m)}{\sigma_m}$$

Power law of deformation. Let us assume that the stress–strain relationship is described by a power law of the type given by eq. 11-24 for shearing strains and by eq. 4-41 for volumetric strains:

$$\tau_i = A(t)\gamma_i^{m_1} + B(t)\sigma_m\gamma_i^{m_2} \qquad \sigma_m = D(t)\varepsilon_m^\chi \tag{13-5}$$

The time functions are, however, of the form given by eq. 11-43:

$$A(t) = \frac{A_0}{1 + \delta_1 t^{\alpha_1}} \qquad B(t) = \frac{B_0}{1 + \delta_2 t^{\alpha_2}} \qquad D(t) = \frac{D_0}{1 + \delta_3 t^{\alpha_3}} \tag{13-6}$$

Then, the functions χ and χ^* in relationships 13-1 may be written as:

$$\chi = \frac{\gamma_i^{1-m_2}}{2[A(t)\gamma_i^{m_1-m_2} + B(t)\sigma_m]} \qquad \chi^* = \frac{\varepsilon_m^{1-\varepsilon}}{D(t)} \tag{13-7}$$

In a special case, for $m_1 = m_2$ and $A(t)/B(t) = H_s$, the χ function is simplified:

$$\chi = \frac{\gamma_i^{1-m}}{2\{A(t)[1 + \sigma_m/H_s]\}} = \frac{\tau_i^{-m)/m}}{2\{A(t)[1 + \sigma_m/H_s]\}^{1/m}} \tag{13-8}$$

13.2 AXI-SYMMETRICAL PROBLEM

Formulation of the problem

Consider the author's solution of the problem of deformation suffered by a thick-walled hollow cylinder (of infinite length) due to an external radial pressure p. The problem stated may arise in analysing ice-soil retaining structures formed by freezing the ground around mine shafts and other underground workings of cylindrical shape to protect them in the course of sinking or driving.

The solution is reduced to determining the optimum thickness of the ice-soil cylinder, on the condition that the radial displacement u_r of the wall under the rock pressure p does not exceed an allowable limit Δ during the time t_w (elapsed between breaking ground and the instant of setting up the support), and that the stability of the ice-soil retaining structure is ensured during this period.

Referring to Fig. 13-1, a is the inner radius of the ice-soil cylinder, b is the outer radius, and r is the current radius; we adopt the polar system of coordinates r, θ, y.

Let the initial equations of state be eqs. 11-47 and 11-43:

$$\tau_i = A(r)\gamma_i^m \left(1 + \frac{\sigma_m}{H_s}\right) \qquad A(t) = \frac{A_0}{1 + \delta t^\alpha}$$

The volumetric strains are disregarded:

$$\varepsilon_m = \frac{\varepsilon_r + \varepsilon_\theta + \varepsilon_y}{3} = 0 \tag{13-9}$$

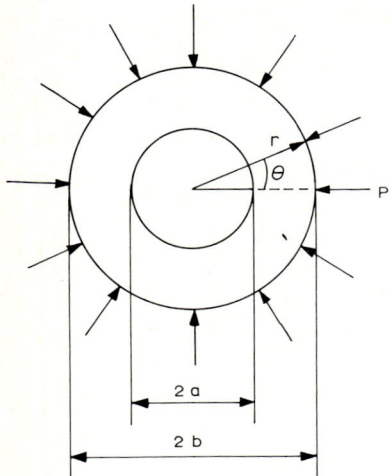

Fig. 13-1. Schematic diagram used in analysing ice-soil retaining structures.

Assuming that the cylinder is deformed in two dimensions, $\varepsilon_y = 0$, $\sigma_y = (\sigma_r + \sigma_\theta)/2$, and noting that according to the conditions of symmetry $\gamma_{r\theta} = \gamma_{ry} = \gamma_{\theta y} = 0$ and $\tau_{r\theta} = \tau_{ry} = 0$ (the y-axis is at right angles to the plane of the drawing), we obtain:

$$\gamma_i = \sqrt{\tfrac{2}{3}}\sqrt{(\varepsilon_r - \varepsilon_\theta)^2 + \varepsilon_r^2 + \varepsilon_\theta^2} \qquad \tau_i = \frac{\sigma_r - \sigma_\theta}{2} \qquad \sigma_m = \frac{\sigma_r + \sigma_\theta}{2} \qquad (13\text{-}10)$$

Hence, Hencky's relations take the form:

$$\varepsilon_r = \chi(\sigma_r - \sigma_m) \qquad \varepsilon_\theta = \chi(\sigma_\theta - \sigma_m) \qquad \varepsilon_y = \chi(\sigma_y - \sigma_m) = 0 \qquad (13\text{-}11)$$

where $\chi = \gamma_i^{1-m} : 2\{A(t)[1 + \sigma_m/H_s]\}$.

Assessing the deformation. The problem is reduced to finding the relation between the radial displacement u_r of the wall and the dimensions a and b of the cylinder under a given outside pressure p coming on the wall, and for known rheological properties of the frozen soil, m, $A(t)$, H_s.

The equation of equilibrium for the problem under consideration is:

$$\frac{d\sigma_r}{dr} - \frac{\sigma_r - \sigma_\theta}{r} = 0 \qquad (13\text{-}12)$$

Caushy's deformation-displacement relations take the form:

$$\varepsilon_r = \frac{du_r}{dr} \qquad \varepsilon_\theta = \frac{u_r}{r} \qquad \varepsilon_y = 0 \qquad (13\text{-}13)$$

Substituting eq. 13-13 into equation of incompressibility 13-9, we obtain a differential equation of displacements:

$$\frac{du_r}{dr} + \frac{u_r}{r} = 0 \qquad (13\text{-}14)$$

Solving eq. 13-14 with the aid of eqs. 13-13, 13-12 and 13-11 for the boundary conditions $\sigma_{r=a} = 0$ and $u_{r=a} = u_a$, we arrive at the following values of stress components:

$$\left.\begin{array}{c}\sigma_r \\ \sigma_\theta\end{array}\right\} = \sigma_m \left(1 \mp \frac{N}{r^{2m}}\right) \mp \frac{NH_s}{r^{2m}} \qquad (13\text{-}15)$$

where:

$$\sigma_m = \frac{(r^{2m} - N)^{(1-m)/m}}{r^{2(1-m)}} \left\{\frac{1}{H_s}\left[\frac{a^{2(1-m)}}{(r^{2m} - N)^{(1-m)/m}}\right.\right.$$

$$\left.\left. - \frac{r^{2(1-m)}}{(r^{2m} - N)^{(1-m)/m}} + \frac{NH_s a^{2(1-m)}}{a^{2m} - N^{1/m}}\right]\right\}$$

$$N = A(t)[2au_0]^m H_s^{-1}$$

Using a second boundary condition $\sigma_{r=b} = p$, we obtain an equation correlating p and u_a:

$$u_a = \frac{b^2}{2a\left[\dfrac{A(t)}{H_s}\right]^{1/m}} \left\{\frac{\left[\dfrac{p}{H_s} + 1\right]^m - 1}{\left(\dfrac{b}{a}\right)^{2m}\left[\dfrac{p}{H_s} + 1\right]^m - 1}\right\}^{1/m} \qquad (13\text{-}16)$$

Equalizing the displacement $u_a = \Delta$, we may find the safe thickness $\delta = b - a$ of the ice-soil wall, from the displacement Δ tolerable during the time t_w, using the solution by inspection method. The time t_w is taken account of in this case by introducing into eq. 13-16 the quantity:

$$A(t_w) = \frac{A_0}{1 + \delta t_w^\alpha}$$

In the case of linear deformation, $m = 1$, eq. 13-16 is simplified:

$$u_a = \frac{ap}{2A(t)\left[\dfrac{p}{H_s} + 1 - \left(\dfrac{a}{b}\right)^2\right]} \qquad (13\text{-}17)$$

If the mean pressure ($H_s = \infty$) is disregarded, eq. 13-16 takes the form:

$$u_a = \frac{a}{2} \left\{\frac{mp}{A(t)} \frac{1}{\left[1 - \left(\dfrac{a}{b}\right)^{2m}\right]}\right\}^{1/m} \qquad (13\text{-}18)$$

For $m = 1$, eq. 13-18 is transformed into the known Lamé formula:

$$u_a = \frac{ap}{2A(t)\left[1 - \left(\frac{a}{b}\right)^2\right]} \qquad (13\text{-}18')$$

As can be seen, the allowance for the mean normal stress significantly reduces the allowable value of soil displacement.

Determining the long-term strength of ice-soil cylinder. Consider the way the long-term strength of the above ice-soil cylinder is determined. The problem is reduced to determining the thickness δ of the cylinder wall which will prevent failure of the ice-soil retaining structure during the given time t_w.

The condition of the limiting state is adopted in the form given by eq. 11-77:

$$\tau_{i(s)} = [H_s + \sigma_m]\tan\psi_s \qquad (13\text{-}19)$$

or, taking into account eq. 13-10, in the form:

$$\pm\frac{\sigma_r - \sigma_\theta}{2H_s + \sigma_r + \sigma_\theta} = \tan\psi_s \qquad (13\text{-}20)$$

where the "plus" sign corresponds to the tearing apart of the cylinder and the "minus" sign indicates the crushing of the cylinder.

Substituting eq. 13-21 into the equilibrium eq. 13-12, we obtain after manipulation:

$$\pm\frac{(1 \pm \tan\psi_s)\,d\sigma_r}{2\tan\psi_s(\sigma_r + H_s)} + \frac{dr}{r} = 0 \qquad (13\text{-}21)$$

The solution of this equation under the boundary condition $\sigma_{r=b} = p$ yields the value of the ultimate load in the case of crushing of the cylinder:

$$P_{ult} = \left[1 - \left(\frac{b}{a}\right)^{2\tan\psi_s/(1-\tan\psi_s)}\right]\frac{\tau_s^0(t)}{\tan\psi_s} \qquad (13\text{-}22)$$

where $\tau_s^0 = H_s \tan\psi_s$ is the ultimate strength in pure shear.

A similar formula, which fails to take account of the time factor, has been suggested by Stroganov (1961, referenced in Section 11.8). In eq. 13-22, the time factor is allowed for by the time-dependent shear strength $\tau_s^{(0)}(t)$ decided by eq. 11-75. The value of $\tau_s^{(0)}$ which corresponds to the set time t_w, i.e. $\tau_s^0(t_w) = \beta : \ln(t_w/T)$ is used in computation.

For $\psi_s = 0$, expression 13-22 provides the well-known solution to the problem of plastic equilibrium of a thick-walled cylinder:

$$P_{ult} = 2\tau_s^0(t)\ln(b/a) \qquad (13\text{-}23)$$

13.3 APPROXIMATE METHODS OF ALLOWING FOR NON-LINEAR STRESS–STRAIN RELATION IN SETTLEMENT COMPUTATIONS

Formulation of the problem

One of the main problems in soil mechanics is that of the stressed and strained state of the soil below a foundation which sustains an external load applied at the surface. The soil is treated either as a half-space, if its deformations are likely to progress in three directions (x, y, z), or as a half-plane if the development of the deformation is possible in two directions only (this is experienced, for example, below a strip foundation).

The above problem has been solved in terms of the theory of linear deformation for many different cases. However, solutions based on the theory of non-linear deformation taking account of creep, the effect of mean normal stress, dilatancy, etc. are obtainable in the general form by recourse to the computer. Recent years have seen certain successes in this field.

The chance of solving a non-linear problem analytically is limited, being possible with respect to the functions χ and χ^* of certain types only. Therefore, it is necessary either to simplify the functions or make use of approximations in their most simple form.

Simple approximation

Non-linearity is allowed for in determining the settlement of the soil below a foundation by a simple empirical procedure wherein the pattern of stress distribution in the soil is determined in terms of elasticity theory and the settlement is computed taking into account non-linearity and creep (see Section 1.5: Vyalov, 1959). In this case, the settlement may be determined by the known method of elementary summation:

$$S = \sum_{j=1}^{n} \frac{\beta}{E_j} p_j h_j \tag{13-24}$$

where p_j is the pressure in the j-th layer of the soil having a depth h_j, as computed by formulae known from the theory of elasticity or by tables; E_j is the modulus of deformation regarded in this case as a variable influenced by load and time, $E_j = E(p, t)$. In the general form, E_j may be expressed by the relation:

$$E(p, t) = \frac{p}{f(p)} \Phi(t) \tag{13-25}$$

Power law of soil deformation. If we proceed from the power law of deformation eq. 5-17, eq. 13-25 takes the form:

$$E(p, t) = \frac{A_z^{1/m} p^{(m-1)/m}}{1 + \delta t^\beta} \tag{13-26}$$

where A_z and m are the coefficients of deformation and hardening for hypothetically instantaneous loading, respectively; δ and β are parameters determined from creep tests.

For $t = 0$, the value of hypothetically instantaneous modulus of deformation is determined from the expression (Cherkasov, 1958):

$$E(p, 0) = A_z^{1/m} p^{(m-1)/m} \tag{13-27}$$

For $t \to \infty$, we have $E(p, \infty) \to 0$.

Thus, corresponding to each value of p_j there will be a certain value of E_j also varying with time as the remainder of the values. The value of E_j used in computations corresponds to the service life of the structure.

The value of $E(p, t)$ is determined from an uniaxial compression test or from a field test, using a settlement plate. The effect of sidewise expansion on the settlement is allowed for by a known coefficient:

$$\beta = \frac{1 - 2v^2}{1 - v} \tag{13-28}$$

Linear–fractional law of deformation. If the law of deformation is adopted in the form of linear–fractional relation 5-22, eq. 13-25 takes the form:

$$E(p, t) = \frac{E_0[T(1 - p/p_s) + t(1 - \delta p/p_s)]}{T + \delta t} \tag{13-29}$$

For $t = 0$ and $t \to \infty$, eq. 13-29 defines the hypothetically instantaneous and ultimate long-term values of the modulus of deformation, respectively:

$$E(p, 0) = E_0(1 - p/p_0) \quad \text{and} \quad E(p, \infty) = E_\infty(1 - p/p_\infty) \tag{13-30}$$

In eqs. 13-29 and 13-30, $p_s = p_0$ is the ultimate load in the case of hypothetically instantaneous load application (for $t \to \infty$); E_0 is the modulus of deformation for $p \to 0$ and $t \to 0$; $E_\infty = E_0/\delta$ is the modulus for $p \to 0$ but at $t \to \infty$; T and δ are parameters. It will be recalled that the ultimate loads p_0 and p_∞ are understood to be the loadings causing unconfined deformations at the initial moment (p_0) or after a long period of load application (p_∞). For details, turn to Fig. 5-8 and the explanatory text.

Flaw law eq. 5-45

The behaviour of the soil below a foundation, obeying either the linear law of Bingham flow or its modification given by eq. 5-45, can be analysed in the case of both vertical and horizontal loadings by a method suggested by Maslov (1968b, referenced in Section 1.5).

Let the yield limit entering into eq. 5-45 be:

$$\tau_y = \tau_{\text{lim}} = \sigma_n \tan \phi_\omega + c_c \tag{13-31}$$

where $\sigma_n = p + \gamma z$; p is the external loading; γ is the soil mass per unit volume; z is the current coordinate.

Thus, we obtain an expression defining the limit below which $\tau < \tau_{lim}$ and no flow is induced. This establishes the zone of soil flow and determines the manner in which the development of shearing deformation takes place with time in soils. Similar reasoning may be employed to determine the rate of soil displacement down a natural or artificial slope.

Use of Schleicher's formula

Another approximate method of allowing for the non-linear behaviour and creep of the soil below a foundation is that the settlement is determined by Schleicher's formula derived for an elastic half-space; the modulus of deformation is, as before, variable in the form given by eq. 13-25. Then:

$$S = \frac{(1 - v^2)}{E(p, t)} \omega b p \tag{13-32}$$

where ω is a coefficient varying with the shape of the footing and foundation rigidity (the values of ω are tabulated, for example, in Tsytovich, 1976, referenced in Section 1.5); b is the width of the footing (or the radius if the foundation is circular).

Taking E in the form given by eq. 13-26, we obtain (Section 1.5: Vyalov, 1959):

$$S = \frac{(1 - v^2)\omega b}{A_z^{1/m}} p^{1/m}(1 + \delta t^\beta) \tag{13-33}$$

If, however, E is adopted as given by eq. 13-25, we will have:

$$S = \frac{(1 - v^2)\omega b}{E_0} \frac{p(T + \delta t)}{T(1 - p/p_s) + t(1 - \delta p/p_s)} \tag{13-34}$$

Eq. 13-34 in the form omitting the time factor has been suggested by Popov (1950). Zaretsky (1972, referenced in Section 7.7) has introduced the time factor into the formula.

Vyalov and Mirenburg (1980) have suggested another settlement equation providing for a better agreement with the experimental data than other equations. It includes a load function in the linear–fractional form and time functions in the power and logarithmic forms:

$$S = \frac{(1 - v^2)nb}{E_0} \frac{p(1 + t/T_1)^\alpha}{1 - \dfrac{p \ln[(t + T_1)/T_2]}{p_0 \ln(1/T_2)}} \tag{13-35}$$

For $t = 0$, eqs. 13-33, 13-34 and 13-35 serve to determine the initial settlement given, respectively, by:

$$S_0 = \frac{(1 - v^2)\omega b}{A_z^{1/m}} p^{1/m} \qquad S_0 = \frac{(1 - v^2)\omega b p}{E_0(1 - p/p_0)} \qquad S_0 = \frac{(1 - v^2)nbp}{E_0(1 - p/p_0)} \tag{13-36}$$

For $t \to \infty$, eq. 13-33 indicates an unconfined increase in the settlement (in computations by the formula, a finite value of t equal to the service life of the structure is adopted) and eqs. 13-34 and 13-35 (for $t = t_\infty$) yield ultimate long-term settlements equalling, respectively:

$$S_\infty = \frac{(1 - v^2)\omega b p}{E_\infty(1 - p/p_\infty)} \qquad S_\infty = \frac{(1 - v^2) n b p}{E_\infty(1 - p/p_\infty)} \qquad (13\text{-}37)$$

Eqs. 13-36 and 13-37 have been checked by a number of researchers. Some of the results are considered in Section 13.6 (see Fig. 13-6).

Allowing for variable loading. The time-dependent changes in a loading increasing slowly and monotonously may be approximated either directly by eq. 13-22 or by its particular cases eqs. 13-33 to 13-37 if a quantity $p = p(t)$ is introduced thereinto. The above changes may be taken into account more accurately by introducing the integral equation from the theory of hereditary creep into eq. 13-32 instead of the term $p(t)/E(p, t)$. Adopting this equation in the form given by eq. 7-43, we obtain:[*1]

$$S = (1 - v^2)\omega b \left\{ f_0[p(t)] + \int_0^t Q(t - \zeta) f[p(\zeta)] \, d\zeta \right\} \qquad (13\text{-}38)$$

Taking $Q(t - \zeta) = (\beta\delta/T^\beta)(\tau - \zeta)^{\beta-1}$ and $f_0(p) = f(p) = (p/A_z)^{1/m}$, we will have:

$$S = \frac{(1 - v^2)\omega b}{A_z^{1/m}} \left\{ p(t)^{1/m} + \frac{\beta\delta}{T^\beta} \int_0^t [p(\zeta)]^{1/m}(t - \zeta)^{\beta-1} \, d\zeta \right\} \qquad (13\text{-}39)$$

For $p = $ const. and $T = 1$, eq. 13-39 is transformed into eq. 13-33. However, inserting $Q(t - \zeta) = T(\delta - 1)/[T + (t - \zeta)]^2$, and writing eq. 13-38 in the form of eq. 7-60, we arrive at:

$$\frac{\bar{E}_0 p_s}{p_s + \bar{E}_0 S} S = p(t) + T(\delta - 1) \int_0^t \frac{p(\zeta) \, d\zeta}{[T + (t - \zeta)]^2} \qquad (13\text{-}40)$$

where $\bar{E}_0 = E_0/(1 - v^2)\omega b$.

For $p = $ const., this expression, if solved for S, is transformed into eq. 13-34.

On coefficient of lateral expansion

In accordance with eq. 4-4, the value of v may be expressed in terms of a relation between the modulus of linear strain E or shear modulus G and the bulk modulus k, provided the moduli are regarded as variables:

$$v = \frac{k(\sigma_m, t) - E(\sigma_z, t)}{2k(\sigma_m, t)} = \frac{k(\sigma_m, t) - 2G(\tau_i, t)}{2[k(\sigma_m, t) + G(\tau_i, t)]} \qquad (13\text{-}41)$$

[*1] Here, the integration variable is denoted by ζ instead of v, as in Chapter 7, to avoid confusion with Poisson's ratio.

It was already pointed out that the coefficient v will be constant in the case when $k(\sigma_m, t)/G(\tau_i, t) = $ const. This is possible when the curves of shearing strains and those of volumetric strains are similar.

In fact, the patterns of these strains are different (see Fig. 1-1) and, in consequence, the coefficient v will be a variable, $v = v(\tau_i, \sigma_m, t)$. However, the changes in the coefficient are in general not very large, the factor $(1 - v^2)$ changing even to a lesser degree. For example, for $0.30 \leqslant v \leqslant 0.5$, the value of $(1 - v^2)$ changes between 0.91 and 0.75. Accordingly, when use is made of eqs. 13-33 to 13-40, we may assume that $v = $ const. This is equivalent, as it was already said, to the assumption that the ratio k/G is constant. Experimental verifications of this assumption will be considered in Section 13.6 (see Fig. 13-6).

However, if necessary, the variability of v may be allowed for by introducing the quantity $v(p, t)$, as determined by eq. 13-41, into eq. 13-32. Another way of allowing for the difference between the patterns of linear and volumetric strains in soil is examined by Mustafaev (1976, referenced in Section 2.9).

13.4 EFFECT OF CONCENTRATED FORCE ON THE SOIL BELOW A FOUNDATION

Boussinesq's problem

As is known, the problem of assessing the stress–strain state of the soil below a foundation sustaining a local load is reduced to (a) determining the components of stress and strain induced at any point M of the half-space by a concentrated force P, and (b) summing up the displacement and stresses arising at the given point due to the action of elemental concentrated forces distributed over the surface in accordance with a given law.

The problem of a concentrated force acting on an elastic half-space was first solved by Boussinesq in 1885; Flamant (1892) examined this problem for a half-plane. In either case the solution was based on a simple radial distribution of the stresses in the half-space. Consequently, it was assumed that the stresses on the planes parallel to the radius are zero.

We shall now discuss the problem of plane deformation ($\varepsilon_y = 0$), i.e. one arising due to the action of a concentrated load per unit length of the half-plane. We shall adopt a cylindrical system of coordinates (r, θ, y) in which the y-axis is at right angles to the plane of the drawing (Fig. 13-2). In such a case we have $\sigma_\theta = 0$; $\tau_{r\theta} = \tau_{yr} = \tau_{\theta y} = 0$; $\sigma_r = \sigma_r(r, \theta)$; $\sigma_y = \sigma_y(r, \theta)$, and the differential equation of equilibrium 3-78 is reduced to the equation:

$$\frac{\partial \sigma_r}{\partial r} + \frac{\sigma_r}{r} = 0 \tag{13-42}$$

Solving this equation simultaneously with the equations of deformation continuity and Hencky's equations, we obtain

$$\sigma_r = \frac{2P}{\pi r} \cos \theta \tag{13-43}$$

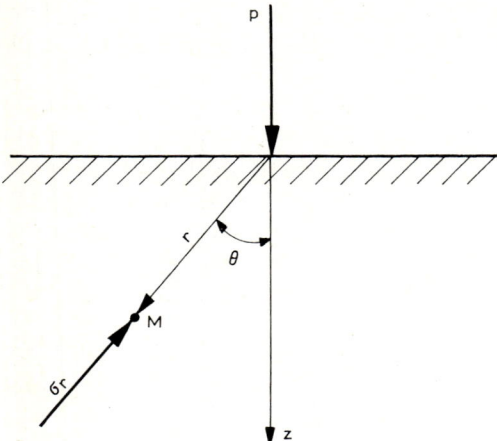

Fig. 13-2. Concentrated force acting on half-space.

Nonlinear problem

In accordance with eq. 11-6, the generalized equation of deformation allowing for nonlinear stress–strain relation, time-dependent deformation and mutual effect of the first and second stress tensor invariants (for $\mu_\sigma = $ const.) is of the form:

$$\gamma_i = f(\tau_i, \sigma_m, t) \qquad \varepsilon_m = f^*(\sigma_m, \tau_i, t) \tag{13-44}$$

In this case, the relations $\chi = \gamma_i/2\tau_i$ and $\chi^* = \varepsilon_m/\sigma_m$, which are the functions of plasticity, enter into Hencky's eqs. 11-13.

Relations 13-44, whatever their form, provide, in conjunction with numerical methods and the computer, solutions to problems of the state of stress and strain arising from local loadings in the soil below a foundation.

Analytical solution of non-linear problem

The stressed and strained state of the soil below a foundation is amenable to analytical solution with allowance for non-linearity, creep and other factors with certain limitations only. Postponing the consideration of these limitations, we review available solutions.

The problem of a concentrated force P which exerts its action on a half-plane and causes soil deformation obeying the power law $\gamma_i = (\tau_i/A_0)^{1/m}$ was first solved by Freilich in 1938. Presented in the form $\sigma_r = (\alpha/r)P\cos^m\theta$, this solution appeared to be valid only for $m = v(1 - v)$.

A more general solution for the case $0 \leqslant m \leqslant 1$ limited, however, by $v = 0.5$ (incompressible medium) was given by Sokolovsky (1950). Arutyunyan (1959a, b) provided a similar solution with reference to the strip foundation and took account of creep in soil. These works have set up the basis for further research into the stressed and strained state of the half-plane in the light of the non-linear stress–strain relation.

At a later date, the problem of concentrated force and power law was considered by Malyshev (1963) for the compressible medium. He has shown that solutions may be obtained only if the patterns of volumetric and shearing strains are similar.

Zaretsky (1965) examined the problem in its most general form, obtaining solutions for concentrated and strip-wise loadings inducing shearing and volumetric strains obeying the power law with allowance for the mutual effect of the first and second stress tensor invariants and the inhomogeneity of the soil below a foundation with depth. A different form of the power law with allowance for creep was examined by the present author and M.E. Slepak (see Table 13-I).

The problem of shearing and volumetric strains obeying the power law with allowance for creep and dilatancy was considered by the present author and Zaretsky (Vyalov and Zaretsky, 1966). The pattern of deformation in the soil below a foundation was derived by taking into account the difference in the resistance of the soil to compression and tension. The deformations were defined by expressions similar to eq. 12-6:

$$\gamma_i = f_1(\tau_i, t) - f_2(\tau_i, \sigma_m, t) \quad \text{and} \quad \varepsilon_m = f_1^*(\sigma_m, t) - f_2^*(\sigma_m, \tau_i, t)$$

A summary of the initial patterns of deformations rendering the problem solvable by analytical means is given in Table 13-I.

Conditions deciding the possibility of an analytical solution

The problem of stress and displacement distribution on a non-linearly deformed half-plane can be solved analytically on the assumption that the stress distribution is of a radial pattern. However, the question then arises: in which cases may it be assumed that the stresses are distributed radially?

According to Zaretsky (1965) this assumption is possible if two conditions are fulfilled: firstly, the functions of eq. 13-44 $\gamma_i = f(\tau_i, \sigma_m, t)$ and $\varepsilon_m = f^*(\sigma_m, \tau_i, t)$ are homogeneous, the degree of their homogeneity being the same, and, secondly, the functions $\chi = \gamma_i/2\tau_i$ and $\chi^* = \varepsilon_m/\sigma_m$ change in the r–θ plane but so with respect to r and θ as to be represented as a product of two functions changing one with the vectorial angle θ only and the other with the radius r, i.e. $\chi = \Phi_1(\theta)\Phi_2(r)$ and $\chi^* = \Phi_1^*(\theta)\Phi_2^*(r)$ in which $\Phi_2(r) = \Phi_2^*(r) = \Phi(r)$ and $\Phi_1^*(\theta) = \alpha_1\Phi_1(\theta)$.

The condition of radial stress distribution may be expressed in a more simple form, taking into account the time factor. Thus, the first condition may be set forth in the form originating from the very concept of similarity:

$$\gamma_i^m = \tau_i\phi\left(\frac{\sigma_m}{\tau_i}\right)[1 + \tilde{L}] \tag{13-45}$$

where $\tilde{L} = \tilde{L}(t)$ is the integral operator which represents the effect of time and is defined by eq. 13-3; $\tilde{L}_0 = 0$.

TABLE 13-I

Patterns of initial soil deformation providing for analytical solution of the problem of concentrated force or stripwise loading sustained by a half-plane

Pattern of shearing strains	Pattern of volumetric strains	Reference
$\gamma_i = \left[\dfrac{\tau_i}{A_0}\right]^{1/m}$	$\varepsilon_m \neq 0, \quad v = \dfrac{m}{1+m}$	Freilich, 1938
$\gamma_i = \left[\dfrac{\tau_i}{A_0}\right]^{1/m}$	$\varepsilon_m = 0$	Sokolovsky, 1950
$\gamma_i = \left[\dfrac{\tau_i}{A(t)}\right]^{1/m}$	$\varepsilon_m = 0$	Arutyunyan, 1959a, b
$\gamma_i = \left[\dfrac{\tau_i}{A_0}\right]^{1/m}$	$\varepsilon_m = \left[\dfrac{\sigma_m}{D_0}\right]^{1/m}$	Malyshev, 1963
$\gamma_i = \dfrac{\tau_i^{1/m}\sigma_m^{1/n}}{N(r,\theta)}$	$\varepsilon_m = \dfrac{\sigma_m^{1/q}\tau_i^{1/\kappa}}{M(r,\theta)}$	Zaretsky, 1965
$\gamma_i = \left[\dfrac{\tau_i}{A(t)} - \dfrac{\sigma_m}{C(t)}\right]^{1/\eta}$	$\varepsilon_m = \left[\dfrac{\sigma_m}{D(t)}\right]^{1/m}$	Vyalov and Slepak, cf. p. 453
$\gamma_i = \left[\dfrac{\tau_i}{A(t)}\right]^{1/m} - \left[\dfrac{\sigma_m}{C(t)\tau_i^{1-\eta/m}}\right]^{1/m}$	$\varepsilon_m = \left[\dfrac{\sigma_m}{D(t)}\right]^{1/m} - \left[\dfrac{\tau_i}{E(t)\sigma_m^{1-\mu/m}}\right]^{1/\mu}$	Vyalov and Zaretsky, 1966

where:
$1/q = 1/m + 1/n - 1/\kappa$,
$M(r, \theta) = 2N(r, \theta)/a(a^2 + a + 1)^\beta$,
$\beta = (\kappa - m - \kappa^m)/2\kappa m$

Note, that the function $f(x; y; \ldots)$ is referred to as a homogeneous one if $f(\lambda x; \lambda y; \ldots) = \lambda^n f(x; y; \ldots)$ where n is the degree of homogeneity.

The second condition may be expressed in terms of a constant coefficient of lateral strain v. This needs to be considered. Assuming that the time functions of the shearing strain $\tilde{L}$ and the volumetric strain $\tilde{L}^*$ are of the type providing for $[1 + \tilde{L}^*] = \alpha_2[1 + \tilde{L}]$, the second condition may be written in the form:

$$\frac{\chi^*}{\chi} = \frac{\Phi^*(\theta)\,\Phi^*(r)[1+\tilde{L}^*]}{\Phi(\theta)\,\Phi(r)[1+\tilde{L}]} = \frac{\alpha_1\alpha_2\Phi(\theta)\,\Phi(r)[1+\tilde{L}]}{\Phi(\theta)\,\Phi(r)[1+\tilde{L}]} = \alpha = \text{const.} \quad (13\text{-}46)$$

where $\alpha = \alpha_1\alpha_2$.

At the same time, the functions χ^* and χ may be expressed as $\chi = 1/2\bar{G}$ and

$\chi^* = 1/\bar{k}$, where $\bar{G} = G(\tau_i, \sigma_m, t)$; $\bar{k} = \bar{k}(\sigma_m, \tau_i, t)$ are the variable shear modulus and bulk modulus, respectively. Hence:

$$\alpha = \frac{\chi^*}{\chi} = \frac{2\bar{G}}{\bar{k}} \tag{13-47}$$

From relationship 4-4 we determine that the coefficients α and ν are interconnected by the expression $\alpha = (1 - 2\nu)/(1 + \nu)$. Since, according to eq. 13-46, $\alpha = $ const. this means that:

$$\nu = \frac{1 - \alpha}{2 + \alpha} = \text{const.} \tag{13-48}$$

This implies that if the shearing strain is described by the relation of the form given by eq. 13-45 and the coefficient of lateral strain $\nu = $ const., we may assume that the stresses on a half-plane sustaining a concentrated loading are distributed radially and, consequently, solve the problem analytically.

Analysing eqs. 13-45 and 13-48. Eq. 13-45 may be fulfilled quite easily. To that end, firstly, the "stress–strain" curves must lend themselves to approximation by a power relation — a possibility coming true for most soils. Secondly, the effect of mean normal stress on the shearing strain must be taken care of by introducing a multiplier $f(\sigma_m/\tau_i)$ into the equation of state; fulfilling this condition is eq. 11-46 and other similar expressions corresponding to many practical cases.

Finally, the time factor must be allowed for by introducing the multiplier $[1 + \tilde{L}]$. This corresponds to the assumption that the isochrones are similar at all moments (see Fig. 5-6) which holds for soils as already discussed.

The condition 13-48 setting forth that the coefficient ν is constant invites difficulties. It was said earlier that ν will be constant only when the curves of shearing and volumetric strains are similar; this applies to the "stress–strain" curves and creep curves as well. However, in fact, the γ_i–τ_i and ε_m–σ_m curves are of different configurations (see Fig. 1). The creep curves representing shearing strains and volumetric strains are also of different shape, the former being of either the attenuating or non-attenuating type and the latter of the attenuating type only.

Nevertheless, the assumption that the coefficient ν is constant has little bearing on the results of an assessment of the behaviour of the soil below a foundation. In Section 13.6 we shall show, using experimental evidence, that in spite of the different laws which changes in shape and volume obey, the resulting "settlement versus loading" curve is of a shape identical with the curve of shearing deformation rather than with that of volumetric changes. Apparently, the changes in volume produce a noticeable effect only at an early stage of loading, during compaction, whereas at a later stage — as shearing gains ground — the pattern of shearing deformation prevails.

All the relationships given in Table 13-I satisfy eqs. 13-56 and 13-48. Let us prove that on the basis of a formula tabulated:

$$\gamma_i = \left[\frac{\tau_i}{A(t)} - \frac{\sigma_m}{C(t)}\right]^{1/m} \quad \varepsilon_m = \left[\frac{\sigma_m}{D(t)}\right]^{1/m} \tag{13-49}$$

Eq. 13-49 describes the pattern of deformation of a cohesive soil with an angle of internal friction $\tan \psi = A(t)/C(t) = $ const.; it corresponds to eqs. 11-46 and 8-5. In a particular case, $C(t) = \infty$ and $\psi = 0$, the soil has no internal friction.

The functions of χ and χ^* are of the form:

$$\chi = \frac{\gamma_i}{2\tau_i} = \frac{1}{2[A(t)]^{1/m}} \left[\tau_i^{1-m} - \frac{A(t)}{C(t)} \tau_i^{-m} \sigma_m \right]^{1/m} \qquad \chi^* = \frac{\varepsilon_m}{\sigma_m} = \left[\frac{\sigma_m^{1-m}}{D(t)} \right]^{1/m} \quad (13\text{-}50)$$

Hence:

$$\alpha = \frac{\chi^*}{\chi} = 2 \left[\frac{A(t)}{D(t)} \right]^{1/m} \left[\left(\frac{\sigma_m}{\tau_i} \right)^{1/m} - \frac{A(t)}{C(t)} \left(\frac{\sigma_m}{\tau_i} \right)^{-m} \right]^{1/m} \quad (13\text{-}51)$$

The condition $v = (1 - \alpha)/(2 + \alpha) = $ const. is satisfied if $\sigma_m/\tau_i = $ const., which is the case of simple loading, and $A(t)/C(t) = $ const. and $A(t)/D(t) = $ const. One may assume for example:

$$A(t) = \frac{A_0}{1 + \delta(t/T)^a} \qquad C(t) = \frac{C_0}{1 + \delta(t/T)^a} \qquad D(t) = \frac{D_0}{1 + \delta(t/T)^a} \quad (13\text{-}52)$$

Then:

$$v = \frac{1 - 2\left(\frac{A_0}{D_0}\right)^{1/m} \left[\left(\frac{\sigma_m}{\tau_i}\right)^{1-m} - \frac{A_0}{C_0}\left(\frac{\sigma_m}{\tau_i}\right)^{-m} \right]^{1/m}}{2\left\{ 1 + \left(\frac{A_0}{D_0}\right)^{1/m} \left[\left(\frac{\sigma_m}{\tau_i}\right)^{1-m} - \frac{A_0}{C_0}\left(\frac{\sigma_m}{\tau_i}\right)^{-m} \right]^{1/m} \right\}} \quad (13\text{-}53)$$

In eqs. 13-49, the time factor is allowed for in the parametric form as prescribed by ageing theory. However, if time is taken account of by means of the integral relations of the equations of hereditary creep, we shall obtain in accordance with eqs. 13-3 and 7-102:

$$\frac{\tau_i(t)}{A(t)} = \frac{1}{A_0} \tau_i(t)[1 + \tilde{L}] = \frac{1}{A_0} \left[\tau_i(t) + \int_0^t K_1(t - \zeta)\tau_i(\zeta) \, d\zeta \right]$$

$$\frac{\sigma_m(t)}{C(t)} = \frac{1}{C_0} \sigma_m(t)[1 + \tilde{L}^*] = \frac{1}{C_0} \left[\sigma_m(t) + \int_0^t K_2(t - \zeta)\sigma_m(\zeta) \, d\zeta \right] \quad (13\text{-}54)$$

Eqs. 13-49 then take the form:

$$(\gamma_i)^m = \frac{1}{A_0} \left[\tau_i(t) + \int_0^t K_1(t - \zeta)\tau(\zeta) \, d\zeta \right] - \frac{1}{C_0} \left[\sigma_m(t) + \int_0^t K_2(t - \zeta)\sigma_m(\zeta) \, d\zeta \right] \quad (13\text{-}55)$$

$$(\varepsilon_m)^m = \frac{1}{D_0} \left[\sigma_m(t) + \int_0^t K_3(t - \zeta)\sigma_m(\zeta) \, d\zeta \right]$$

The condition $v = $ const. will be fulfilled if $K_1(t - \zeta) = K_2(t - \zeta) = K_3(t - \zeta) = K(t - \zeta)$.

For example, this condition is fulfilled when:

$$K(t - \zeta) = \frac{a\delta}{T^a}(t - \zeta)^{a-1} \tag{13-56}$$

which is an expression corresponding to eq. 13-52. Condition 13-48 will thus be fulfilled under the limitations referred to above. Condition 13-45 is fulfilled automatically under the same limitations. In fact, taking into account eq. 13-54, we obtain:

$$\gamma_i^m = \frac{\tau_i}{A(t)} - \frac{\sigma_m}{C(t)} = [1 + \tilde{L}]\left[\frac{1}{A_0} - \frac{1}{C_0}\frac{\sigma_m}{\tau_i}\right]\tau_i \tag{13-57}$$

which corresponds to eq. 13-45.

General solution of the problem

The problem of the stressed and strained state of a half-plane sustaining a concentrated loading is amenable to general solution under the conditions fulfilling eqs. 13-45 and 13-48. This needs to be considered.

The equation of equilibrium 13-42, when solved, takes the form:

$$\sigma_r = \frac{f(\theta)}{r} \tag{13-58}$$

The conditions of compatibility in eq. 3-80 will assume the form:

$$\frac{\partial^2 \varepsilon_r}{\partial \theta^2} = r^2 \frac{\partial^2 \varepsilon_\theta}{\partial r^2} + 2r \frac{\partial \varepsilon_\theta}{\partial r} - r\frac{\partial \varepsilon_r}{\partial r} = 2\left[\frac{\partial^2 \gamma_{r\theta}}{\partial r \partial \theta} - \frac{\partial \gamma_{r\theta}}{\partial \theta}\right] \tag{13-59}$$

and Hencky's relationships 11-13 will be represented by:

$$\varepsilon_r(t) = \tfrac{1}{3}[\sigma_r(2\chi + \chi^*) + \sigma_y(\chi^* - \chi)]$$

$$\varepsilon_\theta(t) = \frac{\sigma_r + \sigma_y}{3}(\chi^* - \chi) \tag{13-60}$$

$$\varepsilon_y(t) = \tfrac{1}{3}[\sigma_y(2\chi + \chi^*) + \sigma_r(\chi^* - \chi)]$$

$$\gamma_{r\theta} = \gamma_{\theta y} = \gamma_{ry} = 0$$

It follows from equality $\varepsilon_y(t) = 0$ that:

$$\sigma_y = \frac{\chi - \chi^*}{2\chi + \chi^*}\sigma_r = \frac{1 - \alpha}{2 + \alpha}\sigma_r = v\sigma_r \tag{13-61}$$

In this case, the stress and strain tensor invariants assume the following meanings:

$$\tau_i = \sqrt{\tfrac{1}{3}}\sqrt{\sigma_r^2 + \sigma_y^2 - \sigma_r\sigma_y} = \sqrt{\tfrac{1}{3}}\sqrt{(1 - v + v^2)}\sigma_r$$

$$\sigma_m = \tfrac{1}{3}(\sigma_r + \sigma_y) = \tfrac{1}{3}(1 + v)\sigma_r \tag{13-62}$$

$$\gamma_i = 2\sqrt{\tfrac{1}{3}}\sqrt{\varepsilon_r^2 + \varepsilon_\theta^2 - \varepsilon_r\varepsilon_\theta} \qquad \varepsilon_\perp = \tfrac{1}{3}(\varepsilon_r + \varepsilon_\theta)$$

Note that eqs. 13-61 and 13-53, if considered jointly, establish a connection between the parameters A_0, C_0, D_0, entering into eqs. 13-49 and 13-52, for the case when $v = $ const.:

$$(1 - 2v)^2 \left(\frac{1}{A_0} \frac{\sqrt{3}\sqrt{1 - v + v^2}}{1 + v} + \frac{1}{C_0} \right) = \frac{1}{D_0} (\sqrt{3}\sqrt{1 - v + v^2})^m$$

Taking into account eq. 13-61, Hencky's relations 13-60 are reduced to:

$$\varepsilon_r(t) = (1 - v)\chi\sigma_r, \quad \varepsilon_\theta(t) = -v\chi\sigma_r, \quad \varepsilon_y = \gamma_{r\theta} = \gamma_{\theta y} = \gamma_{yz} = 0 \quad (13\text{-}63)$$

where $\chi = \chi(t)$.

This statement corresponds to the theory of ageing. However, if we proceed from the theory of hereditary creep, we obtain:

$$\varepsilon_r(t) = (1 - v)\chi \left[\sigma_r(t) + \int_0^t \sigma_r(\zeta) K(t - \zeta) \, d\zeta \right]$$

$$\varepsilon_\theta(t) = -v\chi \left[\sigma_r(t) + \int_0^t \sigma_r(\zeta) K(t - \zeta) \, d\zeta \right] \quad (13\text{-}64)$$

Distribution of stresses. The problem of stress distribution on a half-plane sustaining a concentrated force $P(t)$ is reduced to establishing the type of the function $f(\theta)$ in eq. 13-58. Solving simultaneously eqs. 13-58, 13-59, 13-63 and the equation of the equilibrium of the forces acting along the vertical axis, $\int_{-\pi/2}^{\pi/2} \sigma_r \cos \theta \, d\theta = P$, we shall obtain, as has been shown by Sokolovsky (1950) and Arutyunyan (1959a, b), that:

$$\sigma_r(t) = \frac{2P(t)[\omega'(\lambda\theta)]^m}{rI} \qquad \sigma_y(t) = 2v\sigma_r(t) \quad (13\text{-}65)$$

where:

$$I = 4 \int_0^{\pi/2} \cos \theta [\omega'(\lambda\theta)]^m \, d\theta$$

$$\omega(\lambda\theta) = \begin{cases} \dfrac{1}{\lambda} \sin(\lambda\theta) & \text{for } \lambda^2 > 0 \\ \dfrac{1}{\lambda} \sinh(|\lambda|\theta) & \text{for } \lambda^2 < 0 \\ 1 & \text{for } \lambda^2 = 0 \end{cases}$$

$$\omega'(\lambda\theta) = \begin{cases} \cos(\lambda\theta) & \text{for } \lambda^2 > 0 \\ \cosh(|\lambda|\theta) & \text{for } \lambda^2 < 0 \\ 0 & \text{for } \lambda^2 = 0 \end{cases} \quad (13\text{-}66)$$

$$\lambda^2 = \frac{1}{m} \left[1 + \frac{v}{1 - v} \left(1 - \frac{1}{m} \right) \right]$$

For $m = 1$, eq. 13-65 becomes the solution obtained in terms of elastic theory (eq. 13-43).

The stresses on a half-plane induced by a loading applied to the surface and distributed in accordance with an arbitrary pattern $p = p(x, t)$, are determined by summing up the stresses at a given point M which arise due to the action of concentrated elemental forces.

The vertical stress component produced by a concentrated elemental force $P_j(t)$ may be determined from the expression $\sigma_z = \sigma_r \cos^2 \theta$ on substituting the value of σ_r from eq. 13-65 and taking into account that $z = r \cos \theta$, $r = \sqrt{z^2 + (x\varrho)^2}$, $x - \varrho = r \sin \theta$, where ϱ is the distance of the point under consideration from the z-axis. Thus we obtain:

$$\sigma_z = \frac{2P_i(t)}{I} \left[\cos\left(\lambda \arctan \frac{x-\varrho}{z}\right) \right]^m \frac{z^2}{\sqrt{[z^2 + (x-\varrho)^2]^3}} \quad (13\text{-}67)$$

This expression defines the manner in which the vertical stresses change with the depth of a soil mass in response to a concentrated force varying with time in accordance with a given law.

As can be seen, the stress distribution is affected neither by the type of the stress function $f(\sigma_m/\tau_i)$ nor by the type of the creep function $(1 + \tilde{L})$ entering into eq. 13-45. In other words, eqs. 13-65 through 13-68 appear to be general for all laws satisfying conditions 13-45 and 13-48.

13.5 DETERMINING THE SETTLEMENT OF THE SOIL BELOW A FOUNDATION AND THE REACTION PRESSURE OF SOIL

Determination of displacements

To determine the settlement of the soil below a foundation, one needs to determine the displacement of a given point M on the half-plane acted upon by a concentrated force $P(t)$. To that end, we substitute the values of σ_r and c_θ, as determined by eqs. 13-65, into Hencky's eqs. 13-63 and find the strain components ε_r and ε_θ. The displacement components are obtained by integrating Cauchy's relations:

$$\varepsilon_r = \frac{\partial u_r}{\partial r} \qquad \varepsilon_\theta = \frac{1}{r}\frac{\partial u_\theta}{\partial \theta} + \frac{u_r}{r} \quad (13\text{-}68)$$

taking into account the symmetry of the state of stress and the condition that at infinity the displacements tend toward zero. After some manipulations, slightly modified compared with those found in the work of Arutyunyan (1959a, b), we obtain the following solution:

$$\begin{aligned} u_r(t) &= M[(1 + \tilde{L})P(t)]^{1/m} r^{(m-1)/m} \\ u_\theta(t) &= N[(1 + \tilde{L})P(t)]^{1/m} r^{(m-1)/m} \end{aligned} \quad (13\text{-}69)$$

where N and $M = [(1 - v)/(1 - v/m)] dN/d\theta$ are functions established from the solution of each particular problem; these functions will be discussed below.

Determination of settlement

The settlement of the soil below a foundation, i.e. the displacement of the surface of the half-plane under a loading $P(x, t)$ arbitrarily distributed over a strip of width b (Fig. 13-3), may be determined by summing up the displacements of the free surface of the half-plane due to the action of a system of elemental forces $P_i(x)$. The possibility of such a summation for a material deforming non-linearly in accordance with a power law follows from the principle of superposition of some generalized displacements of the half-plane's boundaries, $u^*(x) = [u_{\theta/\theta} = \pm \pi/2]^m$, as has been shown by N.Kh. Arutyunyan. In fact, since these generalized displacements vary linearly with load, the value of $u^*(x)$ may be determined as the sum of the displacements $u_i^*(x)$ caused by the elemental forces $P_i(x)$. This approach, although an approximate one, yields solutions quite acceptable in engineering practice. By using integration instead of summation and taking the value of $u_{\theta/\theta} = \pm \pi/2$ in the form of the second formula of eq. 13-69 (for $\theta = \pm \pi/2$), we obtain, after some manipulation, the settlement of the surface:

$$S(x, t) = N \left\{ \int_{-b/2}^{b/2} [(1 + \tilde{L})p(\varrho, t)] \frac{d\varrho}{|\varrho - x|^{1-m}} \right\}^{1/m} \qquad (13\text{-}70)$$

In the case of a uniformly distributed loading $p = p(t)$, we will have:

$$S(x, t) = N \left\{ [(1 + \tilde{L})p(t)] \int_{-b/2}^{b/2} \frac{d\varrho}{|\varrho - x|^{1-m}} \right\}^{1/m} \qquad (13\text{-}70')$$

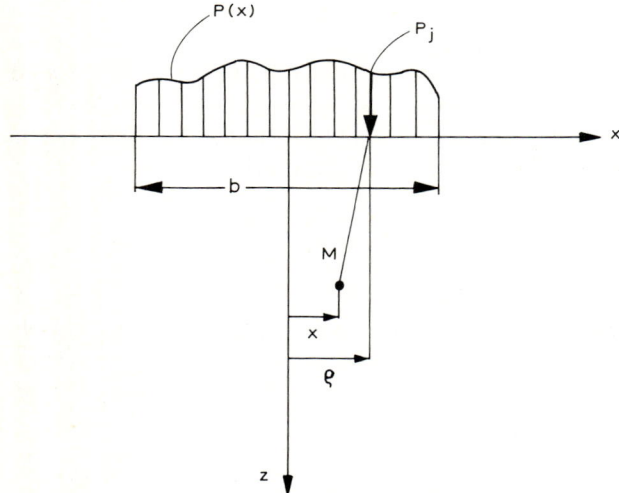

Fig. 13-3. Schematic diagram of local (strip) loading.

where, if we put $\bar{x} = 2x/b$:

$$\left[\int_{-b/2}^{b/2} \frac{d\varrho}{|\varrho - x|^{1-m}}\right]^{1/m} = \frac{b}{2m^{1/m}}[(1 - \bar{x})^m + (1 + \bar{x})^m]^{1/m} \quad \text{for } -1 < \bar{x} < 1$$

$$\left[\int_{-b/2}^{b/2} \frac{d\varrho}{|\varrho - x|^{1-m}}\right]^{1/m} = \frac{b}{2m^{1/m}}[(1 + \bar{x})^m - (\bar{x} - 1)^m]^{1/m} \quad \text{for } \bar{x} > 1$$

$$\left[\int_{-b/2}^{b/2} \frac{d\varrho}{|\varrho - x|^{1-m}}\right]^{1/m} = 2^{(1-m)/m} \frac{b}{m^{1/m}} \quad \text{for } \bar{x} = 0$$

The value of the factor N entering into the equations of displacement 13-69, 13-70 and 13-70' is:

$$N = \kappa \frac{\omega(\lambda\theta)}{I^{1/m}} F(m, v) \tag{13-71}$$

where $\kappa = +1$ or -1 depending on the axes of the coordinates selected; I, $\omega(\lambda\theta)$, λ are the functions of v, m and θ, respectively, as determined by eqs. 13-66; $F(m, v)$ is a factor dependent upon the type of function $f(\sigma_m/\tau_i)$ entering into initial relationship 13-45.

The value of $F(m, v)$ is established in each particular case depending on the form of the relation used to obtain a solution. The values of $F(m, v)$ for some of the initial patterns of soil deformation considered above are given in Table 13-II along with the values of v computed by eqs. 13-47 and 13-48 for the same deformation patterns.

The integral operator $\tilde{L}$ entering into eqs. 13-69, 13-70 and 13-70' is determined from eq. 13-3. Thus:

$$[1 + \tilde{L}]p = p(t) + \int_0^t p(\xi)K(t - \xi)\,d\xi \tag{13-72}$$

When $p = $ const., we will have:

$$[1 + \tilde{L}]p = p\left[1 + \int_0^t K(t)\,dt\right] = p\frac{A_0}{A(t)} \tag{13-73}$$

where $K(t)$ is the creep kernel of the type examined in Section 7.4. If creep is disregarded, $K(t - \xi) = K(t) = 0$.

The integral operator $\tilde{L}$ introduced into eq. 13-70 makes this equation suitable for determining the settlement under a load increasing with time in any manner. Let us assume, for example, that for:

$$(1 + \tilde{L}) = \left[1 + \int_0^t K(t)\,dt\right] = 1 + \delta(t/T)^a$$

TABLE 13-II

Values of v and $F(m, v)$ coefficients depending on the equation of state of soil

Equation of state	Coefficient $v = \dfrac{\frac{1}{2}\sigma_m \gamma_i - \tau_i \varepsilon_m}{\sigma_m \gamma_i + \tau_i \varepsilon_m}$	Coefficient $F(m, v)$
$\gamma_i = \left[\dfrac{\tau_i}{A(t)}\right]^{1/m}$, $\varepsilon_m = 0$	0.5	$\kappa \dfrac{2m-1}{2(1-m)}$, where $\kappa = \pm 1$
$\gamma_i = \left[\dfrac{\tau_i}{A(t)}\right]^{1/m}$, $\varepsilon_m = \left[\dfrac{\sigma_m}{D(t)}\right]^{1/m}$	$\dfrac{1 - 2\left(\dfrac{A_0}{D_0}\right)^{1/m}\left(\dfrac{\sigma_m}{\tau_i}\right)^{(1-m)/m}}{2\left[1 + \left(\dfrac{A_0}{D_0}\right)^{1/m}\left(\dfrac{\sigma_m}{\tau_i}\right)^{(1-m)/m}\right]}$	$\kappa \dfrac{(m-v)\left(\dfrac{2}{\sqrt{3}}\sqrt{1-v+v^2}\right)^{(1-m)/m}}{(1-m)A_0^{1/m}}$
$\gamma_i = \left[\dfrac{\tau_i}{A(t)} - \dfrac{\sigma_m}{C(t)}\right]^{1/m}$, $\varepsilon_m = \left[\dfrac{\sigma_m}{D(t)}\right]^{1/m}$	$\dfrac{1 - 2\left(\dfrac{A_0}{D_0}\right)^{1/m}\left(\dfrac{\sigma_m}{\tau_i}\right)^{(1-m)/m}\left[\left(\dfrac{\sigma_m}{\tau_i}\right)^{1-m} - \dfrac{A_0}{C_0}\left(\dfrac{\sigma_m}{\tau_i}\right)^{-m}\right]^{1/m}}{2\left\{1 + \left(\dfrac{A_0}{D_0}\right)^{1/m}\left(\dfrac{\sigma_m}{\tau_i}\right)^{1-m} - \dfrac{A_0}{C_0}\left(\dfrac{\sigma_m}{\tau_i}\right)^{-m}\right]^{1/m}\right\}}$	$\kappa\left(\dfrac{2}{\sqrt{3}}\right)^{(1-m)/m}\dfrac{(m-v)(\sqrt{1-v+v^2})^{(1-m)/m}}{(1-m)A_0^{1/m}} \times \left(1 - \dfrac{A_0}{C_0}\dfrac{1+v}{\sqrt{3}\sqrt{1-v+v^2}}\right)^{1/m}$
$\gamma_i = \left[\dfrac{\tau_i}{A(t)}\right]^{1/m}$, $\varepsilon_m = \left[\dfrac{\sigma_m}{C(t)\tau_i^{1-\eta}}\right]^{1/\eta} - \left[\dfrac{\tau_i}{E(t)\sigma_m^{1-\mu/m}}\right]^{1/\mu}$	$\dfrac{1 - 2\left(\dfrac{A_0}{D_0}\right)^{1/m}\left(\dfrac{\sigma_m}{\tau_i}\right)^{(1-m)/m}\left[1 - \left(\dfrac{D_0\tau_i}{E_0\sigma_m}\right)^{1/\mu} - \left(\dfrac{A_0\sigma_m}{C_0\tau_i}\right)^{1/\eta}\right]}{2\left\{1 - \left(\dfrac{A_0}{D_0}\right)^{1/m}\left(\dfrac{\sigma_m}{\tau_i}\right)^{(1-m)/m}\left[1 - \left(\dfrac{D_0\tau_i}{E_0\sigma_m}\right)^{1/\mu} - \left(\dfrac{A_0\sigma_m}{C_0\tau_i}\right)^{1/\eta}\right]\right\}}$	$\kappa\left(\dfrac{2}{\sqrt{3}}\right)^{(1-m)/m}\dfrac{(m-v)(\sqrt{1-v+v^2})^{(1-m)/m}}{(1-m)A_0^{1/m}} \times \left\{1 - \left[\dfrac{A_0^{\eta/m}\sqrt{3}(1+v)}{C_0\sqrt{1-v+v^2}}\right]^{1/\eta}\right\}$

we must determine the settlement at the centre of a foundation ($\bar{x} = 0$) in the case when during the interval between 0 and t_1 (construction period) the load increases linearly, $p(t) = nt$, and after t_1 it is constant, $p = p_1 =$ const. Separating the integral in eq. 13-70 into two integrals with the limits between 0 and t_1 and between t_1 and t_w, where t_w is the service life of the structure, we arrive at the following value of the settlement:

$$S(t) = 2^{(1-m)/m} N \frac{bp_1^{1/m}}{m^{1/m}} \left[1 + \frac{\delta}{T^a} \left(\frac{a}{a+1} t_1^a + t^a \right) \right]^{1/m}$$

Assuming that the load was constant from the initial moment of time, we will have:

$$S(t) = 2^{(1-m)/m} N \frac{bp^{1/m}}{m^{1/m}} \left[1 + \delta \left(\frac{t}{T} \right)^a \right]^{1/m}$$

For the case $m = 1$ and $p =$ const. and taking into account that the variable ϱ under the integral is raised to the power -1, the settlement problem acquires a solution corresponding to the known formula from the theory of elasticity:

$$S(x) = \frac{b(1-v^2)}{\pi E} \left[\bar{x} \ln \frac{|1-\bar{x}|}{1+\bar{x}} - \ln|1-\bar{x}^2| + 2 \right] p \tag{13-74}$$

Reaction pressures. The way in which the reaction (contact) pressures of the soil are distributed over the footing of a rigid strip foundation may be assessed from the condition that the points within the area of contact between the foundation and soil displace through the same distance:

$$u_{\theta(1)} + u_{\theta(2)} = \delta - f(x) \tag{13-75}$$

where $u_{\theta(1)} = u_{\theta(2)} = u_\theta$ are the vertical displacements of the interface between soil and footing regarded as an absolutely rigid strip-wise settlement plate; δ is the closing in of the settlement plate footing and soil in the z-direction; $f(x)$ is the equation of the area of contact of the plate; for a flat settlement plate, $f(x) = 0$.

The displacement u_θ, in its turn, is connected with the yet unknown contact pressure $p(x, t)$ by the relationship:

$$u_\theta(t) = N[1 - \tilde{L}] \left[\int_{-b/2}^{b/2} \frac{p(\varrho, t) \, d\varrho}{|\varrho - x|^{1-m}} \right]^{1/m} \tag{13-76}$$

Substituting eq. 13-76 into eq. 13-75 we obtain:

$$W(x, t) - \int_0^t W(x, \zeta) K(t - \zeta) \, d\zeta = [\Phi(t)]^m \tag{13-77}$$

where:

$$W(x, t) = \int_{-b/2}^{b/2} \frac{p(\varrho, t) \, d\varrho}{|\varrho - x|^{1-m}} \tag{13-78}$$

The solution of the problem — establishing the unknown function $p(x, t)$, is reduced to solving the integral eqs. 13-77 and 13-78 with due regard for the equilibrium equation $\int_{-b/2}^{b/2} p(x, t) \, dt = P(t)$. Thus, solving eq. 13-77, we will have:

$$W(t) = [\Phi(t)]^m - \int_0^t [\Phi(\zeta)]^m R(t - \zeta) \, d\zeta \tag{13-79}$$

where $R(t - \zeta)$ is the resolvent of the kernel $K(t - \zeta)$. Applying Arutyunyan's method, we obtain:

$$W(t) = \frac{2P(t)\Gamma\left(\frac{3-m}{2}\right)\Gamma\left(\frac{m}{2}\right)}{\sqrt{\pi}(b/2)^{1-m}} \tag{13-80}$$

Finally, the expression for determining the reaction pressure of the soil on the footing of a rigid foundation, will take the form:

$$p(x, t) = \frac{\Gamma\left(\frac{3-m}{2}\right)\Gamma\left(\frac{m}{2}\right)\sin\left(\frac{\pi m}{2}\right)}{\pi^{3/2}(b/2)\sqrt{(1 - \bar{x}^2)^m}} P(t) \tag{13-81}$$

where $\bar{x} = 2x/b$; $\Gamma(z)$ is the gamma function; its values are tabulated.

As can be seen, the factor N, influenced by the type of the initial pattern of soil deformation (eq. 13-45), is eliminated from the final formula. Thus, this formula holds for any power law satisfying conditions 13-45 and 13-48. For the linear law ($m = 1$), the distribution of reaction pressures is described by the formula from the theory of elasticity:

$$p(x) = \frac{P}{\pi(b/2)\sqrt{1 - \bar{x}^2}} \tag{13-82}$$

Examining formulae

The exponent m appearing in all the formulae considered defines the non-linearity of the pattern of deformation. It also significantly influences the distribution of stresses and displacements in the soil below a foundation which, consequently, differs from the patterns obtained in terms of elasticity theory.

Compare eq. 13-70', determining the superficial settlement of the soil below a foundation under an uniformly distributed loading for $m = 1$, with the corresponding eq. 13-74 from the theory of elasticity. As can be seen from the structure of the formula, superficial settlements of an elastic soil must increase unconfined so that $S \to \infty$ for $x \to \infty$. Apparently this contradicts the physical meaning.

Another point arising in connection with eq. 13-74 is that an arbitrary constant, whose value cannot be determined, enters into the solution after integration. Since this constant is always summed up with the settlement value, an absolute settlement value cannot be obtained. Therefore, eq. 13-74 from the theory of elasticity enables the determination of only the relative settlement, i.e. the displacement of one point at the soil surface with respect to another. However, eq. 13-70′ from the theory of non-linear deformation is free from this drawback. In accordance with the formula, the superficial displacements asymptotically die out and $S \to 0$ for $x \to \infty$; as a result, this formula gives the absolute value of settlement.

The time factor has a direct bearing on the amount of displacement, which increases directly with the creep function $[1 + \int_0^t K(t - \zeta) \, d\zeta]$. However, creep as follows from eqs. 13-68 and 13-81, does not affect the state of stress of the soil below a foundation and the stress distribution varies there with time depending only on changes in external loading also varying with time.

The above follows from the type of creep equation (eq. 13-45) adopted, which corresponds to the case that occurs quite frequently in practice: the pattern of deformation at any moment is similar to the initial pattern. This is represented graphically in Fig. 5-6(c). Thus, under the pattern of deformation given by eq. 13-45, the state of stress will not change with time and the deformation will remain in a condition similar to the initial one.

Assuming, however, that the pattern of deformation varies with time (isochrones are dissimilar, as shown in Figs. 5-6(a) and (b)), the state of stress will then change from an initial to a steady condition. Thus, if the initial state is described by the linear law (at $t > 0$ the power law becomes effective), the stress distribution at the initial moment will correspond to the solution obtained in terms of elasticity theory. However, in nature at a later stage, the stress pattern will be decided by the non-linear formulae considered above. The arrangement of stress and strain will change also in the case when the shearing deformations and volume alterations vary with time in accordance with different laws (e.g., shearing deformation is non-attenuating whereas volumetric changes attenuate) and also in all other cases when the creep curves are dissimilar. Consequently, the coefficient v will everywhere change with time. The state of stress remains unchanged only when $v = $ const. as was shown above.

It is appropriate to point out that the state of stress and strain may change with time also in the cases when e.g. given deformations set up forces in the soil, or the limiting and subliming states are described by different laws, or when the pressure is redistributed between pore water and soil skeleton in the course of consolidation, etc.

Let us assess, by way of illustration, the behaviour of a strip foundation of width $b = 200$ cm sustaining a uniformly distributed loading of an intensity $p = 1.8 \cdot 10^5$ Pa. The pattern of deformation of the soil, as determined from triaxial tests, is defined by eq. 13-49:

$$\gamma_i = \left[\frac{\tau_i}{A(t)} - \frac{\sigma_m}{C(t)}\right]^{1/m} \qquad \varepsilon_m = \left[\frac{\sigma_m}{D(t)}\right]^{1/m}$$

where $m = 0.5$.

The patterns of creep, as established from the same triaxial tests, are described by:

$$A(t) = \frac{A_0}{1 + \tilde{L}} \qquad C(t) = \frac{C_0}{1 + \tilde{L}} \qquad D(t) = \frac{D_0}{1 + \tilde{L}}$$

where $1 + \tilde{L} = 1 + \int_0^t K(t)\,dt = [1 + \delta(t/T)^a]$, provided $\sigma_m \leq [C(t)/A(t)]\tau_i$.

The numerical values of the parameters are as follows: $A_0 = 60.3 \cdot 10^5$ Pa; $C_0 = 10^7$ Pa; $D_0 = 2.23 \cdot 10^7$ Pa; $a = 0.06$; $T = 1$ month; $\delta = 1.3$. The value of the coefficient v, as computed by the formula in line 3 of Table 13-II, is $v = 0.35$.

Determine the superficial settlement of the soil below the foundation, assuming that it behaves absolutely elastically. To that end, use eq. 13-70' and compute the value of N from eq. 13-71:

$$N = \kappa \frac{\omega(\lambda\theta)}{I^{1/m}} F(m, v)$$

In accordance with eq. 13-66 we have:

$$\lambda = \sqrt{\frac{1}{0.5} + \frac{0.35}{1 - 0.35}\left(1 - \frac{1}{0.5}\right)} = 0.96$$

$$\omega\left(\lambda\frac{\pi}{2}\right) = \frac{1}{0.96} \sin\left(0.96\frac{\pi}{2}\right) = 1.04$$

$$I = 4 \int_0^{\pi/2} [\cos(0.96\theta)]^{0.5} \cos\theta\,d\theta = 0.35 \text{ approx.}$$

Then, according to line 3 of Table 13-II, we obtain:

$$F(m, v) = \frac{2(0.5 - 0.35)\sqrt{1 - 0.35 + 0.35^2}}{(0.35)^2 \sqrt{3}(1 - 0.5)60.3^2}\left[1 - \frac{60.3 \cdot 10^5(1 + 0.35)}{10^7\sqrt{3}\sqrt{1 - 0.35 + 0.35^2}}\right]^2$$

$$= 7.78 \cdot 10^{-4}$$

Hence:

$$N = \frac{1.04}{(0.35)^2} 0.916 \cdot 10^{-4} = 7.78 \cdot 10^{-4} \left(\frac{1}{10^5 \text{ Pa}}\right)^2$$

Substituting this value into eq. 13-70', we have for $|\bar{x}| < 1$:

$$S(x, t) = 7.78 \cdot 10^{-4} \times 100 \times 1.8^2 \times 4[\sqrt{1 - \bar{x}} + \sqrt{1 + \bar{x}}]^2 [1 + \tilde{L}]^2$$

for $\bar{x} > 1$:

$$S(x, t) = 1.0[\sqrt{1 + \bar{x}} - \sqrt{\bar{x} - 1}]^2 [1 + \tilde{L}]$$

for $\bar{x} = 0$, $S(x, t) = 1.0$.

Denoting $\bar{x} = x/(b/2)$ as before, we compute the settlement in the dimensionless form, $\bar{S} = S/(b/2)$. For comparison, we also determine the settlement by eq. 13-74 from the theory of elasticity. To enable the comparison, we assume that the settlement

TABLE 13-III

Relative settlements ($\bar{S} \cdot 10^2$) computed in terms of non-linear and linear deformations

$\bar{x} = \dfrac{2x}{b}$	0	0.2	0.4	0.6	0.8	1.0	1.2	1.4	1.6	1.8	83.0
By non-linear eq. 13-70'	1.0	1.0	0.96	0.90	0.80	0.50	0.27	0.21	0.173	0.147	0.875
By linear eq. 13-74	1.0	0.99	0.92	0.805	0.635	0.31	−0.14	−0.225	−0.396	−0.520	

at the centre of the foundation is the same in both cases, $\bar{S} = 1 \cdot 10^{-2}$. Then, the modulus of linear strain may be determined, for $x = 0$, using eq. 13-74:

$$E = \frac{4}{\pi} \frac{b/2}{S}(1 - v^2)p = \frac{4}{\pi} 0.01(1 - 0.35^2)/1.8 = 200 \cdot 10^5 \text{ Pa}$$

Substituting this value into eq. 13-74, we obtain the expression deciding the relative settlement of elastic half-plane:

$$\frac{S(x, t)}{b/2} = \frac{2(1 - 0.35^2)}{200\pi} 1.8 \left[\bar{x} \ln \frac{|1 - \bar{x}|}{1 + \bar{x}} - \ln |1 - \bar{x}^2 + 2| \right]$$

The results of the computations by the two formulae are tabulated in Table 13-III.

The curves of settlement distribution over the surface of the soil below the foundation are depicted in Fig. 13-4(a). It can be seen that the superficial settlement, as computed in terms of the theory of elasticity, assumes a negative value and increases without limit. We have already mentioned that a settlement developing on these lines contradicts reality. However, the settlement computed in terms of non-linear deformation, attenuates from a point closely approaching the load limit. This corresponds to the "cone" of real surface displacements.

The above computations refer to the initial moment $t = 0$. Eventually, the settlement will develop in proportion to the $(1 + \tilde{L})$ function, i.e.:

$$S(x, t) = S_0(x)[1 + 1.3 t^{0.06-1/0.5}]$$

where S_0 are the initial settlements indicated in Table 13-III.

The settlement vs time graph is depicted in Fig. 13-4(b).

Determining reaction pressures. The reaction pressures may be computed by eq. 13-81, assuming that the force applied is $P = pb \times 1.0 = 7.6 \cdot 10^3$ N. From the initial data adopted, we obtain:

$$p(x) = \frac{\Gamma(1.25)\Gamma(0.25) \sin (0.25\pi)}{(\sqrt{\pi})^3 100(1 - \bar{x}^2)^{0.25}} 7.5 \cdot 10^2 (10^5 \text{ Pa})$$

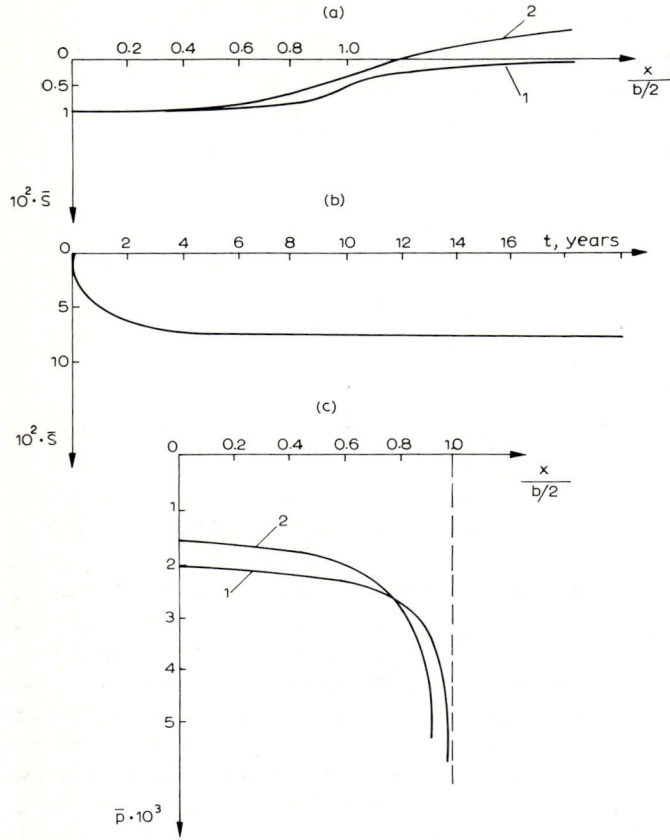

Fig. 13-4. (a) Settlement curves. (b) Curve depicting the development of settlement with time. (c) Curves of reaction pressures. 1 = non-linear deformation; 2 = linear deformation.

To enable comparison, we shall also compute the pressure by eq. 13-82 from the theory of elasticity:

$$p(x) = \frac{7.6 \cdot 10^2}{\pi \times 100 \sqrt{1 - \bar{x}^2}} (10^5 \text{ Pa})$$

The results of computations are tabulated in Table 13-IV where the load is given in a dimensionless form, $\bar{p} = (p \times 1 \times 1)/P$, for convenience.

The curves of reaction pressure distribution are depicted in Fig. 13-4(c). The non-linear curve appears to be of a smooth configuration, exhibiting a closeness of fit with the experimental data for clay soils. However, the extreme values of reactions increase unrestricted by analogy with the solutions obtained in terms of the theory of elasticity. This occurs due to the type of relationship 13-49 adopted, satisfying the condition of an infinite increase in shearing stresses with an increase in deformation.

TABLE 13-IV

Computed reaction pressures ($\bar{p} \cdot 10^3$)

$\bar{x} = \dfrac{2x}{b}$	0	0.1	0.2	0.3	0.4	0.5	0.6	0.7	0.8	0.9	0.99
By non-linear eq. 13-81	2.11	2.115	2.13	2.18	2.20	2.20	2.36	2.50	2.72	3.19	5.60
By linear eq. 13-82	1.59	1.60	1.62	1.67	1.74	1.84	1.99	2.23	2.66	3.65	11.30

To obtain finite values of reaction pressure, it is necessary to introduce boundary conditions ($\tau_i = \tau_{i(s)}$ for $\gamma_i = \gamma_{i(s)}$) into the initial equation or adopt a relationship (e.g., linear–fractional) into which enters the yield limit $\tau_{i(s)}$. However, solutions are obtained in this case by recourse to the computer.

Fig. 13-5 depicts a solution of this kind, as obtained by Shirokov et al. (1970) for the linear–fractional deformation pattern. Curves are shown presenting the distribution of vertical displacements u_z and vertical stresses σ_z (on dimensionless coordinates) under the centre of a rigid strip foundation. For comparison, the curves obtained in terms of the theory of elasticity are given. As can be seen, the non-linear displacements and stresses attenuate much faster with depth than predicted by the theory of elasticity.

Some conclusions drawn from the comparison

If we compare the results of the computations in terms of non-linear and linear deformation theories, we can see that, by virtue of allowing for non-linearity and for the effect of mean normal stress, we obtain analytical solutions approaching the real behaviour of soil.

The assets of the solutions arrived at by employing the non-linear deformation pattern are as follows.

(a) Vertical displacements and vertical stresses in the soil below a foundation, as determined by non-linear formulae, attenuate faster than is the case when the theory of elasticity is used, the explanation being that the soil mass subjected to compression is of limited depth. The notion that the soil below a foundation performs on these lines (compressible layer of limited depth) corresponds to modern concepts and has been implemented into the Building Codes and Regulations II-15-74.

(b) Soil surface settlements attenuate next to the foundation boundaries and tend to zero with an increase in the distance from the foundation. This indicates that a soil deforming non-linearly occupies, in terms of its distributive capacity, a position between the Winkler model (settlement only under a load) and the model of elastic half-space (unconfined deformation at the surface in response to excessive loads). The

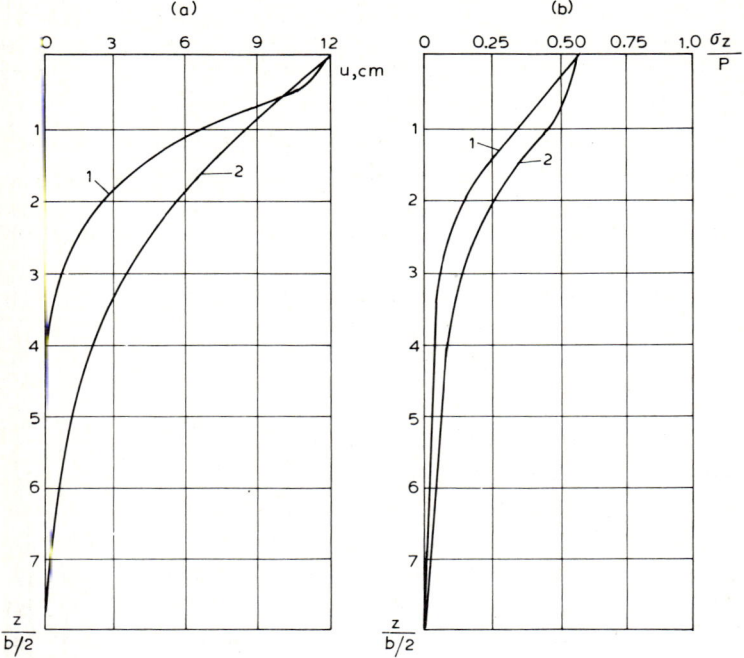

Fig. 13-5. (a) Curves of vertical displacement, and (b) stress curves below the centre of strip foundation. 1 = non-linear deformation; 2 = linear deformation. (Data from V.N. Shirokov and others.)

model of such intermediate status reflects modern concepts and agrees with the experimental data more closely than anything else.

(c) Reaction pressure curves display a configuration which is smoother than the shape of the curves presenting solutions in terms of elasticity theory. This is an indication of a decrease in the magnitude of the design bending moments applied to the foundation.

13.6 EXPERIMENTAL DATA

Conditions of the experiment

Below we shall compare the data on the distribution of stresses and displacements in a non-linearly deforming soil mass below a strip foundation with the solution in terms of the theory of elasticity.

The data were acquired by Vyalov and Mindich (1977; see also Section 11.8: Vyalov and Mindich, 1974) by way of a test using a strip-like settlement plate, a layer of clay soil (kaolin, $W = 36\%$, $W_L = 46.5\%$; $W_p = 28.6\%$) contained in a tray over a rigid substratum. The stresses and displacements at various points in the soil were measured by strain gauges strategically placed in different planes, three at each point, so as to determine all stress components. The soil displacements were registered on two transparent plates, a fixed and a movable one, using a special device. Thus, the displacement path and strain components were determined.

The experiments were conducted at various ratios $\lambda = b/h$, where h is the depth of the soil layer and b is the width of the settlement plate; the ratio varied between 0.25 and 3.0. The stepped load was applied in increments of $0.15 \cdot 10^5$ Pa during a period sufficiently long to obtain deformation stabilization at each stage; the time factor was not considered.

The triaxial compression tests revealed the following stress–strain relation for the soil given:

$$\sigma_m = k\varepsilon_m \qquad \tau_i = \bar{G}\gamma_i \qquad \bar{G} = \frac{(\tau_s^*/\sigma_0)\sigma_m}{(\tau_s^*/G_0^*) + \gamma_i} \tag{13-83}$$

where $\tau_s^* = \tau_s(\sigma_0/\sigma_m)$; $G_0^* = G_0(\sigma_0/\sigma_m)$; τ_s is the yield limit; $\sigma_0 = 10^5$ Pa. According to the experimental data, $G_0^* = 38.8 \cdot 10^5$ Pa, $\tau_s^* = 0.68 \cdot 10^5$ Pa.

Load-settlement relation. This relation, as obtained from experiments, is expressed by eq. 13-37:

$$S_\infty = \frac{(1 - v^2)nbp}{E_\infty(1 - p/p_\infty)} \tag{13-84}$$

where n is a coefficient allowing for the effect of the rigid substratum; $p_\infty = \alpha\lambda/(\lambda - \beta)$ is the ultimate load.

Eq. 13-84 may be reduced to the form:

$$S/Ap = \frac{1}{(1 - p/p_\infty)}$$

where $A = (1 - v^2)nb/E_0$. It follows, if this equation holds, that all the test points for any value of $\lambda = h/b$ must fall on a single curve of a generalized graph presenting S/Ap against p/p_∞.

In Fig. 13-6 we can see such a graph depicting experimental data in a tray along with field test data. The results of testing a layer of clay soil with depth h by a strip-like settlement plate with width b are denoted by points corresponding to various values of $\lambda = h/b$. The solid curve represents the results of analytical computations by eq. 13-84. As can be seen, the test points are closely grouped along the analytical curve (correlation coefficient $r = 0.93$), indicating that eq. 13-84 fits the test data.

Note the following point. As follows from eq. 13-83, the patterns of the changes in volume and shape are materially different for the soil given. The volumetric changes obey the linear law and the shearing deformations are non-linear. This implies that the resulting settlement versus load curve should be described by some other law combining the patterns of shearing and volumetric deformations. However, from Fig. 13-6 one can see that the curve is described by the linear–fractional relation of eq. 13-84, i.e. by the relation which also describes shearing deformation (of course, with different parameters). This is evidence that the process of settlement is influenced

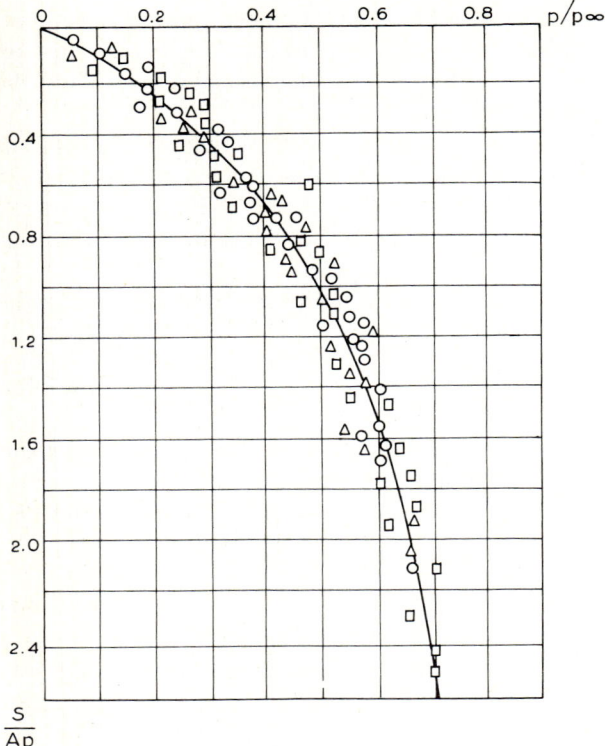

Fig. 13-6. Comparison of settlement data obtained from experiments with the results of computations with eq. 18-84.

predominantly by shearing deformations, and not by volumetric changes. Thus, in determining the settlement of the soil below a foundation it may be assumed, as the first approximation, that the pattern of shearing deformation is the initial one, provided, as was postulated above, that $v = $ const.

Stress and strain state

Ellipses of stress and lobes of strain are of assistance in visualizing the pattern of stressed and strained states. The former show the magnitude and direction of the principal stresses σ_1 and σ_2 at various points of the soil mass and the latter represent the amount and direction of the principal strains ε_1, ε_2.

It will be recalled that the equation of the deformation curve, i.e. of the locus of the points at the end of a line drawn — in proportion to the amount of deformation — from a given point in a given direction, is written in the parametric form as follows:

$$\alpha_1 = \arctan \sqrt{\left|\frac{\varepsilon_1}{\varepsilon_2}\right|} \sin t \qquad \bar{r}_1 = \frac{|\varepsilon_1 \varepsilon_2| \cos^2 t}{|\varepsilon_2| + |\varepsilon_1| \sin^2 t}$$

$$\alpha_2 = \arctan \sqrt{\left|\frac{\varepsilon_2}{\varepsilon_1}\right|} \sin t \qquad \bar{r}_2 = \frac{|\varepsilon_1 \varepsilon_2| \cos^2 t}{|\varepsilon_1| \pm |\varepsilon_2| \sin^2 t}$$

where α_1 and $\bar{r}_1$ are the parameters of the equation of the lobes drawn in the ε_1-direction and defined by positive deformations (compression); α_2 and $\bar{r}_2$ are the same parameters for negative deformations (tension).

Graphs of this kind, plotted from experimental data, are depicted in Fig. 13-7. To the left of the axis of symmetry we find the results of experiments and to the right the data computed in terms of the theory of elasticity. The primed values of ε, σ, α correspond to the loadings $p = 0.3 \cdot 10^5 \text{Pa} = 0.21 p_\infty$ in Fig. 13-7(a) and $p = 0.15 \cdot 10^5 \text{Pa} = 0.1 p_\infty$ in Fig. 13-7(b); the double-primed values correspond to the loading $p = 0.9 \cdot 10^5 \text{Pa} = 0.55 p_\infty$ in both graphs.

Computation of the theory of elasticity, presented for the sake of comparison, was carried out by formulae following Burmister (1956). From the comparison we can visualize the effect of non-linearity on deformation. In contrast to an increase in deformation directly with load as postulated by the elasticity theory, the actual deformation increase in response to a threefold load buildup is 1.5 to 10 times the initial value. Moreover, the statement from the elasticity theory that the modulus of compressive strain ε_1 is greater than that of tensile strain ε_2, at all points, is in fact correct only for the points along the axis of symmetry; at the points along the settlement plate edge we have $|\varepsilon_1| < |\varepsilon_2|$.

On rotation of the axes of principal stresses and principal strains. According to the elasticity theory, the axes of principal stresses and principal strains must not change their direction with an increase in loading. Practically, as can be seen from comparing the angles α' and α'' in Fig. 13-7, an increase in loading causes the axes to rotate. However, the angles of rotation differ but little so that the axes of principal stresses and principal strains are roughly of the same orientation. For example, the angles α_σ and α_ε which the principal axes of stresses and strains, respectively, make with the vertical axis z at various depths under a load $p = 0.55 p_\infty$ are as follows (Fig. 13-7):

$z = 0.25 H \qquad \alpha''_\sigma = 30° \qquad \alpha''_\varepsilon = 30°30'$

$z = 0.5 H \qquad \alpha''_\sigma = 30° \qquad \alpha''_\varepsilon = 28°20'$

$z = 0.75 H \qquad \alpha''_\sigma = 24° \qquad \alpha''_\varepsilon = 26°20'$

This means that the condition of the alignment of stress and strain deviator components (see Section 12.3) is fulfilled with some degree of approximation. However, the difference between α_σ and α_ε increases with depth.

Distribution of stresses with depth. The σ_z–p curves are load-dependent and, as a result, the magnitude of σ_z is not proportional to p, as follows from solutions in terms of the theory of elasticity.

The configuration of the curves decided by the relative thickness of the compressed soil layer, $\lambda = h/b$, are appreciably different, when compared with analytical curves. The stress concentrations over the width of the settlement plate, and along the central axis in particular, cause the curves to assume a basically curvilinear configuration with

474

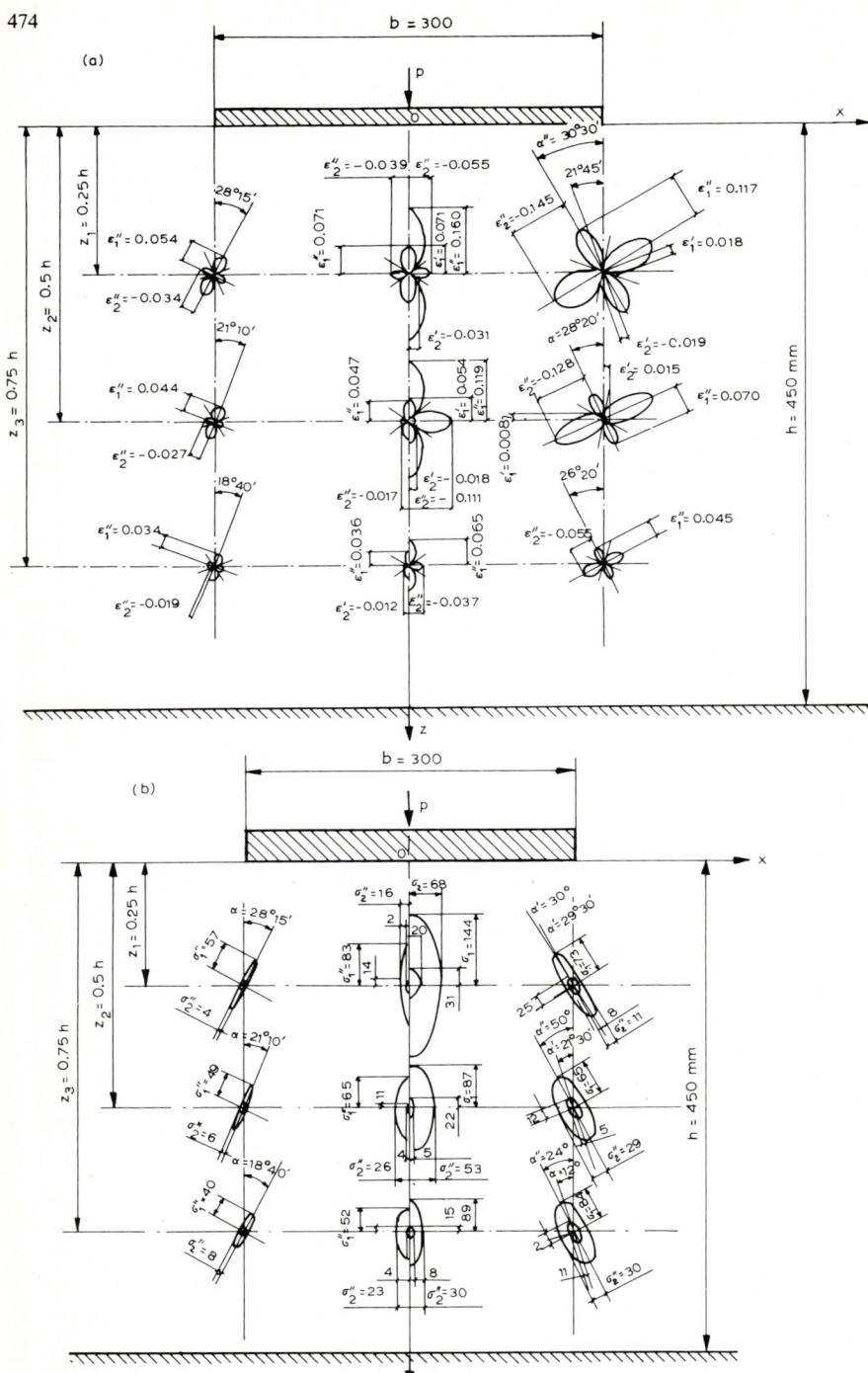

Fig. 13-7. (a) Lobes of strains and (b) ellipses of stresses in soil (σ in 10^5 Pa).

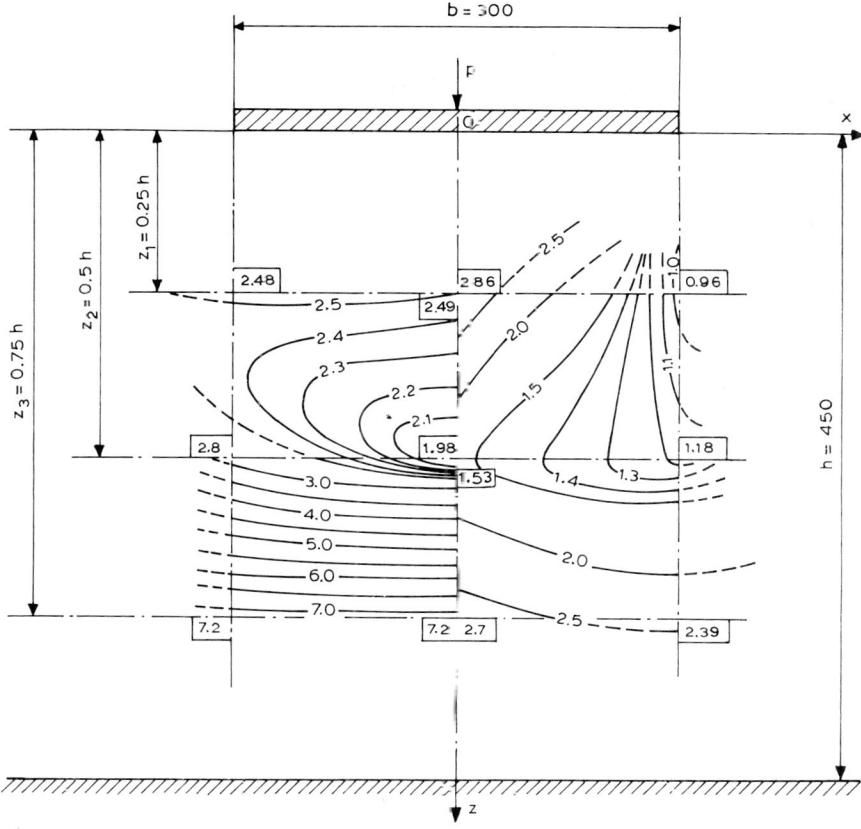

Fig. 13-8. Lines of equal values of modified shear modulus $\bar{G}$ (Pa) for loading $p = 0.3 \cdot 10^5$ Pa $= 0.2p_\infty$ to the left of the axis of symmetry and for $p = 0.9 \cdot 10^5$ Pa $= 0.55p_\infty$ to the right of the axis.

maximum ordinates either at the mid-depth (for small λ) or within the upper third of the layer (for large λ). The ordinates in the example under consideration appeared to be greater than the computed values by 75–100%.

Setting up of zones of limiting stress state. According to eq. 13-83, a soil appears to be in the state of limiting stress, defined by the condition $\tau_i = \tau_s$, when the shearing deformation becomes infinitely large $\gamma \to \infty$ and, consequently, $\bar{G} \to 0$. Thus, at the points where the modified shear modulus $\bar{G}$ is at a minimum, the soil will be in a state of stress approaching the limiting stress more closely than anywhere else.

Referring to Fig. 13-8, we can see the lines of equal values of $\bar{G}$ for a test involving the application of two loadings amounting to 20 and 55% of the ultimate load. Under the load $p = 0.2p_\infty$, the stress concentrations are at a maximum at the centre of the layer where $\bar{G}$ is a minimum. An increase in the loading to $p = 0.55p_\infty$ causes new zones of stress concentration to take shape at the points along an axis running below a settlement plate end face. The value of $\bar{G}$ decreases there to $0.96 \cdot 10^5$ Pa (the maximum value is $7.2 \cdot 10^5$ Pa). The zones thus set up are the cradles of the limiting state.

Contact pressures. Depicted in Fig. 13-9 are the experimental curves and also those plotted from computations by a formula from elasticity theory which show the way contact (reaction) pressures are distributed over the area of a rigid settlement plate and the rigid bottom of a tray containing a layer of clay soil. The curves have been plotted for various values of the average loading on the settlement plate ($p = P/bx$), the maximum loading being $p_\infty = 4.2 \cdot 10^5$ Pa for Fig. 13-9(a) and $p_\infty = 1.6 \cdot 10^5$ Pa for Fig. 13-9(b).

We can see that the shape of contact pressure curves varies with the load p/p_∞ and the relative thickness of the compressed layer $\lambda = h/b$. For small values of λ, the initial shape of curves, when the load is low, approaches the configuration obtained in terms of the theory of elasticity (except the ordinates at the settlement plate edges). With an increase in load, the experimental curves depart in shape from the analytical ones — the stresses under the plate centre increasing more rapidly than those at the edges and, as a result, the curve changing its shape from concave to convex. A further load increase transforms the shape into a kind of a triangular one, corresponding to the curve of the contact pressures as set up in a plastic strip compressed between two pistons. This indicates that a rigid kernel has formed below the settlement plate.

When a layer of comparatively large depth is subjected to compression, the experimental curves of the contact pressures under the settlement plate depart from the analytical curves in point of shape even with low loadings. At an early stage (when λ is small) the curve is saddle-shaped, becoming wave-like eventually. The extreme ordinates do not increase indefinitely whatever the case (as is predicted by the theory of elasticity), but have finite values. This is in accordance with the type of the initial eq. 13-83 setting up a limit for the shearing stress, $\tau_i \leq \tau_s$.

The curve of the contact pressures at the surface of a rigid underlayer is of a shape approaching that of the curve obtained by Gorbunov-Posadov (1946) in terms of the theory of elasticity.

The experimental data referred to above provide evidence that by allowing for a non-linear deformation pattern, we obtain a model of the soil below a foundation which ranks higher in terms of reliability than any other.

Field experiments

To study the development of long-term settlement, Vyalov (1978, 1979) tested compact clay soils (frozen) in the field for 19 years, using three settlement plates each of a diameter $d = 70.5$ cm. The plates, placed at a depth of 3 m in 1951, were loaded in a stepwise manner by loads p equalling $2.5 \cdot 10^5$ Pa, $3.75 \cdot 10^5$ Pa and $4.0 \cdot 10^5$ Pa, i.e. close to the ultimate value $P_{ult} = 5.55 c_{equiv} + \gamma h$, where c_{equiv} was the ultimate long-term cohesion as determined experimentally by eq. 9-40. The last loading was maintained for the rest of the experimental period up to 1971. The settlements resulting from each stage of loading during the 19-year period were described by the power law (eq. 13-33):

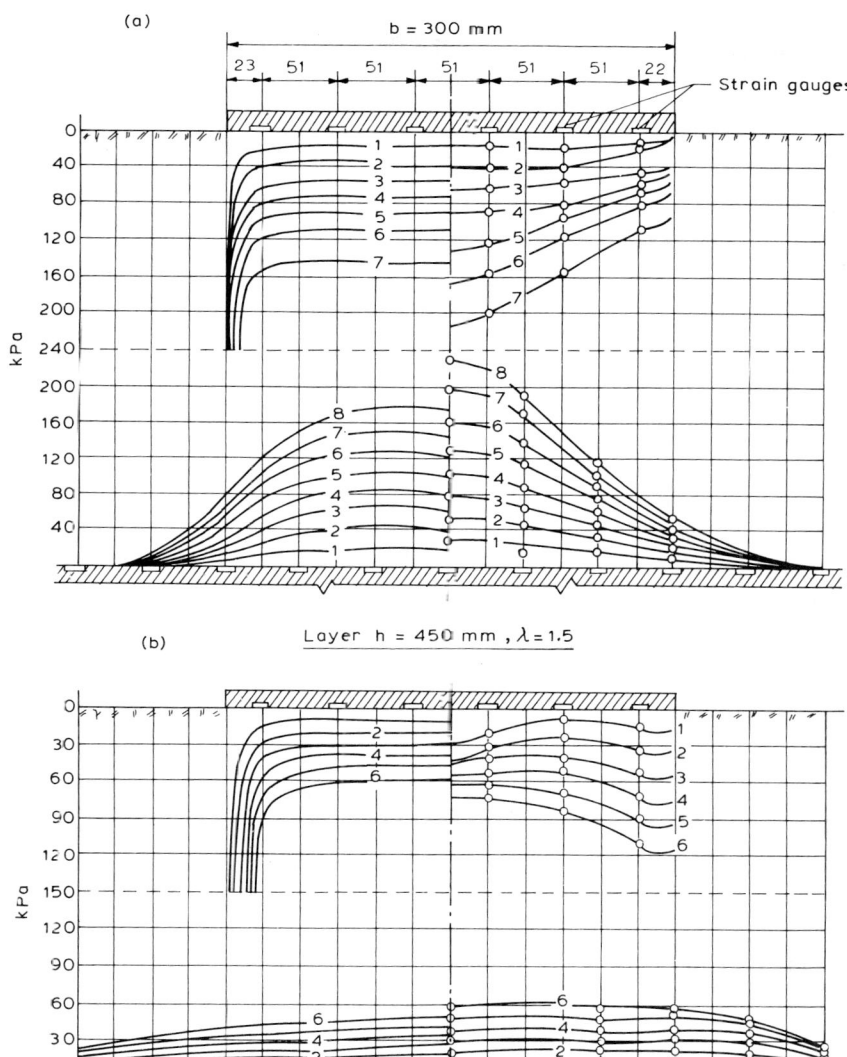

Fig. 13-9. Curves of contact pressures as plotted from test data (on the right) and computed in terms of elasticity theory (on the left).
(a) For relative thickness of layer $h/b = 0.25$ and the load p as follows: $1 = 0.2 \cdot 10^5$ Pa; $2 = 0.4 \cdot 10^5$ Pa; $3 = 0.6 \cdot 10^5$ Pa; $4 = 0.8 \cdot 10^5$ Pa; $5 = 1.0 \cdot 10^5$ Pa; $6 = 1.2 \cdot 10^5$ Pa; $7 = 1.4 \cdot 10^5$ Pa; $8 = 1.6 \cdot 10^5$ Pa.
(b) For relative thickness of layer $h/b = 1.5$ and the load p as follows: $1 = 0.15 \cdot 10^5$ Pa; $2 = 0.3 \cdot 10^5$ Pa; $3 = 0.37 \cdot 10^5$ Pa; $4 = 0.45 \cdot 10^5$ Pa; $5 = 0.75 \cdot 10^5$ Pa; $6 = 0.9 \cdot 10^5$ Pa.

$$S = \frac{(1-v^2)\omega d}{A_z^{1/m}} p^{1/m}(1+\delta t^\beta)$$

The estimated settlements in 25 and 50 years found from extrapolating the experimental data appeared to be $S/d = 0.063$ and $S/d = 0.069$, respectively, under a loading $p = 3.3 \cdot 10^5$ Pa.

Schwab and Broms (1979), experimenting with a soft normally compacted clay (W is 90–120%) at a depth of 4 m, also determined that the settlement produced by plates with a diameter of 16 and 30 cm obeyed a power law. The final loading, amounting to between 67 and 82% of the ultimate load, was attained by a stepwise increase. The settlement equation is of the form:

$$S = I(p/p_{\text{ult}})^{-1/m}(t_1/t)^a t$$

and the time before failure is given by the formula:

$$t_f = \{S_{\text{ult}}(p/p_{\text{ult}})^{-1/m} t_1^{-a}\} \exp \frac{1}{1-a}$$

where p/p_{ult} is the level of stress; t_1 is unit time; t_f and S_g are the time and settlement corresponding to the instant of failure; I, m, a are constants.

The results of the tests for settlement due to creep in hard clay by means of plates with a diameter (d) of 8–100 cm pushed into the soil were reported by Hartlen and Pusch (1973). A stepwise increasing load was applied in increments obeying a log law, $S \sim \ln t$. The data acquired provided the basis for determining the relation between settlement due to creep and load and establishing the ultimate settlement for plates of various diameter. The dependence of p_{ult} on the plate diameter was of a hyperbolic nature, p_{ult} decreasing with an increase in d. A most pronounced decrease in p_{ult} values was associated with the plate diameters of 8–30 cm; values of d between 30 and 100 cm produced a less significant effect.

The way in which the settlement due to creep developed in soft rock (sandstone) in response to the action of settlement plates ($d = 150$ cm) under a constant load $p = 1900$ kN/m^2 applied for 3 months was described by Meigh et al. (1973). Simultaneously, the specimens were exposed to triaxial compression. The development of settlement with time was approximated by a log relation. The tests provided the basis for estimating the possible settlement of the structures erected in the soil examined.

13.7 REFERENCES

Arutyunyan, N.Kh., 1959a. Planar contact problem of creep. Appl. Math. Mech., v. 23, no. 5 (in Russian).
Arutyunyan, N.Kh., 1959b. Planar contact problem of plasticity theory with power hardening of the material. News, Acad. Sci. Armenian SSR, Phys. Math. Sci. Section, issue 2 (in Russian).
Burmister, D.M., 1956. Stress and displacement characteristics of a two-layer rigid base soil system. Natl. Acad. Sci., Highway Res. Board, Proc. 35th Annual Meeting, Washington.
Cherkasov, I.I., 1958. Mechanical Properties of Soil Foundations. Avtostroiizdat, Moscow (in Russian).
Freilich, O.K., 1938. Distribution of Pressure in Soils. Narkomkhoz RSFSR Publ. House, Moscow (in Russian).

Gorbunov-Posadov, N.I., 1946. Settlements of Foundations in a Layer of Soil Resting on Rock. Gosstroiizdat, Moscow (in Russian).

Hartlen, J. and Pusch, R., 1973. Interpretation of creep measurements of stiff clay. Proc. 8th ICSMFE, Moscow, v. 1.1.

Malyshev, M.V., 1963. On the effect of intermediate principal stress on soil strength and on slip planes. SMFE, no. 1 (in Russian).

Meigh, A.C., Skipp, B.O. and Halls, N.B., 1973. Field and laboratory creep of weak rocks. Proc. 8th ICSMFE, Moscow, v. 1.2.

Popov, B.P., 1950. Application of Dimensional Analysis to Experiments with Test-Loads. In collective volume: Engineering Geology Survey for Hydropower Projects. Moscow, v. 2 (in Russian).

Schwab, E.F. and Broms, B.B., 1979. Pressure–settlement–time relationship by screw plate tests in situ. Proc. 9th ICSMFE, Tokyo, v. 1.

Shirokov, V.N., Solomin, V.I., Malyshev, M.V. and Zaretsky, Yu.K., 1970. Stress state and the displacement of a ponderable nonlinearly deformable semi-space below a rigid circular settlement plate. SMFE, no. 1 (in Russian).

Sokolovsky, V.V., 1950. Plasticity Theory. Gostekhteorizdat, Moscow-Leningrad (in Russian).

Vyalov, S.S., 1978. Test plate settlements in plastic frozen soils. SMFE, no. 5 (in Russian).

Vyalov, S.S., 1979. Long-term settlements of foundations in frozen soils. Proc. 3rd Int. Conf. On Permafrost, Canada, 1979.

Vyalov, S.S. and Mindich, A.L., 1977. Experimental studies of the state of stress and strain in a layer of weak soil resting on an insignificantly compressible stratum. SMFE, no. 1 (in Russian).

Vyalov, S.S. and Mirenburg, Yu.S., 1980. Settlements and bearing capacity of foundations formed from weak soils with allowance for nonlinearity and creep. Danube–Europ. Conf. Soil Mech. Found. Eng., Sect. I^a, 6th, Varna, 1980.

Vyalov, S.S. and Zaretsky, Yu.K., 1966. Questions of the theory of deformation of rocks taking into account different tensile and compressive strength. Int. Congr. Rock Mech., 1st, Lisboa, 1966.

Zaretsky, Yu.K., 1965. On radial distribution of stresses on a half-plane in response to concentrated force. In collective volume: Research into the Mechanics of Rocks. Kazakh SSR Acad. Sci., Alma-Ata (in Russian).

Chapter 14

NON-LINEAR SOIL MODELS AND NUMERICAL SOLUTIONS OF PROBLEMS

14.1 ORIGINAL RELATIONSHIPS

Mathematical models

A model of a real body, process or phenomenon is thought of as a system possessing the properties of the modelled entity and simulating the behaviour of the prototype which is being studied. A physical model is a system composed of real material which is similar to the material of the prototype or replaces it. The mechanical models examined in Sections 7.1 and 7.2 simulate the properties of the prototype with the aid of mechanical elements (springs, dashpots with perforated pistons, etc.).

In the mathematical models discussed in this chapter, the properties of the prototype are represented by means of mathematical symbols and operations. For example, elastic properties are expressed in terms of Hooke's equation, viscous properties are simulated by Newton's equations, etc. The behaviour of a modelled body in response to a load is represented in terms of an equation of state. As far as soils are concerned, distinction is made between a model of the soil proper (e.g., an elastic Hookean body) and a model of the soil below a foundation (elastic semi-space, Winkler model, etc.).

Particulars of a soil system

As was said in Chapter 2, soil is a discrete three-component system composed of solid particles interlinked by water-colloidal bonds. An elastic compression of solid particles and air bubbles trapped in the soil, along with a recoverable shear of the particles, defines an elastic deformation of soil. Irrecoverable displacements of particles and expelling of the water and air contained in soil voids leads to plastic deformations.

It was shown in Chapter 10 that the process of soil deformation is associated with structural changes such as the repacking of solid particles, breaking and reforming of interparticle bonds and development and healing of structural defects (micro-cracks, etc.). These changes either strengthen or weaken soil structure and, owing to the viscous nature of the bonds, the processes mentioned are time-dependent. If the soil structure gains in strength, the process of deformation attenuates: the deterioration in strength leads to the non-attenuating deformation ending in failure. This takes place when the defect density reaches a certain critical value.

Particulars of soil deformation

The properties of loose rocks referred to above and the way these properties change with deformation are factors defining the behaviour of soil in response to load. Although examined above, these factors must be given due consideration in deriving models. Therefore, it is appropriate to refer to the following aspects.

(1) Elastic (recoverable) deformations coexist with plastic (irrecoverable) deformations, either type being evident during both shear and volumetric changes associated with a changing porosity.

(2) Plastic deformations are induced almost at the very beginning of loading and attain considerable magnitudes amounting to a considerable percentage (see Fig. 5-4).

(3) The stress–strain relation is non-linear and different for shear and volumetric changes, the shearing deformations increasing to infinity with an increase in stress and the changes in volume tending to a certain limit (see Fig. 1-1).

(4) The state of stress and strain alters with time (the evidence is creep and consolidation, relaxation, long-term strength), the patterns of shear and volumetric changes being different. The time-dependent nature of shear is attributed to creep and depends on the stress, assuming, in consequence, an attenuating or non-attenuating character (see Fig. 5-1). The time-dependent nature of volumetric changes is influenced mainly by the changes in porosity due to percolation as well as by the creep of the soil skeleton. Volumetric changes are of attenuating character (see Fig. 8-3).

(5) Resistance to deformation varies depending on whether it is tension or compression, and the internal friction manifests itself both in the limiting and sublimiting states (see Fig. 11-5). The result is that the deformation in the sublimiting state and the limiting state proper are influenced by all three stress tensor invariants, as is shown by eqs. 11-1 and 11-3.

(6) The effect of the first invariant manifests itself in the value of $I_1(T°)$, i.e. the mean normal stress $\sigma_m = 1/3 I_1(T°)$ (or all-around pressure) affects not only the volumetric strain ε_v, but the shearing strain γ_i as well (see Fig. 11-2).

(7) The effect of the second invariant is evident because of the fact that the value of $I_2(D)$, i.e. the shearing stress intensity $\tau_i = \sqrt{I_2(D)}$, has a bearing not only on the shearing strain γ_i but also on the volumetric strain ε_v. The changes in volume caused by the shearing stress are referred to as dilatancy which, as far as soils are concerned, may be either positive (extra hardening) or negative (softening).

(8) The effect of the third invariant is important in that the processes of deformation and failure are influenced by the value of $I_3(T)$, i.e. by the arrangement of stress, as defined by the parameters μ_σ and ω_σ (see Fig. 12-5).

(9) The coefficient of lateral strains varies with the stress and period of its application owing to the different patterns of shearing and volumetric deformations, $v = v(\tau_i, \sigma_m, t)$ (see Fig. 5-23).

(10) The process of deformation varies with the history and path of loading (see Fig. 12-9).

(11) The stress and strain tensors may fail to be similar and coaxial in some cases.

(12) The deterioration of strength with the period of load application (the evidence is long-term strength) is an outcome of a non-attenuating creep developing at an increasing rate (see Fig. 9-1); as a result, the limiting state is dependent on a time factor, as is shown by eq. 11-9.

On soil models

By taking into account the factors referred to above, we can confidently create models capable of describing the actual behaviour of soil in response to load. Consequently, we can utilize the engineering properties of soil in a much better way. However, the models thus obtained appear to be rather complex. Therefore, the main task facing any researcher is to evaluate the effect of every factor and to select those which are of practical value. In what follows, we will discuss the principles of devising soil models and examine various models which are instrumental in taking into account certain peculiarities of soil behaviour.

Original experimental data

Soil models must be based on experimental data. These are represented by stress–strain curves for shearing and volumetric deformations and a curve defining the limiting state (the so-called failure curve). Any of the loading patterns of Fig. 11-1 are suitable for plotting the curves, but preference is commonly given to triaxi-symmetrical compression (Fig. 11-1(a)).

Configuration of experimental curves

Typical experimental curves are shown in Fig. 14-1. Of this figure (a) depicts the curves of the relation between the stress and the shearing (γ_i) and volumetric ($\varepsilon_m = \varepsilon_v/3$) strains on which the elastic and plastic (e and p) constituents are singled out (this was possible because the test involved unloading). Thus, the strains and strain increments obtained may be represented as the sums:

$$\gamma_i = \gamma_i^e + \gamma_i^p \quad \text{and} \quad \varepsilon_m = \varepsilon_m^e - \varepsilon_m^p \tag{14-1}$$

$$d\gamma_i = d\gamma_i^e + d\gamma_i^p \quad \text{and} \quad d\varepsilon_m = d\varepsilon_m^e + d\varepsilon_m^p$$

In the general form, we can write:

$$\varepsilon_{ij} = \varepsilon_{ij}^e + \varepsilon_{ij}^p \quad d\varepsilon_{ij} = d\varepsilon_{ij}^e + d\varepsilon_{ij}^p \tag{14-1'}$$

where:

$$\varepsilon_{ij} = 1/2 \left(\frac{\partial U_i}{\partial x_i} + \frac{\partial U_j}{\partial x_j} \right)$$

are the components of strain tensors ($i, j = 1, 2, 3$). It will be recalled that the stress tensor components are denoted by σ_{ij}. In substituting $i, j = 1, 2, 3$, the values of σ_{ij} and ε_{ij} become $\sigma_{11} = \sigma_x$, $\sigma_{12} = \tau_{xy}$, $\varepsilon_{11} = \varepsilon_x$, $\varepsilon_{12} = \gamma_{xy}$ and so on.

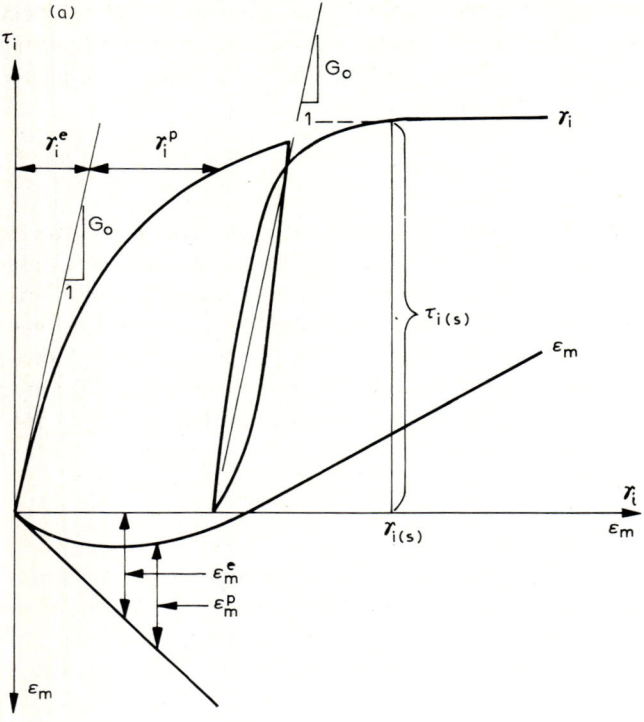
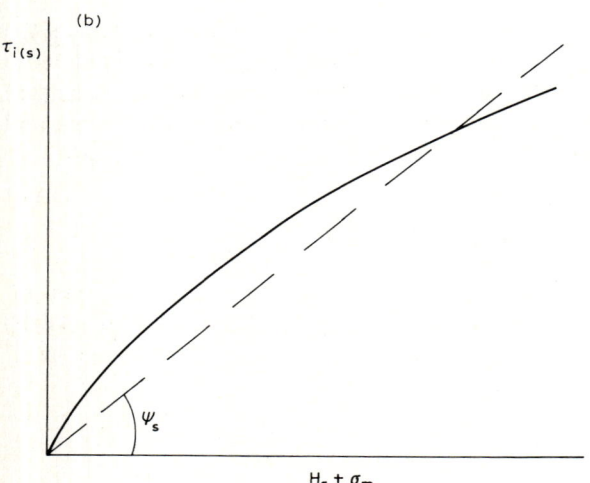

Fig. 14-1. (a) Stress versus shearing strain γ_i and stress versus volumetric strain ε_m curves. (b) Curve representing the limiting state of a soil.

The curves in Fig. 14-1(a) all correspond to a given value of σ_m = const. In general, the process of deformation will be represented by a family of similar curves for various values of σ_m (see Figs. 11-2 and 12-2). To reduce these curves to a single curve, we must insert the value $\tau_i/\Omega(\sigma_m)$ along the stress axis (see Fig. 11-4). The value of $\Omega(\sigma_m)$ is given in eqs. 11-45 to 11-49.

Let us try to reveal other particulars of soil deformation. Dilatancy is represented in Fig. 14-1 by a relationship between the volumetric strain ε_m and the shearing strain γ_i (for γ_i, in its turn, is a function of τ_i). One can see that both a positive and a negative dilatancy are present, i.e. extra hardening and softening, respectively, take place.

The effect of the loading path can be determined by employing various paths during the test (see Fig. 12-9). To find out the effect of the arrangement of stress, one must use various values of the parameter μ_σ or ω_σ during the experiment, the test loading pattern of Fig. 11-1(d) being helpful in this. Similar tests for various values of μ_σ and, correspondingly, ω_σ are conducted in order to check the stress and strain tensors for similarity and alignment. These two conditions will be satisfied if it appears that $\mu_\sigma = \mu_\varepsilon$ and $\alpha_\sigma = \alpha_\varepsilon$, where $\alpha_{\sigma\varepsilon}$ are the angles of inclination of the axes of principal stresses and strains, respectively.

To take into account the time factor, all the tests referred to above should be conducted under conditions of a time-dependent deformation; when examining not only the subliminiting but also the limiting state, the deformation must be brought to the progressive flow stage.

Equations of deformation

Experimental data also provide the basis for establishing the type of relation between the strain (or strain rate), stress and time:

$$\varepsilon_{ij} = f(\sigma_{ij}, t) \quad \text{or} \quad \dot{\varepsilon}_{ij} = f(\sigma_{ij}, t) \tag{14-2}$$

Setting forth these expressions for the shearing and volumetric deformations separately, we obtain equations of the form given by eqs. 11-7 and 11-8.

The equation of volumetric deformation (eq. 11-8) is regarded quite frequently as the product of some functions:

$$\tau_i = \phi(\gamma_i)\Omega(\sigma_m)Z(I_3)F(t) \tag{14-3}$$

where the first function defines the stress–shearing strain relation, the second function determines the effect of the mean normal stress, the third function characterises the effect of stress arrangement and the fourth function indicates the way the process is developing with time.

The function $\phi(\gamma_i)$ is used most frequently in the form of the power law (eq. 4-28) or the linear–fractional relation (eq. 4-36):

$$\phi(\gamma_i) = A\gamma_i^m \qquad \phi(\gamma_i) = \frac{G_0\tau_s\dot{\varrho}_i}{\tau_s + G_0\gamma_i} \tag{a}$$

The function $\Omega(\sigma_m)$ is commonly used in the form given by eq. 11-47:
$$\Omega(\sigma_m) = (1 + \sigma_m/H_s)^\lambda \tag{b}$$
Alternatively, a particular case of the above expression for $\lambda = 1$ is employed.

The $Z(I_3)$ function is adopted according to expression 12-16 as:
$$Z(I_3) = (1 + k\cos 3\omega_\sigma) \tag{c}$$

The most widely used form of the time function is eq. 5-12:
$$\frac{d\psi}{dt} = \left(\frac{T_i}{T_1 + t}\right)^n \tag{d}$$
from which we obtain $F(t) = 1/\psi(t)$, where $\psi(t)$ is in the form given by eqs. 5-13, 5-14 and 5-15.

The statement in the form of eq. 14-3 corresponds to the theory of ageing. However, a similar statement (as a product of functions) may be obtained for other creep theories, proceeding on the assumption that creep curves are similar under all conditions of loading.

The equation of volumetric deformation may be presented in the form given by eq. 12-2:
$$\varepsilon_m = \varepsilon_m^0 \pm \varepsilon_m^D \tag{14-4}$$
where $\varepsilon_m^0 = f^*(\sigma_m, t)$ is the volumetric strain induced by the all-around pressure $\varepsilon_m^D = \Lambda\gamma_i$ is the volumetric strain resulting from the shearing stress (dilatancy) and Λ is the dilatancy coefficient.

Equation of limiting state

This equation can be arrived at, e.g. from the first relation 14-3, by satisfying the condition $\gamma_i = \gamma_{i(s)} = $ const. Hence, we obtain eq. 11-9 which can be set forth, taking into account $\phi(\gamma_{i(s)}) = \tau_s^0$, as:
$$\tau_{i(s)} = \tau_s^0 \Omega(\sigma_m) Z(I_3) F(t) \tag{14-5}$$
or, in the general form, as:
$$\Phi(\sigma_{ij}, t) = 0 \tag{14-5'}$$

The Φ-function, in all its forms for the classical conditions of strength, was considered in Section 4.2; the forms corresponding to the modified conditions will be examined in the following section.

14.2 LOADING SURFACE AND LIMITING SURFACE

Loading surface

In what follows, the concept of loading surface is frequently used; therefore, it is appropriate to consider it.

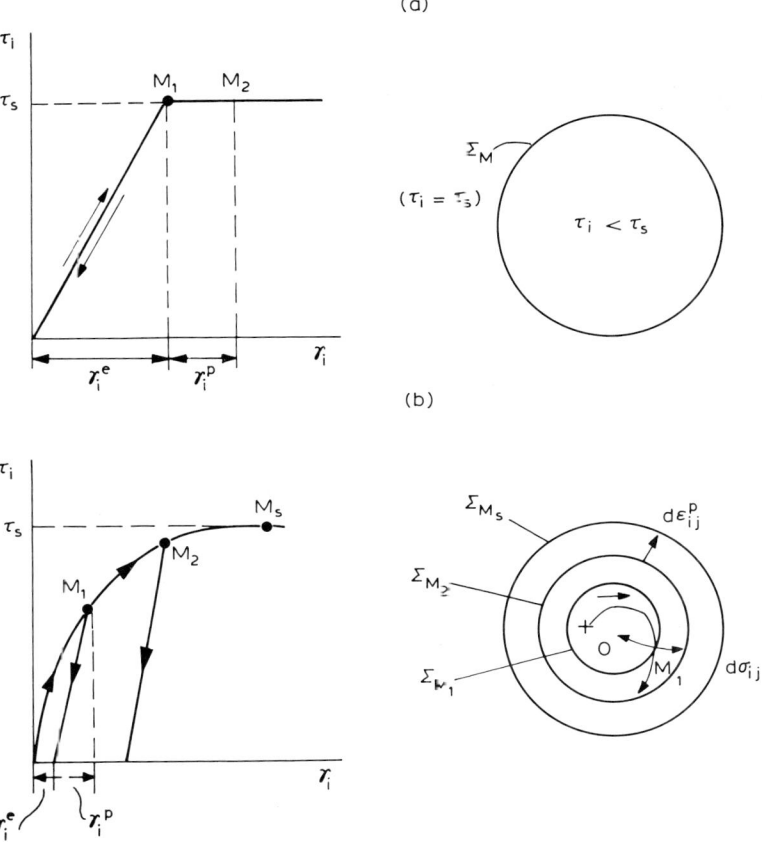

Fig. 14.2. Stress–strain graphs and loading surfaces for (a) ideally elastoplastic medium, and (b) hardening medium.

Referring to eq. 14-1', it was stated that a body subjected to loading suffers from both recoverable (elastic) and irrecoverable (plastic) deformations. Thus, in the σ_{ij} space one can construct a surface Σ separating the regions of elastic and plastic deformation from one another; it is referred to as the loading surface and the analytical expression describing the surface is termed the loading function $f(\sigma_{ij})$.

Loading surface of elastoplastic medium

For an ideal elastoplastic medium, the loading surface remains in a fixed position corresponding to a stress equalling the yield limit τ_s on the $\tau_i - \gamma_i$ graph of Fig. 14-2(a). The loading surface proper, Σ, coincides with a yield surface (of the limiting state), i.e.:

$$f(\sigma_{ij}) = \Phi(\sigma_{ij}) = 0 \tag{14-6}$$

The Φ-function is defined by one of the yield conditions (or conditions of limiting state) discussed in Section 4.2. It can be, for example, the St. Venant–Tresca or von Mises condition for ideally plastic media or the Mohr–Coulomb or von Mises–Botkin condition for plastic media possessing friction. In the $(\sigma_1-\sigma_2-\sigma_3)$ space, the loading surface will thus correspond to the surface of limiting state shown in Figs. 4-3 and 4-4. Note that in literature, along with the term loading surface, the terms yield surface and surface of limiting state (or limiting surface) also occur. Apparently, for an ideally elastic–plastic medium having a fixed yield limit, the loading surface coincides with the yield surface. Therefore, either of the two terms may be used (the term yield surface is used most). For a hardening medium (Fig. 14-2(b)) whose yield limit changes with deformation, it is appropriate to use the term loading surface. The limiting surface is the surface at failure (see Section 4.2) separating the zone of sublimiting and limiting conditions. For an ideal elastic–plastic medium, this surface coincides with the yield surface but appears to be an ultimate position of the loading surface in the hardening-medium case.

Loading surface for hardening medium

Consider the deformation of an elastic–plastic medium subjected to non-linear hardening.

Referring to b in Fig. 14-2, we load this medium from 0 to a point M_1 corresponding to a stress $\tau_i(M_1)$ and then unload along a line $M_1 N_1$. Consequently, an elastic γ_i^e and a plastic γ_i^p strain will appear on the $\tau_i-\gamma_i$ graph in accordance with eq. 14-1. A secondary loading also from 0 to M_1 will induce a strain represented by the same straight line $M_1 N_1$, indicating that the strain is just elastic. However, if we exceed the stress $\tau_i(M_1)$ by an amount $d\tau_i$, the increment in the stress will induce an increment in both the elastic $d\gamma_i^e$ and plastic $d\gamma_i^p$ strains. It follows that the stress $\tau_i(M_1)$ can be considered as a yield limit separating the elastic and plastic strains in the course of loading from 0 to M_1. If we now proceed with the loading up to a point M_2 which is followed by an unloading $(M_2 N_2)$ and another loading $(N_2 M_2)$, we see that the above discussion also holds in this case. The stress $\tau_i(M_2) > \tau_i(M_1)$ can be regarded as a new and higher yield limit. It thus follows that an increment in the plastic strain $d\gamma_i^p$ leads to an increase in the yield limit, i.e. to the hardening of the medium. Interpretation of the process on these lines gives the reason why it is called elastic–plastic deformation with hardening. Each yield strength is represented by a corresponding loading surface $\Sigma(M_1)$, $\Sigma(M_2)$ and so on, separating the region of purely elastic strains induced during unloading and subsequent loading from the region in which the additional loading $d\tau_i$ gives rise to an increment in both the elastic $d\gamma_i^e$ and plastic $d\gamma_i^p$ strains.

Thus, it stands to reason that the position of the loading surface of an elastic–plastic hardening medium is not fixed and changes (the surface expands) with hardening (Fig. 14-2(b)). Ultimately, when the plastic strain reaches a critical value and the medium changes into the limiting state, the loading surface is transformed into the surface of limiting state $\lim \Sigma(M_i) = \Sigma(M_s)$, i.e. eq. 14-6 appears to be satisfied in the limit.

Loading and unloading

The concept of a loading surface appears to be helpful in defining more exactly the concept of loading and unloading referred to in Section 3.6. Apply a stress $d\sigma_{ij}$ to a point M_1 (Fig. 14-2(b)) on the loading surface. If the vector $d\vec{\sigma}_i$ points inside the surface ($d\vec{p}_{ij} < 0$)*¹, this is a case of unloading causing an elastic deformation of the medium; if the vector is in the opposite direction, an extra loading takes place and the deformation is plastic (or, to be precise, elastoplastic). When the $d\vec{p}_{ij}$ vector is tangent to the surface ($d\vec{p}_{ij} = 0$) it corresponds to a neutral state. For $d\vec{p}_{ij} \leq 0$ the surface will not change, for $d\vec{p}_{ij} > 0$ it will expand. A uniform expansion (the surfaces $\Sigma(M_1)$, $\Sigma(M_2)$, ..., $\Sigma(M_S)$ are similar) is evidence of the medium's behaviour referred to as isotropic hardening.

Loading function of hardening medium

In the case of a hardening medium, the loading function f may be set as $f(\sigma_{ij}, t) = F(G)$. Or, in an expanded form:

$$f(\tau_i, \sigma_m, \mu_\sigma, t) = F(G)$$

or:

$$f(\tau_i, \sigma_m, \mu_\sigma, \Gamma, t) = 0 \tag{14-7}$$

where Γ is a measure (parameter) of hardening.

A statement in the form of eq. 14-7 indicates that the shape and position of the loading surface are influenced by the stress deviator, spherical tensor, arrangement of stress, hardening parameter and time. Note that in general the history (or path) of the loading should also be taken into consideration. From eq. 14-7 we obtain that a transition into the limiting state occurs for $\tau_i = \tau_{i(s)}$.

The hardening measure, Γ, used may have different values. In most cases, Γ is adopted as the magnitude of the plastic shearing strain:

$$\Gamma = \gamma_i^p \tag{14-8}$$

Alternatively, either the volumetric plastic strain is used as the hardening measure:

$$\Gamma = \varepsilon_v^p \tag{14-8'}$$

or the combination of γ_i^p and ε_v^p:

$$\Gamma = \Gamma(\gamma_i^p, \varepsilon_v^p) \tag{14-8''}$$

Other suitable forms of Γ are the accumulated plastic strain in shear (the Odqvist parameter):

$$\Gamma = \int \sqrt{\sum_{ij} d\varepsilon_{ij} d\varepsilon_{ij}} \tag{14-9}$$

---

*¹ $\vec{p} = \dfrac{\partial f}{\partial \sigma_{ij}} d\sigma_{ij}$

or the work of plastic strain:

$$\Gamma = \int \sigma_{ij} \, d\varepsilon_{ij}^p \tag{14-9'}$$

Apparently, the choice of the hardening measure will have a bearing on the type of equation of flow. This was demonstrated in the work of Huang et al. (1981). The loading surfaces for all Γ-values given by eqs. 14-8 and 14-9, as constructed on the basis of experimental data according to the models suggested by the authors, all appeared to be different.

Special forms of loading function

Adopting the hardening measure as $\Gamma = \gamma_i$, the equation of loading surface 14-7 is chosen in the form of eq. 11-8. Eqs. 11-45 to 11-52 appear to be special cases of eq. 11-8 solved for τ_i when $\mu_\sigma = $ const.

Consider, as an example, eq. 11-47, rewriting it as:

$$\tau_i = \tan \psi(\gamma_i, t)[H_s + \sigma_m] \tag{14-10}$$

where H_s is a soil constant; $\tan \psi(\gamma_i, t) = A\gamma_i^m/H_s$; ψ is the angle of obliquity at the octahedral plane which varies with the amount of the deformation γ_i.

Adopting the hardening function in the form $F(\Gamma) = A\gamma_i^m = \tan \psi(\gamma_i, t)$, we obtain an equation of the loading surface, and of the limiting state surface, which is the limit for the loading surface (for $A\gamma_i^m = A\gamma_s^m = \tau_s$), as follows:

$$f(\sigma_{ij}) = \tau_i - \tan \psi(\gamma_i, t)[H_s + \sigma_m] = 0 \tag{14-11}$$

$$\Phi(\sigma_{ij}) = \tau_s - \tan \psi_s(t)[H_S + \sigma_m] = 0 \tag{14-11'}$$

where $\Phi(\sigma_{ij}) = \lim f(\sigma_{ij})$.

The loading surface satisfying eq. 14-11, if constructed on the $\sigma_1, \sigma_2, \sigma_3$ coordinates, is represented by a cone with an axis $\sigma_1 = \sigma_2 = \sigma_3 = p$ (the so-called hydrostatic axis). The cone expands with the parameter $\tan \psi(\gamma_i)$, i.e. with an increase in the deformation γ_i (Fig. 14-3). The contours of the loading surface (eq. 14-11) on the deviatoric plane give rise to a family of corresponding circles, and the contours of this surface on the τ_i-σ_m plane yield a family of straight lines all emanating from the same point, $\sigma_m = -H_s$.

Level of stress

To determine the position of the loading surface, it is expedient to use a level of stress which is understood to be the ratio $\bar{\tau}_i = \tau_i/\tau_s$. Applying this to eq. 11-47, we obtain:

$$\bar{\tau}_i = \frac{\tau_i}{\tau_s} = \frac{\tan \psi(\gamma_i)}{\tan \psi_s} \tag{14-12}$$

This ratio changes over the range $0 \leqslant \Delta\bar{\tau}_i \leqslant 1$ and, consequently, the angle of obliquity, ψ, varies between $\psi = 0$ and $\psi = \psi_s$ (Fig. 14-3), thus defining the degree of mobilization of the internal forces of resistance to shear.

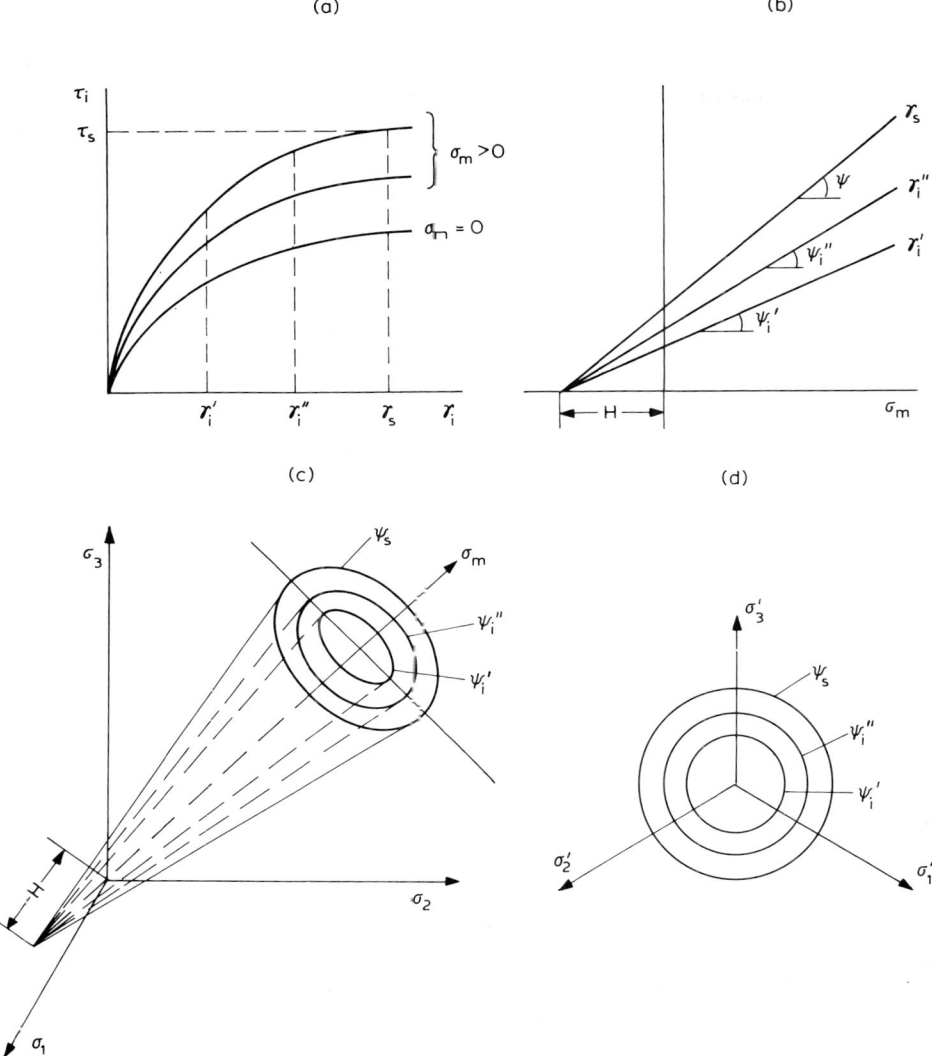

Fig. 14-3. Stress–strain curves and loading surfaces corresponding to eq. 14-11.

Various forms of the equation of limiting state

The equation of limiting state, $\Phi(\sigma_{ij}) = 0$, is one of the principal characteristics of a soil model. The classical equations of limiting state 4-12, 4-14, 4-19 and 4-23 were considered in Section 4.2, and the corresponding surfaces at failure in the σ_1–σ_2–σ_3 space and the contours of these surfaces on the deviatoric plane (one making the same angles with the σ_1-, σ_2-, and σ_3-axes) are depicted in Figs. 4-3 and 4-4. The Tresca–St.

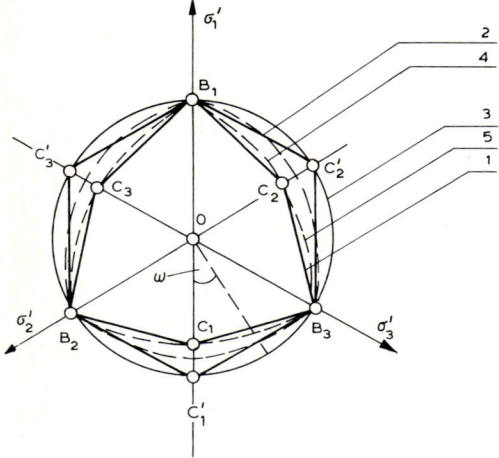

Fig. 14-4. Contours of limiting surface on the deviatoric plane: 1 = Mohr–Coulomb limiting condition, eq. 4-12; 2 = Tresca limiting condition, eq. 14-11; 3 = von Mises–Botkin limiting condition, eq. 4-23; 4, 5 = limiting conditions allowing for the arrangement of stress.

Venant equation (eq. 4-14) and the von Mises equation (eq. 4-19) describe the behaviour of ideally cohesive bodies whose limiting condition is not influenced by the all-around pressure σ_m. The surfaces corresponding to these equations take the form of a hexagon-based or circular cylinder, both of which are symmetrical with respect to the hydrostatic axis $\sigma_1 = \sigma_2 = \sigma_3 = p$ (see Fig. 4-3). The Mohr–Coulomb equation (eq. 4-12) and the von Mises–Schleicher–Botkin equation (eq. 4-23), which is sometimes referred to as the modified von Mises equation, describe the behaviour of cohesive media. These possess both cohesion and internal friction, and their limiting state is dependent upon σ_m. Add the modified Tresca equation to the above expressions:

$$\frac{\sigma_1 - \sigma_3}{2} = c + f\sigma_m \tag{14-13}$$

where f is the friction coefficient.

The surfaces of limiting condition (or surfaces at failure) described by eqs. 4-12, 4-23 and 4-13 are hexagonal-based pyramids for eqs. 4-12 and 4-13 and a cone for eq. 4-23 having the space diagonal as the axis does. In the case of cohesionless soils ($c = 0$), the pyramid and cone emanate from the origin and for cohesive soils ($c \neq 0$), the vertexes of the surfaces are a distance H from the origin along the p-axis (see Fig. 4-4). The contours of the sections of the limiting surfaces on the deviatoric plane are an irregular hexagon for the Mohr–Coulomb equation (eq. 4-12), proceeding from the relation of stresses on the slip plane, a regular hexagon for the Tresca equation (eq. 14-3), considering the maximum tangential stress, and a circle for the von Mises–Botkin equation (eq. 4-23), treating the tangential stress at the octahedral plane.

Fig. 14-4 shows all the contours obtained at the same time on the graph for comparison. The angle $\omega = \omega_\sigma$ is the angle of stress arrangement (eq. 3-50). It will be recalled that this angle is connected to the Lode parameter (eq. 3-51) by eq. 3-52. The effect of the arrangement of stress on the condition of strength is frequently expressed in terms of the Bishop parameter μ_1 (also denoted by b) correlated with μ_σ by the linear relation:

$$\mu_1 = (\sigma_2 - \sigma_3)/(\sigma_1 - \sigma_3) = 1/2(1 + \mu_\sigma) \tag{14-14}$$

providing the values $\omega_\sigma = \pi/3$, $\mu_\sigma = -1$, $\mu_1 = 0$ correspond to compression and $\omega_\sigma = 0$, $\mu_\sigma = \mu_1 = 1$, to tension.

The points $B_{1,2,3}$ in Fig. 14-4 represent the state of uniaxial compression and the points $C_{1,2,3}$ refer to uniaxial tension. As can be seen, the conditions of Tresca and von Mises–Botkin assume that the ultimate strength in both states is the same ($OB_{1,2,3} = OC_{1,2,3}$), whereas according to the Mohr–Coulomb condition the strength in compression is higher than that in tension ($OB_{1,2,3} > OC_{1,2,3}$). Note that in the case of an isotropic material all the surfaces are symmetrical with respect to the σ_1-, σ_2- and σ_3-axes on the deviatoric plane, and the difference between the ultimate strengths in compression and tension inherent in eq. 4-12 does not interfere with the isotropic condition. The real section of the limiting surface (eq. 4-12) changes with angle ϕ, becoming the Tresca condition of eq. 4-14 for $\phi = 0$ and an isosceles triangle for $\phi = 90°$.

In conclusion, it is appropriate to point out that a detailed analysis of strength conditions has been made by Paul (1968).

Experimental data

To make sure that the conditions of limiting equilibrium referred to above are applicable, various authors tested soils under different conditions of stress. Tests in a triaxi-symmetrical compression apparatus (see Fig. 11-1(a)) gave rise to two different states corresponding to $\mu_c = -1$ for $\sigma_1 > \sigma_2 = \sigma_3$ and $\mu_\sigma = 1$ for $\sigma_1 = \sigma_2 > \sigma_3$; tests with apparatus providing independently varying stresses σ_1, σ_2 and σ_3 (see Fig. 11-1(e)) were conducive to any state of stress, $-1 \leq \mu_\sigma \leq 1$.

Numerous experimental results prove that the experimental points fail to fit any of the equations of limiting state discussed; they fall on the irregular hexagons located between the Mohr–Coulomb hexagon and the von Mises circle. In some cases, the points circumscribe the Mohr–Coulomb hexagon, approaching it rather closely and coinciding with all six vertexes (curve 4 of Fig. 14-4). In other cases, the test points depart from the hexagon and coincide with it only at three vertexes, B_1, B_2 and B_3, which correspond to compression. In curve 4, the values of angle ϕ, as determined from tests in compression ($\mu_\sigma = -1$) and tension ($\mu_\sigma = 1$), are the same, increasing, however, for other stress arrangements ($-1 < \mu_\sigma < 1$). In curve 5, which is of a more general nature, the values of ϕ, as determined by the equation of Mohr–Coulomb from compression and tensile tests, are different, i.e. are higher in tension. Likewise, for all other tests ($\mu_\sigma > -1$) the values of ϕ will also be higher than those obtained from compression tests.

Reference to a contour of limiting surface in the form of curve 4 of Fig. 14-4 is found in the works of Shibata and Karube (1965), who experimented with normally compacted clay. This is mentioned in Section 12.2. The same reference is met in the works of L. Barden and his colleagues (Barden and Khayatt, 1966; Barden et al., 1969) who generalize a multitude of experiments with sand. A limiting surface of the same shape was suggested by Kirkpatrick at an earlier date (1957).

A contour of limiting surface in the form of curve 5 of Fig. 14-4 was obtained during the experiments with loose and compact sands, mentioned in Section 12.2, conducted by G.M. Lomize and A.L. Krizhanovsky (1966, 1967). The findings of Ko and Scott (1968), Green and Bishop (1969), Proctor and Barden (1969) and Lade and Duncan (1976) were similar. Henkel (1960) came to the conclusion that curve 5 in Fig. 14-4 is applicable to clays.

Some of the researchers (Bishop, 1966; Green and Bishop, 1969) reported that the contours of limiting surfaces for various soils obtained during various tests were in the form of both curves 4 and 5 of Fig. 14-4. The contours noted by other researchers experimenting with sand (Habib, 1953; Haythornwaite, 1960) were of the form of a triangle inscribed into the Mohr–Coulomb hexagon; the angle ϕ, as determined from the tensile tests, appeared to be smaller in this case than the values obtained from the compression tests. Finally, the contour of the limiting surface obtained by Lade and Musante (1978) experimenting with normally compacted clay of disturbed structure under the undrained conditions approached the Tresca–St. Venant regular hexagon (see Fig. 4-3(a)).

Thus, we come to the conclusion that the actual behaviour of both cohesive and cohesionless soils cannot be accurately described by any of the classical equations of limiting state discussed and that the experimental points on the deviatoric plane representing the contour of the section of limiting surface will fall on a curve (an irregular hexagon) located between the theoretical contours satisfying eqs. 4-12 and 4-23.

14.3 ALLOWING FOR ARRANGEMENT OF STRESS, LOADING PATH AND GEOMETRICAL NON-LINEARITY

Effect of stress arrangement

The lack of fit between the equations of limiting state and experimental data considered above originates mainly from the fact that the equations fail to take into account the arrangement of stress in the soil.

As far as the Mohr–Coulomb equation (eq. 4-12) is concerned this failure is quite obvious, for the intermediate normal stress σ_2 is omitted from the equation.

The effect of σ_2 was examined by many researchers. A particularly exhaustive treatment of the effect of σ_2 on the limiting condition of soil was given in the works of Hvorslev (1960), Bishop (1972) and Malyshev (1980).

The conclusion reached by various authors from the investigations is that the strength parameters, as determined from the Mohr–Coulomb condition (eq. 4-12), will vary with the arrangement of stress, increasing for $\mu_\sigma > -1$. This is demonstrated by Fig. 14-4. If the angle $\omega = \omega_c$ changes from 90° to 60° (this corresponds to an alteration of μ_σ from -1 to 1), the strength of soil obtained from experiments is higher than the theoretical value computed by eq. 4-12.

The question which of the curves 4 or 5 in Fig. 14-4 corresponds to the actual conditions requires further investigation. Apparently, the answer will depend on the type of soil; however, curve 5 appears to be a more general one.

Thus, the Mohr–Coulomb condition will satisfy only the case of planar deformation ($\varepsilon_2 = 0$). Naturally, the strength parameters used during the analysis should be determined in this case from the tests for a given value of μ_σ (or recalculated if the tests were conducted at another value of the parameter). However, the difference between the values of ϕ as predicted in accordance with the Mohr–Coulomb condition and obtained from the experiments for various values of σ_2 is not very significant.

With the von Mises–Botkin equation (eq. 4-23), the dependence of the strength parameters on the arrangement of stress is quite definitely traceable. The experimental and theoretical data coincide only in those cases when the arrangement of stress is the same as used during the test. This is because the von Mises limiting surface extends into the zone of negative values of σ_1, σ_2 and σ_3.

In fact, a regular circle on the deviatoric plane (curve 3 in Fig. 14-4) is described by an equation of the type $\Phi[I_2(D)] = 0$, i.e. by the von Mises equations. If the arrangement of stress is not allowed for, we obtain an equation of the type $\Phi[I_2(D), I_3(D)] = 0$. A surface having a section in the form of a curvilinear figure, the points of which lie at different distances from the centre, corresponds to this equation. Thus, by allowing for the arrangement of stress, we invariably obtain a limiting surface different from the von Mises–Botkin circular cone.

Modified equations of limiting state

To allow for the arrangement of stress, some authors have evolved modified equations of strength into which the third stress tensor invariant I_3 or one of its functions ω_σ, μ_σ, μ_1 defined by eqs. 3-50, 3-51, 14-14, respectively, is entered. Since the equations we deal with are expressed in terms of stress tensor invariants (or stress deviator invariants), it is expedient to present the von Mises–Schleicher–Botkin equation (eq. 14-23) also in the same form:

$$\sqrt{I_2(D)} = \alpha I_1(T) + R \qquad (14\text{-}15)$$

In a more general form, the relation between $\sqrt{I_2(D)}$ and $I_1(T)$ may be assumed to be non-linear, say a parabolic one. Then, we shall have:

$$\sqrt{I_2(D)} = [\alpha I_1(T) + R^2]^{1/2} \qquad (14\text{-}16)$$

The expressions $I_2(D)$ and $I_1(T)$ in the above formulae are the second stress deviator invariant and the first stress tensor invariant, respectively, as follows from eqs. 3-41 and 3-39; $R = \tau_s^0$ is the resistance to pure shear; $3\alpha = \tau_s^0/H_s$, where H_s is the cohesion, and in eq. 14-16, $3\alpha = \tan \psi_s$, where ψ_s is the angle of friction at the octahedral plane.

A modification of eq. 14-15, which allows for the arrangement of stress, has been represented by eq. 12-21 of Geniev (1977). Put in the invariant form, this equation may be written as:

$$\sqrt{I_2(D)} = \alpha[I_1(T) + R]Z(I_3) \qquad (14\text{-}17)$$

where $Z(I_3) = 1 - k \cos 3\omega_\sigma = 1 + (9 - \mu_\sigma^2)/(3 + \mu_\sigma^2)^{3/2}$.

Proceeding, however, from eq. 14-16, we obtain after Geniev:

$$\sqrt{I_2(D)} = \{[\alpha I_1(T) + R^2]Z(I_3)\}^{1/2} \qquad (14\text{-}17')$$

An equation similar to eq. 14-17, except that the $Z(I_3)$ function takes another form, has been suggested by Nagaraj and Somashekar (1977) for cohesionless soils. Modifying Bishop's (1966) equation for sands $(\sigma_1' - \sigma_3')/(\sigma_1' + \sigma_3') = f(\phi, \mu_1)$, the authors arrived at the following expression:

$$\sqrt{I_2(D)} = I_1(T)Z(I_3) \qquad (14\text{-}18)$$

where:

$$Z(I_3) = \frac{2\sqrt{2} \sin \phi}{2\mu_1 \sin \phi - \sin \phi + 3}$$

Eq. 14-18, by analogy with eq. 14-17, gives rise to a contour of limiting surface circumscribing the Mohr–Coulomb hexagon (curve 4 in Fig. 14-4). The contour of the limiting surface corresponding to the equation of strength for sand suggested by Goldsheider and Gudehus (1973) inscribes the Mohr–Coulomb hexagon rather than circumscribes it:

$$\frac{C_1 I_1(T)}{[I_2(D)]^2} + \frac{C_2 I_3(D)}{[I_2(D)]^{3/2}} = 0 \qquad (14\text{-}19)$$

A new hypothesis of cohesive soil strength has been advanced by Matsuoka (1976), Matsuoka and Nakai (1977), and Matsuoka and Ichizaki (1981). It assumes that soil failure occurs neither at the plane of maximum shearing stress (Mohr–Coulomb's hypothesis) nor at the octahedral plane (von Mises' hypothesis) but at the plane of the most probable shear which is arranged in a specific way with respect to the axes of principal stresses and referred to by the authors as a spatial mobilization plane. The condition of strength is represented in this case by a relation between the three stress tensor invariants:

$$I_1(T)I_2(T)/I_3(T) = \text{const.} \qquad (14\text{-}20)$$

This equation, like eqs. 14-17, 14-18 and 14-20, yields a contour of limiting surface in the form of a curvilinear hexagon circumscribing the Mohr–Coulomb hexagon and touching it at all six vertexes (curve 4 in Fig. 14-4).

Consider equations in which the contour of limiting surface acquires a configuration located between the von Mises circle and the Mohr–Coulomb hexagon, touching it at only three points (curve 5 in Fig. 14-4). Some of them, such as eqs. 12-22 and 12-23, were examined in Section 12.2.

Adopting a contour of limiting surface in the form of curve 5 in Fig. 14-4, Lade and Duncan (1975) and Lade (1979) suggested their version of the condition of limiting state. Accordingly, a certain ultimate value of the ratio of the first and third stress tensor invariants, $[I_1(T)]^3/I_3(T) = k_L$, is regarded as the criterion of failure. Hence, the condition of limiting state is:

$$[I_1(T)]^3 - k_L I_3(T) = 0 \tag{14-21}$$

Although the authors have shown that eq. 14-21 satisfactorily fits the experimental data, its physical meaning remains obscure because the second invariant stress deviator is omitted.

The contour of limiting surface in the form of curve 5 of Fig. 14-4 is also obtained if we proceed from the empirical equation of Ergun (1977) for the "peak" strength of sands:

$$(\sigma_1' - \sigma_3')^2 + (\sigma_2' - \sigma_3')^2 = \beta^2 [I_1(T)]^2 \tag{14-22}$$

Note, that the essence of the left-hand side of this equation is not clear, for its form corresponds to none of the stress tensor invariants. Nevertheless, the equation fits the experimental data.

A generalized equation of limiting state which gives rise to various contours of limiting surface, depending on the adopted values of the equation parameters, was derived by Ohmaki (1979). Developing Matsuoka's idea for the case of cohesive soils and assuming, by analogy with this author, that failure takes place at a plane normal to N which intersects the axes of principal stresses at the points $(\sqrt{\sigma_1 + \sigma_0})$, $(\sqrt{\sigma_2 + \sigma_0})$ and $(\sqrt{\sigma_3 + \sigma_0})$, where σ_0 is cohesion, at the instant when the ratio of the tangential and normal stresses reaches its limit $(\tau_N/\sigma_N = k = \text{const.})$, Ohmaki obtained the following equation of limiting state:

$$\frac{I_3 + \sigma_0 I_2 + \sigma_0^3}{(3I_3 + 2\sigma_0 I_2 + \sigma_0^2 I_1)^2} [2\sigma_0(I_1^2 - 3I_2) + I_1 I_2 - 9I_3] = k \tag{14-23}$$

The surface described by eq. 14-23 is a cone whose axis is the space diagonal P. For $\sigma_0 = 0$, this surface is transformed into the Matsuoka surface (eq. 14-20) circumscribing the Mohr–Coulomb hexagon; as $\sigma_0 \to \infty$, it becomes the von Mises–Botkin surface (eq. 14-15). The intermediate values of σ_0 are conducive to surfaces located between the two extremes.

Loading path during standard triaxial tests

The effect of the loading path on the results of soil tests was mentioned in Section 12.3. We shall begin by examining the shape of the loading paths commonly obtained during routine triaxi-symmetrical compression tests, using the loading pattern of Fig. 11-1(a).

As a rule, the notation used during such tests is $q = 1/2(\sigma_z - \sigma_r)$ and $p =$

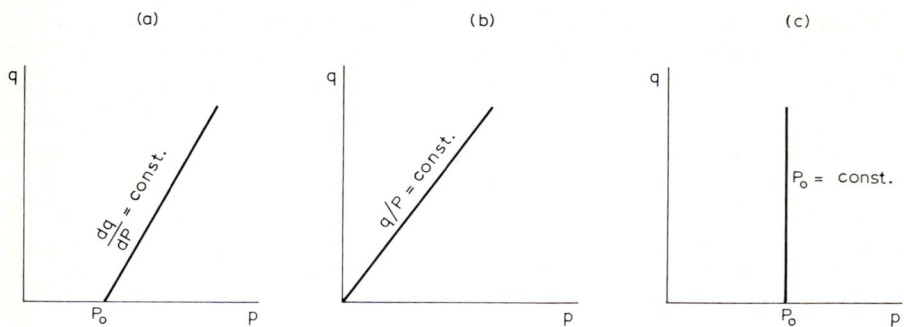

Fig. 14-5. Loading patterns of triaxial compression tests.

$1/3(\sigma_z + 2\sigma_r)$, where the quantity q is the deviatoric part of the stress tensor; according to Table 3-I, $q = \sqrt{3}\tau_i$ and $p = \sigma_m$ is the all-around pressure. During a triaxial compression test ($\sigma_z > \sigma_r$, $\mu = -1$) we have $q > 0$ and during a triaxial tensile stress ($\sigma_z < \sigma_r$, $\mu = 1$), $q < 0$. Note that since the case $\sigma_z = \sigma_1 > \sigma_r = \sigma_2 = \sigma_3$ is the principal one, it is frequently adopted that $q = 1/2(\sigma_1 - \sigma_3)$ and $p = 1/3(\sigma_1 + 2\sigma_3)$.

Typical loading paths employed during triaxi-symmetrical tests are as follows.

(1) An initial reduction of the specimen by hydrostatic (isotropic) pressure $p' = p_0^0$ is succeeded by an increase in axial pressure whereas the radial pressure $\sigma'_r = $ const. so that the ratio of the stress increments $dq' = d\sigma'_z$ and $dp' = 1/3\, d\sigma'_z$ is maintained constant, $dq'/dp' = 3$ (Fig. 14-5(a)). This is the loading path of the standard drained tests with pore pressure measurements to obtain the values of effective pressure.

(2) Proportional increase in q' and p', i.e.:

$$q'/p' = \frac{3}{2}\frac{(\sigma'_z - \sigma'_r)}{(\sigma'_z + 2\sigma'_r)} = \eta = \text{const.}$$

employed in drained tests (Fig. 14-5(b)).

(3) A specimen reduction by a hydrostatic pressure $p = p_0$ is followed by an increase in q for $p = p_0 = $ const., the magnitude of σ_r being decreased with the increase in σ_z (Fig. 14-5(c)). This is the standard undrained test.

The loading paths in use during the crushing tests under a radial pressure $\sigma_r > \sigma_z$, are of the same type as in Fig. 14-5, yet on the $(-q)$–p plane.

The test data are commonly presented in the form of curves of shearing and volumetric strains plotted on the q–ε_z and ε_v–ε_z coordinates and a q–p graph of the resistance to shear.

Since we operated above on effective stresses $\sigma'_1 = \sigma_1 - U$, etc. (see Section 8.1), it is appropriate to point out once more that the concept of effective stresses, which has been formulated by K. Terzaghi, is one of the keystones of soil mechanics. The phenomena of changes in the state of stress and strain — alterations in volume and distortion of shape, variable strength characteristics, etc. — are attributed to changing effective stresses. Therefore, whenever the behaviour of a totally or partially water-saturated soil is concerned, it should be estimated in the light of the response to the effective stresses σ_1', σ_2' and σ_3'. However, for simplicity the primes will be omitted in what follows.

Effect of the loading path

Investigations carried out by various authors have shown that the effect of the loading path on deformation depends on the type of path, being significant in one case and negligible in another. No loading path effect is observed in those cases when the increment in stress level along the entire length of the path is $d\bar{\tau}_i > 0$. Should the stress alter within a segment so that $d\bar{\tau} < 0$, the path of the lowermost value of $\bar{\tau}_i$ will lead to a minimum finite deformation.

To illustrate this, consider different loading paths on a $\tau_i - \sigma_m$ plane (Fig. 14-6) obtained by loading a medium from point M to point N as done by Lade and Duncan (1976). The initial level of stress passing through M is $\bar{\tau}_i'$; the minimum stress level during the experiment in question is denoted by $\bar{\tau}_i''$ and the level corresponding to the limiting condition by $\bar{\tau}_s$. A loading along path 1 ($d\tau_i > 0$, $\sigma_m = $ const.) followed by an unloading and a repeated loading along the same path will have no bearing on the amount of finite deformation. Loading along paths 2 and 3 (both τ_i and σ_m are changeable in this case) give rise to the same finite deformations at the final point N defined by the stress at this point. In other words, the different loading paths 1, 2 and 3 do not influence the finite deformation because $d\bar{\tau}_i > 0$ at all points of the paths. However, in loading along path 4, the level of stress $\bar{\tau}_i$ in a segment Ma becomes lower than the initial value and then decreases to a minimum at a. Although eventually increasing, the level of stress remains within the segment ab lower than the initial values. As a result, the finite deformation at point N will be less than the deformations induced by the loadings along paths 1, 2 and 3.

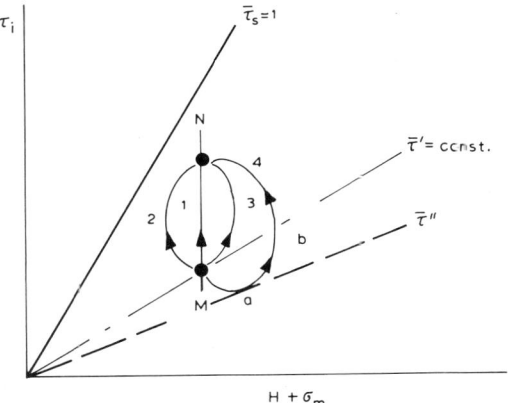

Fig. 14-6. Effect of various loading paths on finite deformation.

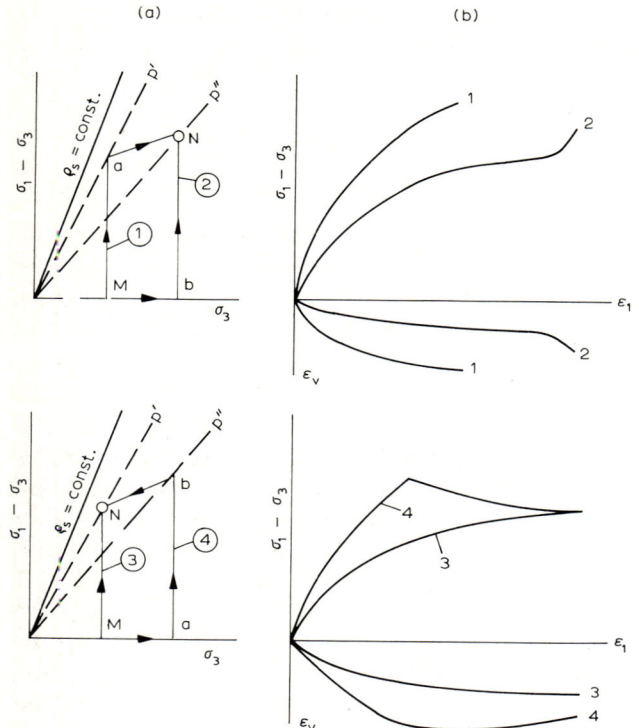

Fig. 14-7. (a) Various loading paths, and (b) corresponding axial and volumetric deformations (Lade and Duncan, 1976). Loading paths: $1 = MaN$; $2 = MeN$; $3 = MNi$; $4 = MabN$.

Consider as an example Fig. 14-7 based on the data obtained by Lade and Duncan (1976). Various loading paths on the $(\sigma_1 - \sigma_3)$–σ_3 plane, used in testing sands by means of a triaxial apparatus capable of applying independently controlled stresses, are depicted. Together with the paths, the corresponding axial ε_1 and volumetric ε_v strains are shown. The level of stress adopted by the authors in accordance with the condition of strength (eq. 14-21) is $\varrho = [I_1(T)^3/I_3(T)]$, varying over the range $0 \leqslant \varrho \leqslant k_L$.

The loadings from point M to point N along paths 1 and 2 gave rise to different finite deformations which were greater along path 2 than path 1, because in this latter case the stress along segment aN decreased compared with the initial level ($\varrho' > \varrho''$), whereas along the entire path 2 $d\varrho > 0$. Loading paths 3 and 4 led to finite axial strains of the same amounts for $d\varrho > 0$ at all points along them.

The effect of the loading path on the strength of the soil is a problem requiring further clarification. One may just say that, according to early experiments of Bishop and Eldin (1953) with sand, this effect is felt for $d\bar{\tau}_i > 0$ within the usual spread of experimental data.

Loading path and creep

Interesting data about the effect of the loading path on the rate of deformation in creep were obtained by Yudhbir and Mathur (1977) from triaxial tests ($\sigma_1 > \sigma_2 = \sigma_3$) of soil in the drained condition, using various loading paths. An initial load $q/p =$ const. applied to all specimens constituting a batch was maintained until given stresses $\sigma_{1(0)}$ and $\sigma_{3(0)}$ were achieved. Next, the specimens wre loaded along different paths defined by various values of the ratio $\bar{\eta} = \Delta\sigma_1/\Delta\sigma_3$. Measurements of the rate of axial (ε_1) and volumetric (ε_v) strains during the test were made in terms of the measures of deformation adopted, on the assumption that creep obeys the logarithmic law of eq. 5-21, in the form:

$$C_{(\varepsilon_1)} = \frac{\Delta\varepsilon_1}{\Delta(\lg t)} \quad \text{and} \quad C_{(\varepsilon_v)} = \frac{\Delta\varepsilon_v}{\Delta(\lg t)}$$

It appeared that the rate of creep was significantly influenced by the axial-to-radial loading relation, i.e. by the loading path. For $\bar{\eta} \to \pm 0$, the rate of axial strain is at a minimum, for all-around pressure prevails in this case. For $\bar{\eta} \to \infty$, the rates acquire maximum values because the deviatoric stress dominates. These data establish the dependence of the logarithmic rate of the shearing strains $C_{(\varepsilon_1)}$ and the volumetric strains $C_{(\varepsilon_v)}$ on the value of $\bar{\eta}$. An increase in $\bar{\eta}$ leads to a linear increase in $C_{(\varepsilon_1)}$ which is evidence that the rate of creep increases uniformly. When $\bar{\eta}$ reaches a certain ultimate value, $\bar{\eta}_s$, the strain rate $C_{(\varepsilon_1)}$ increases to infinity and progressive flow is formed culminating in failure. The rate of volumetric strain appears to be independent of the loading path for $\bar{\eta} < \bar{\eta}_s$ but sharply decreases for $\bar{\eta} = \bar{\eta}_s$.

According to the author, the development of the axial deformation of creep is described by the equation:

$$dC_{(\varepsilon_1)} = d\left[\frac{\Delta\varepsilon_1}{\Delta(\lg t)}\right] = \frac{\partial C_{(\varepsilon_1)}}{\partial p}dp + \frac{\partial C_{(\varepsilon_1)}}{\partial q}dq = A + \frac{a}{q_s - a}dq$$

where $A =$ const. and q_s is the ultimate value of q.

Numerical solutions

Consider two problems of the state of stress and strain of soil solved numerically, with allowance for the loading path, by means of the computer.

One of them dealt with the stability of an earth dam at a reservoir. Computations carried out when the reservoir was filled with water at the time of construction and on completion of the dam showed that the displacements of its core were different (Zaretsky et al., 1980).

The other problem consisted of assessing the behaviour of the soil below a strip foundation sustaining a vertical and a horizontal loading (Bugrov, 1980). Two loading patterns were scrutinized; simultaneous application of the vertical and horizontal loadings and the application of the horizontal load in the wake of the vertical one. It was found that in the former case the sidewise displacement of the flexible foundation was 125% of the displacement computed in the latter case. The average settlement appeared to be almost the same in both cases but was more uniform when the successive loading was applied. At the beginning, plastic deformations were induced in zones of different configuration but this difference became insignificant at the final stages of loading.

The solution of a similar problem for a rigid foundation revealed that the contact stresses differed by 10 to 30%.

The examples presented also prove the effect of the loading path. However, the difference between the stresses and strains produced by a load amounting to 80% of the ultimate value was not greater than 30% for both loading paths. Such a discrepancy is not regarded as very significant for soils.

To make an allowance for the loading path is a difficult task when the deformation theory of plasticity is used, because the cases of combined loading — including the application of a successive external loading — are outside the scope of the theory. The theory of plastic flow, on the other hand, takes account of all types of loading. Consequently, one may follow any loading path by integrating the deformation increments. However, in solving engineering problems with allowance for successive loading in terms of both theories, we obtain final results which, in general, do not differ very much. Therefore, special restriction on the use of deformation theory as a means of allowing for the loading path are hardly necessary. Of course, in those cases when the loading path is very intricate and its effect needs special evaluation, recourse to the theory of plastic flow is necessary.

Fulfilling the condition of similarity and alignment of stress and strain tensors

This aspect was examined in Section 12.8. Here we shall point out that the lack of alignment between the tensors of stress and strain is caused by the anisotropic properties of soil. As shown by De Jong (1967), the condition of alignment can be fulfilled in ideally isotropic media only.

A number of researchers have evolved non-coaxial soil models capable of taking into account the difference between the directions of the stress vector and strain (or rate of strain) vector. These models were examined in the works of Mandel et al. (1970) and Nikolaevsky (1972).

At the same time, the experiments of Roscoe (1970) and the investigations of the author outlined in Section 13.6 have shown that the difference between the angles through which the vertical axes of the stress and strain vectors rotate, is significant only in the regions where the state of stress approaches its limiting condition; in the zones of subliminiting condition the difference is small and can be neglected.

As far as the tensors of stress and strain increment are concerned, the problem of misalignment does not exist at all, for, as is shown in the work of Roscoe referred to above, these tensors are coaxial. The same conclusion was reached by H. Poorooshasb et al. (1966, 1967). Their experimental findings showed that the direction of the displacement vector at the fringe of the surface of limiting state was decided by the stress tensor irrespective of the loading path.

Thus, the question whether or not non-coaxial models should be developed may arise only if use is made of deformation theory based on the condition of similarity and alignment of the tensors of stress and strain. No such question exists in conjunction with the theory of plastic flow which stipulates the alignment of the tensors of stress and strain increment (not total strain). This condition is fulfilled in the case under consideration.

Effect of geometrical non-linearity

Among the particulars of soil deformation referred to in Section 14.1, the non-linearity of the geometrical relations 3-13 attributed to big (finite) deformations was mentioned. These are induced in the zones of maximum stress which are in a state approaching the limiting one; for example, as happens at the edges of a rigid foundation.

The significance of the effect of geometrical non-linearity is demonstrated by the results of the numerical computer-assisted calculations made by Kopeikin and Solomin (1977) in connection with assessing the state of stress and strain of the sand below a rigid strip foundation. The problem was solved in terms of the deformation theory of plasticity. Consequently, the physical equations were adopted in the form given by eqs. 12-16 and 12-12 — in the last named formula both the positive and negative dilatancy was taken into account, and the equations of equilibrium and geometrical relations were obtained from the statements of strain and stress as presented in eqs. 3-13 and 3-18, respectively.

According to eq. 3-13, the strain components are the sum of linear (ε') and non-linear (ε'') constituents, $\varepsilon_x = \varepsilon'_x + \varepsilon''_x, \ldots$. The results of the computations showed that the deformation at the edges of the foundation was high. Thus, the values of ε_x induced in response to the loadings $p/p_{\lim}$ equalling 0.29, 0.67 and 0.35 were $-0.011 + 0.001 = -0.010$, $-0.0030 + 0.013 = -0.017$ and $-0.250 + 0.350 = 0.100$, respectively. The allowance for geometrical non-linearity revealed changes in the configuration of the reaction pressure curves, their ordinates somewhat increasing at the foundation centre and sharply decreasing at the edges. Geometrical non-linearity also influenced settlement. So, a relative settlement, $S/b = 0.1$, obtained by making an allowance for this non-linearity, was caused by a load which was only half of the load bringing about the same settlement when geometrical non-linearity was disregarded; the ultimate load in the former case was half that computed for the latter case.

This effect of geometrical non-linearity has appeared to be surprisingly significant, calling for further studies of the problem.

14.4 EQUATIONS OF DEFORMATION BASED ON THE DEFORMATION THEORY OF PLASTICITY AND THE THEORY OF PLASTIC FLOW

In developing models, one may proceed from either the deformation theory of plasticity or the theory of plastic flow. Consequently, we shall deal with the real deformations in the former case and with the deformation increments (or deformation rates) in the latter one (see Section 3.7).

Deformation theory

The equation defining the theory and establishing the relation between the components of stress and strain tensors may be set forth in the form given by eq. 11-13. Alternatively, it may be presented in the tensor form as:

$$\varepsilon_{ij} = \chi(\sigma_{ij} - \delta_{ij}\sigma_m) + \chi^*\delta_{ij}\sigma_m \tag{14-24}$$

where δ_{ij} is the Kronecker symbol ($\delta_{ij} = 0$ for $i \neq j$ and $\delta_{ij} = 1$ for $i = j$) and $\chi = \gamma_i/2\tau_i$ and $\chi^* = \varepsilon_m/\sigma_m$ are scalar functions defining the relation between the stress and shearing and volumetric strains (including those associated with dilatancy), respectively; their value is determined by expression 11-14. Eq. 14-24 incorporates, according to eq. 14-1, both the elastic and plastic components of the shearing and volumetric strains.

With the transition into the limiting state, eq. 14-24 takes the form given in eq. 14-5'.

It is thus evident that the models based on deformation theory are capable of allowing for the non-linearity of the deformation process, creep and the effect of all stress tensor invariants. However, as pointed out above, the use of deformation theory is justified only for a simple loading, or one close to that, and also in cases when the condition of alignment and similarity of the stress and strain tensors is fulfilled (see Section 3.7).

Theory of plastic flow

The constitutive equations of theory establish a connection (referred to as an incremental defining one) between the components of stress tensor and strain increment tensor (or strain rate tensor). Both the elastic and plastic deformations are taken into account, as in eq. 14-1. The relation between stress and increment in the elastic strain component obeys Hooke's law (eq. 3-101):

$$d\varepsilon_{ij}^e = \frac{1}{2G}\left(d\sigma_{ij} - \frac{3v}{1+v}d\sigma_m\delta_{ij}\right) \tag{14-25}$$

and the relation between stress and increment in the plastic strain component is described by the equation:

$$d\varepsilon_{ij}^p = (\sigma_{ij} - \delta_{ij}\sigma_m)\,d\lambda + \delta_{ij}\sigma_m\,d\xi \tag{14-26}$$

where $d\lambda$ and $d\xi$ are the functions (scalar) establishing a connection between the components of the stress tensors and the tensors of the increments in plastic shearing and plastic volumetric strains:

$$d\lambda = \frac{d\gamma_i^p}{2d\tau_i} = \frac{f(\tau_i, \sigma_m, \mu_\sigma, t)}{2\tau_i}$$

$$d\xi = \frac{d\varepsilon_m^p}{\sigma_m} = \frac{f^*(\sigma_m, \tau_i, \mu_\sigma, t)}{\sigma_m} \qquad (14\text{-}27)$$

The limiting condition is described by eq. 14-5' as in deformation theory. Moreover, to allow for the phenomenon of dilatancy entailing — with the onset of the plastic state — an irrecoverable change in the volume of soil $d\varepsilon_n^F$ in response to an increment in the shearing strain $d\gamma_i$, a condition of the type given in eq. 12-4 is sometimes introduced into the theory of plastic flow:

$$d\varepsilon_m^p = |\dot{\Lambda}\, d\gamma_i^p| \qquad (14\text{-}28)$$

where $\dot{\Lambda}$ is the dilatancy rate.

Loading function and plasticity potential

The concept of plasticity potential was introduced in Section 3.7. Denoted here as f_p, the plasticity potential is connected to the increment in plastic strain by eq. 3-103. This equation may be put in general form as:

$$d\varepsilon_{ij}^p = d\lambda_p \frac{\partial f_p}{\partial \sigma_{ij}} \qquad (14\text{-}29)$$

where $d\lambda > 0$ is an infinitesimal scalar factor. Note that when the loading surface is not smooth, e.g. piecemeal linear, its angles (and faces) form an assembly of special points. In the case of a singular surface, the law of eq. 14-29 is expressed in the form (Koiter, 1960):

$$d\varepsilon_{ij}^p = \sum_r d\lambda_p^r \frac{\partial f_p^r}{\partial \sigma_{ij}} \qquad (14\text{-}30)$$

Distinction is made between the associated (after von Mises) and non-associated laws of plastic flow. The associated law presumes that the plasticity potential coincides with the loading function $f(\sigma_{ij})$ and, taking into account eq. 14-6, also with the failure function $\Phi(\sigma_{ij})$ but in the limiting state only, i.e. $f_p = f = \Phi$. Under the non-associated law, the coincidence between the plasticity potential and the loading (and failure) function does not exist, $f_p \neq f$.

Associated law of plastic flow

Drucker and Prager (1952) have applied the associated law to an elastoplastic (non-hardening) medium. The loading function for the medium, which is also the yield function, is defined by the condition:

$$f = \Phi[I_1(T), I_2(D)] \qquad (14\text{-}31)$$

The Φ-function in the Drucker–Prager model is adopted by proceeding from the von Mises–Schleicher–Botkin condition of strength (eq. 14-15). Note that this condition is sometimes referred to as the Drucker–Prager condition because it was applied in the model of these authors.

Substituting $f_p = f$ into eq. 14-29, we arrive at the following expression establishing connection between the plastic strain increment and stress:

$$d\varepsilon_{ij}^p = d\lambda \left\{ \alpha \delta_{ij} + \frac{1}{2\sqrt{I_2(D)}} \left[\sigma_{ij} - \frac{I_1(T)}{3} \delta_{ij} \right] \right\} \qquad (14\text{-}32)$$

Hence, we obtain the relation:

$$d\varepsilon_{ii}^p = d\varepsilon_v^p = 3\alpha \, d\lambda \qquad (14\text{-}33)$$

which defines the unlimited increases in the volumetric strains during an unlimited plastic flow (in shear).

From eqs. 14-32 and 14-33 we can obtain the equation of dilatancy 14-28. To that end, we find the rate of plastic strain $\dot{J}_2(D)$ and substitute it into eq. 14-33, adopting $6\alpha = \dot{\Lambda}$.

The associated law of plastic flow has several distinctions. Mainly, the product of the increments in plastic strains and stress is, according to Drucker's postulate, always non-negative $d\sigma_{ij} \, d\varepsilon_{ij}^p \geq 0$ (where the symbol "equal to" refers to the case of ideal plasticity, $\tau_s = $ const.). Hence, the loading surface Σ will always be convex. Another distinctive feature is that the vector of plastic strain increment $\overrightarrow{d\varepsilon_{ij}^p}$ is orthogonal relative to the loading surface Σ; this implies that a plastic flow must always progress perpendicularly to the yield surface (see Fig. 14-2).

Under the associated law, the dilatancy rate $\dot{\Lambda}$ in eq. 14-28 must also be related in a strict way with the strength parameter entering the yield condition adopted, e.g. the Mohr–Coulomb or the von Mises–Botkin condition:

$$|\dot{\Lambda}| = \sin\phi \quad \text{and} \quad |\dot{\Lambda}| = \tan\psi \qquad (14\text{-}34)$$

Finally, it follows from eq. 14-32 that in the Drucker–Prager model the volumetric changes take place only in the form of an increase in the volume of the soil (dilatancy of softening); no dilatancy of extra hardening is observed.

Experimental verification of the condition of perpendicularity

The condition of perpendicularity of the plastic strain increment vector $\overrightarrow{d\varepsilon_{ij}^p}$ to the various loading surfaces was tested by a number of authors. However, the results obtained were contradictory. When testing soil samples in an apparatus capable of applying independently controlled stresses, Ivastchenko and Zakharov (1973), came to the conclusion that the vector $\overrightarrow{d\varepsilon_{ij}^p}$ satisfactorily coincides with a normal to the loading surface in the form of an ellipse extending along the σ_1-axis and expanding symmetrically with an increase in the strain. An experimental verification of the perpendicularity of the plastic strain increment vector to a family of closed surfaces by using the von Mises–Botkin cone as a limit, was also obtained by Didukh and Ioselevich (1971).

At the same time other authors (see Section 12.4: Nikolaevsky, 1972) assert that the $\overrightarrow{d\varepsilon_{ij}^p}$ vectors fail to be orthogonal to the traditional loading surfaces Σ. Experimenting with sand, Lade and Musante (1978) showed that the vectors of plastic strain increments were perpendicular to the contour of the failure surface (eq. 14-21) on the octahedral plane but made an acute angle with a normal to the loading surface constructed in the space of effective stresses.

Proof that soil fails to satisfy the condition of perpendicularity was obtained by Poorooshasb et al. (1966, 1967) from measurements of plastic strain vectors during the triaxial compression tests of soil at various loading paths. The plotted equipotential lines, i.e. a family of lines making right angles with the vectors, took the form of ellipses which failed to coincide with the Mohr–Coulomb straight line. This was the reason for the conclusion about the lack of coincidence of the Mohr–Coulomb yield surface with the plastic equipotential lines (see Section 12.4: Nikolaevsky, 1972).

Experimental verification of the fulfillment of eqs. 14-34

This verification, carried out by Roscoe (1970) and other researchers, has shown that the dilatancy rate $\dot\Lambda$ is by far smaller than the friction parameters $\tan\psi$ and $\sin\phi$. In other words, eqs. 14-34 are not fulfilled in real soils. To eliminate this incongruity, some authors have suggested to correct the value of plasticity potential f_p in eq. 14-29, without rendering it inconsistent with the loading function f, by correcting the value of α in the condition of strength of eq. 14-15. According to Stroganov (1966), this can be done by substituting the value of $\dot\Lambda$, as determined from special dilatancy tests, for $\alpha = 1/3 \tan\psi$ in eq. 14-15. Negre and Shutz (1870) also thought it advisable to correct the value of the angle ϕ in the equation of strength, thus bringing the condition of strength into agreement with the plasticity potential.

Thus, it can be taken for granted that eqs. 14-34 fail to be fulfilled. However, the impact of this failure on the solution of engineering problems needs clarification. We will now look at the numerical solutions obtained by Bugrov (1980) in assessing the stressed and strained behaviour of the soil below a strip foundation. It was assumed that the soil (compact sand, $\phi = 22.8°$) obeyed the elastoplastic law and that the Mohr–Coulomb condition (eq. 4-12), as well as the law of plastic flow (eq. 14-29), were applicable in the plastic region. The maximum load sustained by the soil was adopted at 80% of the ultimate strength. Two cases were examined, one with the dilatancy rate value $\dot\Lambda = \sin\phi$ (associated law) and the other with $\dot\Lambda = 0.5 \sin\phi$ (non-associated law). It appeared that even a significant difference between the values of $\dot\Lambda$ used during the calculations did not materially influence the behaviour of the soil. For $\dot\Lambda = 0.5 \sin\phi$, settlement increased by only 10%, the vertical stresses σ_z differed by 12% and the contact stresses τ_{xz} by 8% compared with $\dot\Lambda = \sin\phi$. The curves of contact stresses were of a more non-uniform configuration in the former case than in the latter.

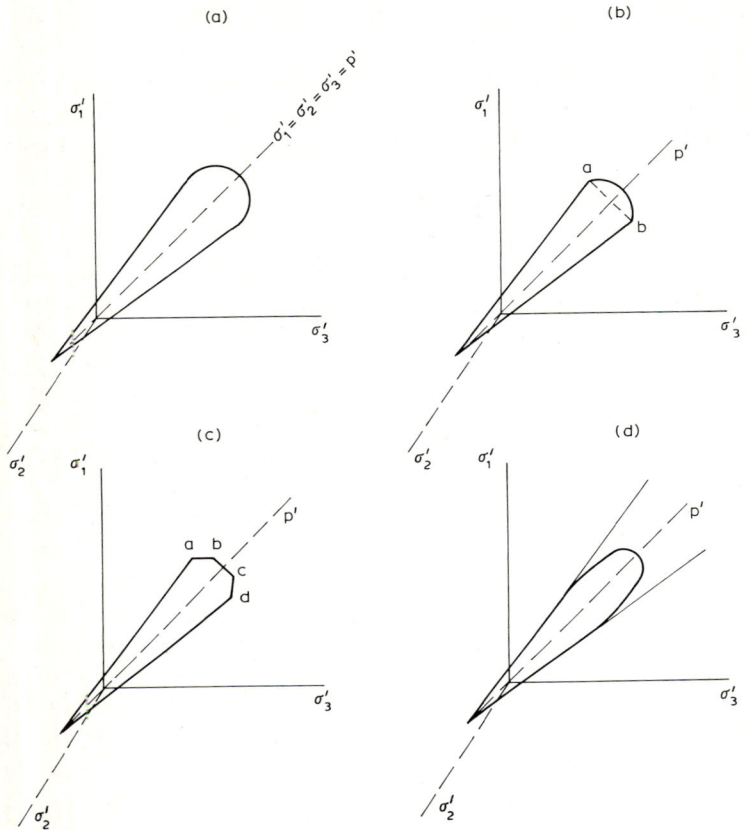

Fig. 14-8. Closed loading surfaces. (a) Cap joining the von Mises–Botkin cone flush with its side surface. (b) Cap joining the cone at an angle to its surface. (c) Cap in the form of a piecemeal linear surface. (d) Loading surface of smooth closed form.

Allowance for extra compacting

An essential limitation of the Drucker–Prager model examined above is the fact, already referred to, that eq. 14-33 describes only the process of increase in soil volume due to the dilatancy of softening and is incapable of defining volumetric shrinking caused by the dilatancy of extra compacting. This is obviously inconsistent with the behaviour of soil, for, as has been proved by numerous experiments, dilatancy may have two different signs; the phenomenon of extra compacting especially occurs in loose soils.

To eliminate the limitation, Drucker et al. (1957) suggested to supplement the von Mises–Botkin cone, representing the surface of limiting state in the $(\sigma_1-\sigma_2-\sigma_3)$ space by a convex cap arranged to complement the limiting surface, i.e. to bridge the gap therein (Fig. 14-8(a)). It was assumed that the cap fitted the cone flush with its side surface. When pressure p increased, the cap shifted along the hydrostatic axis. An increment in the plastic strains directed towards the cap brought about extra compacting of the soil and an increment towards the side surface of the cone resulted in softening. However, in general, the vectors of strain increments were at right angles to the surface. This implies that the associated law holds for both the side surface and the cap.

Later, a number of authors evolved caps of other shapes to complement the limiting cone (see Section 12.4: Nikolaevsky, 1972). In some of the works, these caps were presented in the form of a spherical surface, as in the Drucker-model. However, instead of being flush with the side surface, these caps intersected it so that singular points were formed at the intersections (Fig. 14-8(b)). Among models of this kind there were those capable of representing the hardening of soil as, for example, the model by Sandler and Di Magio (1976). In this model, the loading surface consists of two parts: a cone of the von Mises–Botkin type, having, however, a curvilinear side surface, and an ellipsoid cap. A forward shift of the cap corresponds to volumetric hardening (extra compacting), softening takes place at the side surfaces and the vertexes are the points of incompressibility. The associated law holds for this entire surface.

A modification of the above model is described in the work of Balady and Rohani (1973). It will be noted that the well-known Cam-Clay model of K. Roscoe and fellow workers, which is discussed below, is also based on a closed loading surface.

A. Jenike and R. Shield and also D.D. Ivlev and V.V. Dodukalenko (see Section 12.4: Nikolaevsky, 1972) supplemented the von Mises–Coulomb limiting surface with a flat cap located at right angles to the hydrostatic axis (shown by the dotted line in Fig. 14-8(b)). Zaretsky et al. (1980) suggested supplementing the cone with a singular, piecemeal smooth surface (Fig. 14-8(c)), which meets the requirement of perpendicularity at all points including the sides.

At the same time there a models (Lade, 1979; Tanaka, 1979) which obey the non-associated law along the side surface and the associated law eat the cap.

Finally, it is appropriate to mention the loading surface in the form of the smooth curvilinear and closed figure of Fig. 14-8(d). A surface of this kind represented by a lemniscate of revolution (rotating about the hydrostatic axis) expanding with hardening of the soil was employed by Suh (1969). A similar surface in the form of a family of closed surfaces transformed into the von Mises–Botkin cone with the onset of limiting state was used by Ioselevich et al. (1975). In each case the associated law was obeyed.

Non-associated law of plastic flow

As already mentioned, the non-associated law is based on the difference between the types of the loading function $f = \Phi$ and the plasticity potential f_p. Relevant problems are solved either by specifying a plasticity potential different from the loading function (the potential is specially determined experimentally) or by employing — in addition to the constitutive eq. 14-36 — two mutually independent conditions: the condition of limiting state (eq. 14-15) and that of dilatancy (eq. 14-28). In this latter case, the parameter $\dot{\Lambda}$ in eq. 14-28 needs additional experimental determination. Soil models based on the non-associated law were suggested by V.N. Nikolaevsky, P. Lade, J. Duncan and others. They are the subject of subsequent discussions.

Assessing the assets and limitations of the models which employ the associated and non-associated laws of plastic flow, we come to the following conclusion. Although a general and precise law, the non-associated law is complex and involves labour-consuming determination of additional parameters. The associated law is less general but quite suitable for solving most of the problems, especially those in the engineering field, albeit with due regard to the reservations mentioned.

Numerical methods of solving nonlinear problems

Numerical solutions of problems are commonly obtained in soil mechanics by using the finite difference method or the finite element method.

The finite difference method is based on substituting a system of algebraic equations in the finite difference form for the differential equations. Referring to special literature (Vinokurov and Karamyshev, 1972), we note that the iterative method developed by Vinokurov to solve non-linear problems by the finite difference method is based on Il'yushin's method of elastic solution and makes use of equations from the deformation theory of plasticity. Presenting eqs. 3-78 and 3-14, or eqs. 7-78 and 3-80, in the finite difference form, we examine them in conjunction with the physical eq. 3-90, which is reduced to a hypothetically linear form. To this end, we either introduce variable values of the parameters $\bar{G} = \tau_i/\gamma_i = G(x, y, z)$ and $k = \sigma_m/\gamma_i = k(x, y, z)$ into eq. 3-90 or use Il'yushin's function ω in accordance with eq. 4-27. Solving thus the problem in every nth approximation in terms of elasticity theory with the variable parameters and determining from these approximations the stress and strain at every point of the medium, we find the values of the parameters which are used as the original parameters in the successive approximation and so on.

The finite element method (Zienkiewicz, 1967), given preference nowadays, proceeds from the variational principle on the assumption that the expression for the total potential energy should satisfy the minimum condition:

$$Э = Э(U, V, W) - Э(P, Q) \tag{a}$$

where $Э(U, V, W)$ is the potential energy of system deformation and $Э(P, Q)$ is the force potential. In examining this expression, we obtain the constitutive equation of the finite element method in the matrix form:

$$\{F\} = [k]\{U\} \tag{b}$$

where $\{F\}$ and $\{U\}$ are the matrices (vectors) of the generalized forces and generalized displacements, respectively, and $[k]$ is the generalized matrix of system rigidity correlating the forces and displacements. Expressing eq. b as a set of linear algebraic equations which, on solving, yield the displacements of the node points of each grid element (provided the problem is solved in terms of displacements), we find the corresponding strains and stresses. The generalized matrix $[k]$ in eq. b is then determined as the combination of the matrices of rigidity $[D]$ of the grid elements which break down the region under consideration, proceeding from a given law interconnecting the stresses and strains:

$$\{\sigma\} = [D]\{\varepsilon\} \quad \text{or} \quad \{\Delta\sigma\} = [D^{ep}]\{\Delta\varepsilon\} \tag{c}$$

It is adopted in this case that:

$$\{\Delta\varepsilon\} = \{\Delta\varepsilon^e\} + \{\Delta\varepsilon^p\} \tag{d}$$

The matrix $[D^{ep}]$ defining elastoplastic deformation can be found from Hooke's law and equations of the type given in eq. 3-98*[1] or 3-103:

$$\varepsilon^p = \Delta\lambda_p S \quad \text{or} \quad \Delta\varepsilon^p = \Delta\lambda_p \frac{\partial F}{\partial \sigma}$$

In this latter case, both the associated and non-associated laws hold.

The problem-solving methodology based on the finite element method was developed by many authors. It is appropriate to mention the Rosalie programme developed at Laboratoire Central des Ponts et Chaussées between 1968 and 1976 and reported at the French–Soviet Colloque de Saint-Maximen, Paris, 1978. Employing the finite method, it enables the solution of various problems in continuum mechanics including those associated with the theories of elasticity, elastoplasticity, viscoelasticity, thermoelasticity, dynamics, discontinuous medium, percolation in porous and cracked media and two- and three-dimensional consolidation. Some other methods used in solving boundary-value problems of soil mechanics are mentioned later in discussing the obtained solutions.

14.5 SOIL MODELS SURVEY*[2]

Models based on the deformation theory of plasticity

The preceding sections were devoted to the study of principles and constitutive relationships used in constructing non-linear soil models. Taking into account other factors, models were suggested by various authors. Models providing analytical solutions in the final form (proceeding under certain limitations from the power law) were discussed in Chapter 13. This section deals with more complex models paving the way for solving boundary-value problems on a computer. Let us start with models based on deformation theory.

Available at present is a wealth of experimental data providing the basis for constructing such models. In the U.S.S.R., one of the most comprehensive experimental studies has been carried out at the Moscow Civil Engineering Institute (see

---

*[1] Unlike the matrix $[D^p]$, the symbol D_d^p in eq. 3-98 means the stress deviator.
*[2] V.N. Razbegin took part in compiling the survey.

Section 11.8: Lomize et al., 1966, 1968; Section 12.4: Lomize et al., 1967, 1969; Lomize and Sukhanov, 1973, 1974; Krizhanovsky et al., 1975). They were devoted to testing sands and clays in the state of combined stress (loading patterns of Fig. 11-1, (a), (d) and (e)), with allowances for the dilatancy, the arrangement of stress, the loading path, the rotation of the principal stress axes, creep, etc.

The data thus acquired were conducive to tackling the problem of non-linear behaviour of a soil in the state of stress and strain below a rigid circular or strip foundation in terms of the deformation theory of plasticity. This problem has been solved at the Research Institute for Foundations and Underground Structures and the Chelyabinsk Polytechnic Institute, using the computer and the finite element methods in conjunction with the iteration method and the method of variable parameters (Shirokov et al., 1970; Malyshev et al., 1973). The original equations of state employed in this case were as follows.

(a) The law of the distortion of shape was adopted in the form of a linear–fractional relation:

$$\tau_i = \frac{\tau_{i(s)} \gamma_i}{B + \gamma_i} \tag{14-35}$$

Here:

$$\tau_{i(s)} = (\tau_s^0 + \sigma_m \tan \psi) Z(I_3) \tag{14-36}$$

where $\tau_{i(s)}$ and τ_s^0 are the ultimate shearing strengths in the state of combined stress and pure shear, respectively:

$$Z(I_3) = [1 + \alpha(1 + \cos 3\omega)^n]^{-1} = \left\{1 + \alpha\left[1 + \frac{\mu_\sigma(9 - \mu_\sigma)^2}{(3 + \mu_\sigma^2)^{3/2}}\right]^n\right\}^{-1} \tag{14-37}$$

Eq. 14-36 is the von Mises–Botkin condition of strength in which the factor $Z(I_3)$ allowing for the arrangement of stress is introduced. Note that parameter B equals, in accordance with eq. 4-36, $\tau_{i(s)}/G_0$ so that function $Z(I_3)$ must also be introduced into the expression for B.

(b) The equation of volumetric deformation was adopted in its simplest form without allowance for dilatancy:

$$\varepsilon_m = \frac{1}{d} \sigma_m^{1-\kappa} \tag{14-38}$$

where d and κ are parameters.

Thus, the problem took account of non-linearity, the effect of all-around pressure on the shearing deformation, the variability of Poisson's ratio (attributed to different laws which the distortions in shape and volumetric changes obey) and the arrangement of stress; the dead weight of the soil was also taken into consideration but dilatancy and loading path were disregarded. The condition of similarity and alignment of the stress and strain tensors were postulated, as prescribed by the deformation theory. In solving the problem, the values of the parameters of eqs. 14-36–14-38 were varied to find out their effect.

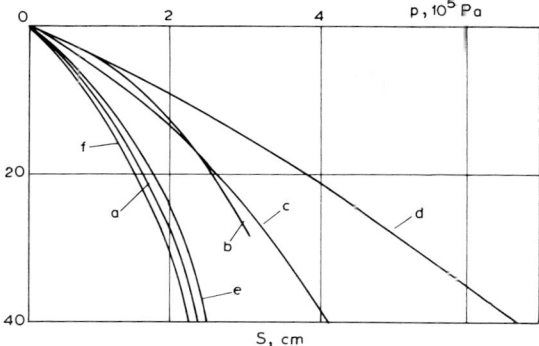

Fig. 14-9. Settlement–load relationships calculated by eqs. 14-35–14-38, using the following values of parameters: (a) $\tan \psi = 0$, $\tau_s^0 = 10^5$ Pa; (b) $\tan \psi = 0$, $\tau_s^0 = 2 \cdot 10^5$ Pa; (c) $\tan \psi = 0.45$, $\tau_s^0 = 10$ Pa; (d) $\tan \psi = 0.96$, $\tau_s^0 = 10^5$ Pa; (e) $\tan \psi = 0.96$, $\tau_s^0 = 0$; (f) $\tan \psi = 0.96$, $\tau_s^0 = 0$; $Z(I_3) = f(\mu_\sigma)$ according to eq. 14-37 (Malyshev et al., 1973).

The results, in general coinciding with the experimental data considered in Section 13.6, have shown the importance of taking into account the non-linear mechanism of soil deformation with due allowance for the mean normal stress σ_m. The settlement versus load curves for a circular foundation, as computed by eqs. 14-35–14-38 for various values of $\tan \psi$ and τ_s^0, are given in Fig. 14-9. In all cases a constant parameter B was used. The effect of $Z(I_3)$ was allowed for only in the case of curve f, in the rest of the computations $Z(I_3) = 1$ was adopted, implying that $\mu_\sigma = -1$.

From Fig. 14-9 we can see that neglecting to take into consideration friction (curves a and b) or cohesion (curves e and f) results in a significant overestimation of the settlement. The arrangement of stress $Z(I_3)$, if disregarded, has no bearing in this case, because the adopted value $\mu_\sigma = -1$ corresponds to the condition of loading of the soil by a circular settlement plate. For a planar problem (strip foundation), the effect of μ_σ appears to be significant.

The computations have shown that the settlement varies inversely with the width and vertical displacement of the settlement plate. The relation between the settlement and foundation width adopted in elasticity theory is not observed, and the depth of the soil layer compressed below the foundation appears to be much less in this case than computed by the theory of elasticity. The stresses attenuate with depth at a quicker rate, particularly when the effect of σ_m is taken into account. Referring to curve f of Fig. 14-9, the depth of the compressed layer (accounting for 94% of the settlement) was, for example, three times the radius of the settlement plate. The curve of the reaction pressures under the footing materially differs from that corresponding to elasticity theory, the edge ordinates having finite values and the curve proper being significantly smoothed (Fig. 14-10).

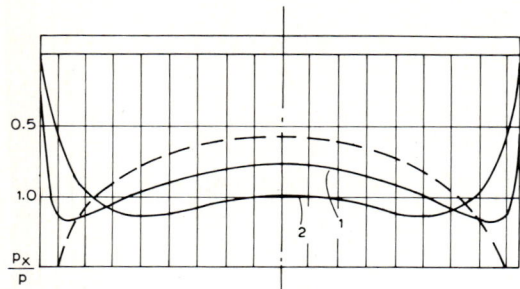

Fig. 14-10. Curves of the reaction pressures below a rigid circular settlement plate resting on sand and sustaining the loads: (1) $p = 10^5$ Pa; (2) $p = 2 \cdot 10^5$ Pa. The results of calculations by elasticity theory are given by the dotted line. Data due to Shirokov et al. (1970) computed by eqs. 14-35 and 14-36 for $\tau_s^0 = 0$; $Z(I_3) = 1$.

Modification of deformation model

A modification of the model based on the linear–fractional pattern of deformation has been suggested by A.L. Krizhanovsky (see Section 12.4: Krizhanovsky et al., 1975). The equations describing the model are of the following types.

(a) The equation of shearing strain:

$$\gamma_i^m = \frac{\tau_{i(s)}^{\alpha_1} \tau_i}{G_0(\tau_{i(s)} - \tau_i)} \tag{14-39}$$

(b) The condition of strength:

$$\tau_{i(s)} = \tau_s^0 + \sigma_m^\lambda Z(I_3) \tag{14-40}$$

where $Z(I_3) = \alpha_2 + \alpha_3 \mu_\sigma + \alpha_4 \mu_\sigma^2$.

(c) The equation of volumetric strain:

$$\varepsilon_m = \varepsilon_m^0 + \varepsilon_m^D \tag{14-41}$$

where ε_m^0 is the volumetric strain induced by the all-around pressure, $\varepsilon_m^0 = \beta_1 \sigma_m^{1/\kappa}$ and ε_m^D is the volumetric strain produced by the stress deviator (dilatancy). The value of ε_m^D may be positive (extra hardening) or negative (softening) depending on the level of stress $\bar{\tau}_i = \tau_i/\tau_{i(s)}$. The curves representing deformation of this kind were depicted in Fig. 12-2(a). The author has approximated the $\varepsilon_m^D - \bar{\tau}_i$ relation by a piecemeal linear function (Fig. 14-11):

$$\varepsilon_m^D = B \frac{\bar{\tau}_i}{\bar{\tau}_i'} \quad \text{(for } \bar{\tau}_i \leq \bar{\tau}_i'\text{)}$$

and:

$$\varepsilon_m^D = B - \frac{B + C}{1 - \bar{\tau}_i'} (\bar{\tau}_i - \bar{\tau}_i') \quad \text{(for } \bar{\tau}_i > \bar{\tau}_i'\text{)}$$

where $B = \beta_2 \sigma_m^{\kappa_2}$, $C = \beta_3 \sigma_m^{1/\kappa_3}$ and $\bar{\tau}_i = \beta_4 + \beta_5 \sigma_m^{\kappa_1}$.

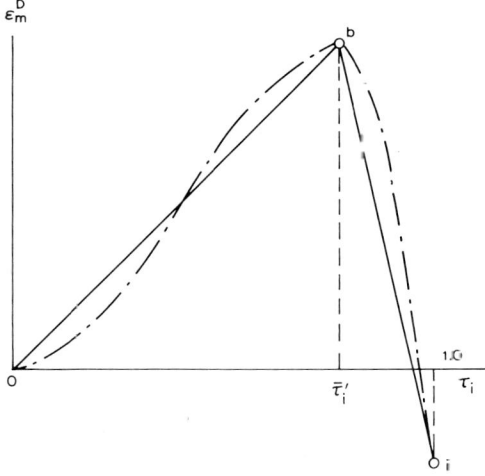

Fig. 14-11. Development of volumetric strains depending on the value of relative shearing stress $\bar{\tau}_i = \tau_i/\tau_{i(s)}$. Experimental curve is shown by the dotted line and the approximated one is given by the solid line (Section 12.4: Krizhanovsky et al., 1975).

Thus, the model incorporates all three stress tensor invariants, i.e. allows for the effect of σ_m on the shearing strain, the variability of Poisson's ratio, the arrangement of stress, and dilatancy of either sign.

It will be noted that numerous parameters inherent in many deformation models are a limitation. However, this may be partially offset if the relationships obtained during the experiments are introduced directly into the computer-aided calculations.

The model considered was employed to solve the problem of non-linearly behaving soil in the state of stress and strain below a strip foundation, using the finite difference method, the iterative method, variable parameters and a computer. The way the non-linearity and other factors allowed for by the model influenced the state of stress and strain of the soil also appeared to be as significant as in the preceding example. Particular attention was paid to evaluating the effect of dilatancy. Comparative studies (Fig. 14-12) showed that this effect became stronger directly with the load, the settlement materially increasing (curves 2 and 3 in Fig. 14-12) if either the dilatancy

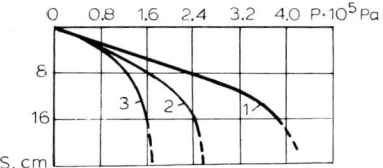

Fig. 14-12. Settlement of the soil below foundation calculated by eqs. 14-35–14-37. (*1*) Dilatancy of hardening and softening taken into account in accordance with eq. 14-36. (*2*) Dilatancy only partly taken into account. (*3*) Dilatancy of softening neglected (Section 12.4: Krizhanovsky et al., 1975).

was partially taken into account (for $\bar{\tau}_i \geq \bar{\tau}'_i$ the role of dilatancy was diminished) or, all the more, if softening was totally excluded.

The finite difference method enabled some authors to solve the problem of stability of a soil mass. So, e.g. at the 5th European Conference on Soil Mechanics and Foundation Engineering (Madrid, 1972) solutions were presented of the problem of earth pressure on retaining walls (Serrano, 1972; Wroth, 1972) and of the state of stress and strain of soil adjacent to a retaining wall (James et al., 1972). Krizhanovsky and Kulikov (1977) arrived at a solution of the problem of stability of the slopes of an embankment by means of eqs. 14-39–14-41, using, however, eq. 14-39 in a modified form:

$$\gamma_i = \beta \tau_{i(s)}^{\alpha_1} \tau_i + \frac{\tau_{i(s)}^{\alpha_2} \tau_i}{G_0(\tau_{i(s)} - \tau_i)}$$

It was found that the stability of a slope should be analysed by considering its state of stress and strain as a whole rather than the condition of limiting equilibrium.

Some other problem-solving methods

Fedorovsky and Koganovskaya (1975) suggested solving non-linear problems of soil deformation by means of a combined variational-difference method. The equations were arranged in the finite difference form proceeding from the variational energy-related principle analogous to the one used in the finite element method and conducive to obtaining a stable solution. Also studied was the impact of a rigid strip foundation on the non-linear behaviour of the underlaying soil obeying the law of eq. 14-35.

A unique method of solving the problem of the response of a non-linear medium to the action of a settlement plate, using the finite element method, was employed by Desai (1971). He approximated an empirical law of the stress–strain relation by means of the so-called spline functions, describing curves drawn through a number of points by means of flexible strips known as splines.

Rheological models

Rheological soil models allowing for non-linearity, the effect of all-around pressure, etc., were discussed in Chapter 11, and analytical solutions of the problem of deformation of the soil below a foundation in the state of non-linear creep obeying the power law of deformation was the subject matter of Chapter 13.

Consider another rheological model of a non-linearly viscous and rigid medium examined by Stroganov (1966). The equation of shearing deformation is:

$$\tau_i = \frac{\tilde{G} \tan \psi_s}{\tan \psi_s + \tilde{G} \gamma_i} (H_s + \sigma_m) \gamma_i + \eta \dot{\gamma}_i \tag{14-42}$$

where $\tilde{G}$, ψ_s and H_s are parameters, the values of which are given in eq. 11-52, and η is the coefficient of viscosity.

Volumetric deformations are described by eq. 14-41, in which the strains ε_m^0 and ε_m^D are determined from the equations $\sigma_0 + \sigma_m = D(\varepsilon_0 + \varepsilon_m^0)^\kappa + \eta_v \varepsilon_m^0$ and $\varepsilon_m^D = -\Lambda \gamma_i$, where σ_0, D, κ are constants; η_v is the coefficient of bulk viscosity and Λ is the dilatancy coefficient (only taken into account during softening of the ground).

For $\gamma_i \to \infty$, eq. 14-42 is transformed into eq. 11-61 describing rigid plastic deformation.

The above model provided the basis for solving some of the boundary-value problems (including those of soil transport down a slope, etc.).

Models based on the theory of plastic flow. The Cam-clay model

One of the most widely used models, the Cam-clay model, was developed by a team of scientists from Cambridge University: K. Roscoe, A. Schofield, C Wroth, J. Burland, H. Poorooshasb, A. Thurainajah and others (Roscoe et al., 1958; Roscoe, 1968; Schofield and Wroth, 1968). Unlike its predecesor, the Granta-gravel model based on the mechanism of the rigid-plastic state (Granta and Cam are the names of the upper and lower parts of the river on which Cambridge is situated), the Cam-clay model treats soil as an elastoplastic hardening medium whose volumetric deformation consists of both recoverable and irrecoverable strains, $\varepsilon_v = \varepsilon_i^e + \varepsilon_i^p$, and the shearing strain is of the irrecoverable kind only, $\gamma_i = \gamma_i^p$. The dilatancies of both signs are taken into account, i.e. the hardening and softening of soil in response to shearing stresses. The model was constructed for isotropic cohesionless soils (sands, water-saturated silty and clayey soils devoid of cohesion). Originally, the model was developed for the case of axisymmetrical compression ($\sigma_1 > \sigma_2 = \sigma_3$, $q = \sigma_1 - \sigma_3$, $p = 1/3(\sigma_1 + 2\sigma_3)$) but then it was raised to generality for the condition $\sigma_1 \neq \sigma_2 \neq \sigma_3$.

The model is based on the concept of critical state formulated by Hvorslev (1937) and Casagrande (1936). In this context, the authors have assumed that an increment in shearing strain is induced by a constant stress with no change in volume provided the porosity (density) and the ratio q/p have certain values; this state is referred to as critical.

Consider the original relationships of the model. It is based on the assumption that existing in the p–q–v space (where v is specific volume) are points which define the position of a line (not, as it will be noted, of a surface) which is the line of the critical state in the sense of the term referred to above. Accordingly, the loading surface is constructed on the p–q–v coordinates as well. This fact is the main difference between the Cam-clay model and most of the other models in which the loading surface is built on the $\sigma_1, \sigma_2, \sigma_3$ coordinates (see Fig. 14-8), being transformed (obviously on the same coordinates) into the surface (not line) of limiting state.

The testing of soil for all-around compression (virgin compression, $q = 0, p > 0$) yields a curve of isotropic hardening which may be described, following Terzaghi, by the logarithmic law (Fig. 14-13(a)):

$$v = v_0 - \lambda_v \ln p \qquad (14\text{-}43)$$

In the case of unloading and reloading, we obtain curves of softening and re-

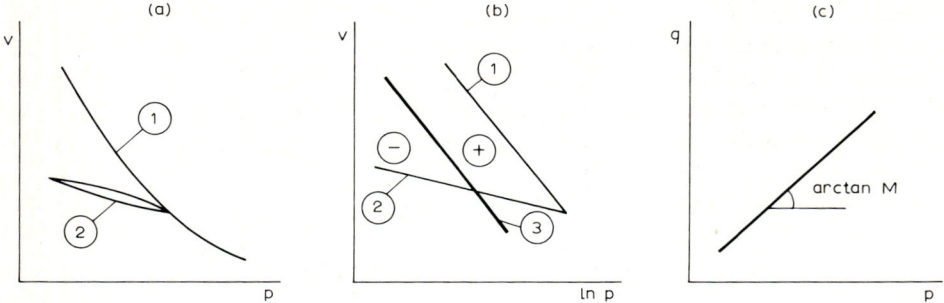

Fig. 14-13. Cam-clay model. (a) Virgin (all-around) compression: (*1*) without unloading, and (*2*) with unloading. (b) Lines of (*1*) virgin compression, (*2*) virgin compression with unloading and (*3*) critical state. (c) *q–p* relationship in the critical state.

hardening which may be described, neglecting a slight hysteresis, by the same logarithmic law:

$$v = v_0 \pm k_v \ln p \qquad (14\text{-}43')$$

where $v = 1 - e$ is the specific volume of the soil, e is the void ratio, $v_0 = 1 - e_0$ is the original volume (for $p = 1$) and λ_v and k_v are soil constants.

In testing under conditions of triaxial compression ($q/p > 0$), we obtain a family of $v-p$ logarithmic curves shown in Figs. 14-13, (a) and (b)[*1]. For a certain value of the ratio q/p, the curve in question (shown by line 3 in Fig. 14-13(b)) will represent the critical state of soil, i.e. a condition giving rise to the plastic flow of soil at a constant volume:

$$v = \Gamma_v - \lambda_v \ln p \qquad (14\text{-}44)$$

where Γ_v is a soil constant corresponding to the specific volume in the critical state for $p = 1$.

The resistance to shear q, at which the critical state is taking shape, is determined from the relation (Fig. 14-13(c)):

$$q = Mp \qquad (14\text{-}45)$$

where M is a soil constant (friction coefficient).

It can be seen that eq. 14-45 corresponds to the von Mises–Botkin condition for a cohesionless medium. However, the critical state of the Cam-clay model differs from the limiting condition (state at failure) in the generally recognized sense of the word and $M \neq \tan \psi$. The shearing stress giving rise to the critical state is slightly higher than the residual strength but lower than the peak one (a fact remaining somewhat obscure if studied from the standpoint of limiting condition).

The associated law of plastic flow is adopted for the model. The loading function

---

[*1] For $q/p = $ const., the v–ln p straight lines are parallel.

is written as $f(p, q, \Gamma) = 0$, where Γ is the parameter of hardening[*1]. This parameter is adopted as the increment in the plastic strains $\partial \Gamma / \partial v^p = 1$ and $\partial \Gamma / \partial \gamma_i^p = 1$. According to the associated law, the loading function is identical with the plasticity potential of eq. 14-29 and is determined on the assumption that the plastic constituent of work (dissipation) is constant and equals the dissipation in the critical state:

$$\frac{dA^p}{v} = p\frac{dv^p}{v} + q \, d\gamma_i^p = Mp|d\gamma_i^p| \tag{14-46}$$

Hence we arrive at an expression for the loading function in the form:

$$f = f_p = \frac{|q|}{Mp} + \ln p - \phi(\Gamma_v) = 0 \tag{14-47}$$

From this we obtain:

$$\frac{|q|}{Mp} + \ln \frac{p}{p_u} - 1 = 0 \qquad p_\perp = \exp\left(\frac{\Gamma_v - v}{\lambda_v}\right) \tag{14-47'}$$

Eq. 14-47 is represented on the p–$(\pm q)$ plane as a closed loading curve (curve 1 in Fig. 14-14) and there is an ith curve corresponding to each value of v_i. The curves emanate from the origin and have zero derivatives at points $C_1, C_2 \ldots$ corresponding to the critical state $\eta = |q|/p = M$. Thus, these points define the position of the line of critical state (eq. 14-45; curve 2 in Fig. 14-14). The rest of the tangents to the family of curves (eq. 14-47) shown by dotted lines in Figs. 14-14(a) and (b) represent the values $\eta \geqslant M$. As already pointed out, for $|q|/p < M$ the soil shrinks in volume (hardening takes place) and for $|q|/p > M$, the volume of the soil increases (softening occurs). For $|q|/p = M$ (critical state) the soil volume remains unchanged.

A projection of the line $|q|/p = M$, i.e. of segment $C_1 C_2 C_3$, on the v–p plane (Fig. 14-14(b), curve 3) gives a logarithmic curve of the critical state (eq. 14-44) and a projection of the lines $|q|/p < M$ and $q|/p > M$ yields logarithmic curves (eq. 14-43).

Loading surface for Cam-clay model

By adding eqs. 14-44 and 14-47 we obtain the equation of loading surface:

$$\Phi = \frac{|q|}{Mp} + \frac{\lambda_v \ln p + v - \Gamma_v}{\lambda_v - k_v} - 1 = 0 \tag{14-48}$$

The upper part of the surface (for $+q$) is depicted in Fig. 14-14(c). Note that eq. 14-48 is not the surface of critical state but the surface confining the states of equilibrium of all possible kinds. The critical state is represented by the line $|q|/p = M$.

This interpretation of the loading surface differs from the traditional concepts according to which the loading surface of a hardening medium is transformed into a limiting surface representing the state of soil at failure.

[*1] Not to be confused with the constant Γ_v in eq. 14-44.

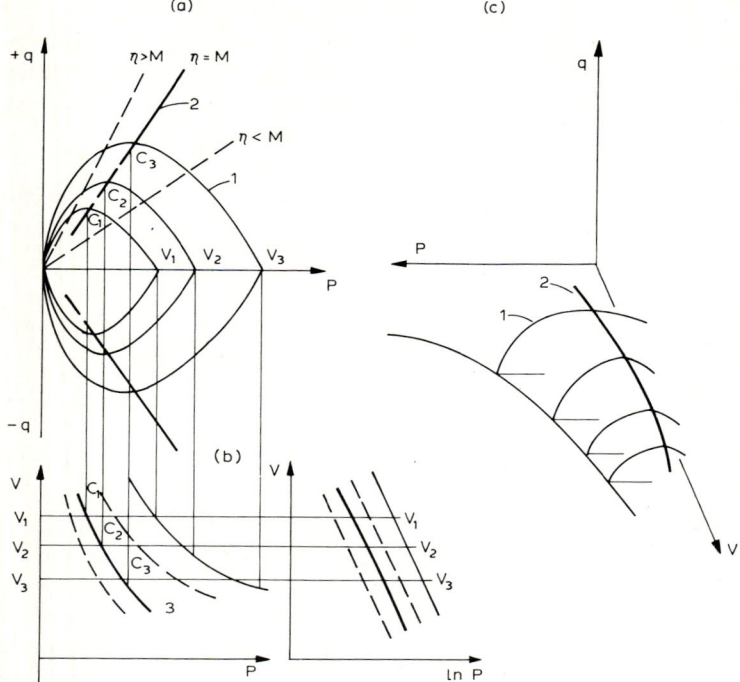

Fig. 14-14. (c) Loading surface and its contours on: (a) the q–p plane; and (b) on the v–p plane for the Cam-clay model of Roscoe et al.

Fig. 14-14(a) is the contour of the surface (eq. 14-48) on the q–p plane, and (b) is the contour of the same surface on the v–p plane.

Note that for $q = 0$ eq. 14-48 is transformed into the expression of the so-called virgin compression line:

$$v_e = v + \lambda_v \ln p = \Gamma_v - \lambda_v - k_v$$

The strain-increment–stress relationship may be obtained either by differentiating the equation of loading surface (eq. 14-48), provided it is taken into account that, in accordance with the associated law,

$$d\gamma_i = \frac{\partial \Phi}{\partial q} d\Phi$$

or (Roscoe and Burland, 1968), from giving due consideration to the equation of dilatancy,

$$\frac{d\gamma_i^p}{d\varepsilon_v^p} = -\left(\frac{dp}{dq}\right) = \frac{1}{\Lambda}$$

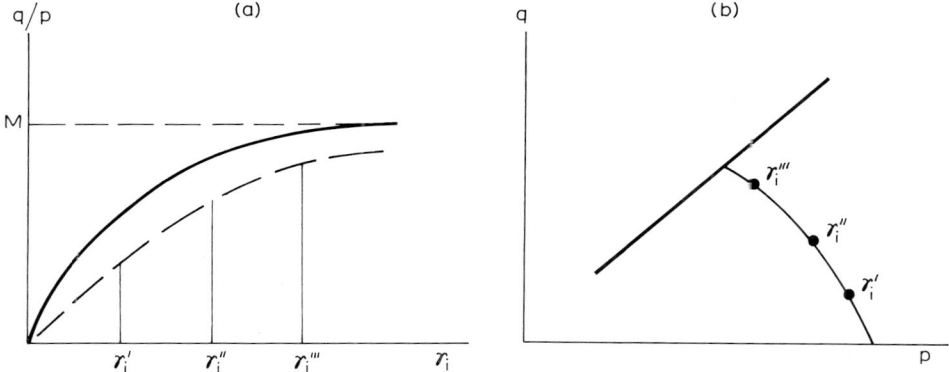

Fig. 14-15. (a) Shearing strain, and (b) loading path observed during undrained tests of the Cam-clay model.

where, taking into account the energy-related condition, $\dot{\Lambda} = M - \eta$ in which $\eta = q/p$:

$$d\varepsilon_v = \frac{1}{1+e}\left(\frac{\lambda_v - k_v}{M} d\eta + \lambda_v \frac{dp}{p}\right)$$

$$d\gamma_i = \frac{\lambda_v - k_v}{1+e}\left(\frac{p\, d\eta + M\, dp}{Mp(M-\eta)}\right)$$

The above equations are solvable in the closed form for a certain stress arrangement only, for example, for the tests $p = $ const. and for undrained tests. Relevant solutions can be found in a work by Schofield and Wroth (1968). A q/p–γ_i curve obtained from undrained tests is depicted in Fig. 14-15(a), being the exponential configuration with an asymptote $q/p \to M$ for $\gamma_i \to \infty$. Corresponding to this curve is the loading path represented in Fig. 14-15(b). In the case of a drained test, the q/p–γ_i curve acquires another configuration, shown in Fig. 14-15(a) by the dotted line, which reflects the dependence of deformation on the loading history.

The model incorporates only four parameters, M, λ_v, k_v and Γ_v, which can be determined from standard triaxial tests. The values of λ_v and k_v are found from all-around compression tests (with unloading), using eqs. 14-43 and 14-43′, and the parameters M and Γ_v are obtained in accordance with eqs. 14-45 and 14-44, proceeding from routine triaxial tests. The value of Γ_v may be verified by eq. 14-49.

A limited number of parameters which all have a clear physical meaning and are unaffected by the initial state of stress and strain is an asset of the Cam-clay model to which it owes its popularity. However, one must point out that some of the aspects involved need a critical examination, and the model itself fails to take into account some real effects.

Numerical solutions based on the Cam-clay model

These were obtained by some authors employing the finite element method. Consider the solution arrived at by Alexandrovich and Fedorovsky (1979).

Although the problem solved was that of a circular rigid settlement plate on a layer of soil whose properties were represented by the Cam-clay model, the model adopted took into account the additionally recoverable shearing strains (the Cam-clay model proper assumes that the shearing strains are irrecoverable). This additional consideration was suggested at an earlier date by Balasubramaniam (1975) who experimentally obtained $\gamma_i - \beta \ln |q| = $ const. Hence $\bar{G} = |q|/\beta$, where $\beta = \alpha |q_s|/p_s$. Introducing a correction into the value of the variable shear modulus, Alexandrovich and Fedorovsky adopted $\bar{G} = pV/g$. A similar expression was adopted for the bulk modulus $\bar{K} = pV/k$ on the basis of eq. 14-43'. Thus, the model used had five constants, λ_v, M, Γ_v, $\bar{K}$ and $\bar{G}$.

A solution was sought for two different initial values of soil porosity, $v_0 = 1.7$ and $v_0 = 1.8$. It appeared that the initial porosity materially influenced the settlement. So, the settlement of 13 cm was produced by a load $p = 0.75 \cdot 10^5$ Pa for $v_0 = 1.8$ and by a load $p = 1.9 \cdot 10^5$ Pa for $v_0 = 1.7$, i.e. an alteration of v_0 from 1.8 to 1.7 called for increasing the load 2.5 times. Correspondingly, in the former case the same load induced a more intensive development of plastic zones at the edges of the settlement plate than in the latter one.

B. Sympson, at Cambridge, carried out a series of comparative numerical calculations for the Cam-clay and other non-linear models by the finite element method (Wroth, 1973).

Some experimental data

Experimental verification of the Cam-clay model was undertaken by many an author. Balasubramaniam et al. (1977) tested hard clays for triaxial compression and tension under the drained and undrained condition, using various loading paths. The deformation obtained during the drained tests corresponded to the Cam-clay model, beginning from a quite significant value of preshearing hardening. Good agreement between the experimental data and the model was obtained by Larsson and Sällfors (1981) from testing soft clay. The loading surface they found during the tests coincided with the surface adopted in the Cam-clay model. Mayne (1980), analysing test data pertaining to 96 clays for various kinds acquired by various authors, came to the conclusion that the connection between the shearing strength and the degree of overhardening corresponds to the Cam-clay model.

However, most of the comparisons referred to above were of a qualitative nature. Also noted, along with the coincidences, were disagreements, which gave rise to suggestions intended to complement and refine the model.

Modifications of the Cam-clay model

Burland (1965) slightly modified the equation of dissipated work increment (eq. 14-46), adopting it in the form:

$$A^p = p\sqrt{(dv^p/v)^2 + M^2(d\gamma_1^p)} \tag{14-49}$$

From this equation, we may obtain equations of loading surface replacing eqs. 14-47 and 14-48:

$$f = f_p = \frac{1}{p_0} - \frac{M^2 p^2}{M^2 p^2 + q^2} = 0 \tag{14-50}$$

$$\Phi = \frac{p}{p_0} - \left[\frac{M^2 p^2}{M^2 p^2 + q^2}\right]^{(\lambda_v - k_v)/\lambda_v}$$

where p_0 and p_e are the initial value of p and that for a given void ratio e, respectively. The equation for f yields an ellipse on the q–p surface.

The equation of the Cam-clay model derived for the triaxi-symmetrical state of stress, was raised to generality by Roscoe and Burland (1968) for an arbitrary three-dimensional state of stress by introducing the parameters $r = \sqrt{2I_2(D)}$, $p^* = 1/3(\sigma_1 + \sigma_2 + \sigma_3)$, $\eta^* = r/p^*$ and M^*. It has been obtained by analogy with the modified Burland model:

$$d\varepsilon_m = \frac{1}{(1 + e)}\left[(\lambda_v - k_v)\frac{2\eta^* \, d\eta^*}{(M^*)^2 + (\eta^*)^2} + \frac{\lambda \, dp}{p}\right]$$

$$d\gamma_i = \frac{\lambda_v - k_v}{(1 + e)}\left[\frac{2\eta^*}{(M^*)^2 + (\eta^*)^2}\right]\left[\frac{2\eta^* \, d\eta^*}{(M^*)^2 + (\eta^*)^2} + \frac{dp}{p}\right] \tag{14-51}$$

Employing the Cam-clay model in their experiments with clay, Tavenas and Leroueil (1977) also obtained an elliptical configuration of the loading surface. Proving that the experimental data generally coincide with the Cam-clay model, the authors extended the field of the model's application to soils with undisturbed structure and introduced an allowance for a time factor.

The applicability of the Cam-clay model as a means of describing the behaviour of sand under triaxial compression was further analysed by Poorooshasb et al. (1966, 1967). The applicability of the Cam-clay model in the case of pure shear of sand was considered by Wroth (1965), and Wroth and Bassett (1965).

Generalized model

Following the suggestion by Hill and Rice (1973), who introduced the history of loading as a parameter of the stress–strain equations and adopted the degree of soil compaction as a similar parameter, Herrera et al. (1977) evolved the theory of deformation of precompacted clays. The loading surface was constructed in a three-dimensional space p, q, v by analogy with the Cam-clay model so that this model appeared to be a particular case of the Herrera model. At the same time it was shown that the postulate concerning the absence of recoverable shearing strains (adopted for the Cam-clay model) is not only irrelevant but contradicts other postulates.

The Rowe dilatancy model

Developed by Rowe (1962, 1972) for a cohesionless medium, this model proceeds from considering physical properties of sand particles — friction between the particles and their elastic behaviour in compression on the slip plane. The model establishes a connection between these properties and the parameters of deformation. The aggregate deformation is regarded as being composed of two constituents, an irrecoverable deformation of slip ε_{ij}^p and an elastic (recoverable) deformation ε_{ij}^e of the contacting particles.

The elastic properties of the model are determined by examining the spatial arrangement of the contacting solid particles which varies with porosity, shape and grading of the particles. The expressions for the strain components in the 1-, 2-, 3-directions are set forth in terms of the Hertz relationship for two contacting spheres:

$$\varepsilon_i^e = S_i \left[\frac{\sigma_i}{E}\right]^n \quad i = 1, 2, 3 \tag{14-52}$$

where S_i and n are parameters.

The plastic properties of the model are determined by considering the slip of the irregularly shaped contacting particles relative to one another along a certain surface (the disintegration of particles in the course of deformation is neglected).

The constitutive relations for slip deformation are the stress–dilatancy ones for an assemblage of solid particles. They may be obtained from the minimum value of the energy-related expression:

$$\bar{E} = \frac{\text{Increment in input of work due to force applied at contact}}{\text{Increment in output of work to counteract extreme forces}}$$

Work input is taken as the energy supplied per unit volume in the 1-direction and work output is defined by the work done per unit volume in the 2- and 3-directions. Hence:

$$\bar{E} = \frac{\sigma_1 \, d\varepsilon_1^p}{-(\sigma_2 \, d\varepsilon_2^p + \sigma_3 \, d\varepsilon_3^p)}$$

Thus, for an axi-symmetrical case, $\sigma_1 > \sigma_2 = \sigma_3$, we find:

$$\bar{E} = \frac{\sigma_1}{\sigma_3(1 - d\varepsilon_v^p/d\varepsilon_1^p)} \tag{a}$$

The minimum value of $\bar{E}$ is given by:

$$\bar{E}_{\min} = K_2 = \tan^2(45° + \phi_f/2) \tag{b}$$

in which ϕ_f is the equivalent angle between the particles which allows for the obliquity of the particle slip with respect to the average direction and varies between the limits $\phi_\mu \leq \phi_f \leq \phi_{cv}$, where ϕ_μ is the mean angle of friction between the surfaces of the particles and ϕ_{cv} is the Coulomb angle of friction in shear in the critical state. Note

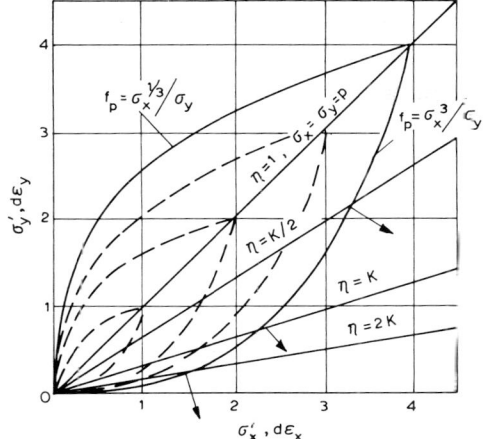

Fig. 14-16. Surfaces of plasticity potential for the Rowe dilatancy model (planar strain).

that $\phi_f = \phi_{cv}$ will be the case during a triaxial compression of both a loose sand and a close one but subjected to considerable deformation, and $\phi_f = \phi_\mu$ will be observed in testing close sand at a peak value of σ_1/σ_3.

Equating eqs. a and b, we obtain the law of flow correlating the stress and increment in slip strain (for $\sigma_1 > \sigma_2 = \sigma_3$):

$$\bar{\eta} = D_R K_R \qquad (14\text{-}53)$$

where:

$$\bar{\eta} = \frac{\Delta\sigma_1}{\Delta\sigma_3}; \qquad D_R = [1 - (d\varepsilon_v^p/d\varepsilon_1^p)]$$

Considering eqs. 14-53 and 14-29 simultaneously, we can determine the plasticity potential f_p. In the case of planar deformation, eq. 14-53 takes the form $d\varepsilon_3^p/d\varepsilon_1^p = -\sigma_1/K_R\sigma_3$. Satisfying this relation and eq. 14-29 is the potential in the form $f_p = \sigma_1^{K_R}/\sigma_3$. The corresponding loading surface on the σ_1–σ_2 plane for a particular case when $K_2 = 3$ is depicted in Fig. 14-16.

To compare the Rowe dilatancy model with the Cam-clay models of Roscoe and Burland, we equate the expressions for dissipated work: $dA^p = \sigma_1\,d\varepsilon_1^p + 2\sigma_3\,d\varepsilon_3^p$. For the Rowe model, we obtain the value of dA^p from eq. 14-53 and for the models of Roscoe and Burland we use expressions 14-46 and 14-49 to find this value. It may be shown that the parameters of critical state $M = |q|/p$ for the models by Rowe, Roscoe and Burland are all identical if the specimen as a whole is in the critical state.

The Rowe theory with its inherently physical approach to the processes of shear was eventually used by some authors as a component of the models they were developing. Some of them are discussed below.

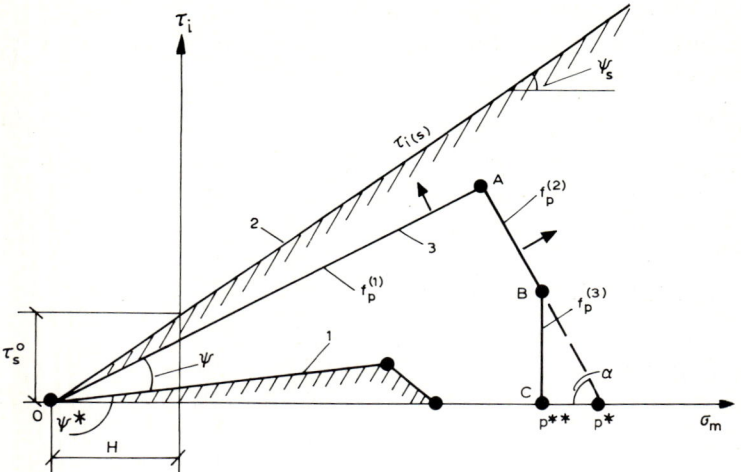

Fig. 14-17. Loading surface for the model of Zaretsky et al.: 1 = original position of the surface $\tau_i = \tau_{i(0)}$; 2 = limiting surface $\tau_i = \tau_{i(s)}$; 3 = surface in variable position $\tau_{i(0)} \leq \tau_i \leq \tau_{i(s)}$.

Model of a hardening elastoplastic soil with piecemeal smooth loading surface

Zaretsky and Lombardo (1979) and Zaretsky et al. (1980) developed a model of clay soil which treated soil as a hardening elastoplastic medium having a loading surface of the piecemeal linear configuration shown in Fig. 14-8(c). The equation of the surface in question was adopted in the form:

$$f(\tau_i, \sigma_m, \gamma_i^p, \varepsilon_m^p) = 0 \tag{14-54}$$

This form indicates that the plastic deformations of shear and volume change are used as the measure of hardening. The loading surface (on the τ_i–σ_m plane) is represented in Fig. 14-17.

Taking into account that the loading surface is of the singular kind, the increment in the components of plastic strain tensor is determined by eq. 14-30.

The values of plastic potential $f_p^{(r)}$ entering eq. 14-30 and describing each of the segments of the loading surface, which all obey the associated law of flow, are as follows.

For segment OA:

$$f_p^{(1)} = \tau_i - [\tau_{i(s)}^* + (\tau_{i(s)} - \tau_{i(s)}^*)\phi_1(\Gamma)] = 0 \tag{a}$$

For segment AB:

$$f_p^{(2)} = \tau_i + (\sigma_m - p^*)\tan\alpha \tag{b}$$

$$\tan\alpha = \tan\alpha^*[1 + \phi_2(\Gamma)] = 0$$

For segment BC:

$$f_p^{(3)} = \sigma_m - p^{**} = 0 \tag{c}$$

Here, the deformation of soil within segment OA of the loading surface is accompanied by the hardening of the soil and in segments AB and BC by the softening.

In eqs. a through c we adopt:

$$\tau_{i(s)} = \tau_s^0 + \sigma_m \tan \psi \tag{d}$$

which is the von Mises–Botkin condition of strength (eq. 14-15) defining the position of the limiting straight line 2 in Fig. 14-17:

$$\tau_{i(s)}^* = \tau_s^{0*} + \sigma_m \tan \psi^* \qquad \tan \alpha^* \tag{e}$$

which are the parameters of the virgin loading surface 1 in Fig. 14-17:

$$\phi_1(\Gamma) = \frac{\Gamma/\Gamma^*}{(1-B) + B(\Gamma/\Gamma^*)} \qquad \phi_2(\Gamma) = \frac{\alpha(\Gamma/\Gamma^*)}{1 - \Gamma/\Gamma^*}$$

$$p^{**} = p^* - \tau_i \cot \alpha \qquad p^* = \left[A_1 + A_2(\Gamma)p^* - \frac{A_3(\Gamma)}{1 + A_4(\Gamma)p^*} \right] \varepsilon_m^p \tag{f}$$

which are the hardening parameters, Γ being the measure of hardening after Odqvist (eq. 14-9).

Although corresponding to each value of the plastic strain γ_i^p, ε_m^p is a particular loading surface, all the surfaces emanate from the same pole $\sigma_m = H_s = \tau_s^0 \tan \psi =$ const. This means that the loading surface is expanding with the development of strain, and for $\psi = \psi_s$ segment OA of the surface is transformed into the limiting straight line 2 of Fig. 14-17.

The equation of loading surface in the adopted form is based on the assumption that soil failure results from a shear at the octahedral planes so that, accordingly, the von Mises–Botkin equation is adopted as the condition of strength. However, experimental evidence indicates that the Hill condition, according to which failure occurs at the plane of maximum shear (with a normal v), is more suitable in the case of planar deformation. The condition of strength then takes the form $\tau_v = \tau_{max} = 1/2(\sigma_1 - \sigma_3)$ and $\sigma_v = 1/2(\sigma_1 + \sigma_3)$, and the loading function appears to be independent of the intermediate principal stress. For a spatial problem, the state of stress and strain of a soil depends on the third invariant I_3, which is not allowed for by the model.

The model in question takes account of the shearing and volumetric hardenings of the soil, the dilatancy of both signs (softening and hardening) and also allows for any loading path.

The model was used by its authors in evaluating the dynamic stability of non-saturated and saturated slopes. In this latter case, eq. 14-54 was considered simultaneously with the equation of consolidation:

$$\sigma'_{ij,j} + \delta_{ij} U_{,j} = F_i$$

$$\left[\frac{k_f}{\gamma(W)} U_{,j} \right]_{,i} = -\dot{\varepsilon}_m + \frac{n}{K(W)} \dot{U}$$

where σ'_{ij} are the components of effective stresses; U is the pore pressure; F_i are the components of the resultant of external forces; k_f is the percolation coefficient; ε_m is the volumetric strain of soil skeleton; $K(W)$ is the modulus of volume compressibility of the "liquid-gas" mixture; $,j$ is the symbol of differentiation with respect to the j coordinate and $\dot{\varepsilon}_m$ and $\dot{U}$ indicate differentiation with respect to time.

Numerical solutions of the problem of equilibrium and stability of the slopes of a high sandy loam embankment, obtained by employing the finite element method, have shown that:

– dilatant changes in the volume of the embankment are predominantly of the hardening type and, consequently, any soil model used should give a closed loading surface representing the dilatancy of both signs (the model suggested by the authors satisfies this condition);

– the rate of loading, as decided by the rate of filling work, significantly influences the stability of slopes – the quicker an embankment is being filled the more probable is a slope failure due to the accumulated plastic deformations.

The loading path is also of significance because, as pointed out earlier, filling the reservoir with water at the time of erecting the dam produces more pronounced horizontal displacement of its core than when the water is admitted after the completion of the dam.

Model with smooth closed loading surface

It was said above that some authors made use of soil models having a smooth closed loading surface of the configuration shown in Fig. 14-8(d). A similar model for cohesive soil was suggested following from the associated law of plastic flow by Ioselevich and Didukh (1970), Didukh and Ioselevich (1971), and Ioselevich et al. (1975). The model describes the behaviour of a soil treated as an elastoplastic medium in which the increment in deformation consists, according to eq. 14-1', of elastic and plastic constituents and the loading function $f(\sigma_{ij}, \Gamma)$ is used in the form:

$$f = y + a(x_0 - x) - (x_0^m - x^m)^{1/m} \tag{14-55}$$

where:

$$y = \sqrt{2}\sqrt{I_2(D)} = \sqrt{2}\sqrt{\tau_i};$$

$$x = \frac{1}{\sqrt{3}\alpha}[\sqrt{I_2(D)}]_s = \tau_{i(s)}/\sqrt{3}\tan\psi = \sqrt{3}(\sigma_m + H) \tag{a}$$

The parameters a, m and x_0 of eq. 14-55 are each a function of the measure of hardening; adopted as this measure are the plastic shearing and volumetric strains:

$$U = \frac{1}{\sqrt{3}}J_1^p(T) = \frac{1}{\sqrt{3}}\varepsilon_v^p = \sqrt{3}\varepsilon_m^p$$

$$V = \sqrt{2}\sqrt{J_2^p(D)} = \frac{1}{\sqrt{2}}\gamma_i^p \tag{b}$$

For cohesive soils, $a = a(V)$, $m = m(V)$; for sandy soils, $a = $ const., $m = m(V)$, and for both soils, $x_0 = x_0(U, V)$. The type of these functions is determined from experiments satisfying the conditions $0 \leqslant a \leqslant \sqrt{\frac{2}{3}} \tan \psi$, $1 < m \leqslant \infty$ and $x_0 \geqslant \sqrt{3} H$. Accordingly, in the limiting state we find:

$$f = \Phi = \tau_i - \tan \psi (\sigma_m + H_s) = 0 \tag{c}$$

which is the von Mises–Botkin condition of limiting state (eq. 14-15).

The relation between stress and increment in plastic strain is determined by employing the plasticity potential f_p in the form frequently used for a hardening medium instead of eq. 14-29 (cf. Section 1.5: Sedov, 1973):

$$d\varepsilon_{ij} = g \frac{\partial f_p}{\partial \sigma_{ij}} d'f_p \tag{d}$$

where:

$$g = -\left(\frac{\partial f_p}{\partial \varepsilon_{ij}^p} \frac{\partial f_p}{\partial \sigma_{ij}}\right)^{-1}$$

is the hardening function and:

$$d'f_p = \frac{\partial f_p}{\partial \sigma_{ij}} d\sigma_{ij}$$

is the increment in the f_p function (for $\varepsilon_{ij}^p = $ const.).

Since the associated law has been adopted, the surface $f_p = f = 0$ is the loading surface. The increment in the plastic strains at this surface is defined by the relations:

$$dU = g \frac{\partial f_p}{\partial x} d'f_p \qquad dV = g \, d'f_p \tag{e}$$

Eq. 14-55 thus correlates the values of x, y, U and V (i.e. τ_i, σ_m, γ_i^p, ε_m^p), and the four parameters mentioned satisfactorily define the state of stress and strain of the soil in accordance with the model. It fails, however, to take into account the arrangement of stress. Accordingly, the loading surface described by eq. 14-55 is constructed in the four-dimensional space x, y, U, V. Its contour on the x, y surface (i.e. on the τ_i–σ_m surface) is depicted in Fig. 14-18.

With an increase in the plastic strains U and V the loading surface (eq. 14-55) expands and there is a particular curve corresponding to each value of U and V (curves OA and OBD' in Fig. 14-18). The contour of the surface on the τ_i–σ_m plane in proceeding to the limit approaches a straight line, and the surface proper degenerates into the von Mises–Botkin cone.

A vast experimental programme was carried out of testing sandy and sandy loam soils under conditions of axi-symmetrical compression ($\mu_\sigma = -1$), using various loading paths. The loading surfaces represented in Fig. 14-18 have been constructed from the data thus acquired. Only some of the loading paths used during the tests are shown. The curves of the γ_i–τ_i and ε_m–σ_m relations constructed from the

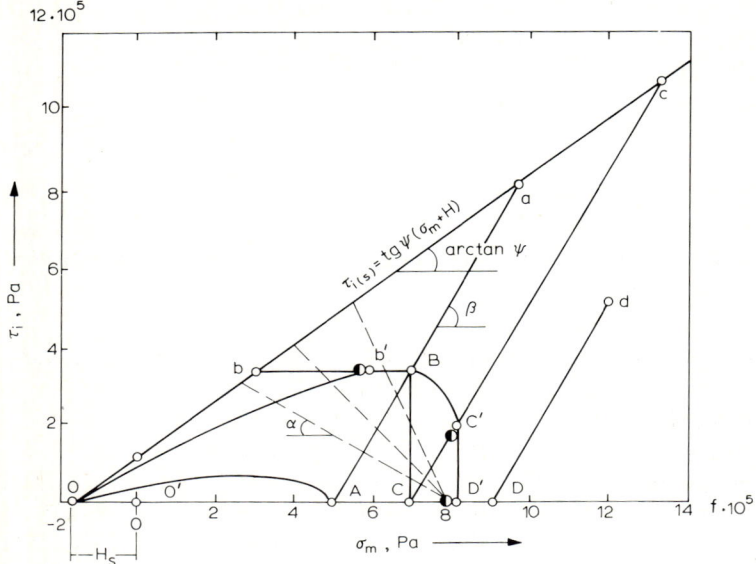

Fig. 14-18. Loading surface for the model after Ioselevich et al.

test data appear to be influenced by the angle $\alpha = -(\Delta\tau_i/\Delta\sigma_m)$; the smaller α is the more intensive is the increase in γ_i.

The above experiments enabled the authors to arrive at the following conclusions.

(a) The vectors $\overrightarrow{d\varepsilon^p_{ij}}$ and $\overrightarrow{df/d\sigma_{ij}}$ are almost parallel to one another, thus corroborating Drucker's postulate about the perpendicularity of the $\overrightarrow{d\varepsilon^p_{ij}}$ vector to the loading surface and, consequently, proving the applicability of the associated law of plastic flow.

(b) The state of stress and strain of a soil is influenced by the history of loading under certain conditions.

(c) The elastic strain is extremely small, if compared with the plastic one, so that one may adopt $\gamma^p_i = \gamma_i$ approximately and $\varepsilon^p_m = \varepsilon_m$ approximately.

(d) Transition into the limiting state takes place when the shearing strain reaches a certain critical value irrespective of the loading path, $\gamma_i = \gamma_{i(s)}$ approx. = const.

These conclusions are generally in agreement with known data.

The parameters of the model discussed may be determined from triaxial compression tests.

Elastoplastic model with fixed loading surface

The Drucker–Prager model examined in Section 14.4 proceeds from the concept of soil as an ideal elastoplastic medium with a fixed loading (yield) surface (see Fig. 14-2(a)). Its application to problems arising in connection with the state of stress and strain of a soil mass means that the zones of both elastic and plastic states are

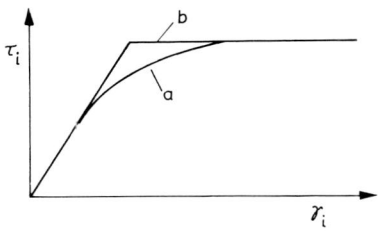

Fig. 14-19. (a) Actual graph of soil deformation, and (b) graph obtained by approximation in terms of elastoplastic behaviour.

assumed to exist at a time, the plastic zones expanding with an increase in the load until some region turns completely into the plastic state.

The problem of the elastic–plastic state of a medium is referred to as a mixed problem; it is stated and analysed by Gorbunov-Posadov (1962).

In formulating this problem, it is assumed that in an elastic zone the soil deforms linearly and that in a plastic one the condition of limiting state (Mohr–Coulomb or von Mises–Botkin) is fulfilled. Accordingly, the stress–strain graph is of the type depicted in Fig. 11-6(b) or, in the case of an ideally cohesive soil, in Fig. 11-6(a). However, generally it is appropriate to proceed from Fig. 11-6(c) assuming that the strains in the sublimiting zone may include both the recoverable and irrecoverable constituents and be affected by all three stress tensor invariants.

The solution of a mixed problem is thus reduced to determining the state of stress and strain of the soil mass in which zones of sublimiting and limiting state exist. For the zones, the equations of equilibrium and geometrical equations will be the same but the physical equations will be different. It goes without saying that the concept of soil as an ideal elastic–plastic body is a certain schematization (Fig. 14-19).

The finite element method, according to which plastic deformations taking shape at any point of a medium may be treated as a sequence of infinitesimal strain increments (Jamada et al., 1968; Hoëg, 1972), creates the prospect of arriving at numerical solutions of many mixed problems. In running a problem on a computer, the method of step-by-step loading is to be used.

Numerical solution of a mixed elastoplastic problem of the state of stress and strain of the soil below a foundation

Consider as an example the technique of solving the above problem developed by Bugrov (1974) and Bugrov and Zarkhi (1978), along with the solutions thus obtained.

The Drucker–Prager model was adopted as the original model, i.e. an elastoplastic medium (see Fig. 11-6(b)) with a fixed yield surface (see Fig. 14-2(a)), subjected to consideration. The relation between the stress and increments in elastic strains was described by Hooke's law and that between the stress and increments in plastic strains by the associated law of plastic flow. Thus, dilatancy was taken into account, but the negative one only (softening).

The constitutive equations, written in matrix form, are as follows:

$$\{\Delta\varepsilon\} = \{\Delta\varepsilon^e\} + \{\Delta\varepsilon^p\} \quad \{\Delta\sigma\} = [D^e]\{\Delta\varepsilon^e\} \quad \{\Delta\sigma\} = [D^p]\{\Delta\varepsilon^p\} \quad (14\text{-}56)$$

where $[D^{e,p}]$ are the matrices defining the relation between the elastic and plastic strains. The $[D^e]$ matrix is adopted in a generalized form of Hooke's law and the $[D^p]$ matrix is determined from the law of plastic flow (eq. 14-29):

$$[D^p] = [D^e] - [D^e][A] \quad [A] = \frac{\left\{\frac{\partial f_p}{\partial \sigma}\right\}\left\{\frac{\partial f_p}{\partial \sigma}\right\}^T [D^e]}{\left\{\frac{\partial f_p}{\partial \sigma}\right\}^T [D^e] \left\{\frac{\partial f_p}{\partial \sigma}\right\}} \quad (14\text{-}57)$$

where f_p is the plasticity potential and T is the symbol of transposition.

In the basic computations, the authors proceeded from the associated law, taking $f = f_p$. Since plane deformation was subjected to consideration, the Mohr–Coulomb condition was adopted as the yield function. Consequently, the yield surface was of the type shown in Fig. 4-4(a).

One of the problems encountered was of the comparative kind, because the original data were similar to the data adopted in the experiment by the present author and others, described in Section 13.6, and the results were compared with the results of this experiment (Fig. 14-20). As can be seen, the curves of the relationship between the load and relative settlement S/b (b is the width of strip foundation) obtained in solving the mixed elastoplastic problem fits, more or less accurately, the experimental curve. The relationship obtained is definitely a non-linear one in spite of adopting the linear law for the sublimiting state. The explanation is that the presence of plastic zones was allowed for in solving the problem (this makes sense of the term "mixed" problem).

The way the zones of plastic deformations spread with an increase in load is depicted in Fig. 14-20(b), representing the solution of the mixed problem in the case of clay loam loaded by an elastic settlement plate ($b = 10$ m). The problem was solved for two values of the coefficient of lateral pressure, $\xi_{LP} = v/(1 - v)$, 0.43 and 1.0. The configuration of the plastic zones appeared to be quite different in the two cases. For $\xi_{LP} = 1$, the zones arise at the edges of the foundation, expanding, extending and joining each other so that a closed region of the limiting state is produced with an increase in load. At the same time, an elastic core, observed more than once in experiments by various authors, is formed below the foundation. The picture described satisfactorily fits the analytical suggestions made by Freilich (referenced in Section 13.7) and K. Terzaghi (see Section 1.5: Terzaghi, 1943), and advanced in the works of Gorbunov–Posadov (1962, 1971) and others.

A different pattern is observed for $\xi_{LP} = 0.43$. The plastic zones come into being first inside the soil mass, along an axis through the foundation centre, and then appear at the edges. An increase in the load below the foundation gives rise to an elastic core whereas the plastic zones join each other, forming, however, a more elongated pattern than is the case for $\xi_{LP} = 1$.

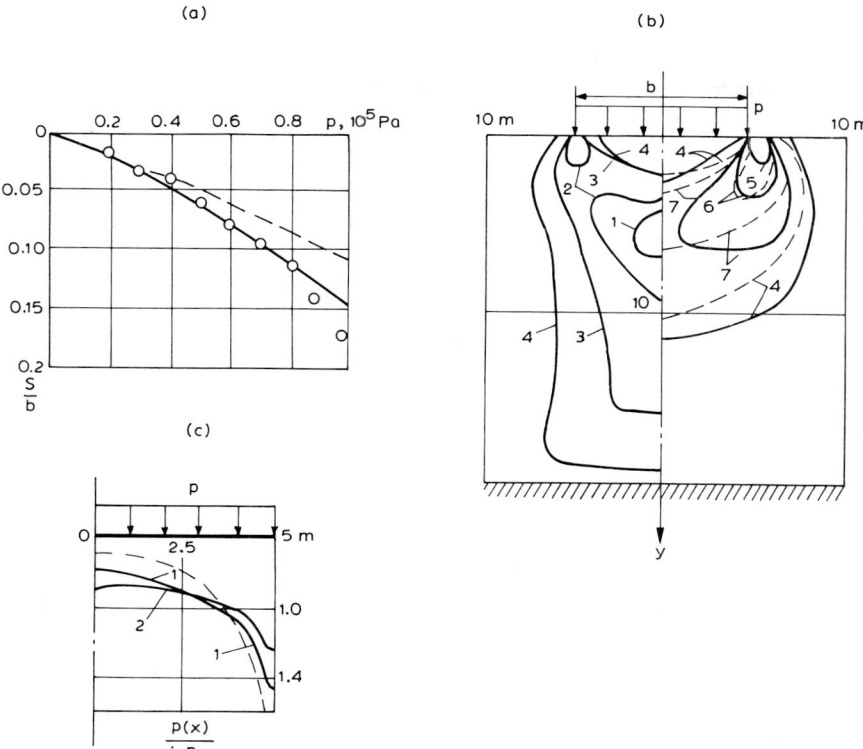

Fig. 14-20. Solutions of the combined elastoplastic problem (after Bugrov). (a) Settlement–load relationship (computer-run solution shown by the solid line, experimental data indicated by the dotted line). (b) Development of plastic zones in the soil below a foundation for $\xi_{LP} = 0.43$ (to the left) and $\xi_{LP} = 1$ (to the right) under the loads: $1 = 1.2 \cdot 10^5$ Pa; $2 = 1.5 \cdot 10^5$ Pa; $3 = 2.7 \cdot 10^5$ Pa; $4 = 7.0 \cdot 10^5$ Pa; $5 = 2.0 \cdot 10^5$ Pa; $6 = 3.0 \cdot 10^5$ Pa; $7 = 4.2 \cdot 10^5$ Pa. (c) Curves of reaction pressures for: $1 = p = 3.0 \cdot 10^5$ Pa; $2 = p = 7.5 \cdot 10^5$ Pa. Solution by elasticity theory shown by the dotted line.

The distribution of the reaction pressures (Fig. 14-20(c)) below the footing corresponds in essence to the curves obtained in the experiments by the present author and others (see Fig. 13-9(b)).

It is interesting to note that the problem referred to above was also solved, for the sake of comparison, using the deformation theory and the finite element method (Bugrov and Grebnev, 1977). The Botkin equation of deformation in a form similar to eqs. 14-35 and 14-36 was adopted as the original equation, the $Z(I_3)$ function, which allows for the arrangement of stress, being in the form $Z(I_3) = 0.7 + 0.0075 (\mu_\sigma - 1)^2$. The outcome appeared to be similar to the results discussed in Section 14.5 (see Figs. 14-9 through 14-12).

Note that variations in the parameters of the law of volumetric changes had a relatively small effect on the results, whereas changes in the parameters of shearing deformations influenced the solution in an appreciable way. Therefore, for example, the difference between the values of settlement computed with the cohesion being ignored and taken into account was 200%.

As far as comparison of the results obtained by solving the problem on the basis of deformation theory and the theory of plastic flow is concerned, the coincidence arrived at is quite acceptable.

Non-associated models of elastoplastic medium

Soil models based on the non-associated law of plastic flow have attracted somewhat less attention of researchers than models relying on the associated law. Apparently, this is because non-associated models are more complex. Consider some of them, beginning with the model suggested by Nikolaevsky (1972, 1979).

The equation defining the relation between the components of plastic strain rate $\dot{\varepsilon}_{ij}^p$ and those of stress σ_{ij} is adopted in the model in a form similar to eq. 14-26:

$$\dot{\varepsilon}_{ij}^p = (\sigma_{ij} + \delta_{ij}\sigma_m)\dot{\lambda} + (H + \sigma_m)\delta_{ij}\dot{\xi} \tag{a}$$

where $\dot{\lambda}$ and $\dot{\xi}$ are unknown functions.

To determine the coefficients $\dot{\lambda}$ and $\dot{\xi}$ and substantiate eq. a in the case of the associated law, it is sufficient to find experimentally the type of the loading function $f = \Phi$ coinciding with the plasticity potential f_p. Simultaneous consideration of the equation $f = 0$ and eq. a enables the constitutive relations in their final form to be obtained.

Since in the case of the non-associated law $f = \Phi \neq f_p$, both equations should be available. In the model under consideration use is made of the equation of limiting state $\Phi_\sigma = (\sigma_{ij}, \alpha, \Gamma) = 0$ and the dilatancy equation $\Phi_\varepsilon(\varepsilon_{ij}^p, \dot{\Lambda}, \Gamma) = 0$, where α is the friction coefficient, $\dot{\Lambda}$ is the dilatancy rate and Γ is the measure of hardening. Note that $\alpha = \alpha(\Gamma)$ and $\dot{\Lambda} = \dot{\Lambda}(\Gamma) = \dot{\Lambda}(\alpha)$. The volumetric strain $\Gamma = \varepsilon_v^p$ is adopted as the measure of hardening, reflecting the fact that the strength of soil depends on its density.

The equation of limiting state is adopted in the von Mises–Botkin form (eq. 14-15):

$$\Phi_\sigma = \sqrt{I_2(D)} - 3\alpha I_1(T) - R = 0 \tag{b}$$

where $\sqrt{I_2(D)} = \tau_i$, $I_1(T) = 3\sigma_m$, $3\alpha = \tan\psi$ and $R = 3\alpha H_s = \tau_s^0$ is cohesion.

The dilatancy equation is adopted in a form similar to eq. 12-4 but expressed in terms of strain rates:

$$\Phi_\varepsilon = \dot{J}_1^p(T) - 2\dot{\Lambda}\sqrt{\dot{J}_2^p(D)} = 0 \tag{c}$$

where:

$$\dot{J}_1^p(T) = \dot{\varepsilon}_v^D \quad \text{and} \quad \sqrt{\dot{J}_2^p(D)} = 1/2\dot{\gamma}_i^p$$

It will be recalled that for the associated law $\dot{\Lambda} = \tan\psi = 3\alpha$ so that only one parameter (commonly α) is determined experimentally. For the non-associated law, both parameters (α and $\dot{\Lambda}$) must be determined.

Considering eqs. a, b and c simultaneously, we find the values of $\dot{\Lambda}$ and $\dot{\xi}$. Substituting them into eq. a, we transform this equation into eq. 14-29, where f_p takes the form:

$$f_p = \tau_i^2 - \dot{\Lambda} \tan \psi (\sigma_m + H)^2 = \text{const.} \tag{14-58}$$

For $\dot{\Lambda} = \tan \psi$, eq. 14-58 is transformed into eq. b. Here it will also be recalled that $\dot{\Lambda} < 0$ corresponds to soil in the soft state which hardens in the course of deformation, $\dot{\Lambda} > 0$ represents a hard soil subjected to softening in deforming and $\dot{\Lambda} = 0$ is the case when the shear is not accompanied by volume changes. For $\dot{\Lambda} < 0$, the plastic equipotential lines (eq. 14-58) are ellipses, for $\dot{\Lambda} > 0$ they are hyperboles and for $\dot{\Lambda} = 0$ the straight line $\tau_i^2 = \text{const.}$ becomes the equipotential line.

The author shows that the model suggested eliminates many misconceptions inherent in associated models.

A non-associated model of another type was proposed by Pender (1978). Assuming that the state of stress and strain of a soil is accurately defined by a point in the $(q-p-e)$ space (where $e = 1 - v$) and adopting the concept of the critical state of a soil defined by eqs. 14-44 and 14-45, he proceeded from the principles of the Cam-clay model. However, unlike this model, it is assumed that the loading function does not coincide with the plasticity potential. Consequently, the plastic flow*[1] is described in terms of the non-associated law displayed in the form (see Section 3.8: Hill, 1950):

$$d\varepsilon_{ij}^p = g \frac{\partial f_p}{\partial \sigma_{ij}} df \tag{14-59}$$

where g is the hardening function, f_p is the plasticity potential and f is the loading function ($f_p \neq f$).

Note, that for $f = f_p$ we have:

$$d\varepsilon_{ij}^p = g \frac{\partial f_p}{\partial \sigma_{ij}} df_p$$

where:

$$df_p = d'f_p = \frac{\partial f_p}{\partial \sigma_{ij}} d\sigma_{ij}$$

(see Section 1.5: Sedov, 1973) and, as a result, eq. 14-59 is transformed into $d\varepsilon_{ij}' = g(\partial f_p / \partial \sigma_{ij}) d'f_p$ which was given above. In its turn, assuming — as is commonly done — that $f = f(\sigma_{ij}, \Gamma)$ and $\Gamma = \Gamma(\varepsilon_{ij}^p)$, we obtain the following expression for the hardening function:

$$g = -\left(\frac{\partial f_p}{\partial \varepsilon_{ij}^p} \frac{\partial f_p}{\partial \sigma_{ij}} \right)^{-1}$$

---

*[1] It will be recalled once more that here the term plastic flow is used in the context of plasticity theory, denoting an increment in irrecoverable strain without allowance for the time factor. This makes all the difference between the plastic flow and the viscous or visco-plastic flow.

Denoting:

$$g\, df_p = -\frac{\partial f_p}{\partial \sigma_{ij}} d\sigma_{ij} \left(\frac{\partial f_p}{\partial \varepsilon_{ij}} \frac{\partial f_p}{\partial \sigma_{ij}}\right) = d\lambda$$

we have, for $f = f_p$, eq. 14-59 transformed into eq. 14-29:

$$d\varepsilon_{ij}^p = d\lambda \frac{\partial f_p}{\partial \sigma_{ij}}$$

To render eq. 14-59 more concrete, we must determine the functions of g, f_p and f. To that end, we assume that:

(1) the f function is specified by the expression:

$$f = q - \eta_i p = 0 \tag{a}$$

where $\eta_i = q/p$ is a certain fixed value $q/p = \text{const.}$;

(2) the loading paths on the q–p plane are parabolas:

$$\left(\frac{\eta}{M}\right)^2 = \frac{p_{es}}{p}\left(\frac{1 - p_0/p}{1 - p_0/p_{es}}\right) \tag{b}$$

where p_0 is the value of p at the point of intersection of the loading path with the p-axis and p_{es} is the value of p on the line of critical state at a point for a given value of e;

(3) the dilatancy rate $\Lambda = d\gamma_i^p/d\varepsilon_v^p$ obeys the relation:

$$d\gamma_i^p/d\varepsilon_v^p = \{[(p_0/p_{es}) - 1][M - (p/p_{es})\eta]\}^{-1} \tag{c}$$

As already pointed out in the definition of the plasticity potential, the vectors of the increments in plastic strains are perpendicular to the surface of the potential at any of its points, i.e.:

$$d\gamma_i^p/d\varepsilon_v^p = -(dp/dq) \tag{d}$$

Considering relations d, c and a simultaneously with eq. 14-59, we obtain the equations of flow (14-50) for the Pender model in the form:

$$d\varepsilon_v^p = \frac{2k_v}{1+e}\left[\frac{(p_0/p_{es} - 1)(p/p_{es})\eta\, d\eta}{M^2(2p_0/p - 1)}\right]$$
$$d\gamma_i^p = \frac{2k_v}{1+e}\left[\frac{(p/p_{es})\eta\, d\eta}{M^2(2p_0/p - 1)(M - \eta p/p_{es})}\right] \tag{14-60}$$

These equations are integrated numerically, being suitable for describing both undrained and drained processes.

Combined non-associated–associated model

It was stated, in analysing loading surfaces of various kinds (Fig. 14-8), that the associated law is adopted by some authors for only a portion of the surface (cap) and the non-associated law is applied to the remaining part (side surface) of the models.

An example of this kind of models is given by Vermeer (1978). Using Rowe's theory to describe shearing deformations and adopting eq. 14-53 as the original equation to describe the plastic flow in shear, the author determined the value of the plasticity potential f_p in shear. However, since it appeared that the value so determined failed to coincide with the loading surface obtained from experiments, $f_p \neq f$, he adopted the non-associated law for shear, describing, however, the volumetric deformations in terms of the associated law. Using the amount of plastic deformation as the parameter of hardening, $\Gamma = \varepsilon_1^p - \varepsilon_3^p$ — for the loading surface appeared to be also the surface of constant plastic deformations in shear — the author obtained several relations establishing the connection between the increments in shearing and volumetric strains and the stress.

Non-associated and combined models due to Lade and Duncan

Describing the behaviour of cohesionless soil (Lade and Duncan, 1973, 1975, 1976; Lade and Musante, 1977; Lade, 1979) these models proceeded from a new criterion of failure, $k_L = I_1^3/I_3$, which was suggested by the authors and defined the type of the equation of strength (14-21) given above:

$$\Phi = \{[I_1(T)]^3 - k_L I_3(T)\} = 0 \tag{14-61}$$

The equation of the loading surface for the models, which described the elasto-plastic behaviour of a hardening soil, was adopted in the form:

$$f = \{[I_1(T)]^3 - \varrho I_3(T)\} = 0 \tag{14-62}$$

The parameter ϱ in eq. 14-62 is considered as a level of stress varying between $\varrho = 27$ for all-around pressure and $\varrho = k_L$ in the limiting state; in this latter case, eq. 14-62 is transformed into eq. 14-61.

The loading surface of eq. 14-62 is represented in the $(\sigma_1 - \sigma_2 - \sigma_3)$ space as an expanding noncircular cone tending to the configuration of eq. 14-61 as to the limit (Fig. 14-21). The contour of this surface on the deviatoric plane is represented by a family of curvilinear hexagons each corresponding to a value of $\varrho = $ const.

It was assumed, by analogy with other models, that the increment in strain is the sum of an elastic and a plastic constituent. The elastic constituent obeys the nonlinear law of elasticity with a modulus of deformation defined by the expression:

$$\bar{E} = k_u p_a \left(\frac{\sigma_3}{p_a}\right)^n \tag{14-63}$$

where p_a is the atmospheric pressure and k_u is a constant.

The increment in plastic deformation is expressed by eq. 14-29 connecting said increment and the plasticity potential f_p.

However, taking into account their own experimental findings and the data acquired by other researchers, the authors assumed in their original model that the plasticity potential f_p did not coincide with the loading function of eq. 14-61. This implied that the law of flow was the non-associated one.

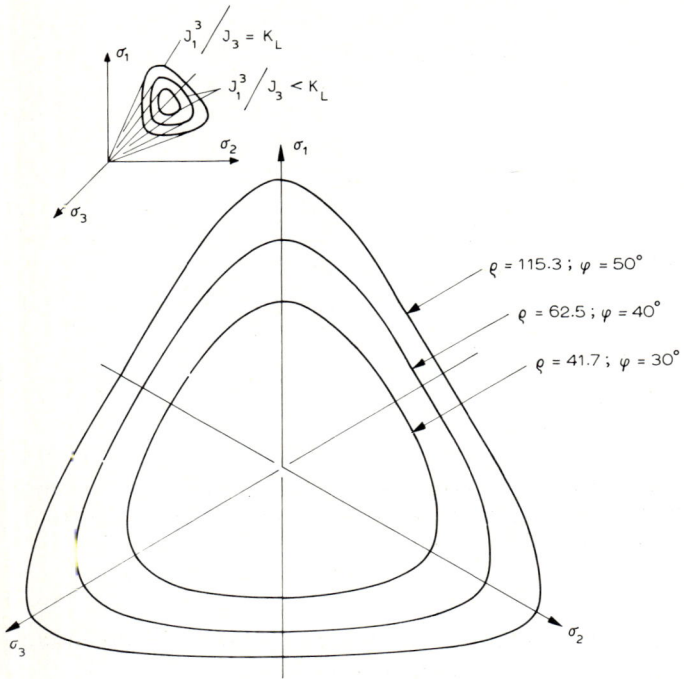

Fig. 14-21. Loading surface for the Lade and Duncan model. (a) In $(\sigma_1-\sigma_2-\sigma_3)$ space. (b) Contours on the deviatoric plane.

The expression for the plasticity potential was obtained in the form:

$$f_p = [I_1(T)]^3 - k_p I_3(T) \tag{14-64}$$

where k_p is not equal to k_L in eq. 14-61.

Substituting eq. 14-64 into eq. 14-29, we obtain the equation of plastic flow. The components of the increments in plastic strains will then be defined by the expressions:

$$d\varepsilon_x^p = d\lambda k_p \left\{ \frac{3}{k_p} [I_1(T)]^2 - \sigma_y \sigma_z + \tau_{yz}^2 \right\} \tag{14-65}$$

$$d\gamma_{xy} = d\lambda k_p (\sigma_x \tau_{xy} - \tau_{yz} \tau_{zx})$$

Two unknown parameters, k_p and $d\lambda$, determined experimentally enter these equations. The value of k_p is given by the formula:

$$k_p = \frac{3[I_1(T)]^2[1 + (d\varepsilon_3/d\varepsilon_1)^p]}{\sigma_3[\sigma_1 + (d\varepsilon_3/d\varepsilon_1)^p \sigma_3]} = B\varrho + 27(1 - B) \tag{14-66}$$

In developing the original model, Lade applied it to normally compacted clays under the condition that the experimental data were reduced to the effective stresses.

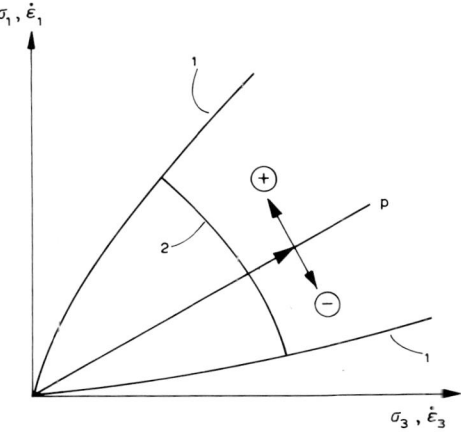

Fig. 14-22. Modified (closed) loading surface after Lade and Duncan: 1 = side surface; 2 = closing sphere. The zones of compressive and tension stresses are marked by the signs + and −, respectively; the directions of the strain increment vectors are shown by arrows.

At the same time, he examined the cases in which the angle of internal friction in soils (both clayey and sandy) varied with the mean normal stress σ_m and when both compacting and softening were likely to occur. The function of plasticity potential f_p and of limiting state Φ are then determined by the expressions:

$$\Phi = \{[I_1(T)]^3/I_3(T) - 27\}[I_1(T)/p_a]^n - n_1 \tag{14-67}$$

$$f_p = [I_1(T)]^3 - \left\{27 + n_2\left[\frac{p_a}{I_1(T)}\right]^m\right\} I_3(T) \tag{14-68}$$

where $n_1 \neq n_2$ are constants.

As can be seen from eq. 14-67, the limiting surface is represented, with allowance for a changing angle of internal friction, by a cone having a convex side surface (Fig. 14-22).

The limiting surface is bound by a cap of the type of Fig. 14-8(b) defined by the function:

$$f_c = [I_1(T)]^2 + 2I_2(T) \tag{14-69}$$

Surface $f_c = 0$ is a sphere (Fig. 14-18) and the vectors of the increment in plastic strain are directed at right angles thereto. Thus, the closing portion of the limiting surface obeys the associated law of flow, whereas the law holding for the side surface is the non-associated one.

The model discussed was also applied to the case of cyclic loads.

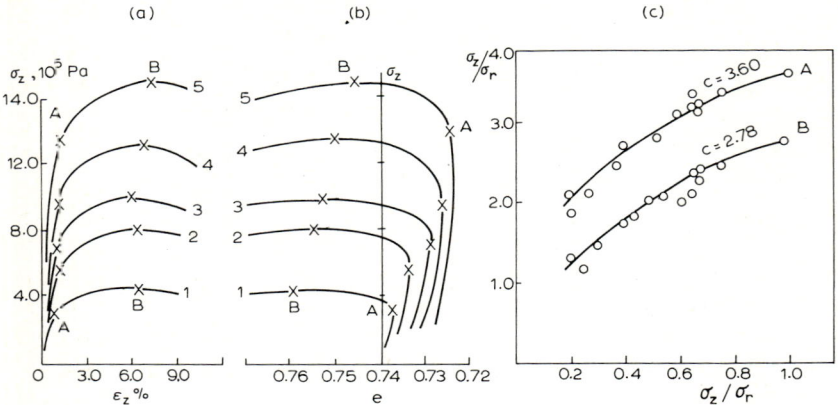

Fig. 14-23. Experimental stress–strain curves after Arnold and Mitchell. (a) Axial strains ε_z. (b) Volumetric strains e. (c) Curves of (A) limiting state and (B) critical state.

The models of Arnold and Mitchell

Describing the behaviour of sandy soil, these models make use of three functions, viz. loading, hardening and plasticity potential (Arnold and Mitchell, 1973). Their type was established from tri-axial tests of hollow specimens by a load $\sigma_z > \sigma_r \geqslant \sigma_t$. Some of the experimental findings are depicted in Figs. 14-23, (a) and (b), by curves of the relationship between the axial stress σ_z, the axial strain ε_z and the volumetric strain (as defined by the changes in void ratio e) for given values of σ_r and σ_t. Two points on the σ_z–ε_z curves are singled out by the authors. Point A, which may be regarded as the conditional yield limit, corresponds to the reversing of the sign of dilatancy: hardening turns into softening with a sharp increase in the value of e. Point B, which may be treated as the ultimate strength represents failures. Corresponding to each of the points there are constant values of strain, $\varepsilon_{z(A)}$ = const., $\varepsilon_{z(B)}$ = const. The two states are represented on the (σ_z/σ_r)–(σ_t/σ_r) plane by a curve of failure and a curve of critical state (Fig. 14-23(c)) referred to by the authors as contraction–dilatation reversion (CDR). This interpretation is close to the principles of the Cam-clay model except that the critical state of Arnold is accompanied by an increase in the shearing strain under conditions of rising stress, i.e. by hardening in shear and not by an unlimited flow in shear.

In the models discussed, the curves of limiting and critical states are described by the same empirical equations:

$$\sigma_z/\sigma_r = C_{1,2} - 2.24(1 - \sigma_t/\sigma_r)^{1.55} \qquad (14\text{-}70)$$

where different values of C_1 and C_2 are used for the limiting and critical curves: $C_1 > C_2$ (the values used by this author in experiments were $C_1 = 3.60$, $C_2 = 2.78$). In other words, the transition from hardening to softening takes place under a stress which is around 75% of the stress at failure.

Note that for $\sigma_t = \sigma_r$ the curve (eq. 14-70) is transformed into a straight line and eq. 14-70 coincides with the equation of critical state (eq. 14-45) by Roscoe and others.

The equation of hardening represents, by analogy with the Cam-clay model, the relation between the void ratio and all-around pressure. However, unlike the Cam-clay model, the p–e relation is adopted in the Arnold model in the power, not the logarithmic, form[*1]:

$$e = e_0 - 0.0000794\, p^{0.76} \qquad (14\text{-}71)$$

where e_0 is the original void ratio.

The connection between the increment in plastic shearing strain and the stress is established by constructing equipotential lines from experimental data, using eq. 14-29. The usual practice is to specify a function of the plasticity potential f_p and experimentally verify the perpendicularity of the vectors of plastic strain increments. However, one may proceed in the opposite way — as was done in the case under consideration — by determining, experimentally, the direction of the strain increment vectors and constructing a family of curves (equipotential lines) at right angles to the vectors. The surface thus obtained will be the surface of plasticity potential. If it coincides with the loading surface, the associated law is in force; if not, the non-associated law holds. This latter case is obtained in the model discussed.

The f_p curves obtained on the σ_z–σ_r plane (for $\sigma_r = \sigma_t$) in the model under consideration have a bullet-like configuration (Fig. 14-24) bound by the line of limiting state (eq. 14-70). The contours of the cross-section of such a surface on the deviatory plane form a family of Mohr circles.

Eqs. 14-71 and 14-29 considered simultaneously enable the determination of the strains ε_r, ε_z and ε_t induced by the stresses σ_z, σ_r and σ_t. However, the work makes no reference to any equations establishing a relevant connection.

Other models for elastoplastic soil subjected to hardening

Let us briefly describe some other soil models based on the theory of plastic flow and treating soil as an elastoplastic medium capable of hardening.

Huang et al. (1981) described the behaviour of soil on establishing experimentally (by analogy with the preceding Arnold model) the connection between three functions — loading function f, plasticity potential f_p and hardening modulus A — the measure of hardening Γ being a parameter of each. Putting the connection between these functions into the matrix form $\{\Delta\sigma\} = [D^{ep}]\{\Delta\varepsilon\}$, the authors suggested the following expression for determining the elastoplastic matrix:

$$[D^{ep}] = [D] - \frac{[D]\left\{\dfrac{\partial f_p}{\partial \sigma}\right\}\left\{\dfrac{\partial f_p}{\partial \sigma}\right\}^T [D]}{A + \left\{\dfrac{\partial f}{\partial \sigma}\right\}^T [D]} \qquad (14\text{-}72)$$

[*1] The numerical values of the parameters in eqs. 14-70 and 14-71 refer to the test in question.

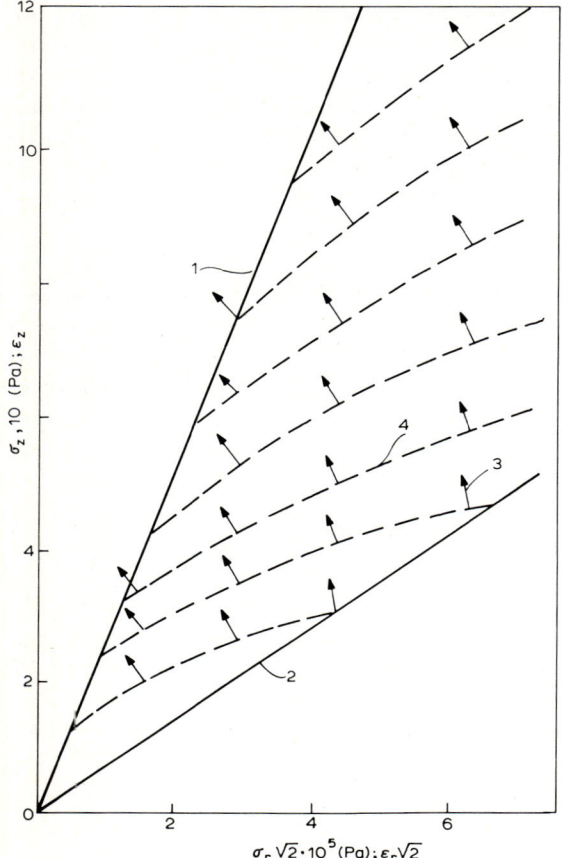

Fig. 14-24. Contour of the surface of plasticity potential on the σ_z–σ_r plane (for $\sigma_r = \sigma_t$) after Arnold and Mitchell: *1* = failure line; *2* = hydrostatic axis; *3* = direction of the spread of the segment of plastic strain increment; *4* = line of plasticity potential.

where:

$$f = f(p, q, \Gamma) \qquad f_p = f_p(p, q, \Gamma) \qquad A = \frac{\partial f}{\partial \Gamma} d\Gamma$$

By comparing its various kinds with the experimental data the measure of hardening Γ was arrived at in the following form:

$$\Gamma = \Gamma(\varepsilon_v^p, \gamma_i^p) = [(\varepsilon_v^p)^2 + 0.5(\gamma_i^p)^2] \cdot 10^4 + 2\sqrt{\gamma_i^p} \tag{a}$$

The loading function and plasticity potential appeared to be identical (associated law) and equal:

$$f = f_p = q^2 + 2(p - 7\Gamma)^2 - 78\Gamma^2 = 0 \tag{b}$$

The elastic constants acquired the following values:

$$G = 217 p_a \left(\frac{p - 0.33q}{p_a}\right)^{0.6} \qquad k = \frac{dp}{d\varepsilon_v^e} = 350p \qquad (c)$$

The Wilde model. Reporting at the French–Polish colloquium on soil mechanics, Wilde suggested a model of sandy soil (Wilde, 1980; Wilde and Zawiszà, 1980). In the papers by his colleagues at the same colloquium, methods of solving certain boundary value problems by using the model were presented.

Although having much in common with the Cam-clay model (or, to be precise, with the Granta-gravel model), the Wilde model differs from this model in that the hardening of soil is attributed not only to the changes in volume but to the shearing deformations as well. The relations which define the Wilde model were obtained in this case in three dimensions and expressed in terms of the stress and strain tensor invariants.

The loading function was adopted in the form:

$$f = f[I_1(T), I_2(D), \varepsilon_v^p, \Gamma]$$

where ε_v^p was regarded as the measure of volumetric hardening and $\Gamma = \gamma_v^p$ was treated as the measure of shear hardening. Accordingly, the equation of the loading surface takes the following form, substituting eq. 14-48:

$$\Phi = \sqrt{I_2(D)} + MI_1(T)\left[\ln \frac{I_1(T)}{I_1^*(T)} - \frac{k_0}{M}\varepsilon_v^p - g(\Gamma)\right] \qquad (14\text{-}73)$$

where $g(\Gamma) = Gk_0/M(M^* - M)\gamma_1^p$.

The constants k_0 and I_1^* are the constants of the original state of the soil (for $\varepsilon_v^p = 0$) and M and M^* are the parameters of strength, where M defines the condition of critical state similar to eq. 14-46 and M^* determines the condition at failure:

$$M = \frac{I_1(T)}{\sqrt{I_2(D)}} \qquad M^* = \alpha M, \text{ where } \alpha > 1 \qquad (14\text{-}74)$$

Although the difference between the critical state and the state at failure is also noted in the Cam-clay model, in the Wilde model it is not only noted but introduced into the equation of the loading surface. The contour of this surface on the $\sqrt{I_2(D)} - I_1(T)$ plane is depicted in Fig. 14-25; similarity with Fig. 14-4(a) is obvious, albeit with due regard for the specific features referred to above.

Proceeding from the associated law of plastic flow, the authors of the model arrived at expressions for the increments in the strains by considering eqs. 14-73 and 14-29 simultaneously:

$$d\varepsilon_{ij}^p = d\lambda \left[\frac{MI_1(T) - \sqrt{I_2(D)}}{I_1(T)}\delta_{ij} + \frac{1}{2\sqrt{I_2(D)}}S_{ij}\right]$$

$$d\varepsilon_v^p = 3\, d\lambda\, \frac{MI_1(T) - \sqrt{I_2(D)}}{I_1(T)} \qquad (14\text{-}75)$$

where $S_{ij} = \sigma_{ij} - \sigma_m$.

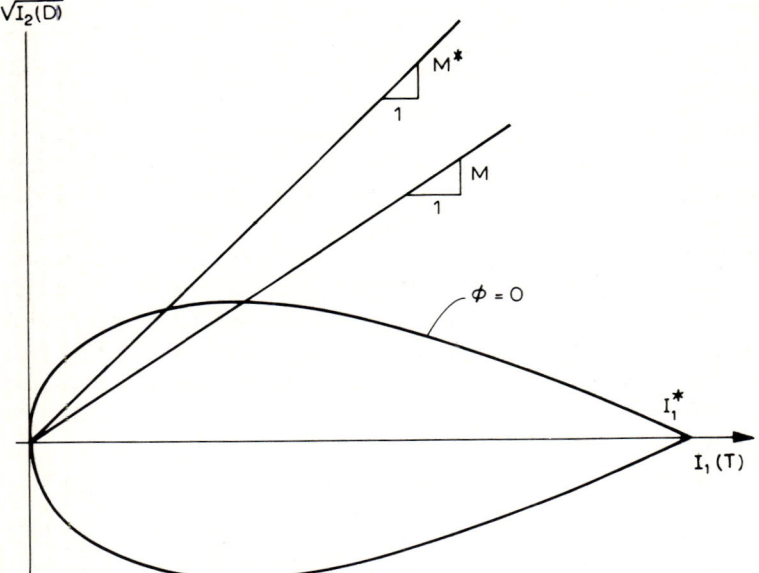

Fig. 14-25. Loading surface for the Wilde model.

The elastic constituent of the strain increment is expressed in terms of Hooke's law.

According to the model suggested by Korhonen (1977) from tri-axial tests of "soft" clay, the yield surface changes its shape depending on the state of deformation. The author singles out the stages of plastic flow and failure. At the former stage, the yield surface is described by an exponential curve and at the latter by an ellipse after Burland. It appears that the relation between the plastic shearing strain and stress at any stage of an undrained tri-axial compression test may be approximated by the linear–fractional relation:

$$q = \frac{\gamma_i^p}{1/G_0 + \gamma_i^p/q_s} \tag{14-76}$$

where q_s is the stress at failure.

The Matsuoka model referred to earlier proceeds from the hypothesis of a "spatial mobilization plane" (cf. eq. 14-2a). In a later work by Matsuoka and Ichizaki (1981) the model was adapted for use with anisotropic soils, the "mobilization plane" being spatially arranged to that end so that the value of angle ϕ can be taken into account during shear in any direction.

The Sandler–Di Maggio–Yamada model was also mentioned previously (the cap model) having the loading surface of the shape shown in Fig. 14-8(a). The model was subsequently modified so as to take account of dynamics, changes in pore pressure, etc. (Sandler and Di Magio, 1976; Sandler, 1979; Yamada, 1979).

Model based on the theory of plastic flow with allowance for time factor

The models discussed above and based on the theory of plastic flow were concerned (as was the theory itself) with the increments in plastic deformations rather than with the time-dependent development of these deformations. Let us consider now a soil model also based on the theory of plastic flow but treating soil as an elastoplastic-viscous (creeping) medium and taking, consequently, into account the time factor. Such a model, suggested by Akai et al. (1977), was derived from triaxial tests of soft sedimentary rock (with a compression strength of $57 \cdot 10^5$ Pa), using various loading paths and various (albeit constant for a given test) strain rates. The tests revealed a connection between the deviatoric stress $q = 1/2(\sigma_1 - \sigma_3)$, the axial strain ε_1 and the volumetric strain ε_v. It was also found that in loading according to the pattern of Fig. 14-5(a), the ln q–ln e_1 relation (where $e_1 = \varepsilon_1 - \varepsilon_m$) was linear but the logarithmic straight line changed into a straight line parallel to the ln e_1-axis when the stress reached the value $q = q_s$. This is an indication that the strain increases in an unconfined way when the stress is constant, $q_s = $ const. Note that corresponding to the value of q_s on the ε_v curve there is a moment when the volume starts to increase unceasingly (i.e., softening occurs). Murayama and Shibata (see Section 1.5: IUTAM Symposium, 1966) suggested this method of determining q_s at an earlier date.

Thus, in the model under consideration there are three limiting states represented on the q–p plane by three curves: (1) curve at failure (for $q = q_s$); (2) yield curve; (3) curve of residual (long-term) strength.

The function of plasticity potential coinciding with the loading function is adopted according to Drucker's associated law of plastic flow and experimental data. Assuming that the plastic flow is not influenced by $I_3(T)$, it may be written as:

$$d\varepsilon_v^p / \sqrt{2J_2(D)} = - d[\sqrt{2I_2(D)}]/d\sigma_m \tag{14-77}$$

The loading (yield) function is obtained by integrating eq. 14-77 for a known law of flow. This law has been established experimentally in the form:

$$\sqrt{2I_2(D)}/\sigma_m = \alpha^*[-d\varepsilon_v^p/\sqrt{2J_2(D)}] + C^* \tag{14-78}$$

where $d\varepsilon_v^p/\sqrt{2J_2(D)} = -(d\varepsilon_v^p/d\varepsilon_1^p)$ is the rate of plastic dilatancy linearly influenced, as can be seen, by the quantity $\sqrt{2I_2(D)}/\sigma_m = q/p$.

Hence, the loading function is obtained in the form:

$$f = C^*\sigma_m \left\{ \left[\frac{C^* + (\alpha^* - 1) \frac{\sqrt{2I_2(D)}}{C_n}}{\alpha^* C^*} \right]^{\alpha/(\alpha^* - 1)} \right\} = K^* \tag{14-79}$$

where $K^* = \sqrt{2I_2(D)}$ and $C^* = [\sqrt{2I_2(D)}]/\sigma_m$.

The applicability of the associated law is proved by the fact that the vectors $\overrightarrow{d\varepsilon_v^p}$ are perpendicular to the loading surface on the q–p coordinates.

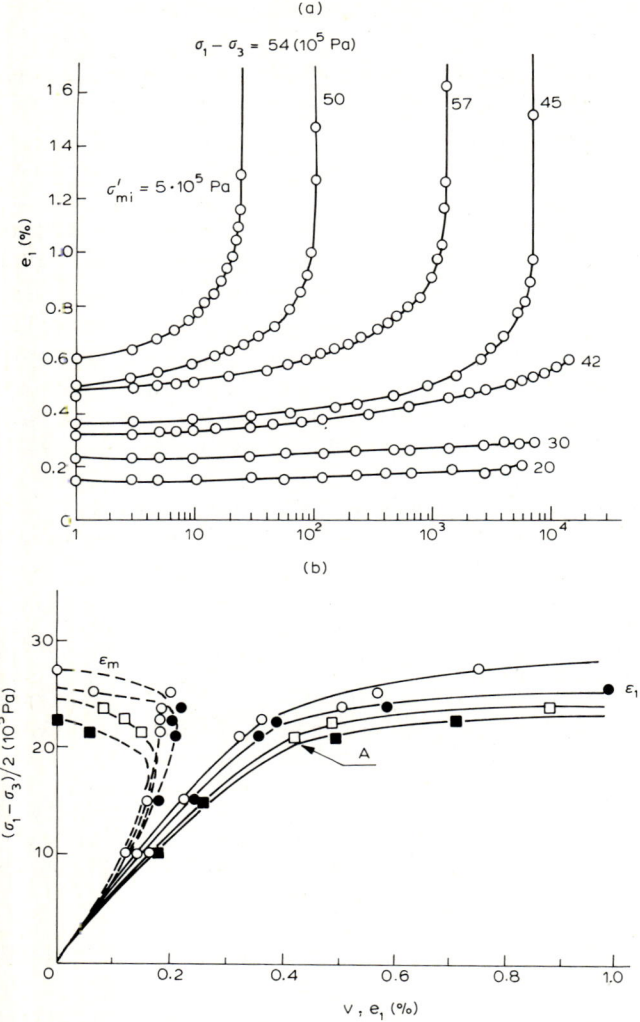

Fig. 14-26. Triaxial compression tests under the conditions of creep (Akai et al., 1977). (a) Creep curves obtained for various values of q. (b) Isochrones of axial, e, and volumetric, ε_m, strains at various times: $1 = 10$ min; $2 = 10^2$ min; $3 = 10^3$ min; $4 = 5 \cdot 10^3$ min; A = point of inflection of the curve corresponding to $q = 1/2(\sigma_1 - \sigma_3) = q_s$.

Experiments indicated that the position of the loading surface was influenced by the rate of strain in loading. To take into account this fact, a family of stress–strain (both shearing and volumetric) curves were plotted for various times (Fig. 14-26(b)), using the data acquired from creep tests (Fig. 14-26(a)). Assuming that the soil was an ideal (not subjected to hardening) elastoplastic-viscous medium, Akai et al. adopted the loading surface in the form of a fixed yield surface (see Fig. 14-2(a)), treating it as

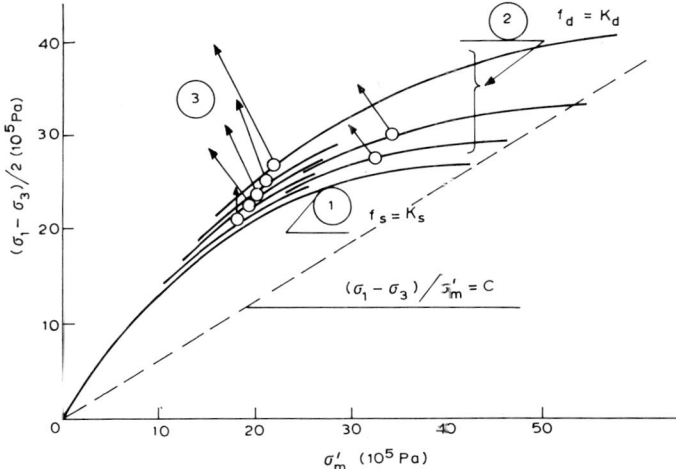

Fig. 14-27. Movable yield surfaces influenced by the strain rate during loading (Akai et al., 1977): 1 = static yield surface $f_s = K_s$; 2 = movable yields $f_d = K_d$ for various strain rates; 3 = vectors of plastic strain rates.

a static one, f_s, in the case of an infinitely slow loading and as movable, dynamic one, $f_d = K_k$, when allowance was made for the strain rate. Accordingly, the loading function was determined from the relation $\Phi = f_d/f_s - 1$, and the connection between the stress and strain rate was described by the expression:

$$\dot{\varepsilon}_{ij} = \eta^* F(\Phi) \partial f_d / \partial \sigma_{ij}$$

where $\eta^* = 1/\eta$.

Assuming that f_s is determined by eq. 14-79 and that η^* therein is equal to K_d, we obtain the following expression for the rate of visco-plastic flow:

$$\dot{\varepsilon}_{ij}^{vp} = \eta^* F(\Phi) \phi[\sqrt{2I_2(D)}/\sigma_m] \left\{ \left[\frac{C^* - \sqrt{2I_2(D)}/\sigma_m}{\alpha^*} \right] \frac{\delta_{ij}}{3} + \frac{S_{ij}}{\sqrt{2I_2(D)}} \right\} \quad (14\text{-}80)$$

where:

$$\phi[\sqrt{2I_2(D)}/\sigma_m] = \{C^* + (\alpha^* - 1)[\sqrt{2I_2(D)}/\sigma_m]\} \alpha^{*1/(\alpha^*-1)}$$

It was experimentally obtained that for conditions of triaxial compression:

$$\Phi = F^{-1}[\eta(\dot{\varepsilon}_1^{vp}/\sqrt{2/3}) \phi(q/p)]$$

$$F(\Phi) = C_2 \exp(\delta_2 \Phi)$$

Yield surfaces of the movable (dynamic) type are depicted in Fig. 14-27. Experiments have shown that the vectors of the rates of steady-state creep (viscous flow) appear to be close to the normals to these surfaces.

Thus, in the elastoplastic–viscous model discussed, the static yield function (loading function) f_s is adopted according to the classical Drucker–Prager theory of plastic flow. When the state of stress exceeds f_s, the flow of soil takes place with an increase in volume due to creep, which leads to a long-term failure. The stress bringing about this failure depends on the strain rate and is determined by the dynamic yield function $f_d = K_d$. Long-term strength, however, is defined by the value $f_s = K_s$ and may be adopted to be roughly equal to the residual strength.

In concluding this chapter we shall try to systematize the various mathematical soil models which have been examined.

Their general tendency towards representing the real properties of soil and taking into account all particulars of its behaviour in response to load is quite obvious. However, the models materially differ from one another by: the type of the plasticity theory employed; the factors deciding the behaviour of a soil under load; the type of constitutive equation and the functions entering.

Original theories of plasticity

In constructing models, either of the two plasticity theories is commonly used: the deformation theory or the theory of plastic flow. In the former case the deformations proper are operated upon; in the latter case we deal with their increments, applying the law of plastic flow in all its three modifications — associated, non-associated and combined (associated–non-associated).

To give preference to any of the theories as the only one deemed acceptable is hardly feasible, all the more because at present we lack experimental proof that calculations carried out on the basis of a certain theory give the closest fit with field performance. Conversely, the comparative calculations referred to in Section 14.5 indicate that the discrepancy between the results obtained is small, being practically within the limits of accuracy observed in determining the original parameters. Therefore, one may say that any of the theories mentioned may be used to construct a soil model, but there is a need to establish the field of application for each model.

The deformation theory, in which many of the limitations mentioned above are inherent, offers maximum simplicity, particularly in assessing experimental data. Therefore, this theory cannot be unconditionally rejected, as is done by some authors; it will always remain a fairly reliable tool in solving engineering problems, especially when the condition of loading is close to the simple one. The theory of plastic flow is more general, allowing for taking into account a greater number of factors. Consequently, the models based on this theory display greater potentialities. As far as the choice between the associated and non-associated laws is concerned, the last-named law, correct as it is from the standpoint of theory, yields models which are practical for use in research only, as the necessity of determining additional parameters renders them complex.

Factors taken into account by various models

The factors which determine the behaviour of soil in response to load and are taken into account by various models are as follows.

(a) Non-linearity of the deformation process and the predominance of the plastic strain component therein. All models represent these peculiarities, but the ways in which they take account of non-linearity and of the development of plastic deformations are different. In most of the models, non-linearity is allowed for in the form of a stress–strain relationship with the plastic strain component manifesting itself from the very beginning of load application (see Fig. 14-1(a)). Such models are, consequently, referred to as elastoplastic models with hardening[*1]. Other models provide for non-linearity by approximating the stress–strain curve with two straight lines (Fig. 14-19) corresponding to the sublimiting and limiting states, the plastic strain developing only during transition into the limiting state. The relevant models are termed ideally elastoplastic ones. The models with hardening make use of both the deformation theory of plasticity (applies to the Shirokov–Solomin, Malyshev–Zaretsky, and Krizhanovsky models, and to the models of some other authors) and the theory of plastic flow (the Cam-clay model and its modifications, the Rowe dilatancy model, the models due to Zaretsky, Ioselevich, Nikolaevsky, Lade and Duncan, Pender, Arnold, Matsuoka and others). The ideal elastoplastic models employ the theory of plastic flow only (the Drucker–Prager model and its modifications, the Bugrov model, etc.).

(b) Dependence of the shearing and volumetric strains upon the three stress tensor invariants. Let us recall that taking into account the effect of the first stress tensor invariant on the shearing strain is equivalent to allowing for the internal friction in soil in both the limiting and sublimiting states, and giving due regard for the influence of the second stress tensor invariant on the volumetric strain is equivalent to taking into consideration the phenomenon of dilatancy; allowing for the effect of the third invariant is equivalent to allowing for the arrangement of stress. The effect of the first and second invariants is taken into account by most models, but that of the third only by some of them. In addition, many models also allow for the loading path.

(c) Lack of alignment and lack of similarity between the state of stress and strain along with the non-linearity of the geometrical relations between the strains and displacements. This factor is taken into account by some models but neglected by most of them without much objection.

(d) Time-dependent changes in the state of stress and strain of soil due to creep, consolidation and deterioration of long-term strength. This aspect of soil behaviour is taken into account only in some models, although, as far as clay soils are concerned, an allowance for the time factor is obligatory.

Type of constitutive equations and of the functions entering

The most significant difference between the models is the type of constitutive equations and the functions entering the equations

---

[*1] It will be remembered that the term hardening, as used in the theory of plasticity, means an increase in the yield limit resulting from an accumulation of the plastic deformation which is likely to progress only with an increase in load (see Fig. 14-2(b)). In the theory of creep, the term hardening is used to denote the process of a decrease in the rate of the plastic strain with an accumulation thereof. A physical interpretation of the hardening process has been given in Section 10.2.

Eq. 14-24 is adopted as the constitutive equation in the deformation models. The models differ from one another by the type of χ and χ^* functions characterizing the relation between the stress tensor invariants and strain tensors (and time). The models based on the theory of plasticity employ eqs. 14-25 to 14-28, or eqs. 14-29 and 14-28 as their constitutive equations. In the former case, when eqs. 14-25–14-28 are used, the models differ by the type of function $d\lambda$, $d\xi$ or $\dot\Lambda$; in the latter case, i.e. when eqs. 14-29 and 14-28 are employed, the difference is the type of function f, f_p or $\dot\Lambda$. As well as this, the models are distinguished by the type of function of the limiting state $\Phi(\sigma_{ij})$.

Most models are based on eq. 14-29 expressing the relation between the increment in plastic strain and stress in terms of the plasticity potential. The greatest diversification exists between the models of this class. Let us consider this in detail.

The models proceeding from eq. 14-29 differ from one another mainly by: (a) the type of the loading function f and that of the function of limiting state Φ; (b) the connection existing between the plasticity potential f_p and the loading function f; (c) the type of connection (eq. 14-28) between the increments in the shearing and volumetric strains.

(a) The loading functions f differ in type depending on the kind of loading surface they describe. It will be recalled that the loading function is a function of stress $f(\sigma_{ij})$ separating during the process of loading and unloading the region of elastic strains from that of plastic strains; if interpreted geometrically in the $(\sigma_1-\sigma_2-\sigma_3)$ space, the loading function represents the loading surface $f(\sigma_{ij}) = 0$. The shape of this surface varies with the type of function f. Distinction is made between fixed and "expanding", nonclosed and closed surfaces. A fixed loading surface corresponds to an ideally elastoplastic model, coinciding with the surface of ultimate loading $f(\sigma_{ij}) = \Phi(\sigma_{ij})$ shown in Fig. 14-2(a) (the Drucker–Prager model and its modifications). An "expanding" loading surface is inherent in elastoplastic models allowing for hardening, in which an increase in plastic strain is accompanied by an increase in the yield limit (Fig. 14-2(b)). In the limit, this surface is transformed into the surface of limiting state, $\lim f = \Phi$ (Fig. 14-3). Thus, for an ideal model it is sufficient to know the type of function $f = \Phi$, whereas the hardening models require knowledge of the type of functions f and Φ and of the law of hardening establishing the relation between the functions.

In most of the models the loading surface and the surface of limiting state are constructed in the $(\sigma_1-\sigma_2-\sigma_3)$ space. An exception is the Cam-clay model and its modifications in which the loading surface is constructed on the $(p-q-e)$ coordinates, where e is the void ratio. However, the critical state is described by two conditions, $\Phi_\sigma(q, p) = 0$ and $\Phi_e = (p, e) = 0$ (Fig. 14-14).

The appearance of the loading surface, which may be closed or not, indicates the type of dilatancy. In the former case account is taken of the dilatancy of softening only, in the latter case allowance is made for both softening and hardening. Nonclosed loading surfaces were employed in the early models of Drucker–Prager, closed surfaces are used in most other models, including the modified models of the two last-named authors. Closed surfaces of various types are depicted in Fig. 14-8.

As far as the type of limiting function $\Phi(\sigma_{ij})$ is concerned, in many models it is of the form corresponding to the classical conditions of limiting state formulated by Mohr–Coulomb and von Mises–Botkin. However, some of the models adopt the Φ function differing from these conditions (Fig. 14-4) mainly because they take into account the arrangement of stress (the models by Shibata–Karube, Lade, Matsuoka and others as well as the deformation models by Lomize and Krizhanovsky, Geniev and others).

(b) The connection between the plasticity potential f_p and the loading function f defines the type of law of plastic flow adopted in the model – associated, non-associated or associated–non-associated. It will be recalled that the plasticity potential is a function f_p, describing in the $(\sigma_1-\sigma_2-\sigma_3)$ space a surface fulfilling the condition of perpendicularity of the vectors of plastic strain increments. In the associated models (the Cam-clay model and its modifications, the Rowe, Zaretsky, Ioselevich–Didukh, Huang, and Wilde models, etc.), the plasticity potential coincides with the loading function, $f = f_p$, and therefore only the value of f is specified in such models. In the non-associated models (Nikolaevsky, Pender, Lade–Duncan models, etc.) $f \neq f_p$ and, consequently, the types of both functions must be specified. In the associated–non-associated models (the modified Lade–Duncan model, the Vermeer and Arnold models, etc.), some segments of the loading surface obey the associated law and others do not.

(c) The type of connection between the increments in plastic shearing strains and plastic volumetric strains defines, according to eq. 14-28, the value of the dilatancy rate Λ. The difference in the type of connection mainly affects the associated and non-associated models. In the former case, the value of Λ is obtained automatically from the condition $f = f_p = \Phi$ so that no extra determination is necessary. In the latter case, Λ is an additional parameter related to the function of plasticity potential f_p and is determined either directly from experiments (the Nikolaevsky and Pender models, etc.) or from the value of the plasticity potential f_p (the Vermeer, Lade–Duncan and Arnold–Mitchell models, etc.).

Outlook

As can be seen from the above survey, a great variety of models is available at present. One may ask, is enthusiasm about the models of any value? In the author's opinion this tendency is quite reasonable and practical, for as long as the problem of non-linear soil mechanics is at the development stage the competing versions of the various models will pave the way to its solution. Dwelling upon the modern soil models in this chapter is aimed at providing the reader with the most exhaustive knowledge of the tenor of the research in the field of non-linear soil mechanics. The problem of selecting the right model will be solved by the practice itself; however, at present it is evident that striving for theoretical correctness of a model should, without prejudice, be aimed at simplicity and clearcut physical meaning. Exactly these factors should govern the selection of a model. Extreme devotion to theoretical construction and complication of soil models is, as is known from experience, of no avail.

Finally some further lines of inquiry in the field of non-linear soil mechanics. As has already been pointed out, the object of developing non-linear soil models is to represent in theoretical constructions the real properties of soil as fully as possible. However, the more factors a model takes account of, the more parameters are incorporated, which all add to the model's complexity. In the light of the capabilities of the computer, this complexity poses no insuperable difficulty in arriving at a solution, but it is quite obvious that if a growing number of soil characteristics must be taken into account, the limits of expediency may be overstepped, not only because of the difficulties experienced in determining them but due to errors which may be introduced owing to the spread in the values of parameters resulting from the inhomogeneity of the soil.

At the same time, because various factors produce different effects, the results of taking into account some factors are felt quite vividly, whereas the role of other factors may be insignificant. Therefore, the object of further studies should be the evaluation of various factors and the effect they produce on the existing conditions rather than unearthing new factors influencing soil behaviour. In other words, the problem of developing non-linear soil models is to be reduced, in our opinion, to selecting or creating models which incorporate a minimum number of design variables and represent the real properties of soil with an adequate closeness of fit. The criterion of the closeness of fit should be the experiment and its comparison with the field performance. In this respect top priority must be given to the development of reliable methods of determining the engineering properties of soil, particularly in the field, along with methods of monitoring the behaviour of the soil used as foundation or construction material, both during the period of construction and in service. A large-scale programme of field experiments and the monitoring of structures are other aspects requiring attention.

14.6 REFERENCES

Akai, K., Adachi, T. and Nishi, K., 1977. Mechanical properties of soft rocks. Proc. 9th ICSMFE, Tokyo, v. 1.

Alexandrovich, V.F. and Fedorovsky, V.G., 1979. Experimental and Analytical Studies of Nonlinear Problems in Foundation Engineering. Novocherkassk Polytechnic Publ. House, Novocherkassk (in Russian).

Arnold, M. and Mitchell, P.W., 1973. Sand deformation in three-dimensional stress state. Proc. 8th ICSMFE, Moscow, v. 1.1.

Balady, G. and Rohani, B., 1973. In: J. Geotech. Eng. Div., Proc. ASCE, v. 105, NGT4.

Balasubramaniam, A.S., 1975. Recoverable strains in triaxial specimens of saturated soil. Proc. 1st Baltic. Conf. Soil Mech. Found. Eng., Gdansk, v. 2.

Balasubramaniam, A.S., Brenner, K.P. and Hazan, Z., 1977. Stress–strain behaviour of stiff Bangkok clay. Proc. 9th ICSMFE, Tokyo, v. 1.

Barden, L. and Khayatt, A.J., 1966. Increment strain rate ratios and strength of sand in triaxial tests. Géotechnique, 16(4).

Barden, L., Khayatt, A.J., et al., 1969. General Deformation Behaviour of a Range of Particulate Materials under a Variety of Test Conditions. Univ. of Manchester Report.

Bishop, A., 1966. The strength of soils as engineering materials. 6th Renkine Lecture. Géotechnique, v. 14, no. 2.

Bishop, A., 1972. Shear strength parameters for undisturbed and remoulded soil specimens. In:

R.H.G. Parry (Editor), Stress–Strain Behaviour of Soils. Proc. Roscoe Memorial Symp. Cambridge Univ., Cambridge.

Bishop, A. and Eldin, A.K.G., 1953. The effect of stress history on the relation between ϕ and porosity of sand. Proc. 3rd ICSMFE, v. 1.

Bugrov, A.K., 1974. On the solution of the combined problem of elasticity theory and plasticity theory. SMFE, no. 6 (in Russian).

Bugrov, A.K., 1980. On the effect of loading path on the stress–strain state of soil. SMFE, no. 2 (in Russian).

Bugrov, A.K. and Grebnev, K.K., 1977. Numerical solutions of physically nonlinear problems for soil foundations. SMFE, no. 3 (in Russian).

Bugrov, A.K. and Zarkhi, A.A., 1978. Some results of combined problems of elasticity theory and plasticity of soil foundations. SMFE, no. 3 (in Russian).

Burland, J.B., 1965. The yielding and dilatation of clay, Correspondence. Géotechnique, v. 15, no. 2.

Casagrande, A., 1936. Characteristics of cohesionless soils affecting the stability of slopes and earth fills. In: Contributions to Soil Mechanics. Boston Soc. Civ. Eng., 1925–1940.

Desai, Ch.S., 1971. Nonlinear analysis using spline function. J. Soil Mech. Found. Div. Proc. ASCE, v. 97, NSM 10.

Didukh, B.N. and Ioselevich, V.A., 1971. On constructing the theory of plastic hardening of soil. News, USSR Acad. Sci., Mechanics of Solid, no. 2 (in Russian).

Drucker, D.C., Gibson, R. and Henkel, D., 1957 Soil mechanics and work-hardening theories of plasticity. Trans. ASCE, v. 122.

Drucker, D.C. and Prager, W., 1952. Soil mechanics and plastic analysis of limit design. Q. Appl. Math., 10(2).

Ergun, M.U., 1977. Evaluation of three-dimensional shear testing. Proc. 10th ICSMFE, Stockholm, v. 1.

Fedorovsky, B.G. and Koganovskaya, S.E., 1975. Rigid settlement plate on a nonlinearly deforming foundation. SMFE, no. 1 (in Russian).

Geniev, G.A., 1977. A variant of strength condition for cohesive soils and rocks. In collective volume: Proc. Res. Inst. for Structures, Investigation of the Stress–Strain State of Structures (in Russian).

Goldsheider, M. and Gudehus, G., 1973. Rectilinear extension of dry sand: testing apparatus and experimental results. Proc. 8th ICSMFE, Moscow, v. 1.1.

Gorbunov-Posadov, M.I., 1962. Stability of Foundations on a Sandy Soil. Gosstroiizdat, Moscow (in Russian).

Gorbunov-Posadov, M.I., 1971. Method for solving the combined problem of elasticity theory and plasticity theory for soils. SMFE, no. 2 (in Russian).

Green, G.E. and Bishop, A.W. 1969. A note on the drained strength of sand under generalized strain conditions. Géotechnique, v. 19, no. 1.

Habib, P., 1953. Influence de la variation de contrainte principal moyenne sur la résistance au cisaillement soiles. Proc. 3rd ICSMFE, Zurich, v. 1.

Haythornwaite, R.M., 1960. Mechanics of the triaxial test for soils. J. Soil Mech. Found. Div. ASCE, 86, no. SM5.

Henkel, D.J., 1960. The shear strength of saturated remoulded clays. ASCE Conf. on Shear Strength of Cohesive Soils, Univ. of Colorado, 1960.

Herrera, J., Leon, J.L. and Fernandes del Omo, R., 1977. Preconsolidation and its rheological implications. Proc. 9th ICSMFE, Tokyo, v. 1.

Hill, R. and Rice, J.R., 1973. Elastic potential and the structure of inelastic constitutive laws, SIAM. J. Appl. Math., v. 25, no. 3.

Höeg Kaar, 1972. Finite element analysis of strain-softening clay. J. Soil Mech. Found. Div. Proc. ASCE, v. 98, no. SMI.

Huang Wen-Xi, Pu Yia-Liu and Chen Yu-Jiong, 1981. Hardening rule and yield function for soils. Proc. 10th ICSMFE, Stockholm, v. 1

Hvorslev, M.J., 1937. Über die Festigkeitseigenschaften gestörter bindiger Böden. Ingvidensk. Skr., A, no. 45. (English transl. No. 69-5, Waterways Experimental Station, Vicksburg, Miss. 1950.)

Hvorslev, M.J., 1960. Physical components of the shear strength of saturated clays. ASCE Res. Conf. Shear Strength of Cohesive Soils, Univ. Colorado, 1960.

Ioselevich, V.I. and Didukh, B.I., 1970. On application of plasticity hardening theory to description of soil deformability. In collective volume: Problems of Soils Mechanics and Construction in Loess Soils. Checheno-Ingush Publ. House, Grozny (in Russian).

Ioselevich, V.I., Zuev, V.V. and Chakhtauri, G.A., 1975. On effects of plastic hardening of soft soils. Proc. Mech. Inst., Moscow State Univ., no. 2, Moscow (in Russian).

Ivastchenko, I.N. and Zakharov, M.N., 1973. On application of flow theories for soils. Proc. 8th ICSMFE, v. 4.3.

Jamada, J., Jashimura, N. and Sakurai, T., 1968. Plastic stress–strain matrix and its application for the solution of elasto-plastic problems by the limit element method. Int. J. Mech. Science.

James, B.G., Smith, I.A. and Bransby, P.L., 1972. The prediction of stress and deformations in a sand mass adjacent to a retaining wall. Proc. 5th Europ. CSMFE, Madrid, v. I.

Josselin de Jong, G. de, 1967. Free discussion. Proc. Geotech. Conf., Oslo, v. 2.

Kirkpatrick, W.M., 1957. The condition of failure for sands. Proc. 4th ICSMFE, v. 1.

Ko, Hon-Yim and Scott, R.F., 1968. Deformation of sand at failure. J. Soil Mech. Found. Div., Proc. ASCE, v. 94, no. SM4.

Koiter, W.T., 1960. General Theorems for Elastic–Plastic Solids. Progress in Solid Mechanics. Amsterdam, v. 1.

Kopeikin, V.S. and Solomin, V.I., 1977. Sand foundation analysis by physically and geometrically non-linear equations. SMFE, 1 (in Russian).

Korhonen, K., 1977. Stress and strain in undrained test. Proc. 9th ICSMFE, Tokyo, v. 1.

Krizhanovsky, A.L. and Kulikov, O.V., 1977. To the analysis of slope stability. Hydr. Engin. Projects, no. 5 (in Russian).

Lade, P.V., 1979. Three-Dimensional Stress–Strain Behaviour and Modelling of Soils. Ruhr-Universität, Bochum.

Lade, P.V. and Duncan, J.M., 1973. Cubical triaxial tests on cohesionless soil. J. Soil Mech. Found. Div., Proc. ASCE, v. 99, no. SM10.

Lade, P.V. and Duncan, J.M., 1975. Elastoplastic stress–strain theory for cohesionless soil. J. Geotech. Eng. Div., ASCE, v. 101, no. GT10.

Lade, P.V. and Duncan, J.M., 1976. Stress path-dependent behaviour of cohesionless soil. J. Geotech. Eng. Div., Proc. ASCE, v. 102, no. GT1.

Lade, P.V. and Musante, H.M., 1977. Failure conditions in sand and remoulded clay. Proc. 9th ICSMFE, Tokyo, v. 1.

Lade, P.V. and Musante, H.M., 1978. Three-dimensional stress–strain behaviour of remoulded clay. J. Geotech. Eng. Div., Proc. ASCE, 104, no. GT2.

Larsson, R. and Sällfors, G., 1981. Hypothetical yield envelope at stress rotation. Proc. 10th ICSMFE, Stockholm, v. 1.

Lomize, G.M. and Krizhanovsky, A.L., 1966. Basic relations of stress state and strength of sandy soils. SMFE, no. 3 (in Russian).

Lomize, G.M. and Krizhanovsky, A.L., 1967. On the strength of sand. Proc. Geotechn. Conf., Oslo, 1967.

Malyshev, M.V., 1980. Strength of soils and stability of foundations. Moscow (in Russian).

Malyshev, M.V., Zaretsky, Yu.K., et al., 1973. Interaction of rigid foundations with a base that deforms nonlinearly. Proc. 8th ICSMFE, Moscow, v. 1.3.

Mandel, G. and Fernandez, Luqul, 1970. Fully developed plastic shear flow of granular materials. Géotechnique, London, v. 20, no. 3.

Matsuoka, H., 1976. On the significance of the spatial mobilized plane. Soil and Foundation. Jap. Soc. Soil Mech. Found. Eng., v. 16, no. I.

Matsuoka, H. and Ichizaki, 1981. Deformation and strength of anysotropic soil. Proc. 10th ICSMFE, Stockholm, v. 1.

Matsuoka, H. and Nakai, T., 1977. Stress–strain relationship of soil based on the SMP. Proc. 9th ICSMFE, Spec. Sect., Tokyo.

Mayne, P.W., 1980. Cam-clay prediction of undrained strength. J. Geotech. Eng. Div., Proc. ASCE, v. 106, no. GT11.

Nagaraj, T.S. and Somashekar, B.V., 1977. Shear strength of soil under general stress field. Proc. 9th ICSMFE, Tokyo, v. 1.
Negre, M.R. and Shutz, P., 1870. Contribution à l'études des fondations de révolution dans l'hypothèse de la plasticité parfaite. Int. J. Solids Struct., v. 1, no. 1.
Nikolaevsky, V.N., 1972. Mechanical properties of soils and plasticity theory. Rev. Sci. Technol., Mech. Deformable Solids. All-Union Inst. Sci. Techn. Inf., Moscow, v. 6 (in Russian).
Nikolaevsky, V.N., 1979. Dilatancy and laws of repeated soil deformation. SMFE, no. 5 (in Russian).
Ohmaki, S., 1979. Strength and Deformation Characteristics of Overconsolidated Cohesive Soil. 3rd Conf. of Num. Math., in Geotechn., Aachen, Rotterdam, v. 1.
Paul, B., 1968. In: Fracture. Acad. Press, New York, London, v. 2.
Pender, M.J., 1978. Model for the behaviour of overconsolidated soil. Géotechnique, v. 28, no. 1.
Poorooshasb, H.B., Holubec, J. and Sherborne, A.N., 1966. Yielding and flow of sand in triaxial compression. Can. Geotech. J., v. 3, no. 4.
Poorooshasb, H.B., Holubec, J. and Sherborne, A.N., 1967. Yielding and flow of sand in triaxial compression. Can. Geotech. J., v. 4, no. 4.
Proctor, D.C. and Barden, L., 1969. Correspondence on Green and Bishop: A Note on the Drained Strength... Géotechnique, v. 19, no. 3.
Roscoe, K.H., 1968. Engineering Plasticity. Cambridge Univ. Press, Cambridge.
Roscoe, K.H., 1970. 12th Rankine Lecture: The Influence of Strains in Soils Mechanics. Géotechnique, v. 20, no. 2.
Roscoe, K.H. and Burland, J.B., 1968. On the generalized behaviour of wet clay. In: Engineering Plasticity. Camb. Univ. Press, Cambridge.
Roscoe, K.H., Schofield, A.N. and Wroth, C.P., 1958. On the yielding of soils. Géotechnique, v. 8, no. 1.
Rowe, P.W., 1962. The stress–dilatancy relation for static equilibrium of an assembly of particles in contact. Proc. R. Soc., A269.
Rowe, P.W., 1972. Theoretical meaning and observed values of deformation parameters for soil. Proc. Roscoe Mem. Symp., Cambridge Univ., 1972.
Sandler, I.S., 1979. In: Proc. 3rd Int. Conf. Num. Method Geomech., Aachen, v. 1.
Sandler, I.S. and Di Magio, F.L., 1976. Generalized cap model for geological materials. J. Geotech. Eng. Div., Proc. ASCE, v. 102, no. GT7.
Schofield, A.N. and Wroth, C.P., 1968. Critical State Soil Mechanics. London.
Serrano, A.A., 1972. The method of associated fields of stress and velocity and its application to earth pressure problems. Proc. 5th Europ. CSMFE. Madrid, v. 1.
Shibata, T. and Karube, D., 1965. Influence of the variation of the intermediate principal stress on the mechanical properties of normally consolidated clays. Proc. 6th ICSMFE, Montreal, v. 1.
Shirokov, V.N., Solomin, V.I., Malyshev, N.V. and Zaretsky, Yu.K., 1970. Stress state and displacement of ponderable non-linearly deformable semi-space below a rigid circular settlement plate. SMFE, no. 1 (in Russian).
Stroganov, A.S., 1966. Sur certains problèmes rhéologiques de la méchanique des sols. Rheology and Soil Mechanics, IUTAM Symp., Grenoble, 1964.
Suh, N.P., 1969. A yield criterion for plastic frictional work-hardening granular materials. Int. J. Powder Metallurgy, 5(1).
Tanaka, T., 1979. In: Bull. Nat. Res. Inst. Agric. Eng., no. 18.
Tavenas, F. and Leroueil, S., 1977. Effects of stress and time on yielding of clays. Proc. 9th ICSMFE, Tokyo, v. 1.
Vermeer, P.A., 1978. A double hardening model for sand. Géotechnique, v. 28, no. 4.
Vinokurov, E.F. and Karamyshev, A.S., 1972. Iterative Method of Foundation Analysis by Means of the Computer. Minsk (in Russian).
Wilde, P., 1980. Principes mathématiques et physiques des modèles élastoplastiques des matériaux granulaires. Coopération Franco-Polonaise Mécan. des Sols. Colloque de Paris, 1979. Lab. Centr. des Ponts et Chauss., Paris.
Wilde, P. and Zawiszà, W., 1980. Détermination expérimentale des paramètres du modèle élasto-plastique

des sols pulvérulents. Coopération Franco-Polonaise Mécan. des Sols. Colloque de Paris, 1979. Lab. Centr. des Ponts et Chauss., Paris.

Wroth, C.P., 1965. The prediction of shear strains in triaxial tests on normally consolidated clays. Proc. 6th ICSMFE, Montreal, v. 1.

Wroth, C.P., 1972. General theories of earth pressure and deformations. Proc. 5th Europ. ICSMFE, v. 2, General Report, Session 1, Madrid.

Wroth, C.P., 1973. Contribution. Spec. Session 2. Proc. 8th ICSMFE, Moscow, v. 4, 3.

Wroth, C.P. and Bassett, R., 1965. A stress–strain relationship for the shearing behaviour of a sand. Géotechnique, v. 15, no. 1.

Yamada, S., 1979. Model Guided by Energy Concept. J. Geotech. Eng. Div., Proc. ASCE, v. 105.

Yudhbir and Mathur, S., 1977. Path-dependent creep of clays. Proc. 9th ICSMFE, Tokyo, v. 1.

Zaretsky, Yu.K. and Lombardo, V.P., 1979. Plastic flow of soil masses. College news "Construction and Architecture", no. 2 (in Russian).

Zaretsky, Yu.K., Lombardo, V.N., Groshev, M.E. and Olimpiev, D.N., 1980. Stability of soil slopes. SMFE, no. 1 (in Russian).

Zienkiewiez, O.C. 1967. The Finite Element Method in Structural Continuum Mechanics. McGraw-Hill, New York, N.Y.

APPENDIX

Some SI units and their conversion into units of other systems

Quantity	SI units	Units in other systems	Equivalents
Length	Metre (m) $1\,m = 10^3\,mm$ $= 10^6\,\mu m$ (micrometres)	Micron (μm) Ångström (Å)	$1\,\mu m = 10^{-6}\,m$ $1\,\text{Å} = 10^{-4}\,\mu m = 10^{-10}\,m$
Force, loading	Newton (N) $1\,N = 10^{-3}\,kN$ $= 10^{-6}\,MN$	Dyne Kilogramme-force (kg)	$1\,\text{dyne} = 10^{-5}\,N$ $1\,kg = 9.81\,N = 10\,N$ approx. $1\,g = 9.81 \cdot 10^{-3}\,N = 10\,mN$ approx.
Stress	Pascal (Pa) ($1\,Pa = 1\,N/m^2$) $1\,Pa = 10^{-3}\,kP$ $= 10^{-6}\,MPa$	Kilogramme-force per square centimetre (kg/cm²)	$1\,kg/cm^2 = 9.81 \cdot 10^4\,Pa$ $= 10^5\,Pa$ approx. $= 10^{-1}\,MPa$ $1\,kg/mm^2 = 9.81 \cdot 10^6\,Pa$ $= 10^7\,Pa$ approx. $= 10\,MPa$
Load per unit area	Newton per square metre (N/m²) $1\,N/m^2 = 1\,Pa$	kg/m²	$1\,kg/m^2 = 9.81\,N/m^2 = 10\,N/m^2$ approx.
Pressure	Pascal (Pa)	Bar, technical atmosphere (at), physical atmosphere (atm), dyne/cm²	$1\,\text{bar} = 10^5\,Pa$ $1\,\text{at} = 1\,kg/cm^2 = 0.98 \cdot 10^5\,Pa$ $= 10^5\,Pa$ approx. $1\,\text{atm} = 76\,cm\,Hg = 1.01 \cdot 10^5\,Pa$ $1\,\text{dyne}/cm^2 = 9.81 \cdot 10^{-2}\,Pa$ $= 10^{-1}\,Pa$ approx.
Work, energy	Joule (J)	Kilogramme-force metre (kgm) Erg	$1\,kgm = 9.81\,J = 10\,J$ approx. $1\,\text{erg} = 10^{-7}\,J$
Heat energy	Joule (J)	Calorie (cal.) Kilocalorie (kcal.)	$1\,\text{cal.} = 4.19\,J$ $1\,\text{kcal.} = 10^3\,\text{cal.}$ $= 4.19 \cdot 10^3\,J = 4.19\,kJ$
Power	Watt (W)	kg m/s cal./s	$1\,kg\,m/s = 9.8\,W = 10\,W$ approx. $1\,\text{cal.}/s = 4.19\,W$ $1\,\text{kcal.}/s = 4.19\,kW$
Viscosity (coefficient of viscosity)	Newton-second per square metre (N s/m²)	Poise (P) Dyne s/cm²	$1\,P = 1\,\text{dyne}\,s/cm^2 = 1.2 \cdot 10^3\,g\,s/cm^2$ $= 10^{-1}\,N\,s/cm^2$
Weight	Newton (N) ($1\,N = 1\,kg\,m/s^2$)	kg (kgf)	$1\,kg = 9.81\,N = 10\,N$ approx.

SUBJECT INDEX

Activation energy, 258, 260, 261, 341
—, changes due to deformation, 370, 371
After-effect of stress, 7, 130
Aggregates, 28, 46, 51
Aggregation factor, 44
All-around compression, 63
Allowance for unloading, 254–257
Angle of obliquity, 102
— — strain arrangement, 79, 82
Angular distortions, 61, 63
Anisotropic strength of soil, 51
Arithmetical mean, 196
Arrangement of stress, angle of, 76, 82
—, effect of, 494
—, generalized stresses and strains for (table), 77

Bach's power expression, 111
Bingham body, 127
— and Schwedoff bodies, 226
Blocks, 28, 46, 51
Body forces, 84, 85
—, method of, 276
Bound water, 31
Boussinesq's problem, 451
Brittleness index, 305
Bulk modulus, 98

Capillary forces, 37
Clay soils, comparative strength of bonds in (table), 38
—, dilatancy of, 422
—, rheological studies of, 10
Clay minerals, classification of, 30
Closeness of fit, 200
Coagulation, 28
—, structure of, 44
Coagulation–thixotropy structure, 43

Coefficient of dilatancy, 424
— — lateral expansion, 116, 272, 456
— — — strain, 98
— — — strength, 75
— — variation, 196
Coefficients of plastic viscosity, connection between, 186
Colloidal clay particles, 30
Components of displacement rates, 71
Condensation–crystallization structure, 43
Confidence interval, 196
Confined compression, 76
Condition of perpendicularity, 506
— — plasticity, Tresca–St.-Venant, 99
Consolidation, degree of, 275
—, percolation theory of, 13
—, primary and secondary, 13
Contraction–dilation reversion, 540
Correlation of plastic strain rates with stress, 93
Crack, classification of, 344
Creep of sand, 184
Creep of soil, attenuating, 166
—, defined, 8
—, dilatancy during, 427
—, deterioration of strength due to, 288
—, exponential relation for, 160
—, hereditary, 13
—, heat-activated process of, 264, 265
—, limit of, 288
—, linear–fractional relation for, 158
—, logarithmic relation for, 157
—, pattern of, 373
—, power relation for, 154
—, secular, 147
—, structural changes due to, 344–347
—, time function for, 154
—, generalized, 159

—, threshold of, 164
—, volumetric deformation due to, 268–270
Critical stress, 162, 376
Curve(s)
    invariant, 153
    creep, 152, 214, 238, 268, 272, 300, 307, 377
    deformation, 390
    finite deformation, 158
    long-term strength, 285, 300
    relaxation of stresses, 255
    settlement versus load, real, 197

Deformation–displacement relation, Cauchy's, 444
Deformation of dams and bridges, 15
— — pile walls, 18
Deformation of soil, analysis of equations, 403
—, approximate formula for, 377
—, description in terms of thermodynamics, 147
—, elastic, 9
—, —, work of, 139
—, elastoplastic, work of, 141
—, equations for, 377, 431
—, finite, 66, 158
—, generalized equation for, 385
—, instantaneous, 158
—, irrecoverable, work of, 140
—, kinetic, nature of, 340
—, linear–fractional law of, 158, 448
—, measures of, 65
—, mechanism of, 350
—, particulars of, 482
—, physical theories of, 257, 258
—, power law of, 156, 447
—, recoverable, 135
—, residual, 150
—, soil-ice retaining structures, 444, 445
—, viscous, 137
—, volumetric, 123
Deformation theory, 388, 504
Deformation, work of, 138
Deviatoric plane, 79, 105
Dispersion, 28
Domains, 28

Elastic after-effect, 7, 10, 225
— constants, 98
Elastoplastic bodies, 129
— Prandtl body, 226
Equation, compatibility, 85
—, differential, of model, 228
—, kinetic, of strain rate, 372
—, quadratic, Volterra, 236

Equations of:
    consolidation, 272
    consolidation and after-effect, 275
    consolidation, single-dimensional problem, 273
    creep, 157, 238, 253
    deformation, 168, 283, 370, 375
    —, in integral form, 400
    engineering theories of creep, 247, 248
    envelope, 395
    equilibrium, plasticity theory, 84
    flow, 161
    flow, Eyring's, 261, 264, 265
    limiting state, 419
    —, von Mises–Botkin, 105
    —, modified, 495–497
    long-term soil strength, 353
    —, linear-fractional, 314, 315
    —, logarithmic, 312
    non-linear viscoplastic flow, 410
    non-linear viscous flow, 409
    plastic potential line, 424
    relaxation, 238, 252, 253
    steady flow, 262
    viscous liquid flow, 118
    —, Navier–Stokes, 119
    volumetric creep, 268
    volumetric strain, 423
Euclidian solid, 97
Exponential relation, 160

Factor of safety, 197
Failure of soil, activation energy of, 358
—, criterion of, 306, 340
—, long-term, effect of temperature on, 360
—, —, kinetic nature of, 355
—, —, pattern of, 352
—, time before, 359
Falling ball viscometer, 122
Fields of slip, 102, 103
Finite difference and finite element methods, 510
Finite volumetric strain, 67
Flow, associated law of, 94
—, defined, 8
— law, 448
— of dispersed medium, 263
—, progressive, cause of, 174
—, steady, 410
—, viscoplastic, 148
—, —, mechanism of, 162
Fluidity, 118
Forces of chemical nature, 35
Forces, wedging, 38
Friction, internal, allowing for in limiting and sub-

limiting states, 395
—, kinematic, 116
— of rest, 116
Frozen soils, rheological studies of, 9
Function, creep, 153, 155, 251
—, deformation, 154
—, linear-fractional, 113, 154, 158
—, loading, 489, 505
—, power, 111, 154
—, strain, Ilushin's, 111
—, stress, 362
—, time, 154, 159, 167, 186, 254

Gas in soil, 31
Generalized stress, 75
— — and strain for various arrangements of stress (table), 77
— viscoplastic body, 225
Graphs, invariant, 406, 410
Green's measure, 66

Hardening measure, 489
Hencky's equations, 441
— measure, 67
— relation, 444
Hookean solid, 97
Hooke's law, 112, 114
Hypothetically instantaneous strength, 9

Ice-cementation adhesion, 43
Increments in elastic and plastic strains, 92
Intensity of rate of shear strain, 83
Invariants of stress deviator, 73
— — — tensor, 77
Ionic-electrostatic forces, 36
Irrecoverable volumetric changes, 13
Isochrones, 152, 153, 158, 214, 216, 238, 248, 249, 268, 272

Kaoline, structural changes in (table), 346
Kelvin body, 133
Kelvin-Voigt body, 222, 230
Kernel, binomial, 245
—, combined exponential power, 243, 244
— of equation of creep, 312
— — exponential function, 235
— — power function, 239
— with variable viscosity, 245–247
Köhn's rule, 32

Lamé constant, 98
— formula, 446
Law, Boltzmann distribution, 259, 262, 263, 340

—, exponential, 340
—, linear-fractional, 115
— of after-effect, 130
— — relaxation, Maxwell's, 132
— — time-dependent changes in deformation, 223
—, power, 250
—, —, of deformation, 443
—, Schwedoff–Bingham, 127
Level of stress, 490
Limiting conditions of soil, comparison of, 432
—, due to Mohr–Coulomb, 102, 103
—, due to von Mises–Schleicher–Botkin, 105, 398
Limiting state, condition of, 384
—, for various values of μ_σ, 433
Linear and non-linear deformation theories, comparison of, 469
Loading and unloading, 489
Loading, effect of, 499
—, path, 436
—, surface, 486
—, typical, 489
Lode–Nadai parameter, 82, 83
Lode parameter, 76
—, as definer of stressed state, 430
Logarithmic relation, 157
Lyosphere, 33

Magnetic forces, 37
Maxwell body, 132, 133, 233
Mean linear strain, 70
— normal stress, 69, 73
— rate of linear strain, 71
Mesotexture, 46
Method of curve rectification, 197
— of isotachs, 14
Micelle, 33
Micro-aggregate composition of soil, 30
Microtexture, 46
Mixed coagulation–crystallization structure, 44
Model of elastoplastic hardening, 526
— — primary and secondary consolidation, 279
Models:
    Abdel-Hardy, 233
    Akai, 545
    Anagosti, 230
    Arnold and Mitchell, 540
    based on deformation theory of plasticity, 511
    Budin, 233, 234
    Burges, 227
    Cam-clay, 517
    Christiansen, 233
    combined non-associated/associated, 536
    dilatancy, due to Rowe, 524

elastoplastic, of hardening, 526
elastoplastic with fixed loading surface, 530
Fedder–Bredth, 232
Folque, 232
Gibson–Lo, 230
Goldstein, 232
Jeffrey, 227
Kisiel, 232
mathematical, 481
Matsuoka, 544
mechanical, allowing for mean normal stress, 397
Murayama–Shibata, 232
non-associated, 534, 535
non-associated and combined, due to Lade and Duncan, 537
Pointing–Thompson, 227
rheological, 516
Rowe, 54
Sandler–Di-Magio–Yamada, 544
Tan, 230
Taylor, 230
Terzaghi–Gersevanov, 229
Vyalov, 231
Wilde, 543

Moduli of elasticity of rocks (table), 99
— — linear deformation (table), 99
Mohr diagram, 102
Moisture content of soil, 31
Multi-component systems, 28

Neutral stress tensor, 267
Newtonian perfectly viscous fluid, 117
Non-linear problems, numerical solution of, 510
Non-linearity, approximation of, 447
—, geometrical effect of, 503
Non-Newtonian liquid, 124
— medium, theory of flow of, 260
Normal stress, 60
—, effect of, 163

Octahedral plane, stress and strain on, 62
— stress, 62, 74, 79
Organic matter, 30

Particle mobility, 259
Pascalian liquid, 97
Phenomenological approach, 6
Plastic deformation with strain hardening effect, 108
Plastic flow, 8, 389
—, associated law of, 505
—, generalized equation of, 385

—, non-associated law of, 509
—, theory of, 504
Plasticity theory, active and passive deformation, 86
—, basic assumptions of, 88
—, boundary conditions of, 85
—, rheological equation of state, 87
—, simple and combined loading, 86
Plasticization–coagulation structure, 44
Potential, elastic, 140
—, plasticity, 93, 505
—, thermodynamic, 143
Pore water pressure, 295, 296
Pores, classification of, 136
Porosity, 31
Power relation, 156
Prandtl elastoplastic body, 100
Principal stresses, 73, 78, 103, 104
— tangential stresses, 61
Pure shear, 64, 75, 78, 82, 111, 113, 118, 124

Rates of unit strain, 72
Reciprocal resolvents, 239
Recoverable strain, 254
Relation between instantaneous, long-term, peak and residual strengths, 305, 306
— — particle reorientation and loading, 47
Relative error, 196
Relaxation of stresses, 130, 226
— test, 132
— time, 259
Repulsive forces, 39
Resistance of soil to tension and compression, difference between, 395
Rheological processes, 4
— characteristics, determination of, 380
— equation of state, 124
— — generalized, 383
Rock minerals, classification of, 29
Rock, rheological properties of, 14

Saint-Venant's rigid–plastic body, 99
Schwedoff body, 227
—, equation of state, 132
Second invariant of rate of strain, 83
— — — strain deviator, 81
Semiaggregates, 28
Shear modulus, 98
Simple error, 196
— shear, 63
— stress, 78
Soil, bond energy in, 42
—, components of stress and strain in, 58

—, criteria of failure, 286, 287, 288
—, critical density of, 127
—, cryptoplastic, 164
—, distortion of shape, 63
—, elements of structure, 28
—, failure of, brittle and viscous, 287
—, —, thermodynamic criterion of, 311
—, fields of energy in, 35
—, forces of attraction and repulsion in, 38
—, geometrical representation of stressed state, 78
—, granulometric composition of, 30
—, hardening of, 330
—, interaction of particles in, 40–42
—, ion exchange in, 32
—, mechanism of structural changes in, 348
—, microtexture and mesotexture of, 46
—, numerical solutions of stress–strain state problems, 501
—, principal stresses and principal strains in, 60
—, recent theories of consolidation of, 281
—, resistance of, 303
—, settlement of, computations of, 447–451
—, —, long-term under load, 20
—, similarities between stress and strain states in, 437
—, structure and texture of, factors deciding, 29
—, typical structural bonds in, 43
Soil below foundation, determining settlement of, 459, 460
—, distribution of vertical stresses in, 457
—, field experiments with, 476
—, load-settlement relation in, 471
—, non-linear stress–strain behaviour of, 453
—, numerical solution of mixed elastoplastic problem, 531
—, patterns of deformation solvable analytically (table), 454
—, reaction pressures in, 463, 467
Soil-ice retaining structure, analysis of, 444–446
Soil skeleton, 267
—, creep of, 182, 278
—, discontinuity of, 342
—, equation of deformation of, 280
Space diagonal, 78
Spatial mobilization plane, 496
Spherical tensor, 267
Stabilization structure, 44
Stacks of particles, 28
Standard deviation, 196
— error, 200
Strain deviator, 71
Strain rate, flow equations allowing for changes in, 167
—, time-dependent changes in, 166

Strain superposition, 195, 235
Strain tensor, 68
—, rate of, 68
—, resolution of, 70
—, spherical, 70, 71
Strength of soil, augmentation due to creep, 328
—, cohesive, new hypothesis of, 496
—, deterioration of, 330
—, effect of density and moisture content on, 305
—, — of history and texture on, 45, 46
—, — of rate of load application on, 330–332
—, — of texture on, 52
—, hypothetically instantaneous, 285
—, in tension and compression, difference between, 352
—, long-term, condition of, 300, 306
—, —, equation of, 418
—, —, field data on, 297–299
—, —, initial equation of, 311
—, —, methods of determining, 332–336
—, —, of dense clays, 292
—, —, of fluid clays, 293
—, —, of frozen soil (table), 316
—, —, of plastic clays, 291
—, —, of rocks, 296
—, —, power equation of, 313
—, —, testing formulae for closeness of fit, 315, 317
Strength of soil-ice retaining structures, determining of, 446
Stress and strain tensors, conditions of similarity and alignment of, 502
— circles, 77
— deviator, 70, 73
— intensity, 74
Stressed state, combined, 267
—, methods of testing for, 386, 387
Stress–strain relation, linear–fractional form of, 113
—, power form of, 111
Stress tensor, 68, 297
—, effective and neutral, 267
—, effect of, 264
—, resolution of, 69
—, spherical, 73
Stress tensor invariants, effect of, 383
Stress–volumetric-strain relation, 115
Structural bonds, dependency of rheological properties on, 181
— damage, degree of, 343, 344, 346, 349
Surface of plastic potential, 94
Surface(s) $(\gamma_i-\tau_i-\sigma_m)$, 391
— $(\tau_i-\gamma_i-\sigma_m)$ and $(\tau_i-\gamma_i-t)$, 399

Tangential stress, 60, 74

Temperature of frozen soil, allowing for changes in, 246
Test apparatus, direct shear, 122
—, ring shear, 11
—, torsion, 122
—, triaxial compression, 122
Tests of soil, accelerated, 334
—, at constant deformation rate, 302
—, by constant and stepwise-increasing loading, 193
—, drained and undrained, 295
Theory, kinetic, applicability to soil. 339
— of ageing, 246, 248–251
— of creep, applicability of, 257
— of flow, 251, 252, 256
— — — of non-Newtonian medium, 260
— of hereditary creep, 256
— of small viscoplastic deformation, 89
— of strain hardening, 253, 254, 256
—, percolation, 276
Third invariant of stress deviator, 76
Time before failure, 262
—, forecasting of, 310
Time of after-effect, 223
— of consolidation, 282
Total displacement, 64
Triaxi-symmetrical compression, 76

Ultimate load sustained by foundation, 336
— long-term strength defined, 9
Uniaxial compression, 75, 78, 82, 111, 114

— — and tension, 103
— stress, state of, 398
— tension, 78, 82
Uniform all-around compression, 76
Unit elongation, 63, 65

Van der Waals–London forces, 36
Variable bulk modulus, 115
Variable-load problems, solving of, 373
Variable loading, 364
—, allowing for, 450
Variable moduli of deformation, 111
Virgin compression line, 520
Viscoelasticity, 126, 129
Viscoplastic bodies, studies of, 11
Viscosity, anomalous, 124
—, apparent, 128
—, bulk, 124
—, effective, 126, 263
—, kinetic theory of, 259
—, methods of determining, 121
—, non-linear, 124
—, structural, 125
Volumetric strain, 115, 127

Water in soil, 31
Work of dissipation, 141

Yield limit, conventional, 110
Yield point in shear, 99